Biomarkers of Brain Injury
and
Neurological Disorders

Biomarkers of Brain Injury and Neurological Disorders

Editors

Kevin K.W. Wang
Zhiqun Zhang
Firas H. Kobeissy

Center for Neuroproteomics and Biomarkers Research
Department of Psychiatry
The Evelyn F. and William L. McKnight Brain Institute
University of Florida
Gainesville, FL
USA

CRC Press
Taylor & Francis Group
Boca Raton London New York

CRC Press is an imprint of the
Taylor & Francis Group, an **informa** business

A SCIENCE PUBLISHERS BOOK

Front Cover Art Work: The Art work is designed by the Graphic designer Hussein Mokdad, B.S. Email: hussein-almokdad@hotmail.com, Lebanon, Beirut

CRC Press
Taylor & Francis Group
6000 Broken Sound Parkway NW, Suite 300
Boca Raton, FL 33487-2742

First issued in paperback 2020

ISBN-13: 978-1-4822-3982-9 (hbk)
ISBN-13: 978-0-367-73941-6 (pbk)

Library of Congress Cataloging-in-Publication Data

Biomarkers of brain injury and neurological disorders / editors, Kevin K.W. Wang, Zhiqun Zhang, Firas H. Kobeissy.
   p. ; cm.
   Includes bibliographical references and index.
   ISBN 978-1-4822-3982-9 (hardcover : alk. paper)
   I. Wang, Kevin K. W., editor. II. Zhang, Zhiqun, editor. III. Kobeissy, Firas H., editor.
   [DNLM: 1. Brain Injuries. 2. Biological Markers. 3. Nervous System Diseases. WL 354]

   RC451.4.B73
   617.4'81044--dc23   2014026023

Visit the Taylor & Francis Web site at
http://www.taylorandfrancis.com

and the CRC Press Web site at
http://www.crcpress.com

Dedication

To my family, for the love and patience they provide me.

Kevin K.W. Wang

For my husband Jianghui Chao, our wonderful children Ryan and Arnall, my parents, sisters and loved ones who supported and believed in me.

Zhiqun Zhang

To my mom Kawsar; may her soul rest in peace, my dear dad Hosni, and Sister Maha whose love and patience gave me hope in my path. To my mentors, Dr. Kevin K. Wang & Dr. Mark S. Gold whose experience and advice drove my success....

Firas H. Kobeissy

Foreword

The central nervous system, including the brain and spinal cord, is arguably the most complex human organ system in terms of structure and function. This book stands out in describing the latest biomarker studies in brain disorders relevant to its mechanism, advanced technologies, experimental methods, and clinical trial studies. Diagnostic tests based on biomarkers have already demonstrated proven clinical diagnostic utility in acute care environments. For example, in the area of cardiac injury, cardiac troponin proteins (T and I) and various forms of brain natriuretic peptide (BNP), often in combination with other biomarkers, are routinely used to facilitate accurate diagnosis of congestive heart failure and myocardial infarction, in patients presenting with chest pain. Recently, there has been broad recognition that biomarkers can also play a critical role in drug discovery and development. There are several areas in which biomarkers can facilitate brain injury drug development and eventually personalized medicine. Biomarker research methods may improve our knowledge of the pathobiology that ensues after CNS disorders, and unlock the connections between this pathology, individual variability, and the heterogeneous outcomes experienced by those with central nervous system disorders.

The aim of this book is to draw together the work of leading experts in the fields of neurological disorders, brain injury, and drug abuse and to highlight what is known on a broad range of topics pertaining to biomarkers in central nervous system disorders ranging from traumatic brain injury, spinal cord injury, multiple sclerosis, and alcohol abuse. This elegant volume begins with chapters discussing basic mechanisms and methods of biomarker identification including neuroproteomics analysis, followed by surveying different textures of currently novel biomarkers such as microRNA, proteins as well as genetic fingerprints and their applications in different neurological disorders. This section is then followed by elaborate chapters written by world-renowned experts, who have tackled several neurodegenerative diseases specifically, along with their mechanisms and their associated "known" biomarkers. This work flows smoothly, discussing novel methods such as Deep Brain Stimulation, the endpoint of which can be used as a neurotherapeutic marker for a number of neuropsychiatric disorders. The book concludes by emphasizing the implications and importance of biomarker research on several fields, such as brain trauma, drug addiction and the need for theranostic (therapeutic and diagnostic) biomarker approaches.

It is my great honor that three of our faculty at the Department of Psychiatry (Wang, Zhang and Kobeissy), who are experts in the areas of biomarkers in brain injury studies and drug abuse–associated neurotoxicity, have gathered in this respected work to deliver one of the most updated presentations in biomarker studies. This collection will have direct implications on neurotherapeutic management, treatment guidance, as well as indicators of injury severity and prognosis.

The book will lay out a foundation for scholars interested in several neurological disorders and related biomarker research. This is currently a unique and timely volume that will provide a comprehensive review for basic scientists, graduate students, medical students, clinical researchers and medical professionals with interest in translational research in the area of brain injury and other brain disorders.

Mark S. Gold, MD
University of Florida Alumni Distinguished Professor (2011–2015)
University of Florida Distinguished Professor
Donald R. Dizney Eminent Scholar
University of Florida College of Medicine and McKnight Brain Institute
Departments of Psychiatry, Neuroscience, Anesthesiology
Community Health & Family Medicine
Chairman, Department of Psychiatry

Preface

Studies in the field of central nervous system (CNS) biomarkers have been evolving at quite a rapid pace especially with the introduction of novel high throughput screening techniques such as proteomics and microarrays coupled with systems biology with its predictive potential of missed targets and their dynamic alteration in respect to disease progression. This in turn has led to the identification of a new generation of biomarkers including signature microRNA biomarkers and even autoantibodies serving as indirect specific indicators of certain brain injury disorders. This exciting advance in biomarker research has encouraged new lines of funding from national and private sectors aiming at identifying novel treatment targets driven via biomarker identification. Among the most challenging and highly demanding fields for biomarker research is the CNS and particularly those related to brain trauma pathology. The CNS, including the brain and spinal cord, is arguably the most complex human organ system in terms of structure and function. Of interest, biomarker identification of CNS disorders is considered the *Holy Grail* that is in constant pursuit by scientists and medical doctors. Identifying these markers in body fluids represents a major challenge due to several factors including their minute levels, turn over/clearance rate, dynamic range in serum/cerebrospinal fluid, etc. which necessitates the use of high throughput technologies such as genomics, neuroproteomics and recently microRNA assessment to decipher these markers dynamics.

In the area of conflicts worldwide, brain injury has been designated the "signature injury" of current military conflicts. CNS trauma primarily from regular accidents or sport associated injury as well as battlefield represents one of the most common causes of morbidity and mortality in our society with enormous societal and economic burdens. Each year, approximately three million people sustain traumatic brain injury; these are often present with neuropsychological deficits that may develop later at chronic phases; coupled with altered emotional, and/or cognitive impairments, exhibiting decreased executive function, depression, and significant drug abuse problem. Such presentation in some TBI patients has been termed as altered psychological health which overlaps with some symptoms of post-traumatic stress disorders (PTSD) leading to its misdiagnosis and consequently, its treatment. This has driven to the increased funding to identify markers that can distinguish brain injury-induced psychological health problems from traditional PTSD symptoms. Currently, the need for biomarkers identification is of high interest and demand to distinguish neuropsychological disorders due to brain trauma as a pipeline to identify novel effective neurotherapeutics that can guide in treatment management for the cure of these devastating injuries.

In this work we have assembled 21 expert contributors renowned in their scientific work of neurological and neuropsychological disorders and the implication of biomarker research in these disorders as chapter lead authors. The book is divided into three sections **(Biomarker technology, CNS injury biomarkers and Other CNS disorder biomarkers)**.

The first section **(Biomarker technology)** describes experimental concepts and molecular mechanisms of biomarker genesis and biomarker discovery and detection methods such as proteomics applications and microRNAs assessment and the role of the protease systems in generating different potential markers. In the following section **(CNS injury biomarkers)**, the authors describe the utility of different biomarkers in the fields of brain and spinal cord trauma and their use as neurotherapeutic recovery endpoints, diagnostic signatures and rehabilitation markers of different brain insult scenarios such as mild brain injury, spinal cord injury acknowledging the different types of markers including inflammatory and protein biomarkers identified.

In the final section **(Other CNS disorder biomarkers)**, biomarkers of several neurological disorders such as Multiple Sclerosis, Alzheimer's disease, Charcot-Marie-Tooth disease, Parkinson's disease as well as alcohol abuse related markers are fully investigated. These chapters represent a valuable addition in the field of biomarker research and to those interested in the experimental studies for identifying advanced neurotherapeutic treatment for brain injury and other neuropsychological disorders. In this initiative, we would like to thank colleagues and contributors who were enthusiastic about this project and dedicated their precious time and expertise in finalizing this wonderful project. A special recognition goes to the CRC Press/Taylor & Francis editorial team who bear with us all the challenges in delivering this book. Our recognition goes to our talented graphic designer Mr. Hussein Mokdad, who had to go through a lot of enjoyable unending requests from us—the editors—to reach a final cover art that is decorating our book. To the readers of this book, we hope that you will find this book both informative and stimulating in your own research or clinical area. We also welcome your feedback to us.

Gainesville, FL, USA, 2014
Kevin K.W. Wang
Zhiqun Zhang
Firas Hosni Kobeissy

Acknowledgements

There are many dear people to thank and acknowledge in the development of this book. Our most abundant gratitude goes to the authors of the chapters in this book. They ultimately made this work possible by providing their top quality manuscripts and comments. Without their extremely valuable and "prompt" contributions, this book would not have been possible. We would like to thank Miss Zeinab Dalloul, MS and Miss Emilia Amrou, BS, for the editing, proof reading and commentary on various components; their help is highly valued.

We would like to take this opportunity to thank our colleagues from the University of Florida, Departments of Psychiatry and Neuroscience, whose help and encouragement have inspired us in completing this book. We want to offer my special thanks to Professor Mark S. Gold, our chairman, who offered his wise advice and guidance that lead to the completion of this book. We feel highly privileged that he agreed to write the Foreword for this book.

Contents

Section B: CNS Injury Biomarkers

Section C: Other CNS Disorder Biomarkers

Section A
BIOMARKER TECHNOLOGY

1

Neuro-proteomics and Neuro-systems Biology in the Quest of TBI Biomarker Discovery

Ali Alawieh,[1] Zahraa Sabra,[3] Zhiqun Zhang,[2] Firas Kobeissy[2] and Kevin K.W. Wang[2],*

INTRODUCTION

Traumatic Brain Injury (TBI) is the leading cause of mortality and disability among the young population in the developed countries, and its worldwide prevalence is sharply increasing (Feigin et al., 2010; Ghajar, 2000; Maas et al., 2008). TBI affects all ages with highest incidence rates among children, young adults and the elderly (Faul et al., 2010; Hemphill III et al., 2012; Koepsell et al., 2011). TBI is associated with increased incidence of disability and premature death along with heightened medical and socioeconomic burden on individuals, families and societies (Leibson et al., 2011). The average annual death from TBI in the US is 53,014, mostly of the young age group (Coronado et al., 2011). This value is only the tip of the iceberg as TBI accounts annually for up to 275,000 hospitalization and 1,365,000 emergency department visits despite those who receive no care or donot appear at the emergency setting (Faul et al., 2010; McCrea et al., 2004)

---

[1] Neuroscience Institute, Department of Neurosciences, Medical University of South Carolina, Charleston, SC 29425, USA.

[2] Center for Neuroproteomics & Biomarkers Research, Departments of Psychiatry & Neuroscience, McKnight Brain Institute, University of Florida, Gainesville, FL 32611, USA.

[3] Department of Electrical and Computer Engineering, Faculty of Engineering and Architecture, American University of Beirut, Lebanon.

* Corresponding author: kwang@ufl.edu

(Figure 1.1). The World Health Organization (WHO) predicts that TBI will rise to the third leading cause of global mortality and disability by 2020 (Organization, 2009). The long-term prevalence of disability secondary to TBI in the US is estimated to be 1–2% of the population (Zaloshnja et al., 2008). Even if the incidence of TBI is much less than strokes, but given its early age incidence, the long-term effects and socio-economic costs of TBI can be as high (van Baalen et al., 2003). TBI accounts for about 10% of the health care budget in the US with an estimated annual cost to society of US$ 30 billion (Hoyt et al., 2004).

TBI is often referred to as a "silent epidemic" since the different complications of the disease is not readily apparent, and the general public has limited awareness about this disease (Faul et al., 2010). There are currently no diagnostic techniques that can confirm whether a blow to the brain has resulted in brain injury or not. Clinical symptoms of brain injury may resolve within one or two months, yet axonal injury may persist for years (Johnson et al., 2012; Williams et al., 2010). Around 43.3% of all Americans having TBI had residual disability one year after injury (Corrigan et al., 2010) despite the fact that up to 90% of TBI cases are classified as mild TBI (mTBI) (McCrea, 2007). Therefore, mTBI does not imply a benign or self-limiting condition since it could be associated with neuronal swelling, axonal energy and disconnection of the white matter (Blumbergs et al., 1994; Kirov et al., 2012). Eventually, mTBI patients may have long-term disabling consequences like dizziness, fatigue, headaches and delayed recall of memory (Heltemes et al., 2011).

Traumatic brain injury is a brain injury caused by an external mechanical force like a blow, concussive force or a bullet (Stergiou-Kita et al., 2012). This injury is a dynamic process that starts with a primary injury and initiates a cascade of biochemical and cellular changes of repair and injury (Ottens

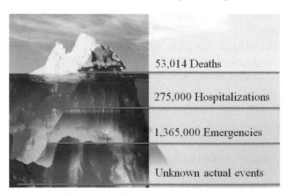

Figure 1.1. Deaths from TBI are only the tip of the iceberg compared to the actual incidence of mild TBI not drawn to medical attention.

Color image of this figure appears in the color plate section at the end of the book.

et al., 2006). These changes contribute to cumulative neuronal death over time resulting in secondary injury and long term complications (Loane et al., 2009). Evidence from pathological studies supported the involvement of several immunological and apoptotic pathways in the progress of this neuronal damage (Raghupathi, 2004) including inflammatory responses (Edwin et al. 2011; Loane and Byrnes, 2010; Ziebell et al., 2012), autophagy and activation of proteases (Clark et al., 2008; Knoblach and Faden, 2005), mitochondrial dysfunction (Lifshitz et al., 2004; Mazzeo et al., 2009), oxidative stress, neurotransmitter release, excitotoxicity and changes in intracranial pressure and cerebrovascular circulation (Cernak and Noble-Haeusslein, 2009; Ghajar, 2000; Maas et al., 2008). Since the early manifestation of these changes is biochemical and molecular in nature; it is in the hands of biochemical and molecular testing to detect and assess the severity of TBI as well as to predict the outcome.

Secondary to the trauma-induced neuronal degeneration, TBI is associated with long-term cognitive deficit (Patterson and Holahan, 2012) that can affect up to 15% of mTBI patients (Røe et al., 2009). Eventually, TBI is considered a risk factor for many neuropsychiatric and neurodegenerative disorders including Alzheimer's disease (Jellinger et al., 2001; Lye and Shores, 2000) where neurofibrillary tangles were detected in the brains of ex-boxers who were subject to mTBI (Tokuda et al., 1991). There is also high comorbidity between TBI and several neuropsychiatric disorders like anxiety, depression, dementia and others (Deb et al., 1999; Rao and Lyketsos, 2000; van Reekum et al., 2000; Whelan-Goodinson et al., 2010).

The aforementioned occult complications of TBI, in the absence of any FDA approved treatment for TBI (Narayan et al., 2002), necessitate the detection of diagnostic and therapeutic biomarkers to improve the quality of life and decrease mortality among patients with TBI. In this chapter we will emphasize the need of biomarker discovery in TBI and highlight the major advances in the field of proteomics as applied to biomarker quest in TBI.

Available Classification and Diagnostic Techniques in TBI

Early classification of acute TBI is of critical importance in the accurate diagnosis, prediction of outcomes (Masel and DeWitt, 2010; Zhu et al., 2009), and; eventually, the clinical workup of patients. TBI is a heterogeneous condition of variable clinical behavior, and a specific targeted therapy for the different subcategories of the disease is essential. In this context, accurate classification is mandatory to discover patients to whom intensive rehabilitation programs are needed and beneficial (van Baalen et al., 2003). The identification of those patients among the heterogeneous population of TBI patients is one of the major challenges in clinical practice (Saatman et al., 2008) especially that proper management of TBI can significantly alter the clinical progression in the first hours or days after injury (Lee and Newberg, 2005). Therefore, an ideal TBI classification includes a rapid assessment of initial severity that is in accordance with the long-term clinical outcome. Noteworthy, the hyper-metabolic state of the brain post-mTBI may render it susceptible to repetitive mTBI that will have dismal, even fatal, consequences on the outcome (McCrory and Berkovic, 1998). Eventually, there is a sincere need to identify and diagnose those patients for medical and occupational management post-mTBI.

Among the traditional classification modalities, computed tomography (CT) and the assessment of severity of injury by the Glasgow Coma Scale are considered the cornerstones of assessment in neuro-traumatology (Marshall, 2000). Other neuroimaging techniques have been also incorporated including Magnetic Resonance Imaging (MRI), Single Photon Emission Tomography (SPECT), and Positron Emission Tomography (PET) (Bigler, 2001; Le and Gean, 2009).

Glasgow Coma Scale (GCS)

The Glasgow Coma Scale (GCS) is the most common modality for TBI classification among clinicians today. GCS is a 15-point index of neurological injury severity that assesses the level of consciousness after TBI. The scale involves three components; assessment of eye opening, best motor response and best verbal response (Scale, 2001). According to this scale, TBI patients are classified into three broad categories: severe TBI (GCS 3–8), moderate TBI (GCS 9–13) or mild TBI (GCS 14–15) (Maas et al., 2008).

The use of GCS as a diagnostic tool is subject to several limitations. Many confounders may obscure the level of consciousness in patients including medical sedation, paralysis, distracting injuries or intoxication due to alcohol or recreational drugs (Green, 2011; Maas et al., 2008). With the increasing use of early sedation, intubation and ventilation in trauma patients, the value of the Glasgow Coma Scale is decreasing (Zhu et al., 2009); the neurological exam is difficult and the severity maybe over-estimated (Stocchetti et al., 2004). Several epidemiological studies have shown that the prevalence of sedation, drug or alcohol abuse and intoxication among patients with TBI is highly increasing (Lindenbaum et al., 1989). The European Brain Injury Consortium Survey of Head Injuries has shown that the GCS was only fully testable on 77% of the TBI patients admitted (Murray et al., 1999). Other populations of patients also are difficult to assess using GCS including infants, young children and patients with pre-existing neurological impairment (Saatman et al., 2008). GCS performs best for severe TBI; however, for mild TBI cases that constitute 80–90% of all TBI, GCS has poor performance. Eventually, GCS is strongly associated with acute morbidity and mortality, but not with long-term outcome (Zasler, 1997).

Aside from clinical management, the GCS does not provide any clue about the underlying pathophysiological underlying the neurological deficits and provides less raw material for targeted therapy (Zasler, 1997).

Neuroimaging Techniques

Neuroimaging techniques were implemented in diagnosis and classification of TBI as a supplement to GCS in order to evade some of the limitations of the GCS especially those associated with cases of early sedation and intubation of patients (Maas et al., 2008). Neuroimaging techniques classify patients based on morphological brain changes that can be detected in TBI patients, and the most common used modality is Computed Tomography (CT).

A neuroimaging descriptive system using CT scan was described by Marshall et al. (1991) and includes criteria such as the presence of mass lesions, and diffuse brain injury assessed by signs of elevated intracranial pressure (ICP) like compression of basal cisterns or midline shift. However, these criteria suffer several limitations. The sensitivity of CT scans to diffuse brain damage is very low and the absence of abnormal finding on CT does not rule out the presence of brain damage especially in case of mild TBI (Güzel et al., 2009; Haydel et al., 2000; Yuh et al., 2012). CT scan, similar to other neuroimaging techniques, can only capture momentary changes in the brain and could not account for the dynamic changes that start at the microscopic level after TBI (Maas et al., 2008). Even though several recommendations have suggested criteria for use of CT scans for high risk patients (Smits et al., 2007) the physicians' lack of confidence in available diagnostic tools have led to the practical consideration of routine use of CT in mTBI patients eventually leading to higher costs and radiation exposure (Stiell et al., 2005).

MRI provided an additional sensitivity to CT scans in terms of detecting diffuse brain damage (Mittl et al., 1994); however, MRI would be impractical to use in the acute phase of the trauma due to limited availability and the physically unstable status of TBI patients. Patients who have metallic items in their body, a common incidence in emergency care, are not candidates for this imaging modality (Wang et al., 2005). Some studies have also associated limited outcome prediction with the use of MRI in TBI (Hughes et al., 2004). The use of other modalities like SPECT and PET also suffers major limitations. Regional blood flow abnormalities as detected by SPECT imaging does not necessarily correlate with a structural brain damage (Wang et al., 2005). SPECT has also low sensitivity in detecting small brain lesions, and the association between abnormalities detected by SPECT and the neurocognitive outcome is still weak (Hofman et al., 2001). The assessment of regional glucose metabolism in TBI by PET scan is also nonspecific due to the heterogeneous nature of TBI where hyper-metabolism and hypo-metabolism in the same regions across different TBI patients have been reported (Le and Gean, 2009).

Other limitations also circumscribe the use of neuroimaging in TBI including their availability, high cost and the inability to carry them in a repeated manner due to inconvenience and radiation exposure. Therefore, they cannot be used to monitor the occurrence of secondary lesions that may occur in a short period of time.

The challenges associated with a reliable and efficient classification of TBI along with the need for such classification have led the National Institute of Neurological Disorders and Stroke (NINDS) to convene a workshop to study the steps needed for a new valid classification system for TBI (Saatman et al., 2008). In October 2011, the collaborative efforts of the European Commission, the Canadian Institutes of Health Research and the National Institutes of Health set the "International Initiative for Traumatic Brain Injury Research" (InTBIR) to advance clinical traumatic brain injury (TBI) research, treatment and care (European-Commission). An expected alternative to imaging based classification is a biomarker-based classification that can overcome many of those limitations.

Needs for Biomarkers

The unapparent progression of TBI consequences and the rapid resolution of signs and symptoms post-mTBI along with the irreversible disabling complications emphasize the need for accurate specific biomarkers for diagnosis, classification and monitoring of the disease and its progression. These biomarkers once discovered and validated will represent definite diagnostic criteria of TBI and reliable outcome predictors added to the current clinical examination and neuroimaging results.

A biomarker has been defined as a *"characteristic that is objectively measured and evaluated as an indicator of normal biological processes, pathogenic processes, or pharmacologic responses to a therapeutic intervention"* (Biomarkers Definitions Working Group, 2001). In the clinical context of TBI, the traumatic injury to the brain may cause a series of cellular changes including degeneration, protease activation, oxidative stress, metabolic disturbances, etc…, and these changes result in shedding specific proteins into the CSF or serum that can be identified and studied for their association with the disease presence, outcome and progression. These biomarkers reflect the earliest changes that occur in the cell before the evidence of injury appear on imaging techniques. Therefore, the use of biomarkers could offer a rapid, noninvasive and cost-effective tool for the diagnosis of TBI and subsequent classification and triage. These diagnostic biomarkers would also help monitor disease progression and assess outcomes of therapeutic interventions. Prognostic biomarkers in TBI may help promote early and effective treatment measures.

Biomarkers are of essential importance for successful decisions in the context of serious situations as with TBI. In these conditions, the rapid evaluation of the severity and future progression of disease is very critical especially in the first hours and days after injury where irreversible damage starts to accumulate (Selassie et al., 2008; Stocchetti et al., 2008).

Away from the bedside, biomarkers have a critical bench role of providing an insight into the underlying pathological processes of TBI associated brain damage. Valid and specific biomarkers are believed to be key players in the cascade of events leading to the pathological manifestations and could provide evidence of the involved pathways in the context of TBI and other diseases (Wagner, 2002). Biomarkers can also act as a surrogate endpoint to substitute a clinical endpoint reducing the cost and risks associated with clinical trials (Woods et al., 2012). The role of biomarkers can be summarized into reflecting disease *traits* (indicating susceptibility, predisposition and risk factors of disease), *states* (assisting in disease diagnosis) and *rates* (providing information about the progression and pathophysiology of the disease) (Fox and Growdon, 2004).

Therefore, the aforementioned limitations of current diagnostic and classification techniques in prediction of occult brain injury in TBI can be best surpassed by the discovery, validation and utilization of diagnostic biomarkers in serum or cerebrospinal fluid (CSF) that can allow for rapid assessment, minute-to-minute monitoring of disease progression, and reliable prediction of the outcome. This temporal profile of disease changes that can be reflected by these biomarkers will be essential in the chronic therapy of TBI to identify the treatment target and timing, and to early detect worsening of neurological status before microscopic lesions become apparent. This has been illustrated in Figure 1.2.

Eventually, an ideal biomarker would (1) diagnose TBI with high sensitivity/specificity before neuroimaging manifestations, (2) measure disease extent and severity, (3) predict the outcome, (4) allow for monitoring of treatment and disease progression, (5) give insight about the underlying pathophysiological mechanisms, (6) uncover new targets for therapy, and (7) this marker should be present in detectable amounts in serum and CSF.

Major advances in biomarker research have taken place in the context of several diseases like troponins in cardiovascular disease, C-reactive protein for inflammation, and creatinine for renal failure. These biomarkers,

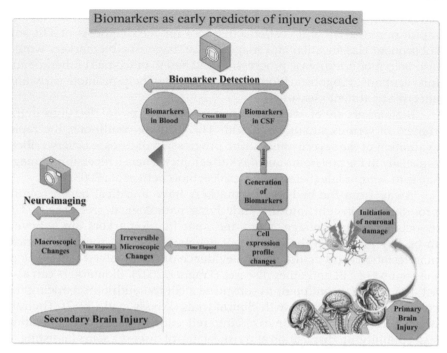

Figure 1.2. The advantage of tissue specific biomarker discovery over current imaging and diagnostic tools is that it can allow for the detection of injury early on before irreversible damage to the brain tissue ensues.

Color image of this figure appears in the color plate section at the end of the book.

among others across the medical disciplines, paved the way for therapeutic revolutions in the corresponding fields. However, in the context of neurological disorders like TBI, the quest of biomarker discovery is still in its early phases and awaits profound advances in discovery and verification strategies and techniques (Maas et al., 2008; Wang et al., 2005). Several areas are suggested as fields for the biomarker quest in TBI and other CNS disorders including proteomics (Cadosch et al., 2010; Conti et al., 2004; Wu et al., 2012b; Yang et al., 2009), transcriptomics (Di Pietro et al., 2010), epigenomics (Conley and Alexander, 2011), lipidomics (Bayir et al., 2007; Sparvero et al., 2010), metabolomics (Keller et al., 2011; Viant et al., 2005; Yang et al., 2012), and microRNA analysis (Lei et al., 2009; Redell et al., 2009; Redell et al., 2010). Later we will illustrate a major area of biomarker quest in TBI; namely, proteomics, illustrating the techniques used and the current standpoint of research along with limitations and future challenges.

Proteomics in TBI

The early definition of a proteome was the entire complement of expressed proteins in a biological system. However, the study of proteomics is not limited to the identification of the expressed protein sets. It rather involves the study of protein abundance, activity, localization, isoforms and modifications, as well as protein-protein interactions and functioning within higher complexes (Tyers and Mann, 2003). This is a golden aim for current proteomic practice that is limited to certain aspects of proteins status and function and to a limited subset of the protein complement (MacBeath, 2002).

The study of proteomics involves two major strategies that could be separate or complementary; (1) a discovery-oriented strategy and (2) a hypothesis-driven strategy. The major emphasis in the research community is on the unbiased discovery-oriented strategy. This methodology studies differential global expression of the proteome across varied conditions to discover new proteins and pathways. The hypothesis-driven strategy assesses the behavior of certain subset of candidate proteins to confirm their implication in certain pathophysiological or physiological effects. Both strategies attempt at the ultimate aim of discovering new disease specific biomarkers that can make their way into clinical practice (Blonder et al., 2011; Mehan et al., 2012; Schiess et al., 2009; Sjödin, 2012). In both aspects, proteomics constitute an important tool for biomarker discovery in TBI through its application on brain tissue, body fluids (serum and CSF) as well as through its utilization in different animal models of TBI as described later. In addition to the discovery of few biomarkers, proteomics can help detect expression profiles that could be disease or stage specific and allow monitoring the progression of disease and assessing response to therapy. Table 1.1 illustrates some advantages of the use of proteomics in TBI biomarker discovery.

Techniques

Traditionally, proteomics made use of two major methodologies: (1) 2-D Difference Gel electrophoresis (2D-DiGE) method, and (2) non-gel based Liquid Chromatography coupled/mass spectrometry (LC/MS) method (Becker et al., 2006; English et al., 2011; Rodriguez-Suarez and Whetton, 2013) (Figure 1.3). 2D gel electrophoresis (2D-GE) involves the separation of proteins based on molecular weight and isoelectric point. With the DiGE modification, the new technique allows for the detection of deferentially expressed fluorescent spots across samples, and then the obtained image is quantified and spots are excised to be identified by MS. In the LC/MS technique, proteins are first digested into fragments, separated by LC,

Table 1.1. Advantages and limitations of the use of proteomics in TBI biomarker discovery.

Status of proteomics in TBI biomarker discovery	
Advantages	**Challenges and limitations**
1. Allow for proteome differential analysis in serum, CSF and tissue specimens	1. Limited sensitivity of current techniques and failure to characterize entire proteome
2. Suggest putative diagnostic and prognostic biomarkers for further validation	2. Dyamic range of protein abundance in biofluids spanning 10 orders of magnitude
3. Detection of protein profile signature that adds to the criteria of patient classification	3. Non-reproducibility of the results of several studies
4. Detection of protein profile signature that can act as an outcome predictor post-injury	4. Emphasis on discovery rather than validation of findings
5. Give an idea about the involved cellular pathways in the pathogenesis of TBI	5. Inability to demonstrate the specificity of findings to TBI
6. Allow for the study of protein degradation patterns to monitor disease progression	6. Paucity of well-controlled studies that adjust for age and other confounders
7. Permit an un-biased discovery-oriented strategy for biomarker discovery	7. Non-uniformity of sample acquisition when it comes to CSF samples
8. Adjuvant protein saturation techniques help detect low abundance proteins	8. Absence of translational studies that can translate results in animals to human
9. Detection of location specific differential protein expression in tissue specimens	9. Low sample size reducing the power and reliability of the studies

and then the eluting solvent is introduced into a mass spectrometer for analysis. Hereby, mass analyzers such as Matrix-Assisted Laser Desorption/Ionization-Time-Of-Flight (MALDI-TOF) or others are used in tandem (MS/MS) to achieve higher degrees of ion separation (Lai et al., 2012). 2D-GE is more quantitative, robust and more reproducible, especially with the use of fluorescent dyes in 2D-differential-in-gel-electrophoresis (2D-DiGE) (Chen et al., 2007). However, 2DiGE is labor intensive and can cover a small range of high abundance proteins. However, LC/MS has wider protein coverage, but is less reproducible and does not support quantification unless coupled to an isotope labeling technique such as iTRAQ (Isobaric tag for relative and absolute quantitation) (Wiese et al., 2006). Despite these limitations, these two techniques are continuously refined and still play a vital role in the study of proteomics (Angel et al., 2012; Rabilloud, 2012; Sabido et al., 2012). Techniques that are more powerful have been also studied for application in high-throughput proteomics like protein microarrays or high-throughput Immunoblotting analysis (Fung et al., 2001; Hall et al., 2007; Talapatra et al., 2002). A protein microarray chip is made of thousands of different affinity reagents like antibodies, aptamers, recombinant peptides or others (for review (Espina et al., 2004)). These can interact specifically for proteins and allow for detection of large numbers of proteins with high

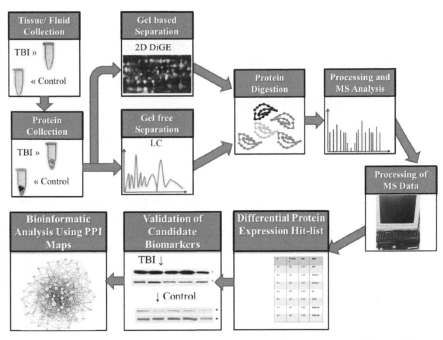

Figure 1.3. Overview of Proteomics-Based Biomarker Discovery: In brief, samples are separated by gel-based or gel-free techniques, digested and then processed into MS that allows for identification of differentially expressed proteins with the help of online databases. The obtained list of differentially expressed proteins provide an insight into candidate biomarkers that can be further validated by traditional molecular biology tools like Western blot. Further use of systems biology protein interaction analysis tools allows for analysis and visualization of resulting data in terms of interacting proteins and protein networks.

Color image of this figure appears in the color plate section at the end of the book.

fidelity through ELISA-style sandwich assays (MacBeath, 2002; Mitchell, 2002). These microarrays can also facilitate the detection of the very high number of proteins, a major limitation of previous techniques, and allows for an extensive study of protein-protein interaction (Melton, 2004; Zhu and Snyder, 2001). Future advances in the utilization and enhancement of protein microarray chips is foreseen with the incorporation of nanotechnology producing miniaturized nano-arrays that can enhance the specificity and sensitivity of current detection technique (Gonzalez-Gonzalez et al., 2012; Mitchell, 2002; Wingren and Borrebaeck, 2007; Yeates, 2011). Linked to the study of proteomics, TBI degradomics play an important role in TBI biomarker detection since protease over activation is a major aspect in brain injury (Knoblach and Faden, 2005; Raghupathi, 2004). Degradomics involves the application of high-throughput genomic and proteomic techniques to identify proteases and the protease-substrate repertoire on organismal scale (López-Otín and Overall, 2002). This allows the study of protein degradation products of CNS-specific proteins in the serum and CSF that could be important indicators of protease activation upon injury; therefore, these degradation products can be important biomarker candidates for detecting and monitoring the progression of brain injury post-trauma.

Target Sources for Proteomics

The quest for biomarkers involves serum and cerebrospinal fluid (CSF) protein pools as well as brain tissue samples. It also involves intact and proteolytic products of proteins.

For brain tissue proteomics, human brain tissue is not readily available for proteomics tools; alternatively, animal models of brain injury have been used as initial evidence to be translated into human studies. The reason behind the use of proteomic in animal models is that brain tissue is the most abundant source of protein biomarkers (Hergenroeder et al., 2008). Several animal models have been proposed like Controlled Cortical Impact (CCI) (Edward Dixon et al., 1991), closed-head Projectile Concussive Impact (PCI) (Chen et al., 2012), fluid percussion injury (McIntosh et al., 1989), penetrating ballistic-like injury (Boutte et al., 2012; Guingab-Cagmat et al., 2012), etc... These models provide a tissue milieu to investigate relevant protein biomarkers at the site of injury or sometimes at distant sites. The studies not only investigate neurons as the master cells of the brain, but also supporting cells like microglia, astrocytes and oligodendrocyte for biomarkers as they also can be involved in major pathologies in the brain. In addition to direct tissue investigation, animal neuronal and glial cell cultures can also be used to monitor the alteration in protein expression profile post-TBI. Tissue studies as well as cell cultures can allow for spatial study of differential protein expression across different cellular locations through studying synapse-enriched or axoplasm-enriched samples (Garland et al., 2012; Wishart et al., 2012).

Akin to tissue sources that provide information about the pathological brain changes, diagnostic biomarkers suggested for the routine use in emergent conditions are to be sought in the serum, or preferentially, the CSF. Although CSF sampling is with lower convenience, CSF sampling is being more preferred for several reasons. The reasons include its proximity to the tissue of origin, higher abundance of potential biomarkers especially if blocked by the Blood-Brain-Barrier (BBB), and less contamination of other proteins present in the plasma (Alawieh et al., 2012). However, an important limitation of CSF sampling is the dynamic protein gradient between the ventricular and the lumbar cisterns (for review (Reiber, 2001)). This acts as an important confounder while comparing protein samples from patients of severe, moderate and mild TBI. Serum is another more convenient source, yet the possibility to fish a CNS specific biomarker in the serum is quite low given the dynamic range of abundance in serum proteins and its contact with all body tissues (Alawieh et al., 2012).

These proteomics techniques were also couples to ontologies, databases and bioinformatics tools that can analyze relationships between discovered proteins and relate several proteins to a certain network, pathway or pathophysiological process (Alawieh et al., 2012). We will illustrate recent examples of the application of neuroproteomics tools to research in TBI.

Application of Different Techniques in TBI

2D-DiGE-MS

2-Dimensional differential gel electrophoresis (2D-DiGE) is one the most commonly used tool in proteomics and neuro-proteomics. Several studies have used 2D-DiGE to investigate diagnostic and therapeutic biomarkers for TBI as well as to monitor differential protein profile expression across time post-injury to monitor damage or response to therapy. The majority of reported studies used animal models to simulate TBI.

One of the most recent studies done by Yang et al. (2013) used impact accelerated model of TBI in rats, and used 2D-DIGE to detect the differential abundance of serum proteins between injured rats and controls (Yang et al., 2013). Five hundred protein spots were detected of which five proteins were identified by Matrix-Assisted Laser Desorption/Ionization-Time of Flight (MALDI-TOF). Of interest was Haptoglobin, a serum protein produced by the liver, whose levels were elevated post TBI and confirmed by mRNA-PCR to be over-expressed by mediation of serum IL-6 that is also elevated post-TBI. As such, serum Haptoglobin is putative biomarker for monitoring damage post-TBI especially that mediated by cytokines and acute phase reactants. Boutte et al. (2012) used a rat model of penetrating ballistic-like brain injury to investigate changes in brain tissue protein expression profile post-TBI and search for CSF biomarkers for brain injury (Boutte et al., 2012). The group used 2D-DIGE coupled to MS to discover 321 differentially expressed proteins in brain tissue that were analyzed by systems biology approach using the Ingenuity Pathway Analysis (IPA) tool or the biological functions. These proteins were involved in common cellular pathways including protein metabolism, signal transduction and cell development. Three proteins were noted to be significantly elevated in CSF and brain tissue, namely ubiquitin carboxyl-terminal hydrolase isozyme L1 (UCH-L1), tyrosine hydroxylase, and syntaxin-6 (UCH-L1) was suggested as a potential biomarker for TBI in this model given that it has been suggested to be elevated in serum and CSF of human TBI patients by several previous studies (Berger et al., 2012; Brophy et al., 2011b; Mondello et al., 2012b). Opii et al. (2007) used CCI model of TBI in young adult rats to investigate the role of mitochondrial oxidation abnormalities in TBI (Opii et al., 2007). Mitochondria were isolated and differential proteomics analysis was performed using 2D-DIGE-MS. The group identified several proteins that were oxidatively modified in the hippocampi and cortices of rats post-TBI including pyruvate dehydrogenase, voltage-dependent anion channel, fumaratehydratase 1, ATP synthase, and prohibitin in the cortex and cytochrome C oxidase Va, isovaleryl coenzyme A dehydrogenase, enolase-1, and glyceraldehyde-3-phosphate dehydrogenase in the hippocampus. Kochanek et al. (2006) studied the change in protein expression 2 weeks post-TBI in rats subject to CCI using 2D-DIGE-MS. Protein identification and function analysis using bioinformatics tools showed significant changes in proteins involved in glial and neuronal stress, oxidative metabolism, calcium uptake and neurotransmitter function. These proteins were further investigated for their possible implication in hippocampal plasticity and cognitive dysfunctions post-TBI (Kochanek et al., 2006). In an attempt to investigate pathways implicated in neuronal damage and degeneration post-TBI, Jenkins et al. (2002) used gel-based MS proteomics in a rat model of moderate CCI studying the changes in protein expression profile of hippocampal neurons after TBI (Jenkins et al., 2002). Their investigation using conventional and functional genomics revealed the implication of protein kinase B (PKB) signal transduction pathway in the pathogenesis of TBI as the substrates of PKB showed altered levels of phosphorylation after injury.

Knowing that aging is an important factor that influences neurodegeneration (Williams, 1995), it can act as a confounder in the study of patients' response to TBI and account for the variable response to TBI with age. Mehan et al. (2011) used a rat model of CCI using rats of all age groups (juvenile, adult and geriatric), and performed proteomics investigation for differential protein expression using 2D-DIGE-MS on neocortical tissue (Mehan and Strauss, 2011). Results have shown the involvement of 13 gene products in the age-related response to TBI including T-kininogen 1, β-actin higher in the geriatric group, collapsin response mediating protein-2 (CRMP2) that were higher in adults than the elderly, and serine protease inhibitors that were higher in the adults than juvenile and elderly. However, apolipoprotein E was upregulated post-TBI in all age groups. These results could provide an insight into the underlying mechanism of differential vulnerability of age groups to TBI, as well as to investigate universal markers of TBI that are not age-related.

The implication of oxidative stress in the pathogenesis of TBI brought up the possibility of anti-oxidant therapy using gamma-glutamylcysteine ethyl ester (GCEE) (Reed et al., 2009). Proteomics tool, namely 2D-DIGE-

MS, was used by Reed et al. (2009) to assess the therapeutic value of GCEE in the treatment of TBI through assessing the effect of GCEE on the protein expression profile difference between rats subject to brain injury with or without GCEE treatment. Results showed that the untreated group showed 19% increase in protein nitration in the brain; however, protein nitration was reduced to below controls with the administration of GCEE. Proteins protected from nitration included synapsin 1, gamma enolase, guanosine diphosphate dissociation inhibitor 1 (GDP), phosphoglyceratemutase 1, heat shock protein 70, ATP synthase, and α-spectrin. This suggests that GCEE could be of a potential therapeutic value in TBI and warrants further investigations.

Rat cortical and glial primary cultures were also used to study the changes in the cellular proteome after TBI (He et al., 2010; Siman et al., 2004). He et al. (2010) used primary astrocyte cultures from rats subject to fluid percussion injury and identified through 2D-DIGE-MS identified five proteins to exhibit significant dynamic changes after injury. These proteins were cofilin 1, destrin, phosphoglyceratemutase 1, NADH dehydrogenase (ubiquinone) 1 alpha subcomplex 10, annexin 1.

In addition to animal models of TBI that suffer the limitations of validation and translation to man, 2D-DIGE-MS was also used in human studies of protein expression changes in both tissue and bio-fluids after TBI. Yang et al. (2009) studied the differential expression of human brain sample from the prefrontal cortex (PFC) of 11 injured patients compared to glioma patients (Yang et al., 2009). Seventy one proteins were identified and investigated for their functions. The main functions attributes to these discovered proteins were cell cytoskeleton, metabolism and oxidative stress, protein turnover, signal transduction and electron transport. Further investigation of these detected proteins can reveal major pathways involved in the pathogenesis of TBI and can give an insight about the identity of corresponding serum or CSF degradation products. Despite this study by Yang et al. 2009, the use of brain tissue from patients with TBI for research investigation is a rare occasion in the literature. Invariably, human neuroproteomics, in contrast to other tissue proteomics, targets patients' bio-fluids (serum and CSF) for protein biomarker discovery. Cadosch et al. (2010) studied the serum and CSF of patients with TBI compared to controls using 2D-DIGE that showed differential expression of serum and CSF proteins that included proteins that can bind to human osteo-progenitor cells (Cadosch et al., 2010). The study identified unique proteins in the serum and CSF of TBI patients, and demonstrated that some of these proteins are osteo-inductive and may be involved in fracture healing. Gao et al. (2007) used 2D-DiGE-MS to investigate the difference between inflicted and non-inflicted injury in pediatric TBI (Gao et al., 2007). Comparison of the CSF from samples of the two groups revealed a four-fold increase in acute phase reactants and Haptoglobin in the non-inflicted group; however, the levels of prostaglandin D2 synthase and cystatin C were 12-fold more elevated in the inflicted group. Moreover, Conti et al. (2004) investigated markers of severe TBI using 2D gel electrophoresis followed by MS, and found that two proteolytic degradation products of fibrinogen beta were present only in TBI patients (Conti et al., 2004).

Gel Free Mass Spectrometry

As mentioned previously, Gel free MS is usually associated with LC separation for protein detection and identification. As 2D-DIGE, this proteomics tool has been widely used in animal models, cell cultures and human samples. In a model of mouse cortical lesion, Wishart et al. (2012) used gel-free MS with iTRAQ on synapse-enriched brain tissue samples to detect differential synaptic protein expression in TBI (Wishart et al., 2012). The investigation revealed 47 proteins six of which are potential regulators of synaptic and axonal degeneration *in vivo* and maybe involved in the pathophysiology of the disease. Protein enrichment techniques are important adjuvants to proteomics tools to overcome the limitations of masking low abundance proteins. Garland et al. (2012) also used this concept to prepare axoplasm-enriched samples of rodent optic nerve post-injury (Garland et al., 2012). Sample was analyzed by MS/MS assisted by iTARQ for quantification and results showed altered expressions of more than 300 proteins. The highest frequency of increased expression was for actin cytoskeleton proteins that showed increased expression 24 and 48 hours post-injury introducing actin cytoskeleton as a novel point of regulation in axon degeneration. Analysis of these proteins also incriminated RhoA pathway that regulates actin cytoskeleton in pathogenesis of brain injury.

MS/MS coupled to iTRAQ was also used by Crawford et al. (2012) on a mouse CCI model of mild and severe TBI (Crawford et al., 2012), and results were further analyzed for their function using Ingenuity Pathway Analysis (IPA). Results showed a temporal difference in plasma protein levels between mild and severe TBI with acute elevation of serum protein levels in mild TBI possibly indicating a reparatory process. Further investigation compared transgenic mice with favorable outcome, i.e., has human APOE3 genotype and mice with unfavorable outcome, i.e., has human APOE4 genotype. This technique allows for the detection of prognostic biomarkers for TBI that can distinguish the two outcomes.

Since different areas of the brain can be the site of TBI sequel as already mentioned, hippocampal tissue was investigated in addition to tissues from neocortex. Wu et al. (2012a) investigated the altered regulation of proteins in the CA3 sub-region of the hippocampus using a fluid percussion model of TBI in rats. They used LC-MS/MS and discovered 1002 dysregulated proteins. An interesting finding was the involvement of calcineurin regulatory subunit, CANB1, and its catalytic binding partner PP2BA, that were decreased in TBI without changes in other calcineurin subunits. Tandem cation-anion exchange chromatography–gel electrophoresis, followed by reversed-phase LC-MS/MS was used by Ottens et al. (2010) to detect common protein markers of traumatic and ischemic brain injury. Comparison of protein profiles between a CCI and middle cerebral artery occlusion rat models revealed common upregulated proteins like albumin, degraded spectrin, and fetuin β, and common downregulated proteins like enolase α, GAPDH, aconitase 2, transgelin-3, aldolase, and MAP2 (Ottens et al., 2010). Less recently, Haskins et al. (2005) used the same technique applied to ipsilateral hippocampal tissue of rats following CCI (Haskins et al., 2005). Haskins et al., 2005 identified a protein profile for TBI that contains more than 100 differentially regulated proteins to be investigated for biomarker discovery. They also confirmed previously established biomarkers like alphaII-spectrin, brain creatine kinase, and neuron-specific enolase. Putative biomarkers like CRMP-2, synaptotagmin, and alphaII-spectrin were also suggested by (Kobeissy et al., 2006) while studying differentially expressed proteins in CCI rat model using the same technique.

Gel free MS techniques were applied to different types of CNS primary cell cultures in search for differential protein expression in response to cytotoxic and apoptotic challenges that are hypothesized to occur post-TBI. For example, Guingab-Cagmat et al. (2012) subjected rat neuronal glial culture to excitotoxic and apoptotic challenges, and identified, using MS, several proteins to be differentially expressed post injury. Of note was GFAP that was proposed as a surrogate biomarker which is degraded post-injury and its degradation products can be detected in serum. In addition, Kim et al. (2012) used gel-free LC-MS/MS to assess the changes in protein profile in response to moderate hypothermia. In their study, moderate hypothermia appeared to have a significant influence on protein profile expression of brain glial cells especially those related to inflammation. Identified proteins may have a possible role in the protective mechanism of hypothermia in response to brain injury.

Studies using Gel free MS on human samples and fluids were also reported. For instance, Lakshmanan et al. (2010) analyzed the microdialysate of patients with severe TBI 96 hours after their injury. Proteins were enriched by magnetic beads and phospho-peptides and analyzed by LC-MS/MS. Differential protein expression was detected among those patients where higher levels of cytoskeletal proteins were expressed in patients with severe metabolic distress. Alternatively, Haqqani et al. (2007) studied the serum protein profile of pediatric patients with severe TBI using ICAT followed by tandem MS. Ninety five differentially expressed proteins were identified, and many of which were compared to the expression pattern of S100B, an established biomarker for brain injury. Eventually, several proteins were identified to exhibit comparable expression profile to S100B such as β-2 Glycoprotein I precursor and Neurofilament triple H protein.

Protein Arrays

The application of protein arrays and high-throughput immunoblotting (HTPI) in neuroproteomics is still limited to few studies. Shu et al. (2011) studied a closed brain injury rate model to identify differential protein expression in serum and hippocampus post-injury. Shu et al. (2011) utilized two types of protein arrays; a weak cationic exchanger (WCX2) chips and immobilized metal affinity capture arrays-Cu (IMAC-Cu) chip. WCX2 chips allowed the identification of 10 differentially expressed proteins and the IMAC-Cu identified 13 differentially expressed proteins; those proteins are suggestive putative biomarkers for TBI. Similarly, Kwon et al. (2011) used protein microarrays applied to both serum and brain tissue to blast injury rat model that were subjected to stress. The study revealed a protein expression profile that allows distinguishing stressed samples from samples with both TBI and stress. An interesting study done by Liu et al. (2006) aimed to compare rat hippocampal lysate in rat post-TBI to the *in vitro* calpain-2 and caspase-3 degradation profile. Liu used HTPI with 1000 monoclonal antibodies and identified 48 proteins to be downregulated in TBI, 42 of which overlapped with the calpain/caspase degradation profile. These identified proteins may serve as proteolytic targets for proteases post-TBI. To compare the difference in protein expression profile between traumatic and ischemic brain injury, Yao et al. (2008) utilized HTPI to identify the differential expression of proteins across two models of acute brain injury, ballistic-like brain injury that mimics TBI and middle cerebral artery occlusion that mimics stroke. Nine hundred and ninety eight proteins were screened; of which, 23 proteins were different across the two groups, and one was differentially expressed. The group verified the results of five proteins through Western blot, namely STAT3, Tau, PKA RIIβ, 14-3-3ε and p43/EMAPII.

Kovesdi et al. (2011) used protein microarrays to assess the effect of physically and socially enriched environment on the neurogenesis after TBI. The group used a blast injury model of TBI, and results have shown that such an environment was successful to ameliorate the long-term effects of TBI; reducing the levels of IL-6 and IFNγ in ventral hippocampus and normalizing the levels of tau and VEGF in dorsal hippocampus.

Nanoproteomics

With the increasing interest in incorporating nanotechnology into "-omics" approaches, nanoproteomics was incorporated in the study of TBI. Sjödin et al. (2010) used hexapeptide ligand libraries (HLL) for the enrichment of low abundant proteins then Shotgun proteomics, in combination with isoelectric focusing (IEF) and nano-LC-MS/MS, to characterize the CSF proteome of two TBI patients. The group of 339 proteins, 130 had overlap in the two studied TBI patients, and 45% of the proteins that had been previously associated with changes post-TBI were also verified. These included NSE, GFAP, S100B, and CK-B. Similarly, Hanrieder et al. (2009) used shotgun proteomics based on nano-LC in conjugation with MALDI-TOF-MS/MS to analyze the temporal profile of protein expression in three patients post-TBI. A series of ventricular CSF samples were analyzed and detected the upregulation of GFAP and NSE in addition to acute phase reactants post-injury.

In addition, new techniques have been studied more recently for the analysis of protein contents of CSF microdialysate for biomarker detection. In the study done by Dahlin et al. (2012) proteins were adsorbed to a surface-modified catheter, followed by on-surface enzymatic digestion in order to identify and quantitate proteins using isobaric tags (iTRAQ) and nanoLC-MS/MS analysis.

Evaluation of the Results of Proteomics Studies

As has been extensively reviewed before in the literature (Lista et al., 2012; Reinders et al., 2004), proteomics techniques still lag behind the promising aims that proteomics wishes to attain. General imitations include (1) inability of these techniques to identify the entire proteome complement in the body and create a unified proteome database (Perez-Riverol et al., 2013), (2) the dynamic range of protein abundance in body fluids and tissue (Zubarev, 2013), (3) interference of various other analytes like lipids, electrolytes and others, (4) absence of robust validation techniques for the findings of these studies, and non-reproducibility of findings (Alawieh et al., 2012). Table 1.1 illustrates the different challenges in proteomics application in TBI biomarker research.

Nevertheless, studies utilizing proteomics in neuro-trauma suffer additional limitations. Many of the reported studies above did not validate the proteomics findings via Western blot or other techniques. Despite the need for validation, these studies still provide a pool of putative biomarkers for others to assess and validate. Another major limitation is the inability of the majority of reported studied to demonstrate that the findings are specific for TBI and not associated with other sequelae like inflammation and bone fracture. Human subjects that were studied for CSF and tissue samples were not always compared to healthy controls but to subjects with other diseases (Yang et al., 2009). There is also a non-uniformity of sample acquisition when it comes to CSF studies. Samples can be withdrawn from ventricles or lumbar puncture, the former usually used for patients with severe injury and the latter for mTBI patients. However, the CSF has differential protein abundance between ventricles and the lumbar levels (Hu et al., 2005). Therefore, this differential abundance if not controlled or corrected for may be the source of the difference in protein profiles and not the disease itself. For example, tau protein was reported by several studies to be differentially expressed comparing ventricular CSF samples of severe TBI and lumbar puncture CSF of control groups; however, this could be due to the normal difference in Tau levels between ventricular samples and lumbar puncture as reported by other studies (Zetterberg et al., 2013). In addition, many of these studies did not account for age-related response where age can act as an important determinant of response to injury post-trauma (Williams, 1995). Once applied to animal models, most of these studies did not demonstrate the translation potential of the findings to human; consequently, a huge effort still resides in animal-to-human translation. Statistically, the majority of these studies has lower sample size due to availability and cost eventually resulting in reduced power of the study and increased incidence of type one errors.

Despite that, studies tend to report differential protein expression in patients with TBI vs. controls; however, few of these studies dwell on the clinical value of the results and their possible prognostic or diagnostic applications in clinical practice. This scientific enthusiasm in putative biomarker discovery rather than validation is still a major limitation in proteomics research specifically as well as in the entire field of "-omics" research (Alawieh et al., 2012).

Data Mining and Neurosystems Biology Analysis

Systems biology is an approach to build a holistic, systematic and unbiased understanding (Feala et al., 2013) of the structural and behavioral model pertaining to biological networks. The exploration and analysis of the different models are based on bioinformatics data collected from genomics, proteomics, transcriptomics and metabolomics studies. The collected data cannot be manually treated and studied due to its abundance, thus data mining using computational tools is a must in order to handle the huge amount of the collected datasets. From the exhaustive work on the collected data, dynamic models are proposed to reflect the genetic regulatory networks, the protein-protein interaction networks, the metabolic networks and canonical biological networks (Alawieh et al., 2012). The developed model is suggestive of expected cellular physiology, with candidate components and networks that can optimize experimental search and investigation and provide new data to be further incorporated in the model and holistic approach (Alawieh et al., 2012).

In vitro experiments on brain biopsy to simulate TBI or performing clinical trials for developing treatments for TBI is not an option. An ideal biomarker would be a perfect solution, but since such a biomarker has not yet been proven to exist, the best of what we can do is to identify a panel or signature of markers (Feala et al., 2013) and try to find out accurate networks that can describe and simulate the pathophysiology of TBI.

The challenge of treating and anticipating secondary damage in TBI can be handled by systems biology through constructing a system scale model to detect and treat TBI, to handle the observations and come up with hypothesis that can be tested and fed back to the system to optimize the results of the model and get insight into the underlying pathophysiology and possible diagnostic and therapeutic modules.

In the context of proteomics, systems biology tools can be incorporated in several ways to overcome many of the limitations of simple proteomics studied. Discovered proteins through common proteomics approaches can be mapped, using protein databases. Many online systems biology resources are available to help building the proteomics and canonical networks of systems biology model into the associated pathways in the cell including cell division, apoptosis, energy metabolism, oxidative stress and others. For instance, Boutte et al. (2012) used IPA software to map 321 differentially expressed proteins obtained by mass spectrometry into relevant cellular pathways.

The importance of the integration that systems biology apply to form hypothesis and propose candidate biomarkers is that the system level would give value to some biomarkers that a human would find of no value at a small scale and may be proven to be of essential role at a system scale. Such a finding could not be realized without application of systems biology. Furthermore, the complexity of the pathways and protein-protein interactions that occurs in the secondary stage, along with the presence of a cascade of injury over a wider area of the brain (diffuse axonal damage). All of these secondary phase TBI diseases are difficult to be studied and interpreted to suggest hypotheses and new biomarkers without a systematic model that takes into accounts the dynamicity of the system along with the different interactions among components. It is through systems biology network analysis also that we can find major players in relevant pathogenic pathways to be identified as putative targets of novel biomarker based theranostics.

Clinical Relevance of Suggested Biomarkers

The immense investigation of protein biomarkers in TBI has resulted in several putative diagnostic and prognostic markers; however, none of these biomarkers is yet approved by FDA for adoption in clinical practice (Robinson et al., 2009). Most of the application of putative biomarkers is restricted to clinical trials, and several algorithms were proposed to incorporate their use in clinical practice. The clinical status of a group of significant biomarkers for TBI will be reviewed here and are summarized in Table 1.2.

Table 1.2. The clinical relevance of major TBI biomarkers.

Prominent TBI biomarkers in scientific literature			
Biomarker	**Location**	**Findings**	**Process involved**
S100B	Serum, CSF	Increased in TBI patients, high negative predictive value toward CT findings, yet not specific for TBI	Astroglial injury/BBB damage
Neuron Specific Enolase (NSE)	Serum, CSF	Increased in serum and CSF of TBI patients; low sensitivity and specificity; limited utility due to effect of hemolysis	Neuronal Injury
Glial Fibrilllary Acidic Protein and BDPs	Serum, CSF	Increased in serum and CSF of TBI patients; controversy toward specificity as biomarker; CNS specific protein; best if used in combination with NSE and S100B	Astroglial injury
Tau Protein	Serum, CSF	Good outcome prediction in patients with severe TBI; limited utility in mTBI	Axonal Injury
AlphaII-Spectrin-SBDP	CSF, Serum	Possible predictor of outcome in severe TBI; not studied in mild TBI	Axonal Injury/ Cell death
UCH-L1	CSF, Serum	Useful in assessment of patients with severe TBI; better outcome prediction if measured with α-II spectrin; limited studies in mTBI	Neuronal Injury
MAP2a/2b	Serum	Recently studied; limited evidence; reported to increase in the sera of patients with severe TBI	Neuronal Injury

S100B

S100B is a low molecular weight calcium-binding protein important in intracellular calcium regulation. It was thought to be specific to astrocytes but later discovered to be present in oligodendrocyte and other extra-cerebral cell types such as adipocytes, chondrocytes, skeletal muscles and bone marrow cells (Berger et al., 2006; Donato, 2001; Olsson et al., 2011). S100B is the earliest and most extensively studied biomarker for TBI, and most of published studies examined its increased level in serum as a putative marker of TBI. However, S100B donot cross the Blood Brain Barrier (BBB), and its presence in the serum is dependent on disruption of the integrity of BBB (Herrmann et al., 2000). Elevated levels of S100B have been highly linked to astroglial injury. The first study to emphasize the role of serum S100B in mTBI patients was done by Ingebrigtsen et al. (1995). The study showed that elevated serum S100B levels in patients with negative CT findings is associated with the occurrence of post-concussive symptoms (Ingebrigtsen et al., 1995). Several other studies, since then, have investigated the clinical prognostic value of elevated serum S100B levels in TBI patients with conflicting evidence (Bazarian et al., 2006; De Kruijk et al., 2001; Egea-Guerrero et al., 2012; Ingebrigtsen et al., 2000; Ingebrigtsen et al., 1999; Mercier et al., 2013; Muller et al., 2007; Schiavi et al., 2012; Spinella et al., 2003; Unden et al., 2007; Vos et al., 2010; Vos et al., 2004). Noteworthy, (Undén and Romner, 2010), did a meta-analysis of articles studying mild head injury comparing CT findings and S100B in the acute phase of injury. The group found 12 eligible articles with a total 2466 patients, and discovered a high sensitivity of low levels of S100B in the prediction of negative CT findings. They suggested that a low serum S100B level (<0.10 µg/L) in the first three hours after injury has more than 90% negative predictive value of the presence of clinically relevant CT findings. Similar findings were also reported by other studies using large samples of patients suggesting the use of serum S100B as a substitute for CT in assessment of mTBI patients (Biberthaler et al., 2006; Zongo et al., 2012).

Even if those studies demonstrate the sensitivity of the use of S100B as a biomarker for TBI, the main limitation towards its use is the lack of specificity to brain trauma especially that S100B can be released by cells other than astrocytes (Berger et al., 2006). Several studies have demonstrated the elevation of S100B in bone fractures without head injury (Anderson et al., 2001; Routsi et al., 2006; Undén et al., 2005).

Despite the abundance of studies reporting serum S100B elevation, studies of CSF levels of S100B in TBI is still limited (Zetterberg et al., 2013). In the study of Neselius et al. (2012) brain injury was found to trigger the release of some biomarkers into the CSF including S100B and other proteins. This; however, does not defy the use of S100B as a screening agent due to its high sensitivity.

Neuron Specific Enolase

Neuron-Specific Enolase (NSE) is an isozyme of glycolytic enzyme enriched in neuronal cell body (Olsson et al., 2011). After its isolation in brain tissue and peripheral neurons, NSE was found to be also expressed in erythrocytes, platelets neuroendocrine cells and oligodendrocyte (Kövesdi et al., 2010). Yet, it was investigated as a biomarker indicating neuronal damage and a possible predictor of TBI outcome.

The levels of NSE were studied in both serum and ventricular CSF. NSE was found to be a predictor of outcome after TBI, especially in patients with severe injury, yet unsatisfactory specificity and sensitivity have been reported (Böhmer et al., 2011; Fridriksson et al., 2000; Geyer et al., 2009; Meric et al., 2010; Skogseid et al., 1992; Topolovec-Vranic et al., 2011). In the studies reporting the utility of NSE as a biomarker for brain trauma, lower sensitivities and specificities than S100B were detected (Herrmann et al., 2000; McKeating et al., 1998; Meric et al., 2010; Topolovec-Vranic et al., 2011). Eventually, it is proposed that NSE is not to be used as a standalone screening biomarker for brain injury (Topolovec-Vranic et al., 2011). The limited utility of NSE as a biomarker of brain trauma may also be related to the high sensitivity of NSE to hemolysis (Ramont et al., 2005).

Glial Fibrillary Acidic Protein

Glial Fibrillary Acidic Protein (GFAP) is an intermediate filament that is believed to be exclusively expressed by astroglia (Olsson et al., 2011). GFAP was studied in both CSF and sera of patients with TBI (Böhmer et al., 2011; Honda et al., 2010; Nylen et al., 2006; Vos et al., 2010; Zetterberg et al., 2013).

Several studies have reported a range of predictive value and specificity. This is likely due to different in ELISA methods used (Honda et al., 2010; Metting et al., 2012; Vos et al., 2010). More recently, GFAP have been reported to be processed into breakdown products (Guingab-Cagmat et al., 2012; Zoltewicz et al., 2012). Detection of GFAP and its breakdown products with a new ELISA format has been reported to detect both mild-moderate TBI (Papa et al., 2012a) and the full spectrum of TBI (TRACK-TBI cohort) (Okonkwo et al., 2013) in two independent studies. Another recent follow up paper with the TRACK-TBI cohort shows that the combination of UCH-L1 with GFAP/BDP further improves its diagnostic utilities (Diaz-Arrastia et al., 2013).

Tau is an axonally enriched microtubule associated protein and one of the best established CSF biomarkers of axonal damage (Zetterberg et al., 2013). TBI was reported to cause the cleavage of tau protein and elevation of levels of cleaved-tau (C-tau) in CSF and serum (Gabbita et al., 2005). The level of C-tau protein in serum and CSF was studied similar to other biomarkers and was associated with both disruption of BBB and cleavage of tau protein post injury (Gabbita et al., 2005). Studies by Zemlan et al. (2002) and Franz et al. (2003) have demonstrated the significance of C-tau in prediction if outcome in patients with severe TBI (Öst et al., 2006). Similarly, other studies reported the utility of C-tau in the prediction of outcome in mTBI (Bulut et al., 2006; Wuthisuthimethawee et al., 2013). However, other studies have reported the poor ability of tau protein to predict outcome and post-concussion syndrome in mTBI (Bazarian et al., 2006; Ma et al., 2008). This limits the utility of C-tau in the outcome prediction of patients with mTBI.

AlphaII-spectrin Breakdown Products (SBDPs)

αII-spectrin degradation products is among the novel biomarkers studied for their clinical relevance in TBI (Zetterberg et al., 2013). αII-spectrin is a cytoskeletal protein enriched in neuronal axons and presynaptic terminals (Berger et al., 2012). While αII-spectrin is present in various nucleated cells, and most tissues, but its high abundance and enrichment of brain still make it a candidate biomarker, especially if used in combination with another more brain-specific marker (Zhang et al., 2011). The breakdown products of αII-spectrin (SBDPs) is thought to be due to the activation of calpain and caspase in the brain after TBI, and thus reflects axonal damage (Pike et al., 2004). SBDP150 and SBDP145 are characteristics of calpain activation, often associated in acute necrotic neuronal cell death while SBDP120 is generated by action of caspase-3 and is affiliated with delayed apoptotic neuronal death (Wang, 2000; Zhang et al., 2009).

Mainly, αII-spectrin was studied in the context of severe rather than mild TBI. Elevation of levels of αII-spectrin degradation products SBDP150 and/or SBDP145 in CSF was reported as a possible outcome predictor in patients with severe TBI (Cardali and Maugeri, 2006; Mondello et al., 2010; Pineda et al., 2007).

Ubiquitin C-terminal Hydrolase (UCH-L1)

Ubiquitin C-terminal Hydrolase (UCH-L1) is a deubiquitinase highly expressed in neuronal cells that is another recently discovered candidate biomarker from a rat TBI model-based differential proteomic study

(Kobeissy et al., 2006). In addition, its high brain specificity and abundance in brain tissue makes it an attractive candidate marker (Brophy et al., 2011a). Similar to αII-spectrin, UCH-L1 CSF and serum levels were found to be elevated in patients with severe TBI correlating with the severity and outcome of injury (Brophy et al., 2011a; Mondello et al., 2012c; Papa et al., 2010; Siman et al., 2009). The elevation of levels of UCH-L1 post-TBI is proposed to be secondary to BBB dysfunction (Blyth et al., 2011). In addition, several recent studies also demonstrated the detectability of UCH-L1 in blood following mild TBI (Diaz-Arrastia et al., 2013; Papa et al., 2012b) and mild TBI injury (Siman et al., 2008).

Together with the breakdown products of α-II spectrin, the serum levels of UCH-L1 was found to change in a similar manner to S100B and GFAP post-injury and to be a an important predictor of outcome in patients with moderate to severe brain injury (Berger et al., 2012; Mondello et al., 2011).

The study of UCH-L1 utility in mTBI is still limited; one study by Papa et al. (2012b) reported that UCH-L1 was identified in the sera of patients with mild to moderate TBI, and its levels correlated with traditional clinical assessment. It was reported that using a cutoff level of 0.09 ng/ml, 100% sensitivity was achieved with 21% specificity demonstrating the negative predictive potential of the test. Nevertheless, the utility of UCH-L1 in mTBI still needs further clinical assessment.

Microtubule Associated Protein-2 (MAP-2)

MAP-2 is dendritically enriched neuronal cytoskeletal protein with two isoforms (MAP-2a and MAP-2b) (Conde and Cáceres, 2009). MAP-2 was reported to be degraded in hippocampal tissue post-TBI so that its degradation products would be suggestive biomarkers for TBI (TAFT et al., 1992). Recently, Mondello et al. (2012a) studied the levels of MAP-2 in patients with severe TBI. This study demonstrated the elevation of the levels of MAP-2 degradation products in patients with severe TBI six months after injury suggesting that MAP-2 could be involved in the chronic neuronal changes that occur post-TBI. However, this remains a solitary study that reported elevation of levels of MAP-2; thus, MAP-2 is still early on the road of application in clinical practice.

Conclusion

Currently, the applications of proteomics in the field of neuro-critical care have failed to provide new FDA-approved biomarkers. However, future work aiming at overcoming the aforementioned limitations of the applied techniques together with better statistical regulation of the analysis and reporting of results, and a more reproducible and robust approaches, still hold a lot of capacity to step up our current understanding of TBI, suggest new therapeutic approaches, and provide a bases for personalized medicine. In this context, basic science researchers, clinicians, epidemiologists, biostatisticians, engineers and mathematics experts together with authoritarian agencies should bring hands together in order to provide a new platform for cooperative work that will ultimately lead to new discoveries.

References

Alawieh, A., F.A. Zaraket, J.L. Li, S. Mondello, A. Nokkari, M. Razafsha, B. Fadlallah, R.M. Boustany and F.H. Kobeissy. 2012. Systems biology, bioinformatics, and biomarkers in neuropsychiatry. *Front Neurosci.* 6: 187.

Anderson, R.E., L.-O. Hansson, O. Nilsson, R. Dijlai-Merzoug and G. Settergren. 2001. High serum S100B levels for trauma patients without head injuries. *Neurosurgery.* 48: 1255–1260.

Angel, T.E., U.K. Aryal, S.M. Hengel, E.S. Baker, R.T. Kelly, E.W. Robinson and R.D. Smith. 2012. Mass spectrometry-based proteomics: existing capabilities and future directions. *Chem Soc Rev.* 41: 3912–3928.

Bayir, H., V.A. Tyurin, Y.Y. Tyurina, R. Viner, V. Ritov, A.A. Amoscato, Q. Zhao, X.J. Zhang, K.L. Janesko-Feldman and H. Alexander. 2007. Selective early cardiolipin peroxidation after traumatic brain injury: an oxidative lipidomics analysis. *Ann Neurol.* 62: 154–169.

Bazarian, J.J., F.P. Zemlan, S. Mookerjee and T. Stigbrand. 2006. Serum S-100B and cleaved-tau are poor predictors of long-term outcome after mild traumatic brain injury. *Brain Injury:* [BI]. 20: 759–765.

Becker, M., J. Schindler and H.G. Nothwang. 2006. Neuroproteomics—the tasks lying ahead. *Electrophoresis.* 27: 2819–2829.

Berger, R.P., R.L. Hayes, R. Richichi, S.R. Beers and K.K. Wang. 2012. Serum Concentrations of Ubiquitin C-Terminal Hydrolase-L1 and αII-Spectrin Breakdown Product 145 kDa Correlate with Outcome after Pediatric TBI. *J Neurotraum.* 29: 162–167.

Berger, R.P., K. Hymel and W.-M. Gao. 2006. The use of biomarkers after inflicted traumatic brain injury: Insight into etiology, pathophysiology, and biochemistry. *Clinical Pediatric Emergency Medicine.* 7: 186–193.

Biberthaler, P., U. Linsenmeier, K.-J. Pfeifer, M. Kroetz, T. Mussack, K.-G. Kanz, E.F. Hoecherl, F. Jonas, I. Marzi and P. Leucht. 2006. Serum S-100B concentration provides additional information fot the indication of computed tomography in patients after minor head injury: a prospective multicenter study. *Shock.* 25: 446–453.

Bigler, E.D. 2001. Quantitative magnetic resonance imaging in traumatic brain injury. *J Head Trauma Rehab.* 16: 117–134.

Biomarkers Definitions Working Group. 2001. Biomarkers and surrogate endpoints: preferred definitions and conceptual framework. *Clinical Pharmacology and Therapeutics.* 69: 89–95.

Blonder, J., H.J. Issaq and T.D. Veenstra. 2011. Proteomic biomarker discovery: It's more than just mass spectrometry. *Electrophoresis.* 32: 1541–1548.

Blumbergs, P.C., G. Scott, J. Manavis, H. Wainwright, D. Simpson and A. McLean. 1994. Stalning af amyloid percursor protein to study axonal damage in mild head Injury. *Lancet.* 344: 1055–1056.

Blyth, B.J., A. Farahvar, H. He, A. Nayak, C. Yang, G. Shaw and J.J. Bazarian. 2011. Elevated serum ubiquitin carboxy-terminal hydrolase L1 is associated with abnormal blood–brain barrier function after traumatic brain injury. *J Neurotraum.* 28: 2453–2462.

Böhmer, A.E., J.P. Oses, A.P. Schmidt, C.S. Perón, C.L. Krebs, P.P. Oppitz, T.T. D'Avila, D.O. Souza, L.V. Portela and M.A. Stefani. 2011. Neuron-specific enolase, S100B, and glial fibrillary acidic protein levels as outcome predictors in patients with severe traumatic brain injury. *Neurosurgery.* 68: 1624.

Boutte, A.M., C. Yao, F. Kobeissy, X.C. May Lu, Z. Zhang, K.K. Wang, K. Schmid, F.C. Tortella and J.R. Dave. 2012. Proteomic analysis and brain-specific systems biology in a rodent model of penetrating ballistic-like brain injury. *Electrophoresis.* 33: 3693–3704.

Brophy, G.M., S. Mondello, L. Papa, S.A. Robicsek, A. Gabrielli, J. Tepas III, A. Buki, C. Robertson, F.C. Tortella and R.L. Hayes. 2011a. Biokinetic analysis of ubiquitin C-terminal hydrolase-L1 (UCH-L1) in severe traumatic brain injury patient biofluids. *J Neurotraum.* 28: 861–870.

Brophy, G.M., S. Mondello, L. Papa, S.A. Robicsek, A. Gabrielli, J. Tepas III, A. Buki, C. Robertson, F.C. Tortella and R.L. Hayes. 2011b. Biokinetic analysis of ubiquitin C-terminal hydrolase-L1 (UCH-L1) in severe traumatic brain injury patient biofluids. *J Neurotraum.* 28: 861–870.

Bulut, M., O. Koksal, S. Dogan, N. Bolca, H. Ozguc, E. Korfali, Y. Ilcol and M. Parlak. 2006. Tau protein as a serum marker of brain damage in mild traumatic brain injury: preliminary results. *Adv Ther.* 23: 12–22.

Cadosch, D., M. Thyer, O.P. Gautschi, G. Lochnit, S.P. Frey, R. Zellweger, L. Filgueira and A.P. Skirving. 2010. Functional and proteomic analysis of serum and cerebrospinal fluid derived from patients with traumatic brain injury: a pilot study. *ANZ J Surg.* 80: 542–547.

Cardali, S. and R. Maugeri. 2006. Detection of alphaII-spectrin and breakdown products in humans after severe traumatic brain injury. *J Neurol Sci.* 50: 25.

Cernak, I. and L.J. Noble-Haeusslein. 2009. Traumatic brain injury: an overview of pathobiology with emphasis on military populations. *J Cerebr Blood F Met.* 30: 255–266.

Chen, S.S., W.E. Haskins, A.K. Ottens, R.L. Hayes, N. Denslow and K.K.W. Wang. 2007. Bioinformatics for traumatic brain injury: Proteomic data mining. *Data Mining in Biomedicine.* 363–387.

Chen, Z., L.Y. Leung, A. Mountney, Z. Liao, W. Yang, X.-C.M. Lu, J. Dave, Y. Deng-Bryant, G. Wei and K. Schmid. 2012. A novel animal model of closed-head concussive-induced mild traumatic brain injury: development, implementation, and characterization. *J Neurotraum.* 29: 268–280.

Clark, R.S.B., H. Bay, C.T. Chu, S.M. Alber, P.M. Kochanek and S.C. Watkins. 2008. Autophagy is increased in mice after traumatic brain injury and is detectable in human brain after trauma and critical illness. *Autophagy.* 4: 88–90.

Conde, C. and A. Cáceres. 2009. Microtubule assembly, organization and dynamics in axons and dendrites. *Nat Rev Neurosci.* 10: 319–332.

Conley, Y.P. and S. Alexander. 2011. Genomic, transcriptomic, and epigenomic approaches to recovery after acquired brain injury. *PM&R.* 3: S52–S58.

Conti, A., Y. Sanchez-Ruiz, A. Bachi, L. Beretta, E. Grandi, M. Beltramo and M. Alessio. 2004. Proteome study of human cerebrospinal fluid following traumatic brain injury indicates fibrin (ogen) degradation products as trauma-associated markers. *J Neurotraum.* 21: 854–863.

Coronado, V.G., L. Xu, S.V. Basavaraju, L.C. McGuire, M.M. Wald, M.D. Faul, B.R. Guzman and J.D. Hemphill. 2011. Surveillance for traumatic brain injury-related deaths—United States, 1997–2007. *MMWR Surveill Summ.* 60: 1–32.

Corrigan, J.D., A.W. Selassie and J.A. Orman. 2010. The epidemiology of traumatic brain injury. *J Head Trauma Rehab.* 25: 72–80.

Crawford, F., G. Crynen, J. Reed, B. Mouzon, A. Bishop, B. Katz, S. Ferguson, J. Phillips, V. Ganapathi, V. Mathura, A. Roses and M. Mullan. 2012. Identification of plasma biomarkers of TBI outcome using proteomic approaches in an APOE mouse model. *J Neurotraum.* 29: 246–260.

Dahlin, A.P., K. Hjort, L. Hillered, M.O. Sjödin, J. Bergquist and M. Wetterhall. 2012. Multiplexed quantification of proteins adsorbed to surface-modified and non-modified microdialysis membranes. *Anal Bioanal Chem.* 402: 2057–2067.

De Kruijk, J., P. Leffers, P. Menheere, S. Meerhoff and A. Twijnstra. 2001. S-100B and neuron-specific enolase in serum of mild traumatic brain injury patients A comparison with healthy controls. *Acta Neurol Scand.* 103: 175–179.

Deb, S., I. Lyons, C. Koutzoukis, I. Ali and G. McCarthy. 1999. Rate of psychiatric illness 1 year after traumatic brain injury. *Am J Psychiat.* 156: 374–378.

Di Pietro, V., D. Amin, S. Pernagallo, G. Lazzarino, B. Tavazzi, R. Vagnozzi, A. Pringle and A. Belli. 2010. Transcriptomics of traumatic brain injury: gene expression and molecular pathways of different grades of insult in a rat organotypic hippocampal culture model. *J Neurotraum.* 27: 349–359.

Diaz-Arrastia, R., K.K. Wang, L. Papa, M.D. Sorani, J.K. Yue, A.M. Puccio, P.J. McMahon, T. Inoue, E.L. Yuh and H.F. Lingsma. 2013. Acute Biomarkers of Traumatic Brain Injury: Relationship between Plasma Levels of Ubiquitin C-Terminal Hydrolase-L1 and Glial Fibrillary Acidic Protein. *J Neurotraum.*

Donato, R. 2001. S100: a multigenic family of calcium-modulated proteins of the EF-hand type with intracellular and extracellular functional roles. *Int J Biochem & Cell Biology.* 33: 637–668.

Edward Dixon, C., G.L. Clifton, J.W. Lighthall, A.A. Yaghmai and R.L. Hayes. 1991. A controlled cortical impact model of traumatic brain injury in the rat. *J Neurosci Meth.* 39: 253–262.

Edwin, Y., H. Sarah, B. Bo-Michael, A. Doreen and M.K.M. Cristina. 2011. Post-traumatic hypoxia exacerbates neurological deficit, neuroinflammation and cerebral metabolism in rats with diffuse traumatic brain injury. *J Neuroinflamm.* 8.

Egea-Guerrero, J., J. Revuelto-Rey, F. Murillo-Cabezas, M. Munoz-Sanchez, A. Vilches-Arenas, P. Sanchez-Linares, J. Dominguez-Roldan and J. Leon-Carrion. 2012. Accuracy of the S100 β protein as a marker of brain damage in traumatic brain injury. *Brain Injury.* 26: 76–82.

English, J.A., K. Pennington, M.J. Dunn and D.R. Cotter. 2011. The neuroproteomics of schizophrenia. *Biol Psychiat.* 69: 163–172.

Espina, V., E.C. Woodhouse, J. Wulfkuhle, H.D. Asmussen, E.F. Petricoin III and L.A. Liotta. 2004. Protein microarray detection strategies: focus on direct detection technologies. *J Immunol Methods.* 290: 121–133.

European-Commission. The International Initiative for Traumatic Brain Injury Research (InTBIR).

Faul, M., L. Xu, M. Wald and V. Coronado. 2010. Traumatic brain injury in the United States: Emergency department visits, hospitalizations and deaths 2002–2006. Atlanta, GA: Centers for Disease Control and Prevention, National Center for Injury Prevention and Control.

Feala, J.D., M.D. Abdulhameed, C. Yu, B. Dutta, X. Yu, K. Schmid, J.R. Dave, F.C. Tortella and J. Reifman. 2013. Systems biology approaches for discovering biomarkers for traumatic brain injury. *J Neurotraum.*

Feigin, V.L., S. Barker-Collo, R. Krishnamurthi, A. Theadom and N. Starkey. 2010. Epidemiology of ischaemic stroke and traumatic brain injury. Best Practice & Research. *Clinical Anaesthesiology.* 24: 485–494.

Fox, N. and J.H. Growdon. 2004. Biomarkers and surrogates. Neurotherapeutics. 1: 181–181.

Franz, G., R. Beer, A. Kampfl, K. Engelhardt, E. Schmutzhard, H. Ulmer and F. Deisenhammer. 2003. Amyloid beta 1-42 and tau in cerebrospinal fluid after severe traumatic brain injury. *Neurology.* 60: 1457–1461.

Fridriksson, T., N. Kini, C. Walsh-Kelly and H. Hennes. 2000. Serum Neuron-specific Enolase as a Predictor of Intracranial Lesions in Children with Head Trauma: A Pilot Study. *Acad Emerg Med.* 7: 816–820.

Fung, E.T., V. Thulasiraman, S.R. Weinberger and E.A. Dalmasso. 2001. Protein biochips for differential profiling. *Curr Opin Biotech.* 12: 65–69.

Gabbita, S.P., S.W. Scheff, R.M. Menard, K. Roberts, I. Fugaccia and F.P. Zemlan. 2005. Cleaved-tau: a biomarker of neuronal damage after traumatic brain injury. *J Neurotraum.* 22: 83–94.

Gao, W.-m., M.S. Chadha, R.P. Berger, G.S. Omenn, D.L. Allen, M. Pisano, P.D. Adelson, R.S. Clark, L.W. Jenkins and P.M. Kochanek. 2007. A gel-based proteomic comparison of human cerebrospinal fluid between inflicted and non-inflicted pediatric traumatic brain injury. *J Neurotraum.* 24: 43–53.

Garland, P., L.J. Broom, S. Quraishe, P.D. Dalton, P. Skipp, T.A. Newman and V.H. Perry. 2012. Soluble axoplasm enriched from injured CNS axons reveals the early modulation of the actin cytoskeleton. *PloS One.* 7: e47552.

Geyer, C., A. Ulrich, G. Gräfe, B. Stach and H. Till. 2009. Diagnostic value of S100B and neuron-specific enolase in mild pediatric traumatic brain injury: Clinical article. *J Neuros-Pediatr.* 4: 339–344.

Ghajar, J. 2000. Traumatic brain injury. *Lancet* (London, England). 356: 923–929.

Gonzalez-Gonzalez, M., R. Jara-Acevedo, S. Matarraz, M. Jara-Acevedo, S. Paradinas, J. Sayagües, A. Orfao and M. Fuentes. 2012. Nanotechniques in proteomics: Protein microarrays and novel detection platforms. *Eur J Pharm Sci.* 45: 499–506.

Green, S.M. 2011. Cheerio, Laddie! Bidding Farewell to the Glasgow Coma Scale. *Ann Emerg Med.* 58: 427.

Guingab-Cagmat, J.D., K. Newsom, A. Vakulenko, E.B. Cagmat, F.H. Kobeissy, S. Zoltewicz, K.K. Wang and J. Anagli. 2012. In vitro MS-based proteomic analysis and absolute quantification of neuronal-glial injury biomarkers in cell culture system. *Electrophoresis.* 33: 3786–3797.

Güzel, A., T. Hiçdönmez, O. Temizöz, B. Aksu, H. Aylanç and S. Karasalihoglu. 2009. Indications for brain computed tomography and hospital admission in pediatric patients with minor head injury: how much can we rely upon clinical findings? *Pediatr Neurosci.* 45: 262–270.

Hall, D.A., J. Ptacek and M. Snyder. 2007. Protein microarray technology. *Mech Ageing Dev.* 128: 161–167.

Hanrieder, J., M. Wetterhall, P. Enblad, L. Hillered and J. Bergquist. 2009. Temporally resolved differential proteomic analysis of human ventricular CSF for monitoring traumatic brain injury biomarker candidates. *J Neurosci Meth.* 177: 469–478.

Haqqani, A.S., J.S. Hutchison, R. Ward and D.B. Stanimirovic. 2007. Biomarkers and diagnosis; protein biomarkers in serum of pediatric patients with severe traumatic brain injury identified by ICAT-LC-MS/MS. *J Neurotraum.* 24: 54–74.

Haskins, W.E., F.H. Kobeissy, R.A. Wolper, A.K. Ottens, J.W. Kitlen, S.H. McClung, B.E. O'steen, M.M. Chow, J.A. Pineda and N.D. Denslow. 2005. Rapid discovery of putative protein biomarkers of traumatic brain injury by SDS-PAGE-capillary liquid chromatography-tandem mass spectrometry. *J Neurotraum.* 22: 629–644.

Haydel, M.J., C.A. Preston, T.J. Mills, S. Luber, E. Blaudeau and P.M.C. DeBlieux. 2000. Indications for computed tomography in patients with minor head injury. *New Engl J Med.* 343: 100–105.

He, T., S.W. Yang, Y. Gup, Z.K. Zhao, X.Y. Chen and Y.L. Zhang. 2010. The study in primary cultured astrocytes following fluid percussion injury. Zhongguo ying yong sheng li xue za zhi = Zhongguo yingyong shenglixue zazhi = Chinese. *J Appl Physiol.* 26: 46–50.

Heltemes, K.J., T.L. Holbrook, A.J. MacGregor and M.R. Galarneau. 2011. Blast-related mild traumatic brain injury is associated with a decline in self-rated health amongst US military personnel. Injury.

Hemphill III, J.C., M.A.S.N. Phan, M.J. Aminoff and J.L. Wilterdink. 2012. Traumatic brain injury: Epidemiology, classification, and pathophysiology.

Hergenroeder, G.W., J.B. Redell, A.N. Moore and P.K. Dash. 2008. Biomarkers in the clinical diagnosis and management of traumatic brain injury. *Mol Diagn Ther.* 12: 345–358.

Herrmann, M., S. Jost, S. Kutz, A.D. Ebert, T. Kratz, M.T. Wunderlich and H. Synowitz. 2000. Temporal profile of release of neurobiochemical markers of brain damage after traumatic brain injury is associated with intracranial pathology as demonstrated in cranial computerized tomography. *J Neurotraum.* 17: 113–122.

Hofman, P.A.M., S.Z. Stapert, M.J.P.G. van Kroonenburgh, J. Jolles, J. de Kruijk and J.T. Wilmink. 2001. MR imaging, single-photon emission CT, and neurocognitive performance after mild traumatic brain injury. *Am J Neuroradiol.* 22: 441–449.

Honda, M., R. Tsuruta, T. Kaneko, S. Kasaoka, T. Yagi, M. Todani, M. Fujita, T. Izumi and T. Maekawa. 2010. Serum glial fibrillary acidic protein is a highly specific biomarker for traumatic brain injury in humans compared with S-100B and neuron-specific enolase. *The Journal of Trauma and Acute Care Surgery.* 69: 104–109.

Hoyt, D.B., J. Holcomb, E. Abraham, J. Atkins and G. Sopko. 2004. Working Group on Trauma Research Program summary report: National Heart Lung Blood Institute (NHLBI), National Institute of General Medical Sciences (NIGMS), and National Institute of Neurological Disorders and Stroke (NINDS) of the National Institutes of Health (NIH), and the Department of Defense (DOD). *The Journal of Trauma and Acute Care Surgery.* 57: 410–415.

Hu, Y., J.P. Malone, A.M. Fagan, R.R. Townsend and D.M. Holtzman. 2005. Comparative proteomic analysis of intra-and interindividual variation in human cerebrospinal fluid. *Mol Cell Proteomics.* 4: 2000–2009.

Hughes, D.G., A. Jackson, D.L. Mason, E. Berry, S. Hollis and D.W. Yates. 2004. Abnormalities on magnetic resonance imaging seen acutely following mild traumatic brain injury: correlation with neuropsychological tests and delayed recovery. *Neuroradiology.* 46: 550–558.

Ingebrigtsen, T., B. Romner, P. Kongstad and B. Langbakk. 1995. Increased serum concentrations of protein S-100 after minor head injury: a biochemical serum marker with prognostic value? *J Neurol Neurosur Ps.* 59: 103.

Ingebrigtsen, T., B. Romner, S. Marup-Jensen, M. Dons, C. Lundqvist, J. Bellner, C. Alling and S.E. Borgesen. 2000. The clinical value of serum S-100 protein measurements in minor head injury: a Scandinavian multicentre study. *Brain Injury:* [BI]. 14: 1047–1055.

Ingebrigtsen, T., K. Waterloo, E.A. Jacobsen, B. Langbakk and B. Romner. 1999. Traumatic brain damage in minor head injury: relation of serum S-100 protein measurements to magnetic resonance imaging and neurobehavioral outcome. *Neurosurgery.* 45: 468–475; discussion 475–466.

Jellinger, K.A., W. Paulus, C. Wrocklage and I. Litvan. 2001. Traumatic brain injury as a risk factor for Alzheimer disease. Comparison of two retrospective autopsy cohorts with evaluation of ApoE genotype. *BMC Neurol.* 1: 3.

Jenkins, L., G. Peters, C. Dixon, X. Zhang, R. Clark, J. Skinner, D. Marion, P. Adelson and P. Kochanek. 2002. Conventional and functional proteomics using large format two-dimensional gel electrophoresis 24 hours after controlled cortical impact in postnatal day 17 rats. *J Neurotraum.* 19: 715–740.

Johnson, V.E., W. Stewart and D.H. Smith. 2012. Axonal pathology in traumatic brain injury. *Exp Neurol.*

Keller, M., D. Enot, M. Urban, U. Felderhoff, U. Kiechl-Kohlendorfer, H. Hagberg, H.P. Deigner, E. Griesmaier, S. Sizonenko and C. Mallard. 2011. Biomarkers of Developmental Brain Injury in Preterm Infants: Reporting on Metabolomics Activities of the Neobrain Consortium. *Pediatr Res.* 70: 29–29.

Kim, J.H., Y.E. Cho, M. Seo, M.C. Baek and K. Suk. 2012. Glial proteome changes in response to moderate hypothermia. *Proteomics.* 12: 2571–2583.

Kirov, I.I., A. Tal, J.S. Babb, Y.W. Lui, R.I. Grossman and O. Gonen. 2012. Diffuse axonal injury in mild traumatic brain injury: a 3D multivoxel proton MR spectroscopy study. *J Neurol.* 1–11.

Knoblach, S. and A. Faden. 2005. Proteases in Traumatic Brain Injury. *Proteases In The Brain.* 79–108.

Kobeissy, F.H., A.K. Ottens, Z. Zhang, M.C. Liu, N.D. Denslow, J.R. Dave, F.C. Tortella, R.L. Hayes and K.K. Wang. 2006. Novel differential neuroproteomics analysis of traumatic brain injury in rats. *Mol Cell Proteomics*. 5: 1887–1898.

Kochanek, A.R., A.E. Kline, W.M. Gao, M. Chadha, Y. Lai, R.S. Clark, C.E. Dixon and L.W. Jenkins. 2006. Gel-based hippocampal proteomic analysis 2 weeks following traumatic brain injury to immature rats using controlled cortical impact. *Dev Neurosci-Basel*. 28: 410–419.

Koepsell, T.D., F.P. Rivara, M.S. Vavilala, J. Wang, N. Temkin, K.M. Jaffe and D.R. Durbin. 2011. Incidence and descriptive epidemiologic features of traumatic brain injury in King County, Washington. *Pediatrics*. 128: 946–954.

Kovesdi, E., A.B. Gyorgy, S.-K.C. Kwon, D.L. Wingo, A. Kamnaksh, J.B. Long, C.E. Kasper and D.V. Agoston. 2011. The effect of enriched environment on the outcome of traumatic brain injury; a behavioral, proteomics, and histological study. *Front Neurosci*. 5.

Kövesdi, E., J. Lückl, P. Bukovics, O. Farkas, J. Pál, E. Czeiter, D. Szellár, T. Dóczi, S. Komoly and A. Büki. 2010. Update on protein biomarkers in traumatic brain injury with emphasis on clinical use in adults and pediatrics. *Acta Neurochir*. 152: 1–17.

Kwon, S.-K.C., E. Kovesdi, A.B. Gyorgy, D. Wingo, A. Kamnaksh, J. Walker, J.B. Long and D.V. Agoston. 2011. Stress and traumatic brain injury: a behavioral, proteomics, and histological study. *Front Neurol*. 2.

Lai, Z.W., Y. Yan, F. Caruso and E.C. Nice. 2012. Emerging Techniques in Proteomics for Probing Nano–Bio Interactions. *ACS Nano*.

Lakshmanan, R., J.A. Loo, T. Drake, J. Leblanc, A.J. Ytterberg, D.L. McArthur, M. Etchepare and P.M. Vespa. 2010. Metabolic crisis after traumatic brain injury is associated with a novel microdialysis proteome. *Neurocrit Care*. 12: 324–336.

Le, T.H. and A.D. Gean. 2009. Neuroimaging of traumatic brain injury. *Mt Sinai J Med: A Journal of Translational and Personalized Medicine*. 76: 145–162.

Lee, B. and A. Newberg. 2005. Neuroimaging in traumatic brain imaging. NeuroRx: *The Journal of The American Society for Experimental Neuro Therapeutics*. 2: 372–383.

Lei, P., Y. Li, X. Chen, S. Yang and J. Zhang. 2009. Microarray based analysis of microRNA expression in rat cerebral cortex after traumatic brain injury. *Brain Res*. 1284: 191–201.

Leibson, C.L., A.W. Brown, J.E. Ransom, N.N. Diehl, P.K. Perkins, J. Mandrekar and J.F. Malec. 2011. Incidence of traumatic brain injury across the full disease spectrum: a population-based medical record review study. *Epidemiology*. 22: 836–844.

Lifshitz, J., P.G. Sullivan, D.A. Hovda, T. Wieloch and T.K. McIntosh. 2004. Mitochondrial damage and dysfunction in traumatic brain injury. *Mitochondrion*. 4: 705–713.

Lindenbaum, G.A., S.F. Carroll, I. Daskal and R. Kapusnick. 1989. Patterns of alcohol and drug abuse in an urban trauma center: the increasing role of cocaine abuse. *J Traum*. 29: 1654–1658.

Lista, S., F. Faltraco and H. Hampel. 2012. Biological and methodical challenges of blood-based proteomics in the field of neurological research. *Prog Neurobiol*.

Liu, M.C., V. Akle, W. Zheng, J.R. Dave, F.C. Tortella, R.L. Hayes and K.K. Wang. 2006. Comparing calpain- and caspase-3-mediated degradation patterns in traumatic brain injury by differential proteome analysis. *Biochem J*. 394: 715–725.

Loane, D.J. and K.R. Byrnes. 2010. Role of microglia in neurotrauma. *Neurotherapeutics*. 7: 366–377.

Loane, D.J., A. Pocivavsek, C.E.H. Moussa, R. Thompson, Y. Matsuoka, A.I. Faden, G.W. Rebeck and M.P. Burns. 2009. Amyloid precursor protein secretases as therapeutic targets for traumatic brain injury. *Nat Med*. 15: 377–379.

López-Otín, C. and C.M. Overall. 2002. Protease degradomics: a new challenge for proteomics. *Nat Rev Mol Cell Bio*. 3: 509–519.

Lye, T.C. and E.A. Shores. 2000. Traumatic brain injury as a risk factor for Alzheimer's disease: a review. *Neuropsychol Rev*. 10: 115–129.

Ma, M., C.J. Lindsell, C.M. Rosenberry, G.J. Shaw and F.P. Zemlan. 2008. Serum cleaved tau does not predict postconcussion syndrome after mild traumatic brain injury. *Am J Emerg Med.* 26: 763–768.

Maas, A.I.R., N. Stocchetti and R. Bullock. 2008. Moderate and severe traumatic brain injury in adults. *Lancet Neurol.* 7: 728–741.

MacBeath, G. 2002. Protein microarrays and proteomics. *Nature Genetics.* 32: 526–532.

Marshall, L.F. 2000. Head injury: recent past, present, and future. *Neurosurgery.* 47: 546–561.

Marshall, L.F., S.B. Marshall, M.R. Klauber, M.B. Clark, H.M. Eisenberg, J.A. Jane, T.G. Luerssen, A. Marmarou and M.A. Foulkes. 1991. A new classification of head injury based on computerized tomography. *Special Supplements.* 75: 14–20.

Masel, B.E. and D.S. DeWitt. 2010. Traumatic brain injury: a disease process, not an event. *J Neurotraum.* 27: 1529–1540.

Mazzeo, A.T., A. Beat, A. Singh and M.R. Bullock. 2009. The role of mitochondrial transition pore, and its modulation, in traumatic brain injury and delayed neurodegeneration after TBI. *Exp Neurol.* 218: 363–370.

McCrea, M., T. Hammeke, G. Olsen, P. Leo and K. Guskiewicz. 2004. Unreported concussion in high school football players: implications for prevention. *Clin J Sport Med.* 14: 13–17.

McCrea, M.A. 2007. Mild Traumatic Brain Injury and Postconcussion Syndrome: The New Evidence Base for Diagnosis and Treatment. Oxford University Press, New York, USA.

McCrory, P.R. and S.F. Berkovic. 1998. Second impact syndrome. *Neurology.* 50: 677–683.

McIntosh, T., R. Vink, L. Noble, I. Yamakami, S. Fernyak, H. Soares and A. Faden. 1989. Traumatic brain injury in the rat: characterization of a lateral fluid-percussion model. *Neuroscience.* 28: 233–244.

McKeating, E., P. Andrews and L. Mascia. 1998. Relationship of neuron specific enolase and protein S-100 concentrations in systemic and jugular venous serum to injury severity and outcome after traumatic brain injury. *Acta Neurochir.* Supplement. 71: 117.

Mehan, M.R., R. Ostroff, S.K. Wilcox, F. Steele, D. Schneider, T.C. Jarvis, G.S. Baird, L. Gold and N. Janjic. 2012. Highly Multiplexed Proteomic Platform for Biomarker Discovery, Diagnostics, and Therapeutics. *Complement Therapeutics.* 283–300.

Mehan, N.D. and K.I. Strauss. 2011. Combined age-and trauma-related proteomic changes in rat neocortex: a basis for brain vulnerability. *Neurobiol Aging.*

Melton, L. 2004. Protein arrays: proteomics in multiplex. *Nature.* 429: 101–107.

Mercier, E., A. Boutin, F. Lauzier, D.A. Fergusson, J.-F. Simard, R. Zarychanski, L. Moore, L.A. McIntyre, P. Archambault and F. Lamontagne. 2013. Predictive value of S-100β protein for prognosis in patients with moderate and severe traumatic brain injury: systematic review and meta-analysis. *BMJ: Brit Med J.* 346.

Meric, E., A. Gunduz, S. Turedi, E. Cakir and M. Yandi. 2010. The prognostic value of neuron-specific enolase in head trauma patients. *J Emerg Med.* 38: 297.

Metting, Z., N. Wilczak, L. Rodiger, J. Schaaf and J. Van Der Naalt. 2012. GFAP and S100B in the acute phase of mild traumatic brain injury. *Neurology.* 78: 1428–1433.

Mitchell, P. 2002. A perspective on protein microarrays. *Nat Biotechnol.* 20: 225–229.

Mittl, R.L., R.I. Grossman, J.F. Hiehle, R.W. Hurst, D.R. Kauder, T.A. Gennarelli and G.W. Alburger. 1994. Prevalence of MR evidence of diffuse axonal injury in patients with mild head injury and normal head CT findings. *AJNR. Am J Neuroradiol.* 15: 1583–1589.

Mondello, S., A. Gabrielli, S. Catani, M. D'Ippolito, A. Jeromin, A. Ciaramella, P. Bossù, K. Schmid, F. Tortella and K.K. Wang. 2012a. Increased levels of serum MAP-2 at 6-months correlate with improved outcome in survivors of severe traumatic brain injury. *Brain Injury.* 26: 1629–1635.

Mondello, S., A. Linnet, A. Buki, S. Robicsek, A. Gabrielli, J. Tepas, L. Papa, G.M. Brophy, F. Tortella and R.L. Hayes. 2012b. Clinical utility of serum levels of ubiquitin C-terminal hydrolase as a biomarker for severe traumatic brain injury. *Neurosurgery.* 70: 666.

Mondello, S., A. Linnet, A. Buki, S. Robicsek, A. Gabrielli, J. Tepas, L. Papa, G.M. Brophy, F. Tortella and R.L. Hayes. 2012c. Clinical utility of serum levels of ubiquitin C-terminal hydrolase as a biomarker for severe traumatic brain injury. *Neurosurgery.* 70: 666–675.

Mondello, S., U. Muller, A. Jeromin, J. Streeter, R.L. Hayes and K.K. Wang. 2011. Blood-based diagnostics of traumatic brain injuries. *Expert Rev Mol Diagn.* 11: 65–78.

Mondello, S., S.A. Robicsek, A. Gabrielli, G.M. Brophy, L. Papa, J. Tepas III, C. Robertson, A. Buki, D. Scharf and M. Jixiang. 2010. αII-spectrin breakdown products (SBDPs): diagnosis and outcome in severe traumatic brain injury patients. *J Neurotraum.* 27: 1203–1213.

Muller, K., W. Townend, N. Biasca, J. Unden, K. Waterloo, B. Romner and T. Ingebrigtsen. 2007. S100B serum level predicts computed tomography findings after minor head injury. *J Traum.* 62: 1452–1456.

Murray, G., G. Teasdale, R. Braakman, F. Cohadon, M. Dearden, F. Iannotti, A. Karimi, F. Lapierre, A. Maas and J. Ohman. 1999. The European brain injury consortium survey of head injuries. *Acta Neurochir.* 141: 223–236.

Narayan, R.K., M.E. Michel, B. Ansell, A. Baethmann, A. Biegon, M.B. Bracken, M.R. Bullock, S.C. Choi, G.L. Clifton, C.F. Contant, W.M. Coplin, W.D. Dietrich, J. Ghajar, S.M. Grady, R.G. Grossman, E.D. Hall, W. Heetderks, D.A. Hovda, J. Jallo, R.L. Katz, N. Knoller, P.M. Kochanek, A.I. Maas, J. Majde, D.W. Marion, A. Marmarou, L.F. Marshall, T.K. McIntosh, E. Miller, N. Mohberg, J.P. Muizelaar, L.H. Pitts, P. Quinn, G. Riesenfeld, C.S. Robertson, K.I. Strauss, G. Teasdale, N. Temkin, R. Tuma, C. Wade, M.D. Walker, M. Weinrich, J. Whyte, J. Wilberger, A.B. Young and L. Yurkewicz. 2002. Clinical trials in head injury. *J Neurotraum.* 19: 503–557.

Neselius, S., H. Brisby, A. Theodorsson, K. Blennow, H. Zetterberg and J. Marcusson. 2012. CSF-biomarkers in Olympic boxing: diagnosis and effects of repetitive head trauma. *PloS One.* 7: e33606.

Nylen, K., M. Öst, L.Z. Csajbok, I. Nilsson, K. Blennow, B. Nellgård and L. Rosengren. 2006. Increased serum-GFAP in patients with severe traumatic brain injury is related to outcome. *J Neurol Sci.* 240: 85–91.

Okonkwo, D.O., J.K. Yue, A.M. Puccio, D. Panczykowski, T. Inoue, P.J. McMahon, M.D. Sorani, E.L. Yuh, H. Lingsma and A. Maas. 2013. GFAP-BDP as an Acute Diagnostic Marker in Traumatic Brain Injury: Results from the prospective TRACK-TBI Study. *J Neurotraum.*

Olsson, B., H. Zetterberg, H. Hampel and K. Blennow. 2011. Biomarker-based dissection of neurodegenerative diseases. *Prog Neurobiol.* 95: 520–534.

Opii, W.O., V.N. Nukala, R. Sultana, J.D. Pandya, K.M. Day, M.L. Merchant, J.B. Klein, P.G. Sullivan and D.A. Butterfield. 2007. Proteomic identification of oxidized mitochondrial proteins following experimental traumatic brain injury. *J Neurotraum.* 24: 772–789.

Organization, W.H. 2009. Global health risks: mortality and burden of disease attributable to selected major risks. World Health Organization.

Öst, M., K. Nylén, L. Csajbok, A.O. Öhrfelt, M. Tullberg, C. Wikkelsö, P. Nellgård, L. Rosengren, K. Blennow and B. Nellgård. 2006. Initial CSF total tau correlates with 1-year outcome in patients with traumatic brain injury. *Neurology.* 67: 1600–1604.

Ottens, A.K., L. Bustamante, E.C. Golden, C. Yao, R.L. Hayes, K.K. Wang, F.C. Tortella and J.R. Dave. 2010. Neuroproteomics: a biochemical means to discriminate the extent and modality of brain injury. *J Neurotraum.* 27: 1837–1852.

Ottens, A.K., F.H. Kobeissy, E.C. Golden, Z. Zhang, W.E. Haskins, S.S. Chen, R.L. Hayes, K.K.W. Wang and N.D. Denslow. 2006. Neuroproteomics in neurotrauma. *Mass Spectrom Rev.* 25: 380–408.

Papa, L., L. Akinyi, M.C. Liu, J.A. Pineda, J.J. Tepas III, M.W. Oli, W. Zheng, G. Robinson, S.A. Robicsek and A. Gabrielli. 2010. Ubiquitin C-terminal hydrolase is a novel biomarker in humans for severe traumatic brain injury*. *Crit Care Med.* 38: 138–44.

Papa, L., L.M. Lewis, J.L. Falk, Z. Zhang, S. Silvestri, P. Giordano, G.M. Brophy, J.A. Demery, N.K. Dixit and I. Ferguson. 2012a. Elevated levels of serum glial fibrillary acidic protein breakdown products in mild and moderate traumatic brain injury are associated with intracranial lesions and neurosurgical intervention. *Ann Emerg Med.* 59: 471–483.

Papa, L., L.M. Lewis, S. Silvestri, J.L. Falk, P. Giordano, G.M. Brophy, J.A. Demery, M.C. Liu, J. Mo and L. Akinyi. 2012b. Serum levels of ubiquitin C-terminal hydrolase distinguish mild traumatic brain injury from trauma controls and are elevated in mild and moderate traumatic brain injury patients with intracranial lesions and neurosurgical intervention. *The Journal of Trauma and Acute Care Surgery.* 72: 1335.

Patterson, Z.R. and M.R. Holahan. 2012. Understanding the neuroinflammatory response following concussion to develop treatment strategies. *Frontiers in Cellular Neuroscience.* 6.

Perez-Riverol, Y., H. Hermjakob, O. Kohlbacher, L. Martens, D. Creasy, J. Cox, F. Leprevost, B.P. Shan, V.I. Pérez-Nueno and M. Blazejczyk. 2013. Computational Proteomics Pitfalls and Challenges: HavanaBioinfo 2012 Workshop Report. *J Proteomics.*

Pike, B.R., J. Flint, J.R. Dave, X.M. Lu, K.K. Wang, F.C. Tortella and R.L. Hayes. 2004. Accumulation of Calpain and Caspase-3 Proteolytic Fragments of Brain-Derived & agr; II-Spectrin in Cerebral Spinal Fluid After Middle Cerebral Artery Occlusion in Rats. *J Cerebr Blood F Met.* 24: 98–106.

Pineda, J.A., S.B. Lewis, A.B. Valadka, L. Papa, H.J. Hannay, S.C. Heaton, J.A. Demery, M.C. Liu, J.M. Aikman and V. Akle. 2007. Clinical significance of α ii-spectrin breakdown products in cerebrospinal fluid after severe traumatic brain injury. *J Neurotraum.* 24: 354–366.

Rabilloud, T. 2012. The whereabouts of 2D gels in quantitative proteomics. *Methods Mol Biol.* 893: 25–35.

Raghupathi, R. 2004. Cell death mechanisms following traumatic brain injury. *Brain Pathol.* 14: 215–222.

Ramont, L., H. Thoannes, A. Volondat, F. Chastang, M.-C. Millet and F.-X. Maquart. 2005. Effects of hemolysis and storage condition on neuron-specific enolase (NSE) in cerebrospinal fluid and serum: implications in clinical practice. *Clin Chem Lab Med.* 43: 1215–1217.

Rao, V. and C. Lyketsos. 2000. Neuropsychiatric sequelae of traumatic brain injury. *Psychosomatics.* 41: 95–103.

Redell, J.B., Y. Liu and P.K. Dash. 2009. Traumatic brain injury alters expression of hippocampal microRNAs: potential regulators of multiple pathophysiological processes. *J Neurosci Res.* 87: 1435–1448.

Redell, J.B., A.N. Moore, N.H. Ward III, G.W. Hergenroeder and P.K. Dash. 2010. Human traumatic brain injury alters plasma microRNA levels. *J Neurotraum.* 27: 2147–2156.

Reed, T.T., J. Owen, W.M. Pierce, A. Sebastian, P.G. Sullivan and D.A. Butterfield. 2009. Proteomic identification of nitrated brain proteins in traumatic brain-injured rats treated postinjury with gamma-glutamylcysteine ethyl ester: Insights into the role of elevation of glutathione as a potential therapeutic strategy for traumatic brain injury. *J Neurosci Res.* 87: 408–417.

Reiber, H. 2001. Dynamics of brain-derived proteins in cerebrospinal fluid. *Clin Chim Acta.* 310: 173–186.

Reinders, J., U. Lewandrowski, J. Moebius, Y. Wagner and A. Sickmann. 2004. Challenges in mass spectrometry-based proteomics. *Proteomics.* 4: 3686–3703.

Robinson, G., U. Muller and K.K. Wang. 2009. Translation of neurological biomarkers to clinically relevant platforms. In Neuroproteomics. *Springer.* 303–313.

Rodriguez-Suarez, E. and A.D. Whetton. 2013. The application of quantification techniques in proteomics for biomedical research. *Mass Spectrom Rev.* 32: 1–26.

Røe, C., U. Sveen, K. Alvsåker and E. Bautz-Holter. 2009. Post-concussion symptoms after mild traumatic brain injury: influence of demographic factors and injury severity in a 1-year cohort study. *Disabil Rehabil.* 31: 1235–1243.

Routsi, C., E. Stamataki, S. Nanas, C. Psachoulia, A. Stathopoulos, A. Koroneos, M. Zervou, G. Jullien and C. Roussos. 2006. Increased levels of serum S100B protein in critically ill patients without brain injury. *Shock.* 26: 20–24.

Saatman, K.E., A.C. Duhaime, R. Bullock, A.I. Maas, A. Valadka and G.T. Manley. 2008. Classification of traumatic brain injury for targeted therapies. *J Neurotraum*. 25: 719–738.

Sabido, E., N. Selevsek and R. Aebersold. 2012. Mass spectrometry-based proteomics for systems biology. *Curr Opin Biotech*. 23: 591–597.

Scale, G.C. 2001. Glasgow Coma Scale.

Schiavi, P., C. Laccarino and F. Servadei. 2012. The value of the calcium binding protein S100 in the management of patients with traumatic brain injury. *Acta Bio-medica: Atenei Parmensis*. 83: 5.

Schiess, R., B. Wollscheid and R. Aebersold. 2009. Targeted proteomic strategy for clinical biomarker discovery. *Mol Oncol*. 3: 33–44.

Selassie, A.W., E. Zaloshnja, J.A. Langlois, T. Miller, P. Jones and C. Steiner. 2008. Incidence of long-term disability following traumatic brain injury hospitalization, United States, 2003. *J Head Trauma Rehab*. 23: 123–131.

Shu, Q., Z. Li, L. Li, S. Yang, L. Zhan and Y. Zhang. 2011. Expression of proteins in serum and hippocampus after closed brain injury in rats. Fa yi xue za zhi. 27: 107.

Siman, R., T.K. McIntosh, K.M. Soltesz, Z. Chen, R.W. Neumar and V.L. Roberts. 2004. Proteins released from degenerating neurons are surrogate markers for acute brain damage. *Neurobiol Dis*. 16: 311–320.

Siman, R., V.L. Roberts, E. McNeil, A. Dang, J.E. Bavaria, S. Ramchandren and M. McGarvey. 2008. Biomarker evidence for mild central nervous system injury after surgically-induced circulation arrest. *Brain Res*. 1213: 1–11.

Siman, R., N. Toraskar, A. Dang, E. McNeil, M. McGarvey, J. Plaum, E. Maloney and M.S. Grady. 2009. A panel of neuron-enriched proteins as markers for traumatic brain injury in humans. *J Neurotraum*. 26: 1867–1877.

Sjödin, M.O., J. Bergquist and M. Wetterhall. 2010. Mining ventricular cerebrospinal fluid from patients with traumatic brain injury using hexapeptide ligand libraries to search for trauma biomarkers. *J Chromatogr B*. 878: 2003–2012.

Sjödin, M.O.D. 2012. Advances for Biomarker Discovery in Neuroproteomics using Mass Spectrometry: From Method Development to Clinical Application. Uppsala University.

Skogseid, I., H. Nordby, P. Urdal, E. Paus and F. Lilleaas. 1992. Increased serum creatine kinase BB and neuron specific enolase following head injury indicates brain damage. *Acta Neurochir*. 115: 106–111.

Smits, M., D.W. Dippel, E.W. Steyerberg, G.G. de Haan, H.M. Dekker, P.E. Vos, D.R. Kool, P.J. Nederkoorn, P.A. Hofman and A. Twijnstra. 2007. Predicting intracranial traumatic findings on computed tomography in patients with minor head injury: the CHIP prediction rule. *Ann Intern Med*. 146: 397–405.

Sparvero, L.J., A.A. Amoscato, P.M. Kochanek, B.R. Pitt, V.E. Kagan and H. Bayır. 2010. Mass-spectrometry based oxidative lipidomics and lipid imaging: applications in traumatic brain injury. *J Neurochem*. 115: 1322–1336.

Spinella, P.C., T. Dominguez, H.R. Drott, J. Huh, L. McCormick, A. Rajendra, J. Argon, T. McIntosh and M. Helfaer. 2003. S-100beta protein-serum levels in healthy children and its association with outcome in pediatric traumatic brain injury. *Crit Care Med*. 31: 939–945.

Stergiou-Kita, M., D. Dawson and S. Rappolt. 2012. Inter-Professional Clinical Practice Guideline for Vocational Evaluation Following Traumatic Brain Injury: A Systematic and Evidence-Based Approach. *J Occup Rehabil*. 1–16.

Stiell, I.G., C.M. Clement, B.H. Rowe, M.J. Schull, R. Brison, D. Cass, M.A. Eisenhauer, R.D. McKnight, G. Bandiera and B. Holroyd. 2005. Comparison of the Canadian CT Head Rule and the New Orleans Criteria in patients with minor head injury. *JAMA: J Am Med Assoc*. 294: 1511–1518.

Stocchetti, N., F. Pagan, E. Calappi, K. Canavesi, L. Beretta, G. Citerio, M. Cormio and A. Colombo. 2004. Inaccurate early assessment of neurological severity in head injury. *J Neurotraum.* 21: 1131–1140.

Stocchetti, N., C. Zanaboni, A. Colombo, G. Citerio, L. Beretta, L. Ghisoni, E.R. Zanier and K. Canavesi. 2008. Refractory intracranial hypertension and "second-tier" therapies in traumatic brain injury. *Intens Care Med.* 34: 461–467.

Taft, W.C., K. Yang, C.E. Dixon and R.L. Hayes. 1992. Microtubule-associated protein 2 levels decrease in hippocampus following traumatic brain injury. *J Neurotraum.* 9: 281–290.

Talapatra, A., R. Rouse and G. Hardiman. 2002. Protein microarrays: challenges and promises. *Pharmacogenomics.* 3: 527–536.

Tokuda, T., S. Ikeda, N. Yanagisawa, Y. Ihara and G. Glenner. 1991. Re-examination of ex-boxers' brains using immunohistochemistry with antibodies to amyloid β-protein and tau protein. *Acta Neuropathol.* 82: 280–285.

Topolovec-Vranic, J., M.-A. Pollmann-Mudryj, D. Ouchterlony, D. Klein, J. Spence, A. Romaschin, S. Rhind, H.C. Tien and A.J. Baker. 2011. The value of serum biomarkers in prediction models of outcome after mild traumatic brain injury. *The Journal of Trauma and Acute Care Surgery.* 71: S478–S486.

Tyers, M. and M. Mann. 2003. From genomics to proteomics. *Nature.* 422.

Unden, J., R. Astrand, K. Waterloo, T. Ingebrigtsen, J. Bellner, P. Reinstrup, G. Andsberg and B. Romner. 2007. Clinical significance of serum S100B levels in neurointensive care. *Neurocrit Care.* 6: 94–99.

Undén, J., J. Bellner, M. Eneroth, C. Alling, T. Ingebrigtsen and B. Romner. 2005. Raised serum S100B levels after acute bone fractures without cerebral injury. *Journal of Trauma and Acute Care Surgery.* 58: 59–61.

Undén, J. and B. Romner. 2010. Can low serum levels of S100B predict normal CT findings after minor head injury in adults?: an evidence-based review and meta-analysis. *The J Head Trauma Rehab.* 25: 228–240.

van Baalen, B., E. Odding, A.I. Maas, G.M. Ribbers, M.P. Bergen and H.J. Stam. 2003. Traumatic brain injury: classification of initial severity and determination of functional outcome. *Disabil Rehabil.* 25: 9–18.

van Reekum, R., T. Cohen and J. Wong. 2000. Can traumatic brain injury cause psychiatric disorders? *J Neuropsych Clin N.* 12: 316–327.

Viant, M.R., B.G. Lyeth, M.G. Miller and R.F. Berman. 2005. An NMR metabolomic investigation of early metabolic disturbances following traumatic brain injury in a mammalian model. *NMR Biomed.* 18: 507–516.

Vos, P., B. Jacobs, T. Andriessen, K. Lamers, G. Borm, T. Beems, M. Edwards, C. Rosmalen and J. Vissers. 2010. GFAP and S100B are biomarkers of traumatic brain injury An observational cohort study. *Neurology.* 75: 1786–1793.

Vos, P.E., K. Lamers, J. Hendriks, M. Van Haaren, T. Beems, C. Zimmerman, W. Van Geel, H. De Reus, J. Biert and M. Verbeek. 2004. Glial and neuronal proteins in serum predict outcome after severe traumatic brain injury. *Neurology.* 62: 1303–1310.

Wagner, J.A. 2002. Overview of biomarkers and surrogate endpoints in drug development. *Dis Markers.* 18: 41–46.

Wang, K.K. 2000. Calpain and caspase: can you tell the difference? *Trends Neurosci.* 23: 20–26.

Wang, K.K., A.K. Ottens, M.C. Liu, S.B. Lewis, C. Meegan, M.W. Oli, F.C. Tortella and R.L. Hayes. 2005. Proteomic identification of biomarkers of traumatic brain injury. *Expert Rev Proteomics.* 2: 603–614.

Whelan-Goodinson, R., J.L. Ponsford, M. Schönberger and L. Johnston. 2010. Predictors of psychiatric disorders following traumatic brain injury. *J Head Trauma Rehab.* 25: 320–329.

Wiese, S., K.A. Reidegeld, H.E. Meyer and B. Warscheid. 2006. Protein labeling by iTRAQ: a new tool for quantitative mass spectrometry in proteome research. *Proteomics.* 7: 340–350.

Williams, L. 1995. Oxidative stress, age-related neurodegeneration, and the potential for neurotrophic treatment. *Cerebrovas Brain Met Reviews.* 7: 55.

Williams, W.H., S. Potter and H. Ryland. 2010. Mild traumatic brain injury and Postconcussion Syndrome: a neuropsychological perspective. *J Neurol, Neurosur Ps.* 81: 1116–1122.

Wingren, C. and C.A.K. Borrebaeck. 2007. Progress in miniaturization of protein arrays—a step closer to high-density nanoarrays. *Drug Discov Today.* 12: 813–819.

Wishart, T.M., T.M. Rooney, D.J. Lamont, A.K. Wright, A.J. Morton, M. Jackson, M.R. Freeman and T.H. Gillingwater. 2012. Combining comparative proteomics and molecular genetics uncovers regulators of synaptic and axonal stability and degeneration *in vivo. PLoS Genet.* 8: e1002936.

Woods, A.G., I. Sokolowska, R. Taurines, M. Gerlach, E. Dudley, J. Thome and C.C. Darie. 2012. Potential biomarkers in psychiatry: focus on the cholesterol system. *J Cell Mol Med.* 16: 1184–1195.

Wu, P., Y. Zhao, S.J. Haidacher, E. Wang, M.O. Parsley, J. Gao, R.G. Sadygov, J.M. Starkey, B.A. Luxon, H. Spratt, D.S. Dewitt, D.S. Prough and L. Denner. 2012a. Detection of Structural and Metabolic Changes in Traumatically Injured Hippocampus by Quantitative Differential Proteomics. *J Neurotraum.*

Wu, P., Y. Zhao, S.J. Haidacher, E. Wang, M.O. Parsley, J. Gao, R.S. Sadygov, J.M. Starkey, B.A. Luxon and H. Spratt. 2012b. Detection of Structural and Metabolic Changes in Traumatically Injured Hippocampus by Quantitative Differential Proteomics. *J Neurotraum.*

Wuthisuthimethawee, P., S. Saeheng and T. Oearsakul. 2013. Serum cleaved tau protein and traumatic mild head injury: a preliminary study in the Thai population. *Eur J Trauma Emerg S.* 1–4.

Yang, S., Y. Ma, Y. Liu, H. Que, C. Zhu and S. Liu. 2012. Arachidonic Acid: a Bridge between Traumatic Brain Injury and Fracture Healing. *J Neurotraum.* 30(9): 775–88.

Yang, S., Y. Ma, Y. Liu, H. Que, C. Zhu and S. Liu. 2013. Elevated serum haptoglobin after traumatic brain injury is synthesized mainly in liver. *J Neurosci Res.* 91: 230–239.

Yang, X., S. Yang, J. Wang, X. Zhang, C. Wang and G. Hong. 2009. Expressive proteomics profile changes of injured human brain cortex due to acute brain trauma. *Brain Injury.* 23: 830–840.

Yao, C., A.J. Williams, A.K. Ottens, X.-C. May Lu, R. Chen, K.K. Wang, R.L. Hayes, F.C. Tortella and J.R. Dave. 2008. Detection of protein biomarkers using high-throughput immunoblotting following focal ischemic or penetrating ballistic-like brain injuries in rats. *Brain Injury.* 22: 723–732.

Yeates, T.O. 2011. Nanobiotechnology: Protein arrays made to order. *Nat Nanotechnol.* 6: 541–542.

Yuh, E.L., S.R. Cooper, A.R. Ferguson and G.T. Manley. 2012. Quantitative CT improves outcome prediction in acute traumatic brain injury. *J Neurotraum.* 29: 735–746.

Zaloshnja, E., T. Miller, J.A. Langlois and A.W. Selassie. 2008. Prevalence of long-term disability from traumatic brain injury in the civilian population of the United States, 2005. *The J Head Trauma Rehab.* 23: 394–400.

Zasler, N.D. 1997. Prognostic indicators in medical rehabilitation of traumatic brain injury: A commentary and review. *Arch Phys Med Rehab.* 78: S12–S16.

Zemlan, F.P., E.C. Jauch, J.J. Mulchahey, S.P. Gabbita, W.S. Rosenberg, S.G. Speciale and M. Zuccarello. 2002. C-tau biomarker of neuronal damage in severe brain injured patients: association with elevated intracranial pressure and clinical outcome. *Brain Res.* 947: 131–139.

Zetterberg, H., D.H. Smith and K. Blennow. 2013. Biomarkers of mild traumatic brain injury in cerebrospinal fluid and blood. *Nat Rev Neurol.*

Zhang, Z., S.F. Larner, M.C. Liu, W. Zheng, R.L. Hayes and K.K. Wang. 2009. Multiple alphaII-spectrin breakdown products distinguish calpain and caspase dominated necrotic and apoptotic cell death pathways. *Apoptosis.* 14: 1289–1298.

Zhang, Z., S. Mondello, F. Kobeissy, R. Rubenstein, J. Streeter, R.L. Hayes and K.K. Wang. 2011. Protein biomarkers for traumatic and ischemic brain injury: from bench to bedside. *Translational Stroke Research.* 2: 455–462.

Zhu, G.W., F. Wang and W.G. Liu. 2009. Classification and prediction of outcome in traumatic brain injury based on computed tomographic imaging. *The Journal of International Medical Research*. 37: 983–995.

Zhu, H. and M. Snyder. 2001. Protein arrays and microarrays. *Curr Opin Chem Biol*. 5: 40–45.

Ziebell, J.M., S.E. Taylor, T. Cao, J.L. Harrison and J. Lifshitz. 2012. Rod microglia: elongation, alignment, and coupling to form trains across the somatosensory cortex after experimental diffuse brain injury. *J Neuroinflamm*. 9: 247.

Zoltewicz, J.S., D. Scharf, B. Yang, A. Chawla, K.J. Newsom and L. Fang. 2012. Characterization of Antibodies that Detect Human GFAP after Traumatic Brain Injury. *Biomarker Insights*. 7: 71.

Zongo, D., R. Ribéreau-Gayon, F. Masson, M. Laborey, B. Contrand, L.R. Salmi, D. Montaudon, J.L. Beaudeux, A. Meurin and V. Dousset. 2012. S100-B protein as a screening tool for the early assessment of minor head injury. *Ann Emerg Med*. 59: 209–218.

Zubarev, R.A. 2013. The challenge of the proteome dynamic range and its implications for in-depth proteomics. *Proteomics*. 13: 723–726.

2

Protein Biomarkers in Traumatic Brain Injury: An Omics Approach

Angela Boutte,[1],* Firas Kobeissy,[2] Kevin K.W. Wang,[2] Zhiqun Zhang,[2] Frank Tortella,[1] Jitendra R. Dave[1] and Kara Schmid[1]

INTRODUCTION

Traumatic brain injuries (TBIs) pose a great health concern. In the United States alone, nearly 2 million people per year will have suffered a TBI. In the civilian population, most TBIs result from falls (35.2%), motor vehicle accidents (17.3%) and assault (10%). Both youth and adult athletes may suffer multiple concussions (http://www.cdc.gov/traumaticbraininjury/statistics.html). They are often exposed to a variety of combat traumas. More than 30,000 military personnel suffered a TBI in 2012. Another 13,000 or more people had a TBI in 2013 (http://www.dvbic.org/dod-worldwide-numbers-tbi). TBI is highly variable and characterized by several severities

[1] Branch of Brain Trauma Neuroprotection and Neurorestoration, Center for Psychiatry and Neuroscience, Walter Reed Army Institute of Research, Silver Spring, MD, 20910, USA.
[2] Center for Neuroproteomics & Biomarkers Research, Departments of Psychiatry & Neuroscience, McKnight Brain Institute, University of Florida, Gainesville, FL 32611, USA.
* Corresponding author: angela.m.boutte.civ@mail.mil

(mild, moderate, severe) as well as multiple injury types (concussive, non-penetrating, penetrating). Mild TBIs are of particular interest to the military, as explosions or "blast"-induced injuries are the most common TBI among active duty personnel (Mondello et al., 2013). More recently, mTBI is a key topic among persons involved in athletic programs (Omalu et al., 2010; Stern et al., 2011). Both youth and adult athletes may suffer multiple concussions (http://www.cdc.gov/traumaticbraininjury/statistics.html).

Defining specific mechanisms for which therapeutic approaches would be effective for TBI, which is heterogeneous, remains elusive. In addition, mild TBIs are often difficult to detect with cognitive tests alone and cannot be resolved using clinical imaging techniques, such as Magnetic Resonance Imaging (MRI) (Yuh et al., 2013). Thus, the use of protein biomarkers to define, diagnose, monitor or treat TBI would greatly enhance efforts to manage patient care. Biomarkers may lead to understanding mechanisms of injury and recovery. Basic and clinical research efforts are currently determining the mechanism of TBI with many methods, including analysis of targeted biomarkers coupled to clinical and/or behavioral tests (Diaz-Arrastia et al., 2013; Okonkwo et al., 2013). The main goal of many biomarkers studies is to provide correlations to severity and/or duration of injury. In addition, biomarkers could potentially be used to evaluate the therapeutic response (Mondello et al., 2013). TBI biomarker research is poised to greatly impact wound healing, recovery, and increased survival with improved quality of life.

The Role of Proteomics in TBI Biomarker Discovery

Biomarkers are naturally occurring molecules found in tissues or bio-fluids that uniquely identify an abnormal/pathological state, such as TBI. They are often (and ideally), indicators of very specific processes, events or conditions (Guingab-Cagmat et al., 2013; Martins-de-Souza, 2010). "Omics" technologies encompass any study that considers all parts of a system or group of analytes collectively. Omics may study any macromolecule or metabolite, e.g., DNA or RNA for genomics fatty acids in lipidomics, and metabolites in metabolomics. Proteins are the molecules of interest in proteomics, which has a plethora of methodologies available to TBI research.

The proteome includes all proteins in the cell, tissue, or organ of interest and proteomics is the investigation of all proteins in a particular physiological state and may also encompass metrics such as TBI type, time lapsed or severity. Proteomics has become one of the most sophisticated and sensitive tools employed in biological studies. It has become more popular due to, in part, (1) the ability to transfer studies across species due to redundancy of the mammalian genome, (2) the ability to detect brain-specific proteins in biofluids in TBI, and (3) the need for novel markers of injury that are readily detectable in biofluids. Furthermore, proteomics may be a preferred method, as opposed to genomics. Genomics assays infer changes in protein abundance; the correlation of gene based studies to protein confirmation is low (Pradet-Balade et al., 2001). In addition, messenger RNA studies are not readily translated to bio-fluid analyses. Proteomics analyses have also expanded from beyond studying groups of proteins to understanding the protein complexes, gaining knowledge of isoforms, as well as discerning distinct roles that certain protein-protein interactions and intact pathways may play within the milieu systems biology.

Proteomics is becoming a well-established approach to discover and validate protein biomarkers in TBI. The collective number of published reports and citations utilizing proteomics in brain injuries is increasing: there were more than 500 in 2012 alone compared to less than 50 a decade ago. In addition, there is an increased interest in clinical and/or deployment related TBI, for the number of military TBI publications and citations has more than doubled since 2010 (Thompson Reuters, Web of Knowledge). Both tissue and biofluid-based biomarkers have been used to determine severity of injury (Zoltewicz et al., 2013) and may be able to determine therapeutic response (Zhang et al., 2010).

Proteomics-based technologies could potentially be used in-theatre (on or near the battlefield) or during triage to aid diagnoses, specifically in mild injuries where greater sensitivity and specificity is key. Omics technologies may be used to determine a variety of endpoints in the realm of protein markers: intact proteins, products of proteolysis and post-translational modifications. This chapter presents an overview of current technologies, sample sources (e.g., cell types, brain tissue regions and biofluids), and possible future directions in TBI proteomics. First the role of animal models is briefly introduced.

Animal (e.g., rodent) models can serve as surrogates to understand injury mechanisms and define treatments based on biomarkers and the vast majority of proteomics research is performed with experimental animal models of TBI. Several types of TBIs are explored in animal models (Xiong et al., 2013), some of which are:

1. Controlled Cortical Impact (CCI)—an open skull injury that uses a piston to penetrate the brain at a specific velocity and distance (Dixon et al., 1991).
2. Projectile Concussive Impact (PCI)—a helmet shielded closed skull injury (Chen et al., 2012).
3. Fluid Percussion Injury (FPI) or Lateral Fluid Percussion (LFP)—an open skull injury that rapidly inject fluid into the intracranial cavity (Alder et al., 2011; Dixon et al., 1987).

4. Penetrating Ballistic-like Brain Injury (PBBI)—a skull breaching gunshot waveform projectile model that forms a permanent, but non-lethal, cavity in the brain (Williams et al., 2006).
5. Middle Cerebral Artery occlusion (MCAo)—a model of stroke, ischemia and infarct that produces a brain lesion (Dave et al., 2009; Carmichael, 2005).
6. Blast or Blast Over-Pressure (BOP)—a model of an explosive charge wherein a whole body pressure wave is administered (Ahlers et al., 2012; Elder et al., 2010).

Many of these models are well defined in terms of post-mortem histopathology; however, there are currently few reliable markers that fully characterize these injuries and harbor clinical utility. These studies may allow definition of injury-specific mechanisms among individuals or groups of patients and facilitate or improve therapeutic regimens. The status of protein biomarkers used in TBI research is introduced next and extensively summarizes applicable proteomics approaches for discovery and confirmation.

Protein Biomarkers and Their Origins

Overview

Proteomics studies may be used to discover biomarkers from many sources. The current convention is to perform proteomics discovery or confirmation with brain tissues or cells containing the lesion induced by TBI. Brain proteins are then assayed in cerebral spinal fluid (CSF) or in the blood (serum or plasma) as a consequence of blood brain barrier (BBB) breakdown and leaky cellular membranes of apoptotic/necrotic cells (Boutte et al., 2012; Guingab-Cagmat et al., 2013). These approaches have proven successful as brain specific proteins have been detected across multiple models and clinical studies.

The Milieu of Protein Biomarkers

Intact proteins, Proteolytic Fragments and Post-translational Modifications

A class of neuronal and glial proteins has been reproducibly identified across animal models and clinical cases of TBI. Some models or clinical samples often identify these same proteins within the proteome, particularly proteins from neurons, astrocytes and oligodendrocytes. In both neurons and glia, proteins are released from necrotic and/or apoptotic cells and are detectable in Cerebral Spinal Fluid (CSF) as well as in blood plasma or serum. These biomarkers have been detectable in the CSF and blood of some patients with severe TBI, including those suffering stroke/cerebral ischemia or penetrating injuries (Hergenroeder et al., 2008; Mondello et al., 2012; Morochovic et al., 2009; Rundgren et al., 2012; Stein et al., 2011a). Defining biomarkers with greater sensitivity and specificity is imperative to determine severity of injury, recovery or response to therapy and to differentiate injuries from one another.

The abundance of neuronal proteins like Neuron Specific Enolase (NSE), ubiquitin C-terminal lyase (UCH)-L1, amyloid precursor protein (APP), neurofilament protein (NF), non erythroid α-II spectrin, and Tau are differentially abundant in tissues and biofluids of TBI models including MCAo, PBBI, LFP and CCI (Aikman et al., 2006; Bohmer et al., 2011; Brophy et al., 2009; Bulut et al., 2006; Magnoni et al., 2012; Papa et al., 2012; Park et al., 2007; Zurek and Fedora, 2012). Microglia, astrocytes and oligodendrocytes also play a key role in TBI as neuro-glial inflammation peaks in response to injury (Ramlackhansingh et al., 2011; Yu et al., 2010). These cells express and release intracellular proteins, such as Glial Fibrillary Acidic Protein (GFAP), and Myelin Basic Protein (MBP), respectively. Microglial cells also secrete greater proportions of cytokines and chemokines when activated after TBI (Ghirnikar et al., 1998). Infiltrating immune cells from the periphery also play a role. Derived from infiltrating immune cells/leukocytes, endothelial monocyte-activating polypeptide II precursor (p43/pro-EMAPII), has been identified as a potential biomarker that is differentially regulated in models of hemorrhagic vs. non-hemorrhagic TBI. P43/pro-EMAPII was upregulated following PBBI which is hemorrhagic and downregulated following MCAo which is non-hemorrhagic (Yao et al., 2009).

In addition to intact proteins, fragments or Break-Down Products (BDPs) may be considered markers of TBI. BDPs are greatly increased after TBI as a consequence of cell death pathway activation and increased activity of pro-apoptotic enzymes (like caspase-3) and pro-necrotic enzymes (like calpain-2). In fact, the 2-dimensional proteomic map of protein fragments isolated from the brain after rodent CCI overlapped with that of brain lysates subjected to caspase-2 or calpain-2 degradation (Liu et al., 2006). This "degradome" introduced several potential targets of study, but several specific protein fragments or BDPs have been the focus of TBI. GFAP-BDPs and non-erythroid α-II spectrin BDPs (SBDPs) have been the focus of many animal models and clinical studies. Both have been detected in biofluids and are correlated to injury severity, poor outcome and mortality (Honda et al., 2010; Lumpkins et al., 2008; Okonkwo et al., 2013; Papa et al., 2012). Additional fragments or BDPs of other proteins have been described, including cleaved tau (Gabbita et al., 2005; Huber et al., 2013; Zemlan et al., 2002).

A new venue of proteomics-based biomarkers is global or individual protein post translational modification. Phospho-proteomics, analysis of hyper-phosphorylated proteins in TBI, is an expanding field. For example, phosphorylated neurofilament protein heavy chain (pNF-H) is increased in a rat model of CCI. Detectable as early as 6 hours after injury, pNF-H abundance peaked 24–48 hours after and correlated with lesion severity (Anderson et al., 2008). Increases in serum pNF-H were recapitulated in a model of severe cortical contusion (Shaw et al., 2005). Furthermore, slow and chronic increases of serum pNF-H were associated with poor outcome and larger infarct volumes in stroke patients (Singh et al., 2011). Other amino acid specific Post-Translational Modifications (PTMs) may also be considered biomarkers. Global 3-nitrotyrosine (NT) steadily increased in the hippocampus after CCI in rats (Ahn et al., 2008) and carbonyl levels of plasma proteins are much higher in patients with severe TBI compared to controls (Hohl et al., 2011).

Biomarker Sources and Distribution

Specific Brain Regions and Spatial Resolution

The anatomical region of interest in TBI varies with the nature of the trauma (force, duration, and repetition), the behavioral outcome, the observed diagnoses and the proposed injury site. Acute, severe, penetrating injuries are perhaps the most obvious injury, since a breach in the skull is readily visible and the adjacent lesion is a site of interest. However, the injury site of mild-moderate, blast related or concussive injuries cannot always be readily determined, especially if a breach is not visible. To address these differences in brain injuries, many models study a brain region of interest that correlates to an observed behavioral outcome (Elder et al., 2012). Studying multiple regions, not just those located within or near the impact site, are crucial. For instance, a spatial distribution study using protein arrays indicated blast TBI affected the proteome of the prefrontal cortex as well as the hippocampus (Kwon et al., 2011). In a model of pediatric TBI, 2-dimensonal differential gel electrophoresis (2D-DIGE) was used to specifically investigate hippocampal biomarkers in CCI (Kochanek et al., 2006). Within brain regions, further analysis may extend to cell types.

Specific Cell Types

The vast majority of studies focus on either neurons and synaptic transmission or glial-mediated inflammation due to advances in histology. The vast majority of total brain content consists of glial cells including astrocytes, microglia, and oligodendrocytes and sub-types therein. As such, cellular subtypes are gaining attention. Most biomarkers discovered by proteomics and validated by other methods are often derived from microglia or recruited inflammatory cells, such as MBP derived from oligodendrocytes, GFAP from astrocytes and various cytokines and chemokines from resident and infiltrating cells (Beschorner et al., 2002). Endothelial cells, although small in population compared to total brain mass, may also be a great source of biomarkers in TBIs with BBB alterations or in the case that endothelial cells themselves are a direct target in TBI via signaling or oxidative stress. Vascular Endothelial Growth Factor (VEGF) is released from damaged blood vessels, is correlated to BBB disruption, and was tested as a cytokine biomarker (Kovesdi et al., 2011; Lee and Agoston, 2010). Overall, several cell types may be the source of definitive biomarkers.

Blood and Cerebral Spinal Fluid

An ideal biomarker study conducted with animal models and/or human samples requires great diligence to maintain practices that are likely to be common to any clinical setting. To alleviate this concern, it is also imperative to define biomarkers that have high sensitivity and specificity when isolated from biofluids. Blood, the most easily attainable biofluid, has been reliably used for detection of proteins such as GFAP and its breakdown products (BDPs), as well as neuronal ubiquitin C-terminal hydrolase (UCH)-L1. However, other proteins remain elusive as plasma and serum are difficult bio-specimens to study due to the high content of abundant "contaminating" proteins. Albumin, immunoglobulin, and other highly abundant proteins interfere with antibody based assays often by simply masking the target of interest. In human plasma, the 20 most abundant proteins represent ~99% of total protein mass. In serum, albumin is ~66 μmol/mL, whereas serum GFAP has been detected at 13.4 ng/mL in human TBI (Stein et al., 2012). Cytokines are even more dilute and have been measured at only 1–10 pmol/mL or less (Anderson, 2010; Anderson and Anderson, 2002).

There is some debate whether plasma or serum is best to use for biomarker proteomics. It is clear that plasma and serum have different proteomes (Tammen et al., 2005) and that handling or preparation (particularly when different anti-clotting factors or clotting chemicals are used) has an effect on the protein population. For example, there were numerous differences in the differential display of detectable peptides in plasma isolated with either EDTA or citrate (Tammen and Hess, 2011). Reaching this level of quantification for multiple dilute proteins discovered via proteomics would be a great feat for TBI proteomics studies.

Cerebral Spinal Fluid (CSF) or ventricular fluid is perhaps the most relevant biofluid to study TBI biomarkers. An early study revealed that the dynamic range of peptide detection in CSF with proteomics was enhanced two-fold (Shores et al., 2008). CSF is extremely limited in the amount of analyte that can be collected from animal models or from patients. For example, approximately 100 µL of vCSF may be isolated from rats (Nirogi et al., 2009). Collecting CSF from patients causes great discomfort, which may lead to a reduction in study sample size. Some proteomics studies have bypassed this caveat by collecting post-mortem CSF and extrapolating results to severe TBI of live patients.

Although a plethora of methods allow study of TBI biomarkers derived from tissues, cells and biofluids, the focus herein is an overview of current proteomics-based methods that are readily applicable to animal models and clinical TBI. Generalized methodology is discussed and includes several studies, many of which have resulted in clinically relevant TBI biomarker analysis.

Instrumentation and Approaches to Define TBI

Overview

The most approachable methodologies are to study intact proteins with either targeted capture arrays, which exploit molecular interactions. In this case, protein targets are often captured with an antibody, but aptamer sequences (DNA, RNA, or peptides) may also be used and are discussed later. Many studies use gel based assays, such that separate proteins via molecular weight, isoelectric focusing or both then identify proteins via their peptide sequences elucidated using mass spectrometry. Mass Spectrometry (MS) is an analytical method that measures the mass to charge ratio of ionized or charged peptides generated by *in vitro* protease cleavage. For early gel-based methods, detection of peptides was used. However, as the field has expanded, fragment ions of the peptides are increasingly used as a metric to determine protein abundance. To generate amino acid sequence information, tandem Mass Spectrometry (MS/MS) is employed. In a typical experiment, intact peptide is ionized and fragmented; then, the ion generated that carries the charge is detected in positive (or negative) ion mode. Multiple copies of the same peptide are fragmented at different peptide bonds providing readout of the peptide sequence; the most labile bonds are cleaved with more frequency. The most abundant ionized fragments are b-type (charge is retained on the N terminal fragment) and y-type (charge is retained on the C-terminal side) (Nilsson et al., 2010). Although proteomics methods employ an array of labeling techniques or MS iterations to perform paired or multiplexed analysis of TBI biomarkers, a few methods are commonly used (Table 2.1).

Gel Based Proteomics

Relative Quantification and Differential Dyes

Two-dimensional electrophoresis (2-DE) requires separation of proteins by at least two chemical properties. The most common separation workflow for intact protein mixtures are by (1) isoelectric point (pI) with isoelectric focusing (IEF) and (2) by ion exchange chromatography. Both methods are followed by polyacrylamide gel-electrophoresis (PAGE) to resolve molecular weight. This method requires a relatively large amount of total protein (50–100 µg per sample) and is well suited for tissues, some cell types, and has been adapted to study pooled protein mixtures from plasma, serum and CSF.

Two-DE has been used in a wide range of studies to define biomarkers of TBI. Differential in-gel electrophoresis (DIGE), relies upon dye intensity changes for the same protein "spot" identified by specific pI/ MW coordinates on an x-y plane using either a total protein dye (such as Coomassie blue or Sypro Ruby) or by labeling cysteine or lysine with fluorescent Cy-2, -3, and/or -5. In the former, sample types such as TBI versus control are compared on separate gels. In the latter, the samples can be labeled and mixed prior to loading then compared within the same gel. Images are acquired with specific excitation/emission wavelengths for each dye, one of which may be used to label an internal standard (Arnold and Frohlich, 2012; Viswanathan et al., 2006).

DIGE can typically identify 1.5–2 fold changes with statistical significance of p <0.05 depending on the number of replicate gels (Friedman and Lilley, 2008). Even greater sensitivity of differential protein abundance may be achieved by using principal component analysis (PCA) to seek out trends due specifically to the TBI and not the gel or dye bias (Friedman, 2012). Conventionally, 2-DE methods like DIGE are coupled to mass spectrometry methods to identify proteins of interest. These methods include Peptide Mass Fingerprinting (PMF) or tandem MS.

Many protein or peptide separation techniques, ionization and mass detection methods are compatible with 2-DE methods. The most widely used types of mass spectrometry in TBI biomarker studies are matrix assisted laser desorption ionization tandem time of flight mass spectrometry

Table 2.1. Protein biomarker proteomic method comparison.

Protein/Peptide Method	Mass Spectrometry Tool	Qualitative or Quantitative Tool	Benefits	Caveats
Two-dimensional Polyacrylamide Gel Electrophoresis (2-DE or 2D-PAGE)	MALDI-TOF-PMF MALDI-TOF/TOF	Densitometry of Coomassie blue or fluorescence of Sypro Ruby post-stain	Provides rapid visually interpretable data of total protein	Quantitation is limited to fairly abundant proteins
Difference in gel electrophoresis (DIGE)		Prelableing with fluorescently labeled cyanine dyes	Two samples may be compared in the same gel	Labeling is limited to exposed cysteine or lysine
2-DE Immunoblot		Densitometry or chemiluminescence of detection antibody	Adaptable to global DNPH-OxyBlotting and post-translational modifications	Dependent upon available antibodies and must be paired with a pre-determined 2-DE map
Targeted Quantification	LC-MS/MS	Peptide peak intensity	Adaptable to peptide SRM or MRM which may have 10^{-19} M sensitivity	May be impractical for larger scale *in vivo* projects and requires knowledge of proteotypic peptides
Spectral Counting		Peptide detection frequency and /or number of unique peptides	Amenable to peptide tags such as iTRAQ and ICAT	Often requires sophisticated analysis to identify false positives
Affinity-based Array	N/A	Densitometry or chemiluminescence (cytokine/chemokine array) or fluorescence (protein microarray and aptamers)	Most targets are well characterized and detection is often adaptable to genomic array technologies	Dependent upon available antibodies or aptamer sequences

Note: Summary of Proteomics Methods for Biomarker Discovery and Validation. Gel and non-gel based assays are compared and listed with commonly used mass spectrometry techniques and quantitative tools. Generalized benefits and caveats are listed for each method.

(MALDI-TOF/TOF-MS). MALDI peptide ionization is triggered by laser activation of a solid, crystalline matrix. The ionized peptides maintain kinetic energy within an electric field and travel through a flight tube toward the mass analyzer. TOF instruments are well regarded for their detection at low m/z values allowing the use of Tandem Mass Tags (TMTs), which are discussed later. Time of Flight (TOF) converts the mass analyzer signal from the time an ion takes to reach the detector to an m/z value. Peptide mass fingerprinting is the result of a single round of TOF analyses, whereas peptide parent ions and their product ions from fragmentation are measured in the tandem method, TOF/TOF. Two-DE matching with secondary protein identification is increasingly used to attain a greater number of true-positives. The vast majority of current studies utilize 2-DE to identify differential protein abundance, but peptide/protein sequence identification requires use of complementary methods such as MS. Peptide identification is greatly augmented by detecting peptide ion fragments using tandem methods (MS/MS or TOF/TOF) (Bienvenut et al., 2002; Gogichaeva and Alterman, 2012; Suckau et al., 2003).

Gel based methods rely upon migration patterns and CAX affinities that, ideally, would not be altered by TBI. A subset of proteins that may shift if mass or charge due to Post-Translational Modifications (PTMs) or alternative splicing may be missed using this methodology alone. In addition, the fold change difference is dependent upon the affinity and resolving power of the dye as well as the densitometric software. Besides these caveats, this method has been proven to be robust and identified many proteins that are differentially abundant in different models of CCI, MCAo and PBBI (Kobeissy et al., 2006; Ottens et al., 2010).

One of the earliest TBI studies utilized 2D-PAGE and matched results against a reference dataset using the murine CCI which collectively identified ~50 differentially abundant cytoskeletal and signaling proteins as a consequence of TBI (Agoston et al., 2009; Jenkins et al., 2002). In a more recent CCI study, proteins were first separated by cation exchange chromatography (CAX) using quaternary ammonium-(Q1) modified sepharose. CAX prior to PAGE or RP-LC-MS/MS may be conducted with a variety of resins including sulfonate (sulfonic acids) or carboxylate functional groups often used with silica or sepharose particles. After staining with Coomassie blue, differentially abundant proteins were excised and peptide sequences were analyzed with RP-LC-MS/MS. Inclusive of increased α-II spectrin (a potential biomarker that recurs across multiple models and methods used), 59 proteins were differentially expressed in TBI compared to controls (Kobeissy et al., 2006). These methods have also identified more than 306 unique proteins and revealed 74 proteins that were present only after injury, which were inclusive of known potential biomarkers of TBI, such as spectrin, NSE and brain creatine kinase (Haskins et al., 2005). A recent study utilized 2D-DIGE to determine several PBBI biomarkers in brain tissues that were then confirmed in biofluids (Boutte et al., 2012).

Immunoblotting

2D-PAGE has also been adapted to antibody based arrays. High Throughput Immunoblotting (HTPI) is a Western blotting technique that relies upon qualitative comparison reactivity of more than 800 monoclonal antibodies. This method was has also been used to identify hippocampal protein degradation patterns or substrates of caspase-3 and calpain-2 in models of TBI. Hippocampal proteins before and after *in vitro* calpain-2 or caspase-3 proteolytic digests were systematically compared to the TBI proteome, wherein intact proteins and their BDPs were simultaneously identified by a pool of HTPI monoclonal antibodies. This "degradome" identified calpain and caspase substrates in CCI that are potential TBI biomarkers (Kane et al., 2012; Liu et al., 2006; Malakhov et al., 2003; Ottens et al., 2007).

Not only may proteomics be used to compare TBI normal controls, but different types of TBI models. CAX-PAGE coupled to HTPI was used to differentially define proteins that are differentially abundant in MCAo compared to PBBI. Although LC-MS/MS was used to identify peptide-protein matches, the key to identifying differentially abundant proteins was dependent upon densitometry of proteins resolved in the immunoblot (Yao et al., 2008). HTPI is a very useful approach, particularly because it uses proteomics to investigate not just protein abundance, but cell death mechanisms as a consequence of TBI. Like commercially available antibody arrays, HTPI is multiplexed but targeted. However, the technique relies on a somewhat limited pool of antibodies that represent only a small fraction of proteins in any cell, tissue, or biofluid. The investigator must pre-determine the pool of potential TBI biomarkers.

Gel-Free Proteomics and Unbiased Multi-dimensional Protein Identification Technology

Overview

Liquid chromatography, tandem mass spectrometry (LC-MS/MS) is a very robust method that uses one to two phases of peptide separation from complex mixtures with column chromatography prior to ionization and detection by the mass spectrometer. LC-MS/MS is often referred to as multidimensional protein identification technology (MudPIT) or shotgun proteomics, although any proteomics method that uses two or more properties to identify differences in protein/peptide abundance may be defined by these terms. The number of studies using this method is growing steadily.

Many combinations of chromatography and ionization may be used; however, the most common in TBI biomarker studies separate tryptic peptides by strong cation exchange fractionation (SCX/CAX) then by reversed phase (RP) chromatography. These methods separate peptides by ionic strength and hydrophobicity, respectfully. Separation is either (1) off-line and peptides are collected in fractions or (2) directly coupled to the mass spectrometer. In both cases, eluted peptides undergo desolvation and protonation by electrospray ionization (ESI). Two or more phases of separation dramatically increases resolving power/detection of individual proteins (Mayr and Rabilloud, 2013).

LC-MS/MS has been adapted for use in label-free and labeled assays to define TBI biomarkers. Labels or tags used include amine or sulfhydryl reactive compounds or tandem mass tags (TMT) (Megger et al., 2013; Sandberg et al., 2013). The peptides may be enriched, which is particularly beneficial for study of biofluids like CSF, to improve proteome coverage (Mertins et al., 2013). LC-MS/MS also has the ability to locate and potentially quantitate PTMs on specific amino acids. In general, protein abundance using LC-MS/MS may be conducted in many ways (Dasilva et al., 2012):

1. Relative peak intensity quantification based on the linear correlation between peak area and peptide (thus protein) abundance.
2. Relative spectral count quantification is based on the number of times (frequency) that a particular peptide per protein is detected by the MS/MS. A cumulative count of peptide frequencies is used to extrapolate the relative abundance.
3. Absolute/targeted quantification is achieved by either adding a known concentration of isotopically labeled peptides or when the number of identified peptides divided by the number of the theoretically observable tryptic peptides for each protein is calculated.

Targeted Quantification

Targeted quantification may be used with unlabeled peptides or heavy isotope labeled peptide standards that are synthesized with ^{15}N and or ^{13}C. In both scenarios, the peptide ion and its charged fragment are measured by the mass spectrometer and compared via peak area as the peptide is separated from a complex mixture during LC-MS/MS. A type of targeted, quantitative peptide based mass spectrometry, was pioneered and developed by the Gygi laboratory (Gerber et al., 2007; Kirkpatrick et al., 2005). Similar to classic stable isotope dilution methods, a heavy labeled peptide synthesized with ^{13}C and/or ^{15}N, of the target protein biomarker of is added to the biological sample as an internal standard. This methodology, coined Absolute Quantification of Abundance (AQUA), monitors intact peptide to one product ion in Select Reaction monitoring (SRM) or multiple ionized product ions in Multiple Reaction Monitoring (MRM). Results are displayed as reconstructed chromatograms displaying the peak intensity of the internal standard and the biological sample. AQUA peptides are reported to have greater than nanomolar sensitivity. Ion fragmentation occurs with differing efficiency, and ions have varying limits of quantitation and/or detection. Thus, SRM or MRM of multiple peptides is preferred for precise targeted quantification of proteins.

Optimal peptide detection and subsequent protein analyses occur with "proteotypic" peptides that are most likely to be observed by the mass spectrometer (Mallick et al., 2007). If the peptides of choice are not known from preparative MS/MS or a prior experiment, SRM/MRM peptides may be predicted by several algorithms including the X! P3 Proteotypic Peptide Profiler (the GPM.org) or a support vector machine which uses more than 30 peptide properties (Craig et al., 2005; Webb-Robertson et al., 2010). Both algorithms have yet to be tested for accuracy of brain protein biomarkers in tissues, cells, CSF or plasma/serum of TBI studies.

Often SRM and MRM are used to validate absolute biomarker abundance and can be sensitive in samples with low proteins abundance, especially after enrichment. SRM/MRM is poised to be of great use to multiplexed biomarker TBI studies using biofluids where TBI biomarkers are quite low compared levels found in brain tissue or cells. The human CSF proteome was recently characterized using highly reproducible MRM-MS after albumin depletion and G-10 resin desalting and concentrating. More than 150 proteins were identified with a false discovery rate of <5%, including cystatin C and apolipoprotein 4/4 (Bora et al., 2012). Apolipoprotein 4/4 is reported to be associated with poor clinical outcome in TBI (Zhou et al., 2008) and increased cystatin C is also increased in human CSF after TBI (Hanrieder et al., 2009). Peptide quantitation is poised to be widely used, specifically for biofluid-based biomarker verification.

Spectral Counting with and without Peptide Tags

Protein biomarker abundance may also be determined by measurement of multiple peptides. In biomarker discovery, labeled or label-free peptide-based proteomics are great options, particularly for LC-MS/MS analyses. Peptide abundance has been stated to be linear to protein abundance (Griffin et al., 2010; Lundgren et al., 2010). Peptide quantification with or without derivatization via tags or labels may be conducted with the following methodologies.

The Use of Tagged Peptide to Estimate Protein Abundance

Chemical tags or labels may be used for relative or absolute quantification. Isobaric tags for relative and absolute quantitation (iTRAQ) are reagents contain multiplex reporter ion compounds, that bind to N-terminal amino groups. These isobaric tags are a set of structurally compounds that bind to the amino-terminus and epsilon-amino functional group of lysine residues and fragment during MS/MS. A short linker normalizes the total mass (m/z) added to each peptide and is flanked by the mass tag and an amine reactive group, like N-hydroxysuccinimide (NHS)-ester (Wiese et al., 2007). Detection and analysis of the reporter ion is somewhat analogous to MRM or SRM transitions. The reporter ion is measured for quantification where the peptide m/z data is stored for identification. Peptide, thus protein, quantification is elucidated from the peak intensity of the reporter ion. Tags or ions are detectable in the low mass range of the MS, which is ideal for analysis with MALDI-TOF/TOF or ESI-Quadrupole-TOF (Hultin-Rosenberg et al., 2013).

Similar to iTRAQ, Isotope Coded Affinity Tags (ICAT) reagents may be used to determine relative protein abundance. However, unlike iTRAQ, ICAT labels bind to sulfhydryl groups and remain fixed to the peptide after ionization, resulting in a mass-shifted peptide. ICAT reagents are either "heavy" and contain ^{13}C, or "light" and contain ^{12}C labeled linkers. Each reagent has a reactive iodoacetyl group on one end and biotinylated linker on the other. In a typical experiment, protein mixtures from two different groups are incubated with one of the each ICAT reagents, mixed 1:1, cleaved with trypsin, then isolated by affinity chromatography to generate an enriched pool of labeled peptides prior to LC-MS/MS. Peptides co-elute and the heavy: light ratio is then compared for each peptide. Often, multiple peptides are used to determine the fold change in protein abundance assembled by programs such as the Trans-Proteomic Pipeline (TPP) (Institute for Systems Biology, Seattle, WA) or Scaffold (Proteome Software, Portland, OR) (Keller et al., 2005; Searle, 2010).

In recent studies, the effect of apolipoprotein E (ApoE) in a mouse model of CCI has been studied using these techniques. This elegant and intricate analysis isolated albumin depleted plasma and labeled peptides with 4-plex iTRAQ. LC/MS-MS was performed on an LTQ-Orbitrap (Thermo Fisher Scientific) and data was analyzed with the TPP followed by protein network analysis. Short and long-term analysis of this model identified several proteins in plasma that decreased 24 hours after severe

CCI. Perhaps, most importantly, this study showed the effects on the proteome in reference to time, injury severity and presence of an allelic variant of ApoE, a protein well known in risk for neurodegeneration and Alzheimer's disease (Crawford et al., 2012).

This approach has also been used in biofluids. CSF, the most pertinent biofluid that is relevant to brain injury, is limited due to low sample volume in animal models and small sample size in many clinical studies. Thus, analytical methods, such as labeled proteomics, which exploit limited sample content, are ideal. A recent study using 4-plex iTRAQ reagents and CSF samples collected from severe TBI patients collected as late as 9 days after injury, demonstrated detection of several biomarkers including those with relatively high peptide counts (like GFAP) and those with low counts (like NSE) (Hanrieder et al., 2009). In a related study, human anti-mortem and postmortem CSF were compared in which putative biomarkers like GFAP and S100β were detectable in postmortem samples (Dayon et al., 2008). In addition to defining a proteomic based on biofluids, these methods have been useful in detecting proteins that correlate to clinical features of TBI. ITRAQ and LC-MS/MS proteomics of depleted serum determined that several protein biomarkers, such as serum amyloid A (SAA) and C-reactive protein (CRP), correlated to intracranial pressure (ICP) (Hergenroeder et al., 2008).

In experimental models of TBI, ICAT labeling has also been proven to be useful. For instance, cerebral vessels were characterized in a rodent model of ischemia-reperfusion. Individual protein lysates were compared to pooled sham controls for which more than 500 ICAT pairs were identified across multiple time-points after injury. Changes in the abundance of multiple cytokines were confirmed to be accurate by secondary methods, including quantitative PCR and enzyme linked immuno-sorbent assays (Haqqani et al., 2005). Although these methods can be sensitive, they capture only a partial proteome as the tag rely upon availability of specific amino acids. In a sense, this strategy may be considered an enrichment method. However, labeling efficiency may vary between users or lots. To address these caveats, label-free peptide based approaches may be used.

The Use of Label-free Spectral Counting

Unlabeled proteomics is achieved via spectral count statistics, e.g., counting the number of times a particular peptide is detected by the mass spectrometer. Essentially, this method provides a frequency distribution of peptides in the form of count data and is independent of labels or dyes. Peptide spectral (or spectrum) count has been shown to be proportional to peptide concentration (Hoehenwarter and Wienkoop, 2010; Old et al.,

2005). Peptides may identify multiple distinct proteins as well as several isoforms and related proteins. To simplify this issue and increase confidence in identification, peptide-protein matches may be parsed and re-assembled into grouped lists with novel algorithms, such as IDPicker (Holman et al., 2012; Ma et al., 2009). Data can then be sorted and analyzed by protein or protein group (wherein shared peptides across proteins are made evident), as well as by peptides or spectra. Other tools for spectral counting are included in many commercially available and open source software programs including Scaffold, emPAI Calc, Absolute Protein Expression (APEX), some of which take into account peptide peak area (Braisted et al., 2008; Shinoda et al., 2010; Zhang et al., 2006). These tools may offer different methods for normalizing and weighting or scoring matches as well as error estimation.

Label-free proteomics was recently used to determine changes in the brain proteome in a model of contusion. Here, subcellular fractionation, extensive peptide separation and informatics were performed in conjunction with MS/MS analysis. Compared to gel-based methods, this strategy increased the number of differential protein identifications by more than 8-fold; the number of differentially abundant peptides increased by more than 12-fold. Biomarkers in brain tissue were confirmed by immunoblotting and may be candidates for study in biofluids (Cortes et al., 2012). Although the majority of studies to date have used ICAT, iTRAQ or gel-based methods interest in liquid phase, label free proteomics is gaining momentum in TBI research and may be a key approach to the discovery and validation of novel biomarkers.

Biomarker Enrichment for LC-MS/MS

LC-MS/MS is capable of detecting 10^{-19} to 10^{-15} Molar peptide concentrations. However, target enrichment is often necessary, especially with CSF or plasma/serum. In addition, any biological sample wherein a sub-set of proteins based on PTMs or other chemical properties is desired benefit from purification, enrichment or abundant protein depletion. Common methods of enrichment are (1) albumin/immunoglobulin (Ig) depletion, (2) high abundant protein depletion with IgY column chromatography (Beckman Coulter, Fullerton, CA), YM-3 and YM-10 filter units (Millipore, Billerica, MA), (3) or PTM selection via site specific antibodies. For example, normal (e.g., healthy) CSF has nearly no protein content and the total protein content of serum or plasma are primarily albumin and immunoglobulins. The lack of enrichment of the target protein pool can severely underestimate the detection of proteins and quantitative measurement of potential biomarkers. Even in brain tissues, the number of peptides identified in MS/MS has been reported to increase by at least 25% with enrichment (Wang et al., 2006).

Peptide Selection and Concentration

In a recent and interesting study, hexapeptide ligand libraries were used to enrich peptides from human ventricular CSF which has very low abundance of protein, even in TBI where biomarkers are increased. Hexapeptide libraries are a form of solid phase affinity adsorption chromatography and are known commercially as ProteoMiner reagents (Boschetti and Giorgio Righetti, 2008). Ligand diversity is dependent upon the length of the peptide and the number of amino acids used. Low abundant proteins are retained; whereas, high molecular weight proteins quickly saturate the ligand and are easily washed away. Isolated proteins were first separated by IEF, digested with an enzyme, then analyzed by LC-MS/MS. Nearly half of the proteins detected were previously associated with neurodegeneration or TBI including GFAP, NSE and CKBB (Sjodin et al., 2010). The authors took a further step and listed the proteins detected to specific features in TBI, such as activation of the immune inflammatory response, apoptosis and hemorrhage. Furthermore, the comparative concentrations of potential biomarkers PDG_2 synthetase and 14-3-3, which are two proteins considered to be biomarkers of TBI in other studies (Siman et al., 2009). The use of hexapeptide libraries can increase the number of proteins identifiable in CSF by 2-fold or greater, easing biomarker discovery and quantification (Mouton-Barbosa et al., 2010).

Phospho-protein Purification

Phosphorylation, well studied form of signal transduction, can itself be considered a biomarker in TBI (Bayes and Grant, 2009). However, the stoichiometry of phosphorylation is quite low dephosphorylation is rapid and often transient. Proteomics studies in TBI may offer varying phospho-protein biomarkers depending on the severity and duration of TBI. Rapid enrichment of phospho-proteins or peptides with Immobilized Metal Affinity Chromatography (IMAC) is often imperative to study the phospho-proteome as a means to reduce sample complexity and enrich potential targets. One caveat of IMAC isolation is that the Fe3+ or TiO_2 resins may aberrantly bind carboxyl groups of mono-acidic peptides; although this background can be decreased by methyl-esterification of the sample prior to enrichment (Negroni et al., 2012). These methods can be very powerful, but the caveat of non-specific binding persists and phospho-specific antibodies have poor enrichment specificity/capability. In addition, pre-selection of phospho-proteins paired with the loss of non-phospho-proteins may skew data interpretation. To circumvent these problems, the direct study of global post-translational modifications prior to individual protein isolation may be employed.

Post-translational Modifications

Overview

Post-translational modifications (PTMs) are chemical modifications to the amino-acid primary structure. Compared to a single protein derived from one gene, the diversity of PTMs expands proteome complexity nearly exponentially. Under healthy or normal conditions PTMs have an active role in protein function and regulate protein, DNA/RNA and lipid interactions, enzyme activity, sub-cellular localization, and a plethora of other biological processes. PTMs are dynamic under normal conditions. However, aberrant and static modifications, such as excessive Tau phosphorylation or global oxidative damage, has been detected in TBI and its models (Goldstein et al., 2012; Hawkins et al., 2013; Schmidt et al., 2001; Shao et al., 2006; Tran et al., 2011). PTMs are proposed to be deleterious to proper protein function. Detection of PTMs within specific proteins or throughout the proteome may serve as TBI biomarkers. Some insights as to what to expect in TBI can be traced from the study of other brain diseases with common endpoints, such as cell death, neurodegeneration and gliosis like Alzheimer's Disease (AD). PTMs in TBI have yet to be fully explored; but (as with neurodegenerative diseases and cancer studies) it has the potential to be a great source of TBI biomarkers. Examples of current and future directions that focus on PTM-based proteomics in TBI are presented. Incorporation of modified proteins containing amino acid-specific information as protein biomarkers achieved through multiplex methods is also discussed.

Phosphorylation

Global phospho-protein detection may be performed with gel based assays or during MS analysis. For gel-based assays, the fluorescent dye, ProQ Diamond, allows detection of (hyper) phosphorylated proteins (Jacob and Turck, 2008). IMAC enrichment increases the ratio of phospho-peptides for direct MS analysis. In the latter case, phospho-peptides may be detected by monitoring fragmentation and loss of the neutral phosphate ion (H_3PO_4) or by mining for phospho-adducts with softer ionization methods, manual spectral interpretation and database searching (Schroeder et al., 2004). Phospho-peptide quantitation may also be determined with SRM-MS (Jin et al., 2010).

ProQ Diamond-PAGE was used in a rat model of spinal cord contusion which indicated that phospho-NF-L was increased compared to controls (Chen et al., 2010). NF protein phosphorylation has also been detected in AD (Rudrabhatla et al., 2010) and phospho-GFAP has been reported to be increased in frontotemporal lobar degeneration (FTLD). This study utilized peptide Immobilized Metal Affinity Chromatography (IMAC), LC-MS/MS, and label-free spectral counting of human post-mortem brain tissues. Increased phospho-GFAP was confirmed by neutral loss of the phosphate ion in MS and by Western blotting (Herskowitz et al., 2010). These studies collectively indicate that phosphoproteomics in TBI is feasible for study of severe, and possibly, mild or moderate injuries. For instance, phospho-NF was detected using 2D-DIGE and TOF/TOF in a study of protein interactions in a model of ischemia-reperfusion (Cid et al., 2007). Overall, phospho-proteins may lead to many TBI biomarkers.

Oxidative Stress

Carbonyl Oxidatation, Tyrosine Nitration and Lipid Peroxidation Products

Oxidative stress or damage has long been associated with TBI (Adibhatla and Hatcher, 2010; Bayir et al., 2002). During this process, reactive oxygen or nitrogen species (ROS, RNS) like superoxide ($O_2^{\bullet -}$), nitrite/nitrate (NO_2/NO_3), hydrogen peroxide (H_2O_2), and carbonate radical ($CO_3^{\bullet -}$), become abundant and form adducts on proteins leading to alteration of structure and function (Kontos and Wei, 1986). Carbonyl oxidation is the result of direct oxidation of amino-acids like proline, arginine, histidine, and glutamic acid. Peroxynitrite ($NO3^{\bullet -}$), the unstable isomer of NO_3, is toxic and reported to form adducts on tyrosine residues to form nitrotyrosine (3-NT) (Darwish et al., 2007). ROS also form adducts to lipids, leading to formation of lipid peroxidation products (LPOs), like hydroxyxynonenal (HNE) (Butterfield et al., 2010; Zarkovic, 2003).

Protein carbonyls may be measured by the use of oxyblots wherein the carbonyl is derivatized to 2,4-dinitrophenylhydrazone by reaction with 2,4-dinitrophenylhydrazine (DNPH). After separation by gel electrophoresis and transfer by Western blotting, the reactive (e.g., oxidatively modified) proteins are detected with an anti-DNP antibody. Under normal conditions in control models, DNPH will react with carbonyl-containing functional groups such as carboxylic acids, amides, or esters. However, the stoichiometry is much lower in the normal proteome compared to that of the experimental proteome affected by oxidative damage. Like DNPH-derivatized proteins, assessment of 3-nitro-tyrosine (3-NT), the product of tyrosine nitration, may be conducted with gel electrophoresis and Western blotting (Aksenov et al., 2001).

Both protein carbonyls and nitrotyrosine were increased in rat brains subjected to a weight drop model of TBI. Protein spots with significant intensity changes for either modification in TBI compared to sham were located on replicate gels that were post-stained with Sypro ruby. Tryptic digests and MALDI-TOF-PMF-MS analysis identified several proteins with increased oxidation, such as neuronal synapsin, and proteins with increased nitration, including α-spectrin (Reed et al., 2009). The role of ROS was also investigated in CCI by 2D-oxyblotting. Several oxidized proteins were identified in the cortex and hippocampus in an experimental CCI (Opii et al., 2007). In terms of direct clinical relevance of biomarkers, albumin oxidation was increased in the CSF of stroke patients (Moon et al., 2011).

These efforts indicate that modifications continue to be a rich source of potential TBI biomarkers that may be indicative of greater processes, such as oxidative stress or cellular signaling. Perhaps these biomarkers, considered collectively, may lead to definitive molecular mechanisms that occur during TBI. Of course, details of these mechanisms are likely to be quite divergent depending on the nature of the TBI model.

Targeted Affinity-based Proteomics

Overview

Proteomics encompasses not only MS based methods, but also targeted arrays that depend upon capture or binding of the protein or peptide from biological samples on a fixed surface. The method of capture may vary and includes nitrocellulose or PVDF membranes. Essentially, the biomarker or analyte is captured to the surface directly or binds to an antibody or nucleotide sequence. Next, a detection molecule, such as another antibody or fluorophore, is added to determine biomarker abundance. This method is more often used to detect intact proteins, although the use of peptides is feasible provided specificity is attainable.

Cytokine and Chemokine Arrays

Cytokines and chemokines are small intracellular signaling molecules that are readily secreted during cell signaling to initiate processes including injury site to inflammatory cell recruitment as well as glial activation in TBI (Woodcock and Morganti-Kossmann, 2013). Cytokines and chemokines are crucial mediators of TBI-induced neuro-inflammation. Cytokine arrays, like HTPI, rely upon dozens, if not hundreds, of antibodies fixed to blotting membrane. Many arrays are commercially available and provide highly sensitive relative quantification. This method is semi-quantitative and compares the signal (often derived from chemiluminescence) to a set of internal controls that may then be normalized to the amount of total protein loaded.

A more complex iteration of this method is a multiplexed, 96-well plate, bead array. Luminex assays (Luminex, Austin, TX, USA) are quantitative and can be combined with kits from many vendors. They allow detection of multiple analytes from a sample. Each bead has a distinct spectral signal to differentiate polystyrene beads (microspheres) from one-another. Here each bead type is dyed with distinct proportions of red and near-infrared fluorophores and can be used to perform ~100 different detection reactions. The assay itself is straight-forward. First, capture antibodies are bound to the beads and incubated with the analyte sample. Next, biotinylated detection antibodies are added followed by a streptavidin-conjugated fluorescent protein. The spectral properties of the bead and of the fluorescence are measured simultaneously. Quantification is achieved by the use of a standard for each target (Elshal and McCoy, 2006). Luminex cytokine assays (Invitrogen, Bio-Rad, Qiagen) are adaptable and some have been pre-optimized for plasma/serum as well as cell or tissue extracts.

Cytokine arrays of brain tissue lysates have indicated that PBBI, a model of severe, penetrating injury, leads to dramatic increases in IL-1β, IL-6, Chemokine (C-C motif) ligand 2 (CCL2, also known as monocyte chemo-attractant protein-1, MCP-1), and Chemokine (C-X-C motif) ligand 2 (CXCL2, also known as macrophage inflammatory protein 2-alpha, MIP2-α) (Wei et al., 2009). In a study which focused on patients with severe TBI, potential CSF and serum biomarkers were correlated to each other and to Cerebral Perfusion Pressure (CPP) and intracerebral pressure (ICP). Serum IL-8 and TNF-α had the highest correlations with ICP and intracerebral hemorrhage (ICH). In addition, the median level of IL-8 in serum was correlated with poor outcome (Stein et al., 2011b). A similar study indicated a strong accordance between the multiplexed cytokine assay and ELISA measurements of IL-8 in the CSF of pediatric TBI patients. Other cytokines and chemokines, such as IL-1β, IL-6, and IL-8 were increased after TBI compared to controls. Specifically, IL-1β and MIP-1α steadily increased 3 days after injury and these increases were attenuated with therapeutic hypothermia (Buttram et al., 2007). A direct comparison to adult TBI with varying degrees of severity or to animal models has yet to be determined, but offers a great challenge to identify novel biomarkers.

Protein Microarrays

A Reversed Phase Protein Microarrays (RPPM also termed Reverse Phase Protein Arrays (RPPA) or Reverse Phase Protein Microarray (RPMA)) is a high-throughput antibody based proteomics method in which antibodies are used to detected proteins spotted on to glass slides (Pierobon et al., 2011; Pierobon et al., 2012; Sereni et al., 2013). Much like genomics arrays, RPPMs lead to relative quantification of protein biomarkers based on binding affinity via detection intensity. RPPMs are similar to antibody microarrays (AbMA) in that they both exploit the antibody-antigen interaction. Great sensitivity of Western blotting and of chip array displays can be achieved; sensitivity will vary based on the antibody-biomarker interaction and fluorophore sensitivity. In addition, position effect (spot location) has to be corrected by appropriate background subtraction and signal-to-noise ratios. Although this method is much more high-throughput than a single dot-blot or Western blot, it relies upon pre-selection and potentially extensive optimization of antibodies to detect potential biomarkers. Thus, the investigator must enter the study with forethought of the biomarkers that are expected to change and rationale for studying specific biological samples as the source, e.g., tissue, CSF or blood. Key features of RPPM are the ability to test many samples or replicates on a single slide or membrane and the ability to exploit detection systems developed for nucleotide microarrays.

RPPM is ideal for individual proteins biomarker that may easily be applied to clinical TBI assays (Gyorgy et al., 2010). Blast TBI biomarkers were recently determined in serum from male Yorkshire pigs prior to and after injury at varying time-points (6, 24, 72 hours, and 2 weeks). Increased levels of serum S100B, NSE, MBP and NF-H after differing blast overpressures in each group: <20 psi, 20–40 psi, and <40 psi were detected (Gyorgy et al., 2011). In contrast, an enriched environment identified that blast TBI induced increases in biomarkers (tau and VEGF), was specifically abrogated in the dorsal hippocampus (Kovesdi et al., 2011; Yao et al., 2008).

Aptamer Arrays

Proteomics and mass spectrometry are inextricably linked, but any targeted approach that uses multiplexed arrays is a viable proteomics method. Aptamers were initially short DNA or RNA sequences used to capture specific proteins in assays that used chip arrays or on micro-well plates. Ligand (protein) binding is reported to be comparable to antibody affinity by vendors, but the premise was initially described on the basis that random oligomer sequence to ligand interactions would be low (Ellington and Szostak, 1990). Aptamers have been expanded to include peptides, L-DNA, L-RNA and other short sequences. Aptamer sequences can be custom synthesized and there is an expanding library of protein-nucleotide interactions. An aptamer database has been reported (Lee et al., 2004) and many vendors offer custom services.

As with a typical array, the target biomarker to be measured is predetermined and spotted onto a glass slide; making aptamer assays a targeted, high throughput approach. The density range of aptamer arrays is reported to be 2,000–94,000 individual oligomer sequences (Evans et al., 2008). Fluorescently tagged proteins in the biological sample are then added to the array prior to detection. On-chip aptamer arrays with fluorescent-tagged proteins have been demonstrated in non-TBI studies. A library of aptamers with nanomolar binding affinity was used to capture biotinylated thrombin, which was detected with Cy5 labeled streptavidin after screening against an initial library containing >40,000 oligomers (Platt et al., 2009). If the user chooses, this assay may be accommodated to become quantitative rather than relative. However, the specific protein-nucleotide affinity and specificity, as well as linear range of detection and of quantification must be determined for each potential biomarker.

Similar to RPPM, this array also benefits from fluorescent detection systems already available from DNA/RNA microarray studies. Another benefit of aptamers is that a library can be created for each protein and plated onto chips for a direct readout of differential protein abundance wherein protein identification confirmation is not an absolute necessity. Even with such a vast multiplexed limit and applicability to central nervous systems studies (Yang et al., 2007), aptamers have yet to be applied to a broad biomarker discovery or confirmation platform in TBI.

The Future of TBI Proteomics and Biomarker Discovery

TBI proteomics of biomarkers in tissue, cells, and biofluids will aid in clinical diagnostics of severity, duration and therapeutic response. Biomarkers are also imperative to understand the mechanism of injury. The myriad of proteomics techniques available allow detection and discovery of novel biomarkers with great assay sensitivity. TBI biomarker proteomics is poised to grow exponentially and represent the complexity of a dynamic pathological state which involves study of several cell types (neurons, glia, endothelial cells), injury types (mild–severe), and biological compartments that contain the biomarker analyte of interest (specific brain regions, CSF or blood). A plethora of methodologies is available: (1) unbiased or targeted proteomics with or without labels to decrease complexity of the proteome, (2) enrichment based on chemical properties or PTMs to enhance signal-noise ratios, and (3) bioinformatics pathway analysis and data-mining proteomics based on published reports.

Individual protein biomarkers may soon give way to understanding biomarkers as multiplexed coordinated features, such as a group of cytoskeletal proteins. To date, neuronal α-spectrin and astrocytic GFAP seem to be the most promising co-occurring TBI biomarkers. In addition, calpain-2 or caspase-3 generated break-down products have been detected across several models and clinical studies (Mondello et al., 2010; Pike et al., 2001; Ringger et al., 2004). Most recently, novel biomarkers may be predicted by bioinformatics analysis based on curated databases containing protein-protein interactions that occur *in vitro* or *in vivo* (Feala et al., 2013). Some biomarkers may also be predicted based on primary sequence or tertiary structure similarity analyses with algorithms like hidden Markov modeling (Asai et al., 1993).

The future of TBI biomarkers may not simply be defined by a few proteins. Much like the board game Clue (Hasbro Pawtucket, Rhode Island), biomarkers can be envisioned as a series of variables that must occur simultaneously. For example, protein X (Colonel Mustard) in anatomical region or compartment Y (The Library), at time t_1 (Saturday Evening), with phosphorylation at amino-acid AA (The Candle Stick), must be present to define TBI (Solve the Crime). Furthermore, multiple proteins, as well as their interactions, may be necessary to develop a biomarker panel that fully describes a TBI with appropriate sensitivity and specificity. Proteomic technologies are poised to have a strong impact in TBI research.

Disclosure

This material has been reviewed by the Walter Reed Army Institute of Research. There is no objection to its presentation and/or publication. The opinions or assertions contained herein are the private views of the authors, and are not to be construed as official, or as reflecting true views of the Department of the Army or the Department of Defense.

References

Adibhatla, R.M. and J.F. Hatcher. 2010. Lipid oxidation and peroxidation in CNS health and disease: from molecular mechanisms to therapeutic opportunities. *Antioxidants & Redox Signaling*. 12: 125–169.

Agoston, D.V., A. Gyorgy, O. Eidelman and H.B. Pollard. 2009. Proteomic biomarkers for blast neurotrauma: targeting cerebral edema, inflammation, and neuronal death cascades. *J Neurotrauma*. 26: 901–911.

Ahlers, S.T., E. Vasserman-Stokes, M.C. Shaughness, A.A. Hall, D.A. Shear, M. Chavko, R.M. McCarron and J.R. Stone. 2012. Assessment of the effects of acute and repeated exposure to blast overpressure in rodents: toward a greater understanding of blast and the potential ramifications for injury in humans exposed to blast. *Front Neurol*. 3: 32.

Ahn, E.S., C.L. Robertson, V. Vereczki, G.E. Hoffman and G. Fiskum. 2008. Normoxic ventilatory resuscitation following controlled cortical impact reduces peroxynitrite-mediated protein nitration in the hippocampus. *J Neurosurg*. 108: 124–131.

Aikman, J., B. O'Steen, X. Silver, R. Torres, S. Boslaugh, S. Blackband, K. Padgett, K.K. Wang, R. Hayes and J. Pineda. 2006. Alpha-II-spectrin after controlled cortical impact in the immature rat brain. *Dev Neurosci.* 28: 457–465.

Aksenov, M.Y., M.V. Aksenova, D.A. Butterfield, J.W. Geddes and W.R. Markesbery. 2001. Protein oxidation in the brain in Alzheimer's disease. *Neuroscience.* 103: 373–383.

Alder, J., W. Fujioka, J. Lifshitz, D.P. Crockett and S. Thakker-Varia. 2011. Lateral fluid percussion: model of traumatic brain injury in mice. *J Vis Exp.*

Anderson, K.J., S.W. Scheff, K.M. Miller, K.N. Roberts, L.K. Gilmer, C. Yang and G. Shaw. 2008. The phosphorylated axonal form of the neurofilament subunit NF-H (pNF-H) as a blood biomarker of traumatic brain injury. *J Neurotrauma.* 25: 1079–1085.

Anderson, N.L. 2010. Counting the proteins in plasma. *Clin Chem.* 56: 1775–1776.

Anderson, N.L. and N.G. Anderson. 2002. The human plasma proteome: history, character, and diagnostic prospects. *Mol Cell Proteomics.* 1: 845–867.

Arnold, G.J. and T. Frohlich. 2012. 2D DIGE saturation labeling for minute sample amounts. *Methods Mol Biol.* 854: 89–112.

Asai, K., S. Hayamizu and K. Handa. 1993. Prediction of protein secondary structure by the hidden Markov model. *Comput Appl Biosci.* 9: 141–146.

Bayes, A. and S.G. Grant. 2009. Neuroproteomics: understanding the molecular organization and complexity of the brain. *Nat Rev Neurosci.* 10: 635–646.

Bayir, H., V.E. Kagan, Y.Y. Tyurina, V. Tyurin, R.A. Ruppel, P.D. Adelson, S.H. Graham, K. Janesko, R.S. Clark and P.M. Kochanek. 2002. Assessment of antioxidant reserves and oxidative stress in cerebrospinal fluid after severe traumatic brain injury in infants and children. *Pediatr Res.* 51: 571–578.

Beschorner, R., T.D. Nguyen, F. Gozalan, I. Pedal, R. Mattern, H.J. Schluesener, R. Meyermann and J.M. Schwab. 2002. CD14 expression by activated parenchymal microglia/macrophages and infiltrating monocytes following human traumatic brain injury. *Acta Neuropathologica.* 103: 541–549.

Bienvenut, W.V., C. Deon, C. Pasquarello, J.M. Campbell, J.C. Sanchez, M.L. Vestal and D.F. Hochstrasser. 2002. Matrix-assisted laser desorption/ionization-tandem mass spectrometry with high resolution and sensitivity for identification and characterization of proteins. *Proteomics.* 2: 868–876.

Bohmer, A.E., J.P. Oses, A.P. Schmidt, C.S. Peron, C.L. Krebs, P.P. Oppitz, T.T. D'Avila, D.O. Souza, L.V. Portela and M.A. Stefani. 2011. Neuron-specific enolase, S100B, and glial fibrillary acidic protein levels as outcome predictors in patients with severe traumatic brain injury. *Neurosurgery.* 68: 1624–1630; discussion 1630–1.

Bora, A., C. Anderson, M. Bachani, A. Nath and R.J. Cotter. 2012. Robust Two-Dimensional Separation of Intact Proteins for Bottom-Up Tandem Mass Spectrometry of the Human CSF Proteome. *J Proteome Res.*

Boschetti, E. and P. Giorgio Righetti. 2008. Hexapeptide combinatorial ligand libraries: the march for the detection of the low-abundance proteome continues. *Biotechniques.* 44: 663–665.

Boutte, A.M., C. Yao, F. Kobeissy, X.C. May Lu, Z. Zhang, K.K. Wang, K. Schmid, F.C. Tortella and J.R. Dave. 2012. Proteomic analysis and brain-specific systems biology in a rodent model of penetrating ballistic-like brain injury. *Electrophoresis.* 33: 3693–3704.

Braisted, J.C., S. Kuntumalla, C. Vogel, E.M. Marcotte, A.R. Rodrigues, R. Wang, S.T. Huang, E.S. Ferlanti, A.I. Saeed, R.D. Fleischmann, S.N. Peterson and R. Pieper. 2008. The APEX Quantitative Proteomics Tool: generating protein quantitation estimates from LC-MS/MS proteomics results. *BMC Bioinformatics.* 9: 529.

Brophy, G.M., J.A. Pineda, L. Papa, S.B. Lewis, A.B. Valadka, H.J. Hannay, S.C. Heaton, J.A. Demery, M.C. Liu, J.J. Tepas, 3rd, A. Gabrielli, S. Robicsek, K.K. Wang, C.S. Robertson and R.L. Hayes. 2009. alphaII-Spectrin breakdown product cerebrospinal fluid exposure metrics suggest differences in cellular injury mechanisms after severe traumatic brain injury. *J Neurotrauma.* 26: 471–479.

Bulut, M., O. Koksal, S. Dogan, N. Bolca, H. Ozguc, E. Korfali, Y.O. Ilcol and M. Parklak. 2006. Tau protein as a serum marker of brain damage in mild traumatic brain injury: preliminary results. *Adv Ther.* 23: 12–22.

Butterfield, D.A., M.L. Bader Lange and R. Sultana. 2010. Involvements of the lipid peroxidation product, HNE, in the pathogenesis and progression of Alzheimer's disease. *Biochimica et Biophysica Acta.* 1801: 924–929.

Buttram, S.D., S.R. Wisniewski, E.K. Jackson, P.D. Adelson, K. Feldman, H. Bayir, R.P. Berger, R.S. Clark and P.M. Kochanek. 2007. Multiplex assessment of cytokine and chemokine levels in cerebrospinal fluid following severe pediatric traumatic brain injury: effects of moderate hypothermia. *J Neurotrauma.* 24: 1707–1717.

Carmichael, S.T. 2005. Rodent models of focal stroke: size, mechanism, and purpose. *NeuroRx.* 2: 396–409.

Chen, A., M.L. McEwen, S. Sun, R. Ravikumar and J.E. Springer. 2010. Proteomic and phosphoproteomic analyses of the soluble fraction following acute spinal cord contusion in rats. *J Neurotrauma.* 27: 263–274.

Chen, Z., L.Y. Leung, A. Mountney, Z. Liao, W. Yang, X.C. Lu, J. Dave, Y. Deng-Bryant, G. Wei, K. Schmid, D.A. Shear and F.C. Tortella. 2012. A novel animal model of closed-head concussive-induced mild traumatic brain injury: development, implementation, and characterization. *J Neurotrauma.* 29: 268–280.

Cid, C., L. Garcia-Bonilla, E. Camafeita, J. Burda, M. Salinas and A. Alcazar. 2007. Proteomic characterization of protein phosphatase 1 complexes in ischemia-reperfusion and ischemic tolerance. *Proteomics.* 7: 3207–3218.

Cortes, D.F., M.K. Landis and A.K. Ottens. 2012. High-capacity peptide-centric platform to decode the proteomic response to brain injury. *Electrophoresis.* 33: 3712–3719.

Craig, R., J.P. Cortens and R.C. Beavis. 2005. The use of proteotypic peptide libraries for protein identification. *Rapid Commun Mass Spectrom.* 19: 1844–1850.

Crawford, F., G. Crynen, J. Reed, B. Mouzon, A. Bishop, B. Katz, S. Ferguson, J. Phillips, V. Ganapathi, V. Mathura, A. Roses and M. Mullan. 2012. Identification of plasma biomarkers of TBI outcome using proteomic approaches in an APOE mouse model. *J Neurotrauma.* 29: 246–260.

Darwish, R.S., N. Amiridze and B. Aarabi. 2007. Nitrotyrosine as an oxidative stress marker: evidence for involvement in neurologic outcome in human traumatic brain injury. *J Trauma.* 63: 439–442.

Dasilva, N., P. Diez, S. Matarraz, M. Gonzalez-Gonzalez, S. Paradinas, A. Orfao and M. Fuentes. 2012. Biomarker discovery by novel sensors based on nanoproteomics approaches. *Sensors (Basel).* 12: 2284–2308.

Dave, J.R., A.J. Williams, C. Yao, X.C. Lu and F.C. Tortella. 2009. Modeling cerebral ischemia in neuroproteomics. *Methods Mol Biol.* 566: 25–40.

Dayon, L., A. Hainard, V. Licker, N. Turck, K. Kuhn, D.F. Hochstrasser, P.R. Burkhard and J.C. Sanchez. 2008. Relative quantification of proteins in human cerebrospinal fluids by MS/MS using 6-plex isobaric tags. *Anal Chem.* 80: 2921–2931.

Diaz-Arrastia, R., K.K. Wang, L. Papa, M.D. Sorani, J.K. Yue, A.M. Puccio, P.J. McMahon, T. Inoue, E.L. Yuh, H.F. Lingsma, A.I. Maas, A.B. Valadka, D.O. Okonkwo, G.T. Manley, S.S. Casey, M. Cheong, S.R. Cooper, K. Dams-O'Connor, W.A. Gordon, A.J. Hricik, D.K. Menon, P. Mukherjee, D.M. Schnyer, T.K. Sinha and M.J. Vassar. 2013. Acute Biomarkers of Traumatic Brain Injury: Relationship between Plasma Levels of Ubiquitin C-Terminal Hydrolase-L1 and Glial Fibrillary Acidic Protein. *J Neurotrauma.*

Dixon, C.E., G.L. Clifton, J.W. Lighthall, A.A. Yaghmai and R.L. Hayes. 1991. A controlled cortical impact model of traumatic brain injury in the rat. *J Neurosci Methods.* 39: 253–262.

Dixon, C.E., B.G. Lyeth, J.T. Povlishock, R.L. Findling, R.J. Hamm, A. Marmarou, H.F. Young and R.L. Hayes. 1987. A fluid percussion model of experimental brain injury in the rat. *J Neurosurg.* 67: 110–119.

Elder, G.A., N.P. Dorr, R. De Gasperi, M.A. Gama Sosa, M.C. Shaughness, E. Maudlin-Jeronimo, A.A. Hall, R.M. McCarron and S.T. Ahlers. 2012. Blast exposure induces post-traumatic stress disorder-related traits in a rat model of mild traumatic brain injury. *J Neurotrauma*. 29: 2564–2575.

Elder, G.A., E.M. Mitsis, S.T. Ahlers and A. Cristian. 2010. Blast-induced mild traumatic brain injury. *Psychiatr Clin North Am*. 33: 757–781.

Ellington, A.D. and J.W. Szostak. 1990. *In vitro* selection of RNA molecules that bind specific ligands. *Nature*. 346: 818–822.

Elshal, M.F. and J.P. McCoy. 2006. Multiplex bead array assays: performance evaluation and comparison of sensitivity to ELISA. *Methods*. 38: 317–323.

Evans, D., S. Johnson, S. Laurenson, A.G. Davies, P. Ko Ferrigno and C. Walti. 2008. Electrical protein detection in cell lysates using high-density peptide-aptamer microarrays. *Journal of Biology*. 7: 3.

Feala, J.D., M.D. Abdulhameed, C. Yu, B. Dutta, X. Yu, K. Schmid, J. Dave, F. Tortella and J. Reifman. 2013. Systems biology approaches for discovering biomarkers for traumatic brain injury. *J Neurotrauma*. 30: 1101–1116.

Friedman, D.B. 2012. Assessing signal-to-noise in quantitative proteomics: multivariate statistical analysis in DIGE experiments. *Methods Mol Biol*. 854: 31–45.

Friedman, D.B. and K.S. Lilley. 2008. Optimizing the difference gel electrophoresis (DIGE) technology. *Methods Mol Biol*. 428: 93–124.

Gabbita, S.P., S.W. Scheff, R.M. Menard, K. Roberts, I. Fugaccia and F.P. Zemlan. 2005. Cleaved-tau: a biomarker of neuronal damage after traumatic brain injury. *J Neurotrauma*. 22: 83–94.

Gerber, S.A., A.N. Kettenbach, J. Rush and S.P. Gygi. 2007. The absolute quantification strategy: application to phosphorylation profiling of human separase serine 1126. *Methods Mol Biol*. 359: 71–86.

Ghirnikar, R.S., Y.L. Lee and L.F. Eng. 1998. Inflammation in traumatic brain injury: role of cytokines and chemokines. *Neurochem Res*. 23: 329–340.

Gogichaeva, N.V. and M.A. Alterman. 2012. Amino acid analysis by means of MALDI TOF mass spectrometry or MALDI TOF/TOF tandem mass spectrometry. *Methods Mol Biol*. 828: 121–135.

Goldstein, L.E., A.M. Fisher, C.A. Tagge, X.L. Zhang, L. Velisek, J.A. Sullivan, C. Upreti, J.M. Kracht, M. Ericsson, M.W. Wojnarowicz, C.J. Goletiani, G.M. Maglakelidze, N. Casey, J.A. Moncaster, O. Minaeva, R.D. Moir, C.J. Nowinski, R.A. Stern, R.C. Cantu, J. Geiling, J.K. Blusztajn, B.L. Wolozin, T. Ikezu, T.D. Stein, A.E. Budson, N.W. Kowall, D. Chargin, A. Sharon, S. Saman, G.F. Hall, W.C. Moss, R.O. Cleveland, R.E. Tanzi, P.K. Stanton and A.C. McKee. 2012. Chronic traumatic encephalopathy in blast-exposed military veterans and a blast neurotrauma mouse model. *Science Translational Medicine*. 4: 134ra160.

Griffin, N.M., J. Yu, F. Long, P. Oh, S. Shore, Y. Li, J.A. Koziol and J.E. Schnitzer. 2010. Label-free, normalized quantification of complex mass spectrometry data for proteomic analysis. *Nat Biotechnol*. 28: 83–89.

Guingab-Cagmat, J.D., E.B. Cagmat, R.L. Hayes and J. Anagli. 2013. Integration of proteomics, bioinformatics, and systems biology in traumatic brain injury biomarker discovery. *Front Neurol*. 4: 61.

Gyorgy, A., G. Ling, D. Wingo, J. Walker, L. Tong, S. Parks, A. Januszkiewicz, R. Baumann and D.V. Agoston. 2011. Time-dependent changes in serum biomarker levels after blast traumatic brain injury. *J Neurotrauma*. 28: 1121–1126.

Gyorgy, A.B., J. Walker, D. Wingo, O. Eidelman, H.B. Pollard, A. Molnar and D.V. Agoston. 2010. Reverse phase protein microarray technology in traumatic brain injury. *J Neurosci Methods*. 192: 96–101.

Hanrieder, J., M. Wetterhall, P. Enblad, L. Hillered and J. Bergquist. 2009. Temporally resolved differential proteomic analysis of human ventricular CSF for monitoring traumatic brain injury biomarker candidates. *J Neurosci Methods*. 177: 469–478.

Haqqani, A.S., M. Nesic, E. Preston, E. Baumann, J. Kelly and D. Stanimirovic. 2005. Characterization of vascular protein expression patterns in cerebral ischemia/reperfusion using laser capture microdissection and ICAT-nanoLC-MS/MS. *FASEB J*. 19: 1809–1821.

Haskins, W.E., F.H. Kobeissy, R.A. Wolper, A.K. Ottens, J.W. Kitlen, S.H. McClung, B.E. O'Steen, M.M. Chow, J.A. Pineda, N.D. Denslow, R.L. Hayes and K.K. Wang. 2005. Rapid discovery of putative protein biomarkers of traumatic brain injury by SDS-PAGE-capillary liquid chromatography-tandem mass spectrometry. *J Neurotrauma*. 22: 629–644.

Hawkins, B.E., S. Krishnamurthy, D.L. Castillo-Carranza, U. Sengupta, D.S. Prough, G.R. Jackson, D.S. DeWitt and R. Kayed. 2013. Rapid accumulation of endogenous tau oligomers in a rat model of traumatic brain injury: possible link between traumatic brain injury and sporadic tauopathies. *J Biol Chem*. 288: 17042–17050.

Hergenroeder, G., J.B. Redell, A.N. Moore, W.P. Dubinsky, R.T. Funk, J. Crommett, G.L. Clifton, R. Levine, A. Valadka and P.K. Dash. 2008. Identification of serum biomarkers in brain-injured adults: potential for predicting elevated intracranial pressure. *J Neurotrauma*. 25: 79–93.

Herskowitz, J.H., N.T. Seyfried, D.M. Duong, Q. Xia, H.D. Rees, M. Gearing, J. Peng, J.J. Lah and A.I. Levey. 2010. Phosphoproteomic analysis reveals site-specific changes in GFAP and NDRG2 phosphorylation in frontotemporal lobar degeneration. *J Proteome Res*. 9: 6368–6379.

Hoehenwarter, W. and S. Wienkoop. 2010. Spectral counting robust on high mass accuracy mass spectrometers. *Rapid Commun Mass Spectrom*. 24: 3609–3614.

Hohl, A., J.D. Gullo, C.C. Silva, M.M. Bertotti, F. Felisberto, J.C. Nunes, B. de Souza, F. Petronilho, F.M. Soares, R.D. Prediger, F. Dal-Pizzol, M.N. Linhares and R. Walz. 2011. Plasma levels of oxidative stress biomarkers and hospital mortality in severe head injury: A multivariate analysis. *J Crit Care*.

Holman, J.D., Z.Q. Ma and D.L. Tabb. 2012. Identifying proteomic LC-MS/MS data sets with Bumbershoot and IDPicker. *Curr Protoc Bioinformatics*. Chapter 13:Unit13 17.

Honda, M., R. Tsuruta, T. Kaneko, S. Kasaoka, T. Yagi, M. Todani, M. Fujita, T. Izumi and T. Maekawa. 2010. Serum glial fibrillary acidic protein is a highly specific biomarker for traumatic brain injury in humans compared with S-100B and neuron-specific enolase. *J Trauma*. 69: 104–109.

Huber, B.R., J.S. Meabon, T.J. Martin, P.D. Mourad, R. Bennett, B.C. Kraemer, I. Cernak, E.C. Petrie, M.J. Emery, E.R. Swenson, C. Mayer, E. Mehic, E.R. Peskind and D.G. Cook. 2013. Blast exposure causes early and persistent aberrant phospho- and cleaved-tau expression in a murine model of mild blast-induced traumatic brain injury. *Journal of Alzheimer's Disease: JAD*. 37: 309–323.

Hultin-Rosenberg, L., J. Forshed, R.M. Branca, J. Lehtio and H.J. Johansson. 2013. Defining, comparing, and improving iTRAQ quantification in mass spectrometry proteomics data. *Mol Cell Proteomics*. 12: 2021–2031.

Jacob, A.M. and C.W. Turck. 2008. Detection of post-translational modifications by fluorescent staining of two-dimensional gels. *Methods Mol Biol*. 446: 21–32.

Jenkins, L.W., G.W. Peters, C.E. Dixon, X. Zhang, R.S. Clark, J.C. Skinner, D.W. Marion, P.D. Adelson and P.M. Kochanek. 2002. Conventional and functional proteomics using large format two-dimensional gel electrophoresis 24 hours after controlled cortical impact in postnatal day 17 rats. *J Neurotrauma*. 19: 715–740.

Jin, L.L., J. Tong, A. Prakash, S.M. Peterman, J.R. St-Germain, P. Taylor, S. Trudel and M.F. Moran. 2010. Measurement of protein phosphorylation stoichiometry by selected reaction monitoring mass spectrometry. *J Proteome Res*. 9: 2752–2761.

Kane, M.J., M. Angoa-Perez, D.I. Briggs, D.C. Viano, C.W. Kreipke and D.M. Kuhn. 2012. A mouse model of human repetitive mild traumatic brain injury. *J Neurosci Methods*. 203: 41–49.

Keller, A., J. Eng, N. Zhang, X.J. Li and R. Aebersold. 2005. A uniform proteomics MS/MS analysis platform utilizing open XML file formats. *Mol Syst Biol*. 1: 2005 0017.

Kirkpatrick, D.S., S.A. Gerber and S.P. Gygi. 2005. The absolute quantification strategy: a general procedure for the quantification of proteins and post-translational modifications. *Methods*. 35: 265–273.

Kobeissy, F.H., A.K. Ottens, Z. Zhang, M.C. Liu, N.D. Denslow, J.R. Dave, F.C. Tortella, R.L. Hayes and K.K. Wang. 2006. Novel differential neuroproteomics analysis of traumatic brain injury in rats. *Mol Cell Proteomics.* 5: 1887–1898.

Kochanek, A.R., A.E. Kline, W.M. Gao, M. Chadha, Y. Lai, R.S. Clark, C.E. Dixon and L.W. Jenkins. 2006. Gel-based hippocampal proteomic analysis 2 weeks following traumatic brain injury to immature rats using controlled cortical impact. *Dev Neurosci.* 28: 410–419.

Kontos, H.A. and E.P. Wei. 1986. Superoxide production in experimental brain injury. *J Neurosurg.* 64: 803–807.

Kovesdi, E., A.B. Gyorgy, S.K. Kwon, D.L. Wingo, A. Kamnaksh, J.B. Long, C.E. Kasper and D.V. Agoston. 2011. The effect of enriched environment on the outcome of traumatic brain injury; a behavioral, proteomics, and histological study. *Front Neurosci.* 5: 42.

Kwon, S.K., E. Kovesdi, A.B. Gyorgy, D. Wingo, A. Kamnaksh, J. Walker, J.B. Long and D.V. Agoston. 2011. Stress and traumatic brain injury: a behavioral, proteomics, and histological study. *Front Neurol.* 2: 12.

Lee, C. and D.V. Agoston. 2010. Vascular endothelial growth factor is involved in mediating increased *de novo* hippocampal neurogenesis in response to traumatic brain injury. *J Neurotrauma.* 27: 541–553.

Lee, J.F., J.R. Hesselberth, L.A. Meyers and A.D. Ellington. 2004. Aptamer database. *Nucleic Acids Res.* 32: D95–100.

Liu, M.C., V. Akle, W. Zheng, J.R. Dave, F.C. Tortella, R.L. Hayes and K.K. Wang. 2006. Comparing calpain- and caspase-3-mediated degradation patterns in traumatic brain injury by differential proteome analysis. *Biochem J.* 394: 715–725.

Lumpkins, K.M., G.V. Bochicchio, K. Keledjian, J.M. Simard, M. McCunn and T. Scalea. 2008. Glial fibrillary acidic protein is highly correlated with brain injury. *J Trauma.* 2008 Oct; 65(4): 778–82; discussion 782-4.

Lundgren, D.H., S.I. Hwang, L. Wu and D.K. Han. 2010. Role of spectral counting in quantitative proteomics. *Expert Rev Proteomics.* 7: 39–53.

Ma, Z.Q., S. Dasari, M.C. Chambers, M.D. Litton, S.M. Sobecki, L.J. Zimmerman, P.J. Halvey, B. Schilling, P.M. Drake, B.W. Gibson and D.L. Tabb. 2009. IDPicker 2.0: Improved protein assembly with high discrimination peptide identification filtering. *J Proteome Res.* 8: 3872–3881.

Magnoni, S., T.J. Esparza, V. Conte, M. Carbonara, G. Carrabba, D.M. Holtzman, G.J. Zipfel, N. Stocchetti and D.L. Brody. 2012. Tau elevations in the brain extracellular space correlate with reduced amyloid-beta levels and predict adverse clinical outcomes after severe traumatic brain injury. *Brain.* 135: 1268–1280.

Malakhov, M.P., K.I. Kim, O.A. Malakhova, B.S. Jacobs, E.C. Borden and D.E. Zhang. 2003. High-throughput immunoblotting. Ubiquitiin-like protein ISG15 modifies key regulators of signal transduction. *J Biol Chem.* 278: 16608–16613.

Mallick, P., M. Schirle, S.S. Chen, M.R. Flory, H. Lee, D. Martin, J. Ranish, B. Raught, R. Schmitt, T. Werner, B. Kuster and R. Aebersold. 2007. Computational prediction of proteotypic peptides for quantitative proteomics. *Nat Biotechnol.* 25: 125–131.

Martins-de-Souza, D. 2010. Is the word 'biomarker' being properly used by proteomics research in neuroscience? *European Archives of Psychiatry and Clinical Neuroscience.* 260: 561–562.

Mayr, M. and T. Rabilloud. 2013. Multidimensional separation prior to mass spectrometry: getting closer to the bottom of the iceberg. *Proteomics.* 13: 2942–2943.

Megger, D.A., L.L. Pott, M. Ahrens, J. Padden, T. Bracht, K. Kuhlmann, M. Eisenacher, H.E. Meyer and B. Sitek. 2013. Comparison of label-free and label-based strategies for proteome analysis of hepatoma cell lines. *Biochimica et Biophysica Acta.*

Mertins, P., J.W. Qiao, J. Patel, N.D. Udeshi, K.R. Clauser, D.R. Mani, M.W. Burgess, M.A. Gillette, J.D. Jaffe and S.A. Carr. 2013. Integrated proteomic analysis of post-translational modifications by serial enrichment. *Nature Methods.* 10: 634–637.

Mondello, S., A. Jeromin, A. Buki, R. Bullock, E. Czeiter, N. Kovacs, P. Barzo, K. Schmid, F. Tortella, K.K. Wang and R.L. Hayes. 2012. Glial neuronal ratio: a novel index for differentiating injury type in patients with severe traumatic brain injury. *J Neurotrauma.* 29: 1096–1104.

Mondello, S., S.A. Robicsek, A. Gabrielli, G.M. Brophy, L. Papa, J. Tepas, C. Robertson, A. Buki, D. Scharf, M. Jixiang, L. Akinyi, U. Muller, K.K. Wang and R.L. Hayes. 2010. alphaII-spectrin breakdown products (SBDPs): diagnosis and outcome in severe traumatic brain injury patients. *J Neurotrauma.* 27: 1203–1213.

Mondello, S., K. Schmid, R.P. Berger, F. Kobeissy, D. Italiano, A. Jeromin, R.L. Hayes, F.C. Tortella and A. Buki. 2013. The Challenge of Mild Traumatic Brain Injury: Role of Biochemical Markers in Diagnosis of Brain Damage. *Medicinal Research Reviews.*

Moon, G.J., D.H. Shin, D.S. Im, O.Y. Bang, H.S. Nam, J.H. Lee, I.S. Joo, K. Huh and B.J. Gwag. 2011. Identification of oxidized serum albumin in the cerebrospinal fluid of ischaemic stroke patients. *Eur J Neurol.* 18: 1151–1158.

Morochovic, R., O. Racz, M. Kitka, S. Pingorova, P. Cibur, D. Tomkova and R. Lenartova. 2009. Serum S100B protein in early management of patients after mild traumatic brain injury. *Eur J Neurol.* 16: 1112–1117.

Mouton-Barbosa, E., F. Roux-Dalvai, D. Bouyssie, F. Berger, E. Schmidt, P.G. Righetti, L. Guerrier, E. Boschetti, O. Burlet-Schiltz, B. Monsarrat and A. Gonzalez de Peredo. 2010. In-depth exploration of cerebrospinal fluid by combining peptide ligand library treatment and label-free protein quantification. *Mol Cell Proteomics.* 9: 1006–1021.

Negroni, L., S. Claverol, J. Rosenbaum, E. Chevet, M. Bonneu and J.M. Schmitter. 2012. Comparison of IMAC and MOAC for phosphopeptide enrichment by column chromatography. *J Chromatogr B Analyt Technol Biomed Life Sci.* 891-892: 109–112.

Nilsson, T., M. Mann, R. Aebersold, J.R. Yates, 3rd, A. Bairoch and J.J. Bergeron. 2010. Mass spectrometry in high-throughput proteomics: ready for the big time. *Nature Methods.* 7: 681–685.

Nirogi, R., V. Kandikere, K. Mudigonda, G. Bhyrapuneni, N. Muddana, R. Saralaya and V. Benade. 2009. A simple and rapid method to collect the cerebrospinal fluid of rats and its application for the assessment of drug penetration into the central nervous system. *J Neurosci Methods.* 178: 116–119.

Okonkwo, D.O., J.K. Yue, A.M. Puccio, D.M. Panczykowski, T. Inoue, P.J. McMahon, M.D. Sorani, E.L. Yuh, H.F. Lingsma, A.I. Maas, A.B. Valadka, Manley, R. Transforming, G.T. Clinical Knowledge In Traumatic Brain Injury Investigators Including, S.S. Casey, M. Cheong, S.R. Cooper, K. Dams-O'Connor, W.A. Gordon, A.J. Hricik, K. Hochberger, D.K. Menon, P. Mukherjee, T.K. Sinha, D.M. Schnyer and M.J. Vassar. 2013. GFAP-BDP as an acute diagnostic marker in traumatic brain injury: results from the prospective transforming research and clinical knowledge in traumatic brain injury study. *J Neurotrauma.* 30: 1490–1497.

Old, W.M., K. Meyer-Arendt, L. Aveline-Wolf, K.G. Pierce, A. Mendoza, J.R. Sevinsky, K.A. Resing and N.G. Ahn. 2005. Comparison of label-free methods for quantifying human proteins by shotgun proteomics. *Mol Cell Proteomics.* 4: 1487–1502.

Omalu, B.I., R.P. Fitzsimmons, J. Hammers and J. Bailes. 2010. Chronic traumatic encephalopathy in a professional American wrestler. *Journal of Forensic Nursing.* 6: 130–136.

Opii, W.O., V.N. Nukala, R. Sultana, J.D. Pandya, K.M. Day, M.L. Merchant, J.B. Klein, P.G. Sullivan and D.A. Butterfield. 2007. Proteomic identification of oxidized mitochondrial proteins following experimental traumatic brain injury. *J Neurotrauma.* 24: 772–789.

Ottens, A.K., L. Bustamante, E.C. Golden, C. Yao, R.L. Hayes, K.K. Wang, F.C. Tortella and J.R. Dave. 2010. Neuroproteomics: a biochemical means to discriminate the extent and modality of brain injury. *J Neurotrauma.* 27: 1837–1852.

Ottens, A.K., F.H. Kobeissy, B.F. Fuller, M.C. Liu, M.W. Oli, R.L. Hayes and K.K. Wang. 2007. Novel neuroproteomic approaches to studying traumatic brain injury. *Prog Brain Res.* 161: 401–418.

Papa, L., L.M. Lewis, J.L. Falk, Z. Zhang, S. Silvestri, P. Giordano, G.M. Brophy, J.A. Demery, N.K. Dixit, I. Ferguson, M.C. Liu, J. Mo, L. Akinyi, K. Schmid, S. Mondello, C.S. Robertson, F.C. Tortella, R.L. Hayes and K.K. Wang. 2012. Elevated levels of serum glial fibrillary acidic protein breakdown products in mild and moderate traumatic brain injury are associated with intracranial lesions and neurosurgical intervention. *Ann Emerg Med.* 59: 471–483.

Park, E., E. Liu, M. Shek, A. Park and A.J. Baker. 2007. Heavy neurofilament accumulation and alpha-spectrin degradation accompany cerebellar white matter functional deficits following forebrain fluid percussion injury. *Exp Neurol.* 204: 49–57.

Pierobon, M., C. Belluco, L.A. Liotta and E.F. Petricoin, 3rd. 2011. Reverse phase protein microarrays for clinical applications. *Methods Mol Biol.* 785: 3–12.

Pierobon, M., A.J. Vanmeter, N. Moroni, F. Galdi and E.F. Petricoin, 3rd. 2012. Reverse-phase protein microarrays. *Methods Mol Biol.* 823: 215–235.

Pike, B.R., J. Flint, S. Dutta, E. Johnson, K.K. Wang and R.L. Hayes. 2001. Accumulation of non-erythroid alpha II-spectrin and calpain-cleaved alpha II-spectrin breakdown products in cerebrospinal fluid after traumatic brain injury in rats. *J Neurochem.* 78: 1297–1306.

Platt, M., W. Rowe, D.C. Wedge, D.B. Kell, J. Knowles and P.J. Day. 2009. Aptamer evolution for array-based diagnostics. *Anal Biochem.* 390: 203–205.

Pradet-Balade, B., F. Boulme, H. Beug, E.W. Mullner and J.A. Garcia-Sanz. 2001. Translation control: bridging the gap between genomics and proteomics? *Trends Biochem Sci.* 26: 225–229.

Ramlackhansingh, A.F., D.J. Brooks, R.J. Greenwood, S.K. Bose, F.E. Turkheimer, K.M. Kinnunen, S. Gentleman, R.A. Heckemann, K. Gunanayagam, G. Gelosa and D.J. Sharp. 2011. Inflammation after trauma: microglial activation and traumatic brain injury. *Ann Neurol.* 70: 374–383.

Reed, T.T., J. Owen, W.M. Pierce, A. Sebastian, P.G. Sullivan and D.A. Butterfield. 2009. Proteomic identification of nitrated brain proteins in traumatic brain-injured rats treated postinjury with gamma-glutamylcysteine ethyl ester: insights into the role of elevation of glutathione as a potential therapeutic strategy for traumatic brain injury. *J Neurosci Res.* 87: 408–417.

Ringger, N.C., B.E. O'Steen, J.G. Brabham, X. Silver, J. Pineda, K.K. Wang, R.L. Hayes and L. Papa. 2004. A novel marker for traumatic brain injury: CSF alphaII-spectrin breakdown product levels. *J Neurotrauma.* 21: 1443–1456.

Rudrabhatla, P., P. Grant, H. Jaffe, M.J. Strong and H.C. Pant. 2010. Quantitative phosphoproteomic analysis of neuronal intermediate filament proteins (NF-M/H) in Alzheimer's disease by iTRAQ. *FASEB J.* 24: 4396–4407.

Rundgren, M., H. Friberg, T. Cronberg, B. Romner and A. Petzold. 2012. Serial soluble neurofilament heavy chain in plasma as a marker of brain injury after cardiac arrest. *Crit Care.* 16: R45.

Sandberg, A., R.M. Branca, J. Lehtio and J. Forshed. 2013. Quantitative accuracy in mass spectrometry based proteomics of complex samples: The impact of labeling and precursor interference. *Journal of Proteomics.* 96C: 133–144.

Schmidt, M.L., V. Zhukareva, K.L. Newell, V.M. Lee and J.Q. Trojanowski. 2001. Tau isoform profile and phosphorylation state in dementia pugilistica recapitulate Alzheimer's disease. *Acta Neuropathologica.* 101: 518–524.

Schroeder, M.J., J. Shabanowitz, J.C. Schwartz, D.F. Hunt and J.J. Coon. 2004. A neutral loss activation method for improved phosphopeptide sequence analysis by quadrupole ion trap mass spectrometry. *Anal Chem.* 76: 3590–3598.

Searle, B.C. 2010. Scaffold: a bioinformatic tool for validating MS/MS-based proteomic studies. *Proteomics.* 10: 1265–1269.

Sereni, M.I., M. Pierobon, R. Angioli, E.F. Petricoin, 3rd and M.J. Frederick. 2013. Reverse phase protein microarrays and their utility in drug development. *Methods Mol Biol.* 986: 187–214.

Shao, C., K.N. Roberts, W.R. Markesbery, S.W. Scheff and M.A. Lovell. 2006. Oxidative stress in head trauma in aging. *Free Radical Biology & Medicine.* 41: 77–85.

Shaw, G., C. Yang, R. Ellis, K. Anderson, J. Parker Mickle, S. Scheff, B. Pike, D.K. Anderson and D.R. Howland. 2005. Hyperphosphorylated neurofilament NF-H is a serum biomarker of axonal injury. *Biochem Biophys Res Commun.* 336: 1268–1277.

Shinoda, K., M. Tomita and Y. Ishihama. 2010. emPAI Calc—for the estimation of protein abundance from large-scale identification data by liquid chromatography-tandem mass spectrometry. *Bioinformatics*. 26: 576–577.

Shores, K.S., D.G. Udugamasooriya, T. Kodadek and D.R. Knapp. 2008. Use of peptide analogue diversity library beads for increased depth of proteomic analysis: application to cerebrospinal fluid. *J Proteome Res*. 7: 1922–1931.

Siman, R., N. Toraskar, A. Dang, E. McNeil, M. McGarvey, J. Plaum, E. Maloney and M.S. Grady. 2009. A panel of neuron-enriched proteins as markers for traumatic brain injury in humans. *J Neurotrauma*. 26: 1867–1877.

Singh, P., J. Yan, R. Hull, S. Read, J. O'Sullivan, R.D. Henderson, S. Rose, J.M. Greer and P.A. McCombe. 2011. Levels of phosphorylated axonal neurofilament subunit H (pNfH) are increased in acute ischemic stroke. *J Neurol Sci*. 304: 117–121.

Sjodin, M.O., J. Bergquist and M. Wetterhall. 2010. Mining ventricular cerebrospinal fluid from patients with traumatic brain injury using hexapeptide ligand libraries to search for trauma biomarkers. *J Chromatogr B Analyt Technol Biomed Life Sci*. 878: 2003–2012.

Stein, D.M., J.A. Kufera, A. Lindell, K.R. Murdock, J. Menaker, G.V. Bochicchio, B. Aarabi and T.M. Scalea. 2011a. Association of CSF biomarkers and secondary insults following severe traumatic brain injury. *Neurocrit Care*. 14: 200–207.

Stein, D.M., A. Lindell, K.R. Murdock, J.A. Kufera, J. Menaker, K. Keledjian, G.V. Bochicchio, B. Aarabi and T.M. Scalea. 2011b. Relationship of serum and cerebrospinal fluid biomarkers with intracranial hypertension and cerebral hypoperfusion after severe traumatic brain injury. *J Trauma*. 70: 1096–1103.

Stein, D.M., A.L. Lindell, K.R. Murdock, J.A. Kufera, J. Menaker, G.V. Bochicchio, B. Aarabi and T.M. Scalea. 2012. Use of serum biomarkers to predict cerebral hypoxia after severe traumatic brain injury. *J Neurotrauma*. 29: 1140–1149.

Stern, R.A., D.O. Riley, D.H. Daneshvar, C.J. Nowinski, R.C. Cantu and A.C. McKee. 2011. Long-term consequences of repetitive brain trauma: chronic traumatic encephalopathy. *PMR*. 3: S460–467.

Suckau, D., A. Resemann, M. Schuerenberg, P. Hufnagel, J. Franzen and A. Holle. 2003. A novel MALDI LIFT-TOF/TOF mass spectrometer for proteomics. *Analytical and Bioanalytical Chemistry*. 376: 952–965.

Tammen, H. and R. Hess. 2011. Collection and handling of blood specimens for peptidomics. *Methods Mol Biol*. 728: 151–159.

Tammen, H., I. Schulte, R. Hess, C. Menzel, M. Kellmann, T. Mohring and P. Schulz-Knappe. 2005. Peptidomic analysis of human blood specimens: comparison between plasma specimens and serum by differential peptide display. *Proteomics*. 5: 3414–3422.

Tran, H.T., F.M. LaFerla, D.M. Holtzman and D.L. Brody. 2011. Controlled cortical impact traumatic brain injury in 3xTg-AD mice causes acute intra-axonal amyloid-beta accumulation and independently accelerates the development of tau abnormalities. *J Neurosci*. 31: 9513–9525.

Viswanathan, S., M. Unlu and J.S. Minden. 2006. Two-dimensional difference gel electrophoresis. *Nature Protocols*. 1: 1351–1358.

Wang, H., W.J. Qian, M.H. Chin, V.A. Petyuk, R.C. Barry, T. Liu, M.A. Gritsenko, H.M. Mottaz, R.J. Moore, D.G. Camp Ii, A.H. Khan, D.J. Smith and R.D. Smith. 2006. Characterization of the mouse brain proteome using global proteomic analysis complemented with cysteinyl-peptide enrichment. *J Proteome Res*. 5: 361–369.

Webb-Robertson, B.J., W.R. Cannon, C.S. Oehmen, A.R. Shah, V. Gurumoorthi, M.S. Lipton and K.M. Waters. 2010. A support vector machine model for the prediction of proteotypic peptides for accurate mass and time proteomics. *Bioinformatics*. 26: 1677–1683.

Wei, H.H., X.C. Lu, D.A. Shear, A. Waghray, C. Yao, F.C. Tortella and J.R. Dave. 2009. NNZ-2566 treatment inhibits neuroinflammation and pro-inflammatory cytokine expression induced by experimental penetrating ballistic-like brain injury in rats. *J Neuroinflammation*. 6: 19.

Wiese, S., K.A. Reidegeld, H.E. Meyer and B. Warscheid. 2007. Protein labeling by iTRAQ: a new tool for quantitative mass spectrometry in proteome research. *Proteomics*. 7: 340–350.

Williams, A.J., J.A. Hartings, X.C. Lu, M.L. Rolli and F.C. Tortella. 2006. Penetrating ballistic-like brain injury in the rat: differential time courses of hemorrhage, cell death, inflammation, and remote degeneration. *J Neurotrauma*. 23: 1828–1846.

Woodcock, T. and M.C. Morganti-Kossmann. 2013. The role of markers of inflammation in traumatic brain injury. *Front Neurol*. 4: 18.

Xiong, Y., A. Mahmood and M. Chopp. 2013. Animal models of traumatic brain injury. *Nat Rev Neurosci*. 14: 128–142.

Yang, Y., D. Yang, H.J. Schluesener and Z. Zhang. 2007. Advances in SELEX and application of aptamers in the central nervous system. *Biomol Eng*. 24: 583–592.

Yao, C., A.J. Williams, A.K. Ottens, X.C. Lu, M.C. Liu, R.L. Hayes, K.K. Wang, F.C. Tortella and J.R. Dave. 2009. P43/pro-EMAPII: a potential biomarker for discriminating traumatic versus ischemic brain injury. *J Neurotrauma*. 26: 1295–1305.

Yao, C., A.J. Williams, A.K. Ottens, X.C. May Lu, R. Chen, K.K. Wang, R.L. Hayes, F.C. Tortella and J.R. Dave. 2008. Detection of protein biomarkers using high-throughput immunoblotting following focal ischemic or penetrating ballistic-like brain injuries in rats. *Brain Inj*. 22: 723–732.

Yu, I., M. Inaji, J. Maeda, T. Okauchi, T. Nariai, K. Ohno, M. Higuchi and T. Suhara. 2010. Glial cell-mediated deterioration and repair of the nervous system after traumatic brain injury in a rat model as assessed by positron emission tomography. *J Neurotrauma*. 27: 1463–1475.

Yuh, E.L., P. Mukherjee, H.F. Lingsma, J.K. Yue, A.R. Ferguson, W.A. Gordon, A.B. Valadka, D.M. Schnyer, D.O. Okonkwo, A.I. Maas, G.T. Manley and T.-T. Investigators. 2013. Magnetic resonance imaging improves 3-month outcome prediction in mild traumatic brain injury. *Ann Neurol*. 73: 224–235.

Zarkovic, K. 2003. 4-hydroxynonenal and neurodegenerative diseases. *Molecular Aspects of Medicine*. 24: 293–303.

Zemlan, F.P., E.C. Jauch, J.J. Mulchahey, S.P. Gabbita, W.S. Rosenberg, S.G. Speciale and M. Zuccarello. 2002. C-tau biomarker of neuronal damage in severe brain injured patients: association with elevated intracranial pressure and clinical outcome. *Brain Res*. 947: 131–139.

Zhang, B., N.C. VerBerkmoes, M.A. Langston, E. Uberbacher, R.L. Hettich and N.F. Samatova. 2006. Detecting differential and correlated protein expression in label-free shotgun proteomics. *J Proteome Res*. 5: 2909–2918.

Zhang, Z., S.F. Larner, F. Kobeissy, R.L. Hayes and K.K. Wang. 2010. Systems biology and theranostic approach to drug discovery and development to treat traumatic brain injury. *Methods Mol Biol*. 662: 317–329.

Zhou, W., D. Xu, X. Peng, Q. Zhang, J. Jia and K.A. Crutcher. 2008. Meta-analysis of APOE4 allele and outcome after traumatic brain injury. *J Neurotrauma*. 25: 279–290.

Zoltewicz, J.S., S. Mondello, B. Yang, K.J. Newsom, F.H. Kobeissy, C. Yao, X.C. Lu, J.R. Dave, D.A. Shear, K. Schmid, V. Rivera, T. Cram, J. Seaney, Z. Zhang, K.K. Wang, R.L. Hayes and F.C. Tortella. 2013. Biomarkers Track Damage Following Graded Injury Severity in a Rat Model of Penetrating Brain Injury. *J Neurotrauma*.

Zurek, J. and M. Fedora. 2012. The usefulness of S100B, NSE, GFAP, NF-H, secretagogin and Hsp70 as a predictive biomarker of outcome in children with traumatic brain injury. *Acta Neurochir (Wien)*. 154: 93–103; discussion 103.

3

Molecular Mechanisms and Biomarker Perspective of MicroRNAs in Traumatic Brain Injury

Nagaraja S. Balakathiresan,[1] Anuj Sharma,[1] Raghavendar Chandran,[1,2] Manish Bhomia,[1,2] Zhiqun Zhang,[3] Kevin K.W. Wang[3,4,5] and Radha K. Maheshwari[1,]*

INTRODUCTION

Worldwide, Traumatic Brain Injury (TBI) is one of the leading causes of death and disability. According to the World Health Organization (WHO), TBI will be the third leading cause of global mortality and disability by the year 2020. It is estimated that each year, TBI affects 1.7 million people resulting in 235,000 hospitalizations. In the United States alone, 50,000 deaths occur per year and currently over 3 million people are living with permanent disabilities due to TBI (Summers et al., 2009). The economic burden of TBI in US is estimated to be US$60 billion annually in total lifetime direct medical

[1] Department of Pathology,Uniformed Services University of the Health Sciences, 4301, Jones Bridge Road, Bethesda,Maryland 20814.

[2] Biological Sciences Group, Birla Institute of Technology and Science, Pilani, Rajasthan, India.

[3] Center for Neuroproteomics & Biomarkers Research, Department of Psychiatry, University of Florida, Gainesville, FL 32610, USA.

[4] Department of Neuroscience, University of Florida, Gainesville, FL 32610, USA.

[5] Department of Psychiatry, University of Florida, Gainesville, FL 32610, USA.

* Corresponding author: radha.maheshwari@usuhs.edu

costs and indirect productivity losses (Langlois et al., 2006; Rockhill et al., 2012). Incidences of TBI in the military are significantly higher. According to the Defense Medical Surveillance System and Theater Medical Data Store, 253,330 U.S. forces personnel have been diagnosed for TBI, since the year 2000 (http://www.health.mil). The Congressional Budget Office reported a total of US$418 million expenditure during 2004–2009 for the diagnosis and treatment of TBI by the Veterans Health Administration (http:// www. cbo.gov/sites/default/files/cbofiles/attachments/02-09-PTSD.pdf).

TBI is defined as the damage to the brain caused by a blow or jolt to the head by external forces such as direct impact by/or against a blunt object, rapid acceleration or deceleration, penetrating objects and/or blast waves from an explosion. TBI is classified as mild, moderate and severe depending on the severity and cause of injury to the brain such as brain laceration, hematoma, contrecoup injury, shearing of nerve fibers, intracranial hypertension and subarachnoid hemorrhage (Sharma and Laskowitz, 2012). Glasgow Coma Scale (GCS) is a standard way of measuring the initial severity and prognosis of TBI. Mild TBI (mTBI) is defined as the loss of consciousness up to 30 minutes, Post Traumatic Amnesia (PTA) up to 24 hours and GCS score above 12 after 30 minutes post-injury. Moderate TBI often involves loss of consciousness between 30 minutes and 24 hours, GCS of 9 to 12, and PTA between 1–7 days. Severe TBI cases often result in loss of consciousness for more than 24 hours and GCS of less than 9 (Bryant, 2010; Laker, 2011).

Imaging techniques viz. Computed Tomography (CT) and Magnetic Resonance Imaging (MRI) are used for measuring the structural damage due to TBI. CT has been the method of choice due to its ease of use, speed and accuracy in detecting skull fractures and acute hemorrhages. MRI is used when neurological damage remains unexplained by CT (Bélanger et al., 2007; Niogi et al., 2008). However, both CT and MRI have limitations in detecting mild brain injuries where the brain tissue damage is minute along with Diffuse Axonal Injury (DAI) (Hunter et al., 2012; Yu et al., 2012). Recent advances and the emergence of newer functional imaging techniques such as Diffusion Tensor Imaging (DTI), fluorodeoxyglucose positron emission tomography (FDG-PET), Single Photon Emission Computed Tomography (SPECT) and functional Magnetic Resonance Imaging (fMRI) have shown great potential in detecting brain tissue damage that are invisible in CT and MRI. These newer technologies therefore, may have a significant potential in diagnosing mTBI in future (Brenner, 2011; Matthews et al., 2011). The availability of these imaging tools, however, is very limited and will require further experimental validation and correlation with the neurophysiological paradigms of TBI. For this reason, several efforts have been put forward for developing minimally invasive biochemical and molecular markers to detect brain damage due to TBI, especially mTBI. Factors associated with the pathophysiology of TBI, i.e., cellular, chemical and molecular cascades have been shown as potential biomarkers of brain injury (Le and Gean, 2009; Giacoppo et al., 2012). Some of the proteins that have been studied for their utility as clinical biomarkers for mild to severe TBI are S-100β, Glial Fibrillary Acidic Protein (GFAP) and ubiquitin C-terminal hydrolase-L1 (UCH-L1) (Svetlov et al., 2010; Zhang et al., 2011).

S-100β is a small dimeric calcium binding protein which is abundantly expressed in glial cells of the Central Nervous System (CNS). Several studies have demonstrated the expression of S-100β in the serum post brain injury (Svetlov et al., 2009, 2012). However, the poor correlation of elevated levels of S-100β to TBI in human subjects indicates that it's elevated level alone is not sufficient to diagnose mTBI (Topolovec et al., 2011). GFAP is a filamentous protein found only in astroglial cytoskeleton. GFAP and it's breakdown products (BDP) were reported to be elevated in the serum of TBI subjects and their elevated levels correlated with the severity of TBI (Papa et al., 2012). Similarly, elevated levels of UCH-L1 have been reported in the serum and CSF of animals and human subjects with TBI (Liu et al., 2010; Brophy et al., 2011). UCH-L1 is a neuronal specific protein and accounts for approximately 5% of total soluble brain protein (Berger et al., 2012). Serum levels of UCH-L1 were found to be elevated in patients with mild to moderate TBI, and were high enough to distinguish them from control subjects (Papa et al., 2012). High levels of UCH-L1 were also detected in serum and CSF of rats exposed to severe blast overpressure (BOP) injury (Svetlov et al., 2010). These studies have shown great promise of using protein factors for the diagnosis of moderate to severe TBI, but a biomarker for mTBI remains a challenge. As mTBI population grows rapidly every year, the lack of diagnostic and prognostic mTBI biomarkers for the management of patients with TBI is of great concern (Crawford et al., 2012). The search for clinically reliable mTBI biomarkers is still an ongoing process, and attempts are being made to find more robust molecules such as microRNAs (miRNA) as biomarkers of TBI.

MicroRNA

MiRNAs are small, endogenous, non-coding RNAs of 20–25 nucleotides (nt) in length. Since their discovery in 1993, miRNAs are now believed to control many cellular aspects of an organism by post transcriptional regulation of gene expression (Pritchard et al., 2012). The landmark discovery of the first miRNA, lin-4, was made by two independent investigators, Ambros and Ruvkun, in the nematode, *Caenorhabditis elegans* during the larval-adult stage transition (Lee et al., 1993; Wightman et al., 1993). After nearly a decade of dormancy, another non-protein coding gene, let-7, was discovered again in

C. elegans that expressed a 21 nt small RNA and regulated the expression of a heterochronic lin-41 protein (Reinhart et al., 2000). After these initial studies, a plethora of new miRNAs in almost all species (viruses, plants, nematodes, fly, fish, mouse and human) have been identified (Saba and Schratt, 2010; Cullen, 2011; Huijser and Schmid, 2011; Jin et al., 2012; Mondol and Pasquinelli, 2012). It is estimated that around 60% of total human protein are regulated by miRNAs (Sayed and Abdellatif, 2011). The latest database "miRbase" contains 25,141 mature miRNA sequences from 193 different species including 2,042 human miRNAs (Release 19; August 2012) (Griffiths-Jones, 2004; Griffiths-Jones et al., 2006, 2008; Kozomara and Griffiths-Jones, 2011). In this chapter, we provide an overview of the recent developments in the field of miRNA regulation in CNS injury with a particular emphasis on TBI.

Biogenesis

The multi-step process of miRNA biogenesis begins with transcription of specific miRNA genes. Except for the Y-chromosome, miRNA genes are scattered throughout the genome, frequently in the intergenic regions and have their own promoters for transcription (Bandiera et al., 2010). Some of the miRNA genes are also present in the intragenic region (overlapping introns and exons) and are transcribed as part of the host gene (Bandiera et al., 2010). In a canonical pathway (Figure 3.1), miRNA genes are transcribed by RNA-polymerase II enzyme into long primary miRNAs (pri-miRNA) inside the nucleus (Issabekova et al., 2011, 2012). Pri-miRNA nucleotide ranges from a few hundred to thousands in length and fold imperfectly to form hairpin-shaped structures having a 5' terminal 7mGpppN-cap and a 3' poly(A) tail (Enciu et al., 2012). Pri-miRNA is further processed in the nucleus by a microprocessor complex containing two core components, RNase III enzyme Drosha and a dsRNA binding protein, Di George Syndrome Critical Region 8 (DGCR8). Drosha-DGCR8 complex processes pri-miRNA into 70 nt long precursor miRNA (pre-miRNA) with stem loop structures containing 2 nt 3'overhangs (Newman and Hammond, 2010). The pre-miRNA is transported to cytoplasm through nucleopore by the nuclear export receptor exportin-5 (XPO5) in a Ran-GTP dependent mechanism (Katahira and Yoneda, 2011). In the cytoplasm, the pre-miRNA is further processed by Dicer, an RNase III family endoribonuclease enzyme (Treiber et al., 2012). Dicer slices the stem loop structure of pre-miRNA and forms a 22 nt length short-lived double stranded RNA intermediate. One of the strands from double stranded miRNA is then incorporated into a micro-ribonucleoprotein complex termed as the RNA Induced Silencing Complex (RISC) (Steitz and Vasudevan, 2009).

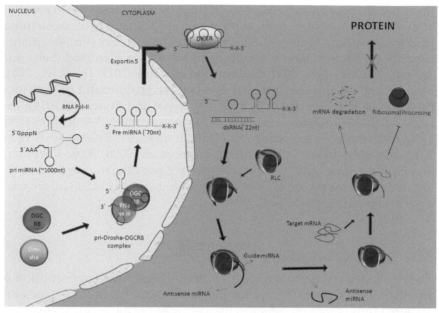

Figure 3.1. MicroRNA biogenesis and functions. MicroRNA (miRNA) genes are transcribed by RNA polymerase II enzyme into long primary miRNAs (pri-miRNA). The pri-miRNA is processed within the nucleus by RNase III enzyme Drosha and the dsRNA binding protein DiGeorge Syndrome Critical Region 8 (DGCR8), into precursor miRNA (Pre-miRNA). Pre-miRNA is then transported to cytoplasm by Exportin-5. In the cytoplasm, the pre-miRNA is cleaved by Dicer and form ~22 bp miRNA duplex. The strand that is less thermodynamically stable at 5' end (guide strand) gets incorporated into RNA induced silencing complex (RISC) whereas the other strand (passenger strand) in most cases is degraded. The mRNA gets degraded when a perfect match between the 2–8 nt from the miRNA 5'end (seed sequence) and 3'UTR target sequence is available. Without a perfect match, mRNA gets either destabilized or repressed.

Color image of this figure appears in the color plate section at the end of the book.

RISC is composed of a Dicer, a catalytic component Argonaut 2 (Ago2), and a double stranded RNA binding protein, TAR RNA-Binding Protein (TRBP). The strand selection into RISC assembly is accomplished by Dicer and TRBP. The strand that is less thermodynamically stable at 5'end, as sensed by the helicase domain of Dicer, is usually incorporated into the RISC, and is termed as the guide strand. The other strand called as the "passenger strand" gets degraded by Ago2 (Krol et al., 2010). The guide strand directs the RISC to specific mRNA targets by binding to 3'UTR of the target mRNA. The interaction between the miRNA and the mRNA requires a near perfect match between the 2–8 nt from the 5'end of the miRNA (seed sequence) and the 3'UTR of the mRNA. Ago mediates post transcriptional silencing of miRNA either by degradation or by destabilization (Krol et al.,

2010). Recent evidence support the notion that destabilization of mRNA via deadenylation or translational repression is the primary mechanism of inhibition of protein synthesis (Pasquinelli, 2012). Non-canonical pathways of miRNA biogenesis have also been reported and are independent of either Drosha or Dicer cleavages as elaborately reviewed by Miyoshi et al. (2010).

Functions

The primary function of miRNAs is the post transcriptional regulation of gene expression. Recent studies indicate that they are also involved in a variety of new functions. These functions are:

(1) MiRNA editing

Conversion of adenosine into inosine (A→I) in pri- or mature miRNA sequence is called as miRNA editing. A→I conversions are mediated by adenosine deaminase, which completely blocks pri-miRNA cleavage by Dicer. Experimental evidence shows that 16% of pri-miRNAs in human brain are edited in this way (Kawahara et al., 2008; Linsen et al., 2010). Edited pri-miRNAs accumulate in the nucleus and editing within the seed regions of the mature miRNA drastically alters their function (Kawahara et al., 2007, 2008; Gommans, 2012). MiRNA editing may either function to reduce the wild type miRNA level or introduce new versions of miRNA that have different target mRNAs (Alon et al., 2012).

(2) Autoregulation

Biogenesis of miRNA can be regulated either by the same or other miRNA(s) inside the nucleus. MiRNAs can directly bind to respective recognition elements that are present on their target pri-miRNA sequences. Such binding prevents the processing of pri-miRNA into pre-miRNA thus suppressing the miRNA expression. MiRNAs can also induce positive feedback regulation, by directly targeting their pri-miRNA, and promoting their downstream processing as seen in the autoregulation of brain enriched miRNA let-7 (Tang et al., 2012; Zisoulis et al., 2012).

(3) Modulation of Splicing Factors

MiRNAs can indirectly modulate the mRNA translation by targeting splicing factors. MiRNAs such as miR-183 and -10a/b target SC35 and SRfamily splicing factors respectively, which in turn changes the translation of their target mRNAs (Meerson et al., 2010; Meseguer et al., 2011).

(4) Brain Development and Aging

A single miRNA exerts diverse functions in the same type of cells or an organ (Ha, 2011). Brain enriched miR-9 has been shown to regulate multiple steps in neurogenesis ranging from neural stem cell proliferation to maturation in a dynamic and context-dependent manner in modulating different downstream mRNA targets (Lang and Shi, 2012). MiRNA expression in the brain changes with aging. Selective upregulation of miR-144 along with two other miRNAs have been reported in the aging brain. MiR-144 regulates expression of ataxin-1 (ATXN-1) and thus may reduce the cytotoxic effect of polyglutamine expanded ATXN-1. MiRNAs may therefore regulate or restrict adverse affects such as neurodegeneration during aging (Fabian et al., 2010; Persengiev et al., 2012).

Regulation

As mentioned before, miRNAs modulate almost all the cellular aspects of biological process and thus play a critical role in normal cellular or tissue processes (Osman, 2012). Normal function of miRNAs can be affected by a wide variety of stimulus, internal or external, to the cell. Some of these factors that have been shown to regulate miRNAs include: (1) translocation of other protein coding genes in the vicinity of miRNA promoters or regulatory regions (Zhou et al., 2012); (2) recombination of miRNA genes close to other regulatory elements (Bisognin et al., 2012); (3) miRNA expression controlled by negative feedback of target mRNA overexpression (Bruning et al., 2011); (4) hypermethylation: Studies in cancer cells have shown that activity of the tumor suppressor miRNAs can be inhibited by hypermethylation of CpG islands in miRNA promoter sequences (Lujambio et al., 2008; Xia and Hu, 2010; Chen et al., 2012). Reintroduction of normal mature miRNAs resulted in inhibition of tumor development thus indicating DNA-methylation mediated miRNA silencing as a potential mechanism of miRNA regulation; (5) other non-coding RNAs: In addition to miRNAs, several other non-coding RNA are transcribed in the cells. These non-coding antisense RNAs bind to the mRNA and inhibit miRNA-mRNA binding resulting in suppression of miRNA activity (Qin et al., 2010). For instance, expression of BACE1, a critical enzyme in Alzheimer's disease pathophysiology, is promoted by a non-coding beta-secretase-1 antisense RNA (BACE1-AS) by preventing the binding of miR-485-5p with BACE 1 mRNA (Faghihi et al., 2010); (6) affected by environmental factors via epigenetic mechanisms (Babenko et al., 2012); (7) genetic polymorphisms of protein or miRNA genes (Saunders et al., 2007; Liu et al., 2010); (8) toxic chemicals from environmental pollution (Hou et al., 2011); and (9) food compounds (Zhang et al., 2011; Vaucheret and Chupeau, 2012).

Alterations in the normal miRNA expression may lead to development of disease and disorders such as cancer, cardiovascular disease, diabetes, obesity, chronic hepatitis, AIDS, schizophrenia, renal function disorders, Tourette's syndrome, psoriasis, primary muscular disorders, Fragile-X mental retardation syndrome and polycythemia vera (Hariharan et al., 2005; Calin and Croce, 2006; Esquela-Kerscher and Slack, 2006; Eisenberg et al., 2007; Hansen et al., 2007; Latronico et al., 2007; Perkins et al., 2007; Sonkoly et al., 2007; van Rooij and Olson, 2007; Abu-Elneel et al., 2008; Bruchova et al., 2008; Li and Jin, 2009; Xie et al., 2009; Heneghanet et al., 2010; Amrouche, et al., 2011; Bravo and Dinan, 2011; Haybaeck et al., 2011; Sun and Rossi, 2011; Chen and Wang, 2012; Kelley et al., 2012; Schneider, 2012). Therefore, without a doubt there exist complex relationships and mechanism(s) between miRNA modulation and disease or disorder development. Human miRNA disease database (HMDD) is a comprehensive collection of published literature for miRNAs and their association with diseases and latest update of HMDD shows about 5,076 miRNA-disease associations, which include 591 miRNA genes and 396 diseases (Lu et al., 2008).

Biomarkers

A biomarker is an "indicator" of specific biological or diseased state and of a response to a treatment. Ideal biomarkers should fit a number of characteristics which includes, non- or minimally invasive access, specificity to disease or pathology, early indicator of disease, it's progression or therapeutic response, and easy to translate from model systems to humans (Etheridge et al., 2011). Altered expression of enzymes, genes, proteins and lipid metabolites in the disease or injury affected tissues may serve as candidate(s) for biomarkers (Jeter et al., 2012). Proteins or their breakdown products have been successfully developed as biomarkers for disease. For example, blood sugar and creatinine levels are used for diagnosis of diabetes and kidney function respectively. Proteins are useful as biomarkers due their diversity and good correlation with a diseased state. However, proteins also have several limitations as biomarkers such as complexity in composition (sulfation, phosphorylation, glycation, etc.), stability in biospecimen, sensitivity to degradation under temperature and pH stress, low abundance affecting detection and sequence variations even within clinically relevant species. On the other hand, miRNAs display many characteristics of an ideal biomarker candidate. MiRNAs are relatively simple molecules, stable, resistant to degradation due to temperature and pH changes, conserved through species, and have tissue or stage specific expression. Detection of miRNAs can be performed using standard, robust molecular biology techniques (Etheridge et al., 2011).

Detection Methods

For measuring the miRNA expression, three major global miRNA profiling platforms, i.e., microarrays, high-throughput real time quantitative PCR (RT-qPCR) and Next Generation Sequencing (NGS) based methods are used (Schmittgen et al., 2004; Bai et al., 2012; Balakathiresan et al., 2012; Huaet et al., 2012). Of these three, qPCR based method of miRNA quantitation provides high specificity and sensitivity (Schmittgen et al., 2008; Gitet et al., 2010; Pritchard et al., 2012). Microarray-based methods require a large amount of starting material for the expression studies. Further, the probe development and hybridization conditions for the simultaneous detection of multiple miRNAs also add to the hardship in the microarray methods (Etheridge et al., 2011). NGS platform has also been used to profile miRNAs by direct sequencing of the study sample (Eacker et al., 2011; Gunaratne et al., 2012). NGS has a great advantage of comprehensive and accurate measurements of miRNAs; however, this platform requires high labor intensive analysis of the readouts and is also very expensive to perform (Morin et al., 2008; Creighton et al., 2009; De Smaele et al., 2010).

MiRNA as Diagnostic Biomarker

Great progress has been made on the identification of miRNAs as biomarkers in many diseases, including that of the brain. The presence of circulating extracellular miRNAs was first demonstrated in the serum of lymphoma patients (Lawrie et al., 2008). Since then many studies in various disease conditions have shown the presence of extracellular miRNAs in various body fluids such as serum/plasma, saliva, urine and CSF. Extracellular miRNAs in the serum or plasma can withstand nuclease activity and harsh conditions such as, extreme pH, repeated freeze-thaw cycles and long-term storage and therefore, have gained importance as non-invasive diagnostic biomarkers (Mitchel et al., 2008). The exact mechanism of origin and stability of extracellular miRNAs in circulatory fluid is not fully understood. Existing evidence indicates that miRNAs are secreted from the injured tissues, enclosed into micro vesicular bodies of endosomal origin known as exosomes. MiRNAs are also found associated with apoptotic bodies, liposome lipoproteins, Argonaute2, as ribonucleoprotein complexes and bound to RNA-binding proteins such as nucleophosmin-1. Such associations of miRNAs are believed to act as a protective shield against the RNase activity (Kosaka et al., 2010; Vickers et al., 2011; Arroyo et al., 2011). Increased expression of extracellular miRNAs in blood has been shown to correlate with the up-regulation of cellular miRNA contents in injured tissues (Kosaka et al., 2010). The altered expression of circulatory miRNAs thus can indicate the pathophysiological status and possibly be used as surrogate biomarkers of injury (Bellingham et al., 2012) (Table 3.1).

MiRNA as Therapeutic Targets

Given the importance of the role that miRNAs play in disease pathology including that of the brain, it may be possible to manipulate miRNA activity for the purpose of treating a disease condition. MiRNA in this way can be used either as a new drug target or as an indicator for the treatment of injury. Few miRNAs have already been used as therapeutic targets and reached pre-clinical (miR-34, -16, -21, -10b and let-7; Broderick and Zamore, 2011) and clinical trials (miR-122; www.santaris.com). MiRNAs can be targeted either to inhibit or increase their activity depending on their modulation in the diseased state. Several strategies of targeting miRNA have been tested as follows:

MiRNA Inhibition

An aberrantly increased activity of miRNA can be inhibited by using synthetic antisense oligonucleotides (ASOs) of 8–25 nt in length. ASOs are synthesized against the target miRNA seed sequence and its neighboring nucleotides. However, ASO in its native form is sensitive to degradation by the endogenous ribonucleases. Therefore, to avoid such degradation *in vivo*, ASO has been modified by adding a methyl group which binds with the 2' oxygen of the ribose (2'-*O*-methyl). 2'-*O*-methyl ASO has been shown to effectively inhibit miRNA activity in *C. elegans*. A more effective *in vivo* inhibition of miRNA activity was noticed with 2'-*O*-methoxyethyl modification instead of 2'-*O*-methyl modification (Broderick and Zamore, 2011; Stenvang et al., 2012).

AntagomiR is another antisense approach that has been used to inhibit the target miRNA activity. AntagomiR is a further modification of 2'-*O*-methyl modified ASO RNA with terminal phosphorothioates (substitution of sulfur to free oxygen to make a phosphorothioate linkage between nucleotides) and a cholesterol group at 3' end. Addition of phosphorothioates and cholesterol conjugation has been shown to respectively improve stability and increase the cellular uptake of antagomiR. The first report on mammalian miRNA knockdown using antagomiRs has been shown to inhibit liver specific

Table 3.1. MiRNA as Biomarker candidates in various brain pathologies

Brain pathology	Altered miRNAs	Biological fluid	Reference
CNS injury			
Traumatic Brain Injury (TBI)	let-7i	Rat serum	Balakathiresan et al. (2012)
	let-7i	Rat CSF	Balakathiresan et al. (2012)
	16, 92a, 765	Human plasma	Redell et al. (2010)
Brain ischemia/stroke	124	Rat plasma	Laterza et al. (2009), Weng et al. (2011)
	126, 144, 16, 21, 223, 320a	Human whole blood	Tan et al. (2009)
	124a, 223, 494	Rat whole blood	Jeyaseelan et al. (2008)
	145	Human whole blood	Gan et al. (2012)
	#298, 333, 7a, 187, 96, 125a-5p, 138, 221, 383, 101a	Rat whole blood	Liu et al. (2010)
Brain hemorrhage - Fresh blood method #	298, 107, Y1, 7a, 145, 342-5p, 130b, 34a, 330, 142-3p	Rat whole blood	Liu et al. (2010)
Brain hemorrhage - Lysed blood model #	298, 333, 340-3p, 505, 107, 330, 322, 125a-5p, 196b, 708	Rat whole blood	Liu et al. (2010)
Brain hemorrhage - Thrombin method #	298, 333, 107, 203, 10b, 342-5p, 383, 130b, 685, 142-3p	Rat whole blood	Liu et al. (2010)
Epileptic seizures (kainate seizure) #	333, 298, 182, 125b-5p, 505, 125a-5p, 342-5p, 350, 685, 421	Rat whole blood	Liu et al. (2010)
Neuropsychiatric diseases			
Schizophrenia	181b, 219-2-3p, 346, 195, 1308, 92a, 17, 103, let-7g	Human serum	Shi et al. (2012)
Bipolar disorder	134	Human plasma	Rong et al. (2011)
Stress	144, 144*, 16	Human whole blood	Katsuura et al. (2012)
Antidepressant treatment in depressed patients #	130b, 505, 26b, 26a, let-7f, 34c-5p, 770-5p	Human whole blood	Bocchio-Chiavetto et al. (2012)
Neurodegenerative disorders			
Alzheimer's disease (AD)	137, 181c, 9, 29a, 29b	Human serum	Geekiyanage et al. (2011)
	let-7b	Human CSF	Lehmann et al. (2012)
	138, 204, 371, 449, 494, 451, 154,	Human CSF	Cogswell et al. (2008)
Huntington's disease (HD)	34b	Human plasma	Gaughwin et al. (2011)
Multiple sclerosis (MS) #	614, 572, 648, 1826, 422a, 22, 1979	Human plasma	Siegel et al. (2012)
	922, 181c, 633	Human CSF	Haghikia et al. (2011)
Parkinson's disease (PD)	1, 22*, 29, 16-2*, 26a2*, 30a	Human whole blood	Margis et al. (2011)
Mild cognitive impairment	134, 323-3p, 382, 128, 132, 874, 491-5p, 370	Human plasma	Sheinerman et al. (2012)
Brain aging	34-a	Mouse plasma	Li et al. (2011)
CNS cancers/tumors			
Primary CNS lymphoma	21, 19b, 92	Human CSF	Baraniskin et al. (2012a)
Primary diffuse large B-cell lymphoma of the CNS	21, 19, 92a	Human CSF	Baraniskin et al. (2011)
Glioblastoma	21	Human Plasma	Ilhan-Mutlu et al. (2012)
Glioma	15b, 21	Human CSF	Baraniskin et al. (2012b)
Astrocytoma	150*, 23a, 133a, 150*, 197, 497, 548b-5p	Human serum	Yang et al. (2012)
CNS infection			
HIV- Associated Neurological Disorders (HAND)	1203, 1224-3p, 182*, 19b-2*, 204, 362-5p, 484, 720, 744*, 934, 937	Human CSF	Pacifici et al. (2012)
Lentivirus-associated CNS disease	125b, 21, 34a, 1233, 130b, 146a	Monkey plasma	Witwer et al. (2011)

- indicates only profiling without qRT-PCR validation

miRNA, miR-122 in mice (Krützfeldt et al., 2005). However, the clinical use of ASOs and antagomiR is restricted as it required higher doses to inhibit target miRNAs (Broderick and Zamore, 2011).

Locked Nucleic Acid (LNA) antisense nucleotide is another modification strategy where a bridge formation is induced with a methylene group in between the 2′ *O* and 4′ *C* atoms of ribose (Li et al., 2009). LNA modification increases the melting temperature by 2–4°C, which prevents degradation from nucleases and also helps to form strong duplex formation with the complementary RNA (Ruberti et al., 2012). LNA-antimiR/ASOs have high affinity binding to target miRNAs and lower dose requirements, making it a more biologically effective inhibition strategy. Systemic delivery of unconjugated LNA-antimiR has been shown to significantly silence miR-122 in liver during chronic hepatitis C virus (HCV) infection (Elmén et al., 2008; Lanford et al., 2010). Currently, the first miRNA-targeted drug, antimiR-122 LNA, is under phase II trial to treat and test the safety and tolerability of the drug in chronic HCV infected patients (Stenvang et al., 2012).

MiRNA sponge is another strategy to inhibit miRNA function. This strategy is based on the principle that a transgene produces RNA molecules that consist of multiple complementary binding sites for a target miRNA (Ebert et al., 2007). Sponge mRNA have bulged sites against complementary sequences 9–12 of target miRNA and have been shown to have more effective inhibition on miRNA activity (Gentner et al., 2009). Improved inhibition of miRNA function has also been shown to increase with the presence of strong promoters and tandem repeats of complementary binding sites in sponge RNAs (Ebert and Sharp, 2010). Targeted delivery of mRNA sponge transgenes both *in vitro* and *in vivo* by viral vectors has been shown to neutralize a family of miRNAs, e.g., miR-17.5p, miR-20 and miR-17-92 within the same seed family (Li et al., 2009; Wang, 2009). MiRNA sponge technology has also been applied to inhibit the activity of miR-9 which regulates the expression of a transcription factor FoxP1. MiR-9-FoxP1 regulation is important in normal spinal cord development. FoxP1 drives specification of lateral motor neuron in developing spinal cord, specific *in vivo* neuronal populations and in primary neuronal cultures (Barbato et al., 2009; Edbauer et al., 2010; Krol et al., 2010; Zhu et al., 2011; Otaegi et al., 2011).

More recently, tough decoy (TuDs) RNA has been shown to have strong and stable suppression of target miRNA activity (Haraguchi et al., 2009). TuD-RNAs have two miRNA-binding sequence (MBS) regions flanking the two stem structures, and are transcribed by RNA polymerase III U6 promoter. Two stem flanking MBS protect TuD-RNA from cellular RNase degradation, and helps to bind tightly to target miRNAs. TuD-RNAs can be expressed both by viral and non-viral vectors, and their inhibitory activity against target miRNAs is higher than that of LNA or sponges. Adenovirus vector-mediated expression of TuD-122a has been shown to significantly reduce HCV replication in Huh-7 cells by efficiently blocking miRNA-122 activity (Sakurai et al., 2012).

MiR-mimetics technology is another inhibition strategy for gene silencing. This approach is based on the exogenous synthetic double-stranded RNA fragments that are similar to endogenous miRNA (Esau and Monia, 2007). MiR-mimic has a 5'end which is designed in such a way to bind partially to a selected sequence in the 3'UTR of the target gene. Therefore, miR-mimics as the name suggest mimic the endogenous miRNA function and repress target genes expression. However, unlike endogenous miRNA, MiR-mimics are designed to target a single gene expression. This strategy is primarily used to study the gene function in mammalian cells (Wang, 2011).

MiRNA Induction

MiR-mimics are now extensively used to rescue under-expressed miRNAs by transiently transfecting double-stranded miRNAs (Stenvang et al., 2012). For instance, miR-34, a master regulator of tumor suppression, is downregulated in different type of cancers. MiR-34 inhibits tumor growth by suppressing the activity of genes that are involved in various oncogenic signaling pathways. Therapeutic delivery of miR-34 mimic has shown to effectively inhibit the tumor growth in various cancers (non-small cell lung cancer, prostate cancer, melanoma, pancreatic cancer and lymphoma) of animal models without affecting the normal tissues. Because of these promising pharmacological features for successful cancer therapy, miR-34 is one of the first miRNA mimics expected to enter clinical trials in 2013 (Bader et al., 2012).

MicroRNAs in Traumatic Brain Injury

High-throughput sequencing experiments have identified the presence of more than 1,000 miRNAs in the human brain (Shao et al., 2010). MiRNAs are differentially expressed in the brain during the developmental stages and also show specific distribution in neural tissue such as cortex, cerebellum, and midbrain. Several miRNAs have been found to be either brain specific or brain enriched (expressed at higher level in brain than any other organ) (Forero et al., 2010). MiRNAs, miR-124, -101, -127, -128, -131, and -132, are predominantly brain specific, of which miR-124 expression dominates and accounts for 25–48% of total miRNAs expressed in the brain (Lagos-Quintana et al., 2002). Studies in human and murine brain have identified six distinct miRNAs namely miR-9, -124a, -124b, -135, -153 and -219 that are expressed in human brain whereas miR-183 that is exclusively expressed in murine brain (Sempere et al., 2004). In addition, miR-125a, -125b, -128, -132, -137, -139 and let-7 family miRNAs were found to be enriched in the brain (Sempere et al., 2004; Nelson et al., 2008). Among the brain enriched miRNAs, let-7 family miRNAs, have been shown to be highly expressed in both mouse and primate brains (Miska et al., 2004).

Stage specific expression of certain miRNAs has been observed during the brain development in rat and mouse. MiRNAs miR-9, -19b -125b, -131 and -178 were found to be expressed at maximum levels only at the time of birth, while miRNA-128 was expressed only during the postnatal period. High levels of expression of miR-124a and -266, was observed throughout embryonic and postnatal stages (Krichevsky et al., 2003; Kosik and Krichevsky, 2005; Mathupala et al., 2007). Differential expression of miRNAs may also play a role in aging. MiR-124 is highly expressed in young mice whereas miR-101, -127, -128, and -132 are expressed in the adult mouse brain (Lagos-Quintana et al., 2002). Some of the aging associated brain miRNAs such as miR-101 and miR-443 have been implicated in neurodegenerative diseases such as spinocerebellar ataxia type 1 and Parkinson's disease (Lee et al., 2008; Wang et al., 2008). MiRNAs have also been found to play a critical role in learning and memory. Bredy et al. (2011) suggested the influence of miRNAs on learning and memory by; (i) controlling dendrite differentiation during growth, (ii) modifying gene function in individual dendrites and (iii) destabilizing old memory to allow new learning and memory. Dendrite spine formations and synaptic plasticity are crucial for memory and learning (Konopka et al., 2011).

In a *dicer* gene mutated mouse, increased dendrite spines are accompanied by enhanced synaptic plasticity, learning and memory, and with long term potentiation of the neurons (Konopka et al., 2010, 2011). MiRNA, miR-125b and -132 regulate dendrite spine formation and synaptic plasticity via regulation of NMDA receptor subunit and have been implicated in learning and memory dysfunctions (Edbauer et al., 2010). MiR-124, the most abundant brain miRNA of all, has also been shown to regulate synaptic plasticity by modulating expression of transcription factor CREB which in turn negatively regulated the serotonin-induced synaptic plasticity (Rajasethupathy et al., 2009). MiRNAs let-7d, miR-132, -124 and -181 mediated regulation of cognition, memory and learning has also been shown in drug addiction models (Chandrasekar and Dreyer, 2009, 2011; Nudelman et al., 2010). Other miRNAs, miR-212, -134, -181a, -324 and -369 have also been shown to regulate learning and memory associated genes (Gao et al., 2010; Spadaro and Bredy, 2012).

As mentioned above, miRNAs are expressed in CNS in a spatially and temporally controlled manner. MiRNAs regulate diverse neurological functions including development, differentiation, growth, proliferation and neural activity (Liu and Xu, 2011). The importance of miRNAs in CNS development was shown by conditional deletion of *dicer* gene which led to a significant reduction in the size of cortex (De Pietri Tonelli et al., 2008). Ablation of Dicer from hippocampus at various phases of mouse development has identified the importance of miRNAs in hippocampus progenitor cell proliferation and differentiation (Li et al., 2011). MiRNAs have also been implicated in the development, differentiation and proliferation of oligodendrocytes and neurons. Deletion of miR-17-92 cluster in the oligodendrocyte genome results in significant reduction in their numbers in brain (Budde et al., 2010). Neural Stem Cell (NSC) proliferation was shown to be regulated by miR-137, which also inhibited neuronal differentiation (Szulwach et al., 2010). Mir-124 was also shown to promote neuronal differentiation along with neurite outgrowth and synaptogenesis (Yu et al., 2008). MiRNAs let-7b, miR-9 and -124 have also been reported to be involved in neuronal differentiation and neurogenesis. As mentioned above, synapse formation and plasticity, an important aspect of learning and memory, is also modulated by miRNAs such as miR-132, -134 and -138 (Liu et al., 2011).

TBI causes a cascade of molecular events that ultimately heal or drive the TBI associated pathology. A variety of cellular events occur at the time of physical injury. During this period, brain triggers the accumulation of substances like neurotransmitters and ions into the neuronal cells and axons, which results in altered homeostasis and subsequent cell death (McAllister, 2011). Secondary sequelae that follow the primary injury is a complex response derived by a mix of inflammatory mediators and innate immune response to damaged tissue, disrupted neurofibers, ischemic damage, bleeding and edema. As described above, miRNAs play a critical role in the normal functions and homeostasis of the brain. So it will be empirical for miRNAs to play a critical role in the tissue repair and pathology of TBI.

Several miRNAs such as miR-107, -130a, -223, -292-5p, -433-3p, -451, -541 and -711 have been found to be modulated in the brains of rats with TBI (Redell et al., 2009). Constitutive upregulation of miR-21 was also observed in hippocampus and cortex of the mouse brain post TBI (Redell et al., 2011; Lei et al., 2009). In a separate miRNA expression kinetics study, a total of 136, 118, 149 and 203 miRNAs were found to be modulated at 6, 24, 48 and 72 hours post TBI respectively in the cerebral cortex of mouse (Lei et al., 2009). We have also shown the modulation of miRNAs in the serum and CSF of rats in which TBI was induced by head-only exposure to blast overpressure waves. Animals were exposed to three repeated blasts and miRNA modulation in serum and CSF was determined at 3 hours and 24 hours after the last blast exposure. Differential expression of miRNAs such as let-7i, miR-122, -340-5p, -200b* and -872 was observed in the serum. Let-7i upregulation was also observed in the CSF of these animals. Pathway analysis of modulated miRNAs suggested their involvement in axon guidance and Wnt signaling pathways (Balakathiresan et al., 2012). Diffuse axonal injury and oxidative stress in the brain following blast TBI has been reported in other studies and therefore, these miRNAs may play an important role in TBI pathology (Readnower et al., 2010; Young, 2010). Let-7 miRNA has been shown to induce neuronal degeneration via TLR activation in neurons, however if a similar role is played by elevated levels of let-7i in brain after the blast injury is not known (Lehman et al., 2012). MiRNAs have also been reported to be differentially modulated in the serum of human TBI patients. Sixty miRNAs were found to be differentially expressed in the serum of the TBI patients of which miR-16, -92a and -765 showed good correlation with the injury (Redell et al., 2010). MiRNAs such as miR-9, -290, -874, -451 and let-7d* are also altered in brains of rats with moderate fluid percussion injury. In the same study, hypothermia, which has been proposed to have protective effect after TBI, was shown to differentially modulate the expression of miRNAs such as miR-874 and -451 post TBI indicating their potential protective role in TBI (Truettner et al., 2011). These studies clearly show that miRNAs are modulated in brain as well as in serum of the animals/humans that have experienced TBI. However, studies to understand the role that miRNAs play in healing after injury or manifestation of TBI induced pathology are lacking.

Chronic TBI pathology must involve the interaction of brain cells resulting in a persistent signaling cascade that ultimately produce the observed brain pathology. Several observed pathologies of TBI such as neuronal degeneration, loss of synaptic plasticity, cognitive dysfunctions, and learning and memory deficits are common with many other neurodegenerative diseases (Figure 3.3). Although, the molecular and biochemical evidences are scant, the epidemiological studies shows that TBI can cause serious and disabling neuropsychiatric disorders. The recent summary of literature published by Australian Centre for Posttraumatic Mental Health indicate that about 32–66% of military personnel who has suffered TBI also suffer from psychiatric disorder (http://www.acpmh. unimelb.edu.au/resources/lit_summary.html/). The high incidences of psychiatric disorders among combat personnel suffering from TBI may be caused by neurological damage along with the stress of battlefield environment.

Clearly, there is an association between the symptoms and pathologies of TBI and neurodevelopmental/degenerative disorders. Therefore, studying the miRNA functions in neurodegenerative disease and their relationship with TBI pathology may help identify and mitigate the risk of neurodegenerative diseases in TBI patients. Such understanding may also help in identifying a TBI biomarker. Some of the diseases where miRNAs have been shown to play a crucial role in deriving the neuropathology and thus, may also be involved in the pathology of TBI are as follows.

TBI, Neurodegenerative Disease and MiRNAs

Neuro-inflammation, -toxicity and -degeneration

Disturbance in brain homeostasis and the accumulation of endogenous toxic factors after TBI may cause neurotoxicity (McAllister, 2011). MiRNA mediated regulation may act to either compensate or exacerbate the neurotoxicity due to TBI (Kaur et al., 2012). As mentioned above, expression levels of miR-107 and let-7i changes in brain and CSF of post TBI, respectively (Redell et al., 2009; Balakathiresan et al., 2012). MiR-107 has been shown to target and suppress expression of progranulin (Wang et al., 2010). Progranulin deficiency has been linked to neurodegeneration and gliosis (Cenik et al., 2012; Martens et al., 2012). Downregulation of miR-107 and subsequent accumulation of progranulin in brain post TBI thus may help toward protecting neurons and a positive outcome of TBI (Wang et al., 2010). Other studies indicate that progranulin is also involved in wound healing and therefore, it may also help in repairing the tissue damage that arises due to TBI (He et al., 2003). Other endogenous factors that have been implicated in neuroinflammation and neurodegeneration are Toll-like receptors (TLR) (Sharma and Maheshwari, 2009; Vezzani et al., 2011; Lehmann et al., 2012).

Lehmann et al. (2012) have shown that miRNA let-7 activates neuronal TLR7 signaling that results in enhanced neurodegeneration. Let-7 expression was also found to be elevated in the CSF of Alzheimer's patients and may therefore, play a critical role in neurodegeneration following TBI. Inflammatory mediators (TNF-α, IL-6, IL-8, IL-10, and TGF-β) are also elevated immediately after TBI. These cytokine molecules alter leukocyte adhesion, breakdown blood-brain barrier (BBB) and increase the cellular stress responses (Das et al., 2012). Inflammatory cytokines are primarily produced to promote neural repair and protect brain from infection after TBI (Timaru-Kast et al., 2012). However, sustained elevated levels of inflammatory cytokines also cause neuronal cell death and drive behavioral and cognitive deficits (Badan et al., 2003; Sandhir et al., 2008). We have shown (Balakathiresan et al., 2012) that several of these cytokines may be regulated by the miRNA let-7i, which was upregulated both in serum and CSF of animals post blast TBI (Figure 3.2). Therefore, understanding the complex interactions of altered cytokines with dysregulated miRNAs may provide a rich array of potential biomarkers and therapeutic targets of TBI.

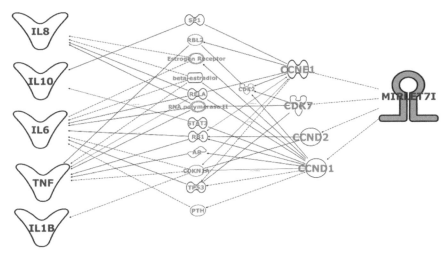

Figure 3.2. Functional interaction of let-7i with known inflammatory markers of TBI. Functional networks of known TBI-related inflammatory molecules predicted to be regulated by miRNA let-7i was carried out by the Ingenuity Pathway Analysis program. Adapted and modified from Balakathiresan et al., 2012.

Brain Tumors

At least one case of post traumatic glioma has been reported where glioma developed after 10 years of injury with the scar of the trauma (Zhou and Liu, 2010). Changes in miRNA expression have been shown in brain tumors. High expression of miR-125b and miR-9 in adult and fetal brain respectively was observed in oligodendrogliomas (Nelson et al., 2006). Mizoguchi et al. (2012) have put forward a comprehensive review of miRNAs in human glioma, which identified four miRNAs, i.e., miR-21, -196, -10b and -128, that were modulated in brain tumors in more than three independent studies. MiR-21, -196 and -10b are upregulated in the glioblastoma and respectively exert anti-apoptotic, malignant-transformative and metastatic effect on tumor cells (Chan et al., 2005; Sasayama et al., 2009; Guan et al., 2010). Currently, there is no evidence that miRNA changes contribute towards the risk of developing glioma post TBI. However, given the role that miRNAs play in brain function and glioma development such a possibility cannot be ruled out until further studies.

Alzheimer's Disease (AD)

AD is the most common of the neurodegenerative diseases. The symptoms of AD usually appear after the age of 60 and an estimated 5.1 million people in the USA may be suffering from AD (http://www.nia.nih.gov/

alzheimers/publication/alzheimers-disease-fact-sheet). Characteristic features of AD are the formation of β-amyloid (Aβ) plaques, neurofibrillary tangles and the loss of connections between neurons. AD and TBI share several common gene expression changes such as amyloid precursor proteins (APP), BACE1, ApoE4 and tau (Johnson et al., 2010; Liliang et al., 2010a,b; Sivanandam and Thakur, 2012). Post mortem analyses of severe TBI patients showed Aβ deposits in the cortical area of the brain in about 30% of patients (Roberts et al., 1994). Aβ is made by proteolytic cleavage of APP by enzyme BACE1. Expression of BACE-1 has been shown to be regulated by miRNAs miR-29a/b-1 and -107 (Hebert et al., 2008; Wanget et al., 2008). Deletion of miR-29a/b-1 cluster was directly proportional to increased BACE-1 activity and in turn increased Aβ deposition (Hebert et al., 2008). Tauopathies are characterized by abnormal intracellular accumulation of microtubulin associated protein "tau" and are also very common in AD. Chronic traumatic encephalopathy, a tau-protein linked neurodegenerative disease has been observed in US veterans exposed to blast TBI and American football players and wrestlers who had suffered concussions (Goldstein et al., 2012). Abnormal splicing of tau gene mediated by a neuronal splicing factor polypyrimidine tract-binding protein 2 (PTBP2), results in tauopathy. The expression of PTBP2 is kept under check by miR-132. Loss of function of miR-132, and thereby increased activity of PTBP2, was proposed to mediate the development of tauopathy in progressive supranuclear palsy patients (Smith et al., 2010).

Parkinson's Disease (PD)

PD is the second most common neurodegenerative disorder in the US and is characterized by resting tremor, rigidity, bradykinesia and loss of postural reflexes. PD patients' brain exhibit Lewy bodies' deposition and progressive loss of dopaminergic neurons. Mutation in or increased expression of α-synuclein gene has been reported to increase the risk of developing PD (Crookson, 2009). The long term effect of damage in the basal ganglia may also contribute to the development of PD (http://www.ninds.nih.gov/disorders/tbi/detail_tbi.htm). PD patients show differential modulation of miR-30b, -30c and -26a miRNAs. These miRNAs modulate glycosphingolipid biosynthesis and the ubiquitin proteosome system by regulating α-synuclein-interacting genes (Martins et al., 2011). Increased sphingolipid levels in the CSF were also reported in TBI patients who died following injury than in the patients who survived (Pasvogel et al., 2010). Differential expression of other miRNAs such as miR-1, -22, -29 and -133 has also been reported in PD (Margis et al., 2011).

Huntington's Disease (HD)

HD is a fatal neurodegenerative disease that is caused by the polyglutamine expansion (CAG repeats) in the gene encoding the Huntington (Htt) protein. HD is characterized by neuronal death in the cortex and striatum, progressive cognitive impairment, neuropsychiatric symptoms, and involuntary choreic movements (Landles and Bates, 2004; Johnson and Buckley, 2009). Along with neurodegeneration, HD and TBI share several cognitive, behavioral and emotional changes such as depression, apathy and irritability (Arciniegas, 2011; Thompson et al., 2012). No direct correlation of miRNA dysregulation and neurodegeneration or cognitive impairments has been shown in HD, however, studies have shown that brain miRNAs are differentially modulated in HD. Aberrant expression of miRNAs such as miR-9/9*, -29b, -124a and -132 has been identified in the brains of both human HD patients and mouse models of HD. These miRNAs may interact with RE1-silencing transcription factor (REST), a master regulator of neuronal genes and Repressor Element 1 (RE1), to derive the molecular etiology of HD (Johnson et al., 2008; Johnson and Buckley, 2009). Aberrant expression of miRNAs such as miR-22, -29c, -128, -132, -138, -218, -222, -344 and -674* have also been reported in the brains of transgenic mouse models of HD (Lee et al., 2011).

Viral Infections

Host immune competence in general prevents the virus infection of the brain. However, viruses have been shown to cause severe neurodegeneration in the brain. Neurodegenerative disease-causing viruses such as retrovirus PVC-211 MuLV and Venezuelan Equine Encephalitis Virus (VEEV) target neuron and glial cells causing neuronal dysfunction and degeneration. Neuroinflammation plays an important role in the viral disease outcome, and is often associated with the secondary neuronal damage in the brain (Berth et al., 2009; Amor et al., 2010). The molecular mechanisms underlying miRNA expression and their role in neurodegeneration in a virus infected brain are not fully understood. We have shown that VEEV infection in mice brain causes modulation of miRNA expression. MicroRNAs such as miR-155, -27a, -381, -154* and -801 were found to be significantly modulated in VEEV infected brain. Pathway analyses showed that majority of these miRNAs regulate neuronal functions and/or inflammation which are hallmarks of virus induced neurodegeneration (Bhomia et al., 2010).

MicroRNA and Spinal Cord Injury

Every year more than 10,000 cases of acute Spinal Cord Injury (SCI) are reported in North America and many more worldwide (Kwon et al., 2010). Diagnosis of acute SCI is easy, however, estimating the severity and the extent of neurological recovery still remains a challenge. Inaccurate assessment of acute SCI can lead to a severe condition such as paralysis, which does not have effective therapeutic intervention. Increased levels of inflammatory molecules in cerebrospinal fluid, such as IL-6, IL-8, MCP-1, tau, S100β, and GFAP have been reported post SCI (Kwon et al., 2010). MiRNA expression studies in acute SCI, although few, have shown that miRNAs may play a major role in the pathophysiology of SCI. For instance, miR-let7a, miR-107 and miR-183 has been shown to play a key role in apoptotic cell death whereas, miR-181b and miR-199 are involved in control of inflammation (Liu et al., 2009; Yunta et al., 2012). MiRNAs have also been tested as therapeutic targets in acute SCI. MiR-486 has been targeted using antagomiRs in a rat model of SCI. MiR-486 was found to be upregulated after SCI and its suppression by using anti-miR486 showed increased recovery from SCI (Jee et al., 2012). Since miRNAs are modulated in SCI, they can be explored as a novel biomarker for diagnosis and therapeutic targets of SCI.

TBI, Neuropsychiatric Disorders and MiRNAs

Post Traumatic Stress Disorder (PTSD)

PTSD, as the name suggests, is an anxiety disorder that develops after exposure to traumatic events. PTSD is characterized by a series of symptoms based on five major criteria: (1) exposure to or witnessing a traumatic event(s); (2) intrusive symptoms; (3) persistence avoidance; (4) negative alterations in cognition and mood and; (5) alteration in arousal (Bottalico and Bruni, 2012). War veterans and soldiers who suffer from PTSD have often been exposed to physical trauma like blast exposure and/or blunt trauma, which could also cause mTBI. Experiencing a traumatic event, either physical or mental or both, is the main cause for development of brain injuries and disorders. This could be the reason that patients with PTSD or TBI show a spectrum of common clinical features such as depression, anxiety, sleep disturbances, irritability, difficulty in concentration and chronic pain (Vasterling et al., 2009). There are evidence that suggest that mTBI can progress into PTSD (Vasterling and Dikmen, 2012). Due to such overlapping symptoms and causative agents, TBI can often be misdiagnosed for PTSD or vice versa. Therefore, it is important to develop biomarker(s) specific for mTBI and PTSD for their accurate diagnosis (Vasterling et al.,

2009). To date, there are no biomarkers that can differentiate PTSD from mTBI. Since miRNAs are specific in their action to injury, these may present an attractive target to be developed as markers that can differentiate between mTBI and PTSD. Altered miRNA expression in the animals exposed to a single or repeated inescapable stress, and also in postmortem brain tissues of different human neuropsychiatric disorders provide some indirect evidence of role of miRNAs in PTSD. Downregulation of 12 miRNAs namely miR-96, -141, -182, -183, -183*, -298, -200a, -200a*, -200b, -200b*, -200c and -429, in the frontal cortex of rats, subjected to repeated inescapable shocks, has been shown to be associated with depression like phenotype (Smalheiser et al., 2011).

Acute and chronic immobilization stress in rats also altered several miRNAs in the hippocampus and amygdala regions (Meerson et al., 2010). Among them, miR-134 and -183 have been shown to be involved in alternative splicing and cholinergic neurotransmission in the stressed brain (Meerson et al., 2010). Another miRNA, miR-34c was found to be upregulated, in the central amygdala of mice, subjected to acute and chronic stress and it correlated with anxiety like behavior (Haramati et al., 2011). MiR-34c has also been shown (Haramati et al., 2011) to reduce the responsiveness of cells to Corticotrophin Releasing Factor (CRF) in neuronal cells, endogenously expressing corticotropin releasing factor receptor type 1 (CRFR1). These studies clearly indicate that miRNA are modulated in response to stress and, therefore, have potential to be explored as biomarkers of PTSD.

Schizophrenia (SCZ)

SCZ is one of the most common forms of psychiatric disorders (Xu et al., 2010). SCZ is characterized by a wide ranging deficit of neuro-cognitive and -physiological functions (Arguello et al., 2010). Studies in the mouse model indicate an association between SCZ and a microdeletion, 22q11.2 on chromosome 22. 22q11.2 deletion causes change in the expression of a number of miRNAs in the brain that may contribute to behavioral and neuronal deficits (Stark et al., 2008). Postmortem brain tissue analyses of SCZ patients have shown a correlation of reduced expression levels of miR-330, -30b/d, -33, -15, -195 and -22, to SCZ pathogenesis (Perkins et al., 2007; Moreau et al., 2011). Genome wide expression profile, in mononuclear leukocytes of SCZ patients, also identified seven significantly altered miRNAs, i.e., miR-34a, -449a, -564, -432, -582, -572 and -652, which correlated with clinical symptoms, neurocognitive performances and neurophysiological functions of SCZ patients (Lai et al., 2011). TBI is known to cause a wide range of psychiatric sequelae but whether TBI can cause schizophrenia is still unclear. However, the meta-analysis of case-controlled population-studies using available literature supports the theory of developing schizophrenia following TBI (Molloy et al., 2011).

Fragile X Syndrome (FXS)

FXS is a form of inherited mental retardation that is caused by the functional loss of Fragile X Mental Retardation Protein (FMRP). FMRP interacts with the Dicer and Ago, two essential components of RNAi machinery. FMRP-Ago interaction is important for normal FMRP function in neuronal development and synaptogenesis (Caudy et al., 2002; Ishizuka et al., 2002; Jin et al., 2004). FMRP-miRNA interaction, specifically with miR-125b and -132 has been shown to regulate FMRP mediated regulation of NMDA receptor subunit NR2A, which in turn affects the spine morphology and synaptogenesis in neurons (Edbauer et al., 2010).

TBI, Neurodevelopmental Disorders and MiRNAs

Tourette's Syndrome (TS)

TS is a neurodevelopmental disorder characterized by the presence of chronic vocal and motor "tics" and behavioral abnormalities. "Tic", an involuntary contraction of skeletal muscle, which is the characteristic feature of TS, has also been reported in at least one adult case of severe TBI, a year after sustaining the injury (Ranjan et al., 2011). Slit and Trk-like family member 1 (SLITRK1) gene is an evolutionarily conserved gene that plays an important role in neuronal circuits. A rare sequence variant of SLITRK1 has been shown to be associated with TS (Stillman et al., 2009). A frame shift mutation identified in the TS SLITRK1 gene, inhibited its binding to miR-189 and also inhibited its activity in enhancing dendrite growth in primary neuronal cultures (Abelson et al., 2005).

Rett Syndrome (RTT)

Girls are exclusively affected by the neurodevelopmental disorder of RTT (Smeets et al., 2011). Mutations in the gene encoding methyl-cpG binding protein 2 (MeCP2) is the main cause of RTT (Wu et al., 2010). Patients have a smaller brain size, and the size and dendrite arborization of individual neuron is also reduced (Stuss et al., 2012). Experimental evidence shows that miR-132 controls MeCP2 expression, which in turn regulates MeCP2 regulated miRNAs (Wu et al., 2010). Therefore, the reduced expression of MeCP2 may lead to the disruption of miRNA regulatory process and leads to clinical RTT phenotypes.

Autism Spectrum Disorders (ASD)

ASD is a neurodevelopmental disorder that is characterized by the abnormalities in reciprocal social interactions and repetitive and stereotyped behavior. Facial emotion recognition that is impaired in TBI patients with frontal lobe injury is also observed in autism patients (Callahan et al., 2011). Genetic studies have shown that microduplication of 22q11.2 sequence in chromosome 22 has potential connection between miRNA abnormalities and ASD phenotypes. DGCR8, which is involved in miRNA biogenesis, and miR-185 genes are present within the 22q11.2 duplication and therefore, micro duplication of 22q11.2 may cause upregulation of DGCR8 gene and duplication of miR-185 (Zhang et al., 2011). MiR-185 has been identified to regulate sarco (endo)plasmic reticulum Ca(2+) ATPase (SERCA2) protein, which is responsible for loading Ca(2+) into the Endoplasmic Reticulum (ER) and contribute to cognitive and psychiatric symptoms (Earls et al., 2012). The upregulation of miR-185 has been shown in ASD samples and their predicted target analysis indicates their involvment in neurological diseases/nervous system development and functions (Sarachana et al., 2010).

Down Syndrome (DS)

DS is caused by triplication of genes located on chromosome 21 (trisomy 21) which causes mild-to-moderate mental retardation (Rachidi and Lopes, 2011). Expression of five miRNAs (let-7c, miR-93a, -125b-2 -155 and -802), which are present on chromosome 21, is increased in DS (Kuhn et al., 2008). MeCP2 expression is suppressed in DS brain specimens and two of the over expressed miRNAs, i.e., miR-155 and -802 have been shown to target and suppress the expression of MeCP2 (Kuhn et al., 2010). Therefore, miRNAs may play a critical role in development of DS symptoms.

MicroRNA as Biomarkers of TBI

The field of development of miRNAs as a biomarker of TBI is still in its infancy. This is mainly because, the TBI pathology is quite complex and its mechanisms are not yet fully understood. In particular, mTBI being an invisible injury often goes undiagnosed (Belanger et al., 2012). Blood based biomarkers are the most ideal markers due to them being minimally invasive and reliable. In relation to brain injury, an ideal clinically usable biomarker should exhibit the following features: (1) high specificity to brain origin and sensitivity to brain injury; (2) released only after the brain tissue damage; (3) detectable in serum immediately after brain injury or should relate to state of injury, i.e., acute vs. chronic; (4) released in a time dependent manner; (5) age and sex based variations should be minimum; and (6) availability of reliable and rapid assays (Bakay and Ward, 1983; Mehta, 2010).

As described above, several investigators including us, have shown the differential modulation of miRNAs in serum and brain following TBI (Lei et al., 2009; Redell et al., 2009, 2010, 2011; Balakathiresan et al., 2012; Truettner et al., 2011). MiRNAs like miR-16, -92a, and -765 have been proposed as good markers for severe TBI (Redell et al., 2010). We have also proposed that let-7i may act as a good acute stage marker of blast induced mTBI (Balakathiresan et al., 2012). Injury severity based screening of miRNA expression pattern may provide specific miRNA signature(s) to classify different categories of TBI. Similarly, based on the state of tissue pathology, acute and chronic TBI may have different miRNAs as biomarkers. Further, the altered expression of miRNAs in blood may not completely reflect the level of miRNA changes in the brain, but a subset of miRNAs in blood may be brain-specific and can serve as diagnostic biomarkers of TBI. Alternatively, CSF which bathes the brain tissue and better reflects the state of tissue pathology may act as a source of reliable miRNA biomarkers. We have described above, several neuro-degenerative, -developmental and -psychiatric disorders, and other CNS injuries/infections, where miRNA mediated disease pathology overlaps with the TBI pathology and symptoms (Figure 3.3). This makes the diagnosis of TBI patients, using miRNAs as biomarkers, a daunting, but not impossible task as the unique miRNA or panel of miRNAs will have to distinguish TBI from all other diseases/conditions that were described above. Prospects of identifying miRNA based biomarker will improve with better understanding of the TBI pathology and the role that the miRNAs play in TBI manifestation and progression.

Conclusion

The brain injury biomarker development is more complex than any other disease or injury. This is due to the complex morphology and heterogeneous nature of the brain. Further, different type of injuries would cause varying degree of damage in brain cells, and BBB regulation would also limit the size and amount of marker in the blood (Giacoppo et al., 2012). Several brain injuries share similar pathological mechanisms. For instance, TBI and cerebral ischemia, in spite of different root causes of injury cause BBB damage, altered blood flow, neuronal and glial damage. It is practically difficult to distinguish these two injuries by traditional imaging techniques. The available imaging techniques and biological markers may be useful for identifying severe TBI, but when it comes to mild-to-moderate TBI the

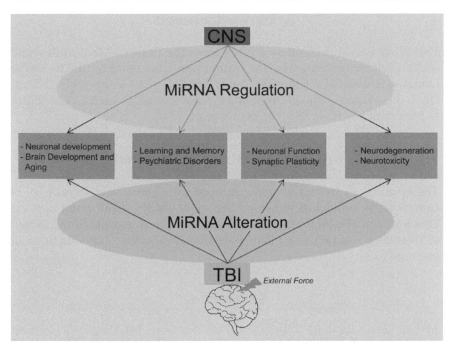

Figure 3.3. Traumatic Brain Injury altered microRNAs in multiple pathophysiological processes. CNS injuries, diseases and disorders altered miRNAs overlap with TBI. TBI—Traumatic Brain Injury; CNS—Central Nervous System.

subtle alterations in the brain and the absence of injury sensitivity delay the process of diagnosis. Neurochemical and metabolic consequences of the initial insult change over time and cause oxidative stress, inflammation, free radical generation, altered calcium influx, apoptosis, axonal injury, degeneration of neurons and synapses. The available studies indicate that miRNAs are altered in different brain diseases and disorders such as chronic pain, neuroinflammation, neurodegenerative diseases, neurotoxicity and neuropsychiatric disorders (Nampiaparampil, 2008; Vaishnavi et al., 2009; Balakathiresan et al., 2012; Kumar and Loane, 2012; Madathil et al., 2011; Kaur et al., 2012). Identification of miRNA based biomarkers for TBI will be a complex task due to overlap of TBI symptoms and pathology with other brain diseases. Probably, it will be a panel or a cohort of miRNAs that only when taken together will specifically diagnose, first the invisible brain injury, and second the causative agent of the injury. Therefore, future studies on miRNAs in the secondary sequelae of TBI and other closely related brain injuries will not only help the identification of biomarkers for varying forms of TBI, but also for its proper diagnosis and treatment.

Acknowledgements

This work was supported by a grant from the Defense Medical Research and Development Program. The authors are also thankful to Dr. Martin Doughty, Assistant Professor, Anatomy, Physiology & Genetics, USUHS and Dr. T. Sreevalsan, Adjunct Professor, Georgetown University, Washington DC for critically reviewing the manuscript.

Authors Disclosure Statement

The opinions expressed here are those of the authors and should not be construed as official or reflecting the views of the Uniformed Services University of Health Sciences, Bethesda, Maryland and the Birla Institute of Technology and Science, Pilani, Rajasthan, India. No competing financial interests exist.

References

Abelson, J.F., K.Y. Kwan, B.J. O'Roak, D.Y. Baek, A.A. Stillman, T.M. Morgan, C.A. Mathews, D.L. Pauls, M.R. Rasin, M. Gunel, N.R. Davis, A.G. Ercan-Sencicek, D.H. Guez, J.A. Spertus, J.F. Leckman, L.S.t. Dure, R. Kurlan, H.S. Singer, D.L. Gilbert, A. Farhi, A. Louvi, R.P. Lifton, N. Sestan and M.W. State. 2005. Sequence variants in SLITRK1 are associated with Tourette's syndrome. *Science.* 310: 317–320.

Abu-Elneel, K., T. Liu, F.S. Gazzaniga, Y. Nishimura, D.P. Wall, D.H. Geschwind, K. Lao and K.S. Kosik. 2008. Heterogeneous dysregulation of microRNAs across the autism spectrum. *Neurogenetics.* 9: 153–161.

Alon, S., E. Mor, F. Vigneault, G.M. Church, F. Locatelli, F. Galeano, A. Gallo, N. Shomron and E. Eisenberg. 2012. Systematic identification of edited microRNAs in the human brain. *Genome Res.* 22: 1533–1540.

Amor, S., F. Puentes, D. Baker and P. van der Valk. 2010. Inflammation in neurodegenerative diseases. *Immunology.* 129: 154–169.

Amrouche, L., R. Bonifay and D. Anglicheau. 2011. MicroRNAs in pathophysiology of renal disease: an increasing interest. *Med Sci (Paris).* 27: 398–404.

Arciniegas, D.B. 2011. Addressing neuropsychiatric disturbances during rehabilitation after traumatic brain injury: current and future methods. *Dialogues Clin Neurosci.* 13: 325–345.

Arguello, P.A., S. Markx, J.A. Gogos and M. Karayiorgou. 2010. Development of animal models for schizophrenia. *Dis Model Mech.* 3: 22–26.

Arroyo, J.D., J.R. Chevillet, E.M. Kroh, I.K. Ruf, C.C. Pritchard, D.F. Gibson, P.S. Mitchell, C.F. Bennett, E.L. Pogosova-Agadjanyan, D.L. Stirewalt, J.F. Tait and M. Tewari. 2011. Argonaute2 complexes carry a population of circulating microRNAs independent of vesicles in human plasma. *Proc Natl Acad Sci USA.* 108: 5003–5008.

Babenko, O., I. Kovalchuk and G.A. Metz. 2012. Epigenetic programming of neurodegenerative diseases by an adverse environment. *Brain Res.* 1444: 96–111.

Badan, I., B. Buchhold, A. Hamm, M. Gratz, L.C. Walker, D. Platt, C. Kessler and A. Popa-Wagner. 2003. Accelerated glial reactivity to stroke in aged rats correlates with reduced functional recovery. *J Cereb Blood Flow Metab.* 23: 845–854.

Bader, A.G. 2012. miR-34—a microRNA replacement therapy is headed to the clinic. *Front Genet.* 3: 120.

Bai, A.H., T. Milde, M. Remke, C.G. Rolli, T. Hielscher, Y.J. Cho, M. Kool, P.A. Northcott, M. Jugold, A.V. Bazhin, S.B. Eichmuller, A.E. Kulozik, A. Pscherer, A. Benner, M.D. Taylor, S.L. Pomeroy, R. Kemkemer, O. Witt, A. Korshunov, P. Lichter and S.M. Pfister. 2012. MicroRNA-182 promotes leptomeningeal spread of non-sonic hedgehog-medulloblastoma. *Acta Neuropathol.* 123: 529–538.

Bakay, R.A. and A.A. Ward, Jr. 1983. Enzymatic changes in serum and cerebrospinal fluid in neurological injury. *J Neurosurg.* 58: 27–37.

Balakathiresan, N., M. Bhomia, R. Chandran, M. Chavko, R.M. McCarron and R.K. Maheshwari. 2012. MicroRNA let-7i is a promising serum biomarker for blast-induced traumatic brain injury. *J Neurotrauma.* 29: 1379–1387.

Bandiera, S., E. Hatem, S. Lyonnet and A. Henrion-Caude. 2010. microRNAs in diseases: from candidate to modifier genes. *Clin Genet.* 77: 306–313.

Baraniskin, A., J. Kuhnhenn, U. Schlegel, A. Chan, M. Deckert, R. Gold, A. Maghnouj, H. Zollner, A. Reinacher-Schick, W. Schmiegel, S.A. Hahn and R. Schroers. 2011. Identification of microRNAs in the cerebrospinal fluid as marker for primary diffuse large B-cell lymphoma of the central nervous system. *Blood.* 117: 3140–3146.

Baraniskin, A., J. Kuhnhenn, U. Schlegel, W. Schmiegel, S. Hahn and R. Schroers. 2012a. MicroRNAs in cerebrospinal fluid as biomarker for disease course monitoring in primary central nervous system lymphoma. *J Neurooncol.* 109: 239–244.

Baraniskin, A., J. Kuhnhenn, U. Schlegel, A. Maghnouj, H. Zollner, W. Schmiegel, S. Hahn and R. Schroers. 2012b. Identification of microRNAs in the cerebrospinal fluid as biomarker for the diagnosis of glioma. *Neuro Oncol.* 14: 29–33.

Barbato, C., I. Arisi, M.E. Frizzo, R. Brandi, L. Da Sacco and A. Masotti. 2009. Computational challenges in miRNA target predictions: to be or not to be a true target? *J Biomed Biotechnol.* 803069.

Belanger, H.G., R.D. Vanderploeg, G. Curtiss and D.L. Warden. 2007. Recent neuroimaging techniques in mild traumatic brain injury. *J Neuropsychiatry Clin Neurosci.* 19: 5–20.

Belanger, H.G., R.D. Vanderploeg, J.R. Soble, M. Richardson and S. Groer. 2012. Validity of the Veterans Health Administration's traumatic brain injury screen. *Arch Phys Med Rehabil.* 93: 1234–1239.

Bellingham, S.A., B.M. Coleman and A.F. Hill. 2012. Small RNA deep sequencing reveals a distinct miRNA signature released in exosomes from prion-infected neuronal cells. *Nucleic Acids Res.* [Epub ahead of print].

Berger, R.P., R.L. Hayes, R. Richichi, S.R. Beers and K.K. Wang. 2012. Serum concentrations of ubiquitin C-terminal hydrolase-L1 and alphaII-spectrin breakdown product 145 kDa correlate with outcome after pediatric TBI. *J Neurotrauma.* 29: 162–167.

Berth, S.H., P.L. Leopold and G.N. Morfini. 2009. Virus-induced neuronal dysfunction and degeneration. *Front Biosci.* 14: 5239–5259.

Bhomia, M., N. Balakathiresan, A. Sharma, P. Gupta, R. Biswas and R. Maheshwari. 2010. Analysis of microRNAs induced by Venezuelan equine encephalitis virus infection in mouse brain. *Biochem Biophys Res Commun.* 395: 11–16.

Bisognin, A., G. Sales, A. Coppe, S. Bortoluzzi and C. Romualdi. 2012. MAGIA(2): from miRNA and genes expression data integrative analysis to microRNA-transcription factor mixed regulatory circuits (2012 update). *Nucleic Acids Res.* 40: W13–21.

Bocchio-Chiavetto, L., E. Maffioletti, P. Bettinsoli, C. Giovannini, S. Bignotti, D. Tardito, D. Corrada, L. Milanesi and M. Gennarelli. 2012. Blood microRNA changes in depressed patients during antidepressant treatment. *Eur Neuropsychopharmacol* (in press).

Bottalico, B. and T. Bruni. 2012. Post traumatic stress disorder, neuroscience, and the law. *Int J Law Psychiatry.* 35: 112–120.

Bravo, J.A. and T.G. Dinan. 2011. MicroRNAs: a novel therapeutic target for schizophrenia. *Curr Pharm Des.* 17: 176–188.

Brenner, L.A. 2011. Neuropsychological and neuroimaging findings in traumatic brain injury and post-traumatic stress disorder. *Dialogues Clin Neurosci.* 13: 311–323.

Broderick, J.A. and P.D. Zamore. 2011. MicroRNA therapeutics. *Gene Ther.* 18: 1104–1110.

Brophy, G.M., S. Mondello, L. Papa, S.A. Robicsek, A. Gabrielli, J. Tepas, 3rd, A. Buki, C. Robertson, F.C. Tortella, R.L. Hayes and K.K. Wang. 2011. Biokinetic analysis of ubiquitin C-terminal hydrolase-L1 (UCH-L1) in severe traumatic brain injury patient biofluids. *J Neurotrauma.* 28: 861–870.

Bruchova, H., M. Merkerova and J.T. Prchal. 2008. Aberrant expression of microRNA in polycythemia vera. *Haematologica.* 93: 1009–1016.

Bruning, U., L. Cerone, Z. Neufeld, S.F. Fitzpatrick, A. Cheong, C.C. Scholz, D.A. Simpson, M.O. Leonard, M.M. Tambuwala, E.P. Cummins and C.T. Taylor. 2011. MicroRNA-155 promotes resolution of hypoxia-inducible factor 1alpha activity during prolonged hypoxia. *Mol Cell Biol.* 31: 4087–4096.

Bryant, R.A., M.L. O'Donnell, M. Creamer, A.C. McFarlane, C.R. Clark and D. Silove. 2010. The psychiatric sequelae of traumatic injury. *Am J Psychiatry.* 167: 312–320.

Budde, H., S. Schmitt, D. Fitzner, L. Opitz, G. Salinas-Riester and M. Simons. 2010. Control of oligodendroglial cell number by the miR-17-92 cluster. *Development.* 137: 2127–2132.

Calin, G.A. and C.M. Croce. 2006. MicroRNA signatures in human cancers. *Nat Rev Cancer.* 6: 857–866.

Callahan, B.L., K. Ueda, D. Sakata, A. Plamondon and T. Murai. 2011. Liberal bias mediates emotion recognition deficits in frontal traumatic brain injury. *Brain Cogn.* 77: 412–418.

Caudy, A.A., M. Myers, G.J. Hannon and S.M. Hammond. 2002. Fragile X-related protein and VIG associate with the RNA interference machinery. *Genes Dev.* 16: 2491–2496.

Cenik, B., C.F. Sephton, B. Kutluk Cenik, J. Herz and G. Yu. 2012. Progranulin: a proteolytically processed protein at the crossroads of inflammation and neurodegeneration. *J Biol Chem.* 287: 32298–32306.

Chan, J.A., A.M. Krichevsky and K.S. Kosik. 2005. MicroRNA-21 is an antiapoptotic factor in human glioblastoma cells. *Cancer Res.* 65: 6029–6033.

Chandrasekar, V. and J.L. Dreyer. 2009. microRNAs miR-124, let-7d and miR-181a regulate cocaine-induced plasticity. *Mol Cell Neurosci.* 42: 350–362.

Chandrasekar, V. and J.L. Dreyer. 2011. Regulation of MiR-124, Let-7d, and MiR-181a in the accumbens affects the expression, extinction, and reinstatement of cocaine-induced conditioned place preference. *Neuropsychopharmacology.* 36: 1149–1164.

Chen, J. and D.Z. Wang. 2012. microRNAs in cardiovascular development. *J Mol Cell Cardiol.* 52: 949–957.

Chen, Y.J., J. Luo, G.Y. Yang, K. Yang, S.Q. Wen and S.Q. Zou. 2012. Mutual regulation between microRNA-373 and methyl-CpG-binding domain protein 2 in hilar cholangiocarcinoma. *World J Gastroenterol.* 18: 3849–3861.

Cogswell, J.P., J. Ward, I.A. Taylor, M. Waters, Y. Shi, B. Cannon, K. Kelnar, J. Kemppainen, D. Brown, C. Chen, R.K. Prinjha, J.C. Richardson, A.M. Saunders, A.D. Roses and C.A. Richards. 2008. Identification of miRNA changes in Alzheimer's disease brain and CSF yields putative biomarkers and insights into disease pathways. *J Alzheimers Dis.* 14: 27–41.

Cookson, M.R. 2009. alpha-Synuclein and neuronal cell death. *Mol Neurodegener.* 4: 9.

Crawford, F., G. Crynen, J. Reed, B. Mouzon, A. Bishop, B. Katz, S. Ferguson, J. Phillips, V. Ganapathi, V. Mathura, A. Roses and M. Mullan. 2012. Identification of plasma biomarkers of TBI outcome using proteomic approaches in an APOE mouse model. *J Neurotrauma.* 29: 246–260.

Creighton, C.J., J.G. Reid and P.H. Gunaratne. 2009. Expression profiling of microRNAs by deep sequencing. *Brief Bioinform.* 10: 490–497.

Cullen, B.R. 2011. Viruses and microRNAs: RISCy interactions with serious consequences. *Genes Dev.* 25: 1881–1894.

Das, M., S. Mohapatra and S.S. Mohapatra. 2012. New perspectives on central and peripheral immune responses to acute traumatic brain injury. *J Neuroinflammation.* 9: 236.

De Pietri Tonelli, D., J.N. Pulvers, C. Haffner, E.P. Murchison, G.J. Hannon and W.B. Huttner. 2008. miRNAs are essential for survival and differentiation of newborn neurons but not for expansion of neural progenitors during early neurogenesis in the mouse embryonic neocortex. *Development*. 135: 3911–3921.

De Smaele, E., E. Ferretti and A. Gulino. 2010. MicroRNAs as biomarkers for CNS cancer and other disorders. *Brain Res*. 1338: 100–111.

Eacker, S.M., M.J. Keuss, E. Berezikov, V.L. Dawson and T.M. Dawson. 2011. Neuronal activity regulates hippocampal miRNA expression. *PLoS One*. 6: e25068.

Earls, L.R., R.G. Fricke, J. Yu, R.B. Berry, L.T. Baldwin and S.S. Zakharenko. 2012. Age-Dependent MicroRNA Control of Synaptic Plasticity in 22q11 Deletion Syndrome and Schizophrenia. *J Neurosci*. 32: 14132–14144.

Ebert, M.S., J.R. Neilson and P.A. Sharp. 2007. MicroRNA sponges: competitive inhibitors of small RNAs in mammalian cells. *Nat Methods*. 4: 721–726.

Ebert, M.S. and P.A. Sharp. 2010. MicroRNA sponges: progress and possibilities. *RNA*. 16: 2043–2050.

Edbauer, D., J.R. Neilson, K.A. Foster, C.F. Wang, D.P. Seeburg, M.N. Batterton, T. Tada, B.M. Dolan, P.A. Sharp and M. Sheng. 2010. Regulation of synaptic structure and function by FMRP-associated microRNAs miR-125b and miR-132. *Neuron*. 65: 373–384.

Eisenberg, I., A. Eran, I. Nishino, M. Moggio, C. Lamperti, A.A. Amato, H.G. Lidov, P.B. Kang, K.N. North, S. Mitrani-Rosenbaum, K.M. Flanigan, L.A. Neely, D. Whitney, A.H. Beggs, I.S. Kohane and L.M. Kunkel. 2007. Distinctive patterns of microRNA expression in primary muscular disorders. *Proc Natl Acad Sci USA*. 104: 17016–17021.

Elmén, J., M. Lindow, S. Schütz, M. Lawrence, A. Petri, S. Obad, M. Lindholm, M. Hedtjärn, H.F. Hansen, U. Berger, S. Gullans, P. Kearney, P. Sarnow, E.M. Straarup and S. Kauppinen. 2008. LNA-mediated microRNA silencing in non-human primates. *Nature*. 452: 896–899.

Enciu, A.M., B.O. Popescu and A. Gheorghisan-Galateanu. 2012. MicroRNAs in brain development and degeneration. *Mol Biol Rep*. 39: 2243–2252.

Esau, C.C. and B.P. Monia. 2007. Therapeutic potential for microRNAs. *Adv Drug Deliv Rev*. 59: 101–114.

Esquela-Kerscher, A. and F.J. Slack. 2006. Oncomirs—microRNAs with a role in cancer. *Nat Rev Cancer*. 6: 259–269.

Etheridge, A., I. Lee, L. Hood, D. Galas and K. Wang. 2011. Extracellular microRNA: a new source of biomarkers. *Mutat Res*. 717: 85–90.

Fabian, M.R., N. Sonenberg and W. Filipowicz. 2010. Regulation of mRNA translation and stability by microRNAs. *Annu Rev Biochem*. 79: 351–379.

Faghihi, M.A., M. Zhang, J. Huang, F. Modarresi, M.P. Van der Brug, M.A. Nalls, M.R. Cookson, G. St-Laurent, 3rd and C. Wahlestedt. 2010. Evidence for natural antisense transcript-mediated inhibition of microRNA function. *Genome Biol*. 11: R56.

Gan, C.S., C.W. Wang and K.S. Tan. 2012. Circulatory microRNA-145 expression is increased in cerebral ischemia. *Genet Mol Res*. 11: 147–152.

Gao, J., W.Y. Wang, Y.W. Mao, J. Graff, J.S. Guan, L. Pan, G. Mak, D. Kim, S.C. Su and L.H. Tsai. 2010. A novel pathway regulates memory and plasticity via SIRT1 and miR-134. *Nature*. 466: 1105–1109.

Gaughwin, P.M., M. Ciesla, N. Lahiri, S.J. Tabrizi, P. Brundin and M. Bjorkqvist. 2011. Hsa-miR-34b is a plasma-stable microRNA that is elevated in pre-manifest Huntington's disease. *Hum Mol Genet*. 20: 2225–2237.

Geekiyanage, H., G.A. Jicha, P.T. Nelson and C. Chan. 2012. Blood serum miRNA: non-invasive biomarkers for Alzheimer's disease. *Exp Neurol*. 235: 491–496.

Gentner, B., G. Schira, A. Giustacchini, M. Amendola, B.D. Brown, M. Ponzoni and L. Naldini. 2009. Stable knockdown of microRNA *in vivo* by lentiviral vectors. *Nat Methods*. 6: 63–66.

Giacoppo, S., P. Bramanti, M. Barresi, D. Celi, V. Foti Cuzzola, E. Palella and S. Marino. 2012. Predictive biomarkers of recovery in traumatic brain injury. *Neurocrit Care*. 16: 470–477.

Git, A., H. Dvinge, M. Salmon-Divon, M. Osborne, C. Kutter, J. Hadfield, P. Bertone and C. Caldas. 2010. Systematic comparison of microarray profiling, real-time PCR, and next-generation sequencing technologies for measuring differential microRNA expression. *RNA.* 16: 991–1006.

Goldstein, L.E., A.M. Fisher, C.A. Tagge, X.L. Zhang, L. Velisek, J.A. Sullivan, C. Upreti, J.M. Kracht, M. Ericsson, M.W. Wojnarowicz, C.J. Goletiani, G.M. Maglakelidze, N. Casey, J.A. Moncaster, O. Minaeva, R.D. Moir, C.J. Nowinski, R.A. Stern, R.C. Cantu, J. Geiling, J.K. Blusztajn, B.L. Wolozin, T. Ikezu, T.D. Stein, A.E. Budson, N.W. Kowall, D. Chargin, A. Sharon, S. Saman, G.F. Hall, W.C. Moss, R.O. Cleveland, R.E. Tanzi, P.K. Stanton and A.C. McKee. 2012. Chronic traumatic encephalopathy in blast-exposed military veterans and a blast neurotrauma mouse model. *Sci Transl Med.* 4: 134ra160.

Gommans, W.M. 2012. A-to-I editing of microRNAs: regulating the regulators? *Semin Cell Dev Biol.* 23: 251–257.

Griffiths-Jones, S. 2004. The microRNA Registry. *Nucleic Acids Res.* 32: D109–111.

Griffiths-Jones, S., R.J. Grocock, S. van Dongen, A. Bateman and A.J. Enright. 2006. miRBase: microRNA sequences, targets and gene nomenclature. *Nucleic Acids Res.* 34: D140–144.

Griffiths-Jones, S., H.K. Saini, S. van Dongen and A.J. Enright. 2008. miRBase: tools for microRNA genomics. *Nucleic Acids Res.* 36: D154–158.

Guan, Y., M. Mizoguchi, K. Yoshimoto, N. Hata, T. Shono, S.O. Suzuki, Y. Araki, D. Kuga, A. Nakamizo, T. Amano, X. Ma, K. Hayashi and T. Sasaki. 2010. MiRNA-196 is upregulated in glioblastoma but not in anaplastic astrocytoma and has prognostic significance. *Clin Cancer Res.* 16: 4289–4297.

Gunaratne, P.H., C. Coarfa, B. Soibam and A. Tandon. 2012. miRNA data analysis: next-gen sequencing. *Methods Mol Biol.* 822: 273–288.

Ha, T.Y. 2011. The Role of MicroRNAs in Regulatory T Cells and in the Immune Response. *Immune Netw.* 11: 11–41.

Haghikia, A., K. Hellwig, A. Baraniskin, A. Holzmann, B.F. Decard, T. Thum and R. Gold. 2012. Regulated microRNAs in the CSF of patients with multiple sclerosis: A case-control study. *Neurology.*

Hansen, T., L. Olsen, M. Lindow, K.D. Jakobsen, H. Ullum, E. Jonsson, O.A. Andreassen, S. Djurovic, I. Melle, I. Agartz, H. Hall, S. Timm, A.G. Wang and T. Werge. 2007. Brain expressed microRNAs implicated in schizophrenia etiology. *PLoS One.* 2: e873.

Haraguchi, T., Y. Ozaki and H. Iba. 2009. Vectors expressing efficient RNA decoys achieve the long-term suppression of specific microRNA activity in mammalian cells. *Nucleic Acids Res.* 37: e43.

Haramati, S., I. Navon, O. Issler, S. Ezra-Nevo, S. Gil, R. Zwang, E. Hornstein and A. Chen. 2011. MicroRNA as repressors of stress-induced anxiety: the case of amygdalar miR-34. *J Neurosci.* 31: 14191–14203.

Hariharan, M., V. Scaria, B. Pillai and S.K. Brahmachari. 2005. Targets for human encoded microRNAs in HIV genes. *Biochem Biophys Res Commun.* 337: 1214–1218.

Haybaeck, J., N. Zeller and M. Heikenwalder. 2011. The parallel universe: microRNAs and their role in chronic hepatitis, liver tissue damage and hepatocarcinogenesis. *Swiss Med Wkly.* 141: w13287.

He, Z., C.H. Ong, J. Halper and A. Bateman. 2003. Progranulin is a mediator of the wound response. *Nat Med.* 9: 225–229.

Hebert, S.S., K. Horre, L. Nicolai, A.S. Papadopoulou, W. Mandemakers, A.N. Silahtaroglu, S. Kauppinen, A. Delacourte and B. De Strooper. 2008. Loss of microRNA cluster miR-29a/b-1 in sporadic Alzheimer's disease correlates with increased BACE1/beta-secretase expression. *Proc Natl Acad Sci USA.* 105: 6415–6420.

Heneghan, H.M., N. Miller and M.J. Kerin. 2010. Role of microRNAs in obesity and the metabolic syndrome. *Obes Rev.* 11: 354–361.

Hou, L., D. Wang and A. Baccarelli. 2011. Environmental chemicals and microRNAs. *Mutat Res.* 714: 105–112.

Hua, D., D. Ding, X. Han, W. Zhang, N. Zhao, G. Foltz, Q. Lan, Q. Huang and B. Lin. 2012. Human miR-31 targets radixin and inhibits migration and invasion of glioma cells. *Oncol Rep.* 27: 700–706.

Huijser, P. and M. Schmid. 2011. The control of developmental phase transitions in plants. *Development.* 138: 4117–4129.

Hunter, J.V., E.A. Wilde, K.A. Tong and B.A. Holshouser. 2012. Emerging imaging tools for use with traumatic brain injury research. *J Neurotrauma.* 29: 654–671.

Ilhan-Mutlu, A., L. Wagner, A. Wohrer, J. Furtner, G. Widhalm, C. Marosi and M. Preusser. 2012. Plasma MicroRNA-21 Concentration May Be a Useful Biomarker in Glioblastoma Patients. *Cancer Invest.* 30: 615–621.

Ishizuka, A., M.C. Siomi and H. Siomi. 2002. A Drosophila fragile X protein interacts with components of RNAi and ribosomal proteins. *Genes Dev.* 16: 2497–2508.

Issabekova, A., O. Berillo, V. Khailenko, S. Atambayeva, M. Regnier and A. Ivachshenko. 2011. Characteristics of Intronic and Intergenic Human miRNAs and Features of their Interaction with mRNA. *World Academy of Academy of Science, Engeinering and Technology.* 59: 63–66

Issabekova, A., O. Berillo, M. Regnier and I. Anatoly. 2012. Interactions of intergenic microRNAs with mRNAs of genes involved in carcinogenesis. *Bioinformation.* 8: 513–518.

Jeyaseelan, K., K.Y. Lim and A. Armugam. 2008. MicroRNA expression in the blood and brain of rats subjected to transient focal ischemia by middle cerebral artery occlusion. *Stroke.* 39: 959–966.

Jee, M.K., J.S. Jung, J.I. Choi, J.A. Jang, K.S. Kang, Y.B. Im and S.K. Kang. 2012. MicroRNA 486 is a potentially novel target for the treatment of spinal cord injury. *Brain.* 135: 1237–1252.

Jeter, C.B., J.B. Redell, A.N. Moore, G.W. Hergenroeder, J. Zhao, D.R. Johnson, M.J. Hylin and P.K. Dash. 2012. Biomarkers of Traumatic Injury. pp. 337–355. *In:* G. Li and S.P. Baker [eds.]. Injury Research: Theories, Methods, and Approaches. Springer, New York, USA.

Jin, H., Y. Yu, W.B. Chrisler, Y. Xiong, D. Hu and C. Lei. 2012. Delivery of MicroRNA-10b with Polylysine Nanoparticles for Inhibition of Breast Cancer Cell Wound Healing. *Breast Cancer (Auckl).* 6: 9–19.

Jin, P., D.C. Zarnescu, S. Ceman, M. Nakamoto, J. Mowrey, T.A. Jongens, D.L. Nelson, K. Moses and S.T. Warren. 2004. Biochemical and genetic interaction between the fragile X mental retardation protein and the microRNA pathway. *Nat Neurosci.* 7: 113–117.

Johnson, R. and N.J. Buckley. 2009. Gene dysregulation in Huntington's disease: REST, microRNAs and beyond. *Neuromolecular Med.* 11: 183–199.

Johnson, V.E., W. Stewart and D.H. Smith. 2010. Traumatic brain injury and amyloid-beta pathology: a link to Alzheimer's disease? *Nat Rev Neurosci.* 11: 361–370.

Katahira, J. and Y. Yoneda. 2011. Nucleocytoplasmic transport of microRNAs and related small RNAs. *Traffic.* 12: 1468–1474.

Katsuura, S., Y. Kuwano, N. Yamagishi, K. Kurokawa, K. Kajita, Y. Akaike, K. Nishida, K. Masuda, T. Tanahashi and K. Rokutan. 2012. MicroRNAs miR-144/144* and miR-16 in peripheral blood are potential biomarkers for naturalistic stress in healthy Japanese medical students. *Neurosci Lett.* 516: 79–84.

Kaur, P., A. Armugam and K. Jeyaseelan. 2012. MicroRNAs in Neurotoxicity. *J Toxicol.* 870150.

Kawahara, Y., M. Megraw, E. Kreider, H. Iizasa, L. Valente, A.G. Hatzigeorgiou and K. Nishikura. 2008. Frequency and fate of microRNA editing in human brain. *Nucleic Acids Res.* 36: 5270–5280.

Kawahara, Y., B. Zinshteyn, T.P. Chendrimada, R. Shiekhattar and K. Nishikura. 2007. RNA editing of the microRNA-151 precursor blocks cleavage by the Dicer-TRBP complex. *EMBO Rep.* 8: 763–769.

Kelley, K., S.J. Chang and S.L. Lin. 2012. Mechanism of repeat-associated microRNAs in fragile X syndrome. *Neural Plast.* 104796.

Konopka, W., A. Kiryk, M. Novak, M. Herwerth, J.R. Parkitna, M. Wawrzyniak, A. Kowarsch, P. Michaluk, J. Dzwonek, T. Arnsperger, G. Wilczynski, M. Merkenschlager, F.J. Theis, G. Kohr, L. Kaczmarek and G. Schutz. 2010. MicroRNA loss enhances learning and memory in mice. *J Neurosci.* 30: 14835–14842.

Konopka, W., G. Schutz and L. Kaczmarek. 2011. The microRNA contribution to learning and memory. *Neuroscientist.* 17: 468–474.

Kosaka, N., H. Iguchi and T. Ochiya. 2010. Circulating microRNA in body fluid: a new potential biomarker for cancer diagnosis and prognosis. *Cancer Sci.* 101: 2087–2092.

Kosik, K.S. and A.M. Krichevsky. 2005. The Elegance of the MicroRNAs: A Neuronal Perspective. *Neuron.* 47: 779–782.

Kozomara, A. and S. Griffiths-Jones. 2011. miRBase: integrating microRNA annotation and deep-sequencing data. *Nucleic Acids Res.* 39: D152–157.

Krichevsky, A.M., K.S. King, C.P. Donahue, K. Khrapko and K.S. Kosik. 2003. A microRNA array reveals extensive regulation of microRNAs during brain development. *RNA.* 9: 1274–1281.

Krol, J., I. Loedige and W. Filipowicz. 2010. The widespread regulation of microRNA biogenesis, function and decay. *Nat Rev Genet.* 11: 597–610.

Krutzfeldt, J., N. Rajewsky, R. Braich, K.G. Rajeev, T. Tuschl, M. Manoharan and M. Stoffel. 2005. Silencing of microRNAs *in vivo* with 'antagomirs'. *Nature.* 438: 685–689.

Kuhn, D.E., G.J. Nuovo, M.M. Martin, G.E. Malana, A.P. Pleister, J. Jiang, T.D. Schmittgen, A.V. Terry, Jr., K. Gardiner, E. Head, D.S. Feldman and T.S. Elton. 2008. Human chromosome 21-derived miRNAs are overexpressed in down syndrome brains and hearts. *Biochem Biophys Res Commun.* 370: 473–477.

Kuhn, D.E., G.J. Nuovo, A.V. Terry, Jr., M.M. Martin, G.E. Malana, S.E. Sansom, A.P. Pleister, W.D. Beck, E. Head, D.S. Feldman and T.S. Elton. 2010. Chromosome 21-derived microRNAs provide an etiological basis for aberrant protein expression in human Down syndrome brains. *J Biol Chem.* 285: 1529–1543.

Kumar, A. and D.J. Loane. 2012. Neuroinflammation after traumatic brain injury: Opportunities for therapeutic intervention. *Brain Behav Immun.* 26: 1191–1201.

Kwon, B.K., A.M. Stammers, L.M. Belanger, A. Bernardo, D. Chan, C.M. Bishop, G.P. Slobogean, H. Zhang, H. Umedaly, M. Giffin, J. Street, M.C. Boyd, S.J. Paquette, C.G. Fisher and M.F. Dvorak. 2010. Cerebrospinal fluid inflammatory cytokines and biomarkers of injury severity in acute human spinal cord injury. *J Neurotrauma.* 27: 669–682.

Lagos-Quintana, M., R. Rauhut, A. Yalcin, J. Meyer, W. Lendeckel and T. Tuschl. 2002. Identification of tissue-specific microRNAs from mouse. *Curr Biol.* 12: 735–739.

Lai, C.Y., S.L. Yu, M.H. Hsieh, C.H. Chen, H.Y. Chen, C.C. Wen, Y.H. Huang, P.C. Hsiao, C.K. Hsiao, C.M. Liu, P.C. Yang, H.G. Hwu and W.J. Chen. 2011. MicroRNA expression aberration as potential peripheral blood biomarkers for schizophrenia. *PLoS One.* 6: e21635.

Laker, S.R. 2011. Epidemiology of concussion and mild traumatic brain injury. *PMR.* 3: S354–358.

Landles, C. and G.P. Bates. 2004. Huntingtin and the molecular pathogenesis of Huntington's disease. Fourth in molecular medicine review series. *EMBO Rep.* 5: 958–963.

Lanford, R.E., E.S. Hildebrandt-Eriksen, A. Petri, R. Persson, M. Lindow, M.E. Munk, S. Kauppinen and H. Ørum. 2010. Therapeutic silencing of microRNA-122 in primates with chronic hepatitis C virus infection. *Science.* 327: 198–201.

Lang, M.F. and Y. Shi. 2012. Dynamic Roles of microRNAs in Neurogenesis. *Front Neurosci.* 6: 71.

Langlois, J.A., W. Rutland-Brown and M.M. Wald. 2006. The epidemiology and impact of traumatic brain injury: a brief overview. *J Head Trauma Rehabil.* 21: 375–378.

Laterza, O.F., L. Lim, P.W. Garrett-Engele, K. Vlasakova, N. Muniappa, W.K. Tanaka, J.M. Johnson, J.F. Sina, T.L. Fare, F.D. Sistare and W.E. Glaab. 2009. Plasma MicroRNAs as sensitive and specific biomarkers of tissue injury. *Clin Chem.* 55: 1977–1983.

Latronico, M.V., D. Catalucci and G. Condorelli. 2007. Emerging role of microRNAs in cardiovascular biology. *Circ Res.* 101: 1225–1236.

Lawrie, C.H., S. Gal, H.M. Dunlop, B. Pushkaran, A.P. Liggins, K. Pulford, A.H. Banham, F. Pezzella, J. Boultwood, J.S. Wainscoat, C.S. Hatton and A.L. Harris. 2008. Detection of elevated levels of tumour-associated microRNAs in serum of patients with diffuse large B-cell lymphoma. *Br J Haematol.* 141: 672–675.

Le, T.H. and A.D. Gean. 2009. Neuroimaging of traumatic brain injury. *Mt Sinai J Med.* 76: 145–162.

Lee, R.C., R.L. Feinbaum and V. Ambros. 1993. The C. elegans heterochronic gene lin-4 encodes small RNAs with antisense complementarity to lin-14. *Cell.* 75: 843–854.

Lee, S.T., K. Chu, W.S. Im, H.J. Yoon, J.Y. Im, J.E. Park, K.H. Park, K.H. Jung, S.K. Lee, M. Kim and J.K. Roh. 2011. Altered microRNA regulation in Huntington's disease models. *Exp Neurol.* 227: 172–179.

Lehmann, S.M., C. Kruger, B. Park, K. Derkow, K. Rosenberger, J. Baumgart, T. Trimbuch, G. Eom, M. Hinz, D. Kaul, P. Habbel, R. Kalin, E. Franzoni, A. Rybak, D. Nguyen, R. Veh, O. Ninnemann, O. Peters, R. Nitsch, F.L. Heppner, D. Golenbock, E. Schott, H.L. Ploegh, F.G. Wulczyn and S. Lehnardt. 2012. An unconventional role for miRNA: let-7 activates Toll-like receptor 7 and causes neurodegeneration. *Nat Neurosci.* 15: 827–835.

Lei, P., Y. Li, X. Chen, S. Yang and J. Zhang. 2009. Microarray based analysis of microRNA expression in rat cerebral cortex after traumatic brain injury. *Brain Res.* 1284: 191–201.

Li, C., Y. Feng, G. Coukos and L. Zhang. 2009. Therapeutic microRNA strategies in human cancer. *AAPS J.* 11: 747–757.

Li, Q., S. Bian, J. Hong, Y. Kawase-Koga, E. Zhu, Y. Zheng, L. Yang and T. Sun. 2011. Timing specific requirement of microRNA function is essential for embryonic and postnatal hippocampal development. *PLoS One.* 6: e26000.

Li, X. and P. Jin. 2009. Macro role(s) of microRNAs in fragile X syndrome? *Neuromolecular Med.* 11: 200–207.

Li, X., A. Khanna, N. Li and E. Wang. 2011. Circulatory miR34a as an RNA based, noninvasive biomarker for brain aging. *Aging (Albany NY).* 3: 985–1002.

Liliang, P.C., C.L. Liang, K. Lu, K.W. Wang, H.C. Weng, C.H. Hsieh, Y.D. Tsai and H.J. Chen. 2010a. Relationship between injury severity and serum tau protein levels in traumatic brain injured rats. *Resuscitation.* 81: 1205–1208.

Liliang, P.C., C.L. Liang, H.C. Weng, K. Lu, K.W. Wang, H.J. Chen and J.H. Chuang. 2010b. Tau proteins in serum predict outcome after severe traumatic brain injury. *J Surg Res.* 160: 302–307.

Linsen, S.E., E. de Wit, E. de Bruijn and E. Cuppen. 2010. Small RNA expression and strain specificity in the rat. *BMC Genomics.* 11: 249.

Liu, D.Z., Y. Tian, B.P. Ander, H. Xu, B.S. Stamova, X. Zhan, R.J. Turner, G. Jickling and F.R. Sharp. 2010. Brain and blood microRNA expression profiling of ischemic stroke, intracerebral hemorrhage, and kainate seizures. *J Cereb Blood Flow Metab.* 30: 92–101.

Liu, N.K., X.F. Wang, Q.B. Lu and X.M. Xu. 2009. Altered microRNA expression following traumatic spinal cord injury. *Exp Neurol.* 219: 424–429.

Liu, N.K. and X.M. Xu. 2011. MicroRNA in central nervous system trauma and degenerative disorders. *Physiol Genomics.* 43: 571–580.

Liu, X.S., M. Chopp, R.L. Zhang, T. Tao, X.L. Wang, H. Kassis, A. Hozeska-Solgot, L. Zhang, C. Chen and Z.G. Zhang. 2011. MicroRNA profiling in subventricular zone after stroke: MiR-124a regulates proliferation of neural progenitor cells through Notch signaling pathway. *PLoS One.* 6: e23461.

Liu, Z., G. Li, S. Wei, J. Niu, A.K. El-Naggar, E.M. Sturgis and Q. Wei. 2010. Genetic variants in selected pre-microRNA genes and the risk of squamous cell carcinoma of the head and neck. *Cancer.* 116: 4753–4760.

Lu, M., Q. Zhang, M. Deng, J. Miao, Y. Guo, W. Gao and Q. Cui. 2008. An analysis of human microRNA and disease associations. *PLoS One.* 3: e3420.

Lujambio, A., G.A. Calin, A. Villanueva, S. Ropero, M. Sanchez-Cespedes, D. Blanco, L.M. Montuenga, S. Rossi, M.S. Nicoloso, W.J. Faller, W.M. Gallagher, S.A. Eccles, C.M. Croce and M. Esteller. 2008. A microRNA DNA methylation signature for human cancer metastasis. *Proc Natl Acad Sci USA.* 105: 13556–13561.

Madathil, S.K., P.T. Nelson, K.E. Saatman and B.R. Wilfred. 2011. MicroRNAs in CNS injury: potential roles and therapeutic implications. *Bioessays.* 33: 21–26.

Margis, R. and C.R. Rieder. 2011. Identification of blood microRNAs associated to Parkinson's disease. *J Biotechnol.* 152: 96–101.

Martens, L.H., J. Zhang, S.J. Barmada, P. Zhou, S. Kamiya, B. Sun, S.W. Min, L. Gan, S. Finkbeiner, E.J. Huang and R.V. Farese, Jr. 2012. Progranulin deficiency promotes neuroinflammation and neuron loss following toxin-induced injury. *J Clin Invest.* 122: 3955–3959.

Martins, M., A. Rosa, L.C. Guedes, B.V. Fonseca, K. Gotovac, S. Violante, T. Mestre, M. Coelho, M.M. Rosa, E.R. Martin, J.M. Vance, T.F. Outeiro, L. Wang, F. Borovecki, J.J. Ferreira and S.A. Oliveira. 2011. Convergence of miRNA expression profiling, alpha-synuclein interaction and GWAS in Parkinson's disease. *PLoS One.* 6: e25443.

Mathupala, S.P., S. Mittal, M. Guthikonda and A.E. Sloan. 2007. MicroRNA and brain tumors: a cause and a cure? *DNA Cell Biol.* 26: 301–310.

Matthews, S.C., I.A. Strigo, A.N. Simmons, R.M. O'Connell, L.E. Reinhardt and S.A. Moseley. 2011. A multimodal imaging study in U.S. veterans of Operations Iraqi and Enduring Freedom with and without major depression after blast-related concussion. *Neuroimage.* 54 Suppl 1: S69–75.

McAllister, T.W. 2011. Neurobiological consequences of traumatic brain injury. *Dialogues Clin Neurosci.* 13: 287–300.

Meerson, A., L. Cacheaux, K.A. Goosens, R.M. Sapolsky, H. Soreq and D. Kaufer. 2010. Changes in brain MicroRNAs contribute to cholinergic stress reactions. *J Mol Neurosci.* 40: 47–55.

Mehta, S.S. 2010. Biochemical serum markers in head injury: an emphasis on clinical utility. *Clin Neurosurg.* 57: 134–140.

Meseguer, S., G. Mudduluru, J.M. Escamilla, H. Allgayer and D. Barettino. 2011. MicroRNAs-10a and -10b contribute to retinoic acid-induced differentiation of neuroblastoma cells and target the alternative splicing regulatory factor SFRS1 (SF2/ASF). *J Biol Chem.* 286: 4150–4164.

Miska, E.A., E. Alvarez-Saavedra, M. Townsend, A. Yoshii, N. Sestan, P. Rakic, M. Constantine-Paton and H.R. Horvitz. 2004. Microarray analysis of microRNA expression in the developing mammalian brain. *Genome Biol.* 5: R68.

Mitchell, P.S., R.K. Parkin, E.M. Kroh, B.R. Fritz, S.K. Wyman, E.L. Pogosova-Agadjanyan, A. Peterson, J. Noteboom, K.C. O'Briant, A. Allen, D.W. Lin, N. Urban, C.W. Drescher, B.S. Knudsen, D.L. Stirewalt, R. Gentleman, R.L. Vessella, P.S. Nelson, D.B. Martin and M. Tewari. 2008. Circulating microRNAs as stable blood-based markers for cancer detection. *Proc Natl Acad Sci USA.* 105: 10513–10518.

Miyoshi, K., T. Miyoshi and H. Siomi. 2010. Many ways to generate microRNA-like small RNAs: non-canonical pathways for microRNA production. *Mol Genet Genomics.* 284: 95–103.

Mizoguchi, M., Y. Guan, K. Yoshimoto, N. Hata, T. Amano, A. Nakamizo and T. Sasaki. 2012. MicroRNAs in Human Malignant Gliomas. *J Oncol.* 732874.

Molloy, C., R.M. Conroy, D.R. Cotter and M. Cannon. 2011. Is traumatic brain injury a risk factor for schizophrenia? A meta-analysis of case-controlled population-based studies. *Schizophr Bull.* 37: 1104–1110.

Mondol, V. and A.E. Pasquinelli. 2012. Let's make it happen: the role of let-7 microRNA in development. *Curr Top Dev Biol.* 99: 1–30.

Moreau, M.P., S.E. Bruse, R. David-Rus, S. Buyske and L.M. Brzustowicz. 2011. Altered microRNA expression profiles in postmortem brain samples from individuals with schizophrenia and bipolar disorder. *Biol Psychiatry.* 69: 188–193.

Morin, R.D., M.D. O'Connor, M. Griffith, F. Kuchenbauer, A. Delaney, A.L. Prabhu, Y. Zhao, H. McDonald, T. Zeng, M. Hirst, C.J. Eaves and M.A. Marra. 2008. Application of massively parallel sequencing to microRNA profiling and discovery in human embryonic stem cells. *Genome Res.* 18: 610–621.

Nampiaparampil, D.E. 2008. Prevalence of chronic pain after traumatic brain injury: a systematic review. *JAMA.* 300: 711–719.

Nelson, P.T., D.A. Baldwin, W.P. Kloosterman, S. Kauppinen, R.H. Plasterk and Z. Mourelatos. 2006. RAKE and LNA-ISH reveal microRNA expression and localization in archival human brain. *RNA.* 12: 187–191.

Newman, M.A. and S.M. Hammond. 2010. Emerging paradigms of regulated microRNA processing. *Genes Dev.* 24: 1086–1092.

Niogi, S.N., P. Mukherjee, J. Ghajar, C. Johnson, R.A. Kolster, R. Sarkar, H. Lee, M. Meeker, R.D. Zimmerman, G.T. Manley and B.D. McCandliss. 2008. Extent of microstructural white matter injury in postconcussive syndrome correlates with impaired cognitive reaction time: a 3T diffusion tensor imaging study of mild traumatic brain injury. *AJNR Am J Neuroradiol.* 29: 967–973.

Nudelman, A.S., D.P. DiRocco, T.J. Lambert, M.G. Garelick, J. Le, N.M. Nathanson and D.R. Storm. 2010. Neuronal activity rapidly induces transcription of the CREB-regulated microRNA-132, *in vivo. Hippocampus.* 20: 492–498.

Osman, A. 2012. MicroRNAs in health and disease—basic science and clinical applications. *Clin Lab.* 58: 393–402.

Otaegi, G., A. Pollock and T. Sun. 2011. An Optimized Sponge for microRNA miR-9 Affects Spinal Motor Neuron Development *in vivo. Front Neurosci.* 5: 146.

Pacifici, M., S. Delbue, P. Ferrante, D. Jeansonne, F. Kadri, S. Nelson, C. Velasco-Gonzalez, J. Zabaleta and F. Peruzzi. 2012. Cerebrospinal fluid miRNA profile in HIV-encephalitis. *J Cell Physiol* (in press).

Papa, L., L.M. Lewis, S. Silvestri, J.L. Falk, P. Giordano, G.M. Brophy, J.A. Demery, M.C. Liu, J. Mo, L. Akinyi, S. Mondello, K. Schmid, C.S. Robertson, F.C. Tortella, R.L. Hayes and K.K. Wang. 2012. Serum levels of ubiquitin C-terminal hydrolase distinguish mild traumatic brain injury from trauma controls and are elevated in mild and moderate traumatic brain injury patients with intracranial lesions and neurosurgical intervention. *J Trauma Acute Care Surg.* 72: 1335–1344.

Pasquinelli, A.E. 2012. MicroRNAs and their targets: recognition, regulation and an emerging reciprocal relationship. *Nat Rev Genet.* 13: 271–282.

Pasvogel, A.E., P. Miketova and I.M. Moore. 2010. Differences in CSF phospholipid concentration by traumatic brain injury outcome. *Biol Res Nurs.* 11: 325–331.

Perkins, D.O., C.D. Jeffries, L.F. Jarskog, J.M. Thomson, K. Woods, M.A. Newman, J.S. Parker, J. Jin and S.M. Hammond. 2007. microRNA expression in the prefrontal cortex of individuals with schizophrenia and schizoaffective disorder. *Genome Biol.* 8: R27.

Persengiev, S.P., Kondova, II and R.E. Bontrop. 2012. The Impact of MicroRNAs on Brain Aging and Neurodegeneration. *Curr Gerontol Geriatr Res.* 359369.

Pritchard, C.C., H.H. Cheng and M. Tewari. 2012. MicroRNA profiling: approaches and considerations. *Nat Rev Genet.* 13: 358–369.

Qin, W., Y. Shi, B. Zhao, C. Yao, L. Jin, J. Ma and Y. Jin. 2010. miR-24 regulates apoptosis by targeting the open reading frame (ORF) region of FAF1 in cancer cells. *PLoS One.* 5: e9429.

Rachidi, M. and C. Lopes. 2011. Mental Retardation and Human Chromosome 21 Gene Overdosage: From Functional Genomics and Molecular Mechanisms Towards Prevention and Treatment of the Neuropathogenesis of Down Syndrome. pp. 21–86. *In*: J.D. Clelland [ed.]. Genomics, Proteomics, and the Nervous System: Advances in Neurobiology. Springer, New York, USA.

Rajasethupathy, P., F. Fiumara, R. Sheridan, D. Betel, S.V. Puthanveettil, J.J. Russo, C. Sander, T. Tuschl and E. Kandel. 2009. Characterization of small RNAs in Aplysia reveals a role for miR-124 in constraining synaptic plasticity through CREB. *Neuron.* 63: 803–817.

Ranjan, N., K.P. Nair, C. Romanoski, R. Singh and G. Venketswara. 2011. Tics after traumatic brain injury. *Brain Inj.* 25: 629–633.

Readnower, R.D., M. Chavko, S. Adeeb, M.D. Conroy, J.R. Pauly, R.M. McCarron and P.G. Sullivan. 2010. Increase in blood-brain barrier permeability, oxidative stress, and activated microglia in a rat model of blast-induced traumatic brain injury. *J Neurosci Res.* 88: 3530–3539.

Redell, J.B., Y. Liu and P.K. Dash. 2009. Traumatic brain injury alters expression of hippocampal microRNAs: potential regulators of multiple pathophysiological processes. *J Neurosci Res.* 87: 1435–1448.

Redell, J.B., A.N. Moore, N.H. Ward, 3rd, G.W. Hergenroeder and P.K. Dash. 2010. Human traumatic brain injury alters plasma microRNA levels. *J Neurotrauma.* 27: 2147–2156.

Redell, J.B., J. Zhao and P.K. Dash. 2011. Altered expression of miRNA-21 and its targets in the hippocampus after traumatic brain injury. *J Neurosci Res.* 89: 212–221.

Reinhart, B.J., F.J. Slack, M. Basson, A.E. Pasquinelli, J.C. Bettinger, A.E. Rougvie, H.R. Horvitz and G. Ruvkun. 2000. The 21-nucleotide let-7 RNA regulates developmental timing in Caenorhabditis elegans. *Nature.* 403: 901–906.

Roberts, G.W., S.M. Gentleman, A. Lynch, L. Murray, M. Landon and D.I. Graham. 1994. Beta amyloid protein deposition in the brain after severe head injury: implications for the pathogenesis of Alzheimer's disease. *J Neurol Neurosurg Psychiatry.* 57: 419–425.

Rockhill, C.M., K. Jaffe, C. Zhou, M.Y. Fan, W. Katon and J.R. Fann. 2012. Health care costs associated with traumatic brain injury and psychiatric illness in adults. *J Neurotrauma.* 29: 1038–1046.

Rong, H., T.B. Liu, K.J. Yang, H.C. Yang, D.H. Wu, C.P. Liao, F. Hong, H.Z. Yang, F. Wan, X.Y. Ye, D. Xu, X. Zhang, C.A. Chao and Q.J. Shen. 2011. MicroRNA-134 plasma levels before and after treatment for bipolar mania. *J Psychiatr Res.* 45: 92–95.

Ruberti, F., C. Barbato and C. Cogoni. 2012. Targeting microRNAs in neurons: tools and perspectives. *Exp Neurol.* 235: 419–426.

Saba, R. and G.M. Schratt. 2010. MicroRNAs in neuronal development, function and dysfunction. *Brain Res.* 1338: 3–13.

Sakurai, F., N. Furukawa, M. Higuchi, S. Okamoto, K. Ono, T. Yoshida, M. Kondoh, K. Yagi, N. Sakamoto, K. Katayama and H. Mizuguchi. 2012. Suppression of hepatitis C virus replicon by adenovirus vector-mediated expression of tough decoy RNA against miR-122a. *Virus Res.* 165: 214–218.

Sandhir, R., G. Onyszchuk and N.E. Berman. 2008. Exacerbated glial response in the aged mouse hippocampus following controlled cortical impact injury. *Exp Neurol.* 213: 372–380.

Sarachana, T., R. Zhou, G. Chen, H.K. Manji and V.W. Hu. 2010. Investigation of post-transcriptional gene regulatory networks associated with autism spectrum disorders by microRNA expression profiling of lymphoblastoid cell lines. *Genome Med.* 2: 23.

Sasayama, T., M. Nishihara, T. Kondoh, K. Hosoda and E. Kohmura. 2009. MicroRNA-10b is overexpressed in malignant glioma and associated with tumor invasive factors, uPAR and RhoC. *Int J Cancer.* 125: 1407–1413.

Saunders, M.A., H. Liang and W.H. Li. 2007. Human polymorphism at microRNAs and microRNA target sites. *Proc Natl Acad Sci USA.* 104: 3300–3305.

Sayed, D. and M. Abdellatif. 2011. MicroRNAs in development and disease. *Physiol Rev.* 91: 827–887.

Schmittgen, T.D., J. Jiang, Q. Liu and L. Yang. 2004. A high-throughput method to monitor the expression of microRNA precursors. *Nucleic Acids Res.* 32: e43.

Schmittgen, T.D., E.J. Lee, J. Jiang, A. Sarkar, L. Yang, T.S. Elton and C. Chen. 2008. Real-time PCR quantification of precursor and mature microRNA. *Methods.* 44: 31–38.

Schneider, M.R. 2012. MicroRNAs as novel players in skin development, homeostasis and disease. *Br J Dermatol.* 166: 22–28.

Sempere, L.F., S. Freemantle, I. Pitha-Rowe, E. Moss, E. Dmitrovsky and V. Ambros. 2004. Expression profiling of mammalian microRNAs uncovers a subset of brain-expressed microRNAs with possible roles in murine and human neuronal differentiation. *Genome Biol.* 5: R13.

Sharma, A. and R.K. Maheshwari. 2009. Oligonucleotide array analysis of Toll-like receptors and associated signalling genes in Venezuelan equine encephalitis virus-infected mouse brain. *J Gen Virol.* 90: 1836–1847.

Sharma, R. and D.T. Laskowitz. 2012. Biomarkers in traumatic brain injury. *Curr Neurol Neurosci Rep.* 12: 560–569.

Sheinerman, K.S., V.G. Tsivinsky, F. Crawford, M.J. Mullan, L. Abdullah and S.R. Umansky. 2012. Plasma microRNA biomarkers for detection of mild cognitive impairment. *Aging (Albany NY).* 4: 590–605.

Shi, W., J. Du, Y. Qi, G. Liang, T. Wang, S. Li, S. Xie, B. Zeshan and Z. Xiao. 2012. Aberrant expression of serum miRNAs in schizophrenia. *J Psychiatr Res.* 46: 198–204.

Siegel, S.R., J. Mackenzie, G. Chaplin, N.G. Jablonski and L. Griffiths. 2012. Circulating microRNAs involved in multiple sclerosis. *Mol Biol Rep.* 39: 6219–6225.

Sivanandam, T.M. and M.K. Thakur. 2012. Traumatic brain injury: a risk factor for Alzheimer's disease. *Neurosci Biobehav Rev.* 36: 1376–1381.

Smalheiser, N.R., G. Lugli, H.S. Rizavi, H. Zhang, V.I. Torvik, G.N. Pandey, J.M. Davis and Y. Dwivedi. 2011. MicroRNA expression in rat brain exposed to repeated inescapable shock: differential alterations in learned helplessness vs. non-learned helplessness. *Int J Neuropsychopharmacol.* 14: 1315–1325.

Smeets, E.E.J., K. Pelc and B. Dan. 2011. Rett Syndrome. *Molecular Syndromology.* 2: 113–127.

Smith, P.Y., C. Delay, J. Girard, M.A. Papon, E. Planel, N. Sergeant, L. Buee and S.S. Hebert. 2011. MicroRNA-132 loss is associated with tau exon 10 inclusion in progressive supranuclear palsy. *Hum Mol Genet.* 20: 4016–4024.

Sonkoly, E., T. Wei, P.C. Janson, A. Saaf, L. Lundeberg, M. Tengvall-Linder, G. Norstedt, H. Alenius, B. Homey, A. Scheynius, M. Stahle and A. Pivarcsi. 2007. MicroRNAs: novel regulators involved in the pathogenesis of psoriasis? *PLoS One.* 2: e610.

Spadaro, P.A. and T.W. Bredy. 2012. Emerging role of non-coding RNA in neural plasticity, cognitive function, and neuropsychiatric disorders. *Front Genet.* 3: 132.

Stark, K.L., B. Xu, A. Bagchi, W.S. Lai, H. Liu, R. Hsu, X. Wan, P. Pavlidis, A.A. Mills, M. Karayiorgou and J.A. Gogos. 2008. Altered brain microRNA biogenesis contributes to phenotypic deficits in a 22q11-deletion mouse model. *Nat Genet.* 40: 751–760.

Steitz, J.A. and S. Vasudevan. 2009. miRNPs: versatile regulators of gene expression in vertebrate cells. *Biochem Soc Trans.* 37: 931–935.

Stenvang, J., A. Petri, M. Lindow, S. Obad and S. Kauppinen. 2012. Inhibition of microRNA function by antimiR oligonucleotides. *Silence.* 3: 1.

Stillman, A.A., Z. Krsnik, J. Sun, M.R. Rasin, M.W. State, N. Sestan and A. Louvi. 2009. Developmentally regulated and evolutionarily conserved expression of SLITRK1 in brain circuits implicated in Tourette syndrome. *J Comp Neurol.* 513: 21–37.

Stuss, D.P., J.D. Boyd, D.B. Levin and K.R. Delaney. 2012. MeCP2 mutation results in compartment-specific reductions in dendritic branching and spine density in layer 5 motor cortical neurons of YFP-H mice. *PLoS One.* 7: e31896.

Summers, C.R., B. Ivins and K.A. Schwab. 2009. Traumatic brain injury in the United States: an epidemiologic overview. *Mt Sinai J Med.* 76: 105–110.

Sun, G. and J.J. Rossi. 2011. MicroRNAs and their potential involvement in HIV infection. *Trends Pharmacol Sci.* 32: 675–681.

Svetlov, S.I., S.F. Larner, D.R. Kirk, J. Atkinson, R.L. Hayes and K.K. Wang. 2009. Biomarkers of blast-induced neurotrauma: profiling molecular and cellular mechanisms of blast brain injury. *J Neurotrauma.* 26: 913–921.

Svetlov, S.I., V. Prima, O. Glushakova, A. Svetlov, D.R. Kirk, H. Gutierrez, V.L. Serebruany, K.C. Curley, K.K. Wang and R.L. Hayes. 2012. Neuro-glial and systemic mechanisms of pathological responses in rat models of primary blast overpressure compared to "composite" blast. *Front Neurol.* 3: 15.

Svetlov, S.I., V. Prima, D.R. Kirk, H. Gutierrez, K.C. Curley, R.L. Hayes and K.K. Wang. 2010. Morphologic and biochemical characterization of brain injury in a model of controlled blast overpressure exposure. *J Trauma.* 69: 795–804.

Szulwach, K.E., X. Li, R.D. Smrt, Y. Li, Y. Luo, L. Lin, N.J. Santistevan, W. Li, X. Zhao and P. Jin. 2010. Cross talk between microRNA and epigenetic regulation in adult neurogenesis. *J Cell Biol.* 189: 127–141.

Tan, K.S., A. Armugam, S. Sepramaniam, K.Y. Lim, K.D. Setyowati, C.W. Wang and K. Jeyaseelan. 2009. Expression profile of MicroRNAs in young stroke patients. *PLoS One.* 4: e7689.

Tang, R., L. Li, D. Zhu, D. Hou, T. Cao, H. Gu, J. Zhang, J. Chen, C.Y. Zhang and K. Zen. 2012. Mouse miRNA-709 directly regulates miRNA-15a/16-1 biogenesis at the posttranscriptional level in the nucleus: evidence for a microRNA hierarchy system. *Cell Res.* 22: 504–515.

Thompson, J.C., J. Harris, A.C. Sollom, C.L. Stopford, E. Howard, J.S. Snowden and D. Craufurd. 2012. Longitudinal evaluation of neuropsychiatric symptoms in Huntington's disease. *J Neuropsychiatry Clin Neurosci.* 24: 53–60.

Timaru-Kast, R., C. Luh, P. Gotthardt, C. Huang and M.K. Schäfer. 2012. Influence of Age on Brain Edema Formation, Secondary Brain Damage and Inflammatory Response after Brain Trauma in Mice. *PLoS One.* 7: e43829.

Topolovec-Vranic, J., M.A. Pollmann-Mudryj, D. Ouchterlony, D. Klein, J. Spence, A. Romaschin, S. Rhind, H.C. Tien and A.J. Baker. 2011. The value of serum biomarkers in prediction models of outcome after mild traumatic brain injury. *J Trauma.* 71: S478–486.

Treiber, T., N. Treiber and G. Meister. 2012. Regulation of microRNA biogenesis and function. *Thromb Haemost.* 107: 605–610.

Truettner, J.S., O.F. Alonso, H.M. Bramlett and W.D. Dietrich. 2011. Therapeutic hypothermia alters microRNA responses to traumatic brain injury in rats. *J Cereb Blood Flow Metab.* 31: 1897–1907.

Vaishnavi, S., V. Rao and J.R. Fann. 2009. Neuropsychiatric problems after traumatic brain injury: unraveling the silent epidemic. *Psychosomatics.* 50: 198–205.

van Rooij, E. and E.N. Olson. 2007. MicroRNAs: powerful new regulators of heart disease and provocative therapeutic targets. *J Clin Invest.* 117: 2369–2376.

Vasterling, J.J. and S. Dikmen. 2012. Mild traumatic brain injury and posttraumatic stress disorder: clinical and conceptual complexities. *J Int Neuropsychol Soc.* 18: 390–393.

Vasterling, J.J., M. Verfaellie and K.D. Sullivan. 2009. Mild traumatic brain injury and posttraumatic stress disorder in returning veterans: perspectives from cognitive neuroscience. *Clin Psychol Rev.* 29: 674–684.

Vaucheret, H. and Y. Chupeau. 2012. Ingested plant miRNAs regulate gene expression in animals. *Cell Res.* 22: 3–5.

Vezzani, A., M. Maroso, S. Balosso, M.A. Sanchez and T. Bartfai. 2011. IL-1 receptor/Toll-like receptor signaling in infection, inflammation, stress and neurodegeneration couples hyperexcitability and seizures. *Brain Behav Immun.* 25: 1281–1289.

Vickers, K.C., B.T. Palmisano, B.M. Shoucri, R.D. Shamburek and A.T. Remaley. 2011. MicroRNAs are transported in plasma and delivered to recipient cells by high-density lipoproteins. *Nat Cell Biol.* 13: 423–433.

Wang, Z. 2009. miRNA Sponge Technology. pp. 153–159. *In*: Z. Wang [ed.]. MicroRNA Interference Technologies. Springer Berlin Heidelberg.

Wang Z. 2011. The guideline of the design and validation of MiRNA mimics. *Methods Mol Biol.* 676: 211–223.

Wang, W.X., B.W. Rajeev, A.J. Stromberg, N. Ren, G. Tang, Q. Huang, I. Rigoutsos and P.T. Nelson. 2008. The expression of microRNA miR-107 decreases early in Alzheimer's disease and may accelerate disease progression through regulation of beta-site amyloid precursor protein-cleaving enzyme 1. *J Neurosci.* 28: 1213–1223.

Wang, W.X., B.R. Wilfred, S.K. Madathil, G. Tang, Y. Hu, J. Dimayuga, A.J. Stromberg, Q. Huang, K.E. Saatman and P.T. Nelson. 2010. miR-107 regulates granulin/progranulin with implications for traumatic brain injury and neurodegenerative disease. *Am J Pathol.* 177: 334–345.

Weng, H., C. Shen, G. Hirokawa, X. Ji, R. Takahashi, K. Shimada, C. Kishimoto and N. Iwai. 2011. Plasma miR-124 as a biomarker for cerebral infarction. *Biomed Res.* 32: 135–141.

Wightman, B., I. Ha and G. Ruvkun. 1993. Posttranscriptional regulation of the heterochronic gene lin-14 by lin-4 mediates temporal pattern formation in C. elegans. *Cell.* 75: 855–862.

Witwer, K.W., S.L. Sarbanes, J. Liu and J.E. Clements. 2011. A plasma microRNA signature of acute lentiviral infection: biomarkers of central nervous system disease. *AIDS.* 25: 2057–2067.

Wu, H., J. Tao, P.J. Chen, A. Shahab, W. Ge, R.P. Hart, X. Ruan, Y. Ruan and Y.E. Sun. 2010. Genome-wide analysis reveals methyl-CpG-binding protein 2-dependent regulation of microRNAs in a mouse model of Rett syndrome. *Proc Natl Acad Sci USA.* 107: 18161–18166.

Xia, M. and M. Hu. 2010. The Role of MicroRNA in Tumor Invasion and Metastasis. *J Cancer Mol.* 5: 33–39.

Xiao, J., B. Yang, H. Lin, Y. Lu, X. Luo and Z. Wang. 2007. Novel approaches for gene-specific interference via manipulating actions of microRNAs: examination on the pacemaker channel genes HCN2 and HCN4. *J Cell Physiol.* 212: 285–292.

Xie, H., L. Sun and H.F. Lodish. 2009. Targeting microRNAs in obesity. *Expert Opin Ther Targets.* 13: 1227–1238.

Xu, B., M. Karayiorgou and J.A. Gogos. 2010. MicroRNAs in psychiatric and neurodevelopmental disorders. *Brain Res.* 1338: 78–88.

Yang, C., C. Wang, X. Chen, S. Chen, Y. Zhang, F. Zhi, J. Wang, L. Li, X. Zhou, N. Li, H. Pan, J. Zhang, K. Zen, C.Y. Zhang and C. Zhang. 2013. Identification of seven serum microRNAs from a genome-wide serum microRNA expression profile as potential noninvasive biomarkers for malignant astrocytomas. *Int J Cancer.* 132: 116–127.

Young, G.B. 2010. Traumatic brain injury: the continued quest for early prognostic determination. *Crit Care Med.* 38: 325–326.

Yu, J., Y. Wang, R. Dong, X. Huang, S. Ding and H. Qiu. 2012. Circulating microRNA-218 was reduced in cervical cancer and correlated with tumor invasion. *J Cancer Res Clin Oncol.* 138: 671–674.

Yu, J.Y., K.H. Chung, M. Deo, R.C. Thompson and D.L. Turner. 2008. MicroRNA miR-124 regulates neurite outgrowth during neuronal differentiation. *Exp Cell Res.* 314: 2618–2633.

Yunta, M., M. Nieto-Diaz, F.J. Esteban, M. Caballero-Lopez, R. Navarro-Ruiz, D. Reigada, D.W. Pita-Thomas, A. del Aguila, T. Munoz-Galdeano and R.M. Maza. 2012. MicroRNA dysregulation in the spinal cord following traumatic injury. *PLoS One.* 7: e34534.

Zhang, L., D. Hou, X. Chen, D. Li, L. Zhu, Y. Zhang, J. Li, Z. Bian, X. Liang, X. Cai, Y. Yin, C. Wang, T. Zhang, D. Zhu, D. Zhang, J. Xu, Q. Chen, Y. Ba, J. Liu, Q. Wang, J. Chen, J. Wang, M. Wang, Q. Zhang, J. Zhang, K. Zen and C.Y. Zhang. 2012. Exogenous plant MIR168a specifically targets mammalian LDLRAP1: evidence of cross-kingdom regulation by microRNA. *Cell Res.* 22: 107–126.

Zhang, Y., Y. Liao, D. Wang, Y. He, D. Cao, F. Zhang and K. Dou. 2011. Altered expression levels of miRNAs in serum as sensitive biomarkers for early diagnosis of traumatic injury. *J Cell Biochem.* 112: 2435–2442.

Zhou, B. and W. Liu. 2010. Post-traumatic glioma: report of one case and review of the literature. *Int J Med Sci.* 7: 248–250.

Zhou, Z.S., H.Q. Zeng, Z.P. Liu and Z.M. Yang. 2012. Genome-wide identification of Medicago truncatula microRNAs and their targets reveals their differential regulation by heavy metal. *Plant Cell Environ.* 35: 86–99.

Zhu, S., Q. Jiang, G. Wang, B. Liu, M. Teng and Y. Wang. 2011. Chromatin structure characteristics of pre-miRNA genomic sequences. *BMC Genomics.* 12: 329.

Zisoulis, D.G., Z.S. Kai, R.K. Chang and A.E. Pasquinelli. 2012. Autoregulation of microRNA biogenesis by let-7 and Argonaute. *Nature.* 486: 541–544.

4

Necrosis, Apoptosis and Autophagy in Acute Brain Injury: The Utilities of Biomarkers

Shankar Sadasivan,[1,]* Kevin K.W. Wang[2,]* and Zhiqun Zhang[2]

INTRODUCTION

Cell death has been classified into three types: a) apoptotic (Type I), b) autophagic (Type II) and c) necrotic/oncotic (Type III). Necrosis and apoptosis have been well studied in neuronal cell death. Despite that we are still witnessing growing evidence for the involvement of autophagy and autophagy-induced cell death in both acute neural injury (such as traumatic brain injury, spinal cord injury and stroke) as well as various neurodegenerative disease conditions, such as Alzheimer's disease, Huntington's disease and Prion disease. Interestingly, there also exists a significant overlap in the biochemical pathways that suggest crosstalk between different forms of cell death. Another common theme appears to be the involvement of proteases in all three forms of cell death. This chapter will focus on our current understanding of all the above mentioned pathways of cell death, and how they together affect the final outcome in the neuron's fate.

[1] Developmental Neurobiology, St Jude Children's Research Hospital, 262 Danny Thomas Place, MS 323 Memphis, TN 38105, USA.
[2] Center for Neuroproteomics & Biomarkers Research, Departments of Psychiatry, University of Florida, Gainesville, FL 32611, USA.
* Corresponding authors: neurostudent@gmail.com; kwang@ufl.edu

Cell death in the nervous systems keeps a conservation model under normal physiological conditions. Developmentally speaking, the neurons in the central nervous system undergo a process of pruning that helps form and refine the neuronal connections and in turn their function developing the neuronal architecture in the brain. Neurons in the central nervous system achieve the process of pruning through controlled cell death. The neurons that are unable to establish neuronal connections are targeted in this form of cell death. Programmed cell death, thus has an important role to play in the neurodevelopment process. In contrast, as the mature neurons age, they are exposed to a multitude of insults during their lifespan which determines their fate and survival. Cell death in brain injury paradigms involve an initial wave of acute necrotizing damage to neurons and tissue followed by a secondary biochemical damage caused by the compromised cellular environment resulting in a more organized or programmed form of cell death. Over the years literature has described, several forms of cell death which occurs in the cell accompanied by distinct phenotypic and molecular markers, depending on the nature of insult experienced by the cell (Galluzzi et al., 2012; Kroemer et al., 2009). This chapter focuses on the role of autophagy and autophagy-induced cell death described to occur following brain injury.

Forms of Cell Death

Programmed cell death in brain injury paradigms has been broadly classified into a) programmed necrosis apoptosis (type III cell death), b) apoptosis (type I cell death) and c) autophagy-induced (autophagic) cell death (type II cell death) (Table 4.1, 4.2). Recently published literature has defined the forms of programmed cell death based on morphological features as proposed by the nomenclature committee on cell death to avoid and in an essence clarify the confusion surrounding the definition of

Table 4.1. Phenotypic characteristics associated with different forms of cell death.

	Apoptosis (type I)	Autophagy (type II)	Necrosis
Morphology	•Cell shrinkage •Cell membrane blebbing to form apoptotic bodies •Nuclear DNA fragmentation and condensation	•Presence of cytosolic double membrane vesicles "autophagosomes" •Apoptotic like nuclear morphology	•Cell swelling •Osmotic & Ionic imbalance •Non-specific nuclear DNA degradation •Cell membrane rupture

Table 4.2. Proteases involved in the execution of different types of cell death.

	Apoptosis (type I)	Autophagic cell death (type II)	Necrosis/ Oncosis (type III)
Proteases	Caspases-3, 6, 7, 8, 9, 12 Calpain-1,2	Atg-4 Cathepsin B (lysosomal) Caspase-3	Calpain-1,2 Calpain-5 Cathepsin D & B (lysosomal)

different forms of programmed cell death (Galluzzi et al., 2012; Kroemer et al., 2009). Based on these recommendations, we can classify the following as programmed cell death:

a) Necrosis

Necrosis has traditionally been considered as an uncontrolled accidental form of cell death with morphological features different from apoptosis or autophagy. A necrotic cell typically undergoes bioenergetic failure due to ATP depletion caused by either exposure to toxins, physical injury, increased Ca^{2+} overload, excitotoxicity or inflammation and is accompanied with swelling of intracellular components and compromise of the plasma membrane (Table 4.1; Figure 4.1). The compromise in cellular integrity results in release of intracellular components into the extracellular space causing inflammation with the ability to cause damage to neighboring cells. Cellular and nuclear lysis resulting from necrosis leads to inflammation,

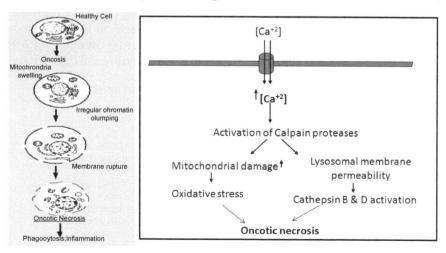

Figure 4.1. Necrotic morphology and cellular signaling pathways.

thus causing further damage in the neighboring environment. Some of the proteases involved in necrosis are calcium-dependent calpains and lysosomalcathepsins that effect most changes within the cell (Table 4.2). Though considered to be accidental, recent studies have evidence suggestive of the fact the necrotic pathway may indeed be a regulated one dependent on the signaling of certain secondary messengers within the cell. One form of "programmed" necrosis also termed *necroptosis* has been demonstrated to be dependent on receptor interacting protein kinase (RIP1)-signaling and is inhibited by RIP1 inhibitors such as necrostatin (Degterev et al., 2008; Degterev et al., 2005; Zhang et al., 2009). Another example of programmed necrosis is the activation of poly-ADP-ribose polymerase1 (PARP-1) due to DNA strand breaks, which have been documented to induce cell death and are observed in certain forms of brain injury.

b) Apoptosis

Apoptosis is one of the well-studied and well characterized forms of programmed cell death. The term was first coined by Kerr and colleagues (Weedon et al., 1979; Wyllie et al., 1980) where they described the morphological features associated with cell death. An apoptotic cell demonstrates morphological features such as chromatin condensation, retraction of psudopodes, nuclear fragmentation, plasma membrane blebbing without release of intracellular contents into the extracellular space (Table 4.1). The cell attains these stages depending on various intracellular as well as extracellular cues that result in the activation of biochemical sequelae of events in the form of proteases and other kinases resulting in cellular signaling leading to cell death (Table 4.2). Depending on the nature of biochemical signaling that the cell is exposed to, apoptosis can be classified as a) caspase-dependent and caspase-independent intrinsic, b) extrinsic. Though these pathways are differentiated based on the biochemical components associated with it, there exists a lot of crosstalk between the pathways through interactions between various proteins involved.

Intrinsic Pathway of Apoptosis: Intrinsic apoptosis can be signaled by multiple signaling modalities including mitochondrial damage, ER stress, DNA damage, oxidative stress and cytoslic Ca+2 overload. A huge volume of research conducted to study intrinsic mechanisms leading to apoptotic cell death has focused on the mitochondria. The loss of mitochondrial membrane potential results forming the Mitochondrial Permeability Transition (MPT) results in the release of mitochondrial proteins such as cytochrome c, Secondary Mitochondria-derived Activator of Caspases (SMAC) and high temperature requirement protein A2 (HTRA2) into the cytosol and also activation of certain pro-apoptotic mitochondrial membrane proteins such as Bid, that trigger a sequence resulting in the activation of pro-apoptotic caspase proteases and also suppression of anti-apoptotic proteins resulting in an environment conducive for the induction of apoptosis (Figure 4.2). Caspase-independent apoptosis involves signaling through proteins released from the Mitochondrial Intermembrane Space (MIS) such as Apoptosis-Inducing Factor (AIF) and endonuclease G that translocate to the nucleus causing irreparable DNA damage.

Extrinsic Pathway of Apoptosis: Extrinsic apoptosis defines a form of cell death propagated by specific transmembrane receptors in response to extracellular stress signals. The signaling is initiated by the docking of the ligand (Fas or TNF) to the death receptors that causes conformational changes in the cytoplasmic domain of the receptors and initiates the formation of the Death-Inducing Signaling Complex (DISC) that activates caspase-8 which in turn activates effector caspases such as caspase-3, -6 and -7 eventually resulting in apoptosis (Figure 4.2).

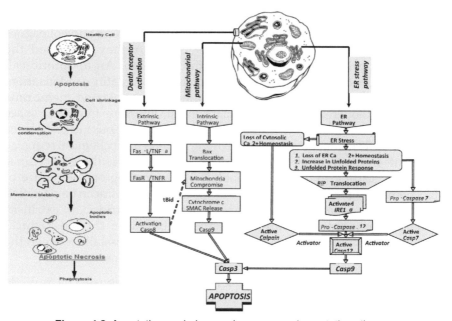

Figure 4.2. Apoptotic morphology and programmed apoptotic pathways.

ER Stress and Unfolded Protein Response Pathway of Apoptosis

A compromise or disturbance in the function of Endoplasmic Reticulum (ER), an organelle involved in protein synthesis, folding and post-translational modification, leads to ER stress. UPR is a cell stress program that is activated due to the accumulation of misfolded proteins in the ER lumen (DeGracia and Montie, 2004). Cell death mediated by endoplasmic reticulum (ER) stress and in turn the Unfolded Protein Response (UPR) has been suggested to be the third form of apoptosis (Figure 4.2). These mechanisms have also been reported to be a contributing factor in chronic neurodegenerative disorders such as Parkinson's Disease and protein aggregation disorders (DeGracia and Montie, 2004; Holtz and O'Malley, 2003; Imai and Takahashi, 2004; Lipinski and Yuan, 2004; Nakka et al., 2010; Ryu et al., 2002). Studies have recognized the cross-talk that exists between the ER and mitochondria ion the execution of cell death. Bcl-2 proteins that are present on the ER membrane are again critical in the sequence of events that play out once ER stress pathway is activated. ER stress has been known to activated caspase-12 which can further activate a host of proteolytic effector caspases downstream leading to cell death (Martinez et al., 2010; Morishima et al., 2002; Rao et al., 2002a,b).

c) Autophagy

Autophagy has been described as an intracellular mechanism that is activated when the cells undergo duress like nutrient deprivation or other cellular insults such as excitotoxicity (Sadasivan et al., 2010). The process involves recycling damaged cellular organelles to sustain cell survival. Autophagy has been characterized by the presence of double membrane vesicles called "autophagosomes", thought to be of endoplasmic reticulum (ER) origin that engulfs damaged organelles such as mitochondria within the cell that finally gets presented to the lysosomes by the fusion of the outer membrane of the autophagosomes with the lysosomal membrane resulting in a single-membraned structure called autolysosomes, wherein the lysosomal hydrolases breakdown the organelles to recycle essential amino acids back into the cellular machinery for cell sustenance and homeostasis. Autophagy has been classified as macroautophagy (referred to autophagy henceforth), microautophagy and chaperone-mediated autophagy. The process of engulfment of cytoplasmic organelles by double membraned autophagosomes followed by fusion with lysosomes where organelle degradation occurs is a hallmark feature of macroautophagy. In microautophagy, the cytoplasmic components are directly presented to the lysosomes for degradation by lysosomalcathepsins (Table 4.2) (Kaminskyy, 2012). Chaperone-mediated autophagy involves the delivery of specific protein substrates containing *KERFQ* motif to the lysosomes by hsc-70 chaperone proteins within the cytoplasm (Figure 4.3).

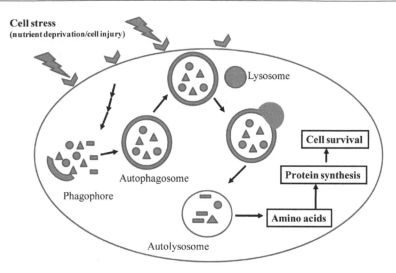

Figure 4.3. Autophagy process.

Autophagy-induced cell death

Autophagy-induced cell death has been termed based on the morphological features presented by the cell during the process of cell death. The cell death is known to occur in instances where apoptosis has been hampered or suppressed (Hou et al., 2010; Li et al., 2011; Sadasivan et al., 2006; Yousefi et al., 2006). Though autophagy is considered to be protective during the initial stages of stress or injury, prolonged induction results in cell death characterized by the presence of vacuoles and autophagosomes within the cell body. Recent studies have linked the interaction between autophagic proteins such as beclin-1 and anti-apoptotic Bcl-2 protein (He and Levine, 2010; Levine et al., 2008) to play an integral role in deciding the fate of cell death following an insult. Since, biochemical machinery involved in autophagic cell death has not been completely established, there is still a cloud of uncertainty in the use of this term.

Evidence for Necrotic and Apoptotic Cell Death in Brain Injury Paradigms and their Biomarkers

Injury in the central nervous system can be classified as acute, chronic or biochemical secondary injury. Depending on the nature of insults, the neurons follow a certain mode of cell death to sustain or terminate it. The range of insults in the brain varies from brain trauma, stroke to neurodegenerative diseases such as Parkinson's disease and Amyotrophic Lateral Sclerosis (ALS). An anomaly to this cell death is the one that occurs in neurons during brain development. This form of programmed cell death is not pathological but is essential as it remodels the structure and connectivity between neurons and in turn affects its functionality. Necrosis appears to be the dominant form of cell death in acute brain insults, such as TBI and ischemic or hemorrhagic brain injury (Wang, 2000). In fact, Wang reviewed that axonally located αII-spectrin breakdown product appear to be excellent neuronal necrosis biomarkers as it is degraded by calpain into two distinct fragments of 150 kDa and 145 kDa (SBDP150, BDP145), sequentially (Table 4.3). Conversely, apoptosis seems to be a smaller component of cell death but appear in a delayed fashion (Pike et al., 2000; Cardali, 2006; Mondello et al., 2010, 2011; Zhang et al., 2011). Again another αII-spectrin breakdown product of 120 kDa (SBDP120) as generated by caspase-3 is now a widely used biomarker for neuronal apoptosis in tissue as well as biofluids such as CSF (Mondello et al., 2010, 2011; Zhang et al., 2011). Similarly, axonally located microtubule associated protein Tau is also alternately degraded by calpain to TauBDP-35K and TauBDP-14K (also called TauBDP-17K) or by caspase into TauBDP-45K under pro-necrosis and pro-apoptosis neural injury challenges (Liu et al., 2011; Canu et al., 1998; Chung et al., 2001; Park et al., 2005).

Table 4.3. Protein Biomarkers for necrosis, apoptosis and autophagy relevant to neuro-diseases or disorders. SBDP: αII-spectrin breakdown product; TauBDP: microtubule associated protein Tau-breakdown product; p62: autophogosome membrane-boundprotein of 62 kDa; LC3-II: microtubule associated light chain 3 form II (truncated and delipidated, 17 kDa).

Protein Biomarker	Pathway	M.W. (Da)	Marker Origin/ Genesis	Detection (tissue, biofluid)
SBDP150, SBDP145	Necrosis	150K 145K	Generated by calpain	Brain tissue, CSF
SBDP120	Apoptosis	120K	Generated by caspase-3	Brain tissue, CSF
TauBDP-35K, TauBDP-14K (Tau-BDP-17K)	Necrosis	35K 14K	Generated by calpain	Brain tissue, CSF
TauBDP-45K	Apoptosis	45K-48K	Generated by caspases	Brain tissue
p62/SQSTM1/A170	Autophagy	62K	Degraded by autophagosome cathepsins	Brain tissue
LC3-II	Autophagy	17K	Generated by ATG-4	Brain tissue

Evidence for Autophagy in Acute Brain Injury and its Biomarkers

Autophagy has been demonstrated to be induced following experimental acute brain injury paradigms in rodent models. Based on reports in literature there is a distinct dichotomy on its purported role in neuronal survival and death. Studies have demonstrated autophagic process to be highly regulated in neurons. Suppression of autophagy and autophagy gene (*atg7*) in the Central Nervous System (CNS) has been found to be result in extensive neurodegeneration (Hara et al., 2006; Komatsu et al., 2006). Hence, autophagy plays a critical role in the survival of post-mitotic neurons in the CNS.

Several papers have now established autophagy induction in different rodent models of traumatic brain injury (Clark et al., 2008; Diskin et al., 2005; Erlich et al., 2007; Erlich et al., 2006; Lai et al., 2008; Sadasivan et al., 2008). Results have demonstrated an upregulation in the expression of autophagy protein beclin-1, a bcl-2 interacting protein following acute traumatic brain injury. Studies from our laboratory suggest an increase in the beclin-1 to bcl-2 ratio at one day post-injury in a Controlled Cortical Impact (CCI) model of TBI (Sadasivan et al., 2008). It has been demonstrated that this ratio is important in maintaining the balance between cell death and cell survival. Increase in the beclin-1 to bcl-2 ratio hence suggests that when autophagy is pushed into overdrive mode following acute brain trauma, it can be detrimental to cell survival. In contrast another study (Erlich et al. 2006), have demonstrated that treatment with rapamycin (an inducer of autophagy), four hours following brain trauma increased beclin-1 levels but also improved neurobehavioral function among rodents subjected to closed head injury. Rapamycin, besides being a pharmacological agent for autophagy induction, is also known to have other cellular targets that suppresses inflammation and has also been used in the clinical setting as an immunesuppressant (Calne et al., 1989; Carlson et al., 1993; Roberge et al., 1995; Sehgal et al., 1975; Warner et al., 1994). Since both the above cellular pathways are influenced singularly or in conjunction with each other in brain injuries, any beneficial neurological effects in TBI due to autophagy-induction following rapamycin administration should be interpreted with caution.

Autophagy has also been demonstrated to be activated in the brain following carotid occlusion and hypoxemia in rodent models (Adhami et al., 2006; Adhami et al., 2007; Degterev et al., 2005; Liu et al., 2010; Qin et al., 2010; Zhu et al., 2005). Autophagy induction in neonatal ischemic models of experimental stroke has been demonstrated to be detrimental to neuronal survival (Puyal and Clarke, 2009; Puyal et al., 2009). In another elegant study, Koike et al. (2008) demonstrated neuroprotective effect among hippocampal neurons, following autophagy gene *atg7* depletion in an experimental model of neonatalacute hypoxic-ischemia injury. These results seem to be applicable only to the neonates as the older *atg7* deficient demonstrate hippocampal degeneration at around 3 weeks of age (Komatsu et al., 2006). Results from experimental ischemia in older wild-type mice (PND 56) demonstrated induction of autophagy, though a detrimental role has not been well established. However, other studies employing pharmacological enhancers of autophagy such as rapamycin have established a beneficial role for autophagy through the activation of Akt/CREB signaling pathway (Carloni et al., 2008; Carloni et al., 2010). These studies further suggest that inhibition of autophagy could result in enhanced cell death (Balduini et al., 2012; Carloni et al., 2012).

As for biomarkers for CNS autophagy, microtubule-associated protein light chain isoform (MAP-LC3, also called *atg8*) is one of the autophagy-linked proteins that are critical in forming the autophagosome double lipid membrane. It appears to be processed at the N-terminal (truncated of 3 amino acid) to expose the new N-terminal glycine that is subsequently lipidated with phosphatidylethanolamine (PE) (LC3-II form) and got incorporated into the forming autophagosome membrane. This lipidated form migrates faster on SDS-gel electrophoresis (17 kDa) versus the full-length non-lipidated counterpart (19 kDa) (Tanida, 2011). Increased LC3 to LC3-II conversion has been observed in neuronal autophagy in both cell culture conditions (Sadasivan et al., 2010) as well as in brain tissue of animal model of TBI (Sadasivan et al., 2008; Lai et al. 2008; Clark et al. 2008) (Table 4.3). Another emerging marker for neuronal autophagy or autophagic cell death is a protein called p62/SQSTM1/A170 (p62). It is a stress-inducible protein that was found to be one of the binding proteins of LC3, resulting in autophagosome formation. Importantly, during active execution of autophagy, levels of p62 protein are diminished (Lou et al., 2011). Thus, p62 is a potential markers for autophagy in CNS injuries and disorders.

Autophagy Regulation and Oxidative Stress and inflammation

TBI has been demonstrated to induce both inflammation and mitochondria-related oxidative stress (Hall et al., 2010; Lucas et al., 2006; Mazzeo et al., 2009; Raghupathi, 2004; Robertson et al., 2006; Ziebell and Morganti-Kossmann, 2010). Studies have reported a complex interplay and feedback mechanisms to exist between autophagy, inflammation and oxidative stress following brain injury (Alirezaei et al., 2009; Levine et al., 2011; Scherz-Shouval et al., 2007). Our studies using cerebellar granule neurons in an NMDA-mediated excitotoxic experimental paradigm (known to induce oxidative stress) suggested a pathological role for prolonged autophagy induction that is alleviated with the use of autophagy inhibitor 3-methyladenine (3-MA) and knockdown of autophagy gene *atg7* (Sadasivan et al., 2010). Similarly, antioxidant therapy has been demonstrated to be neuroprotective, especially in the hippocampus compared to vehicle-treated controls in a controlled cortical impact injury paradigm, possibly due to reduced autophagy induction levels (Lai et al., 2008). Redox imbalance is one of hallmark pathologies following acute brain injury. Mitochondria generated Reactive Oxygen Species (ROS) has emerged as one of the key players that regulate autophagy. Increased ROS levels contributed by dysfunctional mitochondria has been demonstrated to positively regulate autophagy (Lee et al., 2012; Martin, 2012; Sanderson et al., 2012), eventually proving detrimental to the cell.

Recent evidence in literature has suggested that signaling from the peripheral lymphocytes promotes inflammatory response in the brain and signals the microglial response to the site of injury in the central nervous system (Gelderblom et al., 2012; Moreno et al., 2011; Ransohoff and Brown, 2012; Shichita et al., 2012). Autophagy has also been implicated in the development, proliferation and maintenance of peripheral lymphocytes. T and B cell lymphocytes have been demonstrated to play an integral role in the brain contributing to and in certain cases help signal the innate microglia to hone in on the injury in the central nervous system (Ransohoff and Brown, 2012). Autophagy signaling has been demonstrated to alter the viability of T-cells and hence play a key role in controlling the inflammation following brain insults (Brait et al., 2012; Dunkle and He, 2011; Hurn et al., 2007; Pierdominici et al., 2011). Specifically, autophagy gene *atg5* has been identified to play a pivotal role in lymphocyte homeostasis (Jia et al., 2011; Miller et al., 2008; Pua et al., 2007; Pua and He, 2007; Pua and He, 2009). Overexpression of this gene in T-cells has been suggested to contribute to the pathological sequelae in autoimmune diseases especially multiple sclerosis (Alirezaei et al., 2009; Alirezaei et al., 2011).

Cross-talk between Autophagy, Apoptosis and Necrosis

Though programmed cell death has been classified into different forms based on the morphological features presented by the cell, there exists a complex cross-talk at the biochemical level within the cell that makes it difficult to establish the path that the cell might take to undergo cell death. With the advances in molecular biology, it became easier to study the role of each pathway by either genetic manipulation or selective knockdown of secondary messengers using pharmacological inhibitors. Recent studies have demonstrated that intracellular pathways of autophagy and apoptosis are intricately balanced by the interaction of autophagy protein beclin-1, through its BH3 domain with anti-apoptotic proteins such as bcl-2, bcl-xl and mcl-1 (Feng et al., 2007; Oberstein et al., 2007). Increased ratio of beclin-1/bcl-2 directs autophagy-induction which may ultimately if prolonged lead to autophagy-mediated cell death. However, beclin-1 mediated autophagy induction is inhibited by this interaction in nutrient-sufficient conditions.

Thus, the interaction between beclin-1 and bcl-2 supported by other protein components act as a gateway between autophagy and apoptosis (Maiuri et al., 2010; Pattingre et al., 2008; Pattingre et al., 2005; Zhou et al., 2011). Similarly, other reports suggest that the proteolytic function of caspases on beclin-1 inhibits its ability to induce autophagy and also that the C-terminal by-product of the cleaved beclin-1 can amplify mitochondrion-mediated apoptosis (Djavaheri-Mergny et al., 2010; Wirawan et al., 2010). Also, calpain-1 and 2, a non-lysosomal calcium-dependent cysteine protease usually associated with necrotic form of cell death, has been demonstrated by Yousefi et al. (2006) to cleave autophagy protein *atg5*. The N-terminal of the cleaved protein was demonstrated to have the ability to bind to bcl-xl anti-apoptotic protein, thus causing release of cytochrome c leading to apoptosis in neutrophils. Thus, studies reporting the role of proteolytic function of caspase and calpain proteases on autophagy proteins highlight the kind of complex cross-talk that exists among different forms of programmed cell death pathways, within the cell (Cho et al., 2009; Luo et al., 2012; Luo and Rubinsztein, 2010; Rohn et al., 2011; Yousefi and Simon, 2007). Also, Beclin-1 interaction with anti-apoptotic bcl-2 protein in the ER membrane plays a critical role balancing autophagy and apoptosis under nutrient-deprivation conditions.

It has now been demonstrated that stress-induced MAPK JNK signaling phosphorylates residues on bcl-2, thus disrupting its interaction with beclin-1 and induction of autophagy. Additionally, JNK signaling is also responsible for autophagy-induction in response to ER stress, oxidative stress and stimulation through the death receptor Fas (Ogata et al., 2006; Zhang et al., 2008). Also worth noting is the fact that bcl-2-beclin-1 interaction can be disrupted by pro-apoptotic proteins such as bad and bax (Luo and Rubinsztein, 2007; Maiuri et al., 2007). Cleavage of autophagy protein *atg5* or beclin-1 by calpains and caspase proteases respectively has also been demonstrated to inhibit autophagy and enhance apoptosis (Cho et al., 2009; Djavaheri-Mergny et al., 2010; Wirawan et al., 2012; Wirawan et al., 2010; Yousefi et al., 2006). Thus, there is a complex array of interactions between different cell death pathways following injury or perturbation within a cell.

Conclusions

Autophagy has been demonstrated to be important for cellular function and homeostasis. The role of autophagy in brain injury has been extrapolated mainly based on experimental manipulations of autophagy by either genetic or pharmacological avenues. Though the exact physiological role of autophagy in acute brain injury is not well established, it is believed that autophagy though beneficial initially in maintenance of neurons in its compromised state becomes detrimental as prolonged induction of autophagy leads to autophagy-mediated autophagic cell death (Sadasivan et al., 2010). As mentioned earlier, the lack of target specificity by studies employing pharmacological enhancers or inhibitors of autophagy propounds the problem of interpreting results. Disruption of autophagy in the entire central nervous system has not only been demonstrated to be fatal, it also renders the neurons susceptible for neurodegenration (Friedman et al., 2012; Komatsu et al., 2006; Wei et al., 2012). Ultimately, it is the problem of excess as autophagy induction in response to toxic or injury stimulus in the cell is protective, but prolonged activation of this process and the hostile environment due to the injury results in the consumption of too many vital

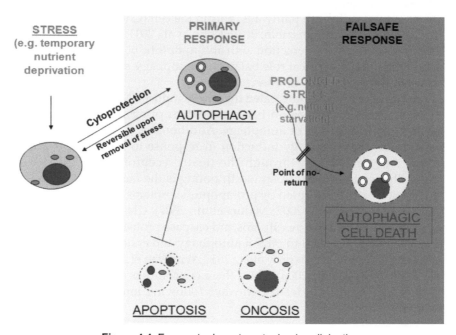

Figure 4.4. From autophagy to autophagic cell death.

Color image of this figure appears in the color plate section at the end of the book.

organelles and other cellular components leading to the demise of the cell (Figure 4.4). Thus further characterization of the autophagy and autophagy-induced cell death would help solve the dilemma on whether to exploit or inhibit the cell process for therapeutic gains in the CNS.

References

Adhami, F., G. Liao, Y.M. Morozov, A. Schloemer, V.J. Schmithorst, J.N. Lorenz, R.S. Dunn, C.V. Vorhees, M. Wills-Karp, J.L. Degen, R.J. Davis, N. Mizushima, P. Rakic, B.J. Dardzinski, S.K. Holland, F.R. Sharp and C.Y. Kuan. 2006. Cerebral ischemia-hypoxia induces intravascular coagulation and autophagy. *Am J Pathol*. 169: 566–83.

Adhami, F., A. Schloemer and C.Y. Kuan. 2007. The roles of autophagy in cerebral ischemia. *Autophagy*. 3: 42–4.

Alirezaei, M., H.S. Fox, C.T. Flynn, C.S. Moore, A.L. Hebb, R.F. Frausto, V. Bhan, W.B. Kiosses, J.L. Whitton, G.S. Robertson and S.J. Crocker. 2009. Elevated ATG5 expression in autoimmune demyelination and multiple sclerosis. *Autophagy*. 5: 152–8.

Alirezaei, M., C.C. Kemball and J.L. Whitton. 2011. Autophagy, inflammation and neurodegenerative disease. *Eur J Neurosci*. 33: 197–204.

Balduini, W., S. Carloni and G. Buonocore. 2012. Autophagy in hypoxia-ischemia induced brain injury. *J Matern Fetal Neonatal Med*. 25 Suppl 1: 30–4.

Brait, V.H., T.V. Arumugam, G.R. Drummond and C.G. Sobey. 2012. Importance of T lymphocytes in brain injury, immunodeficiency, and recovery after cerebral ischemia. *J Cereb Blood Flow Metab*. 32: 598–611.

Calne, R.Y., D.S. Collier, S. Lim, S.G. Pollard, A. Samaan, D.J. White and S. Thiru. 1989. Rapamycin for immunosuppression in organ allografting. *Lancet*. 2: 227.

Carloni, S., G. Buonocore and W. Balduini. 2008. Protective role of autophagy in neonatal hypoxia-ischemia induced brain injury. *Neurobiol Dis*. 32: 329–39.

Carloni, S., G. Buonocore, M. Longini, F. Proietti and W. Balduini. 2012. Inhibition of rapamycin-induced autophagy causes necrotic cell death associated with Bax/Bad mitochondrial translocation. *Neuroscience*. 203: 160–9.

Carloni, S., S. Girelli, C. Scopa, G. Buonocore, M. Longini and W. Balduini. 2010. Activation of autophagy and Akt/CREB signaling play an equivalent role in the neuroprotective effect of rapamycin in neonatal hypoxia-ischemia. *Autophagy*. 6: 366–77.

Carlson, R.P., D.A. Hartman, L.A. Tomchek, T.L. Walter, J.R. Lugay, W. Calhoun, S.N. Sehgal and J.Y. Chang. 1993. Rapamycin, a potential disease-modifying antiarthritic drug. *J Pharmacol Exp Ther*. 266: 1125–38.

Cho, D.H., Y.K. Jo, J.J. Hwang, Y.M. Lee, S.A. Roh and J.C. Kim. 2009. Caspase-mediated cleavage of ATG6/Beclin-1 links apoptosis to autophagy in HeLa cells. *Cancer Lett*. 274: 95–100.

Clark, R.S., H. Bayir, C.T. Chu, S.M. Alber, P.M. Kochanek and S.C. Watkins. 2008. Autophagy is increased in mice after traumatic brain injury and is detectable in human brain after trauma and critical illness. *Autophagy*. 4: 88–90.

DeGracia, D.J. and H.L. Montie. 2004. Cerebral ischemia and the unfolded protein response. *J Neurochem*. 91: 1–8.

Degterev, A., J. Hitomi, M. Germscheid, I.L. Ch'en, O. Korkina, X. Teng, D. Abbott, G.D. Cuny, C. Yuan, G. Wagner, S.M. Hedrick, S.A. Gerber, A. Lugovskoy and J. Yuan. 2008. Identification of RIP1 kinase as a specific cellular target of necrostatins. *Nat Chem Biol*. 4: 313–21.

Degterev, A., Z. Huang, M. Boyce, Y. Li, P. Jagtap, N. Mizushima, G.D. Cuny, T.J. Mitchison, M.A. Moskowitz and J. Yuan. 2005. Chemical inhibitor of nonapoptotic cell death with therapeutic potential for ischemic brain injury. *Nat Chem Biol*. 1: 112–9.

Diskin, T., P. Tal-Or, S. Erlich, L. Mizrachy, A. Alexandrovich, E. Shohami and R. Pinkas-Kramarski. 2005. Closed head injury induces upregulation of Beclin 1 at the cortical site of injury. *J Neurotrauma*. 22: 750–62.

Djavaheri-Mergny, M., M.C. Maiuri and G. Kroemer. 2010. Cross talk between apoptosis and autophagy by caspase-mediated cleavage of Beclin 1. *Oncogene*. 29: 1717–9.

Dunkle, A. and Y.W. He. 2011. Apoptosis and autophagy in the regulation of T lymphocyte function. *Immunol Res*. 49: 70–86.

Erlich, S., A. Alexandrovich, E. Shohami and R. Pinkas-Kramarski. 2007. Rapamycin is a neuroprotective treatment for traumatic brain injury. *Neurobiol Dis*. 26: 86–93.

Erlich, S., E. Shohami and R. Pinkas-Kramarski. 2006. Neurodegeneration induces upregulation of Beclin 1. *Autophagy*. 2: 49–51.

Feng, W., S. Huang, H. Wu and M. Zhang. 2007. Molecular basis of Bcl-xL's target recognition versatility revealed by the structure of Bcl-xL in complex with the BH3 domain of Beclin-1. *J Mol Biol*. 372: 223–35.

Friedman, L.G., M.L. Lachenmayer, J. Wang, L. He, S.M. Poulose, M. Komatsu, G.R. Holstein and Z. Yue. 2012. Disrupted autophagy leads to dopaminergic axon and dendrite degeneration and promotes presynaptic accumulation of alpha-synuclein and LRRK2 in the brain. *J Neurosci*. 32: 7585–93.

Galluzzi, L., I. Vitale, J.M. Abrams, E.S. Alnemri, E.H. Baehrecke, M.V. Blagosklonny, T.M. Dawson, V.L. Dawson, W.S. El-Deiry, S. Fulda, E. Gottlieb, D.R. Green, M.O. Hengartner, O. Kepp, R.A. Knight, S. Kumar, S.A. Lipton, X. Lu, F. Madeo, W. Malorni, P. Mehlen, G. Nunez, M.E. Peter, M. Piacentini, D.C. Rubinsztein, Y. Shi, H.U. Simon, P. Vandenabeele, E. White, J. Yuan, B. Zhivotovsky, G. Melino and G. Kroemer. 2012. Molecular definitions of cell death subroutines: recommendations of the Nomenclature Committee on Cell Death 2012. *Cell Death Differ*. 19: 107–20.

Gelderblom, M., A. Weymar, C. Bernreuther, J. Velden, P. Arunachalam, K. Steinbach, E. Orthey, T.V. Arumugam, F. Leypoldt, O. Simova, V. Thom, M. Friese, I. Prinz, C. Holscher, M. Glatzel, T. Korn, C. Gerloff, E. Tolosa and T. Magnus. 2012. Neutralization of the IL-17 axis diminishes neutrophil invasion and protects from ischemic stroke. *Blood*.

Hall, E.D., R.A. Vaishnav and A.G. Mustafa. 2010. Antioxidant therapies for traumatic brain injury. *Neurotherapeutics*. 7: 51–61.

Hara, T., K. Nakamura, M. Matsui, A. Yamamoto, Y. Nakahara, R. Suzuki-Migishima, M. Yokoyama, K. Mishima, I. Saito, H. Okano and N. Mizushima. 2006. Suppression of basal autophagy in neural cells causes neurodegenerative disease in mice. *Nature*. 441: 885–9.

He, C. and B. Levine. 2010. The Beclin 1 interactome. *Curr Opin Cell Biol*. 22: 140–9.

Holtz, W.A. and K.L. O'Malley. 2003. Parkinsonian mimetics induce aspects of unfolded protein response in death of dopaminergic neurons. *J Biol Chem*. 278: 19367–77.

Hou, W., J. Han, C. Lu, L.A. Goldstein and H. Rabinowich. 2010. Autophagic degradation of active caspase-8: a crosstalk mechanism between autophagy and apoptosis. *Autophagy*. 6: 891–900.

Hurn, P.D., S. Subramanian, S.M. Parker, M.E. Afentoulis, L.J. Kaler, A.A. Vandenbark and H. Offner. 2007. T- and B-cell-deficient mice with experimental stroke have reduced lesion size and inflammation. *J Cereb Blood Flow Metab*. 27: 1798–805.

Imai, Y. and R. Takahashi. 2004. How do Parkin mutations result in neurodegeneration? *Curr Opin Neurobiol*. 14: 384–9.

Jia, W., H.H. Pua, Q.J. Li and Y.W. He. 2011. Autophagy regulates endoplasmic reticulum homeostasis and calcium mobilization in T lymphocytes. *J Immunol*. 186: 1564–74.

Koike, M., M. Shibata, M. Tadakoshi, K. Gotoh, M. Komatsu, S. Waguri, N. Kawahara, K. Kuida, S. Nagata, E. Kominami, K. Tanaka and Y. Uchiyama. 2008. Inhibition of autophagy prevents hippocampal pyramidal neuron death after hypoxic-ischemic injury. *Am J Pathol*. 172: 454–69.

Komatsu, M., S. Waguri, T. Chiba, S. Murata, J. Iwata, I. Tanida, T. Ueno, M. Koike, Y. Uchiyama, E. Kominami and K. Tanaka. 2006. Loss of autophagy in the central nervous system causes neurodegeneration in mice. *Nature*. 441: 880–4.

Kroemer, G., L. Galluzzi, P. Vandenabeele, J. Abrams, E.S. Alnemri, E.H. Baehrecke, M.V. Blagosklonny, W.S. El-Deiry, P. Golstein, D.R. Green, M. Hengartner, R.A. Knight, S. Kumar, S.A. Lipton, W. Malorni, G. Nunez, M.E. Peter, J. Tschopp, J. Yuan, M. Piacentini, B. Zhivotovsky and G. Melino. 2009. Classification of cell death: recommendations of the Nomenclature Committee on Cell Death 2009. *Cell Death Differ*. 16: 3–11.

Lai, Y., R.W. Hickey, Y. Chen, H. Bayir, M.L. Sullivan, C.T. Chu, P.M. Kochanek, C.E. Dixon, L.W. Jenkins, S.H. Graham, S.C. Watkins and R.S. Clark. 2008. Autophagy is increased after traumatic brain injury in mice and is partially inhibited by the antioxidant gamma-glutamylcysteinyl ethyl ester. *J Cereb Blood Flow Metab*. 28: 540–50.

Lee, J., S. Giordano and J. Zhang. 2012. Autophagy, mitochondria and oxidative stress: cross-talk and redox signalling. *Biochem J*. 441: 523–40.

Levine, B., N. Mizushima and H.W. Virgin. 2011. Autophagy in immunity and inflammation. *Nature*. 469: 323–35.

Levine, B., S. Sinha and G. Kroemer. 2008. Bcl-2 family members: dual regulators of apoptosis and autophagy. *Autophagy*. 4: 600–6.

Li, H., P. Wang, J. Yu and L. Zhang. 2011. Cleaving Beclin 1 to suppress autophagy in chemotherapy-induced apoptosis. *Autophagy*. 7: 1239–41.

Lipinski, M.M. and J. Yuan. 2004. Mechanisms of cell death in polyglutamine expansion diseases. *Curr Opin Pharmacol*. 4: 85–90.

Liu, C., Y. Gao, J. Barrett and B. Hu. 2010. Autophagy and protein aggregation after brain ischemia. *J Neurochem*. 115: 68–78.

Lucas, S.M., N.J. Rothwell and R.M. Gibson. 2006. The role of inflammation in CNS injury and disease. *Br J Pharmacol*. 147 Suppl 1: S232–40.

Luo, S., M. Garcia-Arencibia, R. Zhao, C. Puri, P.P. Toh, O. Sadiq and D.C. Rubinsztein. 2012. Bim inhibits autophagy by recruiting beclin 1 to microtubules. *Mol Cell*. 47: 359–70.

Luo, S. and D.C. Rubinsztein. 2007. Atg5 and Bcl-2 provide novel insights into the interplay between apoptosis and autophagy. *Cell Death Differ*. 14: 1247–50.

Luo, S. and D.C. Rubinsztein. 2010. Apoptosis blocks Beclin 1-dependent autophagosome synthesis: an effect rescued by Bcl-xL. *Cell Death Differ*. 17: 268–77.

Maiuri, M.C., A. Criollo and G. Kroemer. 2010. Crosstalk between apoptosis and autophagy within the Beclin 1 interactome. *EMBO J*. 29: 515–6.

Maiuri, M.C., A. Criollo, E. Tasdemir, J.M. Vicencio, N. Tajeddine, J.A. Hickman, O. Geneste and G. Kroemer. 2007. BH3-only proteins and BH3 mimetics induce autophagy by competitively disrupting the interaction between Beclin 1 and Bcl-2/Bcl-X(L). *Autophagy*. 3: 374–6.

Martin, L.J. 2012. Biology of mitochondria in neurodegenerative diseases. *Prog Mol Biol Transl Sci*. 107: 355–415.

Martinez, J.A., Z. Zhang, S.I. Svetlov, R.L. Hayes, K.K. Wang and S.F. Larner. 2010. Calpain and caspase processing of caspase-12 contribute to the ER stress-induced cell death pathway in differentiated PC12 cells. *Apoptosis*. 15: 1480–93.

Mazzeo, A.T., A. Beat, A. Singh and M.R. Bullock. 2009. The role of mitochondrial transition pore, and its modulation, in traumatic brain injury and delayed neurodegeneration after TBI. *Exp Neurol*. 218: 363–70.

Miller, B.C., Z. Zhao, L.M. Stephenson, K. Cadwell, H.H. Pua, H.K. Lee, N.N. Mizushima, A. Iwasaki, Y.W. He, W. Swat and H.W.t. Virgin. 2008. The autophagy gene ATG5 plays an essential role in B lymphocyte development. *Autophagy*. 4: 309–14.

Moreno, B., J.P. Jukes, N. Vergara-Irigaray, O. Errea, P. Villoslada, V.H. Perry and T.A. Newman. 2011. Systemic inflammation induces axon injury during brain inflammation. *Ann Neurol*. 70: 932–42.

Morishima, N., K. Nakanishi, H. Takenouchi, T. Shibata and Y. Yasuhiko. 2002. An endoplasmic reticulum stress-specific caspase cascade in apoptosis. Cytochrome c-independent activation of caspase-9 by caspase-12. *J Biol Chem*. 277: 34287–94.

Nakka, V.P., A. Gusain and R. Raghubir. 2010. Endoplasmic reticulum stress plays critical role in brain damage after cerebral ischemia/reperfusion in rats. *Neurotox Res*. 17: 189–202.

Oberstein, A., P.D. Jeffrey and Y. Shi. 2007. Crystal structure of the Bcl-XL-Beclin 1 peptide complex: Beclin 1 is a novel BH3-only protein. *J Biol Chem*. 282: 13123–32.

Ogata, M., S. Hino, A. Saito, K. Morikawa, S. Kondo, S. Kanemoto, T. Murakami, M. Taniguchi, I. Tanii, K. Yoshinaga, S. Shiosaka, J.A. Hammarback, F. Urano and K. Imaizumi. 2006. Autophagy is activated for cell survival after endoplasmic reticulum stress. *Mol Cell Biol*. 26: 9220–31.

Pattingre, S., L. Espert, M. Biard-Piechaczyk and P. Codogno. 2008. Regulation of macroautophagy by mTOR and Beclin 1 complexes. *Biochimie*. 90: 313–23.

Pattingre, S., A. Tassa, X. Qu, R. Garuti, X.H. Liang, N. Mizushima, M. Packer, M.D. Schneider and B. Levine. 2005. Bcl-2 antiapoptotic proteins inhibit Beclin 1-dependent autophagy. *Cell*. 122: 927–39.

Pierdominici, M., D. Vacirca, F. Delunardo and E. Ortona. 2011. mTOR signaling and metabolic regulation of T cells: new potential therapeutic targets in autoimmune diseases. *Curr Pharm Des*. 17: 3888–97.

Pua, H.H., I. Dzhagalov, M. Chuck, N. Mizushima and Y.W. He. 2007. A critical role for the autophagy gene Atg5 in T cell survival and proliferation. *J Exp Med*. 204: 25–31.

Pua, H.H. and Y.W. He. 2007. Maintaining T lymphocyte homeostasis: another duty of autophagy. *Autophagy*. 3: 266–7.

Pua, H.H. and Y.W. He. 2009. Autophagy and lymphocyte homeostasis. *Curr Top Microbiol Immunol*. 335: 85–105.

Puyal, J. and P.G. Clarke. 2009. Targeting autophagy to prevent neonatal stroke damage. *Autophagy*. 5: 1060–1.

Puyal, J., A. Vaslin, V. Mottier and P.G. Clarke. 2009. Postischemic treatment of neonatal cerebral ischemia should target autophagy. *Ann Neurol*. 66: 378–89.

Qin, A.P., C.F. Liu, Y.Y. Qin, L.Z. Hong, M. Xu, L. Yang, J. Liu, Z.H. Qin and H.L. Zhang. 2010. Autophagy was activated in injured astrocytes and mildly decreased cell survival following glucose and oxygen deprivation and focal cerebral ischemia. *Autophagy*. 6: 738–53.

Raghupathi, R. 2004. Cell death mechanisms following traumatic brain injury. *Brain Pathol*. 14: 215–22.

Ransohoff, R.M. and M.A. Brown. 2012. Innate immunity in the central nervous system. *J Clin Invest*. 122: 1164–71.

Rao, R.V., S. Castro-Obregon, H. Frankowski, M. Schuler, V. Stoka, G. del Rio, D.E. Bredesen and H.M. Ellerby. 2002a. Coupling endoplasmic reticulum stress to the cell death program. An Apaf-1-independent intrinsic pathway. *J Biol Chem*. 277: 21836–42.

Rao, R.V., A. Peel, A. Logvinova, G. del Rio, E. Hermel, T. Yokota, P.C. Goldsmith, L.M. Ellerby, H.M. Ellerby and D.E. Bredesen. 2002b. Coupling endoplasmic reticulum stress to the cell death program: role of the ER chaperone GRP78. *FEBS Lett*. 514: 122–8.

Roberge, F.G., D.F. Martin, D. Xu, H. Chen and C.C. Chan. 1995. Synergism between corticosteroids and Rapamycin for the treatment of intraocular inflammation. *Ocul Immunol Inflamm*. 3: 195–202.

Robertson, C.L., L. Soane, Z.T. Siegel and G. Fiskum. 2006. The potential role of mitochondria in pediatric traumatic brain injury. *Dev Neurosci*. 28: 432–46.

Rohn, T.T., E. Wirawan, R.J. Brown, J.R. Harris, E. Masliah and P. Vandenabeele. 2011. Depletion of Beclin-1 due to proteolytic cleavage by caspases in the Alzheimer's disease brain. *Neurobiol Dis*. 43: 68–78.

Ryu, E.J., H.P. Harding, J.M. Angelastro, O.V. Vitolo, D. Ron and L.A. Greene. 2002. Endoplasmic reticulum stress and the unfolded protein response in cellular models of Parkinson's disease. *J Neurosci*. 22: 10690–8.

Sadasivan, S., W.A. Dunn, Jr., R.L. Hayes and K.K. Wang. 2008. Changes in autophagy proteins in a rat model of controlled cortical impact induced brain injury. *Biochem Biophys Res Commun*. 373: 478–81.

Sadasivan, S., A. Waghray, S.F. Larner, W.A. Dunn, Jr., R.L. Hayes and K.K. Wang. 2006. Amino acid starvation induced autophagic cell death in PC-12 cells: evidence for activation of caspase-3 but not calpain-1. *Apoptosis*. 11: 1573–82.

Sadasivan, S., Z. Zhang, S.F. Larner, M.C. Liu, W. Zheng, F.H. Kobeissy, R.L. Hayes and K.K. Wang. 2010. Acute NMDA toxicity in cultured rat cerebellar granule neurons is accompanied by autophagy induction and late onset autophagic cell death phenotype. *BMC Neurosci*. 11: 21.

Sanderson, T.H., C.A. Reynolds, R. Kumar, K. Przyklenk and M. Huttemann. 2012. Molecular Mechanisms of Ischemia-Reperfusion Injury in Brain: Pivotal Role of the Mitochondrial Membrane Potential in Reactive Oxygen Species Generation. *Mol Neurobiol*.

Scherz-Shouval, R., E. Shvets, E. Fass, H. Shorer, L. Gil and Z. Elazar. 2007. Reactive oxygen species are essential for autophagy and specifically regulate the activity of Atg4. *EMBO J*. 26: 1749–60.

Sehgal, S.N., H. Baker and C. Vezina. 1975. Rapamycin (AY-22,989), a new antifungal antibiotic. II. Fermentation, isolation and characterization. *J Antibiot (Tokyo)*. 28: 727–32.

Shichita, T., R. Sakaguchi, M. Suzuki and A. Yoshimura. 2012. Post-ischemic inflammation in the brain. *Front Immunol*. 3: 132.

Warner, L.M., L.M. Adams and S.N. Sehgal. 1994. Rapamycin prolongs survival and arrests pathophysiologic changes in murine systemic lupus erythematosus. *Arthritis Rheum*. 37: 289–97.

Weedon, D., J. Searle and J.F. Kerr. 1979. Apoptosis. Its nature and implications for dermatopathology. *Am J Dermatopathol*. 1: 133–44.

Wei, K., P. Wang and C.Y. Miao. 2012. A Double-Edged Sword with Therapeutic Potential: An Updated Role of Autophagy in Ischemic Cerebral Injury. *CNS Neurosci Ther*.

Wirawan, E., S. Lippens, T. Vanden Berghe, A. Romagnoli, G.M. Fimia, M. Piacentini and P. Vandenabeele. 2012. Beclin1: a role in membrane dynamics and beyond. *Autophagy*. 8: 6–17.

Wirawan, E., L. Vande Walle, K. Kersse, S. Cornelis, S. Claerhout, I. Vanoverberghe, R. Roelandt, R. De Rycke, J. Verspurten, W. Declercq, P. Agostinis, T. Vanden Berghe, S. Lippens and P. Vandenabeele. 2010. Caspase-mediated cleavage of Beclin-1 inactivates Beclin-1-induced autophagy and enhances apoptosis by promoting the release of proapoptotic factors from mitochondria. *Cell Death Dis*. 1: e18.

Wyllie, A.H., J.F. Kerr and A.R. Currie. 1980. Cell death: the significance of apoptosis. *Int Rev Cytol*. 68: 251–306.

Yousefi, S., R. Perozzo, I. Schmid, A. Ziemiecki, T. Schaffner, L. Scapozza, T. Brunner and H.U. Simon. 2006. Calpain-mediated cleavage of Atg5 switches autophagy to apoptosis. *Nat Cell Biol*. 8: 1124–32.

Yousefi, S. and H.U. Simon. 2007. Apoptosis regulation by autophagy gene 5. *Crit Rev Oncol Hematol*. 63: 241–4.

Zhang, H., C. Zhong, L. Shi, Y. Guo and Z. Fan. 2009. Granulysin induces cathepsin B release from lysosomes of target tumor cells to attack mitochondria through processing of bid leading to Necroptosis. *J Immunol*. 182: 6993–7000.

Zhang, Y., Y. Wu, Y. Cheng, Z. Zhao, S. Tashiro, S. Onodera and T. Ikejima. 2008. Fas-mediated autophagy requires JNK activation in HeLa cells. *Biochem Biophys Res Commun*. 377: 1205–10.

Zhou, F., Y. Yang and D. Xing. 2011. Bcl-2 and Bcl-xL play important roles in the crosstalk between autophagy and apoptosis. *FEBS J*. 278: 403–13.

Zhu, C., X. Wang, F. Xu, B.A. Bahr, M. Shibata, Y. Uchiyama, H. Hagberg and K. Blomgren. 2005. The influence of age on apoptotic and other mechanisms of cell death after cerebral hypoxia-ischemia. *Cell Death Differ*. 12: 162–76.

Ziebell, J.M. and M.C. Morganti-Kossmann. 2010. Involvement of pro- and anti-inflammatory cytokines and chemokines in the pathophysiology of traumatic brain injury. *Neurotherapeutics*. 7: 22–30.

5

Acute, Subacute and Chronic Biomarkers for CNS Injury

Zhiqun Zhang,[1] Ahmed Moghieb[1,3] and Kevin K.W. Wang[1,2,4,]*

INTRODUCTION

Traumatic and ischemic Central Nerve System (CNS) injury is a significant biomedical problem without adequate therapeutic interventions. It includes Traumatic Brain Injury (TBI), ischemic stroke and hemorrhagic stroke (or intracerebral hemorrhage (ICH)), subarachnoid hemorrhage (SAH) and Spinal Cord Injury (SCI). Traumatic brain injury (TBI) is defined as a neurotrauma caused by a mechanical force that is applied to the head. Annually in the United States, there are approximately 1.4 to 2.0 million incidents that involve TBI. Of these, nearly 100,000 patients die, another 500,000 are hospitalized, and thousands of others suffer short and long term effect (Ottens et al., 2007; CDC, 2010). TBI is referred to as a silent epidemic (Siman et al., 2004; Hoffman and Harrison, 2009; Siman et al., 1984; Pike et al., 2004, 2001; Mondello et al., 2010). The Center for Disease Control and Prevention (CDC) reports that approximately 5.3 million Americans live with the effects of TBI. About half of the estimated 1.9 million Americans who experience TBI's each year incur at least some short-term disability. Fifty two thousand people die as a result of their injuries and more than

---

[1] Center for Neuroproteomics & Biomarkers Research, Department of Psychiatry, University of Florida, Gainesville, FL 32611, USA.
[2] Department of Neuroscience, University of Florida, Gainesville, FL 32611, USA.
[3] Department of Chemistry and University of Florida, Gainesville, FL 32611, USA.
[4] Department of Physiological Science, University of Florida, Gainesville, FL 32611, USA.
* Corresponding author: kwang@ufl.edu

90,000 people sustain severe brain injuries leading to debilitating loss of function. Males are 1.6 times more likely than females to suffer TBI until the age of 65 years, when the female rate exceeds the male. The highest overall incidence rate of TBI occurs in children less than 5 years of age, closely followed by seniors more than 85 years old. Falls represent the most common mechanism of TBI injury, followed by motor vehicle-related trauma (Zemlan et al., 2002; Jager et al., 2000; Shaw et al., 2002; Siman et al., 2005). In addition, in the U.S., TBI accounts for 1.3% of all emergency department visits (Dambinova et al., 2012; Jager et al., 2000). The direct medical costs for treatment of TBI in the U.S. have been estimated to be more than US$ 4 billion annually. Unlike other disease processes with later onset such as cancer and cardiovascular disease, survivors of TBI often have many decades of productive life lost, costing themselves, their families, and society the capability for competitive employment and other meaningful roles (Petzold et al., 2005; Elovic et al., 2006; Petzold and Shaw, 2007; Englander et al., 2010; Anderson et al., 2008). Mild TBIs often are under diagnosed and the societal burden grossly underestimated. More Americans are disabled by TBI than Alzheimer's disease.

Clinicians treating those with TBI need better methods to quantify the type and extent of injury to the brain. This is especially critical in the operating theater where timing of care delivery is the key. Neuroimaging (CAT can and MRI) does not capture the full extent of injury, nor does it provide much assistance in tailoring the recovery process. Thus, there exists an urgent need to develop and refine biological measures of acute injury and chronic recovery after TBI. Such measures, "biomarkers", can assist clinicians in developing treatment paradigms for the acutely injured to reduce secondary injury processes. Such markers may also assist in helping to define and refine the recovery process. The absence of biochemical markers of brain injury has severely handicapped therapy development in this field. Equally important are chronic biomarkers that would allow researchers and clinicians to design, refine and monitor innovative therapeutic strategies for the recovery of those with TBI.

Stroke is the second leading cause of death worldwide and the third leading cause of death in the USA with a annual incidence of 750,000 (Kobeissy et al., 2006; Centers for Disease Control and Prevention (CDC), 2012; Liu et al., 2006; Gerberding and Binder, 2003; Yao et al., 2008, 2009; Ottens et al., 2010). A stroke is a sudden interruption in the blood supply of the brain. The most common type of stroke, accounting for almost 80% of all strokes, is caused by a clot or other blockage within an artery leading to the brain (ischemic stroke) (Kobeissy et al., 2008; Carandang et al., 2006). Other strokes are caused by bleeding into brain tissue when a blood vessel bursts (hemorrhagic stroke). Because stroke occurs rapidly and requires immediate treatment, stroke is also called a brain attack. When the symptoms of a stroke last only a short time (less than an hour), this is called a Transient Ischemic Attack (TIA) or mini-stroke. An intracerebral hemorrhage is a type of stroke caused by the sudden rupture of an artery within the brain. Blood is then released into the brain compressing brain structures. Subarachnoid hemorrhage is also a type of brain hemorrhage caused by the sudden rupture of an artery. A subarachnoid hemorrhage differs from an intracerebral hemorrhage in that the location of the rupture leads to blood filling the space surrounding the brain rather than inside of it. It was estimated there are 9 in 100,000 person year of SAH in developed countries with unruptured intracranial aneurysms occurs at much higher rate with 3.2% of the population (Liu et al., 2010; de Rooij et al., 2007; Vlak et al., 2011). Similarly, spinal cord injury (SCI) is considered among the most frequent causes of mortality and morbidity in every medical care system around the world. The incidence of SCI in the United States alone is estimated to be 11,000 new cases each year affecting a total of 183,000 to 230,000 individuals. There are approximately 900 to 1000 cases in a million in the general population (Svetlov et al., 2010; Ravenscroft et al., 2000).

Acute Phase CNS Injury Protein Biomarkers

Significant scientific advances in the last decade have increased our understanding of the pathobiology and biochemical pathways of CNS injury. During the same period, numerous experimental drugs have been shown to be neuroprotective in animal models of brain injury. Unfortunately, these efforts have yet to translate into successful TBI clinical trials (Papa et al., 2010; Saatman et al., 2008). The frustration with failed clinical therapy trials is attributed to the lack of therapeutic intervention-tracking CNS biomarkers. A number of markers have already been identified in the literature. New proteomics methods have also led to the discovery of novel brain injury biomarker candidates (Brophy et al., 2011; Denslow et al., 2003; Kobeissy et al., 2006, 2011). Many agree that there is an unmet medical need for a simple-biofluid-based rapid diagnostic test for the management of TBI patients, whether it be for monitoring severe TBI patients in the intense care unit, or triaging mild and moderate TBI patients in the emergency room. As outlined below, these biomarkers have the potential to revolutionize medical practice and biomedical research of TBI (Siman et al., 2008; Papa et al., 2008).

To aid in the diagnosis and evaluation of acute and chronic TBI, there is an urgent need for biomarkers (Siman et al., 2009; Manley et al., 2010; Saatman et al., 2008). For the acute phase of TBI, a number of brain injury biomarkers have been documented in the literature including neuron specific enolase, glial protein S100β, Glial Fibrillary Acid Protein (GFAP) and Myelin Basic Protein (MBP) (Lewis et al., 2010; Missler et al., 1999; Ross et al., 1996; Yamazaki et al., 1995; Raabe and Seifert, 1999; Romner et al., 2000). Some studies illustrated the diagnostic potential of these brain injury biomarkers, but other studies have provided conflicting results (Kobeissy et al., 2006; Johnsson et al., 2000; Liu et al., 2006; Pelinka et al., 2003; Yao et al., 2008; Pelinka et al., 2005; Yao et al., 2009; Pelinka et al., 2004; Ottens et al., 2010; Berger et al., 2004; Zhang et al., 2011; Berger et al., 2007; Svetlov et al., 2012). NSE, for example, was thought to be strictly neuronal. Assays of serum NSE together with S100β have been valuable in prediction of TBI outcome (Mondello et al., 2010; Berger et al., 2007; Papa et al., 2012, 2010; Mondello et al., 2012). However, additional research found that NSE was also present in red blood cells and platelets decreasing its diagnostic utility as a marker due to possible cross contamination that could occur in blood samples (Zhang et al., 2011; Johnsson et al., 2000). After multiple trauma, increases in NSE levels has been observed, but systemic NSE levels increased correspondingly with and without TBI, limiting its ability to be a discriminator of brain injury magnitude (Allard et al., 2005; Pelinka et al., 2005). The field of genomics and proteomics could be very helpful in identifying specific protein biomarkers that are involved in both degeneration and regeneration for the benefit of patients of acute CNS injury. Several studies have demonstrated the role of proteomics (Suarez et al., 2012; Fountoulakis, 2004; Lubec et al., 2003; Haskins et al., 2005; Kalanu et al., 2006) and genomics (Choi, 2002; Crack et al., 2009; Lei et al., 2009; Balakathiresan et al., 2012; Redell et al., 2009) in providing significant insight into understanding changes, modifications and functions in certain proteins post TBI. Proteomic approaches have been applied to studying rat models to identify novel proteins following SCI (Gordon et al., 2006; Ding et al., 2006; Kang et al., 2006). Another study by Kunz et al. (2005) used proteomics methods to identify novel proteins associated with chronic pain in SCI (Saatman et al., 2008). Different neuroproteomic methods and protocols have been described by Ottens and Wang (Varma et al., 2003; Ottens et al., 2007; Berger et al., 2004; Wagner et al., 2004).

More recently, our group and others have characterized αII-spectrin breakdown products (SBDP150 and SBDP145 produced by calpain in acute necrosis phase, while SBDP120 produced by caspase-3 during delayed apoptosis phase) as potential biomarkers for excitotoxic, traumatic and ischemic brain injury in rat and in human brain trauma (Begaz et al., 2006; Siman et al., 2004, 1984; Pike et al., 2004, 2001; Mondello et al., 2010), while others proposed that the cleaved tau protein (c-tau) (Kobeissy et al., 2011; Zemlan et al., 2002; Shaw et al., 2002; Siman et al., 2005) and a fragmented peptidic form of the glutamate-N-methyl-D-aspartate (NMDA) receptor (NR2A/2B subtype) dam (Kang and Lin, 2012; Dambinova et al., 2012; Dams-O'Connor et al., 2013; Sivanandam and Thakur, 2012; Lee et al.,

2012) might have similar utilities. In addition, studies by Petzold and Shaw (2007) and their colleagues has identified neurofilament-H as a promising axonal injury biomarker for various forms of acute brain damage (Smith et al., 1999; Petzold et al., 2005; Anderson et al., 2008).

Using differential neuroproteomic methods, a systematic assessment was made to identify additional, previously unidentified protein biomarkers for TBI, ischemic and penetrating brain injury with relevant animal models (Duan et al., 2012; Kobeissy et al., 2006; Liu et al., 2006; Yao et al., 2008, 2009; Ottens et al., 2010). As a follow-up, a Systems Biology based approach was applied to select down top candidate markers that represent distinct pathways and hot spots (McKee et al., 2009; Kobeissy et al., 2008; Omalu et al., 2010). One such candidate biomarker identified was Ubiquitin C-terminal hydrolase L1 (UCH-L1). It was recently published by us that UCH-L1 is released into both CSF and blood following both experimental TBI (controlled cortical impact) and ischemic stroke (transient middle cerebral artery occlusion) (Goldstein et al., 2012; Liu et al., 2010) and in a model of blast overpressure wave-induced brain injury in rats (Canu et al., 1998; Svetlov et al., 2010; Park et al., 2007; Liu et al., 2011). Similarly, Papa et al. (2010) and (Haas et al., 2012; Mollenhauer and Zhang, 2012) and Brophy et al. (Ramlackhansingh et al., 2011; Brophy et al., 2011) also reported UCH-L1 released into CSF and blood almost immediately following severe TBI incidents. Siman et al. (2008) and their colleagues described that UCH-L1 (in addition to phospho-neurofilament-H, αII-spectrin breakdown product and 14-3-3 proteins) is elevated in the CSF in humans following surgically-induced circulation arrest (Smith et al., 2012; Jacobowitz et al., 2012; Johnson et al., 2013) as well as in a small severe TBI cohort (Yao et al., 2009; Siman et al., 2009). Independently, UCH-L1 was found released into CSF following in aneurysmal subarachnoid hemorrhage in humans (Ziebell and Morganti-Kossmann, 2010; Lewis et al., 2010; Kumar and Loane, 2012; Maier et al., 2005; Chiaretti et al., 2004; Folkersma et al., 2008; Bonneh-Barkay et al., 2010). We also pioneered in using neuroproteomics/systems biology methods to discover additional unobvious brain injury biomarkers (UCH-L1, αII-spectrin breakdown products (SBDPs), myelin-BDP, Tau-BDP, MAP2, GFAP and its BDP and EMAPII) (Kobeissy et al., 2006; Liu et al., 2006; Yao et al., 2008, 2009; Ottens et al., 2010; Zhang et al., 2011; Svetlov et al., 2012). Many of them (UCH-L1, GFAP and BDP, MAP2, SBDPs) are now verified in clinical studies (Mondello et al., 2010; Papa et al., 2012, 2010; Mondello et al., 2012).

In parallel, other investigative groups have also identified additional protein biomarker candidates for acute TBI, including S100b, NSE (not specific to brain), neurofilament protein NF-H, b- and h-fatty acid binding proteins (Zhang et al., 2011). Allard et al. (2005) identified PARK7 (also called DJ-1) and nucleotide diphosphate kinase A (NDKA) as potential ischemic stroke markers, but their brain specificity and distribution is not completely known.

Thus, the acute phase of CNS injury, candidate biomarker types include: cell body injury markers (UCH-L1, NSE), neurite degeneration markers (SBDP/MAP2, c-Tau, NF-H), demyelination markers (MBP & fragment), gliosis marker (GFAP & BDP, S100b) (Table 5.1).

Table 5.1. Various types of biomarkers for acute, subacute and chronic TBI.

Acute injury markers	**Neurite injury markers: (SBDP150 & SBDP145, MAP2, NF-H, c-Tau)**
	Cell body injury:(UcH-L1, NSE, NMD-R fragment)
	Gliosis/ Glial cell injury (GFAP & BDP; S100b)
Subacute injury markers	**Neuronal apoptosis marker (SBDP120)**
	Demyelination (MBP & fragment)
Neuroinflammatory markers	**Cytokines: IL-6, IL-8 ; TNF-alpha, IL-10**
	Inflammasome markers (caspase-1, NALP1, ASC)
	Microgliosis (iba-1)
Neurodegeneration markers	**AD / CTE markers: P-tau (e.g.T181, S202), Tau , TDP-43**
	AD markers: Aβ 1-40, 1-42
	Alpha-synuclein, Park-7 (DJ-1)
Systemic chronic response markers	**Neuro-endocrine markers (GH, gonadotropin, ACTH/cortisol)**
	Autoimmune (autoantibodies)
Neuro-regeneration markers	**Neurotropic markers (BDNF, NGF)**
	Neuro-stem cell markers: (Nestin, doublecortin)
	Neurite Outgrowth markers: (GAP43, CRMPs)

Need for Biochemical Markers of Subacute and Chronic CNS Injury

With the exception of diuretics, supportive measures and, when appropriate, recombinant tissue plasminogen activator (tPA) there are currently no approved drug treatments for traumatic or ischemic brain injury (Suarez et al., 2012). Historically, most research has focused on closed head brain injury etiologies (e.g., ischemia), with limited study of the implications of overpressure or blast injury. There have been a large number of acute clinical trials studying potential therapies for TBI that have resulted in negative findings with a cost of over US\$ 200 million (Choi, 2002). Post acute care has been plagued with a similar lack of definitive findings and a paucity of evidence based interventions (Gordon et al., 2006). Many investigators have pointed out that the absence of biochemical markers of injury may have contributed to these failures (Saatman et al., 2008). Unlike other organ-based diseases, where rapid diagnosis employing biomarkers (usually involving blood tests), prove invaluable to guide treatment, there are yet no definitive diagnostic tests for traumatic or ischemic brain injury to provide quantifiable neurochemical biomarkers. However, we are currently working on several promising markers to help determine the seriousness of the injury, the anatomical and cellular pathology of the injury, and to guide implementation of appropriate triage and medical management, either short term or long term.

Biomarkers would have important applications in diagnosis, prognosis and clinical research of CNS injuries. While there is currently ongoing research on the acute (1–7 days post injury) time phase of TBI, little research has been done on the subacute (7–30 days) and the chronic (1–3 months) phases after injury. Accurate diagnosis of the subacute and chronic phase care environments can significantly enhance decisions about patient management including decisions whether to admit or discharge or administer other time consuming and expensive tests such as Computer Tomography (CT) and Magnetic Resonance Imaging (MRI) scans. As time progresses neuroimaging plays a minimal role in the care paradigm design and thus biomarkers may be of great assistance in therapeutic design. Post acute care paradigms are needed of markers that could be used to design and track improvement from interventions. Biomarkers would have important prognostic functions; facilitating the development of guidelines for returns to duty or work, and also provide opportunities for counseling of patients suffering from these deficits. Biomarkers would provide major opportunities for the conduct of clinical research including confirmation of injury mechanism(s) and drug target identification. A temporal profile of changes in biomarkers would guide timing of treatment. Biomarkers would provide a clinical trial of early outcome measurements that would be more reliable and economical than conventional neurological assessments. These biomarkers would, thereby, significantly reduce the risks and costs of human clinical trials and ultimately patient treatment and management. Potential gender and developmentally related differences in biomarker profiles (Varma et al., 2003; Berger et al., 2004; Wagner et al., 2004), as well as sensitivity to therapeutic interventions, can be determined in later studies.

Subacute, Chronic CNS Injury Biomarkers

It is important to note that the biofluid (CSF, blood) levels of these acute TBI markers will likely return to baseline level after a matter of days following TBI, especially for those who suffered from mild CNS injury. In fact, A meta-analysis by Begaz et al. (2006) identified that no existing biomarker has consistently demonstrated the ability to predict post-concussive syndrome after mTBI. Thus, we hypothesize that we need another toolkit to approach the diagnosis and management of the chronic effects of TBI (>3 months after injury) (Kobeissy et al., 2011). If there is an unmet medical need for simple biofluid-biomarker tests that can help diagnose and monitor these conditions, we then submit that the chronic effects of TBI are very complex with various CNS and systemic comorbidities and yet-unclear pathway into development into neurodegenerative diseases such as CTE and AD, neuro-endocrine dysfunction and chronic fatigue symptoms, etc., thus it requires multiple types of biomarkers that encompass different neuronal and systemic responses to monitor.

It is also of interest to think of TBI biomarkers as a continuum of biomarkers that might be released at different time points following the initial brain injury event (Figure 5.1). These biomarkers could represent different pathways that can be at play at various time points after the initial injury.

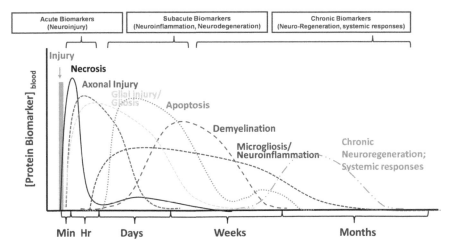

Figure 5.1. CNS Injury Pathophysiology—a continuum of biomarkers. Modified from (118).

Color image of this figure appears in the color plate section at the end of the book.

Post-TBI Neurodegeneration Markers

Recent data indicate that TBI may also be a risk factor for the development of age-associated neurodegenerative disorders including Alzheimer's Disease, Parkinson's Disease (PD), Amyotrophic Lateral Sclerosis and Multiple Sclerosis (Kang and Lin, 2012; Dams-O'Connor et al., 2013; Sivanandam and Thakur, 2012; Lee et al., 2012). Moderate to severe TBI has been shown in autopsy studies to result in increased amyloid deposition in the brain. As we are unable to perform brain biopsies on participants, Aβ 1–40 and Aβ 1–42 are proxies for altered amyloid metabolism and easily followed over time. Tau is a neuronal protein which helps stabilize microtubules in the axon. Moreover, Smith et al. (1999) have show P-tau deposited following TBI. Tau is phosphorylated at many sites potentially by kinases such as casein kinase II, GSK3b, CDk5. Especially of interests are P-tau at both T181 and S202 (Duan et al., 2012). Elevated levels of phosphorylated Tau (P-Tau) are seen in the brain in chronic traumatic encephalopathy years following mild TBI of repeated concussions as a significant tauopathic neurodegenerative disease likely occurring as a result of repeated brain trauma or concussion (McKee et al., 2009; Omalu et al., 2010). Goldstein et al. (2012) demonstrated in a mouse model that single simulated blast exposure resulted in increased deposition of P-Tau. In addition, Tau is post-translationally modified by both calpain and caspase proteases, producing distinct BDPs (Canu et al., 1998; Park et al., 2007; Liu et al., 2011). Tau-BDPs have been identified in both TBI and ischemic brain injury models. Thus, neurodegeneration markers include P-tau (both T181/S202 sites), Tau; Aβ 1–40, 1–42. Additional markers are PD marker candidates DJ-1 and alpha-synuclein (Haas et al., 2012; Mollenhauer and Zhang, 2012) (Table 5.1).

Neuroinflammatory Markers

Ramlackhansigh et al. (2011) demonstrated that microglia, which are turned on in TBI, can remain activated up to 17 years after trauma in 10 patients with moderate to severe TBI or PK11,159 PET ligand. Others have also show sustained microgliosis in both human and animal models using marker such as iba-1 (Smith et al., 2012; Jacobowitz et al., 2012; Johnson et al., 2013). In addition, differential neuroproteomics identified endothelial monocyte-activating polypeptide II precursor (EMAPII) as another possible microglia marker for penetrating brain injury and ischemic brain injury (Yao et al., 2009). Pro-inflammatory cytokines (including IL1β, IL-6 and IL-8, TNF-alpha and YKL-40) are released likely by reactive or over-activated microglia or glia after traumatic brain injury and elevated in biofluids (CSF, blood) or tissue after traumatic brain injury in human and in rat (Ziebell and Morganti-Kossmann, 2010; Kumar and Loane, 2012; Maier et al., 2005; Chiaretti et al., 2004; Folkersma et al., 2008; Bonneh-Barkay et al., 2010). More recently, University of Miami has evaluated a novel innate inflammatory cascade that is initiated by inflammasome activation within glial and neural cells (Abulafia et al., 2008; de Rivero Vaccari et al., 2008, 2009) in various CNS injury conditions. Abnormal inflammasome activation leads to caspase-1 activation and downstream inflammatory cytokines including IL-1b and IL-18 (de Rivero Vaccari et al., 2009). These inflammasome markers can be measured in CSF and plasma following experimental or clinical TBI. These reported changes appear to be sensitive to injury severity, temperature changes and histopathological damage. In a recent study by our group, Adamczak et al. (2012) reported that patients with severe or moderate TBI exhibited significantly higher CSF levels of inflammasome proteins (ASC, caspase-1, NALP1) compared to non-trauma controls. Interestingly, in that study a positive correlation was documented between levels of inflammasome proteins and GOS five months post-injury. Thus we include pro- and anti-inflammatory cytokines (e.g., including interleukin-1β, IL-6 and IL-8) and inflammasomes as TBI neuro-inflammation biomarker (Table 5.1).

Neuroregeneration Markers

Stem cell-based treatment has been applied to experimental CNS injury models and has shown promising results (Daadi, 2011). Yet, little is known about the endogenous neuroregeneration after mild TBI, especially in the chronic stage. It is envisioned that neuroregeneration can occur at two levels: (a) addition proliferation activation and differentiation of neuro-stem cells; (b) factors are released to promote neurite-regeneration and functional recovery (e.g., BDNF, NGF (Chiaretti et al., 2008, 2009; Johanson et al., 2010)). Interestingly, even with NSC due to the small numbers that are available, it is now commonly viewed as not likely to directly make an impact on direct replacement of the loss of functional neurons after acute neural injury. Instead, the current hypothesis is that new neuro-stems cell might release growth factors such as BDNF, GDNF that can help promote neuro-recovery and neuro-remodeling including new neurite outgrowth and nee synapses formation (Kim et al., 2010; Joyce et al., 2010; Feng et al., 2012). Thus both of these two pathways converge onto detecting CNS-relevant growth factors as neuroregeneration biomarkers. In fact, several recent studies (Chiaretti et al., 2008, 2009; Johanson et al., 2010) had examined several growth factors (BDNF, GDNF, NGF) and found that all of them are detectable and elevated in human CSF following TBI, with NGF particularly correlate to outcome measure (GOS) in one study. In addition, stem cell markers such as nestin, dobulecortin and neurite outgrowth markers GAP43 and collapsin response mediator proteins (CRMPs) might be very useful to monitor the status of neuroregeneration (Temple, 2001; Jain, 2010) and neurite outgrowth markers GAP43 (Zakharov et al., 2005) and (CRMPs) (Wang and Strittmatter, 1996; Touma et al., 2007; Zhang et al., 2007). Currently these markers are readily monitored in brain tissue in animal models of CNS injury, but increasing work is beginning to explore their potential in biofluid-detection of such markers. Thus, neuro-regeneration markers include such as neurotropic markers BDNF and NGF, neuro-stem cell markers such as nestin, dobulecortinand neurite outgrowth markers GAP43 and CRMPs (Table 5.1).

Systems Biology-assisted Biomarker Integration

With the rapid growth of high throughput technology in genomic and proteomic studies, a large amount of data has been generated. By using data integration and warehousing techniques, a comprehensive database of brain injury-related information, including both high throughput "omic" datasets (genomics, proteomics, metabolomics, lipidomics, etc.) and "targeted" pathway, and molecular imaging studies from the relevant scientific literature can be developed. Systems modeling and simulation is now considered fundamental to the future development of effective therapeutics. Different model representations have been established to serve these purposes. The graphical diagrams of biological processes such as Pathway Studio, Gene Spring, Ingenuity Pathway and Gene GO give visual presentations of network models by incorporating genome, proteome and metabolome data. However, different formats which incorporate quantitative data generated from or validated with directed biological studies have emerged and have found further use in system simulation and analysis. These bioinformatics software (as Pathway Studio, Gene Spring, Ingenuity Pathway and Gene GO) have been used to construct functional interaction maps of the generated high-throughput data. Functional interaction maps depict altered subsets of genes and/or proteins describing perturbed cellular functions relevant to a specific disorder in question. A major strength of interaction maps lies in their ability to predict and identify certain gene(s) and or protein(s) that have been missed by experimental analysis as well as to provide potential functions of identified proteins with an unknown physiology.

Since CNS injury is a very complex condition with multiple comorbidities, we thus employ a wild range of systems biomarker candidates to match. We submit that it is plausible to use systems biology tools as an unbiased approach to identify non-redundant neuromodulation/

neurotoxicity pathways or hotspots and to further pinpoint candidate diagnostic biomarkers and/or molecular targets for therapy (Kitano, 2002; Zhang et al., 2010; Kobeissy et al., 2011). These tools help organize the available data and knowledge according to the high-level, holistic view required by SB and integrate them with suitable analysis methods (e.g., Support Vector Machines, Bayesian Networks, Cognitive Maps, Dynamic Feature Selection), and they support the iterative, hypothesis-free approach to scientific discovery that is increasingly being adopted in biological research. The resulting high-level views (pathways, maps, biomarkers), together with the data sources will be available through a comprehensive in-house service. Following that, Predictive Systems Responses can also be generated by applying Optimization-based Predictive Model that factor in multi-group multi-stage analysis (Lee, 2008; J and Lee, 2008). Figure 5.2 is an example of using systems biology tools to map the interactome of the various CNS injury biomarker types we described in this chapter.

Although there have been many biomarker studies of brain injury, it continues to be a big challenge to generate the database and functional network map that link those potential biomarker with their corresponding biological effects. A major challenge facing systems biology based protein biomarker approach is the level of statistical confidence related to the personal brain injury or prognosis prediction on those biomarkers based on the profiling data. Therefore, a clinic validation and personalized clinical

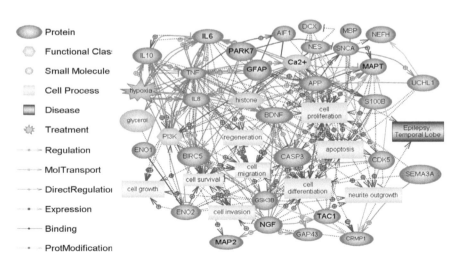

Figure 5.2. Systems Biology mapping of interactome of various TBI biomarker types.

Color image of this figure appears in the color plate section at the end of the book.

medicine could be used in addition to experimental genomic or proteomic analysis.

Conclusion

It is important to note that the biofluid levels of these acute brain injury markers will likely return to baseline level after days following injury, especially for those who suffered from mild brain injury. As noted, a single biomarker cannot present the detailed scenario of the chronic CNS injury which necessitates the use of a panel of biomarkers that may reflect biomarker dynamic changes (Please refer to Figure 5.1). It is possible that the chronic effects of brain injury are very complex with various CNS and systemic comorbidities and yet-unclear pathways contribute to neurodegenerative diseases such as CTE and AD, neuro-endocrine dysfunction and chronic fatigue symptoms, etc. Thus, multiple types of biomarkers will encompass different neuronal and systemic responses as well as various pathways involved that need be monitored (Please refer to Figure 5.2). In summary, the possible ideal neurological injury-associated biomarkers that have been identified and validated with preclinical animal models need to be translated and confirmed in clinical studies as well. Confirmatory testing should be applied in terms of their ability to detect injury magnitude as well as drug-based biomarker level reduction coupled with their sensitivity and specificity levels. A direct comparison of biomarker occurrence between preclinical models and biomarker data from human clinical studies would allow investigators to gain considerable insight into the validity (or challenges to the validity) of the employed preclinical animal models. Finally, systems biology-based biomarker approach will enable researchers to understand the overall underlying pathophysiological theme of the long term sequelae of brain injury. A detailed chronic brain injury biomarker profiling, analysis of correlation to various chronic brain injury comorbidities and neurodegeneration progression analysis would require the use of systems biology platform in order to integrate biomarker data for prediction modeling, and recommend biomarker subset for further clinical utility verification in the near future.

References

Abulafia, D.P., J.P. de Rivero Vaccari, J.D. Lozano, G. Lotocki, R.W. Keane and W.D. Dietrich. 2008. Inhibition of the inflammasome complex reduces the inflammatory response after thromboembolic stroke in mice. *J Cereb Blood Flow Metab.* 29: 534–544.

Adamczak, S., G. Dale, J.P. de Rivero Vaccari, M.R. Bullock, W.D. Dietrich and R.W. Keane. 2012. Inflammasome proteins in cerebrospinal fluid of brain-injured patients as biomarkers of functional outcome. *J Neurosurg.* 117: 1119–1125.

Allard, L., P.R. Burkhard, P. Lescuyer, J.A. Burgess, N. Walter, D.F. Hochstrasser and J.-C. Sanchez. 2005. PARK7 and nucleoside diphosphate kinase A as plasma markers for the early diagnosis of stroke. *Clinical Chemistry.* 51: 2043–2051.

Anderson, K.J., S.W. Scheff, K.M. Miller, K.N. Roberts, L.K. Gilmer, C. Yang and G. Shaw. 2008. The phosphorylated axonal form of the neurofilament subunit NF-H (pNF-H) as a blood biomarker of traumatic brain injury. *J Neurotrauma.* 25: 1079–1085.

Balakathiresan, N., M. Bhomia, R. Chandran, M. Chavko, R.M. McCarron and R.K. Maheshwari. 2012. MicroRNA Let-7i Is a Promising Serum Biomarker for Blast-Induced Traumatic Brain Injury. *J Neurotrauma.* 29: 1379–1387.

Begaz, T.T., D.N.D. Kyriacou, J.J. Segal and J.J.J. Bazarian. 2006. Serum biochemical markers for post-concussion syndrome in patients with mild traumatic brain injury. *J Neurotrauma.* 23: 1201–1210.

Berger, R.P., P.M. Kochanek and M.C. Pierce. 2004. Biochemical markers of brain injury: could they be used as diagnostic adjuncts in cases of inflicted traumatic brain injury? *Child Abuse & Neglect.* 28: 739–754.

Berger, R.P., S.R. Beers, R. Richichi, D. Wiesman and P.D. Adelson. 2007. Serum Biomarker Concentrations and Outcome after Pediatric Traumatic Brain Injury. *J Neurotrauma.* 24: 1793–1801.

Bonneh-Barkay, D., P. Zagadailov, H. Zou, C. Niyonkuru, M. Figley, A. Starkey, G. Wang, S.J. Bissel, C.A. Wiley and A.K. Wagner. 2010. YKL-40 expression in traumatic brain injury: an initial analysis. *J Neurotrauma.* 27: 1215–1223.

Brophy, G.M., S. Mondello, L. Papa, S.A. Robicsek, A. Gabrielli, J. Tepas, A. Buki, C. Robertson, F.C. Tortella, R.L. Hayes and K.K.W. Wang. 2011. Biokinetic analysis of ubiquitin C-terminal hydrolase-L1 (UCH-L1) in severe traumatic brain injury patient biofluids. *J Neurotrauma.* 28: 861–870.

Canu, N., L. Dus, C. Barbato, M.T. Ciotti, C. Brancolini, A.M. Rinaldi, M. Novak, A. Cattaneo, A. Bradbury and P. Calissano. 1998. Tau cleavage and dephosphorylation in cerebellar granule neurons undergoing apoptosis. *J Neurosci.* 18: 7061–7074.

Carandang, R., S. Seshadri, A. Beiser, M. Kelly-Hayes, C.S. Kase, W.B. Kannel and P.A. Wolf. 2006. Trends in incidence, lifetime risk, severity, and 30-day mortality of stroke over the past 50 years. *JAMA.* 296: 2939–2946.

CDC. 2010. Traumatic Brain Injury in the United States: Emergency Department Visits, Hospitalizations and Deaths 2002–2006. 1–74.

Centers for Disease Control and Prevention (CDC). 2012. Prevalence of stroke—United States, 2006–2010. *MMWR Morb Mortal Wkly Rep.* 61: 379–382.

Chiaretti, A., A. Antonelli, R. Riccardi, O. Genovese, P. Pezzotti, C. Di Rocco, L. Tortorolo and G. Piedimonte. 2008. Nerve growth factor expression correlates with severity and outcome of traumatic brain injury in children. *Eur J Paediatr Neurol.* 12: 10–10.

Chiaretti, A., G. Barone, R. Riccardi, A. Antonelli, P. Pezzotti, O. Genovese, L. Tortorolo and G. Conti. 2009. NGF, DCX, and NSE upregulation correlates with severity and outcome of head trauma in children. *Neurology.* 72: 609–616.

Chiaretti, A., O. Genovese, L. Aloe, A. Antonelli, M. Piastra, G. Polidori and C. Rocco. 2004. Interleukin 1? and interleukin 6 relationship with paediatric head trauma severity and outcome. *Childs Nerv Syst.* 21: 185–193.

Choi, D.W. 2002. Exploratory clinical testing of neuroscience drugs. *Nat Neurosci.* 5 Suppl: 1023–1025.

Crack, P.J.P., J.J. Gould, N.N. Bye, S.S. Ross, U.U. Ali, M.D.M. Habgood, C.C. Morganti-Kossman, N.R.N. Saunders and P.J.P. Hertzog. 2009. The genomic profile of the cerebral cortex after closed head injury in mice: effects of minocycline. *Journal of Neural Transmission.* 116: 1–12.

Daadi, M.M. 2011. Novel paths towards neural cellular products for neurological disorders. *Regen Med.* 6: 25–30.

Dambinova, S.A., K. Bettermann, T. Glynn, M. Tews, D. Olson, J.D. Weissman and R.L. Sowell. 2012. Diagnostic potential of the NMDA receptor peptide assay for acute ischemic stroke. *PLoS One.* 7: e42362.

Dams-O'Connor, K., L.E. Gibbons, J.D. Bowen, S.M. McCurry, E.B. Larson and P.K. Crane. 2013. Risk for late-life re-injury, dementia and death among individuals with traumatic brain injury: a population-based study. *J Neurol Neurosurg Psychiatr.* 84: 177–182.

de Rivero Vaccari, J.P., G. Lotocki, A.E. Marcillo, W.D. Dietrich and R.W. Keane. 2008. A molecular platform in neurons regulates inflammation after spinal cord injury. *Journal of Neuroscience.* 28: 3404–3414.

de Rivero Vaccari, J.P., G. Lotocki, O.F. Alonso, H.M. Bramlett, W.D. Dietrich and R.W. Keane. 2009. Therapeutic neutralization of the NLRP1 inflammasome reduces the innate immune response and improves histopathology after traumatic brain injury. *Journal of Cerebral Blood Flow & Amp; Metabolism.* 29: 1251–1261.

de Rooij, N.K., F.H.H. Linn, J.A. van der Plas, A. Algra and G.J.E. Rinkel. 2007. Incidence of subarachnoid haemorrhage: a systematic review with emphasis on region, age, gender and time trends. *J Neurol Neurosurg Psychiatr.* 78: 1365–1372.

Denslow, N., M.E. Michel, M.D. Temple, C.Y. Hsu, K. Saatman and R.L. Hayes. 2003. Application of proteomics technology to the field of neurotrauma. *J Neurotrauma.* 20: 401–407.

Ding, Q.Q., Z.Z. Wu, Y.Y. Guo, C.C. Zhao, Y.Y. Jia, F.F. Kong, B.B. Chen, H.H. Wang, S.S. Xiong, H.H. Que, S.S. Jing and S.S. Liu. 2006. Proteome analysis of up-regulated proteins in the rat spinal cord induced by transection injury. *Proteomics.* 6: 505–518.

Duan, Y., S. Dong, F. Gu, Y. Hu and Z. Zhao. 2012. Advances in the Pathogenesis of Alzheimer's Disease: Focusing on Tau-Mediated Neurodegeneration. *Transl Neurodegener.* 1: 24.

Elovic, E., H.S. Doppalapudi and M. Miller. 2006. Endocrine abnormalities and fatigue after traumatic brain injury. *J Head Trauma Rehab.* 21: 426–427.

Englander, J., T. Bushnik, J. Oggins and L. Katznelson. 2010. Fatigue after traumatic brain injury: Association with neuroendocrine, sleep, depression and other factors. *Brain Inj.* 24: 1379–1388. doi:10.3109/02699052.2010.523041.

Folkersma, H., J.J.P. Brevé, F.J.H. Tilders, L. Cherian, C.S. Robertson and W.P. Vandertop. 2008. Cerebral microdialysis of interleukin (IL)-1beta and IL-6: extraction efficiency and production in the acute phase after severe traumatic brain injury in rats. *Acta Neurochir.* 150: 1277–1284.

Fountoulakis, M. 2004. Application of proteomics technologies in the investigation of the brain. *Mass Spectrom Rev.* 23: 231–258.

Gerberding, J.L. and S. Binder. 2003. National Center for Injury Prevention and Control. Report to Congress on Mild Traumatic Brain Injury in the United States: Steps to Prevent a Serious Public Health Problem. CDC report. http://www.cdc.gov/ncipc/pub-res/mtbi/report.htm.

Goldstein, L.E., A.M. Fisher, C.A. Tagge, X.L. Zhang, L. Velisek, J.A. Sullivan, C. Upreti, J.M. Kracht, M. Ericsson, M.W. Wojnarowicz, C.J. Goletiani, G.M. Maglakelidze, N. Casey, J.A. Moncaster, O. Minaeva, R.D. Moir, C.J. Nowinski, R.A. Stern, R.C. Cantu, J. Geiling, J.K. Blusztajn, B.L. Wolozin, T. Ikezu, T.D. Stein, A.E. Budson, N.W. Kowall, D. Chargin, A. Sharon, S. Saman, G.F. Hall, W.C. Moss, R.O. Cleveland, R.E. Tanzi, P.K. Stanton and A.C. McKee. 2012. Chronic Traumatic Encephalopathy in Blast-Exposed Military Veterans and a Blast Neurotrauma Mouse Model. *Science Translational Medicine.* 4: 134ra60–134ra60.

Gordon, W.A., R. Zafonte, K. Cicerone, J. Cantor, M. Brown, L. Lombard, R. Goldsmith and T. Chandna. 2006. Traumatic brain injury rehabilitation: state of the science. *Am J Phys Med Rehabil.* 85: 343–382.

Haas, B.R., T.H. Stewart and J. Zhang. 2012. Premotor biomarkers for Parkinson's disease—a promising direction of research. *Transl Neurodegener.* 1: 11.

Haskins, W.E., F.H. Kobeissy, R.A. Wolper, A.K. Ottens, J.W. Kitlen, S.H. McClung, B.E. O'Steen, M.M. Chow, J.A. Pineda, N.D. Denslow, R.L. Hayes and K.K.W. Wang. 2005. Rapid discovery of putative protein biomarkers of traumatic brain injury by SDS-PAGE-capillary liquid chromatography-tandem mass spectrometry. *J Neurotrauma.* 22: 629–644.

Hoffman, S.W. and C. Harrison. 2009. The interaction between psychological health and traumatic brain injury: a neuroscience perspective. *Clin Neuropsychol.* 23: 1400–1415.

J, L.G. and E.K. Lee. 2008. Optimization in Medicine And Biology. L.G. J and E.K. Lee, editors. Auerbach Publications, Taylor & Francis Group, Boca Raton, FL, 1 pp.

Jacobowitz, D.M., J.T. Cole, D.P. McDaniel, H.B. Pollard and W.D. Watson. 2012. Microglia activation along the corticospinal tract following traumatic brain injury in the rat: a neuroanatomical study. *Brain Res.* 1465: 80–89.

Jager, T.E., H.B. Weiss, J.H. Coben and P.E. Pepe. 2000. Traumatic brain injuries evaluated in U.S. emergency departments, 1992–1994. *Acad Emerg Med.* 7: 134–140.

Jain, K.K. 2010. The Handbook of Biomarkers. Humana Press, New York, pp. 1–492.

Johanson, C., E. Stopa, A. Baird and H. Sharma. 2010. Traumatic brain injury and recovery mechanisms: peptide modulation of periventricular neurogenic regions by the choroid plexus–CSF nexus. *Journal of Neural Transmission.* 118: 115–133.

Johnson, V.E., J.E. Stewart, F.D. Begbie, J.Q. Trojanowski, D.H. Smith and W. Stewart. 2013. Inflammation and white matter degeneration persist for years after a single traumatic brain injury. *Brain.* 136: 28–42.

Johnsson, P., S. Blomquist, C. Luhrs, G. Malmkvist, C. Alling, J.O. Solem and E. Stahl. 2000. Neuron-specific enolase increases in plasma during and immediately after extracorporeal circulation. *Ann Thorac Surg.* 69: 750–754.

Kang, J.-H. and H.-C. Lin. 2012. Increased risk of multiple sclerosis after traumatic brain injury: a nationwide population-based study. *J Neurotrauma.* 29: 90–95.

Kang, S.K., H.H. So, Y.S. Moon and C.H. Kim. 2006. Proteomic analysis of injured spinal cord tissue proteins using 2-DE and MALDI-TOF MS. *Proteomics.* 6: 2797–2812.

Katano, T., T. Mabuchi, E. Okuda-Ashitaka, N. Inagaki, T. Kinumi and S. Ito. 2006. Proteomic identification of a novel isoform of collapsin response mediator protein-2 in spinal nerves peripheral to dorsal root ganglia. *Proteomics.* 6: 6085–6094.

Kitano, H. 2002. Systems biology: a brief overview. *Science.* 295: 1662–1664.

Kobeissy, F.H., A.K. Ottens, Z. Zhang, M.C. Liu, N.D. Denslow, J.R. Dave, F.C. Tortella, R.L. Hayes, K.K. Wang, F.H. Kobeissy, A.K. Ottens, Z. Zhang, M.C. Liu, N.D. Denslow, J.R. Dave, F.C. Tortella, R.L. Hayes and K.K.W. Wang. 2006. Novel differential neuroproteomics analysis of traumatic brain injury in rats. *Molecular & Cellular Proteomics.* 5: 1887–1898.

Kobeissy, F.H., J.D. Guingab-Cagmat, M. Razafsha, L. O'Steen, Z. Zhang, R.L. Hayes, W.-T. Chiu and K.K.W. Wang. 2011. Leveraging biomarker platforms and systems biology for rehabilomics and biologics effectiveness research. *PM & R.* 3: S139–S147.

Kobeissy, F.H., S. Sadasivan, M.W. Oli, G. Robinson, S.F. Larner, Z. Zhang, R.L. Hayes and K.K.W. Wang. 2008. Neuroproteomics and systems biology-based discovery of protein biomarkers for traumatic brain injury and clinical validation. *Proteomics Clin Appl.* 2: 1467–1483.

Kumar, A. and D.J. Loane. 2012. Neuroinflammation after traumatic brain injury: Opportunities for therapeutic intervention. *Brain, Behaviour, and Immunity.* 1–11.

Kunz, S., I. Tegeder, O. Coste, C. Marian, A. Pfenninger, C. Corvey, M. Karas, G. Geisslinger and E. Niederberger. 2005. Comparative proteomic analysis of the rat spinal cord in inflammatory and neuropathic pain models. *Neurosci Lett.* 381: 289–293.

Lee, E.K. 2008. Optimization-based predictive models in medicine and biology. *Optimization in Medicine.* 12: 127–151.

Lee, P.-C., Y. Bordelon, J. Bronstein and B. Ritz. 2012. Traumatic brain injury, paraquat exposure, and their relationship to Parkinson disease. *Neurology.* 79: 2061–2066.

Lei, P., Y. Li, X. Chen, S. Yang and J. Zhang. 2009. Microarray based analysis of microRNA expression in rat cerebral cortex after traumatic brain injury. *Brain Res.* 1284: 191–201.

Lewis, S.B., R. Wolper, Y.-Y. Chi, L. Miralia, Y. Wang, C. Yang and G. Shaw. 2010. Identification and preliminary characterization of ubiquitin C terminal hydrolase 1 (UCHL1) as a biomarker of neuronal loss in aneurysmal subarachnoid hemorrhage. *Journal of Neuroscience Research.* 88: 1475–1484.

Liu, M.C., F. Kobeissy, W. Zheng, Z. Zhang, R.L. Hayes and K.K.W. Wang. 2011. Dual vulnerability of tau to calpains and caspase-3 proteolysis under neurotoxic and neurodegenerative conditions. *ASN Neuro.* 3: e00051.

Liu, M.C., L. Akinyi, D. Scharf, J. Mo, S.F. Larner, U. Muller, M.W. Oli, W. Zheng, F. Kobeissy, L. Papa, X.-C. Lu, J.R. Dave, F.C. Tortella, R.L. Hayes and K.K.W. Wang. 2010. Ubiquitin C-terminal hydrolase-L1 as a biomarker for ischemic and traumatic brain injury in rats. *Eur J Neurosci.* 31: 722–732.

Liu, M.C., V. Akle, W. Zheng, J.R. Dave, F.C. Tortella, R.L. Hayes, K.K. Wang, M.C. Liu, V. Akle, W. Zheng, J.R. Dave, F.C. Tortella, R.L. Hayes and K.K.W. Wang. 2006. Comparing calpain- and caspase-3-mediated degradation patterns in traumatic brain injury by differential proteome analysis. *Biochem J.* 394: 715–725.

Lubec, G., K. Krapfenbauer and M. Fountoulakis. 2003. Proteomics in brain research: potentials and limitations. *Progress in Neurobiology.* 69: 193–211.

Maier, B., H.-L. Laurer, S. Rose, W.A. Buurman and I. Marzi. 2005. Physiological levels of pro- and anti-inflammatory mediators in cerebrospinal fluid and plasma: a normative study. *J Neurotrauma.* 22: 822–835.

Manley, G.T., R. Diaz-Arrastia, M. Brophy, D. Engel, C. Goodman, K. Gwinn, T.D. Veenstra, G. Ling, A.K. Ottens, F. Tortella and R.L. Hayes. 2010. Common Data Elements for Traumatic Brain Injury: Recommendations From the Biospecimens and Biomarkers Working Group. *YAPMR.* 91: 1667–1672.

McKee, A.C., R.C. Cantu, C.J. Nowinski, E.T. Hedley-Whyte, B.E. Gavett, A.E. Budson, V.E. Santini, H.S. Lee, C.A. Kubilus and R.A. Stern. 2009. Chronic traumatic encephalopathy in athletes: progressive tauopathy after repetitive head injury. *J Neuropathol Exp Neurol.* 68: 709–735.

Missler, U., M. Wiesmann, G. Wittmann, O. Magerkurth and H. Hagenstrom. 1999. Measurement of glial fibrillary acidic protein in human blood: analytical method and preliminary clinical results. *Clin Chem.* 45: 138–141.

Mollenhauer, B. and J. Zhang. 2012. Biochemical premotor biomarkers for Parkinson's disease. *Mov Disord.* 27: 644–650.

Mondello, S., A. Gabrielli, S. Catani, M. D'Ippolito, A. Jeromin, A. Ciaramella, P. Bossù, K. Schmid, F. Tortella, K.K.W. Wang, R.L. Hayes and R. Formisano. 2012. Increased levels of serum MAP-2 at 6-months correlate with improved outcome in survivors of severe traumatic brain injury. *Brain Inj.* 1–7.

Mondello, S., S.A. Robicsek, A. Gabrielli, G.M. Brophy, L. Papa, J. Tepas, C. Robertson, A. Buki, D. Scharf, M. Jixiang, L. Akinyi, U. Muller, K.K.W. Wang and R.L. Hayes. 2010. AlphaII-spectrin breakdown products (SBDPs): diagnosis and outcome in severe traumatic brain injury patients. *J Neurotrauma.* 27: 1203–1213.

Omalu, B.I., R.L. Hamilton, M.I. Kamboh, S.T. DeKosky and J. Bailes. 2010. Chronic traumatic encephalopathy (CTE) in a National Football League Player: Case report and emerging medicolegal practice questions. *Journal of Forensic Nursing.* 6: 40–46.

Ottens, A.K., F.H. Kobeissy, B.F. Fuller, M.C. Liu, M.W. Oli, R.L. Hayes, K.K. Wang, A.K. Ottens, F.H. Kobeissy, B.F. Fuller, M.C. Liu, M.W. Oli, R.L. Hayes and K.K.W. Wang. 2007. Novel neuroproteomic approaches to studying traumatic brain injury. *Prog Brain Res.* 161: 401–418.

Ottens, A.K., L. Bustamante, E.C. Golden, C. Yao, R.L. Hayes, K.K.W. Wang, F.C. Tortella and J.R. Dave. 2010. Neuroproteomics: a biochemical means to discriminate the extent and modality of brain injury. *J Neurotrauma.* 27: 1837–1852.

Papa, L., G. Robinson, M. Oli, J. Pineda, J. Demery, G. Brophy, S.A. Robicsek, A. Gabrielli, C.S. Robertson, K.K. Wang and others. 2008. Use of biomarkers for diagnosis and management of traumatic brain injury patients. *Expert Opin Med Diagn.* 2: 937–945.

Papa, L., L. Akinyi, M.C. Liu, J.A. Pineda, J.J. Tepas, M.W. Oli, W. Zheng, G. Robinson, S.A. Robicsek, A. Gabrielli, S.C. Heaton, H.J. Hannay, J.A. Demery, G.M. Brophy, J. Layon, C.S. Robertson, R.L. Hayes and K.K.W. Wang. 2010. Ubiquitin C-terminal hydrolase is a novel biomarker in humans for severe traumatic brain injury. *Crit Care Med.* 38: 138–144.

Papa, L., L.M. Lewis, J.L. Falk, Z. Zhang, S. Silvestri, P. Giordano, G.M. Brophy, J.A. Demery, N.K. Dixit, I. Ferguson, M.C. Liu, J. Mo, L. Akinyi, K. Schmid, S. Mondello, C.S. Robertson, F.C. Tortella, R.L. Hayes and K.K.W. Wang. 2012. Elevated levels of serum glial fibrillary acidic protein breakdown products in mild and moderate traumatic brain injury are associated with intracranial lesions and neurosurgical intervention. *Ann Emerg Med.* 59: 471–483.

Park, S.Y., C. Tournell, R.C. Sinjoanu and A. Ferreira. 2007. Caspase-3- and calpain-mediated tau cleavage are differentially prevented by estrogen and testosterone in beta-amyloid-treated hippocampal neurons. *NSC.* 144: 119–127. doi:10.1016/j.neuroscience.2006.09.012.

Pelinka, L.E., A. Kroepfl, R. Schmidhammer, M. Krenn, W. Buchinger, H. Redl and A. Raabe. 2004. Glial fibrillary acidic protein in serum after traumatic brain injury and multiple trauma. *J Trauma.* 57: 1006–1012.

Pelinka, L.E., H. Hertz, W. Mauritz, N. Harada, M. Jafarmadar, M. Albrecht, H. Redl and S. Bahrami. 2005. Nonspecific increase of systemic neuron-specific enolase after trauma: clinical and experimental findings. *Shock.* 24: 119–123.

Pelinka, L.E., L. Szalay, M. Jafarmadar, R. Schmidhammer, H. Redl and S. Bahrami. 2003. Circulating S100B is increased after bilateral femur fracture without brain injury in the rat. *Br J Anaesth.* 91: 595–597.

Petzold, A. and G. Shaw. 2007. Comparison of two ELISA methods for measuring levels of the phosphorylated neurofilament heavy chain. *J Immunol. Methods.* 319: 34–40.

Petzold, A., K. Rejdak, A. Belli, J. Sen, G. Keir, N. Kitchen, M. Smith and E.J. Thompson. 2005. Axonal pathology in subarachnoid and intracerebral hemorrhage. *J Neurotrauma.* 22: 407–414.

Pike, B.R., J. Flint, J.R. Dave, X.-C.M. Lu, K.K.K. Wang, F.C. Tortella and R.L. Hayes. 2004. Accumulation of calpain and caspase-3 proteolytic fragments of brain-derived alphaII-spectrin in cerebral spinal fluid after middle cerebral artery occlusion in rats. *J Cereb Blood Flow Metab.* 24: 98–106.

Pike, B.R., J. Flint, S. Dutta, E. Johnson, K.K. Wang, R.L. Hayes, B.R. Pike, J. Flint, S. Dutta, E. Johnson, K.K. Wang and R.L. Hayes. 2001. Accumulation of non-erythroid alpha II-spectrin and calpain-cleaved alpha II-spectrin breakdown products in cerebrospinal fluid after traumatic brain injury in rats. *J Neurochem.* 78: 1297–1306.

Raabe, A. and V. Seifert. 1999. Fatal secondary increase in serum S-100B protein after severe head injury. Report of three cases. *J Neurosurg.* 91: 875–877.

Ramlackhansingh, A.F., D.J. Brooks, R.J. Greenwood, S.K. Bose, F.E. Turkheimer, K.M. Kinnunen, S. Gentleman, R.A. Heckemann, K. Gunanayagam, G. Gelosa and D.J. Sharp. 2011. Inflammation after trauma: Microglial activation and traumatic brain injury. *Ann Neurol.* 70: 374–383.

Ravenscroft, A., Y.S. Ahmed and I.G. Burnside. 2000. Chronic pain after SCI. A patient survey. *Spinal Cord.* 38: 611–614.

Redell, J.B., Y. Liu and P.K. Dash. 2009. Traumatic brain injury alters expression of hippocampal microRNAs: Potential regulators of multiple pathophysiological processes. *Journal of Neuroscience Research.* 87: 1435–1448.

Romner, B., T. Ingebrigtsen, P. Kongstad and S.E. Borgesen. 2000. Traumatic brain damage: serum S-100 protein measurements related to neuroradiological findings. *J Neurotrauma.* 17: 641–647.

Ross, S.A., R.T. Cunningham, C.F. Johnston and B.J. Rowlands. 1996. Neuron-specific enolase as an aid to outcome prediction in head injury. *Br J Neurosurg.* 10: 471–476.

Saatman, K.E., A.-C. Duhaime, R. Bullock, A.I.R. Maas, A. Valadka and G.T. Manley. 2008. Classification of Traumatic Brain Injury for Targeted Therapies. *J Neurotrauma.* 25: 719–738.

Shaw, G.J., E.C. Jauch and F.P. Zemlan. 2002. Serum cleaved tau protein levels and clinical outcome in adult patients with closed head injury. *Ann Emerg Med.* 39: 254–257.

Siman, R., C. Zhang, V.L. Roberts, A. Pitts-Kiefer and R.W. Neumar. 2005. Novel surrogate markers for acute brain damage: cerebrospinal fluid levels corrrelate with severity of ischemic neurodegeneration in the rat. *J Cereb Blood Flow Metab.* 25: 1433–1444. doi:10.1038/sj.jcbfm.9600138.

Siman, R., M. Baudry and G. Lynch. 1984. Brain fodrin: substrate for calpain I, an endogenous calcium-activated protease. *Proc Natl Acad Sci USA.* 81: 3572–3576.

Siman, R., N. Toraskar, A. Dang, E. McNeil, M. McGarvey, J. Plaum, E. Maloney and M.S. Grady. 2009. A Panel of Neuron-Enriched Proteins as Markers for Traumatic Brain Injury in Humans. *J Neurotrauma.* 26: 1867–1877.

Siman, R., T.K. McIntosh, K.M. Soltesz, Z. Chen, R.W. Neumar, V.L. Roberts, R. Siman, T.K. McIntosh, K.M. Soltesz, Z. Chen, R.W. Neumar and V.L. Roberts. 2004. Proteins released from degenerating neurons are surrogate markers for acute brain damage. *Neurobiology of Disease.* 16: 311–320.

Siman, R., V.L. Roberts, E. McNeil, A. Dang, J.E. Bavaria, S. Ramchandren and M. McGarvey. 2008. Biomarker evidence for mild central nervous system injury after surgically-induced circulation arrest. *Brain Research.* 1213: 1–11.

Sivanandam, T.M. and M.K. Thakur. 2012. Traumatic brain injury: a risk factor for Alzheimer's disease. *Neurosci Biobehav Rev.* 36: 1376–1381.

Smith, C., S.M. Gentleman, P.D. Leclercq, L.S. Murray, W.S.T. Griffin, D.I. Graham and J.A.R. Nicoll. 2012. The neuroinflammatory response in humans after traumatic brain injury. *Neuropathol Appl Neurobiol.* 39(6): 654–66.

Smith, D.H., X.H. Chen, M. Nonaka, J.Q. Trojanowski, V.M. Lee, K.E. Saatman, M.J. Leoni, B.N. Xu, J.A. Wolf and D.F. Meaney. 1999. Accumulation of amyloid beta and tau and the formation of neurofilament inclusions following diffuse brain injury in the pig. *J Neuropathol Exp Neurol.* 58: 982–992.

Suarez, J.I., R. Geocadin, C. Hall, P.D. Le Roux, S. Smirnakis, C.A.C. Wijman and O.O. Zaidat. 2012. The neurocritical care research network: NCRN. *Neurocrit Care.* 16: 29–34.

Svetlov, S.I., V. Prima, D.R. Kirk, H. Gutierrez, K.C. Curley, R.L. Hayes and K.K.W. Wang. 2010. Morphologic and biochemical characterization of brain injury in a model of controlled blast overpressure exposure. *J Trauma.* 69: 795–804.

Svetlov, S.I., V. Prima, O. Glushakova, A. Svetlov, D.R. Kirk, H. Gutierrez, V.L. Serebruany, K.C. Curley, K.K.W. Wang and R.L. Hayes. 2012. Neuro-glial and systemic mechanisms of pathological responses in rat models of primary blast overpressure compared to "composite" blast. *Front Neurol.* 3: 15.

Temple, S. 2001. The development of neural stem cells. *Nature.* 414: 112–117.

Touma, E., S. Kato, K. Fukui and T. Koike. 2007. Calpain-mediated cleavage of collapsin response mediator protein(CRMP)-2 during neurite degeneration in mice. *European Journal of Neuroscience.* 26: 3368–3381.

Varma, S., K.L. Janesko, S.R. Wisniewski, H. Bayir, P.D. Adelson, N.J. Thomas and P.M. Kochanek. 2003. F2-isoprostane and neuron-specific enolase in cerebrospinal fluid after severe traumatic brain injury in infants and children. *J Neurotrauma.* 20: 781–786.

Vlak, M.H., A. Algra, R. Brandenburg and G.J. Rinkel. 2011. Prevalence of unruptured intracranial aneurysms, with emphasis on sex, age, comorbidity, country, and time period: a systematic review and meta-analysis. *The Lancet Neurology.* 10: 11–11.

Wagner, A.K., H. Bayir, D. Ren, A. Puccio, R.D. Zafonte and P.M. Kochanek. 2004. Relationships between cerebrospinal fluid markers of excitotoxicity, ischemia, and oxidative damage after severe TBI: the impact of gender, age, and hypothermia. *J Neurotrauma.* 21: 125–136.

Wang, L.H. and S.M. Strittmatter. 1996. A family of rat CRMP genes is differentially expressed in the nervous system. *The Journal of Neuroscience.* 16: 6197–6207.

Yamazaki, Y., K. Yada, S. Morii, T. Kitahara and T. Ohwada. 1995. Diagnostic significance of serum neuron-specific enolase and myelin basic protein assay in patients with acute head injury. *Surg Neurol.* 43: 267–70.

Yao, C., A.J. Williams, A.K. Ottens, X.-C. May Lu, R. Chen, K.K. Wang, R.L. Hayes, F.C. Tortella, J.R. Dave, C. Yao, A.J. Williams, A.K. Ottens, X.-C. May Lu, R. Chen, K.K. Wang, R.L. Hayes, F.C. Tortella and J.R. Dave. 2008. Detection of protein biomarkers using high-throughput immunoblotting following focal ischemic or penetrating ballistic-like brain injuries in rats. *Brain Inj.* 22: 723–732.

Yao, C., A.J. Williams, A.K. Ottens, X.-C.M. Lu, M.C. Liu, R.L. Hayes, K.K. Wang, F.C. Tortella and J.R. Dave. 2009. P43/pro-EMAPII: a potential biomarker for discriminating traumatic versus ischemic brain injury. *J Neurotrauma.* 26: 1295–1305.

Zakharov, V.V., M.N. Bogdanova and M.I. Mosevitsky. 2005. Specific Proteolysis of Neuronal Protein GAP43 by Calpain: Characterization, Regulation, and Physiological Role. *Biochemistry (Moscow).* 70: 897–907.

Zemlan, F.P., E.C. Jauch, J.J. Mulchahey, S.P. Gabbita, W.S. Rosenberg, S.G. Speciale and M. Zuccarello. 2002. C-tau biomarker of neuronal damage in severe brain injured patients: association with elevated intracranial pressure and clinical outcome. *Brain Research.* 947: 131–139.

Zhang, Z., A.K. Ottens, S. Sadasivan, F.H. Kobeissy, T. Fang, R.L. Hayes and K.K.W. Wang. 2007. Calpain-mediated collapsin response mediator protein-1, -2, and -4 proteolysis after neurotoxic and traumatic brain injury. *J Neurotrauma.* 24: 460–472.

Zhang, Z., S. Mondello, F. Kobeissy, R. Rubenstein, J. Streeter, R.L. Hayes and K.K.W. Wang. 2011. Protein biomarkers for traumatic and ischemic brain injury: from bench to bedside. *Translational Stroke Research.* 3: 1–32.

Zhang, Z., S.F. Larner, F. Kobeissy, R.L. Hayes and K.K.W. Wang. 2010. Systems biology and theranostic approach to drug discovery and development to treat traumatic brain injury. *Methods Mol Biol.* 662: 317–329.

Ziebell, J.M. and M.C. Morganti-Kossmann. 2010. Involvement of pro-and anti-inflammatory cytokines and chemokines in the pathophysiology of traumatic brain injury. *Neurotherapeutics.* 7(1): 22–30.

6

Biomarkers for Differential Calpain Activation in Healthy and Diseased Brains: a Systematic Review

Hussam Jourdi

INTRODUCTION

Calpains refer to a family of intracellular non-lysosomal cysteine proteases. They are regulatory rather than digestive proteases because of limited substrate proteolysis. Activation of calpains at neutral pH distinguishes them from other cysteine proteases such as cathepsins. Most calpains have widespread distribution in various tissues of the body. The best-characterized isoforms μ-calpain and m-calpain are the predominant calpains in the central nervous system (CNS). At least *in vitro*, μ-calpain and m-calpain (a.k.a. calpain-1 and calpain-2) are activated by μM and nearly mM Ca^{2+} concentrations, respectively. Considering their numerous important functions in the CNS, identifying plausible biomarkers for calpain activation represents a major advancement for the diagnosis and treatment of brain diseases and for better understanding of their individual physiological roles. This is a challenging and complicated task as μ-calpain and m-calpain have overlapping substrate specificity and because the presence of substrate degradation products cannot distinguish between activation of these two calpains.

New York University, Centre for Neural Sciences, 4 Washington Place, New York, NY 10003.
Current Address: University of Balamand, Souk-El-Gharb, Lebanon.
Email: jourdi@nyu.edu

Moreover, although substrate degradation products have been used as indicators (biomarkers) of calpain activation, it has been difficult to discriminate between those degraded by physiological or pathological processes that involve either one of the two isoforms. Further, the limited substrate cleavage under physiological conditions coupled to little prevalence of degradation products renders their detection more challenging. Indeed, aside from possible quantitative differences in the magnitude of degradation of any single substrate, it remains to be determined whether any specific degradation product can be used as a biomarker to point out physiological substrate cleavage and tell it apart from pathological degradation. In light of exciting new studies revealing phosphorylation-dependent activation of calpains published in the past few years this chapter reviews the pathological and physiological roles of μ-calpain and m-calpain, discusses recent progress in the search for specific upstream and downstream biomarkers for physiological and pathological activation of each calpain and suggests new venues to explore in the brain.

Calpains are ubiquitous intracellular proteases that play critical roles in a wide range of diverse physiological and pathological conditions. More recently, it has become widely acknowledged that calpains contribute to the regulation of shape and motility in numerous cell types by partially truncating a variety of cytoskeletal proteins. In the CNS, calpains are implicated in Long-Term Potentiation (LTP), learning and memory and neurodegeneration. In contrast to other regulators of intracellular processes, calpains, being proteases, execute irreversible modifications of their substrates, making them unidirectional organizers of cellular functions.

Several excellent reviews and chapters have been published in the past decade on calpains and regulation of proteolysis of their substrates in neurons and glial cells (Liu et al., 2008; Yamashima, 2004; Yamashima and Oikawa, 2009). As most of these reviews have focused on calpain activation under pathological conditions of neuronal injury, ischemic death and neurodegeneration, particular attention will be given in this chapter to calpain regulation under physiological conditions. Equal attention will be given to upstream signaling pathways regulating calpains and to downstream pathways regulated by these proteases in neurons in order to point out potential biomarkers.

Structure of Ubiquitous Brain Calpains

Both μ- and m-calpain are heterodimeric proteins sharing a common small regulatory subunit (~28–30 kDa) encoded by the calpain 4 gene and differing in the large catalytic subunit (~80 kDa). The large subunit of calpains is organized into four domains (I, II, III and IV), whereas the small subunit contains two domains (V and VI). The composition and function of each domain is briefly described below:

1. Domain I is the N-terminus of the large subunit. Autolysis has been observed only in this region, and when it occurs, the large and small subunits dissociate.
2. Domain II is rich in cysteine and histidine, can interact with calpain substrates or with the endogenous unique calpain inhibitor, calpastatin. It also includes the catalytic domain of calpains.
3. Domain III may be involved in binding Ca^{2+} and phospholipids.
4. Domain IV is the C-terminus of the large subunit and is involved in dimer formation with domain VI. It also contains EF-hand domains, which are Ca^{2+} binding sites.
5. Domain V is the N-terminus of the small subunit. It is rich in glycine and may function as a membrane anchor.
6. Domain VI, similar to domain IV, contains EF-hand motifs and is the C-terminus of the small subunit.

Expression and Distribution of Brain Calpains

Various calpains have differential distribution in multiple tissues (Simonson et al., 1985; Sorimachi et al., 1994). In the brain, immunohistochemistry studies indicate that μ-calpain is expressed mostly in cortical and hippocampal neurons, while m-calpain expression is largely limited to glia and hippocampal interneurons (Fukuda et al., 1990; Hamakubo et al., 1986; Perlmutter et al., 1990; Siman et al., 1985). In contrast, *in situ* hybridization studies demonstrate diffuse μ-calpain mRNA throughout the entire brain, in both neurons and glia, and reveal that m-calpain mRNA is localized in distinct neuronal populations such as all the hippocampal pyramidal neurons, cortical pyramidal neurons, and cerebellar Purkinje cells (Li et al., 1996). Immunoelectron microscopy studies show that μ-calpain is associated with dendritic spines and postsynaptic densities in hippocampal neurons (Perlmutter et al., 1988). Later studies identify both μ- and m-calpain in the nuclear fraction of rabbit hippocampal homogenates and in the cytoplasm, but mostly associated with membranes of the Endoplasmic Reticulum (ER) and Golgi apparatus (Hood et al., 2003; Ostwald et al., 1994).

Although the apparent distribution of μ-calpain and m-calpain varies depending on the probing method, it is commonly agreed that their distribution is ubiquitous in the brain. More recent studies indicate differential subcellular localization of μ-calpain and m-calpain in neurons, implying distinct roles. Despite all of these studies, and possibly because of them, systematic and thorough reevaluation of m-calpain and μ-calpain distributions and expression patterns in the brain is necessary. Namely, reassessments of the regulation of calpains' mRNA expression, mRNA translation, protein distribution in different brain areas, and calpains' subcellular localization are indispensable and need to be examined in details using modern tools. This should be combined with similar analyses of the expression patterns for the protein and mRNA levels of the endogenous calpain inhibitor, calpastatin, across different brain areas as calpastatin levels may interfere with calpain activation (see next two paragraphs). Such studies are warranted in light of recent findings revealing that μ-calpain is localized to the mitochondrial intermembrane space and that its N-terminal domain acts as a mitochondrial targeting motif (Badugu et al., 2008; Garcia et al., 2005; Kar et al., 2010).

Regulation of Brain Calpain Activation

The traditional view on activation of calpain in the brain proposes two modalities: controlled activation under physiological conditions involving few calpain molecules per cell and excessive activation under pathological conditions comprising sustained Ca^{2+} overload triggering all calpain molecules (Liu et al., 2008). Calpain hyperactivation in the CNS generally correlates with severe cellular impairment and damage whereas regulated calpain activation is discrete as it contributes to various vital cellular functions including synaptic plasticity and memory formation (Figure 6.1). Although limited or physiological calpain activation may involve

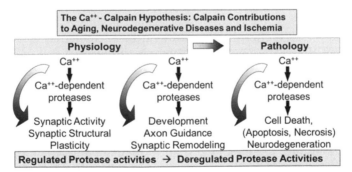

Figure 6.1. Depiction of the classical view of Ca^{++}'s role in regulating brain calpain activities in physiological and pathological conditions.

The early events in deregulation of calpains that lead to neurodegenerative diseases are not well understood. Recent findings indicate that calpains are also regulated by phosphorylation and that m-calpain is bidirectionally regulated by phosphorylation with ERK1/2-dependent phosphorylation leading to m-calpain activation and the CaMKII-PKA pathway inhibiting m-calpain activity (see text). It is possible that phosphorylation-dependent regulation of calpains contributes to pathological brain conditions as in certain brain tumors.

various substrates including structural and scaffolding proteins, enzymes, and glutamate receptor subunits, it remains largely unclear whether this changes the subcellular distribution of degraded substrates and alters their functions (see below).

Figure 6.1 depicts the classical view of physiological (regulated) calpain activation by Ca^{++} and calpain deregulation in pathological brain conditions because of excessive Ca^{++} levels. The roles of Ca^{++} and endogenous calpain inhibitors in the coordination of these proteases in neurons will be discussed in the next two paragraphs.

Calpains and Ca^{2+}

Until recently, it has been assumed that, in response to a stimulus, Ca^{2+} is released from intracellular stores, such as the Endoplasmic Reticulum (ER) and mitochondria, or it enters into the cells through plasma membrane receptors and voltage-gated Ca^{2+} channels, resulting in a rapid 10–100 fold increase of intracellular Ca^{2+} concentration and that such Ca^{2+} increases are responsible for calpain activation. The intracellular (cytosolic) concentration of Ca^{2+} is estimated to be 10^{-7} M in some studies while other reports estimate it to be 50–300 nM, several orders of magnitude lower than the extracellular Ca^{2+} concentration (Goll et al., 2003). A wide range of Ca^{2+} concentrations have been stated to activate these proteases; for half maximal activation *in vitro*, μ-calpain requires 3–50 μM Ca^{2+} and m-calpain needs 300–800 μM. As free cytoplasmic Ca^{2+} is generally maintained at less than 1 μM other mechanisms must contribute to regulate the activation of both calpains *in vivo*.

Indeed, Ca^{2+} levels needed to activate calpains (namely m-calpain) *in vitro* are much higher than would ever be encountered in living cells. Cells must therefore contain mechanisms to reduce this requirement (Friedrich, 2004; Sabatini et al., 2002). Because calpains contain Ca^{2+}-binding EF-hand motifs in domains IV and VI and because domain IV of μ-calpain and domain IV of m-calpain are different, it has been assumed that these domains are responsible for differences in the Ca^{2+} levels needed for differential activation of these two proteases; these same criteria have helped promoting the long-held dogma that calpain activation requires Ca^{2+} influx followed by auto-proteolysis. However, structural data suggest that conformational changes caused by Ca^{2+} occupancy of the EF hands alone are insufficient to properly line up the active site catalytic residues in these proteases. Furthermore, functional studies demonstrate that domain II alone exhibits Ca^{2+}-dependent protease activity and that non-EF-hand Ca^{2+}-binding sites within the protease domain act as a Ca^{2+} switch to align the catalytic subdomains IIa and IIb (Hata et al., 2001; Moldoveanu et al., 2002, 2004).

Subcellular localization of calpains is an important factor in determining their exposure to Ca^{2+} and their individual roles in neuronal cell death, synaptic plasticity and learning and memory formation. Recent findings indicate that μ-calpain is not localized in the cytoplasm or dendritic spines but rather in the intermembrane space of mitochondria implying that m-calpain is likely to play a more prominent physiological role than previously presumed based on its *in vitro* dependence on very high Ca^{2+} concentrations for activation. In the CNS, sudden non-physiological increases in intracellular Ca^{2+} levels are predictably associated with pathological conditions such as stroke. Nerve tissue damage may also result from deregulated calpain activation associated with poor control of intracellular Ca^{2+} levels and new evidence indicates that such rise in cytoplasmic Ca^{2+} only occurs under pathological conditions (Geddes and Saatman, 2010; Higuchi et al., 2005b; Liu et al., 2008). While a mild Ca^{2+} increase preferentially induces apoptotic cell death, a sudden and severe Ca^{2+} increase initiates necrosis. During stroke, the core area is instantly affected by limited blood flow leading to death by necrosis. On the contrary, many neurons undergo apoptosis in the adjacent penumbra area which is usually less affected by oxygen and nutrient deprivation (Broughton et al., 2009).

The diversity of calpain substrates implies that multiple mechanisms can exist with which calpains trigger or contribute to neuronal death (Goll et al., 2003; Liu et al., 2008). However, the precise *in vivo* substrates of calpains that directly contribute to the early events leading to ischemic neuronal necrosis remain unknown. Although technically challenging, the identification of such substrates in neurons *in vivo* is necessary to precisely understand calpain-mediated degradation events and distinguish them from later degradation events that lead to lysosomal disruption and necrotic cell death (Yamashima and Oikawa, 2009).

In summary, the magnitude of change in intracellular Ca^{2+} concentrations is apparently inadequate to elicit significant activation of m-calpain as it is generally agreed that mM Ca^{2+} concentrations cannot be reached within the cell under physiological conditions (Friedrich, 2004; Sabatini et al., 2002). Surely, alternative *in vivo* activation mechanisms for m-calpain have been proposed and demonstrated (see below).

Endogenous Calpain Inhibitors

How endogenous calpain inhibitors control the proteolytic activity of brain calpains remains relatively unresolved. This could be due to interference by other corroborating factors that modulate calpain activation, alter calpain interaction with endogenous inhibitors, or affect substrate susceptibility to degradation by calpains (Bi et al., 1998, 2000; Franco and Huttenlocher, 2005; Rong et al., 2001).

Numerous studies indicate a complex multilayered regulation of calpain activation with most studies focusing on the calpain-calpastatin system since calpastatin stood for a long time as the only known endogenous calpain inhibitor (Mellgren and Carr, 1983; Wendt et al., 2004). More recently, another endogenous inhibitor, rapsyn, has been identified. Rapsyn acts at the neuromuscular junction where it participates in acetyl choline-induced dispersion of acetyl choline receptor clusters (Chen et al., 2007). Calpastatin regulation of calpain activation is much better characterized than that of rapsyn. As calpastatin expression varies with the brain region, it is plausible, but remains to be proven, that the magnitude of calpain activation can be inversely proportional to the calpastatin expression levels but this may not be true (Nakajima et al., 2008). Systematic evaluation of calpastatin protein levels in different brain regions has been missing. Although calpastatin mRNA and protein levels need not be directly correlated, some information can be inferred from the Allen Brain Atlas, an *in situ* hybridization database of mouse brain (Allen Institute for Brain Science, Seattle, WA; http://www.brain-map.org).

Calpastatin mRNA expression levels differ with brain regions, with highest levels found in the medulla, pons and cerebellum (Nakajima et al., 2008). Should a positive correlation exist between calpastatin mRNA and protein levels, higher levels of calpastatin in these brain structures is expected to confer better protection against cell death. Unfortunately, little is known on the regulation of calpain activation and calpain functions in these brain areas. In addition, differences in calpastatin protein expression across various brain regions have not been systematically evaluated. As such, it is difficult to conclude whether calpastatin levels are directly correlated with the magnitude of calpain activation and whether this holds true under physiological conditions too. Calpastatin mRNA levels have been assessed, at least in the mouse brain, and data have been obtained from the Allen Brain Atlas and used to compare the mRNA expression levels of μ-calpain, m-calpain and calpastatin in various brain regions (Nakajima et al., 2008).

Calpastatin can undergo phosphorylation mediated by Protein Kinase A (PKA) and Protein Kinase C (PKC) and these phosphorylations affect calpastatin efficiency in inhibiting calpain and provide an additional mechanism to control calpains (Averna et al., 1999; Salamino et al., 1997). Interestingly, a study that uses a Tat-calpastatin construct to transduce primary rat cortical neurons shows that over-expression of calpastatin does not inhibit calpain activation and suggests that higher calpastatin protein levels are not necessarily associated with reduced calpain activation (Sengoku et al., 2004). In contrast, three other studies that use calpastatin over-expressing or calpastatin knock-out mice indicate that calpastatin protein levels *in vivo* are inversely correlated with calpain activation only under pathological conditions, that calpastatin deficiency exacerbates neuronal injury such as following traumatic brain injury, and that calpain inhibition reduces neuronal excitotoxic injury (Higuchi et al., 2005a,b; Schoch et al., 2012; Takano et al., 2005). Evidently, results from these three *in vivo* studies contradict the findings from the *in vitro* overexpression of Tat-calpastatin (Higuchi et al., 2005a; Takano et al., 2005; Schoch et al., 2012; Sengoku et al., 2004). The cause of this discrepancy is not clear yet, but could be due to the experimental models and methodologies used in the said studies or to dissimilarities in the *in vitro* and *in vivo* regulation of calpastatin and calpain. It is important to mention that μ-calpain and calpastatin expression levels are developmentally regulated in the brain and that this may contribute to well-known differences in brain tissue susceptibility to neurotoxicity and ischemic death at various stages of brain development (Li et al., 2009; Zhou and Baudry, 2006; Zhou et al., 2009).

Importantly, it is not known yet whether calpastatin regulates the dispersion of glutamatergic receptors in neurons. Should this be the case, the calpain-calpastatin system would be expected to regulate the redistribution and maintenance of functional glutamatergic receptors at the synapse. Under pathological conditions such regulation may exacerbate neuronal injury as it would allow continued influx of ions into the cytoplasm. Several studies indicate that this is not the case; calpastatin and various other biochemical calpain inhibitors protect neurons from cell death (Higuchi et al., 2005a,b; Takano et al., 2005).

Calpastatin has recently been found to be a key regulator of axon degeneration during development and following injury and as a downstream convergence point of distinct pathways involved in axon degeneration (Yang et al., 2013). Calpastatin depletion is detected in degenerating axons after physical injury, and maintaining calpastatin expression inhibits degeneration of transected axons *in vitro* and *in vivo*; trophic factor deprivation results in caspase-mediated calpastatin depletion *in vitro* and calpastatin depletion *in vivo* is important for the developmental pruning/elimination of axons from their target fields (Yang et al., 2013).

High molecular weight calmodulin-binding protein (HMWCaMBP) is highly homologous to calpastatin I and calpastatin II, is expressed in the mammalian heart, lung and brain, and has a calpastatin-like activity to inhibit calpains. Most work with this protein has been done in heart tissue showing decreased expression and revealing degradation during ischemia-reperfusion (Parameswaran and Sharma, 2012). This protein is phosphorylated by PKA and dephosphorylated by the phosphatase calcineurin (a.k.a. protein phosphatase 3) (Kakkar et al., 1997). The expression levels of HMWCaMBP are low in the brain. However, it is plausible that it can play significant role in the regulation of neuronal calpains. This has not been explored yet but can be an interesting venue for future research because of this protein's ability to interact with calmodulin and because it is a substrate of two major regulatory proteins involved in synaptic plasticity, PKA and calcineurin.

Degradation Products of Calpain Substrates as Biomarkers in Neurodegenerative Diseases

Proteins released from degenerating neurons can be important biomarkers for acute brain damage (Siman et al., 2004). Degradation products of calpain substrates have been detected and used in the histopathological diagnosis of various neurodegenerative diseases including Alzheimer's disease, Huntington's disease and Parkinson's disease (Cai et al., 2012; Fernández-Shaw et al., 1997; Higuchi et al., 2012; Masliah et al., 1990; Nguyen et al., 1999, 2012; Sihag and Cataldo, 1996; Vanderklsih and Bahr, 2000). Degradation products for spectrin, amyloid-β-precursor protein, α-synuclein, members of the 14-3-3 family of cytoskeletal scaffolding proteins, huntingtin, striatal-enriched protein tyrosine phosphatase (STEP), and the cyclin-dependent kinase 5 (Cdk5) regulatory protein p35 have been studied and used as biomarkers of these brain diseases (Adamec et al., 2002; Bizat et al., 2003; Foote and Zhou, 2012; Gafni and Ellerby, 2002; Gafni et al., 2004; Gladding et al., 2012; Kim et al., 2003a,b; Landles et al., 2010; Nguyen et al., 1999; Schilling et al., 2006; Southwell et al., 2011).

Although these degradation products have been widely used as biomarkers of calpain activation in countless number of studies and discussed in a large number of reviews, there has only been a very limited set of studies that validated their functions in the brain in pathological conditions and even a lesser number of studies that addressed their physiological functions, if any (Singh et al., 2012; Wu et al., 2007a,b; Xu et al., 2007; Zhou et al., 2009). A systematic approach that uses modern proteomics tools is needed to evaluate the accumulation and localization of these degradation products in the brain following injury or physiological activation of calpains. Such an approach may help draw a detailed map of substrate degradation at different points in time and may result in a better understanding of the early events, such as the role of dysfunctional mitochondria, in deregulation of calpains (Esteves et al., 2010) (Figure 6.1).

Autophagy and Calpains

Autophagy is an evolutionarily conserved catabolic process that involves invagination and degradation of cytoplasmic components through an autophagosome-lysosome pathway. It serves physiological functions as a quality control of the intracellular milieu. Impaired autophagy is implicated in a wide variety of neurodegenerative and pathological brain conditions (Foote and Zhou, 2012). When nutrients are plentiful, the PI3K-Akt signaling pathway, stimulated by growth factors, such as BDNF, affects mTORC1 (Liang, 2010). This activates a cascade of anabolic cell growth processes and contributes to local protein synthesis in neurons mainly mediated by the ribosomal S6 kinase (p70S6K) and eukaryotic translation initiation factor 4E (eIF4E)-binding protein (4EBP), which are two direct downstream substrates of mTORC1 (Briz et al., 2013; Jourdi et al., 2009b). Concomitantly, activation of the mTORC1 pathway leads to autophagy suppression.

Overexpression of α-synuclein, one of the factors that contribute to Parkinson's disease, is associated with decreased autophagosome formation (Ravikumar et al., 2010). α-Synuclein is a calpain substrate and its proteolytic processing by μ-calpain leads to formation of aggregated high molecular weight species and adoption of a β-sheet configuration that is resistant to further degradation (Dufty et al., 2007). m-Calpain and the proteasome preferentially cleave and degrade the N-terminal half of α-synuclein but not the C-terminal fragment which resists further degradation by the proteasome leading to its accumulation (Kim et al., 2006). Interestingly, α-synuclein aggregates colocalize with activated calpain in brains of patients with Parkinson's Disease (PD) and dementia with Lewy bodies (Dufty et al., 2007).

A degenerative form of epilepsy called Lafora progressive myoclonic epilepsy (MELF) is caused by mutations in a protein called laforin that has a Dual Specificity Phosphatase domain (DSP) and a Carbohydrate Binding Module (CBM). Laforin dephosphorylates glycogen and mutations in laforin lead to the accumulation of hyperphosphorylated polymers of glucose. MELF, a.k.a. LaFora disease is also associated with decreased autophagosome formation (Minassian, 2001; Turnbull et al., 2011). MELF is the most common teenage-onset neurodegenerative disease and is characterized by accumulation of polyglucosan (glucose polysaccharide), a starch/glycogen-like compound, in inclusion bodies (Minassian, 2001). Considering the regulation of glycogen synthase by glycogen synthase kinase 3 (GSK-3), whose activity is enhanced by Cdk5 via calpain-mediated cleavage of the Cdk5 regulatory/inhibitory protein p35 to p25, which then becomes an activator of Cdk5, it will be interesting to examine whether there is a causative correlation between calpain regulation of GSK-3 and accumulation of phosphorylated glycogen on one side and the inability of mutated laforin to counteract neuronal accumulation of polyglucosan on the other side.

Negative regulation of autophagy implicates μ-calpain where it operates as a switch controlling autophagy or induction of apoptosis (Liang, 2010).

Interestingly, μ-calpain depletion is coupled to increased sensitivity to apoptosis which is triggered by a number of autophagy-inducing stimuli in mammalian cells. Autophagy is impaired in cells deficient in the calpain small regulatory subunit (Capns1; 28 kDa), which interacts with the large subunits Capn1 and Capn2 (80 kDa each) of μ-calpain and m-calpain, respectively. The enhancement of lysosomal activity and long-lived protein degradation, normally occurring upon starvation, are also reduced in the absence of the calpain small regulatory subunit (Demarchi et al., 2007; Demarchi and Schneider, 2007). These studies imply that μ-calpain and m-calpain can both be implicated in autophagy. Conversely, there are indications that disassembling the autophagy machinery in dying cells requires calpains (Corazzari et al., 2012).

Recent results afford important inferences for calpains' contributions to autophagy disruption in neurodegenerative diseases. For instance, the correlation between the mTORC1 pathway and autophagy, namely the recent findings that mTORC1 is regulated by m-calpain imply that deregulation of calpains can contribute to disruption of autophagy by implicating signaling pathways that regulate mTOR (Briz et al., 2013).

Thus, deregulation of calpains to affect upstream regulators of mTORC1 activity, such as PTEN, TSC1 and TSC2, is likely to influence autophagy in neurons. In this regard, autophagy protein 5 (Atg5), an E3 ubiquitin ligase necessary for autophagy due to its role in autophagosome elongation, is a μ-calpain substrate. Atg5 cleavage by μ-calpain disrupts Atg5 interaction with Atg12, and this acts as a switch to abrogate autophagy. m-Calpain's role in autophagy is less well understood despite indications that it also can cleave Atg5 (Gordy and He, 2012). The truncated Atg5 translocates to the mitochondria, binds to the anti-apoptotic protein Bcl-X_L to participate in the induction of apoptosis (Liang, 2010). These studies indicate a critical role for autophagy proteins, namely Atg5, in the crosstalk between autophagy and apoptosis while adding calpains to the growing list of regulators of the autophagy pathway (Liang, 2010; Yousefi et al., 2006).

Peptides that specifically inhibit Atg5 degradation by calpain can be instrumental in abrogating apoptotic cell death and promoting autophagy to protect nerve cells from undergoing apoptosis. Apoptotic cell death that happens in the penumbra of ischemic brain tissue can benefit from such a strategy and significantly reduce tissue damage. Quantitation of Atg5 degradation products may serve as interesting biomarkers for autophagy and apoptosis induction in neurons. Detailed future studies should be directed at investigating this possibility while examining the regulation of Atg5 proteolysis by m-calpain in parallel to degradation of mTOR regulatory proteins.

Calpains and Regulation of the Actin Cytoskeleton

Cytoskeletal proteins and proteins involved in the regulation of actin polymerization such as Focal Adhesion Kinase (FAK), cortactin, α-actinin, and several members of the 14.3.3 family of proteins including 14.3.3ς are calpain substrates. At various places within this chapter, many of these proteins have been discussed with reference to their roles in synaptic plasticity and neurodegenerative diseases as well as their possible contributions to brain cancer metastasis (Frame et al., 2002; Perrin et al., 2006; Potter et al., 1998; Storr et al., 2011; Villa et al., 1998). The role of calpains in brain tumors will be discussed in the next paragraph. This will be followed by descriptions of alternative (i.e., other than Ca^{2+}) regulators of calpains. Together, these deliberations will be used to identify potential venues to explore for brain tumor therapy and for better understanding to the physiological functioning of calpains in neurons.

Calpains, Cell Spreading, Metastasis, and Brain Tumors

Calpains have been clearly implicated in cell adhesion and motility (Chan et al., 2010; Wells et al., 2005). Calpain substrates include several intracellular anchoring proteins and membrane-bound adhesion molecules involved in cell-cell interaction and signaling between adjacent cells; calpain-mediated degradation of these adhesion proteins and their intracellular anchoring partners can facilitate the metastasis of transformed cells (Huang et al., 2004). Calpains have been directly implicated in cancer metastasis and inhibition of calpains has been suggested as a cancer treatment (Leloup et al., 2010; Leloup and Wells, 2011; Storr et al., 2011). The metastatic dissemination of rhabdomyosarcomas, soft-tissue sarcomas commonly encountered in childhood, implicates many proteases and deregulation of μ-calpain and m-calpain activities seems to play prominent roles in the anarchic adhesion, migration and invasiveness of these cancers (Roumes et al., 2010).

Calpains control cell spreading and retraction downstream from integrin and RhoA GTPase signaling (Flevaris et al., 2007). Interestingly, phosphorylation of both μ-calpain and m-calpain regulates their activities in a cellular model of cancer cell migration and dephosphorylation of both calpains by protein phosphatase 2A (PP2A) suppresses cancer cell migration (Xu and Deng, 2004, 2006a,b). Calpain 2, the larger subunit of m-calpain, is upregulated in breast cancer and its expression correlates with increased tumor invasiveness (Libertini et al., 2005). m-Calpain also regulates the activity of non-receptor protein tyrosine phosphatase 1B (PTP1B) to modulate the activity of Src kinase and promote cancer progression and invasiveness of transformed breast cancer cells (Cortesio et al., 2008). Thus, results from several studies imply that activation of calpains can serve as a biomarker of tumor aggressiveness. Similar studies should be conducted for brain tumors to better validate the roles of calpains in their dissemination.

A recent report reveals that, in neurons, m-calpain cleaves PTEN, a phosphatase that is mutated in numerous cancers. However, it is not known whether cancer-related PTEN mutations interfere with its susceptibility to cleavage by m-calpain (Briz et al., 2013; Pianetti et al., 2001; Zhang et al., 2008). Should this be the case, calpain inhibitors may not be useful; rather, inhibitors of signaling pathways downstream from calpains should be inspected.

Recent reports implicate m-calpain in the metastasis of glioblastoma in the brain (Jang et al., 2010; Lal et al., 2012). Inhibiting tumor cell migration and invasion in the brain may reduce dissemination of metastatic cells, lower tumor neovascularization and prohibit tumor expansion. Inhibition of calpain activity and calpain knock-down approaches may become useful strategies to abrogate the development of primary tumors and formation of metastases (Lal et al., 2012; Zadran et al., 2012).

An interesting and simple model to test whether interference with calpain activity can impede brain tumor progression can be done using glioblastoma and neuroblastoma cell lines. In glioblastoma, m-calpain involvement in metastasis is mediated by the extracellular matrix metalloproteinase 2 (MMP2) (Jang et al., 2010). The signaling pathways linking MMP2 with m-calpain await future studies.

N-cadherin (a.k.a. NCAD, CDH2, or Cadherin-2), a Ca^{2+}-dependent adhesion glycoprotein that functions in presynaptic-postsynaptic adhesion is a calpain substrate. N-cadherin has been implicated in brain tumor metastasis, namely neuroblastoma, although the mechanisms involved have not been fully elucidated (Ara and DeClerck, 2006; Ramis-Conde et al., 2009). Calpain also cleaves the cytoplasmic tail of N-cadherin in brain injury (Jang et al., 2009). ErbB1, the receptor for Epidermal Growth Factor (EGF) and other EGF-like cytokines, is mutated and its activity is increased in several types of highly metastatic cancers, including high-grade diffusely infiltrative pediatric brainstem glioma (Gilbertson et al., 2003). Neuroblastomas also express ErbB1 where it is involved in the high proliferation rate of tumor cells (Ho et al., 2005). Aggressive metastatic neuroblastomas (type B neuroblastoma) express high levels of the BDNF receptor TrkB which displays high activity (Brodeur et al., 2009; Hecht et al., 2005; Nakagawara et al., 1994). Higher activation of calpains (possibly m-calpain), possibly because of excessive activation of TrkB and/or ErbB1, may provide the missing link between the high grade invasiveness of these tumors and the higher levels of expression and functioning of these growth factor receptors. We have recently shown that BDNF and EGF treatments activate neuronal m-calpain in a phosphorylation dependent manner via

MAPK/ERK and that activation of the CaMKII-PKA pathway downstream from BDNF-TrkB inhibits m-calpain activity in neurons (Jourdi et al., 2010).

These findings may have important consequences for brain cancer research. For instance, it is conceivable that higher expression levels and activity of TrkB and ErbB1 could be coupled to higher levels of m-calpain activation and degradation of adhesion molecules, leading to facilitated metastasis of transformed neuroblastoma cells (Figure 6.2). A model for m-calpain activation downstream from ErbB1 and TrkB via MAPK and inhibition by CaMKII/PKA downstream from TrkB is shown in Figure 6.2. The use of sensitive FRET assays can be an easy and rapid way to estimate calpain activation levels in neuroblastoma (and glioma) cell lines. Evidently, this should be complemented by the assessment of the extent of calpain-dependent degradation of adhesion molecules in transformed cells. In parallel, studies to estimate invasion and metastasis of the transformed cells could be done using standard *in vivo* and *in vitro* methods (DeClerck et al., 1997; Sohara et al., 2005). Unfortunately, such studies have been lacking. Calpain inhibitors or molecules that activate the PKA pathway can also be tested in a similar experimental paradigm to assess their ability to abrogate metastasis into other tissues and organs, although PKA antagonists have been shown to inhibit tumor progression in other types of cancer.

Thus, a combinatorial therapy can also be envisaged for neuroblastoma with standard anticancer drugs concomitantly administered with inhibitors of ERK1/2 and/or agonists of PKA (Brudvik et al., 2011). Indeed, there are indications that PKA pathway activation leads to differentiation of at least one neuroblastoma cell line and that the PKA pathway is inhibited following ERK activation in a different neuroblastoma cell line (Canals et al., 2005; Davis et al., 2003). Although inhibition and knock-down of calpain can interfere with anti-cancer treatments, as ubiquitous calpains play crucial roles in chemotherapy-induced cell death, it will be interesting to investigate the role of phosphorylation-dependent activation of calpains in brain tumor as this can be a useful biomarker for the aggressiveness of the tumor and its spreading potential. Taken together, there is an evident need for the reassessment of the role of individual calpains and of the signaling pathways that control them in brain tumors.

Lipids and Alternative Regulation of Activation of Calpains

Except under pathological conditions associated with cell death, such as axonal transection, neurodegeneration and tissue ischemia, levels of Ca^{2+} comparable to those required to maximally activate m-calpain *in vitro* do not exist within living cells, implying that m-calpain can only be activated

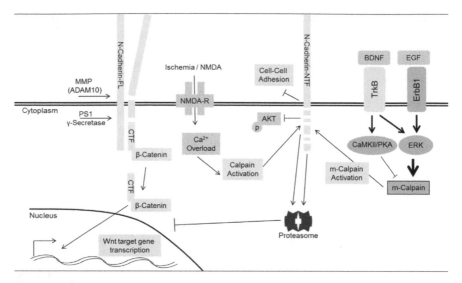

Figure 6.2. Pathways involved in m-calpain regulation and their potential implications in brain tumors exhibiting high expression levels and activation of tyrosine kinase receptors.

N-cadherine is cleaved by MMP/ADAM10 and by PS1/γ-secretase. The resulting C-terminal fragment (CTF) interacts with β-catenin and together they translocate to the nucleus to activate the transcription of genes that are Wnt targets. Ca^{2+}-dependent activation of m-calpain results in proteolysis of the C-terminal domain of N-cadherin into several fragments that ultimately get further degraded by the proteasome. N-cadherin cleavage by calpains is associated with loss of CTF/β-catenin nuclear translocation, AKT phosphorylation, and cell adhesion. It is expected that over-activation of the EGF-ErbB1 and BDNF-TrkB pathways, which have been observed in some brain tumors, can lead to over-activation of ERK and m-calpain, disruption of the CTF/β-catenin complex nuclear translocation, and loss of tumor cell adhesion. Conversely, inhibition and knock-down of ErbB1, TrkB, ERK pathway mediators, and m-calpain are expected to delay the progression of the tumor and reduce its metastasis. Considering new results on TrkB cleavage by calpains, it is plausible that truncated TrkB isoforms may promote neuroblastoma cell survival too (Gomes et al., 2012; Vidaurre et al., 2012). Inhibition of m-calpain has been postulated as a potential treatment for brain cancers. However, inhibition of calpains may also disrupt calpains roles in induction of cell death in transformed cells (see text). Investigations of combinatorial treatment paradigms with activators of the adenylate cyclase—PKA pathway given together with traditional chemotherapy treatments are warranted.

MMP, Matrix metalloproteinase; ADAM10, A disintegrin and metalloproteinase domain-containing protein 10; AKT, product of the c-AKT proto-oncogene; a.k.a. protein kinase B (a kinase similar to a viral oncoprotein originally isolated from a murine T cell lymphoma infected with the acutely-transforming retrovirus AKT8 that infected a mouse strain named Ak mice); CTF, C-terminal fragment; NTF; N-terminal fragment, BDNF, brain-derived neurotrophic factor; EGF, epidermal growth factor; ErbB1, EGF receptor; ERK, extracellular signal-regulated kinase; CaMKII, Ca^{2+}/calmodulin-dependent protein kinase; PKA, protein kinase A; PI3K, phosphatidylinositide 3-kinase; mTOR, mammalian target of rapamycin; PS1, presenilin-1; Wnt, wingless (Wg)-type and integration1 (Int1) morphogenic protein (adapted from Jang et al., 2009; Brodeur et al., 2009).

in pathological conditions. This has been at the basis of an apparent paradox that led researchers to the idea that other regulatory mechanisms can lower this requirement *in vivo* (Franco and Huttenlocher, 2005). Among these, the binding of phospholipids is reported to reduce the Ca^{2+} requirement for *in vitro* calpain activation (Arthur and Crawford, 1996; Melloni et al., 1996; Saido et al., 1992; Suzuki et al., 2004; Tompa et al., 2001). For example, activation of m-calpain is regulated by its binding to phosphatidylinositol 4,5-bisphosphate and by its membrane localization (Leloup et al., 2010).

Until recently, the *in vivo* relevance of phospholipid regulation of m-calpain activity in neurons has remained largely unknown despite indications that it could be involved in degradation of submembrane cytoskeletal proteins and in adhesion complex turnover (Franco and Huttenlocher, 2005). Exciting recent results show that m-calpain, but not μ-calpain, cleaves phosphatase and tensin homolog (PTEN), a tumor suppressor that is mutated at high frequency in a large number of cancers (Briz et al., 2013). PTEN functions as a phosphatidylinositol-3,4,5-trisphosphate (PIP3) 3-phosphatase. Therefore, m-calpain cleavage of PTEN is likely to contribute to increasing PI3K activity in neurons. Having higher levels of PIP3 is then associated with reduced PIP2 levels which reduce activation of m-calpain (Figure 6.3). Figure 6.3 illustrates a model for interactions between PI3K, PTEN and m-calpain in a feedback loop to regulate m-calpain activation. Under such paradigm, and considering the recent evidence that μ calpain is localized to the mitochondria, it is not surprising that μ-calpain does not cleave PTEN (Briz et al., 2013). Interestingly, *in vitro* treatment of total brain homogenates with μ-calpain does not cleave PTEN, suggesting differential susceptibility of PTEN to cleavage by m-calpain but not by μ-calpain (Briz et al., 2013). These results merit independent replication as they hint for the first time for a protein that is differentially cleaved by these two calpains. Should this be the case, PTEN cleavage may become a useful biomarker for differential activation of calpain isoforms in neuronal tissue.

Lipid binding to substrate can affect substrate susceptibility to calpain cleavage. For example, phosphoinositide (both the triphosphorylated and biphosphorylated phosphatidylinositides, PIP3 and PIP2, respectively) binding to α-actinin, a cytoskeletal adhesion protein, increases its susceptibility to proteolysis by μ-calpain and m-calpain. The same study indicates that regulation of α-actinin by calpains plays an important role in metastasis of glioblastoma cells (Sprague et al., 2008).

The PTEN-PI3K-AKT signaling pathway regulates the hamartin-tuberin (a.k.a. TSC1-TSC2) complex whereby AKT phosphorylates TSC2 on 2–5 sites and inhibits the TSC1-TSC2 complex (Huang and Manning,

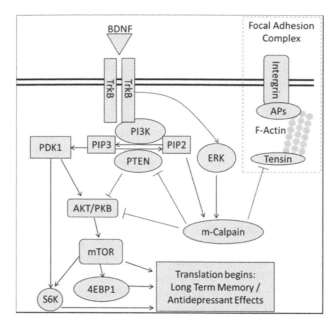

Figure 6.3. Pathways involved in m-calpain regulation of protein translation and synaptic adhesion complexes.

PTEN is a negative regulator of the PI3K/AKT and mTOR pathways and is therefore widely implicated in linking extracellular stimuli to regulation of growth processes through the control of protein synthesis. m-Calpain regulation of focal adhesion complexes in non-neuronal cells involves several proteins, among which are the GTPases RhoA, Rac1 and Cdc42, all of which have been implicated in cell migration and tumor metastasis. RhoA stimulates ROCK that phosphorylates/activates LIM kinase resulting in cofilin phosphorylation/inactivation to promote polymerization of filamentous actin (Stanyon and Bernard, 1999). Rac and Cdc42 activate p21-activated kinases (PAKs). In the brain, PAK3 is the predominant isoform. All of these proteins together with other downstream effectors, such as myosin light chain kinase (MLCK), paxillin, filamin A and cortactin, regulate actin filament stability and branching (Kumar et al., 2009). Regulation of m-calpain by phosphatidylinositol biphosphate (PIP2) is also depicted.

PDK1: (a.k.a PDPK1) Phosphoinositide-Dependent Kinase-1; Tensin: (a.k.a. TNS1) crosslinks F-Actin filaments and is a regulator of Focal Adhesion that is mediated by integrin signaling; Aps: Actin adaptor proteins (including cofilin and talin); PTEN: phosphatase and tensin homolog. PTEN and Tensin are both m-calpain substrates (Adapted from Briz et al., 2013).

2009; Tee et al., 2003). TSC2 phosphorylation seems to facilitate its cleavage by m-calpain to release the inhibitory control exerted by the TSC1-TSC2 complex on mTOR and promote protein translation (Briz et al., 2013).

m-Calpain can also be modified by sumoylation to reduce its activity; small ubiquitin-like modifier (SUMO) proteins are added at lysine residue 390 of m-calpain and this reduces m-calpain elicited functions in enhancing cell motility (Wang et al., 2009). It is possible that regulation of m-calpain sumoylation could impact similar as well as other cellular functions in the brain including those related to synaptic plasticity, synapse formation and synapse elimination (pruning) but this awaits future studies.

Regulation of protein-protein interactions, such as Acyl-CoA-binding protein interaction with calpains, is another element that alters the Ca^{2+} requirements of calpains but detailed characterization of other specific molecules and elucidation of their roles in activation of calpains will hopefully be addressed in future reports (Melloni et al., 1998, 2000a,b,c).

Regulation of Calpains by Phosphorylation

Recent evidence indicates that calpains are phosphorylated and that phosphorylation affects their activities (Glading et al., 2004; Jourdi et al., 2010; Xu and Deng, 2004, 2006a). Proteomic discovery-mode mass spectrometry data indicate that multiple phosphorylation sites exist in mouse, rat, and human calpain 1, the large subunit of μ-calpain (http://www.phosphosite.org/proteinAction.do?id=12597).

Unfortunately however, phosphorylation of μ-calpain has not been investigated in detail in the context of brain diseases and no functions have been ascribed to these phosphorylation sites yet. In addition, site specific phosphorylations of various calpain 1 residues are not well characterized and have not been used as biomarkers for specific signaling pathways, diseases or physiological functions. Based on similar mass spectrometry data, calpain 2, the large subunit of m-calpain is also phosphorylated at multiple sites (http://www.phosphosite.org/proteinAction.do?id=4844). Some of these sites have been more thoroughly investigated and implicated in fibroblast cell migration and healing from tissue injury as well as in synaptic plasticity and regulation of the actin cytoskeleton within the synapse (Glading et al., 2002, 2004; Jourdi et al., 2010; Shao et al., 2006). Phosphorylation of Ser 50, one of the best studied sites, increases m-calpain activation whereas phosphorylation of Ser369/Thr370 stabilizes m-calpain in an inactive state (Glading et al., 2004; Shiraha et al., 1999, 2002). Another report indicates that both μ- and m-calpain have at least six phosphorylation sites; two each phospho-Tyrosine, phospho-Serine, and phospho-Threonine as demonstrated in experiments with phospho-antibodies against these residues (Vazquez et al., 2008). Other mass spectrometry data identify four phosphorylation sites in the large subunit of μ-calpain, eight phosphorylation sites in the large subunit of m-calpain, one phosphorylation site in the small subunit of μ-calpain, and three phosphorylation sites in the small subunit of m-calpain but a follow up study to these findings has been missing (Vazquez et al., 2008). Interestingly, at least four of the identified phosphorylated sites occur within domain II of calpains implying that they modulate the Ca^{2+} requirements for activation of proteolysis (Vazquez et al., 2008).

As stated earlier, m-calpain phosphorylation has been better documented. Figure 6.4 illustrates a model of m-calpain regulation by phosphorylation downstream from BDNF-TrkB signaling.

A recent study has shown that m-calpain can be directly activated following treatment with Epidermal Growth Factor (EGF) via extracellular signal-regulated kinase (ERK)-mediated phosphorylation at Ser50 without intracellular Ca^{2+} elevation and that m-calpain can be equally activated by both ERK and high Ca^{2+} concentrations in a murine fibroblast cell line (Glading et al., 2004). The same is also seen in neurons whereby m-calpain activation in neurons requires ERK-mediated serine phosphorylation (Jourdi et al., 2010).

Remarkably, m-calpain can be activated in the presence of BAPTA-AM, an intracellular Ca^{2+} chelator. In contrast, phosphorylation of other amino acid residues in m-calpain can be inhibitory. For instance, cAMP-mediated activation of protein kinase A (PKA) phosphorylates m-calpain at a consensus PKA site (Ser369-Thr370) found in the m-calpain regulatory domain III to inhibit its activity (Shiraha et al., 2002). PKA and CaMKII have also been shown to inhibit neuronal m-calpain activation (Jourdi et al., 2010). Adenylate cyclase inhibitors are expected to inhibit m-calpain too because they reduce cAMP levels and limit PKA activation. Thus, phosphorylation of the PKA consensus sequence Ser360/Thr370 abrogates the Ca^{2+} dependence of m-calpain in various cell types including neurons (Jourdi et al., 2010; Shiraha et al., 2002).

Studies using amino acid substitutions in m-calpain to produce various m-calpain mutants (Ser50Asp, Ser50Glu, Ser67Glu, and Thr70Glu) show that the negatively charged glutamate and aspartate mutants have the same specific activity and Ca^{2+} requirement as the wild-type enzyme. In contrast, Ser369Asp, Ser369Glu, and Thr370Glu m-calpain mutants are inactive. These data are consistent with the notion that substitutions that mimic phosphorylation at the m-calpain PKA consensus site are inhibitory to m-calpain activity (Smith et al., 2003).

The finding that m-calpain is regulated by phosphorylation is at best unsettling to the long held dogma and arguments that Ca^{2+} is implicated in m-calpain activation under physiological conditions in neurons because m-calpain could still be activated in a phosphorylation dependent way even in the presence of BAPTA-AM, an intracellular Ca^{2+} chelator (Jourdi et al., 2010). In the same report, we have shown that PKA activity inhibits brain m-calpain reproducing similar results seen in fibroblasts (Shiraha et al., 1999, 2002). Together, these results indicate bidirectional phosphorylation-mediated regulation of m-calpain activity in neurons.

Calpains cleave and regulate the activities of several phosphatases (Briz et al., 2013; Janssens et al., 2009; Qian et al., 2011; Shimizu et al., 2007, 2010; Watkins et al., 2012; Wu et al., 2004, 2007b; Xu and Deng, 2006). However,

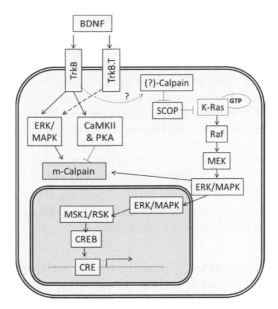

Figure 6.4. m-Calpain interaction with ERK1/2, CaMKII and SCOP.

Because ERK1/2 is critically involved in hippocampal-dependent memory, mechanisms that control ERK1/2 activation are of considerable interest. Calpain activity is also required for memory formation and its underlying synaptic plasticity. Only m-calpain has been studied for its interaction with ERK1/2 but this interaction has been controversial placing calpain upstream or downstream from ERK1/2. Recent findings show that ERK1/2 directly activates m-calpain by phosphorylating its serine 50 residue (Jourdi et al., 2010). Another study shows that an unidentified calpain, indicated as (?-Calpain in Figure 6.4), cleaves SCOP (suprachiasmatic nucleus circadian oscillatory protein, a.k.a. pleckstrin homology leucine-rich repeat protein phosphatase, PHLPP1β); SCOP is a blocker of Ras GTPase which is an upstream activator of ERK1/2. Thus, SCOP is a negative regulator of ERK1/2 signaling pathway and its cleavage leads to the activation of ERK1/2 (Shimizu et al., 2007). The controversy lies in whether ERK1/2 directly activates m-calpain or calpain(s) degrades SCOP to indirectly cause ERK1/2 activation (see text). The calpain isoform activated downstream from BDNF-TrkB that is involved in SCOP degradation has not been identified (Shimizu et al., 2007). The time at which studies have been conducted seems to unravel this controversy (Briz et al., 2013; Jourdi et al., 2010; Shimizu et al., 2007). Early and rapid activation of m-calpain can involve either rapid SCOP degradation or be independent of SCOP degradation. Conversely, delayed ERK1/2 activation is likely to be associated with calpain-mediated degradation of SCOP or other negative regulators of the ERK1/2-m-calpain pathway (Michel et al. 2012, 2013). Noteworthy, m-calpain activation downstream from ERK1/2 has not been associated with calpastatin degradation (Jourdi et al., 2010).

Truncated/short isoforms of TrkB (TrkB.T1 and TrkB.T2) have been recently isolated and they have distinct localization from the full length TrkB with respect to the synapse. Truncated TrkB isoforms seem to activate a subset of the signaling pathways typically activated by full length TrkB (Gomes et al., 2012; Vidaurre et al., 2012). This can have consequences on neuroprotection, neurotoxicity, and synaptic plasticity. Future studies should uncover the role of calpain and the mechanisms that lead to calpain activation to cleave full length TrkB. Similarly, extrasynaptic or parasynaptic localization of distinct NMDA receptor subunits and estrogen receptors may also have consequences on differential calpain activation. Dotted lines indicate pathways that have not been characterized.

dephosphorylation of calpains by phosphatases has not received much attention in the brain. Indeed, very few reports document phosphatase-dependent modulation of calpain activation in pathological brain conditions involving NMDA receptor-mediated excitoxicity, Alzheimer's disease, and brain tumors (Wu et al., 2007a; Yuen et al., 2008). Better characterization of such phosphatases awaits future studies. Such studies should address the regulation of m-calpain and μ-calpain by phosphatases and kinases in the contexts of brain tumors and the newly discovered functions ascribed physiological functions to m-calpain and its implication in synaptic plasticity (regulation of the synaptic actin cytoskeleton, local dendritic protein synthesis via the mTORC1 pathway in neurons) and its anticipated role in the regulation of the actin cytoskeleton via the mTORC2 and ERK pathways (see below; Briz et al., 2013; Huang and Manning, 2009; Huang et al., 2004, 2013; Jourdi et al., 2009b, 2010).

Phosphorylation of m-calpain as a Biomarker of ERK/MAPK Activation

ERK/MAP kinase is a major hub of rapid intracellular signaling downstream from several trophic factors, cytokines and hormones. BDNF, EGF, and estrogen (E2) are among the recently uncovered activators of m-calpain via MAPK/ERK (Jourdi et al., 2010; Zadran et al., 2009). As stated earlier, m-calpain activation has consequences on synaptic plasticity, LTP and learning and memory formation whereas deregulation of m-calpain is more relevant in neurodegenerative diseases, neuronal cell death and possibly in metastasis of brain tumors. The important finding that m-calpain is rapidly activated by MAPK-mediated phosphorylation has significance for monitoring MAPK activity in living cells and tissue.

The Fluorescence Resonance-Energy Transfer (FRET) signal that arises from the degradation of a calpain substrate consisting of a calpain-specific peptide flanked by two fluorophores exhibiting FRET and a hepta-arginine sequence to facilitate the substrate penetration across cell membranes can be used to monitor not only m-calpain activation but also MAPK activity (Jourdi et al., 2010). The advantage of this approach compared to classical methods used to monitor MAPK activity, such as Western blotting, is twofold: it allows for observing the fast activation of MAPK and m-calpain in discrete subcellular domains such in dendritic spines in living cells and it preserves the tissue and cells for further analyses (Jourdi et al., 2010).

Three reports indicate that MAPK is upstream from calpain, including two reports that m-calpain is activated by MAPK, whereas one report indicates that an unidentified calpain is upstream from MAPK signaling (Briz et al., 2013; Jourdi et al., 2010; Shimizu et al., 2007; Zadran et al., 2009).

These reports will be discussed in more details elsewhere in this chapter. Considering that MAPK/ERK signaling can be engaged repeatedly and at different time points during the early, intermediate and late phases of memory formation, monitoring of m-calpain activation using FRET or well-characterized calpain substrates can be useful biomarkers of such ERK activation waves in living neurons (Michel et al., 2012, 2013).

Synaptic Calpain Substrates

Calpain 2, the larger subunit of m-calpain, cleaves several proteins involved in the reorganization of the postsynaptic cytoskeleton including α-actinin, α and β spectrins and dystrophins. Such cleavage is expected to affect the regulation of cytoskeletal proteins and actin filament assembly regulated by focal adhesion kinase and integrins within the synapse. α-Actinin belongs to the spectrin gene superfamily which represents a diverse group of submembrane cytoskeletal proteins, including the α and β spectrins and dystrophins. It is unknown whether the degradation products/fragments of these proteins can alter cellular functions in a 2nd messenger-like fashion (Bloomer et al., 2007; Liu et al., 2008). Recent evidence indicates that this could be the case at least for other calpain substrates; β-catenin and Src kinase are both cleaved by calpains and the cleavage products participate in regulation of gene transcription and in induction of cell death, respectively. Truncation of the N-terminal domain of Src by calpain eliminates the prosurvival functions of Src (Abe and Takeichi, 2007; Guo et al., 2006; Hossain et al., 2013; Liu et al., 2002, 2008; Sridharan et al., 2006).

Various synaptic targets of calpains and focus on regulation of glutamate receptors and their scaffolding proteins in light of m-calpain's regulation by phosphorylation and their relevance to synaptic plasticity and behavioral disorders will be discussed next.

Glutamate Receptors and their Scaffolding Proteins are Calpains Substrates

Ionotropic and metabotropic glutamate receptors and several of their scaffolding proteins and interacting enzymes involved in glutamatergic signaling are calpain substrates. Among them are several subunits of α-amino-3-hydroxy-5-methyl-4-isoxazolepropionic acid (AMPA) and N-methyl d-aspartate (NMDA) receptors, metabotropic glutamate receptor 1α (mGluR1α), postsynaptic density proteins, such as postsynaptic density protein of 95 kDa (PSD-95), glutamate receptor-interacting protein 1 (GRIP1), AMPA receptor binding protein (ABP, a.k.a. GRIP2; Jourdi H., unpubl. data), and synapse associated protein of 97 kDa (SAP97) as well as enzymes, such as calcineurin, and the tumor suppressor proteins p53 and PTEN (Briz et al., 2013; Lakshmikuttyamma et al., 2004; Qin et al., 2009, 2010). These molecules are involved in important aspects of neuronal physiology and growth, such as in growth cone collapse, and in synaptic plasticity and its underlying structural and cytoskeletal reorganizations within the dendritic spine (Bi et al., 1998, 2000; Jourdi et al., 2005a,b, 2013; Rong et al., 2001).

NMDA-type glutamate receptors are critical mediators of synaptic plasticity; depending on their mode of activation and on subunit composition and localization, they can participate in Long-Term Potentiation (LTP) and Long-Term Depression (LTD) induction. These receptors are calpain substrates (Dong et al., 2004, 2006; Doshi and Lynch, 2009; Guttmann et al., 2001; Simpkins et al., 2003; Wu et al., 2007a). Importantly, developmental and cell-selective variations exist in both calpain expression and NMDA receptor subunit distribution, resulting in differential NMDA receptor degradation by calpains and disparity in their contributions to neuronal cell death or survival at least under pathological conditions (Dong et al., 2004, 2006; Doshi and Lynch, 2009; Guttmann et al., 2001; Lu et al., 2000a,b, 2001; Simonson et al., 1985; Simpkins et al., 2003; Wu et al., 2007a; Zhou and Baudry, 2006). NMDA receptor degradation by calpain under physiological conditions has not been documented.

Calpains truncate the C-terminal domains of several AMPA and NMDA glutamate receptor subunits, implying dissociation of the truncated subunits from their scaffolding proteins. These scaffolding proteins belong to a large family of molecules characterized by having multiple PDZ (PSD-95, Dlg-1, ZO1) domains and many of them are calpain substrates too (Lu et al., 2000a, 2001; Nagano et al., 1998). AMPA receptors mediate most of the fast excitatory neurotransmission in the mammalian central nervous system. Over the past two decades, several studies have implicated the trafficking of these receptors into and out of the synaptic compartment in important neurophysiological phenomena such as LTP and LTD in support of the initial hypothesis of Baudry et al. (2011).

As stated earlier, AMPA receptor interactions with distinct scaffolding PDZ proteins regulate intracellular receptor trafficking and localization. For instance, SAP97 and PSD-95 are important for targeting the receptors to synaptic membranes, Pick1 is involved in receptor internalization, and SAP97, Pick1 and GRIP1 contribute to the stabilization of receptor subunits and to receptor trafficking between different subcellular compartments (Ahmadian et al., 2004; Collingridge et al., 2004; Iwakura et al., 2001; Jourdi et al., 2003). Moreover, treatment of neurons with glutamate, NMDA, or BDNF results in translocation of AMPA receptor to the cell surface via a mechanism involving interaction of the AMPA receptor subunit GluR2 with the N-ethylmaleimide Sensitive Factor (NSF) (Braithwaite et al., 2002; Broutman and Baudry, 2001; Collingridge et al., 2004; Narisawa-Saito et al., 2002). PSD-95 and GRIP1 can be degraded by calpain and calpain-mediated proteolysis of AMPA receptor-associated PDZ proteins results in downregulation of AMPA receptor subunits (Jourdi et al., 2005a,b; Lu et al., 2000, 2001).

The finding that m-calpain can be activated in neurons following EGF and BDNF treatments is very important as it has implications for the role of m-calpain in synaptic plasticity and its associated structural changes (structural plasticity) as well as in behavioral disorders (Jourdi et al., 2010; Jourdi and Kabbaj, 2013b). For instance, BDNF is important for LTP induction and maintenance, stabilization of AMPA receptors in synaptic compartments, and for enhancing glutamatergic neurotransmission whereas EGF and members of its family, such as HB-EGF, TGFα and heregulin, reduce glutamatergic synaptic transmission by lowering the expression levels and activity of AMPA receptors and reducing the protein levels of AMPA receptor scaffolding proteins (Jourdi et al., 2003; Jourdi and Kabbaj, 2013a; Namba et al., 2006; Narisawa-Saito et al., 2002; Yokomaku et al., 2003, 2005). As such, it is highly plausible that the degradation of AMPA receptor subunits and/or their scaffolding proteins by calpain could contribute to the observed reduction in AMPA receptor activity and expression seen following treatments with basic Fibroblast Growth Factor (bFGF), EGF and EGF-like cytokines (Jourdi and Nawa, 2002; Jourdi et al., 2005a,b; Yokomaku et al., 2003, 2005). Such findings have implications for neuroprotection elicited with exposure to bFGF, EGF and EGF-related cytokines following traumatic central nervous system injury or induction of neuronal cell death (Boniece and Wagner, 1993; Casper and Blum, 1995; Maiese and Boccone, 1995; Zhou and Besner, 2010).

AMPA and NMDA receptors are phosphorylated by various kinases in their C-terminal domains, sites known to be cleaved by calpains, at least *in vitro* (Barria et al., 1997; Bi et al., 1998, 2000; Malenka et al., 1989; Mammen et al., 1997). Studies have shown that phosphorylation of these sites regulates their susceptibility to calpain-dependent cleavage.

Figure 6.5 illustrates a model where phosphorylation of the AMPA receptor subunit GluR1 by CaMKIIα reduces its cleavage by calpain. This mechanism is conceivable because CaMKII inhibits m-calpain activation and because CaMKII, similar to other kinases, phosphorylates this subunit. As phosphorylation of GluR1 by CaMKII has been associated with trafficking and synaptic targeting of this subunit, it is unlikely that CaMKII-mediated phosphorylation of GluR1 would promote its degradation to facilitate its escape and trafficking away from synaptic sites. Opposing results have also been reported (Yuen et al., 2007a,b).

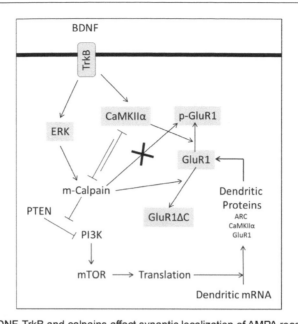

Figure 6.5. BDNF-TrkB and calpains affect synaptic localization of AMPA receptor subunits.
AMPA receptor subunit GluR1 is phosphorylated by CaMKII, PKA and PKC at Serine residues 831 and 845 (see text). GluR1 is also cleaved by calpains, though mostly under pathological conditions which mimic ischemia and excitotoxicity. Since m-calpain is activated by BDNF-TrkB signaling via ERK1/2-dependent phosphorylation, m-calpain may cleave the C-terminal domain and alter the synaptic localization of GluR1. It is likely that phosphorylation of GluR1 alters its susceptibility for cleavage by calpains. Interestingly, phosphorylation of m-calpain by CaMKII-PKA downstream from BDNF-TrkB inhibits/reduces m-calpain activation. GluR2 phosphorylation by PKC seems to have opposite effects rendering it more susceptible for cleavage by calpains (Jourdi, unpubl. results). The diagram depicts additional links between BDNF-TrkB and its regulation of GluR1 protein levels. BDNF, the NMDA receptor antagonist ketamine, and positive AMPA receptor modulators (the latter two not shown in diagram) lead to activation of mTOR-dependent mRNA translation and upregulation of GluR1 protein levels. Translation and protein levels of ARC, CaMKIIα and PSD95 have also been shown to increase in response to BDNF, ketamine and positive AMPA receptor modulator (Briz et al., 2013; Duman et al., 2012; Jourdi et al., 2009b). Timing of phosphorylation of AMPA receptor subunits and the bidirectional dynamics of m-calpain activation/inhibition downstream from BDNF-TrkB and other factors (such as EGF-ErbB1 pathway that does not engage PKA-CaMKII signaling) need to be determined in details as this will have significant impact on synaptic plasticity involving the regulation of AMPA receptor localization and trafficking.

Calmodulin, high molecular weight calmodulin-binding protein (HMWCaMBP), Ca²⁺-calmodulin kinase II (CaMKII) and CaMKIV are calpain substrates (Hajimohammadreza et al., 1997; Kosaki et al., 1983; McClelland et al., 1994; McGinnis et al., 1998; Parameswaran and Sharma, 2012). These proteins play central roles in Ca^{2+} signaling in various cell types and affect the transcription and expression of plasticity-related genes. Calmodulin, CaMKIV and CaMKII are key players in synaptic plasticity and their degradation by calpains can regulate various aspects of neuronal functioning within the nucleus, cytoplasm and synapse (Malenka et al., 1989; Tremper-Wells et al., 2005; Wang et al., 1989; Wu et al., 2007a; Wu and Lynch, 2006; Yoshimura et al., 1996).

Metabotropic glutamate receptor 1 alpha (mGluR1α) is a calpain substrate (LeBeau and Jourdi, unpubl. results) that has recently been found to play an essential role in neuronal cell death (Xu et al., 2007). The calpain-truncated mGluR1α maintains its ability to increase cytosolic Ca^{2+} while it no longer activates the neuroprotective and prosurvival PI3K-AKT signaling pathways (Xu et al., 2007). The use of a peptide that specifically inhibits mGluR1α degradation by calpain has been shown to promote neuronal cell survival (Xu et al., 2007; Zhou et al., 2009). Neuronal death in excitotoxicity has very recently been shown to also be mediated by a calpain-generated fragment of the Src protein tyrosine kinase (Hossain et al., 2013). To abrogate function-specific calpain activation, approaches that use overexpression of substrate-specific inhibitory peptides might be needed. An approach similar to the one used for mGluR1α has the advantage of targeting calpain-mediated cleavage of particular substrates involved in specific types of cellular functions. Such an approach can be tested for apoptosis induced with Atg5 degradation and excitotoxicity induced with Src cleavage (Hossain et al., 2013; Liang, 2010).

BDNF, EGF and Calpains Regulate the AMPA Receptor Compartment

As indicated earlier, m-calpain activation following EGF or BDNF treatments has important implications for synaptic plasticity and the underlying structural changes (Jourdi et al., 2010; Zadran et al., 2010). Since AMPA receptor scaffolding proteins SAP97 and GRIP1 are substrates for calpain, it is plausible that BDNF- and EGF-mediated activation of m-calpain may cause degradation of SAP97 and GRIP1. Earlier studies have proposed that GluR2-GRIP1 binding stabilizes a cytoplasmic pool of AMPA receptor subunits and that NMDA-dependent activation of calpain, which is involved in NMDA-dependent induction of LTP, causes GRIP1 cleavage to release GluR2 from its interaction with GRIP1 and promote its exocytosis and cell membrane insertion of AMPA receptors (Bednarski et al., 1995; Broutman and Baudry, 2001; Lu et al., 2001; Vanderklish et al., 1995, 1996, 2000).

BDNF treatment drives GluR1 to synapses and GluR1-SAP97 interaction is also suggested to participate in GluR1 trafficking to synaptic sites (Fortin et al., 2012; Hayashi et al., 2000). Recent findings indicate that calpain (most likely m-calpain) activation following BDNF treatment and synaptic activity degrades ankyrin repeat-rich membrane-spanning/kinase D-interacting substrate of 220 kDa (ARMS/Kidins220) and releases GluR1 from its binding to ARMS/Kidins220 (Wu and Chao, 2007; Wu et al., 2010).

Although SAP97 and GRIP1 are substrates of m-calpain, which is rapidly activated by BDNF, our published results do not display BDNF-mediated reduction in native SAP97 and GRIP1 protein levels in neurons or in human embryonic cells expressing stable levels of the BDNF receptor TrkB (HEKTrkB), whether expressed individually or together with GluR1, full length GluR2 or a GluR2 mutant (GluR2Δ5) lacking the binding domain that interacts with the PDZ proteins GRIP1, ABP/GRIP2 and Pick1 (Jourdi and Kabbaj, 2013a). Instead, prolonged BDNF treatment increases SAP97 and GRIP1 protein levels rather than causes their degradation (Jourdi et al., 2003).

In addition, our recent results argue against a role for BDNF in increasing AMPA receptor accumulation in the cytoplasmic fraction as BDNF treatment increases AMPA receptor subunits in the cell membrane fraction and at the surface of BDNF-treated neurons as evidenced with biotinylation results (Jourdi and Kabbaj, 2013a). Accordingly, BDNF-mediated stabilization/accumulation of AMPA receptor subunits is better associated with scaffolding proteins interacting with membrane-bound AMPA receptor subunits rather than with AMPA receptor subunits found in the cytoplasmic pool. As GluR1, GluR2, SAP97, PSD-95 and GRIP1 are all substrates of calpain, future studies should address the relationships between BDNF-dependent activation of m-calpain and their susceptibility to degradation by the protease following BDNF treatment. Such studies should investigate whether phosphorylation or other post-translational modifications of all of these proteins alters their susceptibility to degradation by calpain. Future studies should also be directed at determining whether various isoforms of these scaffolding proteins are differentially affected by BDNF treatment and calpain activation (Jourdi and Kabbaj, 2013a; Zheng et al., 2011).

EGF and members of its family, such as HB-EGF, TGFα and heregulin, reduce glutamatergic synaptic transmission by lowering AMPA receptor activity and expression levels (Namba et al., 2006). In agreement, EGF and similar inflammatory cytokines reverse LTP (Kaphzan et al., 2012; Kwon et al., 2005, 2008; Xu et al., 2004, 2008). Further, EGF and other members of the EGF family of cytokines reduce the expression of AMPA receptor-interacting PDZ proteins (Yokomaku et al., 2003, 2005). Since EGF can activate m-calpain and because AMPA receptor-interacting PDZ proteins are calpain substrates, it is highly plausible that EGF-induced calpain activation could result in degradation of AMPA receptor scaffolding proteins to contribute to the observed reduction in AMPA receptor activity and expression seen with EGF and EGF-like cytokines (Jourdi et al., 2005; Yokomaku et al., 2005). It is important to note that prenatal and perinatal exposure to inflammatory cytokines has been linked to the development of schizophrenia-like behavior in experimental animal models and this has been associated with reduced expression and activity of AMPA receptors in discrete brain regions (Abe et al., 2011; Buonanno et al., 2008; Eda et al., 2013; Iwakura and Nawa, 2013; Mizuno et al., 2013; Namba et al., 2006; Nawa and Yamada, 2012; Nawa et al., 2000).

It will be interesting to examine the roles of anti-inflammatory drugs on AMPA receptors and their scaffolding proteins considering that these drugs inhibit calpain activity and membrane localization of the calpain 2 protease (Silver et al., 2010).

Presynaptic Effects of Calpains

Application of leupeptin, a calpain inhibitor, to cultured hippocampal neurons reduces the frequency of miniature excitatory postsynaptic currents (mERSC), indicating a presynaptic effect of calpain on neurotransmission (Di Rosa et al., 2002). Several presynaptic proteins are found to be calpain substrates, including SNAP-25 and amphiphysin I (Ando et al., 2005; Wu et al., 2007b). SNAP-25 is a member of the soluble N-ethylmaleimide-sensitive factor attachment protein receptor (SNARE family) of proteins. Amphiphysin I is a major partner of dynamin, a protein involved in endocytosis and localized at the neck of retrieved recycling synaptic vesicles and dynamin plays a key role in clathrin-mediated endocytosis of synaptic vesicles (Wu et al., 2007b).

Furthermore, synaptic vesicle fusion is subject to calpain regulation; calpain-mediated truncation of SNAP-25 and amphiphysin I affects vesicular exocytosis and endocytosis, respectively, to affect neurotransmitter release (Grumelli et al., 2008; Wu et al., 2007b). Calpain activity partially contributes to downregulation of SNAP-25 in differentiated GABAergic cells and this effect helps in reducing epilepsy-associated hyperexcitability of neurons (Ando et al., 2005; Grumelli et al., 2008). In addition, calpain-dependent amphiphysin I cleavage attenuates kainate-induced epileptic seizures (Wu et al., 2007b).

Calpains and GABAergic Neurotransmission

Calpain activity seems to be distinct in different neuronal populations, and is apparently considerably higher in GABAergic interneurons (Ando et al., 2005). These neurons have been implicated in various brain disorders, including epilepsy, autism and Huntington's disease. Calpains roles in epilepsy and autism are less well understood. Calpain plays a role in the regulation of neuronal hyperexcitability as in epilepsy (Fujikawa, 2005; Wu and Lynch, 2006; Wu et al., 2007b). The roles played by calpains in inhibitory neurotransmission are corroborated by the fact that they cleave several molecules specifically expressed in GABAergic neurons and involved in inhibitory neurotransmission; calpains cleave the C-terminal cytoplasmic domain of GABA transporter GAT1, vesicular GABA transporter (VGAT), glutamic acid decarboxylase 65 (GAD65) and KCl cotransporter KCC2, which plays a crucial role in neuronal chloride regulation in GABAergic neurons (Baliova et al., 2009; Buddhala et al., 2012; Gomes et al., 2011; Monnerie and Le Roux, 2008; Monnerie et al., 2010; Puskarjov et al., 2012).

In addition, calpain-dependent mechanisms have lately been implicated in MAPK- and GSK3β regulation of gephyrin postsynaptic aggregation and GABAergic synaptic function (Tyagarajan et al., 2013). Epilepsy is known to predominantly affect GABAergic neurons and epilepsy induced with various treatments leads to μ-calpain activation and neuronal cell death (Gao and Geng, 2013; Lopez-Meraz and Niquet, 2009; Wang et al., 2008). In summary, most results indicate that activation of μ-calpain is pathological, impairs inhibitory neurotransmission, and seems to be mediated by excessive NMDA receptor activation (Buddhala et al., 2012; Diwakarla et al., 2009; Gomes et al., 2011; Monnerie et al., 2010; Puskarjov et al., 2012; Sha et al., 2008). There is an evident need for better elucidation of the physiological contributions of μ-calpain activation, if any, to normal inhibitory synaptic transmission.

BDNF treatment promotes the maturation of the AMPA receptor compartment in cultured GABAergic neurons, results in enhanced GABAergic neurotransmission and increases GABA release (Jourdi et al., 2003; Nagano et al., 2003; Xiong et al., 2002). Conversely, treatments with EGF receptor (a.k.a. ErbB1) ligands, such as EGF and TGFα attenuate, *in vitro* and *in vivo*, respectively, the maturation of the AMPA receptor compartment in GABAergic neurons (Nagano et al., 2007; Namba et al., 2006). m-Calpain, rather than μ-calpain, is likely to contribute to the observed opposite effects of TrkB and ErbB1 ligands on AMPA receptor expression and functioning in GABAergic neurons; BDNF-dependent activation of m-calpain is mediated by the BDNF receptor TrkB and activation of ERK leads to ERK-dependent phosphorylation of m-calpain at serine50.

The BDNF-induced activation of m-calpain via ERK is opposed by TrkB-mediated activation of the CaMKII-PKA pathway-dependent phosphorylation of m-calpain at serine369/threonine370 residues. In contrast, EGF-mediated activation of m-calpain involves ERK but does not recruit the CaMKII-PKA pathway to constrain m-calpain activity following EGF treatment (Jourdi et al., 2010). ErbB1-dependent reduction of AMPA receptor anchoring proteins is likely to contribute to AMPA receptor downregulation following treatment with ErbB1 ligands (Jourdi et al., 2005a;

Nagano et al., 2007; Yokomaku et al., 2003, 2005). BDNF and EGF treatments do not activate μ-calpain; thus, μ-calpain is unlikely to participate in the regulation of AMPA receptors' functioning and expression at least under physiological conditions.

These findings can be important for autism spectrum disorders and Huntington's disease, both of which have been associated with deficiencies in the BDNF-TrkB signaling and their downstream effectors (Castrén and Castrén, 2013). Since ErbB1 ligands exert a dominant effect over those of TrkB signaling, ErbB1 ligands and inhibition or downregulation of signaling pathways stimulated by BDNF, such ERK inhibition, may be promising tools for treating some pathophysiological aspects of autism (Namba et al., 2006; Xiong et al., 2002). Inhibition of the PI3K pathway exerts the opposite effect and abrogates BDNF-dependent AMPA receptor upregulation, possibly because PI3K is an upstream activator of mTOR signaling which contributes to upregulation of AMPA receptor expression (Jourdi et al., 2009b; Manadas et al., 2009; Xiong et al., 2002).

Intriguingly, fever has been correlated with amelioration of autistic behavior (Good, 2011, 2013). However, this has never been correlated with the possibility that the presence of higher levels of inflammatory cytokines, such as ErbB1 ligands, can have modulatory effects on signaling pathways that are deregulated in autism (Bhattacharya et al., 2012; Darnell and Klann, 2013; Park et al., 2013; Zoghbi and Bear, 2012). Similarly, amelioration of autistic symptoms with fever has not been correlated with the effects exerted by inflammatory cytokines on AMPA receptors, AMPA receptor scaffolding proteins and AMPA receptor-mediated neurotransmission within GABAergic neurons (Nagano et al., 2003, 2007; Namba et al., 2006; Yokomaku et al., 2003, 2005). Huntingtin protein is prominently expressed in GABAergic neurons and is a calpain substrate. Huntingtin is involved in intracellular trafficking of BDNF and BDNF precursors. Deregulation of calpains is thus likely to alter BDNF transport and release resulting in altered activation of downstream signaling pathways that affect synaptic plasticity and neuronal survival (Gafni et al., 2004; Kim et al., 2003; Landles et al., 2010; Reijonen et al., 2010; Schilling et al., 2006; Southwell et al., 2011).

Calpains and Long Term Potentiation

Calpain regulation of cellular motility has been compared to the expansion and retraction of dendritic spines in nerve cells and to processes that are implicated in LTP, LTD and in refining connections between neurons. Indeed, the regulation of calpain activity within the dendritic spine to enlarge and consolidate spine structures, as in LTP or to eliminate, shrink and retract dendritic spines, as in spine pruning and LTD, can be compared biochemically to the regulation of cell migration (Baudry and Bi, 2013; Briz et al., 2013; Jourdi et al., 2010).

Currently, it is widely accepted that LTP maintenance is due to modifications of AMPA receptor function and localization (trafficking) coupled with structural modifications of dendritic spines. In this capacity, and depending on results discussed earlier in this chapter, calpain activation in neuronal spines is anticipated to act as a major molecular switch controlling the number of spines and the number of receptors at each spine to affect neurotransmission. Calpain activation is required for the induction of LTP and occurs following incubation of neurons with NMDA, glutamate, and theta burst stimulation (Bednarski et al., 1995; Muller et al., 1995; Oliver et al., 1989; Vanderklish et al., 1995, 1996, 2000). LTP induction involves degradation of several calpain substrates whereas suppression of calpain expression or activity blocks LTP (Bednarski et al., 1995; Denny et al., 1990; Oliver et al., 1989; Vanderklish et al., 1996; Zadran et al., 2012).

Until very recently, there has been a persistent controversy as to which calpain isoform is more involved in LTP. In fact, a recent study implicating calpain in LTP does not indicate which calpain isoform is more relevant (Shimizu et al., 2007). The long-lasting dogma in this field has been that μ-calpain, because it requires lower Ca^{2+} concentrations to be activated, is the more relevant isoform in LTP. However, μ-calpain knock-out mice do not exhibit deficits in LTP or in fear conditioning memory (Grammer et al., 2005). In addition, μ-calpain localization to the mitochondria is another factor that argues against a role for μ-calpain in synaptic structural changes associated with LTP (Badugu et al., 2008). Two confounding factors contributing to this dogma are the requirement of m-calpain for non-physiological mM Ca^{2+} concentrations to be activated and the fact that such Ca^{2+} concentrations can only occur under pathological conditions of ischemia and cell death (Friedrich, 2004; Sabatini et al., 2002). Our recent finding that m-calpain activity is regulated by phosphorylation is important in that a physiological role has now been ascribed to m-calpain in neurons (Jourdi et al., 2010; Zadran et al., 2009, 2010, 2012). Therefore, m-calpain rather than μ-calpain is the likely candidate in playing a role in LTP.

Newly Characterized Calpain Substrates Play Important Physiological Functions in Neurons

As stated earlier, it has been a challenge to identify calpain substrates that are cleaved under physiological conditions because of the little prevalence of degradation products, because activation of calpain is brief under such conditions, and because most of these substrates have been identified following excitotoxic or ischemic injury. Too many proteins are degraded by calpains in neurons and such proteolytic processes may play critical roles in the reorganization of glutamatergic postsynaptic compartment following a number of stimuli. Recently, significant advances have been reached where several new calpain substrates have been characterized and implicated in LTP and learning and memory formation. Among these new substrates are the suprachiasmatic nucleus oscillatory protein (SCOP), striatal enriched protein tyrosine phosphatase (STEP), ankyrin repeat-rich membrane spanning (ARMS) scaffold protein (a.k.a. kinase D-interacting substrate of 220 kDa, Kidins220), hamartin (a.k.a. tuberous sclerosis protein 1, TSC1) and tuberin (a.k.a. TSC2), two negative regulators of mTOR, and phosphatase and tensin homolog deleted in chromosome 10 (PTEN) (Briz et al., 2013; Gladding et al., 2012; Gurd et al., 1999; Nguyen et al., 1999; Shimizu et al., 2007, 2010; Wu et al., 2010).

SCOP (a.k.a. pleckstrin homology leucine-rich repeat protein phosphatase, PHLPP1β) interacts with and inhibits multiple proteins important for intracellular signaling, either by directly binding to K-Ras or by dephosphorylating AKT and PKC. SCOP/PHLPP1β is one of three protein products of the same gene produced from three different start codons. SCOP is the predominant form in the brain and its structure contains the following domains: a Pleckstrin Homology (PH) domain, a Leucine-Rich Repeat (LRR), a Protein Phosphatase 2C (PP2C)-like domain, a glutamine (Q)-rich region, and a PDZ-binding domain (Shimizu et al., 2010). In the hippocampus, SCOP inhibits ERK1/2 signaling and proteolysis of SCOP by an unidentified calpain leads to activation of ERK1/2. This has been associated with learning, namely in hippocampus-dependent memory tasks (Shimizu et al., 2007). The same study shows that SCOP is cleaved by calpain when hippocampal neurons are treated with BDNF, KCl, or NMDA. Intriguingly, our recent results indicate that BDNF treatment of cortical and hippocampal neurons results in ERK1/2 activation and phosphorylation/activation of m-calpain, but not μ-calpain, in an ERK1/2-dependent way (Jourdi et al., 2010). Evidently, our results oppose those of Shimizu et al. (2007). Very recently, SCOP/PHLLP1 has been shown to be degraded by μ-calpain downstream from activation of synaptic NMDA receptor and this has been implicated in neuroprotection via the activation of prosurvival AKT and ERK1/2 (Wang et al., 2013b).

It is evident that this controversy merits further characterization to determine which event precedes ERK1/2 activation of m-calpain or calpain degradation of SCOP to lead to ERK1/2 activation. One model predicts that calpain inhibition prior to incubation with neuromodulators such as BDNF, estrogen, and AMPAkines (for example CX614) or before LTP induction should eliminate ERK1/2 activation in these experimental paradigms and interfere with early events associated with synaptic plasticity prompted by these modulators and by LTP-inducing paradigms. The same model also predicts that a signaling loop could exist whereby an unidentified calpain truncates SCOP leading to ERK1/2 activation which then phosphorylates m-calpain to further degrade SCOP and promote ERK1/2 and m-calpain activation. Importantly, recent data argue against this model and indicate that chemical inhibition of calpains and knock-down of m-calpain do not interfere with ERK1/2 activation up to 1 hour after treatments with BDNF or with a positive AMPA receptor modulator (CX614), implying that m-calpain is downstream rather than upstream from ERK1/2 (Briz et al., 2013). Considering that BDNF-dependent calpain activation and degradation of SCOP is observed by Shimizu et al. (2007) 24 hours after BDNF treatment, it is plausible to speculate that SCOP may not necessarily be involved in the initial activation of m-calpain following BDNF and CX614 treatments, and possibly soon after LTP induction. LTP induction requires BDNF release and it has been known for a while that calpain inhibition prior to LTP induction eliminates ERK1/2 activation and abrogates LTP induction (Vanderklish et al., 2000).

Conversely, another model predicts that m-calpain could play a role in LTP via a delayed activation of ERK1/2 at later time points by degrading SCOP during LTP maintenance. In this model, m-calpain is expected to play a role in the maintenance of LTP and according to this model, unopposed activation of m-calpain is anticipated to promote saturation of LTP (ceiling effect) and lead to detrimental consequences on homeostatic synaptic plasticity. Considering our previous findings that m-calpain activity can be repressed in a delayed way by BDNF itself but not by Epidermal Growth Factor (EGF), via the CaMKII-PKA pathway and the inhibitory PKA-dependent phosphorylation of m-calpain, it is highly likely that m-calpain activation needs to be restricted during later phases of LTP (such as in LTP maintenance) to counteract LTP saturation. Consequently, PKA and CaMKII inhibitors are expected to facilitate further activation of m-calpain and increase LTP ceiling.

Depending on recent results showing m-calpain as an important regulator of mTOR-dependent protein translation, it is conceivable that the protein levels of a repressor of m-calpain or ERK1/2 could be enhanced creating a negative feedback loop to limit m-calpain activity during late LTP phases. Possible candidates can be SCOP itself or calpastatin. However, previous findings dismiss upregulation of calpastatin following BDNF and CX614 treatments (Jourdi et al., 2010). It is relatively easy to examine whether m-calpain-dependent activation of mTOR and the resulting enhancement of protein translation can increase the protein levels of inhibitors of ERK1/2-m-calpain signaling, including SCOP. Should this be the case, m-calpain regulation of mTOR-dependent translation could lead to a delayed increase of SCOP protein levels or other inhibitors of ERK1/2 signaling. Accordingly, upregulation of expression of SCOP or other downregulators of ERK1/2 activity can appose m-calpain activation to limit the effects of m-calpain on translation and on other parameters of synaptic plasticity, including structural and electrophysiological ones in order to limit the LTP magnitude.

Along these lines of thoughts interesting results have recently emerged pertaining to the long known fact that spaced training is far superior compared to massed training as a fundamental feature of improved learning and memory formation (Kramár et al., 2012a; Lynch et al., 2013). The enhanced LTP seen with strong initial induction of LTP, such as with using tetanization or strong theta burst stimulation protocols to induce LTP, is reported to have detrimental effects on subsequent episodes of LTP induction, limiting the ability to augment LTP magnitude in massed stimulation protocols (up to 40 minutes between stimuli). Under such conditions, substantial and continuous (cumulative) activation of m-calpain is expected to occur because of the short duration between two successive trains (up to 40 minutes) of electrical stimuli and limit the ability of the second train of stimuli to augment LTP magnitude. A prolonged activation of m-calpain would lead to degradation of newly synthesized SCOP and/or other repressors of ERK1/2-m-calpain pathway to mitigate LTP augmentation that is seen in the spaced stimulation protocols. These possibilities warrant future investigations to better define the timing of m-calpain activation following exposure of neurons to neuromodulators (such as BDNF, estrogen and CX614) during early and late phases of synaptic plasticity such as during LTP induction and maintenance.

As stated earlier, SCOP binds and inhibits K-Ras and calpain degradation of SCOP releases this inhibition leading to activation of the Raf-MEK-ERK pathway in neurons (Shimizu et al., 2007). Interestingly, mice expressing a constitutively active H-Ras (G12V mutant mice) have increased degradation products of spectrin, a major cytoskeletal protein and one of the best studied calpain substrates. These mice exhibit a gain of function in signaling pathways involved in synaptic plasticity leading to increased rate of ocular dominance change in the visual cortex in response to monocular deprivation and in accelerated recovery from deprivation by reverse eyelid occlusion (Kaneko et al., 2010). These effects have been linked to decreased baseline presynaptic neurotransmitter release probability and an enhanced form of presynaptically-expressed LTP (Kaneko et al., 2010). The effects of this mutation on postsynaptic activation of ERK and calpain have not been fully studied. However, hippocampal samples from these mice reveal increased phosphorylation of ERK1/2 and increased abundance of calpain-cleaved spectrin degradation products implying enhanced calpain activation in these mice (H. Jourdi, M. Zhou, M. Baudry and A.J. Silva; unpubl. results). Such enhanced calpain activity is expected to alter synaptic plasticity and neuronal cell death. Unfortunately, studies that address these expectations have been lacking but may constitute an interesting venue to explore in order to better reveal the role of endogenously higher m-calpain activation in physiological and pathological conditions.

Striatal-enriched tyrosine phosphatase (STEP) is another calpain substrate that plays important roles in neurodegenerative diseases, behavioral disorders and synaptic plasticity (Braithwaite et al., 2006; Fitzpatrick and Lombroso, 2011; Goebel-Goody et al., 2012; Johnson and Lombroso, 2012; Xu et al., 2009, 2012). STEP is involved in several pathological brain disorders including Alzheimer's disease and Huntington's disease (Gladding et al., 2012; Xu et al., 2012). STEP seems to be cleaved by calpain in a Ca^{++}-dependent manner (Gladding et al., 2012; Gurd et al., 1999).

Recent findings also indicate that m-calpain cleaves STEP downstream from extrasynaptic NMDA receptor activation under *in vitro* experimental conditions that mimic excitotoxicity (Wang et al., 2013b). STEP is cleaved by calpain under pathological conditions. However, the physiological interactions between STEP and m-calpain require further elucidation. Although highly likely, it is not known yet whether physiological activation of m-calpain also leads to STEP cleavage. The same applies to whether calpain-mediated cleavage of STEP is important in synaptic plasticity (Zhang et al., 2011). Being a phosphatase, STEP substrates need further elucidation and the availability of STEP knock-out animals will be instrumental in such studies (Tashev et al., 2009). Importantly, knocking-out STEP leads to increased ERK phosphorylation (Venkitaramani et al., 2009).

It is therefore feasible that calpain-mediated cleavage of STEP, in a manner similar to the cleavage of SCOP, would also result in enhanced activation of ERK. Prospectively therefore, it is reasonable to assume that STEP plays an intermediate role between an unknown calpain and ERK phosphorylation. Future studies should endeavor to elucidate the roles of the two phosphatases STEP and SCOP and their interactions with both calpains. Such studies should give paramount importance to elucidating the timing and sequential activation of both phosphatases and compare them to the dynamics of bidirectional regulation of m-calpain activation by phosphorylation and to the activation of µ-calpain. Attention should also be given to synaptic and parasynaptic localization of m-calpain and µ-calpain and to the coupling of their upstream regulators with downstream substrates such as NMDA and AMPA receptors (Jourdi et al., 2005a; Wang et al., 2013b).

Calpains Cleave Phosphorylated p53 to Regulate Axonal Growth Cones

Axonal outgrowths (growth cones) are important for the proper development of brain circuitry and calpains control their development and motility (Qin et al., 2009, 2010). Calpains seem to regulate this process by mediating the effects of Semaphorin 3A on growth cone collapse (Qin et al.,

2010). Curiously, the proto-oncogene p53 is implicated in this phenomenon; semaphorin 3A-mediated calpain activation and growth cone collapse are associated with m-calpain phosphorylation and truncation of p53 leading to activation of RhoA kinase (ROCK)-mediated cytoskeletal reorganization.

These effects are eliminated by inhibition of MAPKK, ERK, or p38. In this paradigm the inhibitory role of semaphoring 3A to limit or reduce the number of growth cones could be overcome by using calpain inhibitors. These results are important in that they provide a signaling pathway mediated by calpain to regulate growth cones and because they may have value for brain injury (including spinal cord injury) where axonal regrowth is needed for functional recovery (Qin et al., 2010).

Calpain Regulation of Dendritic mRNA Translation and Spine Actin Polymerization

Local dendritic mRNA translation and *de novo* protein synthesis are important elements for synaptic consolidation and for maintenance of LTP. Neurotrophic signaling, namely by BDNF and its receptor TrkB, is required in learning and memory consolidation and in the underlying structural plasticity in dendritic spines (Bramham and Panja, 2014; Panja and Bramham, 2014). Dicer and eukaryotic initiation factor 2c (eIF2c, a member of the Argonaute proteins) are two proteins involved in mRNA regulation as parts of the RNA-Induced Silencing Complex (RISC) and they are enriched at postsynaptic densities in the adult mouse brain and are modified by neuronal activity in a calpain-dependent manner (Lugli et al., 2005). Interestingly, Pick1, a protein involved in AMPA receptor internalization has recently been implicated in the localization and proper functioning of argonaute 2 protein; these findings have implications for an association, and possibly a coupling, between LTD-dependent AMPA receptor internalization and LTD-induced mRNA translation (Antoniou et al., 2013).

Learning triggers BDNF-TrkB signaling and BDNF-TrkB activation facilitates memory formation (Chen et al., 2010; Lynch et al., 2008). Mutant BDNF and TrkB mice have defective LTP and memory formation and flawed BDNF-TrkB signaling impairs LTP and memory (Bramham and Messaoudi, 2005). We have recently shown that BDNF treatment of cultured primary neurons and hippocampal slices increases the activation of a major kinase pathway (the mTOR pathway) involved in new protein synthesis. We have also examined whether CX614, a positive modulator of AMPA receptors (PARM) that facilitates memory formation and enhances LTP, could also increase local dendritic mRNA translation in neurons (Jourdi et al., 2009b). mTOR and its downstream substrates p70S6K and 4EBP1 are all involved in

BDNF- and CX614-mediated stimulation of local mRNA translation. CX614 and BDNF treatments increase the expression of ARC and CaMKIIα, two proteins that play fundamental roles in synaptic plasticity and in various forms of learning and memory formation (Jourdi et al., 2009b). Regulation of actin polymerization is another essential element in LTP maintenance and synaptic consolidation (Jourdi et al., 2009b, 2010; Kramár et al., 2006; Rex et al., 2006, 2007).

In another report, we have shown that the same BDNF treatment paradigm results in m-calpain activation and triggers actin polymerization (Jourdi et al., 2010). As indicated earlier, calpain and mTOR activities are both required for LTP and memory formation (Argüelles et al., 2013; Bramham and Messaoudi, 2005; Gooney et al., 2004; Soulé et al., 2006; Vanderklish et al., 1996). BDNF and CX614 activate m-calpain as well as mTOR-mediated local protein synthesis. mTOR activation is essential in mediating the BDNF-dependent effects on memory consolidation and long term memory formation. m-Calpain regulation of mTOR affects local protein translation (Briz et al., 2013). Although these biochemical studies have fundamental implications for synaptic plasticity and memory formation, they lack electrophysiological proof to indicate that they are indeed required for synaptic plasticity and induction and maintenance of LTP and/or memory (Briz et al., 2013; Jourdi et al., 2009b, 2010).

It has recently been shown that BDNF and CX614 treatments rapidly reduce the protein levels of hamartin and tuberin (a.k.a. TSC1 and TSC2, respectively), two negative regulators of mTOR, in an m-calpain-dependent manner (Briz et al., 2013). m-Calpain activity is required for BDNF- and CX614-stimulated local protein synthesis as they both increase the phosphorylation of AKT, mTOR, p70S6K, 4EBP1 and ERK (Jourdi et al., 2009b; Jourdi unpubl. results). Calpain inhibition abrogates all of these phosphorylations except that of ERK, indicating that m-calpain is upstream from AKT, mTOR, p70S6K and 4EBP1 (Briz et al., 2013; Jourdi unpubl. data). The results also imply that ERK is upstream from calpain rather than downstream in contrast to an earlier report and in agreement with recent findings (Jourdi et al., 2010; Shimizu et al., 2007; Zadran et al., 2010).

Exciting results show that m-calpain, but not μ-calpain, cleaves PTEN, a tumor suppressor that is mutated at high frequency in a large number of cancers (Briz et al., 2013). Interestingly, *in vitro* treatment of total brain homogenates with μ-calpain does not cleave PTEN, suggesting differential susceptibility of PTEN to cleavage by m-calpain but not by μ-calpain (Briz et al., 2013).

These results merit independent replication as they hint for the first time ever for a protein that is differentially cleaved by calpain. Should this be the case, PTEN cleavage may become a biomarker for differential activation of the m-calpain isoform, but not μ-calpain, at least in neuronal tissue.

It is expected that the link between m-calpain and mTOR pathways (structural plasticity and protein translation pathways) may not be in one direction, but rather bidirectional. It has long been known that calpain knock down and application of calpain inhibitors shortly before LTP induction abolish LTP induction (Vanderklisch et al., 1996, 2000). As detailed earlier, the detrimental effects of calpain inhibitors when applied before synaptic potentiation could be mediated by abrogation of structural plasticity and be transient (i.e., they do not extend into the maintenance phase of LTP, l-LTP). PKA (and CaMKII) inhibition promotes BDNF-dependent activation of m-calpain indicating that PKA (and CaMKII) activation counteracts BDNF-dependent, ERK1/2-mediated activation of m-calpain (Jourdi et al., 2010). In view of this BDNF-elicited bidirectional regulation of m-calpain by ERK1/2 and CaMKII-PKA, application of calpain inhibitors during LTP maintenance is expected to enhance late LTP, possibly because prolonged m-calpain activation during the maintenance phase of LTP is likely to cause degradation of newly synthesized negative regulators of proteins and kinases required for LTP maintenance. Importantly, BDNF release has been implicated in long-lasting synapse formation induced by repetitive BDNF-dependent PKA activation (Taniguchi et al., 2006). Since BDNF treatment reduces m-calpain activity via a PKA-dependent phosphorylation of m-calpain, it is expected that m-calpain inhibition by PKA-dependent phosphorylation plays a role in LTP stabilization and in promoting long-lasting synapse formation and in consolidating the underlying synaptic structural changes (a.k.a. structural plasticity) (Taniguchi et al., 2006). With the same logic, the use of mTOR inhibitors such as rapamycin during LTP maintenance is expected to reduce the translation of negative regulators of the ERK1/2-m-calpain pathway. This invites the prediction that mTOR inhibition with rapamycin shortly after LTP induction, would lead to enhancement of LTP. Such experiments are relatively easy to investigate and should provide better understanding of the intricate and convoluted controls exerted by m-calpain (and possibly μ-calpain) during the different phases of LTP.

Negative regulation of protein translation, such as with genetic removal of p70S6K, which abrogates mRNA translation, corrects the molecular, synaptic, and behavioral phenotypes in a mouse model of fragile X syndrome where mRNA translation is revealed to be excessive (Bhattacharya et al., 2012; Osterweil et al., 2010). Taken together, the use of calpain inhibitors is anticipated to also abrogate the various manifestations of the fragile X phenotype.

Calpains and Regulation of Epigenetic Mechanisms

Early studies indicate that Histone 1 is a substrate of m-calpain and μ-calpain (Buki et al., 1997; Kuo et al., 1993; Kuroda et al., 1991; Mellgren, 1991).

Histones H2B, H2A, and H3 are also m-calpain substrates (Sakai et al., 1987). Other reports implicate both calpains in the regulation of H1 kinase in the nucleus and in the regulation of nuclear matrix protein turnover (Magnaghi-Jaulin et al., 2010; Mellgren, 1991; Mellgren and Rozanov, 1990; Panigrahi et al., 2011; Rajendran et al., 2011). Histone deacetylases (HDACs) are also calpain substrates that play a role in cell survival (Artus et al., 2010; Baritaud et al., 2010). HDACs remove acetyl groups from an ε-N-acetyl lysine amino acid on histones, allowing the histones to wrap the DNA more tightly; this mechanism is associated with suppression of gene transcription. Thus, calpain-mediated degradation of HDACs should abrogate HDAC-dependent suppression of gene transcription (Zhang et al., 2011). Importantly, such studies have not been replicated in neurons or in brain tumors (Lopes et al., 2009; Mataga et al., 2012).

Considering the role of epigenetics in addiction, in encoding of long term memory and in various aspects of brain behavior future studies should address in detail histone regulation by calpains in the context of epigenetic mechanisms affecting the brain, namely predispositions to neurodegenerative diseases and behavioral disorders (Duclot et al., 2013; McCarthy et al., 2013; Nestler, 2012; Schmidt et al., 2013; Wang et al., 2013a).

Phosphorylation of Substrate as a Regulator of Calpain-mediated Proteolysis

Phosphorylation of calpain substrates can promote or reduce their susceptibility to degradation by calpain. Calpastatin, the endogenous calpain inhibitor, is also a calpain substrate. Calpastatin is phosphorylated by Protein Kinase A (PKA) and Protein Kinase C (PKC) and these phosphorylations seem to affect calpastatin efficiency in inhibiting calpain and possibly its susceptibility to calpain-dependent cleavage.

These phosphorylations provide additional mechanisms that regulate the calpain-calpastatin system (Averna et al., 1999; Salamino et al., 1997). It seems that phosphorylation of calpain substrates to regulate their susceptibility to degradation by calpains applies to several proteins in neurons, including AMPA receptor subunits and protein kinase M zeta (PKMζ), a prominent player in the maintenance of memories (Hrabetova and Sacktor, 1996). For more details, see the paragraph on glutamate receptors and their scaffolding proteins being calpain substrates. Indeed, many examples have so far been elucidated in various cell types, including neurons (Briz et al., 2013; Canu et al., 1998; Flevaris et al., 2007; Hisanaga and Saito, 2003; Schilling et al., 2006; Zakharov and Mosevitsky, 2007).

Unfortunately however, very little has been done to systematically examine the correlations between substrate phosphorylation and its susceptibility to degradation within dendritic spines under physiological and pathological conditions. Such studies can be very important as they may reveal specific biomarkers of pathological versus physiological calpain activation and unravel the critical early steps involved in the deregulation of calpains (Geddes and Saatman, 2010; Jourdi et al., 2009b, 2010).

Cytoskeletal Proteins are Calpain Substrates

Calpain-mediated degradation of sub-membrane cytoskeletal proteins, namely spectrins, has been the subject of numerous studies (Czogalla and Sikorski, 2005; Jourdi et al., 2005; Liu et al., 2008; Lynch et al., 2007; Saatman et al., 2010; Vanderklish and Bahr, 2000; Yan et al., 2012). Spectrins are likely the best characterized calpain substrates and the generation of calpain-cleaved degradation products of spectrins (SBDP) has repeatedly been used as a "biomarker" of calpain activation (Wang, 2000). Indeed, a more systematic evaluation of spectrin breakdown products has recently emerged, suggesting the use of various SBDPs as biomarkers of various brain pathologies including those induced with trauma and exposure to addictive and psychoactive compounds. Interestingly, some of these studies have characterized new calpain-cleaved spectrin degradation species (Gold et al., 2009; Warren et al., 2005, 2007a,b). Importantly, calpain-mediated spectrin cleavage also occurs under physiological conditions that underlie synaptic plasticity and modulation of synaptic activity by neuromodulators including BDNF, EGF and estrogen as well as by positive modulators of AMPA receptors (Jourdi et al., 2005a,b, 2010). Under explicit experimental conditions calpain-mediated spectrin degradation can serve as a biomarker for specific signaling pathways or receptors. For example, spectrin degradation is mediated by facilitation of AMPA receptor activity and this degradation does not implicate NMDA receptors as NMDA receptor blockers do not abrogate such spectrin degradation (Jourdi et al., 2005a,b).

Under these conditions calpain-mediated spectrin degradation is abrogated by the AMPA receptor antagonist CNQX but not by the NMDA receptor antagonist AP5 (Jourdi et al., 2005b). This observation could be relevant for (and possibly play a role in) the synaptic structural reorganizations that follow glutamate stimulation of AMPA receptors but precede glutamate-dependent activation of NMDA receptors that depends on the removal of Mg^{++} ion from the NMDA receptor pore during the induction phase of LTP (Jourdi et al., 2005a).

Do Calpains Have Neuroprotective Roles?

New calpain substrates and novel regulatory mechanisms that control calpains have been identified. Some of these molecules and mechanisms provide new venues to investigate how they modulate neuronal survival and death. As stated earlier, a large number of reviews and studies provide excellent material to consult on the roles of calpains in traumatic brain injury, excitotoxicity, ischemic cell death and in various neurodegenerative diseases, including Alzheimer's disease, Parkinson's disease, Huntington's disease (Camins et al., 2006; Gladding et al., 2012; Kühl, 2004; Lai et al., 2013; Landles et al., 2010; Lee and Kim, 2006; Liu et al., 2008; Paoletti et al., 2008; Reijonen et al., 2010; Rohn, 2012; Samantaray et al., 2008; Shacka et al., 2008; Southwell et al., 2011; Vosler et al., 2008).

EGF, EGF-like cytokines and bFGF have neuroprotective effects in traumatic central nervous system injury and neuronal cell death. However, the implicated prosurvival signaling pathways have not been well characterized. Some reports indicate that they are independent of PKC, PKA, and protein synthesis while others implicate PKC in the neuroprotection provided by these peptide growth factors (Boniece and Wagner, 1993; Casper and Blum, 1995; Maiese and Boccone, 1995; Zhou and Besner, 2010). Considering that EGF activates m-calpain via ERK1/2-dependent phosphorylation and that this activation is unidirectional, rather than bidirectional as with BDNF, investigating the role of ERK1/2 and its upstream regulators in neuroprotection becomes critical. We have found that, EGF exerts neuroprotective effects to promote neuronal survival of acutely prepared and cultured hippocampal slices in *in vitro* models of ischemia (oxygen glucose deprivation). We have also found that these neuroprotective effects of EGF are completely abrogated by pre-incubation with inhibitors of calpain and ERK1/2 signaling (Jourdi unpubl. data).

Under specific experimental conditions, calpain activity can promote or abrogate neuroprotection in hippocampal and mesencephalic cultured slices depending on whether calpain inhibitor is applied prior or together with neuroprotective or neurotoxic factors (Jourdi et al., 2009a). The critical factor in determining the role of calpain in these experiments is the time of incubation. These data emphasize the role of calpain as a major switch that can determine neuronal death or survival (Jourdi et al., 2009a).

Regulation of calpain activity by the two phosphatases SCOP and STEP provides noteworthy candidates to scrutinize in the regulation of neuronal cell death (Chen et al., 2013; Gladding et al., 2012; Xu et al., 2009, 2012). As indicated earlier, SCOP (a.k.a. PHLPP1) is degraded by calpain. Remarkably, PHLPP1 gene deletion is neuroprotective (Chen et al., 2013). Thus, it is highly plausible that SCOP/PHLPP1 degradation by calpain can be neuroprotective too. STEP degradation by calpain has also been implicated in neuronal cell death (Gurd et al., 1999; Venkitaramni et al., 2009; Xu et al., 2009).

The long known opposite roles of different NMDA receptor subunits in neuronal survival and death is paramount in the field of neuroscience.

NMDA receptor subunits (namely, NR2A and NR2B) have been documented to have opposing effects on neuronal survival and death, at least at various stages of brain development and on synaptic plasticity to either induce LTP or LTD (Zhou and Baudry, 2006).

NR2A-associated activation of calpain is expected to promote survival, whereas NR2B-associated activation of calpain is more likely to promote neuronal injury. Accordingly, a differential (i.e., synaptic vs. parasynaptic) localization of calpain isoforms may play a paramount role in eliciting neuronal survival or death on one hand and LTP or LTD on the other hand. A recent study suggests that μ-calpain is localized to synaptic sites and mediates NMDA receptor-mediated neuroprotection and that m-calpain is localized to parasynaptic sites where it participates in NMDA-mediated neurotoxicity (Wang et al., 2013b). The model proposed in this study suggests that synaptically localized NR2A subunits are coupled to μ-calpain whereby μ-calpain promotes NR2A-associated neuronal survival. The same study proposes that the parsasynaptically localized NR2B subunits are coupled to m-calpain which mediated neuronal death. Unfortunately however, these claims are not supported by any piece of data to reveal the differential localization of both calpains; namely, there is a remarkable lack of any direct and irrevocable evidence for differential localization (i.e., synaptic for μ-calpain versus parasynpatic for m-calpain). It is very important to reiterate that studies that address the subcellular localization of μ-calpain and m-calpin in neurons are much needed, namely because other recent studies indicate that μ-calpain is localized to the mitochondria and has an mitochondrial localization signal peptide and that m-calpain is found in close proximity to the cell membrane (including synaptic membrane). The subcellular distribution of calpain isoforms is expected to govern their individual contributions to neuronal cell death and survival, possibly via differential recruitment of signaling pathways in close proximity to synaptic or parasynaptic sites.

Similarly, differential distribution of full length TrkB (TrkB.FL, synaptic) and truncated TrkB isoforms (TrkB.T, parasynaptic) is likely to yield differences in m-calpain (and possibly μ-calpain) activation and modulate neuronal survival and death. Importantly, excitotoxicity downregulates TrkB.FL and upregulates the truncated TrkB receptors which are found to be neuroprotective in cultured hippocampal and striatal neurons (Gomes et al., 2012). Likewise, estrogen's role in survival and plasticity is anticipated to be governed by the differential distribution of its well characterized receptors, ER2α and ER2β, and other membrane-bound estrogen receptors that have not been fully characterized.

Figure 6.6 illustrates possible convergence points in signaling pathways acutely activated by AMPA receptor potentiators, such as the AMPAkine CX614, and inhibited by NMDA receptor antagonists, such as ketamine

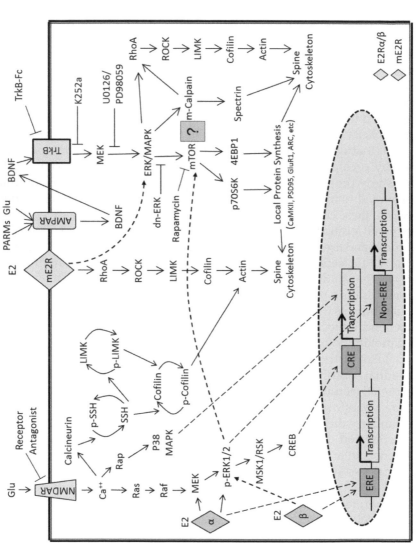

Figure 6.6. contd....

Figure 6.6. Illustration of possible convergence points in signaling pathways acutely activated by NMDA receptor antagonists and positive AMPA receptor modulators (PARMs) and their overlap with those activated downstream from BDNF-TrkB and estrogen (E2)-estrogen receptors (E2R).

Several similarities (and overlaps) exist in the signaling pathways activated following blockade of NMDA receptors and upregulation (positive modulation) of AMPA receptor activity. These pathways are critical for synaptic plasticity and for the underlying activation of the translation machinery and structural changes that occur in dendritic spines and are required for synaptic plasticity. These pathways are highly relevant for various brain diseases and behavioral disorders, including schizophrenia, autism and depression and implicate regulation of gene expression and local (dendritic) protein translation. Similarly, BDNF and E2 also activate the well-defined Ras-Raf-MEK-ERK pathway to engage gene transcription, activate the mammalian target of rapamycin (mTOR) kinase and its downstream substrates, the ribosomal proteins p70S6K and 4EBP1, and stimulate new protein synthesis (Jourdi et al., 2009b; Takei et al., 2004; Yu and Henske, 2006). BDNF and E2 treatments also activate the protease m-calpain to induce degradation of the cytoskeletal protein spectrin, activate the mTOR pathway and stimulate actin polymerization in dendritic spines (Briz et al., 2013; Jourdi et al., 2009b, 2010; Zadran et al., 2009).

E2 and BDNF treatments lead to recruitment of RhoA, ROCK, LIMK, Cofilin and the dual-specificity phosphatase Slingshot (SSH), all of which are important regulators of actin polymerization and stability within the dendritic spine cytoskeleton, so as to participate in "structural" synaptic plasticity underlying learning and memory formation (Carlisle and Kennedy, 2005; Carlisle et al., 2008; Lynch et al., 2011; Kramár et al., 2009, 2013). NMDAR activation increases intracellular Ca^{++} which, among others, leads to activation of the Ca^{++}-dependent phosphatase calcineurin. Calcineurin dephosphorylates/activates SSH which leads to dephosphorylation of two SSH substrates, LIMK and cofilin, and to abrogation of actin cytoskeletal plasticity (inhibition of actin polymerization). Treatments with NMDA receptor antagonists, such as phencyclidine and ketamine, exert fast effects and engage many of the signaling molecules that are activated by BDNF-TrkB and E2-E2R. For instance, the fast antidepressant effects of ketamine, an NMDA receptor antagonist, are mediated by activation of mTOR and some of its downstream substrates, leading to enhanced expression of important proteins required for synaptic plasticity such as the AMPA receptor subunit GluR1 (a.k.a. GluA1) and postsynaptic density scaffolding protein PSD95. The acute blockade of NMDAR by ketamine leads to reduced activation of calcineurin and this in turn increases phosphorylation/inhibition of cofilin to abrogate its actin-severing function and promote actin polymerization; thus, ketamine blockade of NMDAR is expected to promote actin polymerization (Carlisle and Kennedy, 2005; Carlisle et al., 2008). Acute treatment with PARMs leads to increased BDNF release and stimulates the mTOR (and ERK) signaling pathway(s) to increase local protein translation (Jourdi et al., 2009b). In addition, several reports indicate that PARMs stimulate actin polymerization by engaging several of its upstream regulatory proteins (Baudry et al., 2012; Kramár et al., 2012b, 2013; Simmons et al., 2009). The described pathways make for convergence points as the basis of synaptic plasticity and are essential for the proper functioning of the synapse and the brain. Disruption of these pathways has been implicated in various brain diseases and behavioral disorders. The dynamic nature of these pathways merits investigating their roles in real-time conditions as with *in vivo* recordings of spatial learning and cognitive discriminative tasks.

and phencyclidine. Disruption of these pathways has been implicated in numerous brain diseases and behavioral disorders including depression (Jourdi and Kabbaj, 2013b). These pathways are remarkably similar to those activated downstream from BDNF-TrkB and estrogen (E2)-estrogen receptors (E2R) where they play important roles in neuronal survival as well as in synaptic plasticity associated with learning and with antidepressant drug treatments. Recent advances in studying physiological functions of m-calpain place it as prominent hub at the intersection of major intracellular signaling pathways that affect multiple aspects of synaptic physiology and pathology. Interestingly, additional calpain-mediated mechanisms exist and control cell survival. For example, calpain activation by hepatitis C virus proteins impedes the extrinsic apoptotic signaling pathway (Simonin et al., 2009). It will be interesting to examine the modality of virus-dependent calpain activation in the brain and how this can be employed to abrogate neuronal death and alter other aspects of brain physiology and behavior.

Future Directions

New methods and tools have been developed to better evaluate calpains' roles *in vivo* and *in vitro*. Mutant animals have also been made available to evaluate the developmental, pathological and physiological contributions of calpains (Bartoli et al., 2006; Stockholm et al., 2005). Novel tools can now assess calpain activity within seconds of stimulation (Jourdi et al., 2010). Considering the important roles played by calpains in the brain, the utilization of these tools continues to be largely inadequate. It remains unclear how and why the tight control over calpains is lost and such tools could significantly improve our understanding of calpains functioning because factors that contribute to calpain deregulation need to be elucidated in much detail. The role of m-calpain in the regulation of LTD-associated protein translation is yet to be examined and this line of research merits attention because it underlies specific aspects and modalities of learning and contributes to the proper maturation and functioning of neuronal circuits. The role of calpains in GABAergic neurotransmission and the accompanying synaptic plasticity also requires significant elaboration. It is also anticipated that m-calpain-dependent regulation of mTORC1 and mTORC2 could affect behaviors distinct from learning and memory formation such as addiction and depression. Resolution of the role of degradation products is also needed for better appreciation of their possible roles as 2nd messengers that alter intracellular signaling pathways and possibly gene expression too (Liu et al., 2008; Lai et al., 2013).

Recent elucidation of mechanisms that regulate the activation of calpains under physiological conditions constitutes major advancement that should motivate further evaluation of the Ca^{2+}-versus phosphorylation-dependent regulation of calpains. However, stringent control of experimental conditions is essential and much needed to validate claims made depending on limited sets of results; one such example is PTEN degradation by m-calpain but not μ-calpain (Briz et al., 2013). For example, these findings are published without any assessment or monitoring of the levels of PIP2 and PIP3; PIP2 and PIP3 alter m-calpain activation and monitoring of their levels should accompany studies that aim at evaluating the subcellular distribution of these two proteases in designated experimental conditions as well as studies that address differential cleavage of substrates by the two calpain isoforms.

The lack of isoform-specific calpain inhibitors continues to be challenging to endeavors for better understanding of calpains' physiological and pathological roles (Donkor, 2000). Earlier studies have systematically examined and characterized various protease inhibitors and attempted to characterize isoform-specific calpain inhibitors (Powers et al., 2002). Calpain 1- and calpain 2-selective inhibitors have already been identified, showing 3–7 fold differences in specificity for individual isoforms, but very little has been done with them (Ovat et al., 2010). Unfortunately, the *in vivo* usefulness of these inhibitors and their derivatives against brain injury or for elucidation of calpain isoform-specific functions have yet to be done because little attention has been given to them and because they have never been sold or made commercially available for research purposes. It will be very important to revisit these earlier findings and make better use of them in future studies carried out under various experimental paradigms to better evaluate their usefulness in eliciting isoform-specific inhibition of calpain 1 (μ-calpain) and calpain 2 (m-calpain). Most importantly, it will also be essential to evaluate their functions in the living brain and examine their ability to modulate learning, synaptic plasticity and alter the progression of neurodegenerative diseases, cognitive impairment and behavioral disorders.

The utilization of new techniques and tools, such as high-throughput strategies, is expected to generate substantial amount of data. However, the value of such data sets should not be overestimated and must always be coupled with electrophysiological and behavioral studies to confirm their relevance because selective cleavage of various calpain substrates is likely to play pivotal roles in the regulation of various aspects of brain functions under different conditions. Also, a systematic evaluation of major players in synaptic functions, such as AMPA and NMDA receptors, is needed because these receptors are themselves calpain substrates (Baudry et al., 2011; Briz et al., 2013; Jourdi et al., 2010; Oertner and Matus, 2005).

Conclusions

While a large body of work supports a critical role for calpain(s) in LTP formation and in neurotoxicity, it has been extremely difficult to provide definitive evidence and detailed information regarding the mechanisms of calpain activation and the nature of their critical substrate(s) in the cascades of events leading to stable LTP and to neuronal death. As a result, controversies over the physiological involvement of calpains in LTP have persisted for a very long time and these have generally caused calpains to be overlooked in most reviews that discuss synaptic mechanisms and functions. Similarly, these controversies are at the basis of absence of research addressing the pro-survival and neuroprotective roles for calpains. It is evident that much remains to be done despite very encouraging recent strides. Likewise, it is clear that calpains, their regulators and their substrates make for excellent biomarker candidates but better elucidation of their functions and significance is required. For instance, a systematic evaluation of their expression and accumulation profiles under physiological and pathological neuronal conditions merits more attention.

Recent advances in neuroscience and biomedical research ascribes important new functions for calpains in the brain. New tools have been developed, but they have yet to be fully employed in efficient and organized efforts to unravel the intricacies of calpains as regulators of major cellular functions and as substrates of upstream signaling pathways under various physiological and pathological brain conditions. Based on recent advances, it also is clear that m-calpain stands out as a major regulator of important physiological functions in neurons and can be placed at the intersection of major signaling pathways that control structural plasticity and local protein translation in dendrites, both of which are essential elements in synaptic plasticity and in learning and memory. These mechanisms are disrupted or deregulated in various pathological conditions but the mechanisms leading to deregulation of calpains remain poorly understood. Better understanding of mechanisms that lead to such deregulation is therefore essential for future advances in this field and for better elucidation of calpains as important players in neuroprotection and neuronal survival. The same can be extended to the need for better evaluation and utilization of calpains, calpains' regulators and calpains' substrates as biomarkers for brain health and disease.

This chapter constitutes an effort to better situate the physiological roles of calpains depending on recent findings. Several possible research topics and new venues to explore have been alluded to or discussed at numerous instances throughout the chapter. It is evident that research on calpains needs much concerted efforts to better understand the physiological and pathological functions played by these proteases in the brain. It is hoped that the recent advances and findings would be at the start of a new lasting revival of interest in calpains.

References

Abe, Y., H. Namba, T. Kato, Y. Iwakura and H. Nawa. 2011. Neuregulin-1 signals from the periphery regulate AMPA receptor sensitivity and expression in GABAergic interneurons in developing neocortex. *J Neurosci.* 31(15): 5699–709.

Abe, K. and M. Takeichi. 2007. NMDA-receptor activation induces calpain-mediated beta-catenin cleavages for triggering gene expression. *Neuron.* 53: 387–397.

Adamec, E., P. Mohan, J.P. Vonsattel and R.A. Nixon. 2002. Calpain activation in neurodegenerative diseases: confocal immunofluorescence study with antibodies specifically recognizing the active form of calpain 2. *Acta Neuropathol.* 104(1): 92–104.

Ahmadian, G., W. Ju, L. Liu, M. Wyszynski, S.H. Lee, A.W. Dunah, C. Taghibiglou, Y. Wang, J. Lu, T.P. Wong, M. Sheng and Y.T. Wang. 2004. Tyrosine phosphorylation of GluR2 is required for insulin-stimulated AMPA receptor endocytosis and LTD. *EMBO J.* 23(5): 1040–50.

Ando, K., Y. Kudo and M. Takahashi. 2005. Negative regulation of neurotransmitter release by calpain: a possible involvement of specific SNAP-25 cleavage. *J Neurochem.* 94(3): 651–8.

Antoniou, A., L.M.R. Cockbill and J.G. Hanley. 2013. A role for pick1 in argonaute2 localization and function. Program No. 706.10/E12. *Neuroscience 2013 Abstracts.* San Diego, CA: Society for Neuroscience. Online.

Ara, T. and Y.A. DeClerck. 2006. Mechanisms of invasion and metastasis in human neuroblastoma. *Cancer Metastasis Rev.* 25(4): 645–57. Review.

Argüelles, S., S. Camandola, E.R. Hutchison, R.G. Cutler, A. Ayala and M.P. Mattson. 2013. Molecular control of the amount, subcellular location, and activity state of translation elongation factor 2 in neurons experiencing stress. *Free Radic Biol Med.* 61(C): 61–71.

Arthur, J.S. and C. Crawford. 1996. Investigation of the interaction of m-calpain with phospholipids: calpain-phospholipid interactions. *Biochim Biophys Acta.* 1293(2): 201–6.

Artus, C., H. Boujrad, A. Bouharrour, M.N. Brunelle, S. Hoos, V.J. Yuste, P. Lenormand, J.C. Rousselle, A. Namane, P. England, H.K. Lorenzo and S.A. Susin. 2010. AIF promotes chromatinolysis and caspase-independent programmed necrosis by interacting with histone H2AX. *EMBO J.* 29(9): 1585–99.

Averna, M., R. De Tullio, F. Salamino, E. Melloni and S. Pontremoli. 1999. Phosphorylation of rat brain calpastatins by protein kinase C. *FEBS Lett.* 450(1-2): 13–6.

Badugu, R., M. Garcia, V. Bondada, A. Joshi and J.W. Geddes. 2008. N terminus of calpain 1 is a mitochondrial targeting sequence. *J Biol Chem.* 283(6): 3409–17.

Baliova, M., A. Knab, V. Franekova and F. Jursky. 2009. Modification of the cytosolic regions of GABA transporter GAT1 by calpain. *Neurochem Int.* 55(5): 288–94.

Baritaud, M., H. Boujrad, H.K. Lorenzo, S. Krantic and S.A. Susin. 2010. Histone H2AX: The missing link in AIF-mediated caspase-independent programmed necrosis. *Cell Cycle.* 9(16): 3166–73. Review.

Barria, A., V. Derkach and T. Soderling. 1997. Identification of the Ca²⁺/calmodulin-dependent protein kinase II regulatory phosphorylation site in the alpha-amino-3-hydroxyl-5-methyl-4-isoxazole-propionate-type glutamate receptor. *J Biol Chem.* 272(52): 32727–30.

Bartoli, M., N. Bourg, D. Stockholm, F. Raynaud, A. Delevacque, Y. Han, P. Borel, K. Seddik, N. Armande and I. Richard. 2006. A mouse model for monitoring calpain activity under physiological and pathological conditions. *J Biol Chem.* 281(51): 39672–80.

Baudry, M. and X. Bi. 2013. Learning and memory: an emergent property of cell motility. *Neurobiol Learn Mem.* 104: 64–72.

Baudry, M., X. Bi and C. Aguirre. 2013. Progesterone-estrogen interactions in synaptic plasticity and neuroprotection. *Neuroscience.* 239: 280–94. Review.

Baudry, M., X. Bi, C. Gall and G. Lynch. 2011. The biochemistry of memory: The 26year journey of a 'new and specific hypothesis'. *Neurobiol Learn Mem.* 95(2): 125–33.

Bednarski, E., P. Vanderklish, C. Gall, T.C. Saido, B.A. Bahr and G. Lynch. 1995. Translational suppression of calpain I reduces NMDA-induced spectrin proteolysis and pathophysiology in cultured hippocampal slices. *Brain Res.* 694(1-2): 147–57.

Bhattacharya, A., H. Kaphzan, A.C. Alvarez-Dieppa, J.P. Murphy, P. Pierre and E. Klann. 2012. Genetic removal of p70 S6 kinase 1 corrects molecular, synaptic, and behavioral phenotypes in fragile X syndrome mice. *Neuron.* 76(2): 325–37.

Bi, R., X. Bi and M. Baudry. 1998. Phosphorylation regulates calpain-mediated truncation of glutamate ionotropic receptors. *Brain Res.* 797(1): 154–8.

Bi, X., R. Bi and M. Baudry. 2000. Calpain-mediated truncation of glutamate ionotropic receptors. Methods for studying the effects of calpain activation in brain tissue. *Methods Mol Biol.* 144: 203–17.

Bizat, N., J.M. Hermel, F. Boyer, C. Jacquard, C. Créminon, S. Ouary, C. Escartin, P. Hantraye, S. Kajewski and E. Brouillet. 2003. Calpain is a major cell death effector in selective striatal degeneration induced *in vivo* by 3-nitropropionate: implications for Huntington's disease. *J Neurosci.* 23(12): 5020–30.

Bloomer, W.A., H.M. VanDongen and A.M. VanDongen. 2007. Activity-regulated cytoskeleton-associated protein Arc/Arg3.1 binds to spectrin and associates with nuclear promyelocytic leukemia (PML) bodies. *Brain Res.* 1153: 20–33.

Boniece, I.R. and J.A. Wagner. 1993. Growth factors protect PC12 cells against ischemia by a mechanism that is independent of PKA, PKC, and protein synthesis. *J Neurosci.* 13(10): 4220–8.

Braithwaite, S.P., S. Paul, A.C. Nairn and P.J. Lombroso. 2006. Synaptic plasticity: one STEP at a time. *Trends Neurosci.* 29(8): 452–8. Review.

Braithwaite, S.P., H. Xia and R.C. Malenka. 2002. Differential roles for NSF and GRIP/ABP in AMPA receptor cycling. *Proc Natl Acad Sci USA.* 99(10): 7096–101.

Bramham, C.R. and E. Messaoudi. 2005. BDNF function in adult synaptic plasticity: the synaptic consolidation hypothesis. *Prog Neurobiol.* 76(2): 99–125. Review.

Bramham, C.R. and D. Panja. 2014. BDNF regulation of synaptic structure, function, and plasticity. *Neuropharmacology.* 76(Pt C): 601–2. doi:10.1016/j.neuropharm.2013.08.012.

Briz, V., Y.T. Hsu, Y. Li, E. Lee, X. Bi and M. Baudry. 2013. Calpain-2-mediated PTEN degradation contributes to BDNF-induced stimulation of dendritic protein synthesis. *J Neurosci.* 33(10): 4317–28.

Brodeur, G.M., J.E. Minturn, R. Ho, A.M. Simpson, R. Iyer, C.R. Varela, J.E. Light, V. Kolla and A.E. Evans. 2009. Trk receptor expression and inhibition in neuroblastomas. *Clin Cancer Res.* 15(10): 3244–50.

Broughton, B.R., D.C. Reutens and C.G. Sobey. 2009. Apoptotic mechanisms after cerebral ischemia. *Stroke.* 40(5): e331–9.

Broutman, G. and M. Baudry. 2001. Involvement of the secretory pathway for AMPA receptors in NMDA-induced potentiation in hippocampus. *J Neurosci.* 21(1): 27–34.

Brudvik, K.W., J.E. Paulsen, E.M. Aandahl, B. Roald and K. Taskén. 2011. Protein kinase A antagonist inhibits β-catenin nuclear translocation, c-Myc and COX-2 expression and tumor promotion in Apc(Min/+) mice. *Mol Cancer.* 10: 149. doi:10.1186/1476-4598-10-149.

Buddhala, C., M. Suarez, J. Modi, H. Prentice, Z. Ma, R. Tao and J.Y. Wu. 2012. Calpain cleavage of brain glutamic acid decarboxylase 65 is pathological and impairs GABA neurotransmission. *PLoS One.* 7(3): e33002.

Buki, K.G., P.I. Bauer and E. Kun. 1997. Isolation and identification of a proteinase from calf thymus that cleaves poly(ADP-ribose) polymerase and histone H1. *Biochim Biophys Acta.* 1338(1): 100–6.

Buonanno, A., O.B. Kwon, L. Yan, C. Gonzalez, M. Longart, D. Hoffman and D. Vullhorst. 2008. Neuregulins and neuronal plasticity: possible relevance in schizophrenia. *Novartis Found Symp.* 289: 165–77; discussion 177–9, 193–5. Review.

Cai, Y., H.X. Zhu, J.M. Li, X.G. Luo, P.R. Patrylo, G.M. Rose, J. Streeter, R. Hayes, K.K. Wang, X.X. Yan and A. Jeromin. 2012. Age-related intraneuronal elevation of αII-spectrin breakdown product SBDP120 in rodent forebrain accelerates in 3×Tg-AD mice. *PLoS One.* 7(6): e37599.

Camins, A., E. Verdaguer, J. Folch, A.M. Canudas and M. Pallàs. 2006. The role of CDK5/P25 formation/inhibition in neurodegeneration. *Drug News Perspect.* 19(8): 453–60. Review.

Canals, M., E. Angulo, V. Casadó, E.I. Canela, J. Mallol, F. Viñals, W. Staines, B. Tinner, J. Hillion, I. Agnati, K. Fuxe, S. Ferré, C. Lluis and R. Franco. 2005. Molecular mechanisms involved in the adenosine A and A receptor-induced neuronal differentiation in neuroblastoma cells and striatal primary cultures. *J Neurochem.* 92(2): 337–48.

Canu, N., L. Dus, C. Barbato, M.T. Ciotti, C. Brancolini, A.M. Rinaldi, M. Novak, A. Cattaneo, A. Bradbury and P. Calissano. 1998. Tau cleavage and dephosphorylation in cerebellar granule neurons undergoing apoptosis. *J Neurosci.* 18(18): 7061–74.

Carlin, R.K., D.C. Bartelt and P. Siekevitz. 1983. Identification of fodrin as a major calmodulin-binding protein in postsynaptic density preparations. *J Cell Biol.* 96(2): 443–8.

Carlisle, H.J. and M.B. Kennedy. 2005. Spine architecture and synaptic plasticity. *Trends Neurosci.* 28(4): 182–7. Review.

Carlisle, H.J., P. Manzerra, E. Marcora and M.B. Kennedy. 2008. SynGAP regulates steady-state and activity-dependent phosphorylation of cofilin. *J Neurosci.* 28(50): 13673–83.

Casper, D. and M. Blum. 1995. Epidermal growth factor and basic fibroblast growth factor protect dopaminergic neurons from glutamate toxicity in culture. *J Neurochem.* 65(3): 1016–26.

Castrén, M.L. and E. Castrén. 2013. BDNF in fragile X syndrome. *Neuropharmacology.* 76(Pt C): 729–36. doi:10.1016/j.neuropharm.2013.05.018.

Chan, K.T., D.A. Bennin and A. Huttenlocher. 2010. Regulation of adhesion dynamics by calpain-mediated proteolysis of focal adhesion kinase (FAK). *J Biol Chem.* 285(15): 11418–26.

Chen, B., J.A. Van Winkle, P.D. Lyden, J.H. Brown and N.H. Purcell. 2013. PHLPP1 gene deletion protects the brain from ischemic injury. *J Cereb Blood Flow Metab.* 33(2): 196–204.

Chen, F., L. Qian, Z.H. Yang, Y. Huang, S.T. Ngo, N.J. Ruan, J. Wang, C. Schneider, P.G. Noakes, Y.Q. Ding, L. Mei and Z.G. Luo. 2007. Rapsyn interaction with calpain stabilizes AChR clusters at the neuromuscular junction. *Neuron.* 55(2): 247–60.

Chen, L.Y., C.S. Rex, Y. Sanaiha, G. Lynch and C.M. Gall. 2010. Learning induces neurotrophin signaling at hippocampal synapses. *Proc Natl Acad Sci USA.* 107(15): 7030–5.

Collingridge, G.L., J.T. Isaac and Y.T. Wang. 2004. Receptor trafficking and synaptic plasticity. *Nat Rev Neurosci.* 5(12): 952–62. Review.

Corazzari, M., G.M. Fimia and M. Piacentini. 2012. Dismantling the autophagic arsenal when it is time to die: concerted AMBRA1 degradation by caspases and calpains. *Autophagy.* 8(8): 1255–7.

Cortesio, C.L., K.T. Chan, B.J. Perrin, N.O. Burton, S. Zhang, Z.Y. Zhang and A. Huttenlocher. 2008. Calpain 2 and PTP1B function in a novel pathway with Src to regulate invadopodia dynamics and breast cancer cell invasion. *J Cell Biol.* 180(5): 957–71.

Czogalla, A. and A.F. Sikorski. 2005. Spectrin and calpain: a 'target' and a 'sniper' in the pathology of neuronal cells. *Cell Mol Life Sci.* 62(17): 1913–24. Review.

Darnell, J.C. and E. Klann. 2013. The translation of translational control by FMRP: therapeutic targets for FXS. *Nat Neurosci.* 16(11): 1530–6.

Davis, M.I., J. Ronesi and D.M. Lovinger. 2003. A predominant role for inhibition of the adenylate cyclase/protein kinase A pathway in ERK activation by cannabinoid receptor 1 in N1E-115 neuroblastoma cells. *J Biol Chem.* 278(49): 48973–80.

DeClerck, Y.A., S. Imren, A.M. Montgomery, B.M. Mueller, R.A. Reisfeld and W.E. Laug. 1997. Proteases and protease inhibitors in tumor progression. *Adv Exp Med Biol.* 425: 89–97. Review.

Demarchi, F., C. Bertoli, T. Copetti, E.L. Eskelinen and C. Schneider. 2007. Calpain as a novel regulator of autophagosome formation. *Autophagy.* 3(3): 235–7.

Demarchi, F. and C. Schneider. 2007. The calpain system as a modulator of stress/damage response. *Cell Cycle.* 6(2): 136–8. Review.

Denny, J.B., J. Polan-Curtain, A. Ghuman, M.J. Wayner and D.L. Armstrong. 1990. Calpain inhibitors block long-term potentiation. *Brain Res.* 534(1-2): 317–20.

Di Rosa, G., T. Odrijin, R.A. Nixon and O. Arancio. 2002. Calpain inhibitors: a treatment for Alzheimer's disease. *J Mol Neurosci.* 19(1-2): 135–41.

Diwakarla, S., L.D. Mercer, L. Kardashsyan, P.W. Chu, Y.S. Shin, C.L. Lau, M.L. Hughes, P. Nagley and P.M. Beart. 2009. GABAergic striatal neurons exhibit caspase-independent, mitochondrially mediated programmed cell death. *J Neurochem.* 109(Suppl 1): 198–206.

Dong, Y.N., E.A. Waxman and D.R. Lynch. 2004. Interactions of postsynaptic density-95 and the NMDA receptor 2 subunit control calpain-mediated cleavage of the NMDA receptor. *J Neurosci.* 24(49): 11035–45.

Dong, Y.N., H.Y. Wu, F.C. Hsu, D.A. Coulter and D.R. Lynch. 2006. Developmental and cell-selective variations in N-methyl-D-aspartate receptor degradation by calpain. *J Neurochem.* 99(1): 206–17.

Donkor, I.O. 2000. A survey of calpain inhibitors. *Curr Med Chem.* 7(12): 1171–88. Review.

Doshi, S. and D.R. Lynch. 2009. Calpain and the glutamatergic synapse. *Front Biosci* (Schol Ed.). 1: 466–76. Review.

Duclot, F. and M. Kabbaj. 2013. Individual differences in novelty seeking predict subsequent vulnerability to social defeat through a differential epigenetic regulation of brain-derived neurotrophic factor expression. *J Neurosci.* 33(27): 11048–60.

Dufty, B.M., L.R. Warner, S.T. Hou, S.X. Jiang, T. Gomez-Isla, K.M. Leenhouts, J.T. Oxford, M.B. Feany, E. Masliah and T.T. Rohn. 2007. Calpain-cleavage of alpha-synuclein: connecting proteolytic processing to disease-linked aggregation. *Am J Pathol.* 170(5): 1725–38.

Duman, R.S., N. Li, R.J. Liu, V. Duric and G. Aghajanian. 2012. Signaling pathways underlying the rapid antidepressant actions of ketamine. *Neuropharmacology.* 62(1): 35–41. Review.

Eda, T., M. Mizuno, K. Araki, Y. Iwakura, H. Namba, H. Sotoyama, A. Kakita, H. Takahashi, H. Satoh, S.Y. Chan and H. Nawa. 2013. Neurobehavioral Deficits of Epidermal Growth Factor-Overexpressing Transgenic Mice: Impact on Dopamine Metabolism. *Neurosci Lett.* 547: 21–5.

Esteves, A.R., D.M. Arduíno, R.H. Swerdlow, C.R. Oliveira and S.M. Cardoso. 2010. Dysfunctional mitochondria uphold calpain activation: contribution to Parkinson's disease pathology. *Neurobiol Dis.* 37(3): 723–30.

Fernández-Shaw, C., A. Marina, P. Cazorl, F. Valdivieso and J. Vázquez. 1997. Anti-brain spectrin immunoreactivity in Alzheimer's disease: degradation of spectrin in an animal model of cholinergic degeneration. *J Neuroimmunol.* 77(1): 91–8.

Flevaris, P., A. Stojanovic, H. Gong, A. Chishti, E. Welch and X. Du. 2007. A molecular switch that controls cell spreading and retraction. *J Cell Biol.* 179(3): 553–65.

Foote, M. and Y. Zhou. 2012. 14-3-3 proteins in neurological disorders. *Int J Biochem Mol Biol.* 3(2): 152–64.

Fortin, D.A., T. Srivastava, D. Dwarakanath, P. Pierre, S. Nygaard, V.A. Derkach and T.R. Soderling. 2012. Brain-derived neurotrophic factor activation of CaM-kinase kinase via transient receptor potential canonical channels induces the translation and synaptic incorporation of GluA1-containing calcium-permeable AMPA receptors. *J Neurosci.* 32(24): 8127–37.

Frame, M.C., V.J. Fincham, N.O. Carragher and J.A. Wyke. 2002. v-Src's hold over actin and cell adhesions. *Nat Rev Mol Cell Biol.* 3(4): 233–45. Review.

Franco, S.J. and A. Huttenlocher. 2005. Regulating cell migration: calpains make the cut. *J Cell Sci.* 118(Pt 17): 3829–38. Review.

Friedrich, P. 2004. The intriguing Ca^{2+} requirement of calpain activation. *Biochem Biophys Res Commun.* 323(4): 1131–3. Review.

Fukuda, T., E. Adachi, S. Kawashima, I. Yoshiya and P.H. Hashimoto. 1990. Immunohistochemical distribution of calcium-activated neutral proteinases and endogenous CANP inhibitor in the rabbit hippocampus. *J Comp Neurol.* 302: 100–9.

Fujikawa, D.G. 2005. Prolonged seizures and cellular injury: understanding the connection. *Epilepsy Behav.* 7(Suppl 3): S3–11. Review.

Gafni, J. and L.M. Ellerby. 2002. Calpain activation in Huntington's disease. *J Neurosci.* 22(12): 4842–9.

Gafni, J., E. Hermel, J.E. Young, C.L. Wellington, M.R. Hayden and L.M. Ellerby. 2004. Inhibition of calpain cleavage of huntingtin reduces toxicity: accumulation of calpain/caspase fragments in the nucleus. *J Biol Chem.* 279(19): 20211–20.

Gao, H. and Z. Geng. 2013. Calpain I activity and its relationship with hippocampal neuronal death in pilocarpine-induced status epilepticus rat model. *Cell Biochem Biophys.* 66(2): 371–7.

Garcia, M., V. Bondada and J.W. Geddes. 2005. Mitochondrial localization of mu-calpain. *Biochem Biophys Res Commun.* 338(2): 1241–7.

Geddes, J.W. and K.E. Saatman. 2010. Targeting individual calpain isoforms for neuroprotection. *Exp Neurol.* 226(1): 6–7.

Gilbertson, R.J., D.A. Hill, R. Hernan, M. Kocak, R. Geyer, J. Olson, A. Gajjar, L. Rush, R.L. Hamilton, S.D. Finkelstein and I.F. Pollack. 2003. ERBB1 is amplified and overexpressed in high-grade diffusely infiltrative pediatric brain stem glioma. *Clin Cancer Res.* 9(10 Pt 1): 3620–4.

Gladding, C.M., M.D. Sepers, J. Xu, L.Y. Zhang, A.J. Milnerwood, P.J. Lombroso and L.A. Raymond. 2012. Calpain and STriatal-Enriched protein tyrosine phosphatase (STEP) activation contribute to extrasynaptic NMDA receptor localization in a Huntington's disease mouse model. *Hum Mol Genet.* 21(17): 3739–52.

Glading, A., R.J. Bodnar, I.J. Reynolds, H. Shiraha, L. Satish, D.A. Potter, H.C. Blair and A. Wells. 2004. Epidermal growth factor activates m-calpain (calpain II), at least in part, by extracellular signal-regulated kinase-mediated phosphorylation. *Mol Cell Biol.* 24(6): 2499–512.

Glading, A., D.A. Lauffenburger and A. Wells. 2002. Cutting to the chase: calpain proteases in cell motility. *Trends Cell Biol.* 12(1): 46–54. Review.

Goebel-Goody, S.M., M. Baum, C.D. Paspalas, S.M. Fernandez, N.C. Carty, P. Kurup and P.J. Lombroso. 2012. Therapeutic implications for striatal-enriched protein tyrosine phosphatase (STEP) in neuropsychiatric disorders. *Pharmacol Rev.* 64(1): 65–87. Review.

Gold, M.S., F.H. Kobeissy, K.K. Wang, L.J. Merlo, A.W. Bruijnzeel, I.N. Krasnova and J.L. Cadet. 2009. Methamphetamine- and trauma-induced brain injuries: comparative cellular and molecular neurobiological substrates. *Biol Psychiatry.* 66(2): 118–27.

Goll, D.E., V.F. Thompson, H. Li, W. Wei and J. Cong. 2003. The calpain system. *Physiol Rev.* 83(3): 731–801. Review.

Gomes, J.R., J.T. Costa, C.V. Melo, F. Felizzi, P. Monteiro, M.J. Pinto, A.R. Inácio, T. Wieloch, R.D. Almeida, M. Grãos and C.B. Duarte. 2012. Excitotoxicity downregulates TrkB.FL signaling and upregulates the neuroprotective truncated TrkB receptors in cultured hippocampal and striatal neurons. *J Neurosci.* 32(13): 4610–22.

Gomes, J.R., A.C. Lobo, C.V. Melo, A.R. Inácio, J. Takano, N. Iwata, T.C. Saido, L.P. de Almeida, T. Wieloch and C.B. Duarte. 2011. Cleavage of the vesicular GABA transporter under excitotoxic conditions is followed by accumulation of the truncated transporter in nonsynaptic sites. *J Neurosci.* 31(12): 4622–35.

Gooney, M., E. Messaoudi, F.O. Maher, C.R. Bramham and M.A. Lynch. 2004. BDNF-induced LTP in dentate gyrus is impaired with age: analysis of changes in cell signaling events. *Neurobiol Aging.* 25(10): 1323–31.

Good, P. 2011. Does fever relieve autistic behavior by improving brain blood flow? *Neuropsychol Rev.* 21(1): 66–7.

Good, P. 2013. Does infectious fever relieve autistic behavior by releasing glutamine from skeletal muscles as provisional fuel? *Med Hypotheses.* 80(1): 1–12.

Gordy, C. and Y.W. He. 2012. The crosstalk between autophagy and apoptosis: where does this lead? *Protein Cell.* 3(1): 17–27.

Grammer, M., S. Kuchay, A. Chishti and M. Baudry. 2005. Lack of phenotype for LTP and fear conditioning learning in calpain 1 knock-out mice. *Neurobiol Learn Mem.* 84(3): 222–7.

Grumelli, C., P. Berghuis, D. Pozzi, M. Caleo, F. Antonucci, G. Bonanno, G. Carmignoto, M.B. Dobszay, T. Harkany, M. Matteoli and C. Verderio. 2008. Calpain activity contributes to the control of SNAP-25 levels in neurons. *Mol Cell Neurosci.* 39(3): 314–23.

Guo, J., H.W. Wu, G. Hu, X. Han, W. De and Y.J. Sun. 2006. Sustained activation of Src-family tyrosine kinases by ischemia: a potential mechanism mediating extracellular signal-regulated kinase cascades in hippocampal dentate gyrus. *Neuroscience.* 143(3): 827–36.

Gurd, J.W., N. Bissoon, T.H. Nguyen, P.J. Lombroso, C.C. Rider, P.W. Beesley and S.J. Vannucci. 1999. Hypoxia-ischemia in perinatal rat brain induces the formation of a low molecular weight isoform of striatal enriched tyrosine phosphatase (STEP). *J Neurochem.* 73(5): 1990–4.

Guttmann, R.P., D.L. Baker, K.M. Seifert, A.S. Cohen, D.A. Coulter and D.R. Lynch. 2001. Specific proteolysis of the NR2 subunit at multiple sites by calpain. *J Neurochem.* 78(5): 1083–93.

Hajimohammadreza, I., K.J. Raser, R. Nath, R. Nadimpalli, M. Scott and K.K. Wang. 1997. Neuronal nitric oxide synthase and calmodulin-dependent protein kinase IIalpha undergo neurotoxin-induced proteolysis. *J Neurochem.* 69(3): 1006–13.

Hamakubo, T., R. Kannagi, T. Murachi and A. Matus. 1986. Distribution of calpains I and II in rat brain. *J Neurosci.* 6: 3103–11.

Hata, S., H. Sorimachi, K. Nakagawa, T. Maeda, K. Abe and K. Suzuki. 2001. Domain II of m-calpain is a Ca(2+)-dependent cysteine protease. *FEBS Lett.* 501(2-3): 111–4.

Hayashi, Y., S.H. Shi, J.A. Esteban, A. Piccini, J.C. Poncer and R. Malinow. 2000. Driving AMPA receptors into synapses by LTP and CaMKII: requirement for GluR1 and PDZ domain interaction. *Science.* 287(5461): 2262–7.

Hecht, M., J.H. Schulte, A. Eggert, J. Wilting and L. Schweigerer. 2005. The neurotrophin receptor TrkB cooperates with c-Met in enhancing neuroblastoma invasiveness. *Carcinogenesis.* 26(12): 2105–15.

Higuchi, M., N. Iwata, Y. Matsuba, J. Takano, T. Suemoto, J. Maeda, B. Ji, M. Ono, M. Staufenbiel, T. Suhara and T.C. Saido. 2012. Mechanistic involvement of the calpain-calpastatin system in Alzheimer neuropathology. *FASEB J.* 26(3): 1204–17.

Higuchi, M., N. Iwata and T.C. Saido. 2005. Understanding molecular mechanisms of proteolysis in Alzheimer's disease: progress toward therapeutic interventions. *Biochim Biophys Acta.* 1751(1): 60–7.

Higuchi, M., M. Tomioka, J. Takano, K. Shirotani, N. Iwata, H. Masumoto, M. Maki, S. Itohara and T.C. Saido. 2005. Distinct mechanistic roles of calpain and caspase activation in neurodegeneration as revealed in mice overexpressing their specific inhibitors. *J Biol Chem.* 280(15): 15229–37.

Hisanaga, S. and T. Saito. 2003. The regulation of cyclin-dependent kinase 5 activity through the metabolism of p35 or p39 Cdk5 activator. *Neurosignals.* 12(4-5): 221–9. Review.

Ho, R., J.E. Minturn, T. Hishiki, H. Zhao, Q. Wang, A. Cnaan, J. Maris, E.A. Evans and G.M. Brodeur. 2005. Proliferation of human neuroblastomas mediated by the epidermal growth factor receptor. *Cancer Res.* 65(21): 9868–75.

Hood, J.L., B.B. Logan, A.P. Sinai, W.H. Brooks and T.L. Roszman. 2003. Association of the calpain/calpastatin network with subcellular organelles. *Biochem Biophys Res Commun.* 310: 1200–12.

Hossain, M.I., C.L. Roulston, M.A. Kamaruddin, P.W. Chu, D.C. Ng, G.J. Dusting, J.D. Bjorge, N.A. Williamson, D.J. Fujita, S.N. Cheung, T.O. Chan, A.F. Hill and H.C. Cheng. 2013. A Truncated Fragment of Src Protein Kinase Generated by Calpain-Mediated Cleavage is a Mediator of Neuronal Death in Excitotoxicity. *J Biol Chem.* 288(14): 9696–709.

Hrabetova, S. and T.C. Sacktor. 1996. Bidirectional regulation of protein kinase M zeta in the maintenance of long-term potentiation and long-term depression. *J Neurosci.* 16(17): 5324–33.

Huang, C., K. Jacobson and M.D. Schaller. 2004. MAP kinases and cell migration. *J Cell Sci.* 117(Pt 20): 4619–28. Review.

Huang, J. and B.D. Manning. 2009. A complex interplay between Akt, TSC2 and the two mTOR complexes. *Biochem Soc Trans.* 37(Pt 1): 217–22.

Huang, W., P.J. Zhu, S. Zhang, H. Zhou, L. Stoica, M. Galiano, K. Krnjević, G. Roman and M. Costa-Mattioli. 2013. mTORC2 controls actin polymerization required for consolidation of long-term memory. *Nat Neurosci.* doi:10.1038/nn.3351.

Iwakura, Y., T. Nagano, M. Kawamura, H. Horikawa, K. Ibaraki, N. Takei and H. Nawa. 2001. N-methyl-D-aspartate-induced alpha-amino-3-hydroxy-5-methyl-4-isoxazoleproprionic acid (AMPA) receptor down-regulation involves interaction of the carboxyl terminus of GluR2/3 with Pick1. Ligand-binding studies using Sindbis vectors carrying AMPA receptor decoys. *J Biol Chem.* 276(43): 40025–32.

Iwakura, Y. and H. Nawa. 2013. ErbB1-4-dependent EGF/neuregulin signals and their cross talk in the central nervous system: pathological implications in schizophrenia and Parkinson's disease. *Front Cell Neurosci.* 7: 4. Review.

Jang, Y.N., Y.S. Jung, S.H. Lee, C.H. Moon, C.H. Kim and E.J. Baik. 2009. Calpain-mediated N-cadherin proteolytic processing in brain injury. *J Neurosci.* 29(18): 5974–84.

Jang, H.S., S. Lal and J.A. Greenwood. 2010. Calpain 2 is required for glioblastoma cell invasion: regulation of matrix metalloproteinase 2. *Neurochem Res.* 35(11): 1796–804.

Janssens, V., R. Derua, K. Zwaenepoel, E. Waelkens and J. Goris. 2009. Specific regulation of protein phosphatase 2A PR72/B" subunits by calpain. *Biochem Biophys Res Commun.* 386(4): 676–81.

Johnson, M.A. and P.J. Lombroso. 2012. A common STEP in the synaptic pathology of diverse neuropsychiatric disorders. *Yale J Biol Med.* 85(4): 481–90.

Jourdi, H., L. Hamo, T. Oka, A. Seegan and M. Baudry. 2009. BDNF mediates the neuroprotective effects of positive AMPA receptor modulators against MPP+-induced toxicity in cultured hippocampal and mesencephalic slices. *Neuropharmacology.* 56(5): 876–85.

Jourdi, H., Y.T. Hsu, M. Zhou, Q. Qin, X. Bi and M. Baudry. 2009. Positive AMPA receptor modulation rapidly stimulates BDNF release and increases dendritic mRNA translation. *J Neurosci.* 29(27): 8688–97.

Jourdi, H., Y. Iwakura, M. Narisawa-Saito, K. Ibaraki, H. Xiong, M. Watanabe, Y. Hayashi, N. Takei and H. Nawa. 2003. Brain-derived neurotrophic factor signal enhances and maintains the expression of AMPA receptor-associated PDZ proteins in developing cortical neurons. *Dev Biol.* 263(2): 216–30.

Jourdi, H. and M. Kabbaj. 2013. Acute BDNF treatment upregulates GluR1-SAP97 and GluR2-GRIP1 interactions: implications for sustained AMPA receptor expression. *PLoS One.* 8(2): e57124.

Jourdi, H. and M. Kabbaj. 2013. Characterization of CX614, an AMPAkine, as a fast onset antidepressant. Program No. 503.08. *Neuroscience 2013 Abstracts.* San Diego, CA: Society for Neuroscience, 2013. Online.

Jourdi, H., X. Lu, T. Yanagihara, J.C. Lauterborn, X. Bi, C.M. Gall and M. Baudry. 2005. Prolonged positive modulation of alpha-amino-3-hydroxy-5-methyl-4-isoxazolepropionic acid (AMPA) receptors induces calpain-mediated PSD-95/Dlg/ZO-1 protein degradation and AMPA receptor down-regulation in cultured hippocampal slices. *J Pharmacol Exp Ther.* 314(1): 16–26.

Jourdi, H. and H. Nawa. 2002. Basic Fibroblast Growth Factor Modulates the Expression of PDZ Domain-containing Proteins in Cultured Cortical Neurons. *Acta Medica et Biologica.* 50(3): 107–115.

Jourdi, H., K. Rostamiani, Q. Qin, X. Bi and M. Baudry. 2010. BDNF- and EGF-mediated neuronal Calpain activation through MAPK-dependent phosphorylation. *J Neurosci.* 30(3): 1086–95.

Jourdi, H., T. Yanagihara, U. Martinez, X. Bi, G. Lynch and M. Baudry. 2005. Effects of positive AMPA receptor modulators on calpain-mediated spectrin degradation in cultured hippocampal slices. *Neurochem Int.* 46(1): 31–40.

Kakkar, R., S. Taketa, R.V. Raju, S. Proudlove, P. Colquhoun, K. Grymaloski and R.K. Sharma. 1997. *In vitro* phosphorylation of bovine cardiac muscle high molecular weight calmodulin binding protein by cyclic AMP-dependent protein kinase and dephosphorylation by calmodulin-dependent phosphatase. *Mol Cell Biochem.* 177(1-2): 215–9.

Kaneko, M., C.E. Cheetham, Y.S. Lee, A.J. Silva, M.P. Stryker and K. Fox. 2010. Constitutively active H-ras accelerates multiple forms of plasticity in developing visual cortex. *Proc Natl Acad Sci USA.* 107(44): 19026–31.

Kaphzan, H., P. Hernandez, J.I. Jung, K.K. Cowansage, K. Deinhardt, M.V. Chao, T. Abel and E. Klann. 2012. Reversal of impaired hippocampal long-term potentiation and contextual fear memory deficits in Angelman syndrome model mice by ErbB inhibitors. *Biol Psychiatry.* 72(3): 182–90.

Kar, P., K. Samanta, S. Shaikh, A. Chowdhury, T. Chakraborti and S. Chakraborti. 2010. Mitochondrial calpain system: an overview. *Arch Biochem Biophys.* 495(1): 1–7.

Kim, H.J., D. Lee, C.H. Lee, K.C. Chung, J. Kim and S.R. Paik. 2006. Calpain-resistant fragment(s) of alpha-synuclein regulates the synuclein-cleaving activity of 20S proteasome. *Arch Biochem Biophys.* 455(1): 40–7.

Kim, M., J.K. Roh, B.W. Yoon, L. Kang, Y.J. Kim, N. Aronin and M. DiFiglia. 2003. Huntingtin is degraded to small fragments by calpain after ischemic injury. *Exp Neurol.* 183(1): 109–15.

Kim, S.J., J.Y. Sung, J.W. Um, N. Hattori, Y. Mizuno, K. Tanaka, S.R. Paik, J. Kim and K.C. Chung. 2003. Parkin cleaves intracellular alpha-synuclein inclusions via the activation of calpain. *J Biol Chem.* 278(43): 41890–9.

Kosaki, G., T. Tsujinaka, J. Kambayashi, K. Morimoto, K. Yamamoto, K. Yamagami, K. Sobue and S. Kakiuchi. 1983. Specific cleavage of calmodulin-binding proteins by low Ca2+-requiring form of Ca2+-activated neutral protease in human platelets. *Biochem Int.* 6(6): 767–75.

Kramár, E.A., A.H. Babayan, C.M. Gall and G. Lynch. 2013. Estrogen promotes learning-related plasticity by modifying the synaptic cytoskeleton. *Neuroscience.* 239: 3–16. Review.

Kramár, E.A., A.H. Babayan, C.F. Gavin, C.D. Cox, M. Jafari, C.M. Gall, G. Rumbaugh and G. Lynch. 2012. Synaptic evidence for the efficacy of spaced learning. *Proc Natl Acad Sci USA.* 109(13): 5121–6.

Kramár, E.A., L.Y. Chen, N.J. Brandon NJ, C.S. Rex, F. Liu, C.M. Gall and G. Lynch. 2009. Cytoskeletal changes underlie estrogen's acute effects on synaptic transmission and plasticity. *J Neurosci.* 29(41): 12982–93.

Kramár, E.A., L.Y. Chen, J.C. Lauterborn, D.A. Simmons, C.M. Gall and G. Lynch. 2012 BDNF upregulation rescues synaptic plasticity in middle-aged ovariectomized rats. *Neurobiol Aging.* 33(4): 708–19.

Kramár, E.A., B. Lin, C.S. Rex, C.M. Gall and G. Lynch. 2006. Integrin-driven actin polymerization consolidates long-term potentiation. *Proc Natl Acad Sci USA.* 103(14): 5579–84.

Kühl, M. 2004. The WNT/calcium pathway: biochemical mediators, tools and future requirements. *Front Biosci.* 9: 967–74. Review.

Kumar, A., P.R. Molli, S.B. Pakala, T.M. Bui Nguyen, S.K. Rayala and R. Kumar. 2009. PAK thread from amoeba to mammals. *J Cell Biochem.* 107(4): 579–85.

Kuo, W.N., U. Ganesan, D.L. Davis, D.L. Walbey, M.A. Bell, K. Allen, S. Siddeeq, L.K. McCall and N. Carwell. 1993. Regulation of the phosphorylation of histones and glycogen synthase. *Cytobios.* 76(304): 41–8.

Kuroda, T., K. Mikawa, H. Mishima and A. Kishimoto. 1991. H1 histone stimulates limited proteolysis of protein kinase C subspecies by calpain II. *J Biochem.* 110(3): 364–8.

Kwon, O.B., M. Longart, D. Vullhorst, D.A. Hoffman and A. Buonanno. 2005. Neuregulin-1 reverses long-term potentiation at CA1 hippocampal synapses. *J Neurosci.* 25(41): 9378–83.

Kwon, O.B., D. Paredes, C.M. Gonzalez, J. Neddens, L. Hernandez, D. Vullhorst and A. Buonanno. 2008. Neuregulin-1 regulates LTP at CA1 hippocampal synapses through activation of dopamine D4 receptors. *Proc Natl Acad Sci USA.* 105(40): 15587–92.

Lai, T.W., S. Zhang and Y.T. Wang. 2013. Excitotoxicity and stroke: Identifying novel targets for neuroprotection. *Prog Neurobiol.* doi: 10.1016/j.pneurobio.2013.11.006.

Lakshmikuttyamma, A., P. Selvakumar, A.R. Sharma, D.H. Anderson and R.K. Sharma. 2004. *In vitro* proteolytic degradation of bovine brain calcineurin by m-calpain. *Neurochem Res.* 29(10): 1913–21.

Lal, S., J. La Du, R.L. Tanguay and J.A. Greenwood. 2012. Calpain 2 is required for the invasion of glioblastoma cells in the zebrafish brain microenvironment. *J Neurosci Res.* 90(4): 769–81.

Landles, C., K. Sathasivam, A. Weiss, B. Woodman, H. Moffitt, S. Finkbeiner, B. Sun, J. Gafni, L.M. Ellerby, Y. Trottier, W.G. Richards, A. Osmand, P. Paganetti and G.P. Bates. 2010. Proteolysis of mutant huntingtin produces an exon 1 fragment that accumulates as an aggregated protein in neuronal nuclei in Huntington disease. *J Biol Chem.* 285(12): 8808–23.

Lee, S.T. and M. Kim. 2006. Aging and neurodegeneration. Molecular mechanisms of neuronal loss in Huntington's disease. *Mech Ageing Dev.* 127(5): 432–5. Review.

Leloup, L., H. Shao, Y.H. Bae, B. Deasy, D. Stolz, P. Roy and A. Wells. 2010. m-Calpain activation is regulated by its membrane localization and by its binding to phosphatidylinositol 4,5-bisphosphate. *J Biol Chem.* 285(43): 33549–66.

Leloup, L. and A. Wells. 2011. Calpains as potential anti-cancer targets. *Expert Opin Ther Targets.* 15(3): 309–23. Review.

Li, J., F. Grynspan, S. Berman, R. Nixon and S. Bursztajn. 1996. Regional differences in gene expression for calcium activated neutral proteases (calpains) and their endogenous inhibitor calpastatin in mouse brain and spinal cord. *J Neurobiol.* 30(2): 177–91.

Li, Y., V. Bondada, A. Joshi and J.W. Geddes. 2009. Calpain 1 and Calpastatin expression is developmentally regulated in rat brain. *Exp Neurol.* 220(2): 316–9.

Liang, C. 2010. Negative regulation of autophagy. *Cell Death Differ.* 17(12): 1807–15. Review.

Libertini, S.J., B.S. Robinson, N.K. Dhillon, D. Glick, M. George, S. Dandekar, J.P. Gregg, E. Sawai and M. Mudryj. 2005. Cyclin E both regulates and is regulated by calpain 2, a protease associated with metastatic breast cancer phenotype. *Cancer Res.* 65: 10700–10708.

Liu, C., Y. Li, M. Semenov, C. Han, G.H. Baeg, Y. Tan, Z. Zhang, X. Lin and X. He. 2002. Control of beta catenin phosphorylation/degradation by a dual kinase mechanism. *Cell.* 108: 837–847.

Liu, J., M.C. Liu and K.K. Wang. 2008. Calpain in the CNS: from synaptic function to neurotoxicity. *Sci Signal.* 1(14): re1. doi:10.1126/stke.114re1.

Lopes, J.P., C.R. Oliveira and P. Agostinho. 2009. Cdk5 acts as a mediator of neuronal cell cycle re-entry triggered by amyloid-beta and prion peptides. *Cell Cycle.* 8(1): 97–104.

Lopez-Meraz, M.L. and J. Niquet. 2009. Participation of mu-calpain in status epilepticus-induced hippocampal injury. *Brain Res Bull.* 78(4-5): 131.

Lu, X., Y. Rong and M. Baudry. 2000. Calpain-mediated degradation of PSD-95 in developing and adult rat brain. *Neurosci Lett.* 286(2): 149–53.

Lu, X., Y. Rong, R. Bi and M. Baudry. 2000. Calpain-mediated truncation of rat brain AMPA receptors increases their Triton X-100 solubility. *Brain Res.* 863(1-2): 143–50.

Lu, X., M. Wyszynski, M. Sheng and M. Baudry. 2001. Proteolysis of glutamate receptor-interacting protein by calpain in rat brain: implications for synaptic plasticity. *J Neurochem.* 77(6): 1553–60.

Lugli, G., J. Larson, M.E. Martone, Y. Jones and N.R. Smalheiser. 2005. Dicer and eIF2c are enriched at postsynaptic densities in adult mouse brain and are modified by neuronal activity in a calpain-dependent manner. *J Neurochem.* 94(4): 896–905.

Lynch, G., E.A. Kramár, A.H. Babayan, G. Rumbaugh and C.M. Gall. 2013. Differences between synaptic plasticity thresholds result in new timing rules for maximizing long-term potentiation. *Neuropharmacology.* 64: 27–36. Review.

Lynch, G., L.C. Palmer and C.M. Gall. 2011. The likelihood of cognitive enhancement. *Pharmacol Biochem Behav.* 99(2): 116–29.

Lynch, G., C.S. Rex, L.Y. Chen and C.M. Gall. 2008. The substrates of memory: defects, treatments, and enhancement. *Eur J Pharmacol.* 585(1): 2–13.

Lynch, G., C.S. Rex and C.M. Gall. 2007. LTP consolidation: substrates, explanatory power, and functional significance. *Neuropharmacology.* 52(1): 12–23.

Lyons, L.C. 2011. Critical role of the circadian clock in memory formation: lessons from Aplysia. *Front Mol Neurosci.* 4: 52. doi:10.3389/fnmol.2011.0005.

Magnaghi-Jaulin, L., A. Marcilhac, M. Rossel, C. Jaulin, Y. Benyamin and F. Raynaud. 2010. Calpain 2 is required for sister chromatid cohesion. *Chromosoma.* 119(3): 267–74.

Maiese, K. and L. Boccone. 1995. Neuroprotection by peptide growth factors against anoxia and nitric oxide toxicity requires modulation of protein kinase C. *J Cereb Blood Flow Metab.* 15(3): 440–9.

Malenka, R.C., J.A. Kauer, D.J. Perkel, M.D. Mauk, P.T. Kelly, R.A. Nicoll and M.N. Waxham. 1989. An essential role for postsynaptic calmodulin and protein kinase activity in long-term potentiation. *Nature.* 340(6234): 554–7.

Mammen, A.L., K. Kameyama, K.W. Roche and R.L. Huganir. 1997. Phosphorylation of the alpha-amino-3-hydroxy-5-methylisoxazole4-propionic acid receptor GluR1 subunit by calcium/calmodulin-dependent kinase II. *J Biol Chem.* 272(51): 32528–33.

Manadas, B., A.R. Santos, K. Szabadfi, J.R. Gomes, S.D. Garbis, M. Fountoulakis and C.B. Duarte. 2009. BDNF-induced changes in the expression of the translation machinery in hippocampal neurons: protein levels and dendritic mRNA. *J Proteome Res.* 8(10): 4536–52.

Masliah, E., D.S. Iimoto, T. Saitoh, L.A. Hansen and R.D. Terry. 1990. Increased immunoreactivity of brain spectrin in Alzheimer disease: a marker for synapse loss? *Brain Res.* 531(1-2): 36–44.

Mataga, M.A., S. Rosenthal, S. Heerboth, A. Devalapalli, S. Kokolus, L.R. Evans, M. Longacre, G. Housman and S. Sarkar. 2012. Anti-breast cancer effects of histone deacetylase inhibitors and calpain inhibitor. *Anticancer Res.* 32(7): 2523–9.

McCarthy, D.M., A.N. Brown and P.G. Bhide. 2012. Regulation of BDNF expression by cocaine. *Yale J Biol Med.* 85(4): 437–46. Review.

McClelland, P., L.P. Adam and D.R. Hathaway. 1994. Identification of a latent Ca^{2+}/calmodulin dependent protein kinase II phosphorylation site in vascular calpain II. *J Biochem.* 115(1): 41–6.

McGinnis, K.M., M.M. Whitton, M.E. Gnegy and K.K. Wang. 1998. Calcium/calmodulin-dependent protein kinase IV is cleaved by caspase-3 and calpain in SH-SY5Y human neuroblastoma cells undergoing apoptosis. *J Biol Chem.* 273(32): 19993–20000.

Mellgren, R.L. 1991. Proteolysis of nuclear proteins by mu-calpain and m-calpain. *J Biol Chem.* 266(21): 13920–4.

Mellgren, R.L. and T.C. Carr. 1983. The protein inhibitor of calcium-dependent proteases: purification from bovine heart and possible mechanisms of regulation. *Arch Biochem Biophys.* 225(2): 779–86.

Mellgren, R.L. and C.B. Rozanov. 1990. Calpain II-dependent solubilization of a nuclear protein kinase at micromolar calcium concentrations. *Biochem Biophys Res Commun.* 168(2): 589–95.

Melloni, E., M. Averna, F. Salamino, B. Sparatore, R. Minafra and S. Pontremoli. 2000. Acyl-CoA-binding protein is a potent m-calpain activator. *J Biol Chem.* 275(1): 82–6.

Melloni, E., M. Michetti, F. Salamino, R. Minafra and S. Pontremoli. 1996. Modulation of the calpain autoproteolysis by calpastatin and phospholipids. *Biochem Biophys Res Commun.* 229(1): 193–7.

Melloni, E., M. Michetti, F. Salamino, R. Minafra, B. Sparatore and S. Pontremoli. 2000. Isolation and characterization of calpain activator protein from bovine brain. *Methods Mol Biol.* 144: 99–105.

Melloni, E., M. Michetti, F. Salamino and S. Pontremoli. 1998. Molecular and functional properties of a calpain activator protein specific for mu-isoforms. *J Biol Chem.* 273(21): 12827–31.

Melloni, E., R. Minafra, F. Salamino and S. Pontremoli. 2000. Properties and intracellular localization of calpain activator protein. *Biochem Biophys Res Commun.* 272(2): 472–6.

Michel, M., J.S. Gardner, C.L. Green, C.L. Organ and L.C. Lyons. 2013. Protein phosphatase-dependent circadian regulation of intermediate-term associative memory. *J Neurosci.* 33(10): 4605–13.

Michel, M., C.L. Green, J.S. Gardner, C.L. Organ and L.C. Lyons. 2012. Massed training-induced intermediate-term operant memory in aplysia requires protein synthesis and multiple persistent kinase cascades. *J Neurosci.* 32(13): 4581–91.

Minassian, B.A. 2001. Lafora's disease: towards a clinical, pathologic, and molecular synthesis. *Pediatr Neurol.* 25(1): 21–9. Review.

Mizuno, M., H. Sotoyama, H. Namba, M. Shibuya, T. Eda, R. Wang, T. Okubo, K. Nagata, Y. Iwakura and H. Nawa. 2013. ErbB inhibitors ameliorate behavioral impairments of an animal model for schizophrenia: implication of their dopamine-modulatory actions. *Transl Psychiatry.* 3: e252. doi:10.1038/tp.2013.29.

Moldoveanu, T., C.M. Hosfield, D. Lim, J.S. Elce, Z. Jia and P.L. Davies. 2002. A Ca(2+) switch aligns the active site of calpain. *Cell.* 108(5): 649–60.

Moldoveanu, T., Z. Jia and P.L. Davies. 2004. Calpain activation by cooperative Ca²⁺ binding at two non-EF-hand sites. *J Biol Chem.* 279(7): 6106–14.

Monnerie, H. and P.D. Le Roux. 2008. Glutamate alteration of glutamic acid decarboxylase (GAD) in GABAergic neurons: the role of cysteine proteases. *Exp Neurol.* 213(1): 145–53.

Monnerie, H., F.C. Hsu, D.A. Coulter and P.D. Le Roux. 2010. Role of the NR2A/2B subunits of the N-methyl-D-aspartate receptor in glutamate-induced glutamic acid decarboxylase alteration in cortical GABAergic neurons *in vitro. Neuroscience.* 171(4): 1075–90.

Muller, D., I. Molinari, I. Soldati and G. Bianchi. 1995. A genetic deficiency in calpastatin and isovalerylcarnitine treatment is associated with enhanced hippocampal long-term potentiation. *Synapse.* 19(1): 37–45.

Nakagawara, A., C.G. Azar, N.J. Scavarda and G.M. Brodeur. 1994. Expression and function of TRK-B and BDNF in human neuroblastomas. *Mol Cell Biol.* 14(1): 759–67.

Nakajima, R., K. Takao, S.M. Huang, J. Takano, N. Iwata, T. Miyakawa and T.C. Saido. 2008. Comprehensive behavioral phenotyping of calpastatin-knockout mice. *Mol Brain.* 1: 7. doi: 10.1186/1756-6606-1-7.

Nagano, T., H. Jourdi and H. Nawa. 1998. Emerging roles of Dlg-like PDZ proteins in the organization of the NMDA-type glutamatergic synapse. *J Biochem.* 124(5): 869–75. Review.

Nagano, T., H. Namba, Y. Abe, H. Aoki, N. Takei and H. Nawa. 2007. *In vivo* administration of epidermal growth factor and its homologue attenuates developmental maturation of functional excitatory synapses in cortical GABAergic neurons. *Eur J Neurosci.* 25(2): 380–90.

Nagano, T., Y. Yanagawa, K. Obata, M. Narisawa-Saito, H. Namba, Y. Otsu, N. Takei and H. Nawa. 2003. Brain-derived neurotrophic factor upregulates and maintains AMPA receptor currents in neocortical GABAergic neurons. *Mol Cell Neurosci.* 24(2): 340–56.

Namba, H., T. Nagano, Y. Iwakura, H. Xiong, H. Jourdi, N. Takei and H. Nawa. 2006. Transforming growth factor alpha attenuates the functional expression of AMPA receptors in cortical GABAergic neurons. *Mol Cell Neurosci.* 31(4): 628–41.

Narisawa-Saito, M., Y. Iwakura, M. Kawamura, K. Araki, S. Kozaki, N. Takei and H. Nawa. 2002. Brain-derived neurotrophic factor regulates surface expression of alpha-amino-3-hydroxy-5-methyl-4-isoxazoleproprionic acid receptors by enhancing the N-ethylmaleimide-sensitive factor/GluR2 interaction in developing neocortical neurons. *J Biol Chem.* 277(43): 40901–10.

Nawa, H., M. Takahashi and P.H. Patterson. 2002. Cytokine and growth factor involvement in schizophrenia—support for the developmental model. *Mol Psychiatry.* 5(6): 594–603. Review.

Nawa, H. and K. Yamada. 2012. Experimental schizophrenia models in rodents established with inflammatory agents and cytokines. *Methods Mol Biol.* 829: 445–51.

Nestler, E.J. 2012. Transcriptional mechanisms of drug addiction. *Clin Psychopharmacol Neurosci.* 10(3): 136–43.

Nguyen, T.H., S. Paul, Y. Xu, J.W. Gurd and P.J. Lombroso. 1999. Calcium-dependent cleavage of striatal enriched tyrosine phosphatase (STEP). *J Neurochem.* 73(5): 1995–2001.

Nguyen, H.T., D.R. Sawmiller, Q. Wu, J.J. Maleski and M. Chen. 2012. Evidence supporting the role of calpain in the α-processing of amyloid-β precursor protein. *Biochem Biophys Res Commun.* 420(3): 530–5.

Oertner, T.G. and A. Matus. 2005. Calcium regulation of actin dynamics in dendritic spines. *Cell Calcium.* 37(5): 477–82. Review.

Oliver, M.W., M. Baudry and G. Lynch. 1989. The protease inhibitor leupeptin interferes with the development of LTP in hippocampal slices. *Brain Res.* 505(2): 233–8.

Ono, Y. and H. Sorimachi. 2012. Calpains: an elaborate proteolytic system. *Biochim Biophys Acta.* 1824(1): 224–36.

Osterweil, E.K., D.D. Krueger, K. Reinhold and M.F. Bear. 2010. Hypersensitivity to mGluR5 and ERK1/2 leads to excessive protein synthesis in the hippocampus of a mouse model of fragile X syndrome. *J Neurosci.* 30(46): 15616–27.

Ostwald, K., M. Hayashi, M. Nakamura and S. Kawashima. 1994. Subcellular distribution of calpain and calpastatin immunoreactivity and fodrin proteolysis in rabbit hippocampus after hypoxia and glucocorticoid treatment. *J Neurochem.* 63: 1069–76.

Ovat, A., Z.Z. Li, C.Y. Hampton, S.A. Asress, F.M. Fernández, J.D. Glass and J.C. Powers. 2010. Peptidyl alpha-ketoamides with nucleobases, methylpiperazine, and dimethylaminoalkyl substituents as calpain inhibitors. *J Med Chem.* 53(17): 6326–36.

Panigrahi, A.K., N. Zhang, Q. Mao and D. Pati. 2011. Calpain-1 cleaves Rad21 to promote sister chromatid separation. *Mol Cell Biol.* 31(21): 4335–47.

Panja, D. and C.R. Bramham. 2014. BDNF mechanisms in late LTP formation: A synthesis and breakdown. *Neuropharmacology.* 76(Pt C): 664–76.

Paoletti, P., I. Vila, M. Rifé, J.M. Lizcano, J. Alberch and S. Ginés. 2008. Dopaminergic and glutamatergic signaling crosstalk in Huntington's disease neurodegeneration: the role of p25/cyclin-dependent kinase 5. *J Neurosci.* 28(40): 10090–101.

Parameswaran, S. and R.K. Sharma. 2010. High molecular weight calmodulin-binding protein: 20 years onwards-a potential therapeutic calpain inhibitor. *Cardiovasc Drugs Ther.* 26(4): 321–30.

Park, J., I. Al-Ramahi, Q. Tan, N. Mollema, J.R. Diaz-Garcia, T. Gallego-Flores, H.C. Lu, S. Lagalwar, L. Duvick, H. Kang, Y. Lee, P. Jafar-Nejad, L.S. Sayegh, R. Richman, X. Liu, Y. Gao, C.A. Shaw, J.S. Arthur, T.F. Westbrook, J. Botas and H.Y. Zoghbi. 2013. RAS-MAPK-MSK1 pathway modulates ataxin 1 protein levels and toxicity in SCA1. *Nature.* 498(7454): 325–31. doi:10.1038/nature12204.

Perlmutter, L.S., C. Gall, M. Baudry and G. Lynch. 1990. Distribution of calcium-activated protease calpain in the rat brain. *J Comp Neurol.* 296: 269–76.

Perlmutter, L.S., R. Siman, C. Gall, P. Seubert, M. Baudry and G. Lynch. 1988. The ultrastructural localization of calcium-activated protease "calpain" in rat brain. *Synapse.* 2(1): 79–88.

Perrin, B.J., K.J. Amann and A. Huttenlocher. 2006. Proteolysis of cortactin by calpain regulates membrane protrusion during cell migration. *Mol Biol Cell.* 17(1): 239–50.

Pianetti, S., M. Arsura, R. Romieu-Mourez, R.J. Coffey and G.E. Sonenshein. 2001. Her-2/neu overexpression induces NF-kappaB via a PI3-kinase/Akt pathway involving calpain-mediated degradation of IkappaB-alpha that can be inhibited by the tumor suppressor PTEN. *Oncogene.* 20(11): 1287–99.

Potter, D.A., J.S. Tirnauer, R. Janssen, D.E. Croall, C.N. Hughes, K.A. Fiacco, J.W. Mier, M. Maki and I.M. Herman. 1998. Calpain regulates actin remodeling during cell spreading. *J Cell Biol.* 141(3): 647–62.

Powers, J.C., J.L. Asgian, O.D. Ekici and K.E. James. 2002. Irreversible inhibitors of serine, cysteine, and threonine proteases. *Chem Rev.* 102(12): 4639–750. Review.

Puskarjov, M., F. Ahmad, K. Kaila and P. Blaesse. 2012. Activity-dependent cleavage of the K-Cl cotransporter KCC2 mediated by calcium-activated protease calpain. *J Neurosci.* 32(33): 11356–64.

Qian, W., X. Yin, W. Hu, J. Shi, J. Gu, I. Grundke-Iqbal, K. Iqbal, C.X. Gong and F. Liu. 2011. Activation of protein phosphatase 2B and hyperphosphorylation of Tau in Alzheimer's disease. *J Alzheimers Dis.* 23(4): 617–27.

Qin, Q., G. Liao, M. Baudry and X. Bi. 2010. Role of calpain-mediated p53 truncation in semaphorin 3A-induced axonal growth regulation. *Proc Natl Acad Sci USA.* 107(31): 13883–7.

Qin, Q., M. Baudry, G. Liao, A. Noniyev, J. Galeano and X. Bi. 2009. A novel function for p53: regulation of growth cone motility through interaction with Rho kinase. *J Neurosci.* 29(16): 5183–92.

Rajendran, P., B. Delage, W.M. Dashwood, T.W. Yu, B. Wuth, D.E. Williams, E. Ho and R.H. Dashwood. 2011. Histone deacetylase turnover and recovery in sulforaphane-treated colon cancer cells: competing actions of 14-3-3 and Pin1 in HDAC3/SMRT corepressor complex dissociation/reassembly. *Mol Cancer.* 10: 68.

Ramis-Conde, I., M.A. Chaplain, A.R. Anderson and D. Drasdo. 2009. Multi-scale modelling of cancer cell intravasation: the role of cadherins in metastasis. *Phys Biol.* 6(1): 016008.

Ravikumar, B., S. Sarkar, J.E. Davies, M. Futter, M. Garcia-Arencibia, Z.W. Green-Thompson, M. Jimenez-Sanchez, V.I. Korolchuk, M. Lichtenberg, S. Luo, D.C. Massey, F.M. Menzies, K. Moreau, U. Narayanan, M. Renna, F.H. Siddiqi, B.R. Underwood, A.R. Winslow and D.C. Rubinsztein. 2010. Regulation of mammalian autophagy in physiology and pathophysiology. *Physiol Rev.* 90(4): 1383–435. Review.

Reijonen, S., J.P. Kukkonen, A. Hyrskyluoto, J. Kivinen, M. Kairisalo, N. Takei, D. Lindholm and L. Korhonen. 2010. Downregulation of NF-kappaB signaling by mutant huntingtin proteins induces oxidative stress and cell death. *Cell Mol Life Sci.* 67(11): 1929–41.

Rex, C.S., J.C. Lauterborn, C.Y. Lin, E.A. Kramár, G.A. Rogers, C.M. Gall and G. Lynch. 2006. Restoration of long-term potentiation in middle-aged hippocampus after induction of brain-derived neurotrophic factor. *J Neurophysiol.* 96(2): 677–85.

Rex, C.S., C.Y. Lin, E.A. Kramár, L.Y. Chen, C.M. Gall and G. Lynch. 2007. Brain-derived neurotrophic factor promotes long-term potentiation-related cytoskeletal changes in adult hippocampus. *J Neurosci.* 14; 27(11): 3017–29.

Rohn, T.T. 2012. Targeting alpha-synuclein for the treatment of Parkinson's disease. *CNS Neurol Disord Drug Targets.* 11(2): 174–9. Review.

Rong, Y., X. Lu, A. Bernard, M. Khrestchatisky and M. Baudry. 2001. Tyrosine phosphorylation of ionotropic glutamate receptors by Fyn or Src differentially modulates their susceptibility to calpain and enhances their binding to spectrin and PSD-95. *J Neurochem.* 79(2): 382–90.

Roumes, H., L. Leloup, E. Dargelos, J.J. Brustis, L. Daury and P. Cottin. 2010. Calpains: markers of tumor aggressiveness? *Exp Cell Res.* 316(9): 1587–99.

Saatman, K.E., J. Creed and R. Raghupathi. 2010. Calpain as a therapeutic target in traumatic brain injury. *Neurotherapeutics.* 7(1): 31–42.

Sabatini, B.L., T.G. Oertner and K. Svoboda. 2002. The life cycle of Ca(2+) ions in dendritic spines. *Neuron.* 33(3): 439–52.

Saido, T.C., M. Shibata, T. Takenawa, H. Murofushi and K. Suzuki. 1992. Positive regulation of mu-calpain action by polyphosphoinositides. *J Biol Chem.* 267(34): 24585–90.

Sakai, K., H. Akanuma, K. Imahori and S. Kawashima. 1987. A unique specificity of a calcium activated neutral protease indicated in histone hydrolysis. *J Biochem.* 101(4): 911–8.

Salamino, F., M. Averna, I. Tedesco, R. De Tullio, E. Melloni and S. Pontremoli. 1997. Modulation of rat brain calpastatin efficiency by post-translational modifications. *FEBS Lett.* 412(3): 433–8.

Samantaray, S., S.K. Ray and N.L. Banik. 2008. Calpain as a potential therapeutic target in Parkinson's disease. *CNS Neurol Disord Drug Targets.* 7(3): 305–12. Review.

Schilling, B., J. Gafni, C. Torcassi, X. Cong, R.H. Row, M.A. LaFevre-Bernt, M.P. Cusack, T. Ratovitski, R. Hirschhorn, C.A. Ross, B.W. Gibson and L.M. Ellerby. 2006. Huntingtin phosphorylation sites mapped by mass spectrometry. Modulation of cleavage and toxicity. *J Biol Chem.* 281(33): 23686–97.

Schmidt, H.D., J.F. McGinty, A.E. West and G. Sadri-Vakili. 2013. Epigenetics and psychostimulant addiction. *Cold Spring Harb Perspect Med.* 3(3): a012047. Review.

Schoch, K.M., H.N. Evans, J.M. Brelsfoard, S.K. Madathil, J. Takano, T.C. Saido and K.E. Saatman. 2012. Calpastatin overexpression limits calpain-mediated proteolysis and behavioral deficits following traumatic brain injury. *Exp Neurol.* 236(2): 371–82.

Sengoku, T., V. Bondada, D. Hassane, S. Dubal and J.W. Geddes. 2004. Tat-calpastatin fusion proteins transduce primary rat cortical neurons but do not inhibit cellular calpain activity. *Exp Neurol.* 188(1): 161–70.

Sha, D., Y. Jin, H. Wu, J. Wei, C.H. Lin, Y.H. Lee, C. Buddhala, S. Kuchay, A.H. Chishti and J.Y. Wu. 2008. Role of mu-calpain in proteolytic cleavage of brain L-glutamic acid decarboxylase. *Brain Res.* 1207: 9–18.

Shacka, J.J., K.A. Roth and J. Zhang. 2008. The autophagy-lysosomal degradation pathway: role in neurodegenerative disease and therapy. *Front Biosci.* 13: 718–36. Review.

Shao, H., J. Chou, C.J. Baty, N.A. Burke, S.C. Watkins, D.B. Stolz and A. Wells. 2006. Spatial localization of m-calpain to the plasma membrane by phosphoinositide biphosphate binding during epidermal growth factor receptor-mediated activation. *Mol Cell Biol.* 26(14): 5481–96.

Shimizu, K., S.M. Mackenzie and D.R. Storm. 2010. SCOP/PHLPP and its functional role in the brain. *Mol Biosyst.* 6(1): 38–43. Review.

Shimizu, K., T. Phan, I.M. Mansuy and D.R. Storm. 2007. Proteolytic degradation of SCOP in the hippocampus contributes to activation of MAP kinase and memory. *Cell.* 128(6): 1219–29.

Shiraha, H., A. Glading, J. Chou, Z. Jia and A. Wells A. 2002. Activation of m-calpain (calpain II) by epidermal growth factor is limited by protein kinase A phosphorylation of m-calpain. *Mol Cell Biol.* 22(8): 2716–27.

Shiraha, H., A. Glading, K. Gupta and A. Wells. 1999. IP-10 inhibits epidermal growth factor-induced motility by decreasing epidermal growth factor receptor-mediated calpain activity. *J Cell Biol.* 146(1): 243–54.

Sihag, R.K. and A.M. Cataldo. 1996. Brain beta-spectrin is a component of senile plaques in Alzheimer's disease. *Brain Res.* 743(1-2): 249–57.

Silver, K., L. Leloup, L.C. Freeman, A. Wells and J.D. Lillich. 2010. Non-steroidal anti-inflammatory drugs inhibit calpain activity and membrane localization of calpain 2 protease. *Int J Biochem Cell Biol.* 42(12): 2030–6.

Siman, R., C. Gall, L.S. Perlmutter, C. Christian, M. Baudry and G. Lynch. 1985. Distribution of calpain I, an enzyme associated with degenerative activity, in rat brain. *Brain Res.* 347: 399–403.

Siman, R., T.K. McIntosh, K.M. Soltesz, Z. Chen, R.W. Neumar and V.L. Roberts. 2004. Proteins released from degenerating neurons are surrogate markers for acute brain damage. *Neurobiol Dis.* 16(2): 311–20.

Simmons, D.A., C.S. Rex, L. Palmer, V. Pandyarajan, V. Fedulov, C.M. Gall and G. Lynch. 2009. Up-regulating BDNF with an ampakine rescues synaptic plasticity and memory in Huntington's disease knockin mice. *Proc Natl Acad Sci USA.* 106(12): 4906–11.

Simonin, Y., O. Disson, H. Lerat, E. Antoine, F. Binamé, A.R. Rosenberg, S. Desagher, P. Lassus, P. Bioulac-Sage and U. Hibner. 2009. Calpain activation by hepatitis C virus proteins inhibits the extrinsic apoptotic signaling pathway. *Hepatology.* 50(5): 1370–9.

Simonson, L., M. Baudry, R. Siman and G. Lynch. 1985. Regional distribution of soluble calcium activated proteinase activity in neonatal and adult rat brain. *Brain Res.* 327(1-2): 153–9.

Simpkins, K.L., R.P. Guttmann, Y. Dong, Z. Chen, S. Sokol, R.W. Neumar and D.R. Lynch. 2003. Selective activation induced cleavage of the NR2B subunit by calpain. *J Neurosci.* 23(36): 11322–31.

Singh, R.B., L. Hryshko, D. Freed and N.S. Dhalla. 2012. Activation of proteolytic enzymes and depression of the sarcolemmal Na+/K+-ATPase in ischemia-reperfused heart may be mediated through oxidative stress. *Can J Physiol Pharmacol.* 90(2): 249–60.

Smith, S.D., Z. Jia, K.K. Huynh, A. Wells and J.S. Elce. 2003. Glutamate substitutions at a PKA consensus site are consistent with inactivation of calpain by phosphorylation. *FEBS Lett.* 542(1-3): 115–8.

Sohara, Y., H. Shimada and Y.A. DeClerck. 2005. Mechanisms of bone invasion and metastasis in human neuroblastoma. *Cancer Lett.* 228(1-2): 203–9. Review.

Sorimachi, H., S. Hata and Y. Ono. 2011. Calpain chronicle—an enzyme family under multidisciplinary characterization. *Proc Jpn Acad Ser B Phys Biol Sci.* 87(6): 287–327. Review.

Sorimachi, H., S. Hata and Y. Ono. 2011. Impact of genetic insights into calpain biology. *J Biochem.* 150(1): 23–37.

Sorimachi, H., H. Mamitsuka and Y. Ono. 2012. Understanding the substrate specificity of conventional calpains. *Biol Chem.* 393(9): 853–71.

Sorimachi, H., T.C. Saido and K. Suzuki. 1994. New era of calpain research. Discovery of tissue-specific calpains. *FEBS Lett.* 343(1): 1–5. Review.

Soulé, J., E. Messaoudi and C.R. Bramham. 2006. Brain-derived neurotrophic factor and control of synaptic consolidation in the adult brain. *Biochem Soc Trans.* 34(Pt 4): 600–4. Review.
Southwell, A.L., C.W. Bugg, L.S. Kaltenbach, D. Dunn, S. Butland, A. Weiss, P. Paganetti, D.C. Lo and P.H. Patterson. 2011. Perturbation with intrabodies reveals that calpain cleavage is required for degradation of huntingtin exon 1. *PLoS One.* 6(1): e16676.
Sprague, C.R., T.S. Fraley, H.S. Jang, S. Lal and J.A. Greenwood. 2008. Phosphoinositide binding to the substrate regulates susceptibility to proteolysis by calpain. *J Biol Chem.* 283(14): 9217–23.
Sridharan, D.M., L.W. McMahon and M.W. Lambert. 2006. αII-Spectrin interacts with five groups of functionally important proteins in the nucleus. *Cell Biol Int.* 30(11): 866–78.
Stanyon, C.A. and O. Bernard. 1999. LIM-kinase1. *Int J Biochem Cell Biol.* 31(3-4): 389–94. Review.
Stockholm, D., M. Bartoli, G. Sillon, N. Bourg, J. Davoust and I. Richard. 2005. Imaging calpain protease activity by multiphoton FRET in living mice. *J Mol Biol.* 346(1): 215–22.
Storr, S.J., N.O. Carragher, M.C. Frame, T. Parr and S.G. Martin. 2011. The calpain system and cancer. *Nat Rev Cancer.* 11(5): 364–74. Review.
Suzuki, K., S. Hata, Y. Kawabata and H. Sorimachi. 2004. Structure, activation, and biology of calpain. *Diabetes.* 53 Suppl 1: S12–8. Review.
Takano, J., M. Tomioka, S. Tsubuki, M. Higuchi, N. Iwata, S. Itohara, M. Maki and T.C. Saido. 2005. Calpain mediates excitotoxic DNA fragmentation via mitochondrial pathways in adult brains: evidence from calpastatin mutant mice. *J Biol Chem.* 280(16): 16175–84.
Takei, N., N. Inamura, M. Kawamura, H. Namba, K. Hara, K. Yonezawa and H. Nawa. 2004. Brain-derived neurotrophic factor induces mammalian target of rapamycin-dependent local activation of translation machinery and protein synthesis in neuronal dendrites. *J Neurosci.* 24(44): 9760–9.
Taniguchi, N., Y. Shinoda, N. Takei, H. Nawa, A. Ogura and K. Tominaga-Yoshino. 2006. Possible involvement of BDNF release in long-lasting synapse formation induced by repetitive PKA activation. *Neurosci Lett.* 406(1-2): 38–42.
Tashev, R., P.J. Moura, D.V. Venkitaramani, C. Prosperetti, D. Centonze, S. Paul and P.J. Lombroso. 2009. A substrate trapping mutant form of striatal-enriched protein tyrosine phosphatase prevents amphetamine-induced stereotypies and long-term potentiation in the striatum. *Biol Psychiatry.* 65(8): 637–45.
Tee, A.R., B.D. Manning, P.P. Roux, L.C. Cantley and J. Blenis. 2003. Tuberous sclerosis complex gene products, Tuberin and Hamartin, control mTOR signaling by acting as a GTPase-activating protein complex toward Rheb. *Curr Biol.* 13(15): 1259–68.
Tompa, P., Y. Emori, H. Sorimachi, K. Suzuki and P. Friedrich. 2001. Domain III of calpain is a Ca^{2+}-regulated phospholipid-binding domain. *Biochem Biophys Res Commun.* 280(5): 1333–9.
Tremper-Wells, B. and M.L. Vallano. 2005. Nuclear calpain regulates Ca^{2+}-dependent signaling via proteolysis of nuclear Ca^{2+}/calmodulin-dependent protein kinase type IV in cultured neurons. *J Biol Chem.* 280(3): 2165–75.
Turnbull, J., A.A. DePaoli-Roach, X. Zhao, M.A. Cortez, N. Pencea, E. Tiberia, M. Piliguian, P.J. Roach, P. Wang, C.A. Ackerley and B.A. Minassian. 2011. PTG depletion removes Lafora bodies and rescues the fatal epilepsy of Lafora disease. *PLoS Genet.* 7(4): e1002037.
Tyagarajan, S.K., H. Ghosh, G.E. Yévenes, S.Y. Imanishi, H.U. Zeilhofer, B. Gerrits and J.M. Fritschy. 2013. Extracellular signal-regulated kinase and glycogen synthase kinase 3β regulate gephyrin postsynaptic aggregation and GABAergic synaptic function in a calpain-dependent mechanism. *J Biol Chem.* 288(14): 9634–47.
Vanderklish, P., E. Bednarski and G. Lynch. 1996. Translational suppression of calpain blocks long-term potentiation. *Learn Mem.* 3(2-3): 209–17.
Vanderklish, P., T.C. Saido, C. Gall, A. Arai and G. Lynch. 1995. Proteolysis of spectrin by calpain accompanies theta-burst stimulation in cultured hippocampal slices. *Brain Res Mol Brain Res.* 32(1): 25–35.
Vanderklish P.W. and B.A. Bahr. 2000. The pathogenic activation of calpain: a marker and mediator of cellular toxicity and disease states. *Int J Exp Pathol.* 81(5): 323–39. Review.

Vanderklish, P.W., L.A. Krushel, B.H. Holst, J.A. Gally, K.L. Crossin and G.M. Edelman. 2000. Marking synaptic activity in dendritic spines with a calpain substrate exhibiting fluorescence resonance energy transfer. *Proc Natl Acad Sci USA.* 97(5): 2253–8.

Vazquez, R., A. Wendt, V.F. Thompson, S.M. Novak, C. Ruse, J.R. Yates and D.E. Goll. 2008. Phosphorylation of the calpains. *FASEB J.* 22: 793.5.

Venkitaramani, D.V., S. Paul, Y. Zhang, P. Kurup, L. Ding, L. Tressler, M. Allen, R. Sacca, M.R. Picciotto and P.J. Lombroso. 2009. Knockout of striatal enriched protein tyrosine phosphatase in mice results in increased ERK1/2 phosphorylation. *Synapse.* 63(1): 69–81.

Vidaurre, O.G., S. Gascón, R. Deogracias, M. Sobrado, E. Cuadrado, J. Montaner, A. Rodríguez-Peña and M. Díaz-Guerra. 2012. Imbalance of neurotrophin receptor isoforms TrkB-FL/TrkB-T1 induces neuronal death in excitotoxicity. *Cell Death Dis.* 3: e256; doi:10.1038/cddis.2011.143.

Villa, P.G., W.J. Henzel, M. Sensenbrenner, C.E. Henderson and B. Pettmann. 1998. Calpain inhibitors, but not caspase inhibitors, prevent actin proteolysis and DNA fragmentation during apoptosis. *J Cell Sci.* 111 (Pt 6): 713–22.

Vosler, P.S., C.S. Brennan and J. Chen. 2008. Calpain-mediated signaling mechanisms in neuronal injury and neurodegeneration. *Mol Neurobiol.* 38(1): 78–100. Review.

Wang, H., F. Duclot, Y. Liu, Z. Wang and M. Kabbaj. 2013. Histone deacetylase inhibitors facilitate partner preference formation in female prairie voles. *Nat Neurosci.* 16(7): 919–24.

Wang, H.C., Y.S. Huang, C.C. Ho, J.C. Jeng and H.M. Shih. 2009. SUMO modification modulates the activity of calpain-2. *Biochem Biophys Res Commun.* 384(4): 444–9.

Wang, K.K. 2000. Calpain and caspase: can you tell the difference? *Trends Neurosci.* 23(1): 20–6. Review.

Wang, K.K., A. Villalobo and B.D. Roufogalis. 1989. Calmodulin-binding proteins as calpain substrates. *Biochem J.* 262(3): 693–706. Review.

Wang, S., S. Wang, P. Shan, Z. Song, T. Dai, R. Wang and Z. Chi. 2008. Mu-calpain mediates hippocampal neuron death in rats after lithium-pilocarpine-induced status epilepticus. *Brain Res Bull.* 76(1-2): 90–6.

Wang, Y., V. Briz, A. Chishti, X. Bi and M. Baudry. 2013. Distinct Roles for μ-Calpain and m-Calpain in Synaptic NMDAR-Mediated Neuroprotection and Extrasynaptic NMDAR-Mediated Neurodegeneration. *J Neurosci.* 33(48): 18880–92.

Warren, M.W., F.H. Kobeissy, M.C. Liu, R.L. Hayes, M.S. Gold and K.K. Wang. 2005. Concurrent calpain and caspase-3 mediated proteolysis of alpha II-spectrin and tau in rat brain after methamphetamine exposure: a similar profile to traumatic brain injury. *Life Sci.* 78(3): 301–9.

Warren, M.W., S.F. Larner, F.H. Kobeissy, C.A. Brezing, J.A. Jeung, R.L. Hayes, M.S. Gold and K.K. Wang. 2007. Calpain and caspase proteolytic markers co-localize with rat cortical neurons after exposure to methamphetamine and MDMA. *Acta Neuropathol.* 114(3): 277–86.

Warren, M.W., W. Zheng, F.H. Kobeissy, M. Cheng Liu, R.L. Hayes, M.S. Gold, S.F. Larner and K.K. Wang. 2007. Calpain- and caspase-mediated alphaII-spectrin and tau proteolysis in rat cerebrocortical neuronal cultures after ecstasy or methamphetamine exposure. *Int J Neuropsychopharmacol.* 10(4): 479–89.

Watkins, G.R., N. Wang, M.D. Mazalouskas, R.J. Gomez, C.R. Guthrie, B.C. Kraemer, S. Schweiger, B.W. Spiller and B.E. Wadzinski. 2012. Monoubiquitination promotes calpain cleavage of the protein phosphatase 2A (PP2A) regulatory subunit α4, altering PP2A stability and microtubule-associated protein phosphorylation. *J Biol Chem.* 287(29): 24207–15.

Wells, A., A. Huttenlocher and D.A. Lauffenburger. 2005. Calpain proteases in cell adhesion and motility. *Int Rev Cytol.* 245: 1–16. Review.

Wendt, A., V.F. Thompson and D.E. Goll. 2004. Interaction of calpastatin with calpain: a review. *Biol Chem.* 385(6): 465–72. Review.

Wu, H.Y., F.C. Hsu, A.J. Gleichman, I. Baconguis, D.A. Coulter and D.R. Lynch. 2007. Fyn-mediated phosphorylation of NR2B Tyr-1336 controls calpain-mediated NR2B cleavage in neurons and heterologous systems. *J Biol Chem.* 282(28): 20075–87.

Wu, H.Y. and D.R. Lynch. 2006. Calpain and synaptic function. *Mol Neurobiol.* 33(3): 215–36.

Wu, H.Y., K. Tomizawa and H. Matsui. 2007. Calpain-calcineurin signaling in the pathogenesis of calcium-dependent disorder. *Acta Med Okayama.* 61(3): 123–37. Review.

Wu, H.Y., K. Tomizawa, Y. Oda, F.Y. Wei, Y.F. Lu, M. Matsushita, S.T. Li, A. Moriwaki and H. Matsui. 2004. Critical role of calpain-mediated cleavage of calcineurin in excitotoxic neurodegeneration. *J Biol Chem.* 279(6): 4929–40.

Wu, S. and M.V. Chao. 2007. Activity-dependent regulation of ARMS, a Trk receptor substrate. Program No. 572.15. *Neuroscience 2007 Abstracts.* San Diego, CA: Society for Neuroscience.

Wu, S.H., J.C. Arévalo, V.E. Neubrand, H. Zhang, O. Arancio and M.V. Chao. 2010. The ankyrin repeat-rich membrane spanning (ARMS)/Kidins220 scaffold protein is regulated by activity-dependent calpain proteolysis and modulates synaptic plasticity. *J Biol Chem.* 285(52): 40472–78.

Xiong, H., T. Futamura, H. Jourdi, H. Zhou, N. Takei, M. Diverse-Pierluissi, S. Plevy and H. Nawa. 2002. Neurotrophins induce BDNF expression through the glutamate receptor pathway in neocortical neurons. *Neuropharmacology.* 42(7): 903–12.

Xu, J., P. Kurup, A.C. Nairn and P.J. Lombroso. 2012. Striatal-enriched protein tyrosine phosphatase in Alzheimer's disease. *Adv Pharmacol.* 64: 303–25. Review.

Xu, J., P. Kurup, Y. Zhang, S.M. Goebel-Goody, P.H. Wu, A.H. Hawasli, M.L. Baum, J.A. Bibb and P.J. Lombroso. 2009. Extrasynaptic NMDA receptors couple preferentially to excitotoxicity via calpain-mediated cleavage of STEP. *J Neurosci.* 29(29): 9330–43.

Xu, L. and X. Deng. 2006. Protein kinase Ciota promotes nicotine-induced migration and invasion of cancer cells via phosphorylation of micro- and m-calpains. *J Biol Chem.* 281(7): 4457–66.

Xu, L. and X. Deng. 2006. Suppression of cancer cell migration and invasion by protein phosphatase 2A through dephosphorylation of mu- and m-calpains. *J Biol Chem.* 281(46): 35567–75.

Xu, L. and X. Deng. 2004. Tobacco-specific nitrosamine 4-(methylnitrosamino)-1-(3-pyridyl)-1-butanone induces phosphorylation of µ- and m-calpain in association with increased secretion, cell migration, and invasion. *J Biol Chem.* 279(51): 53683–90.

Xu, W., T.P. Wong, N. Chery, T. Gaertner, Y.T. Wang and M. Baudry. 2007. Calpain-mediated mGluR1alpha truncation: a key step in excitotoxicity. *Neuron.* 53(3): 399–412.

Xu, Z., G.D. Ford, D.R. Croslan, J. Jiang, A. Gates, R. Allen and B.D. Ford. 2005. Neuroprotection by neuregulin-1 following focal stroke is associated with the attenuation of ischemia-induced pro-inflammatory and stress gene expression. *Neurobiol Dis.* 19(3): 461–70.

Xu, Z., J. Jiang, G. Ford and B.D. Ford. 2004. Neuregulin-1 is neuroprotective and attenuates inflammatory responses induced by ischemic stroke. *Biochem Biophys Res Commun.* 322(2): 440–6.

Yamashima, T. 2004. Ca^{2+}-dependent proteases in ischemic neuronal death: a conserved 'calpain-cathepsin cascade from nematodes to primates. *Cell Calcium.* 36(3-4): 285–93. Review.

Yamashima, T. and S. Oikawa. 2009. The role of lysosomal rupture in neuronal death. *Prog Neurobiol.* 89(4): 343–58.

Yan, X.X., A. Jeromin and A. Jeromin. 2012. Spectrin Breakdown Products (SBDPs) as Potential Biomarkers for Neurodegenerative Diseases. *Curr Transl Geriatr Exp Gerontol Rep.* 1(2): 85–93.

Yang, J., R.M. Weimer, D. Kallop, O. Olsen, Z. Wu, N. Renier, K. Uryu and M. Tessier-Lavigne. 2013. Regulation of axon degeneration after injury and in development by the endogenous calpain inhibitor calpastatin. *Neuron.* 80(5): 1175–89.

Yokomaku, D., H. Jourdi, A. Kakita, T. Nagano, H. Takahashi, N. Takei and H. Nawa. 2005. ErbB1 receptor ligands attenuate the expression of synaptic scaffolding proteins, GRIP1 and SAP97, in developing neocortex. *Neuroscience.* 136(4): 1037–47.

Yokomaku, D., H. Jourdi, T. Nagano, M. Mizuno, N. Takei and H. Nawa. 2003. Epidermal growth factor (EGF)/ErbB signaling regulates the expression of synaptic proteins in rat cerebral cortex. *Bulletin of the Japanese Society for Neurochemistry.* 42(2/3): 350.

Yoshimura, Y., T. Nomura and T. Yamauchi. 1996. Purification and characterization of active fragment of Ca2+/calmodulin-dependent protein kinase II from the post-synaptic density in the rat forebrain. *J Biochem.* 119(2): 268–73.

Yousefi, S., R. Perozzo, I. Schmid, A. Ziemiecki, T. Schaffner, L. Scapozza, T. Brunner and H.U. Simon. 2006. Calpain-mediated cleavage of Atg5 switches autophagy to apoptosis. *Nat Cell Biol.* 8(10): 1124–32.

Yu, J. and E.P. Henske. 2006. Estrogen-induced activation of mammalian target of rapamycin is mediated via tuberin and the small GTPase Ras homologue enriched in brain. *Cancer Res.* 66(19): 9461–6.

Yu, C.G., Y. Li, K. Raza, X.X. Yu, S. Ghoshal and J.W. Geddes. 2012. Calpain 1 knockdown improves tissue sparing and functional outcomes following spinal cord injury in rats. *J Neurotrauma.* doi:10.1089/neu.2012.2561.

Yuen, E.Y., W. Liu and Z. Yan. 2007. The phosphorylation state of GluR1 subunits determines the susceptibility of AMPA receptors to calpain cleavage. *J Biol Chem.* 282(22): 16434–40.

Yuen, E.Y., Z. Gu and Z. Yan. 2007. Calpain regulation of AMPA receptor channels in cortical pyramidal neurons. *J Physiol.* 580(Pt 1): 241–54.

Yuen, E.Y., Y. Ren and Z. Yan. 2008. Postsynaptic density-95 (PSD-95) and calcineurin control the sensitivity of N-methyl-D-aspartate receptors to calpain cleavage in cortical neurons. *Mol Pharmacol.* 74(2): 360–70.

Zadran, S., G. Akopian, H. Zadran, J. Walsh and M. Baudry. 2012. RVG-Mediated Calpain2 Gene Silencing in the Brain Impairs Learning and Memory. *Neuromolecular Med.* 15(1): 74–81.

Zadran, S., X. Bi and M. Baudry. 2010. Regulation of calpain-2 in neurons: implications for synaptic plasticity. *Mol Neurobiol.* 42(2): 143–50.

Zadran, S., Q. Qin, X. Bi, H. Zadran, Y. Kim, M.R. Foy, R. Thompson and M. 2009. Baudry. 17-Beta-estradiol increases neuronal excitability through MAP kinase-induced calpain activation. *Proc Natl Acad Sci USA.* 106(51): 21936–41.

Zakharov, V.V. and M.I. Mosevitsky. 2007. M-calpain-mediated cleavage of GAP-43 near Ser41 is negatively regulated by protein kinase C, calmodulin and calpain-inhibiting fragment GAP-43-3. *J Neurochem.* 101(6): 1539–51.

Zhang, R., N.L. Banik and S.K. Ray. 2008. Combination of all-trans retinoic acid and interferon-gamma upregulated p27(kip1) and down regulated CDK2 to cause cell cycle arrest leading to differentiation and apoptosis in human glioblastoma LN18 (PTEN-proficient) and U87MG (PTEN-deficient) cells. *Cancer Chemother Pharmacol.* 62(3): 407–16.

Zhang, Y, S.J. Matkovich, X. Duan, A. Diwan, M.Y. Kang and G.W. Dorn 2nd. 2011. Receptor-independent protein kinase C alpha (PKCalpha) signaling by calpain-generated free catalytic domains induces HDAC5 nuclear export and regulates cardiac transcription. *J Biol Chem.* 286(30): 26943–51.

Zheng, C.Y., G.K. Seabold, M. Horak and R.S. Petralia. 2011. MAGUKs, synaptic development, and synaptic plasticity. *Neuroscientist.* 17(5): 493–512.

Zhou, M., W. Xu G. Liao, X. Bi and M. Baudry. 2009. Neuroprotection against neonatal hypoxia/ischemia-induced cerebral cell death by prevention of calpain-mediated mGluR1alpha truncation. *Exp Neurol.* 218(1): 75–82.

Zhou, M. and M. Baudry. 2006. Developmental changes in NMDA neurotoxicity reflect developmental changes in subunit composition of NMDA receptors. *J Neurosci.* 26(11): 2956–63.

Zhou, Y. and G.E. Besner. 2010. Heparin-binding epidermal growth factor-like growth factor is a potent neurotrophic factor for PC12 cells. *Neurosignals.* 18(3): 141–51.

Zoghbi, H.Y. and M.F. Bear. 2012. Synaptic dysfunction in neurodevelopmental disorders associated with autism and intellectual disabilities. *Cold Spring Harb Perspect Biol.* doi:pii: a009886.10.1101/cshperspect.a009886. Review.

Section B
CNS INJURY BIOMARKERS

7

Protein Biomarkers for Mild Traumatic Brain Injury

Linda Papa,[1,] Neema J. Ameli,[1] Ashley Waplinger,[1] Zhiqun Zhang[2] and Kevin K.W. Wang[2,3,4]*

INTRODUCTION

There are an estimated 10 million people affected annually by Traumatic Brain Injury (TBI) across the globe (Hyder et al., 2007). According to the World Health Organization, TBI will surpass many diseases as the major cause of death and disability by the year 2020 (Hyder et al., 2007). Research in the field of TBI has long been dominated by severe brain injury (Narayan et al., 2002). However, of the estimated 1.8 million people in the United States who sustain a TBI each year, over 90% will have either a "moderate" (GCS 9–12) or "mild" (GCS 13–15) injury (Consensus conference 1999; Vollmer and Dacey, 1991; Yealy and Hogan, 1991). The distinction between mild, moderate and severe TBI is initially based on a GCS score and this may be influenced by factors such as perfusion and intoxication from drugs or alcohol, sedative medications and other distracting injuries. The majority of these patients will present to Emergency Departments (ED's) around the country for assessment and treatment (Langlois et al., 2004).

---

[1] Department of Emergency Medicine, Orlando Regional Medical Center, Orlando Florida.
[2] Center for Neuroproteomics & Biomarkers Research, Department of Psychiatry, University of Florida, Gainesville, FL 32611, USA.
[3] Neuroscience and University of Florida, Gainesville, FL 32611, USA.
[4] Physiological Science, University of Florida, Gainesville, FL 32611, USA.
* Corresponding author: lpstat@aol.com

Mild TBI is often difficult to assess clinically during the first hours after injury because neurological examinations are of restricted value. Mild TBI is significantly under-diagnosed and has been referred to by many as the silent epidemic of our time (Hoffman et al., 2010; Vaishnavi et al., 2009). The term "mild TBI" is actually a misnomer. Individuals who incur a TBI and have an initial GCS score of 13–15 are acutely at risk for intracranial bleeding and diffuse axonal injury (Stein et al., 2009). Additionally, a significant proportion is at risk for impairment of physical, cognitive and psychosocial functioning (Alexander, 1995; Alves et al., 1993; Barth et al., 1983; Millis et al., 2001; Rimel et al., 1981). Although some patients with mild TBI may be admitted to the hospital overnight, the vast majority are treated and released from emergency departments with basic discharge instructions. This group of TBI patients represents the greatest challenges to accurate diagnosis and outcome prediction. The lack of clinical tools to detect the deficits that affect daily function, have left these individuals with little or no treatment options. The injury is often seen as "not severe" and subsequently therapies have not been aggressively sought for mild TBI.

The Utility of Serum Biomarkers for TBI

Currently, diagnosis of TBI depends on a variety of measures including neurological examination and neuroimaging. Neuroimaging techniques such as CT scanning and MRI are used to provide objective information. However, CT scanning has low sensitivity to diffuse brain damage and confers exposure to radiation. MRI can provide information on the extent of diffuse injuries but its widespread application is restricted by cost, the limited availability of MRI in many centers, and the difficulty of performing it in physiologically unstable patients. Additionally, its role in the clinical management of TBI patients acutely has not been established (Jagoda et al., 2008; Kesler, 2000).

Diagnostic and prognostic tools for risk stratification of TBI patients are limited in the early stages of injury in the emergency setting for all severities of TBI. Unlike other organ-based diseases where rapid diagnosis employing biomarkers from blood tests are clinically essential to guide diagnosis and treatment, such as for myocardial ischemia or kidney and liver dysfunction, there are no rapid, definitive diagnostic tests for traumatic brain injury. Research in the field of TBI biomarkers has increased exponentially over the last 20 years (Kochanek et al., 2008; Papa, 2012), with most of the publications on the topic occurring in the last 10 years (Papa et al., 2013). Accordingly, studies assessing biomarkers in TBI have looked at a number of potential markers that could lend diagnostic, prognostic, as well as therapeutic information. Despite the large number of published studies, there is still a lack of any FDA-approved biomarkers for clinical use in adults and children (Papa, 2012; Papa et al., 2013). This chapter will review some of the most widely studied biomarkers for TBI in the clinical setting, with an emphasis on those that have been evaluated in mild TBI. Figure 7.1 shows the neuroanatomical locations of the biomarkers that will be reviewed.

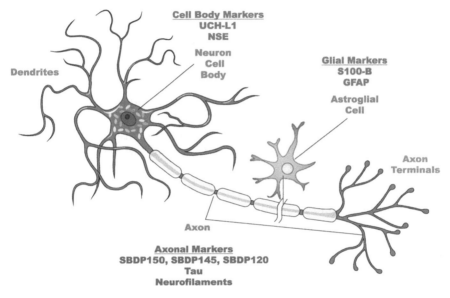

Cell Body Markers
UCH-L1
NSE

Neuron
Cell
Body

Dendrites

Glial Markers
S100-B
GFAP

Astroglial
Cell

Axon
Terminals

Axon

Axonal Markers
SBDP150, SBDP145, SBDP120
Tau
Neurofilaments

Figure 7.1. Neuroanatomic areas where biomarkers are most abundant.

Color image of this figure appears in the color plate section at the end of the book.

Biofluid Biomarkers of Astroglial Injury

S100β

S100β is the major low affinity calcium binding protein in astrocytes (Xiong et al., 2000) that helps to regulate intracellular levels of calcium and it is considered a marker of astrocyte injury or death. It can also be found in non-neural cells such as adipocytes, chondrocytes and melanoma cells (Olsson et al., 2011; Zimmer et al., 1995). S100β is one of the most extensively studied biomarkers (Berger et al., 2002; Haimoto, 1987; Jonsson, 2000; Korfias et al., 2007; Missler, 1997; Raabe et al., 1999; Romner et al., 2000; Usui, 1989; Vos et al., 2010; Woertgen et al., 1997; Ytrebo, 2001). Elevated S100β levels in serum have been associated with increased incidence of post concussive syndrome and problems with cognition (Ingebrigtsen and Romner, 1997; Waterloo et al., 1997). Other studies have reported that serum levels of S-100β are associated with MRI abnormalities and with neuropsychological examination disturbances after mild TBI (Ingebrigtsen and Romner, 1996; Ingebrigtsen et al., 1999). A number of studies have found significant correlations between elevated serum levels of S100β and CT abnormalities (Biberthaler et al., 2006; Ingebrigtsen et al., 2000; Muller et al., 2007). It has been suggested that adding the measurement of S100β concentration to clinical decision tools for mild TBI patients could potentially reduce the number of CT scans by 30% (Biberthaler et al., 2006). However, other investigators have failed to detect associations between S100β with CT abnormalities (Phillips et al., 1980; Piazza et al., 2007; Rothoerl et al., 1998; Bechtel et al., 2009).

Although S100β remains promising as an adjunctive marker, its utility in the setting of multiple trauma remains controversial because it is also elevated in trauma patients without head injuries (Anderson et al., 2001; Pelinka et al., 2004b; Romner and Ingebrigtsen, 2001; Rothoerl and Woertgen, 2001).

Glial Fibrillary Acid Protein (GFAP)

Glial Fibrillary Acidic Protein (GFAP) is a monomeric intermediate protein found in astroglial skeleton that was first isolated by Eng et al. (1971). GFAP is found in white and gray brain matter and is strongly upregulated during astrogliosis (Duchen, 1984). Current evidence indicates that serum GFAP might be a useful marker for various types of brain damage from neurodegenerative disorders (Baydas et al., 2003; Mouser et al., 2006) and stroke (Herrmann et al., 2000) to severe traumatic brain injury (Missler et al., 1999; Mondello et al., 2011; Nylen et al., 2006; Pelinka et al., 2004a,b; van Geel et al., 2002). In 2010, Vos et al. described serum GFAP profile in severe and moderate TBI with GCS <12 and found an association with unfavorable outcome at 6 months. More recently, Metting et al. (2012) found GFAP to be elevated in patients with axonal injury on MRI in patients with mild TBI at 3 months post-injury but it was not predictive of global outcome at 6 months. In a study by Papa et al. in 2012, GFAP was detectable in serum in less than 1 hour after a concussion and was able to distinguish concussion patients from other trauma patients (without head injury) who had orthopedic injuries or who were in motor vehicle crashes. In this same study, serum GFAP was significantly higher in mild TBI patients with intracranial lesions on CT compared to those without lesions and predicted patients who required neurosurgical intervention (Papa et al., 2012a). Similarly, Metting et al., 2012, demonstrated that serum GFAP was increased in patients with an abnormal CT after mild TBI. These studies suggest that GFAP has a good specificity for brain injury acutely after injury.

Biofluid Biomarkers of Neuronal Injury

Neuron Specific Enolase (NSE)

Neuron specific enolase (NSE) is one of the five isozymes of the gycolytic enzyme enolase found in central and peripheral neuronal cell bodies and it has been shown be elevated following cell injury (Skogseid et al., 1992). It is also present in erythrocytes and endocrine cells and has a biological half-life of 48 hours (Schmechel et al., 1978). This protein is passively released into the extracellular space only under pathological conditions during cell destruction. Several reports on serum NSE measurements of mild TBI have been published (Ergun et al., 1998; Fridriksson et al., 2000; Ross et al., 1996; Skogseid et al., 1992; Yamazaki et al., 1995). Many of these studies either contained inadequate control groups or concluded that serum NSE had limited utility as a marker of neuronal damage. Early levels of NSE and MBP concentrations have been correlated with outcome in children, particularly those under 4 years of age (Bandyopadhyay et al., 2005; Berger et al., 2005; Berger et al., 2007; Varma et al., 2003). In the setting of diffuse axonal injury in severe TBI, levels of NSE at 72 hours of injury have shown an association with unfavorable outcome (Chabok et al., 2012). A limitation of NSE is the occurrence of false positive results in the setting of hemolysis (Johnsson et al., 2000; Ramont et al., 2005).

Ubiquitin C-terminal Hydrolase (UCH-L1)

A promising candidate biomarker for TBI currently under investigation is Ubiquitin C-terminal Hydrolase-L1 (UCH-L1). UCH-L1 was previously used as a histological marker for neurons due to its high abundance and specific expression in neurons (Jackson and Thompson, 1981). This protein is involved in the addition and removal of ubiquitin from proteins that are destined for metabolism (Tongaonkar et al., 2000). It has an important role in the removal of excessive, oxidized or misfolded proteins during both normal and pathological conditions in neurons (Gong and Leznik, 2007). Clinical studies in humans with severe TBI have confirmed, using ELISA analysis, that the UCH-L1 protein is significantly elevated in human CSF (Papa et al., 2010; Siman et al., 2009) is detectable very early after injury, and remains significantly elevated for at least 1 week post-injury (Papa et al., 2010). Further studies in severe TBI patients have revealed a very good correlation between CSF and serum levels (Brophy et al., 2011). Increases in serum UCH-L1 have also been found in children with moderate and severe TBI (Berger et al., 2012). Most recently, UCH-L1 was detected in the serum of Mild and Moderate TBI (MMTBI) patients within an hour of injury (Papa et al., 2012b). Serum levels of UCH-L1 discriminated concussion patients from uninjured and non-head injured trauma control patients that had orthopedic injuries or motor vehicle trauma without head injury. Most notable was that levels were significantly higher in those with intracranial lesions on CT than those without lesions as well as those eventually requiring a neurosurgical intervention (Papa et al., 2012b).

Biofluid Biomarkers of Axonal Injury

Alpha-II Spectrin Breakdown Products

Alpha-II-spectrin (280 kDa) is the major structural component of the cortical membrane cytoskeleton and is particularly abundant in axons and presynaptic terminals (Goodman et al., 1995; Riederer et al., 1986). It is also a major substrate for both calpain and caspase-3 cysteine proteases (McGinn et al., 2009; Wang et al., 1998). A hallmark feature of apoptosis and necrosis is an early cleavage of several cellular proteins by activated caspases and calpains. A signature of caspase-3 and calpain-2 activation is cleavage of several common proteins such as cytoskeletal αII-spectrin (Pike et al., 2004; Ringger et al., 2004). Levels of spectrin breakdown products (SBDP's) have been reported in CSF from adults with severe TBI and they have shown a significant relationship with severity of injury and clinical outcome (Cardali and Maugeri, 2006; Farkas et al., 2005; Mondello et al., 2010; Papa et al., 2004; Papa et al., 2006; Papa et al., 2005; Pineda et al., 2007). The time course of calpain mediated SBDP150 and SBDP145 (markers of necrosis) differs from that of caspase-3 mediated SBDP120 (marker of apoptosis). Average SBDP values measured in CSF early after injury have been shown to correlate with severity of injury, CT scan findings and outcome at 6 months post injury (Brophy et al., 2009).

Serum SBDP145 has also been measured in serum in children with TBI. Levels were significantly greater in subjects with moderate and severe TBI than in controls (but not in mild TBI) and were correlated with dichotomized GOS at 6 months. This correlation did not hold true for mild TBI. More recently, however, serum levels of SBDP150 has been examined in patients with mild TBI and has shown a significant association with acute measures of injury severity, such as GCS score, intracranial injuries on CT and neurosurgical intervention (Papa et al., 2012c). In this study, serum SBDP150 levels were much higher in patients with mild TBI/concussion than other trauma patients who did not have a head injury (Papa et al., 2012c).

Tau Protein

Tau is an intracellular, microtubule-associated protein that is highly enriched in axons and is involved with assembling axonal microtubule bundles and participating in anterograde axoplasmic transport (Teunissen et al., 2005). Since Tau is preferentially localized in the axon, tau lesions are apparently related to axonal disruption (Higuchi et al., 2002; Kosik and Finch, 1987). A supposedly cleaved form of tau, c-tau, has been investigated as a potential biomarker of CNS injury. Following TBI tau is proteolytically cleaved (c-tau) and gains access to cerebrospinal fluid (CSF) and serum. In a study by Shaw et al. (2002) CSF levels of c-tau were significantly elevated in TBI patients compared to control patients and these levels correlated with clinical outcome. In a similar study, initial elevated CSF c-tau levels in severe TBI patients were significant predictors of ICP and clinical outcome (Zemlan et al., 2002). However these findings did not hold true when c-tau was measured in peripheral blood or in mild TBI. Though levels of c-tau were also elevated in plasma from patients with severe TBI, there was no correlation between plasma levels and clinical outcome (Chatfield et al., 2002). Two studies assessed whether serum cleaved tau (C-tau) could predict Post Concussion Syndrome (PCS) in adults after mild TBI at 3 months post injury. C-tau was a poor predictor of CT lesions and also a poor predictor of PCS (Bazarian et al., 2006; Ma et al., 2008).

Total tau protein is highly expressed in thin, nonmyelinated axons of cortical interneurons (Trojanowski et al., 1989) thus, may be indicative of axonal damage in gray matter neurons. It has been found to be correlated with severity of injury in severe TBI (Franz et al., 2003; Marklund et al., 2009; Ost et al., 2006; Sjogren et al., 2001). Ost et al. (2006) found that total Tau measured in CSF on days 2 to 3 discriminated between TBI and controls (normal pressure hydrocephalus) and also between good and bad outcome at one year per dichotomized GOS score. However, total tau was not detected in serum throughout the study.

Neurofilaments

Neurofilaments are heteropolymeric components of the neuron cytoskeleton that consist of a 68 kDa light neurofilament subunit (NF-L) backbone with either 160 kDa medium (NF-M) or 200 kDa heavy subunit (NF-H) side-arms (Julien and Mushynski, 1998). Following TBI, calcium influx into the cell contributes to a cascade of events that activates calcineurin, a calcium-dependent phosphatase that dephosphorylates neurofilament side-arms, presumably contributing to axonal injury (Buki and Povlishock, 2006). Phosphorylated NF-H has been found to be elevated in the CSF of adult patients with severe TBI compared to controls (Siman et al., 2009). Similarly, hyperphosphorylated NF-H has also been correlated with severity of brain injury in children (Zurek et al., 2012). In a study by Zurek et al. (2012) NF-H levels taken on the second to fourth day remained significantly higher in patients with poor outcome in comparison to patients with good outcome. Additionally, NF-H was significantly higher in those children with diffuse axonal injury on initial CT scan (Zurek et al., 2012). Accordingly, Vajtr et al. (2013) compared 10 patients with diffuse axonal injury (DAI) to 28 patients with focal injuries and found that serum NF-H was much higher in patients with DAI over 10 days after admission. Serum NF-H levels were highest from the fourth to the tenth day in both groups.

Future Directions

The release of substances and potential biomarkers after an injury is not a static process. Understanding the biokinetic properties of a biomarker will be essential to understanding the release pattern and "optimum" time for measurement. Clinicians and researchers will have to keep in mind that different injury types may demonstrate different kinetic parameters and, thus, may produce different quantities of a marker with different peaks and rates of decay. The following properties will need to be carefully considered when evaluating a biomarker for clinical application. Studies will need to assess: 1) degree of sensitivity and specificity for brain injury; 2) ability to stratify of patients by severity of injury; 3) rate of appearance in accessible biological fluid; 4) association with injury mechanisms; 5) biokinetic properties; 6) ability to monitor progress of disease and response to treatment; and 7) prognostic features of outcome (Papa, 2012; Papa et al., 2008).

Conclusion

Over the last decade there have been significant advances in the field of human biochemistry and physiology which have increased insight into the neurobiology of traumatic brain injury (TBI). Newer proteomic techniques that are more specific and selective than traditional methods have allowed the detection of proteins amid complex high-protein content biofluids such as serum or plasma. These developments have brought neurobiomarkers a step closer to the bedside. Biomarkers could potentially facilitate diagnosis and risk stratification of mild TBI and impact management of these patients.

Moreover, there are major opportunities for improving the conduct of clinical research by providing early information about injury mechanism(s) and possible drug targets.

References

Alexander, M.P. 1995. Mild traumatic brain injury: pathophysiology, natural history, and clinical management. *Neurology*. 45: 1253–60.

Alves, W., S. Macciocchi and J.T. Barth. 1993. Postconcussive Symptoms After Uncomplicated Mild Head Injury. *J Head Trauma Rehabil*. 8: 48–59.

Anderson, R.E., L.O. Hansson, O. Nilsson, R. Dijlai-Merzoug and G. Settergen. 2001. High serum S100B levels for trauma patients without head injuries. *Neurosurgery*. 49: 1272–3.

Bandyopadhyay, S., H. Hennes, M.H. Gorelick, R.G. Wells and C.M. Walsh-Kelly. 2005. Serum neuron-specific enolase as a predictor of short-term outcome in children with closed traumatic brain injury. *Acad Emerg Med*. 12: 732–8.

Barth, J.T., S.N. Macciocchi, B. Giordani, R. Rimel, J.A. Jane and T.J. Boll. 1983. Neuropsychological sequelae of minor head injury. *Neurosurgery*. 13: 529–33.

Baydas, G., V.S. Nedzvetskii, M. Tuzcu, A. Yasar and S.V. Kirichenko. 2003. Increase of glial fibrillary acidic protein and S-100B in hippocampus and cortex of diabetic rats: effects of vitamin E. *Eur J Pharmacol*. 462: 67–71.

Bazarian, J.J., F.P. Zemlan, S. Mookerjee and T. Stigbrand. 2006. Serum S-100B and cleaved-tau are poor predictors of long-term outcome after mild traumatic brain injury. *Brain Inj*. 20: 759–65.

Bechtel, K., S. Frasure, C. Marshall, J. Dziura and C. Simpson. 2009. Relationship of serum S100B levels and intracranial injury in children with closed head trauma. *Pediatrics*. 124: e697–704.

Berger, R.P., P.D. Adelson, M.C. Pierce, T. Dulani, L.D. Cassidy and P.M. Kochanek. 2005. Serum neuron-specific enolase, S100B, and myelin basic protein concentrations after inflicted and noninflicted traumatic brain injury in children. *J Neurosurg*. 103: 61–8.

Berger, R.P., S.R. Beers, R. Richichi, D. Wiesman and P.D. Adelson. 2007. Serum biomarker concentrations and outcome after pediatric traumatic brain injury. *J Neurotrauma*. 24: 1793–801.

Berger, R.P., R.L. Hayes, R. Richichi, S.R. Beers and K.K. Wang. 2012. Serum concentrations of ubiquitin C-terminal hydrolase-L1 and alphaII-spectrin breakdown product 145 kDa correlate with outcome after pediatric TBI. *J Neurotrauma*. 29: 162–7.

Berger, R.P., M.C. Pierce, S.R. Wisniewski, P.D. Adelson and P.M. Kochanek. 2002. Serum S100B concentrations are increased after closed head injury in children: a preliminary study. *J Neurotrauma*. 19: 1405–9.

Biberthaler, P., U. Linsenmeier, K.J. Pfeifer, M. Kroetz, T. Mussack, K.G. Kanz, E.F. Hoecherl, F. Jonas, I. Marzi, P. Leucht, M. Jochum and W. Mutschler. 2006. Serum S-100B concentration provides additional information for the indication of computed tomography in patients after minor head injury: a prospective multicenter study. *Shock*. 25: 446–53.

Brophy, G., S. Mondello, L. Papa, S. Robicsek, A. Gabrielli, J. Tepas Iii, A. Buki, C. Robertson, F.C. Tortella and K.K. Wang. 2011. Biokinetic Analysis of Ubiquitin C-Terminal Hydrolase-L1 (Uch-L1) in Severe Traumatic Brain Injury Patient Biofluids. *J Neurotrauma*.

Brophy, G.M., J.A. Pineda, L. Papa, S.B. Lewis, A.B. Valadka, H.J. Hannay, S.C. Heaton, J.A. Demery, M.C. Liu, J.J. Tepas, 3rd, A. Gabrielli, S. Robicsek, K.K. Wang, C.S. Robertson and R.L. Hayes. 2009. alphaII-Spectrin breakdown product cerebrospinal fluid exposure metrics suggest differences in cellular injury mechanisms after severe traumatic brain injury. *J Neurotrauma*. 26: 471–9.

Buki, A. and J.T. Povlishock. 2006. All roads lead to disconnection?—Traumatic axonal injury revisited. *Acta Neurochir (Wien).* 148: 181–93; discussion 193–4.

Cardali, S. and R. Maugeri. 2006. Detection of alphaII-spectrin and breakdown products in humans after severe traumatic brain injury. *J Neurosurg Sci.* 50: 25–31.

Chabok, S.Y., A.D. Moghadam, Z. Saneei, F.G. Amlashi, E.K. Leili and Z.M. Amiri. 2012. Neuron-specific enolase and S100BB as outcome predictors in severe diffuse axonal injury. *J Trauma Acute Care Surg.* 72: 1654–7.

Chatfield, D.A., F.P. Zemlan, D.J. Day and D.K. Menon. 2002. Discordant temporal patterns of S100beta and cleaved tau protein elevation after head injury: a pilot study. *Br J Neurosurg.* 16: 471–6.

Consensus conference. 1999. Rehabilitation of persons with traumatic brain injury. NIH Consensus Development Panel on Rehabilitation of Persons With Traumatic Brain Injury. *Jama.* 282: 974–83.

Duchen, L.W. 1984. General pathology of neurons and neuroglia. pp. 1–52. *In*: J.A. Adams, J.A.N. Corsellis and L.W. Duchen [eds.]. Greenfield's Neuropathology. Edward Arnold, London.

Eng, L.F., J.J. Vanderhaeghen, A. Bignami and B. Gerstl. 1971. An acidic protein isolated from fibrous astrocytes. *Brain Res.* 28: 351–4.

Ergun, R., U. Bostanci, G. Akdemir, E. Beskonakli, E. Kaptanoglu, F. Gursoy and Y. Taskin. 1998. Prognostic value of serum neuron-specific enolase levels after head injury. *Neurol Res.* 20: 418–20.

Farkas, O., B. Polgar, J. Szekeres-Bartho, T. Doczi, J.T. Povlishock and A. Buki. 2005. Spectrin breakdown products in the cerebrospinal fluid in severe head injury—preliminary observations. *Acta Neurochir (Wien).* 147: 855–61.

Franz, G., R. Beer, A. Kampfl, K. Engelhardt, E. Schmutzhard, H. Ulmer and F. Deisenhammer. 2003. Amyloid beta 1-42 and tau in cerebrospinal fluid after severe traumatic brain injury. *Neurology.* 60: 1457–61.

Fridriksson, T., N. Kini, C. Walsh-Kelly and H. Hennes. 2000. Serum neuron-specific enolase as a predictor of intracranial lesions in children with head trauma: a pilot study. *Acad Emerg Med.* 7: 816–20.

Gong, B. and E. Leznik. 2007. The role of ubiquitin C-terminal hydrolase L1 in neurodegenerative disorders. *Drug News Perspect.* 20: 365–70.

Goodman, S.R., W.E. Zimmer, M.B. Clark, I.S. Zagon, J.E. Barker and M.L. Bloom. 1995. Brain spectrin: of mice and men. *Brain Res Bull.* 36: 593–606.

Haimoto, H.H., S. Kato, K. 1987. Differential distribution of immunoreactive S100-a and S100-b proteins in normal nonnervous human tissues. *Lab Invest.* 57: 489–498.

Herrmann, M., P. Vos, M.T. Wunderlich, C.H. de Bruijn and K.J. Lamers. 2000. Release of glial tissue-specific proteins after acute stroke: A comparative analysis of serum concentrations of protein S-100B and glial fibrillary acidic protein. *Stroke.* 31: 2670–7.

Higuchi, M., V.M. Lee and J.Q. Trojanowski. 2002. Tau and axonopathy in neurodegenerative disorders. *Neuromolecular Med.* 2: 131–50.

Hoffman, S.W., K. Shesko and C.R. Harrison. 2010. Enhanced neurorehabilitation techniques in the DVBIC Assisted Living Pilot Project. *NeuroRehabilitation.* 26: 257–69.

Hyder, A.A., C.A. Wunderlich, P. Puvanachandra, G. Gururaj and O.C. Kobusingye. 2007. The impact of traumatic brain injuries: a global perspective. *NeuroRehabilitation.* 22: 341–53.

Ingebrigtsen, T. and B. Romner. 1996. Serial S-100 protein serum measurements related to early magnetic resonance imaging after minor head injury. Case report. *J Neurosurg.* 85: 945–8.

Ingebrigtsen, T. and B. Romner. 1997. Management of minor head injuries in hospitals in Norway. *Acta Neurol Scand.* 95: 51–5.

Ingebrigtsen, T., B. Romner, S. Marup-Jensen, M. Dons, C. Lundqvist, J. Bellner, C. Alling and S.E. Borgesen. 2000. The clinical value of serum S-100 protein measurements in minor head injury: a Scandinavian multicentre study. *Brain Inj.* 14: 1047–55.

Ingebrigtsen, T., K. Waterloo, E.A. Jacobsen, B. Langbakk and B. Romner. 1999. Traumatic brain damage in minor head injury: relation of serum S-100 protein measurements to magnetic resonance imaging and neurobehavioral outcome. *Neurosurgery.* 45: 468–75; discussion 475–6.

Jackson, P. and R.J. Thompson. 1981. The demonstration of new human brain-specific proteins by high-resolution two-dimensional polyacrylamide gel electrophoresis. *J Neurol Sci.* 49: 429–38.

Jagoda, A.S., J.J. Bazarian, J.J. Bruns, Jr., S.V. Cantrill, A.D. Gean, P.K. Howard, J. Ghajar, S. Riggio, D.W. Wright, R.L. Wears, A. Bakshy, P. Burgess, M.M. Wald and R.R. Whitson. 2008. Clinical policy: neuroimaging and decisionmaking in adult mild traumatic brain injury in the acute setting. *Ann Emerg Med.* 52: 714–48.

Johnsson, P., S. Blomquist, C. Luhrs, G. Malmkvist, C. Alling, J.O. Solem and E. Stahl. 2000. Neuron-specific enolase increases in plasma during and immediately after extracorporeal circulation. *Ann Thorac Surg.* 69: 750–4.

Jonsson, H., J.P., P. Hoglund, C. Alling and S. Blomquist. 2000. The elimination of S-100b and renal function after cardiac surgery. *J Cardiothorac Vasc Aneth.* 14: 698–701.

Julien, J.P. and W.E. Mushynski. 1998. Neurofilaments in health and disease. *Prog Nucleic Acid Res Mol Biol.* 61: 1–23.

Kesler, e.a. 2000. APECT, MR and quantitative MR imaging: correlates with neuropsycholgical. *Brain Injury.* 14: 851–857.

Kochanek, P.M., R.P. Berger, H. Bayr, A.K. Wagner, L.W. Jenkins and R.S. Clark. 2008. Biomarkers of primary and evolving damage in traumatic and ischemic brain injury: diagnosis, prognosis, probing mechanisms, and therapeutic decision making. *Curr Opin Crit Care.* 14: 135–141.

Korfias, S., G. Stranjalis, E. Boviatsis, C. Psachoulia, G. Jullien, B. Gregson, A.D. Mendelow and D.E. Sakas. 2007. Serum S-100B protein monitoring in patients with severe traumatic brain injury. *Intensive Care Med.* 33: 255–60.

Kosik, K.S. and E.A. Finch. 1987. MAP2 and tau segregate into dendritic and axonal domains after the elaboration of morphologically distinct neurites: an immunocytochemical study of cultured rat cerebrum. *J Neurosci.* 7: 3142–53.

Langlois, J.A., W. Rutland-Brown and K.E. Thomas. 2004. Traumatic Brain Injury in the United States: Emergency Department Visits, Hospitalizations, and Deaths. Division of Injury and Disability Outcomes and Programs. National Center for Injury Prevention and Control. CDC, Atlanta. 1–55.

Ma, M., C.J. Lindsell, C.M. Rosenberry, G.J. Shaw and F.P. Zemlan. 2008. Serum cleaved tau does not predict postconcussion syndrome after mild traumatic brain injury. *Am J Emerg Med.* 26: 763–8.

Marklund, N., K. Blennow, H. Zetterberg, E. Ronne-Engstrom, P. Enblad and L. Hillered. 2009. Monitoring of brain interstitial total tau and beta amyloid proteins by microdialysis in patients with traumatic brain injury. *J Neurosurg.* 110: 1227–37.

McGinn, M.J., B.J. Kelley, L. Akinyi, M.W. Oli, M.C. Liu, R.L. Hayes, K.K. Wang and J.T. Povlishock. 2009. Biochemical, structural, and biomarker evidence for calpain-mediated cytoskeletal change after diffuse brain injury uncomplicated by contusion. *J Neuropathol Exp Neurol.* 68: 241–9.

Metting, Z., N. Wilczak, L.A. Rodiger, J.M. Schaaf and J. van der Naalt. 2012. GFAP and S100B in the acute phase of mild traumatic brain injury. *Neurology.* 78: 1428–33.

Millis, S.R., M. Rosenthal, T.A. Novack, M. Sherer, T.G. Nick, J.S. Kreutzer, W.M. High, Jr. and J.H. Ricker. 2001. Long-term neuropsychological outcome after traumatic brain injury. *J Head Trauma Rehabil.* 16: 343–55.

Missler, U., M. Wiesmann, G. Wittmann, O. Magerkurth and H. Hagenstrom. 1999. Measurement of glial fibrillary acidic protein in human blood: analytical method and preliminary clinical results. *Clin Chem.* 45: 138–41.

Missler, U. 1997. S-100 protein and neuron-specific enolase concentrations in blood as indicators of infarction volume and prognosis in acute ischemic stroke. *Stroke.* 28: 1956–1960.

Mondello, S., L. Papa, A. Buki, R. Bullock, E. Czeiter, F. Tortella, K.K. Wang and R.L. Hayes. 2011. Neuronal and glial markers are differently associated with computed tomography findings and outcome in patients with severe traumatic brain injury: a case control study. *Crit Care.* 15: R156.

Mondello, S., S.A. Robicsek, A. Gabrielli, G.M. Brophy, L. Papa, J. Tepas, C. Robertson, A. Buki, D. Scharf, M. Jixiang, L. Akinyi, U. Muller, K.K. Wang and R.L. Hayes. 2010. alphaII-spectrin breakdown products (SBDPs): diagnosis and outcome in severe traumatic brain injury patients. *J Neurotrauma.* 27: 1203–13.

Mouser, P.E., E. Head, K.H. Ha and T.T. Rohn. 2006. Caspase-mediated cleavage of glial fibrillary acidic protein within degenerating astrocytes of the Alzheimer's disease brain. *Am J Pathol.* 168: 936–46.

Muller, K., W. Townend, N. Biasca, J. Unden, K. Waterloo, B. Romner and T. Ingebrigtsen. 2007. S100B serum level predicts computed tomography findings after minor head injury. *J Trauma.* 62: 1452–6.

Narayan, R.K., M.E. Michel, B. Ansell, A. Baethmann, A. Biegon, M.B. Bracken, M.R. Bullock, S.C. Choi, G.L. Clifton, C.F. Contant, W.M. Coplin, W.D. Dietrich, J. Ghajar, S.M. Grady, R.G. Grossman, E.D. Hall, W. Heetderks, D.A. Hovda, J. Jallo, R.L. Katz, N. Knoller, P.M. Kochanek, A.I. Maas, J. Majde, D.W. Marion, A. Marmarou, L.F. Marshall, T.K. McIntosh, E. Miller, N. Mohberg, J.P. Muizelaar, L.H. Pitts, P. Quinn, G. Riesenfeld, C.S. Robertson, K.I. Strauss, G. Teasdale, N. Temkin, R. Tuma, C. Wade, M.D. Walker, M. Weinrich, J. Whyte, J. Wilberger, A.B. Young and L. Yurkewicz. 2002. Clinical trials in head injury. *J Neurotrauma.* 19: 503–57.

Nylen, K., M. Ost, L.Z. Csajbok, I. Nilsson, K. Blennow, B. Nellgard and L. Rosengren. 2006. Increased serum-GFAP in patients with severe traumatic brain injury is related to outcome. *J Neurol Sci.* 240: 85–91.

Olsson, B., H. Zetterberg, H. Hampel and K. Blennow. 2011. Biomarker-based dissection of neurodegenerative diseases. *Prog Neurobiol.* 95: 520–34.

Ost, M., K. Nylen, L. Csajbok, A.O. Ohrfelt, M. Tullberg, C. Wikkelso, P. Nellgard, L. Rosengren, K. Blennow and B. Nellgard. 2006. Initial CSF total tau correlates with 1-year outcome in patients with traumatic brain injury. *Neurology.* 67: 1600–4.

Papa, L. 2012. Exploring the Role of Biomarkers for the Diagnosis and Management of Traumatic Brain Injury Patients. *In:* T.K. Man and R.J. Flores [eds.]. Poteomics—Human Diseases and Protein Functions. In Tech Open Access Publisher.

Papa, L., L. Akinyi, M.C. Liu, J.A. Pineda, J.J. Tepas, 3rd, M.W. Oli, W. Zheng, G. Robinson, S.A. Robicsek, A. Gabrielli, S.C. Heaton, H.J. Hannay, J.A. Demery, G.M. Brophy, J. Layon, C.S. Robertson, R.L. Hayes and K.K. Wang. 2010. Ubiquitin C-terminal hydrolase is a novel biomarker in humans for severe traumatic brain injury. *Crit Care Med.* 38: 138–44.

Papa, L., D. D'Avella, M. Aguennouz, F.F. Angileri, O. de Divitiis, A. Germano, A. Toscano, F. Tomasello, G. Vita, J.A. Pineda, R.L. Hayes, B. Pike and K.K.W. Wang. 2004. Detection of Alpha-II Spectrin And Breakdown Products In Humans After Severe Traumatic Brain Injury [abstract]. *Acad Emerg Med.* 11.

Papa, L., L.M. Lewis, J.L. Falk, Z. Zhang, S. Silvestri, P. Giordano, G.M. Brophy, J.A. Demery, N.K. Dixit, I. Ferguson, M.C. Liu, J. Mo, L. Akinyi, K. Schmid, S. Mondello, C.S. Robertson, F.C. Tortella, R.L. Hayes and K.K. Wang. 2012a. Elevated levels of serum glial fibrillary acidic protein breakdown products in mild and moderate traumatic brain injury are associated with intracranial lesions and neurosurgical intervention. *Ann Emerg Med.* 59: 471–83.

Papa, L., L.M. Lewis, S. Silvestri, J.L. Falk, P. Giordano, G.M. Brophy, J.A. Demery, M.C. Liu, J. Mo, L. Akinyi, S. Mondello, K. Schmid, C.S. Robertson, F.C. Tortella, R.L. Hayes and K.K. Wang. 2012b. Serum levels of ubiquitin C-terminal hydrolase distinguish mild traumatic brain injury from trauma controls and are elevated in mild and moderate traumatic brain injury patients with intracranial lesions and neurosurgical intervention. *J Trauma Acute Care Surg.* 72: 1335–1344.

Papa, L., K.W. Wang, B. G.B., J.A. Demery, S. Silvestri, P. Giordano, J.L. Falk, K. Schmid, F.C. Tortella, R.L. Hayes and C.S. Robertson. 2012c. Serum Levels of Spectrin Breakdown Product 150 (SBDP150) distinguish Mild Traumatic Brain Injury from Trauma and

Uninjured Controls and Predict Intracranial Injuries on CT and Neurosurgical Intervention. *J Neurotrauma*. 29: A28.

Papa, L., S.B. Lewis, S. Heaton, J.A. Demery, J.J. Tepas III, K.K.W. Wang, C.S. Robertson and R.L. Hayes. 2006. Predicting Early Outcome Using Alpha-II Spectrin Breakdown Products In Human CSF After Severe Traumatic Brain Injury [abstract]. *Acad Emerg Med*. 13.

Papa, L., J. Pineda, K.K.W. Wang, S.B. Lewis, J.A. Demery, S. Heaton, J.J. Tepas III and R.L. Hayes. 2005. Levels of Alpha-II Spectrin Breakdown Products in Human CSF and Outcome After Severe Traumatic Brain Injury [abstract]. *Acad Emerg Med*. 12.

Papa, L., M.M. Ramia, J.M. Kelly, S.S. Burks, A. Pawlowicz and R.P. Berger. 2013. Systematic review of clinical research on biomarkers for pediatric traumatic brain injury. *J Neurotrauma*. 30: 324–38.

Papa, L., G. Robinson, M. Oli, J. Pineda, J. Demery, G. Brophy, S.A. Robicsek, A. Gabrielli, C.S. Robertson, K.W. Wang and R.L. Hayes. 2008. Use of Biomarkers for Diagnosis and Management of Traumatic Brain Injury Patients. *Expert Opinion on Medical Diagnostics*. 2: 937–945.

Pelinka, L.E., A. Kroepfl, M. Leixnering, W. Buchinger, A. Raabe and H. Redl. 2004a. GFAP versus S100B in serum after traumatic brain injury: relationship to brain damage and outcome. *J Neurotrauma*. 21: 1553–61.

Pelinka, L.E., A. Kroepfl, R. Schmidhammer, M. Krenn, W. Buchinger, H. Redl and A. Raabe. 2004b. Glial fibrillary acidic protein in serum after traumatic brain injury and multiple trauma. *J Trauma*. 57: 1006–12.

Phillips, J.P., H.M. Jones, R. Hitchcock, N. Adama and R.J. Thompson. 1980. Radioimmunoassay of serum creatine kinase BB as index of brain damage after head injury. *Br Med J*. 281: 777–9.

Piazza, O., M.P. Storti, S. Cotena, F. Stoppa, D. Perrotta, G. Esposito, N. Pirozzi and R. Tufano. 2007. S100B is not a reliable prognostic index in paediatric TBI. *Pediatr Neurosurg*. 43: 258–64.

Pike, B.R., J. Flint, J.R. Dave, X.C. Lu, K.K. Wang, F.C. Tortella and R.L. Hayes. 2004. Accumulation of calpain and caspase-3 proteolytic fragments of brain-derived alphaII-spectrin in cerebral spinal fluid after middle cerebral artery occlusion in rats. *J Cereb Blood Flow Metab*. 24: 98–106.

Pineda, J.A., S.B. Lewis, A.B. Valadka, L. Papa, H.J. Hannay, S.C. Heaton, J.A. Demery, M.C. Liu, J.M. Aikman, V. Akle, G.M. Brophy, J.J. Tepas, K.K. Wang, C.S. Robertson and R.L. Hayes. 2007. Clinical significance of alphaII-spectrin breakdown products in cerebrospinal fluid after severe traumatic brain injury. *J Neurotrauma*. 24: 354–66.

Raabe, A., C. Grolms and V. Seifert. 1999. Serum markers of brain damage and outcome prediction in patients after severe head injury. *Br J Neurosurg*. 13: 56–9.

Ramont, L., H. Thoannes, A. Volondat, F. Chastang, M.C. Millet and F.X. Maquart. 2005. Effects of hemolysis and storage condition on neuron-specific enolase (NSE) in cerebrospinal fluid and serum: implications in clinical practice. *Clin Chem Lab Med*. 43: 1215–7.

Riederer, B.M., I.S. Zagon and S.R. Goodman. 1986. Brain spectrin (240/235) and brain spectrin (240/235E): two distinct spectrin subtypes with different locations within mammalian neural cells. *J Cell Biol*. 102: 2088–97.

Rimel, R.W., B. Giordani, J.T. Barth, T.J. Boll and J.A. Jane. 1981. Disability caused by minor head injury. *Neurosurgery*. 9: 221–8.

Ringger, N.C., B.E. O'Steen, J.G. Brabham, X. Silver, J. Pineda, K.K. Wang, R.L. Hayes and L. Papa. 2004. A novel marker for traumatic brain injury: CSF alphaII-spectrin breakdown product levels. *J Neurotrauma*. 21: 1443–56.

Romner, B. and T. Ingebrigtsen. 2001. High serum S100B levels for trauma patients without head injuries. *Neurosurgery*. 49: 1490; author reply 1492–3.

Romner, B., T. Ingebrigtsen, P. Kongstad and S.E. Borgesen. 2000. Traumatic brain damage: serum S-100 protein measurements related to neuroradiological findings. *J Neurotrauma*. 17: 641–7.

Ross, S.A., R.T. Cunningham, C.F. Johnston and B.J. Rowlands. 1996. Neuron-specific enolase as an aid to outcome prediction in head injury. *Br J Neurosurg*. 10: 471–6.

Rothoerl, R.D. and C. Woertgen. 2001. High serum S100B levels for trauma patients without head injuries. *Neurosurgery.* 49: 1490–1; author reply 1492–3.

Rothoerl, R.D., C. Woertgen, M. Holzschuh, C. Metz and A. Brawanski. 1998. S-100 serum levels after minor and major head injury. *J Trauma.* 45: 765–7.

Schmechel, D., P.J. Marangos and M. Brightman. 1978. Neurone-specific enolase is a molecular marker for peripheral and central neuroendocrine cells. *Nature.* 276: 834–6.

Shaw, G.J., E.C. Jauch and F.P. Zemlan. 2002. Serum cleaved tau protein levels and clinical outcome in adult patients with closed head injury. *Ann Emerg Med.* 39: 254–7.

Siman, R., N. Toraskar, A. Dang, E. McNeil, M. McGarvey, J. Plaum, E. Maloney and M.S. Grady. 2009. A panel of neuron-enriched proteins as markers for traumatic brain injury in humans. *J Neurotrauma.* 26: 1867–77.

Sjogren, M., M. Blomberg, M. Jonsson, L.O. Wahlund, A. Edman, K. Lind, L. Rosengren, K. Blennow and A. Wallin. 2001. Neurofilament protein in cerebrospinal fluid: a marker of white matter changes. *J Neurosci Res.* 66: 510–6.

Skogseid, I.M., H.K. Nordby, P. Urdal, E. Paus and F. Lilleaas. 1992. Increased serum creatine kinase BB and neuron specific enolase following head injury indicates brain damage. *Acta Neurochir (Wien).* 115: 106–11.

Stein, S.C., A. Fabbri, F. Servadei and H.A. Glick. 2009. A critical comparison of clinical decision instruments for computed tomographic scanning in mild closed traumatic brain injury in adolescents and adults. *Ann Emerg Med.* 53: 180–8.

Teunissen, C.E., C. Dijkstra and C. Polman. 2005. Biological markers in CSF and blood for axonal degeneration in multiple sclerosis. *Lancet Neurol.* 4: 32–41.

Tongaonkar, P., L. Chen, D. Lambertson, B. Ko and K. Madura. 2000. Evidence for an interaction between ubiquitin-conjugating enzymes and the 26S proteasome. *Mol Cell Biol.* 20: 4691–8.

Trojanowski, J.Q., T. Schuck, M.L. Schmidt and V.M. Lee. 1989. Distribution of tau proteins in the normal human central and peripheral nervous system. *J Histochem Cytochem.* 37: 209–15.

Usui, A., K.K., T. Abe, M. Murase, M. Tanaka and E. Takeuchi. 1989. S-100ao protein in blood and urine during open-heart surgery. *Clin Chem.* 35: 1942–1944.

Vaishnavi, S., V. Rao and J.R. Fann. 2009. Neuropsychiatric problems after traumatic brain injury: unraveling the silent epidemic. *Psychosomatics.* 50: 198–205.

Vajtr, D., O. Benada, P. Linzer, F. Samal, D. Springer, P. Strejc, M. Beran, R. Prusa and T. Zima. 2013. Immunohistochemistry and serum values of S-100B, glial fibrillary acidic protein, and hyperphosphorylated neurofilaments in brain injuries. *Soud Lek.* 57: 7–12.

van Geel, W.J., H.P. de Reus, H. Nijzing, M.M. Verbeek, P.E. Vos and K.J. Lamers. 2002. Measurement of glial fibrillary acidic protein in blood: an analytical method. *Clin Chim Acta.* 326: 151–4.

Varma, S., K.L. Janesko, S.R. Wisniewski, H. Bayir, P.D. Adelson, N.J. Thomas and P.M. Kochanek. 2003. F2-isoprostane and neuron-specific enolase in cerebrospinal fluid after severe traumatic brain injury in infants and children. *J Neurotrauma.* 20: 781–6.

Vollmer, D.G. and R.G. Dacey, Jr. 1991. The management of mild and moderate head injuries. *Neurosurg Clin N Am.* 2: 437–55.

Vos, P.E., B. Jacobs, T.M. Andriessen, K.J. Lamers, G.F. Borm, T. Beems, M. Edwards, C.F. Rosmalen and J.L. Vissers. 2010. GFAP and S100B are biomarkers of traumatic brain injury: an observational cohort study. *Neurology.* 75: 1786–93.

Wang, K.K., R. Posmantur, R. Nath, K. McGinnis, M. Whitton, R.V. Talanian, S.B. Glantz and J.S. Morrow. 1998. Simultaneous degradation of alphaII- and betaII-spectrin by caspase 3 (CPP32) in apoptotic cells. *J Biol Chem.* 273: 22490–7.

Waterloo, K., T. Ingebrigtsen and B. Romner. 1997. Neuropsychological function in patients with increased serum levels of protein S-100 after minor head injury. *Acta Neurochir (Wien).* 139: 26–31; discussion 31–2.

Woertgen, C., R.D. Rothoerl, M. Holzschuh, C. Metz and A. Brawanski. 1997. Comparison of serial S-100 and NSE serum measurements after severe head injury. *Acta Neurochir (Wien).* 139: 1161–4; discussion 1165.

Xiong, H., W.L. Liang and X.R. Wu. 2000. Pathophysiological alterations in cultured astrocytes exposed to hypoxia/reoxygenation. *Sheng Li Ke Xue Jin Zhan*. 31: 217–21.

Yamazaki, Y., K. Yada, S. Morii, T. Kitahara and T. Ohwada. 1995. Diagnostic significance of serum neuron-specific enolase and myelin basic protein assay in patients with acute head injury. *Surg Neurol*. 43: 267–70; discussion 270–1.

Yealy, D.M. and D.E. Hogan. 1991. Imaging after head trauma. Who needs what? *Emerg Med Clin North Am*. 9: 707–17.

Ytrebo, L.M., N.G., C. Korvald et al. 2001. Renal elimination of protein S-100beta in pigs with acute encephalopathy. *Scand J Clin Lab Invest*. 61: 217–225.

Zemlan, F.P., E.C. Jauch, J.J. Mulchahey, S.P. Gabbita, W.S. Rosenberg, S.G. Speciale and M. Zuccarello. 2002. C-tau biomarker of neuronal damage in severe brain injured patients: association with elevated intracranial pressure and clinical outcome. *Brain Res*. 947: 131–9.

Zimmer, D.B., E.H. Cornwall, A. Landar and W. Song. 1995. The S100 protein family: history, function, and expression. *Brain Res Bull*. 37: 417–29.

Zurek, J., L. Bartlova and M. Fedora. 2012. Hyperphosphorylated neurofilament NF-H as a predictor of mortality after brain injury in children. *Brain Inj*. 25: 221–6.

8

Rehabilomics Concepts: An Overview of Genetic, Proteomic and Hormonal Biomarkers in TBI Recovery

James Crownover[1] and Amy K. Wagner[1,2,3,]*

INTRODUCTION

A leading cause of mortality and long-term disability within the United States is Traumatic Brain Injury (TBI), with an estimated 1.7 million related deaths, hospitalizations and emergency department visits annually (Frieden et al., 2006). At least 5.3 million Americans have long-term needs for assistance with activities of daily living as a result of TBI (Thurman et al., 1999). There are also many unaccounted for incidents of TBI that are managed outside of the emergency department or hospital, or who receive no care (Frieden et al., 2006), and TBI is a contributing factor to a third of all injury-related deaths in the US (Frieden et al., 2006). TBI also carries a heavy financial burden, with associated direct and indirect medical costs estimated to be in the area of $60 billion in the US in the year 2000 (Frieden et al., 2006).

[1] Department of Physical Medicine & Rehabilitation, University of Pittsburgh, School of Medicine, Pittsburgh PA.
[2] Center for Neuroscience, University of Pittsburgh, Pittsburgh, PA.
[3] Safar Center for Resuscitation Research.
* Corresponding author: wagnerak@upmc.edu

Both primary mechanical injury and secondary biochemical cascades contribute to the overall pathophysiology of TBI. There is a great deal of heterogeneity in the type of injury sustained, and the pattern of pathophysiology that ensues, to affect injury severity and potential outcome. Demographic factors, social support, and health-related behavior can all influence both the injury and recovery course. Even among individuals with similar demographic backgrounds, clinical management course, and injury types, there can often be a great deal of variability in treatment outcomes, making the search for effective therapies for TBI particularly challenging.

Until recently, researchers have tended to focus on the development of interventions for providing neuroprotection during the acute stages of TBI by targeting the secondary mechanisms of injury. These acute secondary injury processes include excitotoxicity, inflammation, oxidative injury and cerebral edema, among others. Despite the lack of major breakthroughs in neuroprotective treatments, advances made in the standard of acute care provided for those with significant TBI have reduced mortality resulting in more persons surviving beyond the sub-acute stage.

As a result of these advancements, recent research is shifting to study how TBI pathophysiology evolves over the long-term, temporal spectrum of rehabilitation and recovery. The pathophysiology of plasticity and repair, in the setting of ongoing neurodegeneration, makes understanding the neurobiology of TBI recovery complex. Also, TBI is increasingly viewed as a chronic disease state in which recovery mechanisms interact with ongoing neurodegeneration and other chronic state pathology. Several complications can commonly arise during these later stages of evolving pathology and recovery, and it is critically important that we develop a better understanding of how the acute, initial patho-mechanisms of TBI affect the equilibrium of degeneration and plasticity as well as recovery in the post-acute and chronic periods after TBI.

TBI commonly results in severe functional deficits, and therefore, patients with significant injury often require some type of rehabilitation prior to transitioning back into the community. In addition to the improvements made in the acute management of TBI, promising gains have been made with rehabilitation techniques treating some of the chronic consequences associated with TBI (Arabi et al., 2010; Badjatia et al., 2008; Brain Trauma Foundation et al., 2007; Espinosa-Aguilar et al., 2008). However, it is extremely challenging to establish and validate the biological basis of many rehabilitation therapies because rehabilitation does not lend itself to a singular "protocolized" plan of care or therapy, and the evaluation of each person must include an individualized assessment of physical, cognitive, emotional and social systems, each of which uniquely interacts to affect recovery (World Health Organization et al., 2001). As TBI rehabilitation research continues to move forward, more emphasis has been placed on personalized (theranostic) approaches to the scientific development and implementation of efficacious treatments. This movement aligns well with recent momentum in the area of comparative effectiveness research (Timmons and Toms, 2012; Wagner and Sowa, 2014). Rehabilomics is a conceptual research framework with translational relevance to clinical care that takes a personalized approach to rehabilitation research (Wagner, 2010). Within this framework, a full range of individual biological characteristics is examined to determine which characteristics impact recovery and the nature of this impact. This neurobiological grounded concept personalizes the approach to rehabilitation constructs and interventions in order to optimize individual recovery. Central to the Rehabilomics approach is the utilization of molecular tools to understand the mechanistic underpinnings of current treatments and care algorithms that impact functional recovery.

Biomarker research methods may improve our knowledge of the pathobiology that ensues after TBI, and unlock the connections between this pathology, individual variability, and the heterogeneous outcomes experienced by those with TBI (Conley and Alexander, 2011). Biomarkers are currently used as a major part of clinical care and decision making in many patient populations, but have not yet been fully utilized in TBI (Berger, 2006). There has, however, been an increasing number of studies using multiple biomarkers to understand TBI pathology (Wagner et al., 2004, 2005a, 2007, 2011a,b; Bonneh-Barkey et al., 2010), to predict outcome (Wagner et al., 2011a,b; Hoh et al. 2010; Salonia et al., 2010; Berger et al., 2010), and to understand risk for complications (Darrah et al., 2013; Wagner et al., 2011a, 2012a; Wagner and Zitelli, 2013; Miller et al., 2010) and are therefore are important in characterizing pathophysiology during rehabilitation and recovery. Biomarker research will help capture important elements about heterogeneity in TBI and the factors/therapeutic targets that lead to the varied course of recovery observed with this population (Wagner, 2010, 2011; Wagner and Zitelli, 2013). The purpose of this chapter is to overview a wide variety of biomarkers currently being used in TBI research and clinical management, in the context of a rehabilitation relevant (or Rehabilomics) framework, with the goal of informing the reader and demonstrating the areas that require further research. We will discuss specific Rehabilomics relevant examples of genetic, proteomic, and endocrine biomarker research and how this research has advanced our current understanding of TBI as a complex chronic disease.

Genetic Biomarkers

Genomic biomarkers have tremendous potential to help better understand the molecular neurobiology of brain injury, neuroplasticity and recovery, in addition to revealing the molecular mechanisms driving the effectiveness of many rehabilitative interventions and risks for a range of common complications. Increasing evidence suggests that genetic variability likely plays a critical role in the highly variable outcomes observed amongst patients with TBI despite there being similar mechanisms of injury, demographics and clinical care. Individual genetic variability can also interact with other innate factors, including sex and age, resulting in a unique response to the post-injury environment, and therefore, affecting neuro-recovery.

The advancement of molecular and biotechnology has greatly expanded our ability to study genetic variability and its interaction with the environment following TBI. Techniques such as gene array studies have already contributed to our understanding of the mechanisms of secondary injury in both human and animal models of TBI, and these approaches have been used to characterize response to treatment (Crack, 2009). Epigenetics examines the interface between genetics and the environment, in which observable outcomes or function result from mechanisms of functional genome modification without changing nucleotide sequences. Researchers have increased their discussion on the Epigenome and how chromatin modification affects injury outcomes and recovery. Now that significant portions of the Human Methylome have been mapped, comparative analysis could conceivably be made in relation to pathological conditions (Berman et al., 2009). Because of these techniques, we are beginning to understand how genes and the environment interact and affect each other, and researchers are provided with additional high-impact methods and platforms to study the molecular mechanisms of complex treatments such as cognitive rehabilitation, exercise, stimulants or other rehabilitative techniques commonly used for treating patients with TBI. To fully utilize current research techniques and capacity for studying genetic biomarkers in TBI, it is important to establish rigorous, high quality biorepositories with associated relevant clinical and functional information. Collecting genetic data a priori in high-risk populations such as those participating in contact sports, or members of the military, could also be considered.

In addition to functional variants, Single Nucleotide Polymorphisms (SNPs) are often evaluated for their associations with outcome or health and complication status. Some SNPs are considered functional, and they affect protein, structure, translation or stability, while others are considered tagging SNPs (tSNPs) that represent a unique block of DNA in which other polymorphisms may reside and contribute to associations with these phenotypes.

Cognition, Function, Recovery

Dopamine

One area of post-acute pathology that has evolved (in part) through the introduction of biomarkers into TBI research design is dopamine (DA) dysfunction after TBI. Dopamine pathways have been implicated in secondary injury and outcome (Wagner et al., 2007; McAllister et al., 2008). DA cell terminals begin in the substantia nigra (SN) and project to cerebral forebrain structures, including the hippocampus, prefrontal cortex and striatum, are important for memory, executive function and attention (Cohen and Servan-Schreiber, 1992; Floresco et al., 2006; Alexander and Crutcher, 1990). DAergic neurons also innervate brain structures which function in complex motor behavior and drug reinforcement (Bannon and Roth, 1983; Brozoski et al., 1979; Goeders and Smith, 1993; Tassin et al., 1978). After experimental TBI, regional, time dependent changes in DA systems have been observed. TBI is associated with changes in dopaminergic state that are specific to genetic variation within the DA system pathways (Wagner et al., 2007; Wagner et al., 2014). Also, DA enhancing drugs can improve cognition and recovery (Bales et al., 2009; Neurobehavioral Guidelines Working Group, 2006; Giacino et al., 2012).

The most important regulator of DA neurotransmission is the dopamine transporter (DAT) (Sesack et al., 1998) which plays a central role in determining the duration of action of DA by rapidly taking up extracellular DA into pre-synaptic terminals after release (Horn, 1990; Jones et al., 1998). Regional decreases in DAT expression have been reported after TBI (Wagner et al., 2005c, 2009b), suggesting that the benefits of DAT inhibitors (Goldstein, 2003; Kline et al., 2000) are conferred by increasing extracellular DA. While subject to rapid (Daws et al., 2002) regulation, DAT is a primary target for drugs, like methylphenidate (MPD), used with Attention Deficit Disorder (ADD) and TBI. Following TBI, alterations in striatal DAT expression (Wagner et al., 2005b, 2005c), which may reflect a compensatory response to improve DA neurotransmission in a "low or dysfunctional DA state" that is hypothesized to occur in the chronic phases of TBI recovery. Changes in striatal DA release and DAT modulation may occur through D2 auto-receptors (Wagner et al., 2009b; Liu et al., 2009). This hypothesis is supported by the use of real time DA electrochemical monitoring techniques in the DA rich striatum, that demonstrate reduced DA neurotransmission capacity and dysfunctional DA reuptake (Wagner et al., 2005b, 2009b). Although neurostimulants have discrete on-off effects on behavior, our work shows that daily treatment with the DAT inhibitor and neurostimulant, methylphenidate is neuro-restorative, and can reverse dopaminergic deficits after two weeks of treatment (Wagner et al., 2009a,b).

Clinically, MPD improves attention and motor processing speed (Whyte et al., 2004), and DA agonists also have beneficial effects on performance of spatial learning tasks in experimental TBI (Bales et al., 2009). Amantadine also shows promise in treating TBI (Giacino et al., 2012; Dixon et al., 1999), however, not all patients demonstrate beneficial effects from DA agonist medications. Genetic and temporal biomarker profiles, as well as high resolution, real-time symptom/complication assessments, may provide insight on who derives the largest benefits from these treatments.

There are several variants within DAergic candidate genes associated with a range of cognitive and neuropsychiatric conditions, including ADD, depression, impulsivity and Parkinson's Disease (Swanson, 2000; Dresel et al., 2000; Krause et al., 2000; Rowe et al., 1999; Virgos et al., 2001; Noble, 2000; Maher et al., 2002; Schulze et al., 2000; Manuck et al., 2000; Altar et al., 1992; Momose et al., 2002). The above diseases share many symptoms in common with those resulting from TBI. One of these candidate DAergic genes is the DAT1 gene, which has a variable number of tandem repeats (VNTR) located in the 3′ untranslated region (UTR) located on chromosome 5p15.3. The variation in this region influences gene expression. The two most frequent alleles, the nine repeat and the 10 repeat, can alter transcription of DAT, and therefore, the amount of available transporter (Michelhaugh et al., 2001). The DAT1 VNTR is implicated with ADD, and behaviors and deficits that often characterize TBI such as inattention and impulsivity. We have demonstrated that there are elevated DA levels in cerebral spinal fluid in response to blood pressure supporting agents and innate factors like female sex and genetic variation within the VNTR of DAT1, with 10/10 homozygotes having higher CSF DA levels acutely following severe TBI (Wagner et al., 2007). This work further supports a hypothesis for altered dopamine systems following severe TBI. The DRD2 gene is another DAergic gene of which the Taq1A polymorphism is one of the most studied. This restriction fragment length polymorphism can affect D2 receptor density (Jönsson et al., 1999; Thompson et al., 1997; Pohjalainen et al., 1998), and TaqIA1 carriers can have decreased glucose metabolism in several regions of the brain (Noble et al., 1997). How the genetic variation of DAT1 and DRD2 effects D2 and DAT binding has not been well studied in the context of TBI. Using PET imaging to evaluate adults 1-year after severe TBI, we have demonstrated reductions in striatal DAT binding after severe TBI and have shown that this change is genotype specific, where DAT 9 carriers (DAT1) and D2 A2/A2 (DRD2) had less DAT receptor binding (Wagner 2014).

The enzymes monoamine oxidase (MAO) and catechol-o-methyltransferase (COMT) are the primary enzymes that metabolize DA into its major metabolites 3,4-dihydroxyphenylacetic acid (DOPAC) and homovanilleic acid (HVA) (Oechsner et al., 2002). The COMT gene

Val58Met functional Single Nucleotide Polymorphism (SNP) variant (rs4680) influences COMT enzyme activity. Val/Val homozygotes have a 3–4 fold increase in activity over the Met/Met homozygote (Rujescu et al., 2003), leading to lower DA levels for Val carriers. This SNP modulates DA levels in healthy individuals (Chen et al., 2004) and may moderate executive function test performance in healthy controls (Wishart et al., 2011), but findings are less clear after TBI (Lipsky et al., 2005). However, these data suggest that genetic variation in DA systems may continue to be a fruitful target in delineating possible predictors and mechanisms for TBI outcomes.

APO-E/Neprilysin

Recent reports also suggest that ongoing, chronic mechanisms of pathology associated with TBI features characteristics that overlap with those normally associated with neurodegenerative diseases like Alzheimer's Dementia (AD) and Parkinson's disease. In fact, epidemiological studies suggest that persons with TBI, including those with mild or recurrent injuries, are at increased risk for later development of these neurodegenerative conditions (Masel and DeWitt, 2010). There are striking postmortem histological similarities between TBI and AD, including the presence of extracellular amyloid beta plaques and neurofibrillary tangles (Dziedzic, 2006; Sivanandam and Thakur, 2012). Axonal damage, and the accumulation of amyloid beta substrate proteins within the damaged axons, are common to both TBI and AD (Sivanandam and Thakur, 2012; Johnson et al., 2010). Accumulation of several key proteins, axonal degeneration, and dysfunctional axon transport can persist after TBI (Sivanandam and Thakur, 2012; Johnson et al., 2010). It is now believed that carriers of the apolipoprotein epsilon 4 (APOE4) allele have a higher risk of developing AD, and the APOE4 allele is associated with increased accumulation of amyloid-beta protein (Hashimoto et al., 2012; Youmans et al., 2012). Some studies suggest that APOE4 carriers who sustain a TBI tend to have more severe cognitive deficits, an increased likelihood of developing dementia, and a slower overall recovery rate than individuals with TBI who do not carry this allele (Sivanandam and Thakur, 2012; Johnson et al., 2010; Mauri et al., 2006; Alexander et al., 2006). The largest study conducted in TBI evaluating possession of the APOE4 allele and outcome found a significant age-genotype interaction, where adolescent-young adult patients carrying APOE4 had worse outcomes compared to younger patients without this allele (Teasdale et al., 2005).

Neprilysin is an enzyme that helps metabolize beta-amyloid, and therefore, reduces plaque formation; mutations in the neprilysin gene have been linked to familial forms of Alzheimer's disease (Helisalmi et al.,

2004). In addition, variability within the promoter region of the neprolysin gene impacts the incidence of Alzheimer's type beta-amyloid plaques in a cohort of post-mortem brains from patients with previous history of TBI (Johnson et al., 2009). These histological findings compliment the previous studies implicating the Apolipoprotein E4 genotype in TBI pathology and outcomes (Conley and Alexander, 2011; McAllister et al., 2010), however, more work may be helpful in order to determine if this functional variant within the neprilysin gene is linked to functional recovery.

KIBRA Gene

Recent genetic advances have allowed researchers to identify genes that play a role in shaping human memory capacity. The KIBRA gene, which contains the (SNP) rs17070145, has been implicated as influencing episodic memory. KIBRA is highly expressed in the medial temporal lobe and hippocampus (Kremerskothen et al., 2002; Papassotiropoulos et al., 2006), and it acts as a substrate for the atypical protein kinase C zeta (aPKC zeta), which is important for the occurrence of long-term potentiation (Büther et al., 2004; Johannsen et al., 2008). There are a number of neurological functions that KIBRA may be involved in, including synaptic transmission and plasticity, signal transduction, and memory formation. The results of a genome-wide association study (Papassotiropoulos et al., 2006) suggested a strong association between episodic memory performance and the (SNP) rs17070145, a common T→C exchange located on the ninth intron of the KIBRA gene. Healthy individuals carrying the rs17070145 T allele (CT/TT genotypes) significantly outperformed non-carriers of the allele (CC genotype) on measures of delayed episodic recall in three independent populations including young and older adults. Functional Magnetic Resonance Imaging (fMRI) studies (Papassotiropoulos et al., 2006) further support the relationship between the T-allele and better memory performance in healthy young adults. Some studies have not replicated these findings though (Need et al., 2008; Jacobsen et al., 2009), and the associations with KIBRA genotype have been mixed for populations like those with Alzheimer's type dementia (Burgess et al., 2011; Corneveaux et al., 2010; Rodriguez-Rodriguez et al., 2009). Interestingly, a study involving elderly adults with subjective memory complaints revealed that carriers of the T-allele performed worse on measures of episodic memory than non-carriers, an effect opposite of that typically found in healthy populations (Nacmias et al., 2008). Similarly, our recent study on a population with severe TBI demonstrated that non-carriers of the T-allele performed better on episodic memory measures than T-carriers at both 6 and 12 months post injury (Wagner et al., 2012b). This finding supports the concept that variability in the KIBRA gene may influence episodic memory performance in a population with severe TBI. However, the data suggest that injury and genetic background may interplay to provide unique associations with cognition and outcome.

BDNF

The gene for the neurotophin, brain-derived neurotrophic factor (BDNF) is located on chromosome 11, and it plays an important role in synaptic connectivity (Huang and Reichardt, 2001), synaptic plasticity (McAllister et al., 1995; Lu, 2003), and neuron survival throughout adulthood (Huang and Reichardt, 2001; Poo, 2001; Okada et al., 2006; Leibrock et al., 1989; Binder and Scharfman, 2004). Alzheimer's disease (Huang et al., 2007), obsessive compulsive disorder (Hall et al., 2003), eating disorders (Ribases et al., 2003), and bipolar disorder (Sklar et al., 2002; Neves-Pereira et al., 2002) are neurological diseases demonstrated to have an association with BDNF polymorphism. There is also an association with the BDNF polymorphism and episodic memory (Egan et al., 2003) and hippocampal activity, performance and volume in healthy human populations (Hariri et al., 2003; Pezawas et al., 2004; Szeszko et al., 2005; Bueller et al., 2006). These findings are interesting in the context of experimental TBI studies that show increased hippocampal BDNF expression early following traumatic injury in animal models (Hicks et al., 1997). In a clinical population with severe TBI, both serum and CSF BDNF levels are reduced acutely after injury, and serum BDNF levels remain depressed for at least 1 year after TBI, particularly in those who are clinically depressed and who have impaired functional cognition (Failla et al., 2013). BDNF is important for synaptic remodeling in the adult hippocampus (Heldt et al., 2007). The importance of BDNF in synapse formation and plasticity (Lu, 2003; Frisen et al., 1992; Huang and Reichardt, 2003; Luikart and Parada, 2006) makes it likely that polymorphisms at this gene significantly influence cognitive recovery following TBI.

Within the BDNF gene, a frequent Single-Nucleotide Polymorphism (SNP) results in a valine-to-methionine (Val66Met) substitution in the propeptide of the BDNF molecule (Chen et al., 2004; Chen and Westerlain, 2006), and this SNP affects the regulated secretion and neuroplastic effects of mature BDNF (Egan et al., 2003). It was recently shown that the Met allele promoted recovery of executive functioning following TBI and is considered protective for higher-order executive functions in patients with pre-frontal damage after TBI compared to the Val carriers (Krueger et al., 2011). It has also been reported that BDNF polymorphism has an effect on the recovery of general cognitive intelligence following penetrating TBI in Vietnam combat veterans 10–15 years and 30–35 years after injury (Rostami et al., 2011).

No association has been demonstrated between the BDNF polymorphism and general cognitive intelligence in a healthy population (Raymont et al., 2008; Egan et al., 2003; Tsai et al., 2004), suggesting that the association of the BDNF polymorphism and recovery of general cognitive intelligence is likely secondary to lesion-induced neuroplasticity (Rostami et al., 2011). Identifying mechanisms of how BDNF affects cognitive recovery after TBI will be an important step towards fully understanding and therefore facilitating functional recovery after a TBI as well as help identify higher risk patients.

Post Traumatic Epilepsy

The development of Post-Traumatic Epilepsy (PTE) is a frequent complication after TBI. Higher PTE rates co-occur with severe TBI compared to more mild injuries, and the time from injury to PTE onset can be months to years (Yablon and Dostrow, 2001). However, those who develop PTE often go on to have lower quality of life scores, and have worse function outcomes than those who do not have this complication (Bushnik and Gordon, 2012). Prolonged convulsive seizures in the acute setting can be detrimental to the brain, and non-convulsive seizures can have adverse outcomes in both the acute and sub-acute settings (Chen and Westerlain, 2006; Young et al., 1996). Also, Post Traumatic Seizures (PTS) can significantly exacerbate the structural damage observed in rat models of TBI (Bao et al., 2011). Though common, the neurobiological underpinnings associated with epileptogenesis and subsequent PTE are not well understood, and not every patient with clinical risk factors for PTE goes on to develop PTE. Genetic background has been implicated in some epilepsy populations (Cavalleri et al., 2007; Kwan et al., 2007, 2009; Jamali et al., 2010), and thus, it is plausible that variability in seizure occurrence may be attributable to the unique genetic makeup that each patient brings to post-TBI recovery. Variation in the APOE gene has been studied with regard to PTE risk, the findings of which have been mixed (Miller et al., 2010; Anderson et al., 2009; Diaz-Arrastia et al., 2003). The C677T genotype variant within the methylenetetrahydrofolate reductase (MTHFR) gene also has been associated with epilepsy, including increased risk for development of post-traumatic seizures (Scher et al., 2011). Interestingly, recent studies implicate variation in both GABAergic and Adenosine pathways in PTE.

Glutamic Acid Decarboxylase (GAD) Gene

GABAergic neurotransmission impacts TBI pathophysiology in several ways after injury. Increased GABA levels acutely after injury may represent a compensatory mechanism to counteract excitotoxicity, and therefore, mitigate the development of early seizure activity in the injured brain. A common linkage between the pathophysiology associated with secondary TBI and seizure susceptibility is the gamma-aminobutyric acid (GABA) neurotransmission system. GABA is a primary inhibitory neurotransmitter in the brain, and under physiological conditions, modulates excitatory tone. GABA is synthesized from glutamate by the enzyme Glutamic Acid Decarboxylase (GAD). Mature GAD protein has two isoforms: GAD67 is coded by the GAD1 gene, and GAD65 is coded by the GAD2 gene. Both genes are responsible for catalyzing the production of GABA from glutamate in the brain (Kuo et al., 2009). It is known that GAD1 maps to chromosome 2q31, while GAD2 maps to 10p11 (Bu et al., 1992). A recent study in humans with severe TBI demonstrated that the genetic variability within the GAD1 gene was associated with early PTS and PTE (Darrah et al., 2013). Specifically, SNPs located in the GAD1 gene were associated with seizures during the first week post-injury also PTE onset between 1 week-6 months post-injury (Darrah et al., 2013). The TT genotype for SNP rs3828275 was associated with PTS <1week, and the GG genotype for rs3791878 and the AA genotype for rs769391 were both associated with increased PTE development 1 week-6 months after TBI. Haplotype analysis showed that those with haplotype pairs containing two copies of the haplotype containing both variant risk alleles were the subjects at increased risk for PTS during this time (Darrah et al., 2013). For GAD1, associations with early PTS (<1 week) may be due to GABAergic modulation of acute excitotoxicity that is known to occur after TBI (Neese et al., 2007; Muir et al., 1996; O'Dell et al., 2000; Wagner et al., 2005a; Ruppel et al., 2001; Palmer et al., 1993). These results implicate GAD systems in PTE pathology and suggest that other inhibitory neurotransmitter systems may also be implicated in PTE development. The fact that different SNPs were associated with early seizures versus seizures in the sub-acute phase may imply different biologically plausible mechanisms of susceptibility given that the substrates of early seizures and later development of PTE are likely different.

Adenosine A1 receptor

Genetic variability with the Adenosine A1 receptor (A1AR) gene also is associated with PTE susceptbility (Wagner et al., 2010). CSF Glutamate levels are elevated in adults with severe TBI (Bell et al., 2001), and experimental models suggest a relationship between excitatory amino acid levels and seizure development after cortical impact (Nilsson et al., 1994). Neuronal A1AR's are localized in regions susceptible to both seizures and TBI and often co-localize with NMDA receptors (Deckert and Jorgensen,

1988). Adenosine binding limits both excitatory neurotransmission and excitotoxicity (Rudolphi et al., 1992; Andiné et al., 1990), and is therefore, an important acute neuroprotectant in the setting of TBI and ischemia. Adenosine and its target A1AR are implicated with non-traumatic seizure development. A1AR knockout (KO) mice, when subjected to the Controlled Cortical Impact (CCI) model of experimental TBI, develop profound PTS, leading to lethal status epilepticus (Kochanek et al., 2006). Also, glial scar formation, common to TBI, has been implicated in seizure development by compromising adenosine synthesis (Fedele et al., 2005). Recent reports suggest that genetic variability within the A1AR gene represented by multiple tagging Single Nucleoptide Polymorphisms (tSNPs), was associated with the development of PTS in a human population with severe (Wagner et al., 2010). Specifically, there were significant association between tSNPs rs3766553 and rs10920573 and PTE development. The A1AR can undergo dimerization and heteromerization to modulate neurotransmission (Ciruela et al., 2006; Franco et al., 2000), and genetic variation in the A1AR gene represented by rs3766553 and rs10920573 may influence PTS susceptibility via these mechanisms. The findings implicating A1AR variability in PTE susceptibility, the spatial proximity of A1ARs to glutamatergic receptors (Deckert and Jorgensen, 1988), and the relationship between adenosine and glutamate in human TBI (Robertson et al., 2001) suggests that adenosine and A1AR are important in countering the contribution of excitotoxic injury to immediate/early PTS. Future work needs to focus on identifying the specific portion of DNA underlying these tSNP associations with PTE and how differences in the biological functioning represented by rs3766553 and rs10920573 could contribute to PTE development. Further, targeted multivariate analysis of genes for other proteins and receptors involved in adenosine synthesis and neurotransmission may increase understanding of how genetic variability affects the adenosine neurotransmitter system with regard to PTE susceptibility and treatment, and these findings may lead to a better understanding of this common complication and targeted effective therapies for prevention and treatment.

Post-Traumatic Depression

Major depression is one of the most common neuropsychiatric complications following TBI with prevalence between 14–29% (Gomez-Hernandez et al., 1997; Jorge et al., 1993; Rapoport et al., 2003). The pathogenesis of depression post-TBI remains unclear, yet it presents a significant barrier to rehabilitation and functional recovery, being associated with cognitive deficits (Rapoport et al., 2005), more post-concussive symptoms (Herrmann et al., 2009) and worse overall outcomes (Rapoport et al., 2006). Many psychosocial factors, including pre-morbid psychiatric history, likely contribute to the risk of developing post-TBI depression, however to what degree remains unclear. Anterior frontal lobe injury may have an association with Post-Traumatic Depression (PTD) risk, and the frontal lobes are a major area of projection for the serotonergic system (Silver et al., 2009; Jorge et al., 1993; Fedoroff et al., 1992; Robinson and Szetela, 1981; Pazos et al., 1987). Injury and disruption of serotonergic neurotransmission may contribute to PTD development.

In addition to its general effects on recovery after injury, BDNF gene variation may also influence serotonin system function thus influence PTD risk; specifically, BDNF can affect serotonin transporter function (Daws et al., 2007) and influence neuronal plasticity (Castren et al., 2007; Duman and Monteggia, 2006), via this pathway. Further, the val66met polymorphism for the BDNF gene has demonstrated some association with depression risk and severity, though data has been inconsistent (Hwang et al., 2006; Oswald et al., 2005; Schumacher et al., 2005; Strauss et al., 2005). Another polymorphism involved in the serotonergic system includes the methylenetetrahydrofolate reductase (MTHFR) enzyme. MTHFR is responsible for homocysteine metabolism, which provides the methyl group necessary for the production of serotonin (Bottiglieri et al., 1992). It has thus been theorized that genetic variation in the MTHFR gene could influence depression rates through decreased serotonin levels, however, data supporting this possibility has been inconsistent (Gilbody, 2007; Zintzaras, 2006).

The commonly prescribed Selective Serotonin Reuptake Inhibitors (SSRIs) mainly target the serotonergic system, and it is likely that genetic variability involved in this complex system plays a role in treatment response variability. Several studies have evaluated genetic variability within the serotonergic system, particularly associated with the serotonin transporter protein (5-HTT) (Caspi et al., 2003; Trivedi et al., 2006). The serotonin-transporter-linked promoter region (5-HTTLPR) is adjacent to the gene which codes 5-HTT (Heils et al., 1996), and this region has been the most consistent in predicting depression in the non-TBI population (Eley et al., 2004; Jacobs et al., 2006; Kendler et al., 2005; Wilhelm et al., 2006; Zalsman et al., 2006). It has been demonstrated that the presence of one or two copies of the S-allele of 5HTTLPR is associated with lower 5HTT expression and also increased susceptibility to major depression (Eley et al., 2004; Jacobs et al., 2006; Kendler et al., 2005; Wilhelm et al., 2006; Zalsman et al., 2006), however results in this area have been variable (Gillespie et al., 2005; Surtees et al., 2006). This particular polymorphism was previously evaluated in individuals 1 year post-TBI, and no significant associations were linked with depression risk or severity among those who developed depression (Chan et al., 2008). Our recent work, however, does implicate the 5HTTLPR in PTD risk, particularly in the context of premorbid depression history and the temporal trajectory of PTD onset (Failla et al., 2013). A recent study was completed evaluating several of these genetic variants and their associations with treatment response to citalopram in post-TBI patients with depression. Polymorphisms in the BDNF and MTHFR genes were predictive of response to treatment, and a polymorphism in the 5HTTLPR seemed to predict medication tolerability (Lanctôt et al., 2010). Further studies are certainly required to solidify our understanding of this sub-set of serotonergic genes and how they influence treatment response and medication tolerability.

Protein Biomarkers

Overview

Proteins biomarkers also have extensive potential for utility in the study of mechanisms and management with TBI. Currently proteomic biomarkers are a part of mainstream clinical care used to quantitatively assess and define injury in almost every organ system except the brain (Berger, 2006; Kochanek et al., 2008), yet new markers continue to be reported, suggesting the need for definitive translational studies that identify biomarkers as clinically relevant for specific indications like diagnosis, prognosis, as well as treatment stratification and treatment effects. Several cerebrospinal fluid (CSF) and serum biomarkers have shown promise in predicting long-term outcome after severe TBI. Some of the serum proteomic biomarkers lack specificity for neural tissue injury, complicating the ability to utilize effectively following TBI if additional injuries or systemic medical complications have occurred. However, some markers, including inflammatory mediators, though not necessarily CNS specific, can provide useful information about pathophysiology and outcome. Below we summarize the literature on several biomarkers associated with structural damage and cell death after TBI.

Biomarker Analysis Issues

In TBI, there is a tremendous unmet need to deal effectively with the significant heterogeneity associated with injury response, and it is possible that the individual differences in outcome may be due to unique and dynamic temporal profiles in an injury induced proteome. Traditionally, biomarker data collected over time are summarized into single point estimates, and these point estimates are then used to examine the relationship between biomarker levels and outcome of interest. Some studies have used average biomarker levels to predict TBI outcome (Naeimi et al., 2006; Pelinka et al., 2004a; Ucar et al., 2004; Chatfield et al., 2002). Others have used peak levels (Hayakata et al., 2004; Nylén et al., 2008; Rainey et al., 2009) or levels obtained during the first day after injury (Berger et al., 2007; Townend et al., 2002; Rainey et al., 2009), while others have used arbitrary cut-offs (Raabe et al., 1999; Woertgen et al., 1999; Chatfield et al., 2002; Townend et al., 2002; Leblanc et al., 2006). It is also important to note that demographic, premorbid, and psychosocial factors have been identified that are associated with functional impairments, disability, and community integration for those with TBI (Testa et al., 2005; Niyonkuru et al., 2013), yet it is largely unknown if/how these factors affect the temporal biomarker profiles. We have applied a systematic and practical approach recently to biomarker analysis (group based trajectory analysis) to evaluate temporal trends, to determine factors affecting these trends, and to assess outcome differences associated with these temporal biomarker patterns (Niyonkuru et al., 2013). This group based modeling approach provides the ability to identify qualitatively distinct subpopulations rather than estimating the overall biomarker pattern from a singular population. There are now examples that have demonstrated that biomarker profiles using this technique can be more informative than single point estimates in predicting outcome (Salonia et al., 2010; Wagner et al., 2011a,b). Replicating these studies and generalizing this modeling approach to other populations will be essential to better understand the utility of this technique for outcome prognostication and management across the spectrum of TBI. There may also be some utility for using biomarker based trajectory groups in combination with genetic and/or epigenetic information to further characterize secondary injury and recovery mechanisms as well as to assess prognosis and/or treatment effects using a gene stratification approach. Additionally, there is a growing need for bioinformatics tools and options like structural equations modeling (Wagner et al., 2011a) to generate an understanding of biomarker profiles within a larger biomarker pathway.

S100b

S100b is a protein primarily found in astrocytic glial cells of the central nervous system (Donato, 2003). S100b has significant calcium homeostatic functions and is secreted for neuroprotective and neurotrophic cellular mechanisms (Kleindienst and Bullock, 2006). Previous work has shown increases in both CSF and serum S100b levels following severe TBI, and this body of work likely represents ongoing structural damage and cell death within the central nervous system (Petzold et al., 2003; Ingebrigtsen and Romner, 2002; Missler et al., 1999). It is unclear if elevations of serum S100b levels occur secondary to breached Blood Brain Barrier (BBB) integrity or if they are strictly associated with the extent of brain parenchymal damage, as the S100b protein is too large to pass through an intact and healthy BBB (Dash et al., 2010), and disruption of this barrier after TBI allows extravasation of proteins from the CSF to serum and vice versa. It is also important to note that, to a certain degree, serum S100b protein can also be generated from sources outside of the central nervous system. Elevated serum levels have been demonstrated in association with bone fractures (Pelinka et al., 2003), burns and muscular injury (Anderson et al., 2001). As a result, the specific contributors to S100b levels within the serum are not fully understood when evaluating multitrauma patients (Korfias et al., 2006; Kleindienst et al., 2010).

We recently demonstrated that serum and CSF S100b values in patients following severe TBI had a predictive capacity for both global outcomes and disability (Goyal et al., 2012). These findings were demonstrated by using a TRAJ approach. Acutely, temporal profiles of elevated CSF S100b levels were predictive of mortality and worse global (Goyal et al., 2012). This finding is consistent with additional previous studies (Townend and Ingebrigtsen, 2006; Dimopoulou et al., 2003; Elting et al., 2000; Rainey et al., 2009). Early after brain injury, CSF and serum levels of S100b protein closely correlated with each other. At later time points this correlation weakened, which may be secondary to both dynamic changes in BBB integrity over time and the dynamic contributions of S100b protein from sources outside of the central nervous system. However, distinct longitudinal CSF S100b profiles were predictive of outcome, while mean serum S100b levels predicted acute mortality (Goyal et al., 2012). Taken together, S100b shows promise as an informative post-TBI biomarker if used appropriately in the context of these limitations.

GFAP

Glial Fibrillary Acid Protein (GFAP) is found in astroglial cells and is not synthesized outside of the CNS (Missler et al., 1999; Webster et al., 2001). GFAP is a monomeric intermediate filament protein providing strength and support to the astroglial cytoskeleton (Eng et al., 1971) and is considered highly specific to CNS injury (Missler et al., 1999; Pelinka et al., 2004b). GFAP production increases in response to injury, which is likely involved in managing the function of astroglial cells to optimize their ability to support neurons (Vos et al., 2010). Since GFAP is specific to the CNS and is released when the CNS is damaged, it is considered a promising biomarker.

Serum GFAP levels are related to the severity of brain injury (Pelinka et al., 2004a,b; Vos et al., 2010), acute mortality (Pelinka et al., 2004b), the presence of mass lesions (Lumpkins et al., 2008) and/or elevated intracranial pressures (De Kruijk et al., 2002); however polytrauma without brain injury does not appear to increase serum levels, indicating CNS specificity (Pelinka et al., 2004b). An association between higher serum GFAP levels and worse outcomes following severe TBI has also been reported (Vos et al., 2010; Lumpkins et al., 2008; Nylén et al., 2006), and levels have been shown to be stronger predictors of unfavorable outcomes than age and GCS (Vos et al., 2010). When evaluating serum levels in patients with mild TBI, GFAP was elevated when there was an abnormality on neuroimaging and correlated with return to work time, however these values were not associated with overall severity of injury or outcomes (Metting et al., 2012). GFAP Break Down Products (BDP) are detectable in serum within one hour of brain injury (Papa et al., 2012), found to be elevated in mild TBI (GCS of 15) and highly elevated levels are found when there was a traumatic intracranial lesion observed on head CT scans (Papa et al., 2012). If validated, GFAP BDP has the potential to improve the diagnosis of mild TBI and reduce unnecessary head CT scans in the future. Elevated levels of serum glial fibrillary acidic protein breakdown products in mild and moderate traumatic brain injury also are associated with intracranial lesions and neurosurgical intervention (Papa et al., 2012).

UCHL-1

Ubiquitin C-terminal Hydrolase-L1 protein (UCH-L1) is a neuron specific protein that is involved in the processing of ubiquitin from damaged proteins to be degraded by proteasomes (Tongaonkar et al., 2000) and is therefore important in physiological and pathological conditions (Gong and Leznik, 2007). UCH-L1 levels in CSF are increased in severe TBI, positively correlating with severity levels (Papa et al., 2010), serum levels in the acute setting and outcome predictions (Mondello et al., 2011). More specifically, serum and CSF UCH-L1 levels have been demonstrated to have unique temporal profiles associated with injury type assessed radiologically by Marshall score (Papa et al., 2010; Vos et al., 2004) such as mass lesions versus diffuse axonal injury (Mondello et al., 2011). There is a strong correlation between CSF and serum levels of UCH-L1 that are strongest in the acute period following injury, and serum levels are reported as more likely to be associated with survival in the post-acute period (Brophy et al., 2011).

Bcl-2 and Cytochrome C

Apoptosis is a significant component of secondary injury following TBI. With the intrinsic (mitochondrial mediated) pathway associated with caspase-dependent apoptosis, stress on cellular organelles leads to the initiation of apoptosis via mitochondrial permeability, transition pore formation and subsequent release of cytochrome C (CytoC) into the cytosol. CytoC release initiates apoptosis via the intrinsic caspase-dependent pathway, while Bcl-2 protein modulates mitochondrial permeability transition pore formation through its ability to inhibit the pore forming function of the BH-3 domains of other pro-apoptotic Bcl-2 family proteins and subsequent CytoC release. Cytosolic CytoC promotes "apoptosome" formation, a molecular platform for caspase-9 activation which catalyzes the proteolytic activation of effector caspases including caspase-3 (Boehning et al., 2003). The Bcl-2 family proteins are important apoptosis modulators. They comprise both pro-apoptotic and pro-survival proteins that regulate mitochondrial permeability through transition pore formation (Clark et al., 2005; Liou et al., 2003; Zhang et al., 2005). Bcl-2 is increased in injured neurons and reduces programmed cell death in rats (Clark et al., 1997) and mice (Raghupathi et al., 1998) by inhibiting release of mitochondrial proteins (Zhang et al., 2005). Bcl-2 over expression in experimental models can be neuroprotective and reduce cortical neuron loss (Nakamura et al., 1999; Raghupathi et al., 1998); however, tissue sparing is not necessarily associated with improved behavioral outcomes (Tehranian et al., 2006). While these proteins are critical to cell death and survival pathways after TBI, additional work has focused on evaluating their potential as prognostic biomarkers for TBI outcome.

It has been established that CytoC and Bcl-2 levels are elevated in CSF after pediatric TBI (Clark et al., 2000; Satchell et al., 2005). Increased CSF Bcl-2 levels following TBI have been observed in children and adults with TBI (Clark et al., 1999, 2000; Uzan et al., 2006). Higher CSF levels of Bcl-2 were associated with increased survival in pediatric patients (Clark et al., 1999), and genetic variation in the Bcl-2 gene has been linked to adult TBI outcomes (Hoh et al., 2010). Recently, temporal profiles for CSF Bcl-2 and CytoC have been studied in adults after severe TBI, and these profiles can predict long-term outcomes and capture dynamic changes in these markers after severe TBI that are sensitive to outcome prediction (Wagner et al., 2011b). Specifically, Bcl-2 and CytoC profiles were effective predictors of scores on the Glasgow Outcome Scale and Disability Rating Scale after severe TBI, with subjects having sustained high levels of each biomarker experiencing worse outcomes. Interestingly, using group based TRAJ analysis there was a distinct group of subjects with temporally rising Bcl-2 levels that had excellent outcome (best long-term outcomes). Also, a distinct decliner group for CytoC had unique associations with outcome. Since Bcl-2 modulates CytoC release in experimental studies (Antonsson et al., 1997), these temporal profiles may reflect some regulatory elements of mitochondrial mediated apoptotic pathways. However, further work assessing CytoC and other Bcl-2 family proteins, and also how genetic variation influences these biomarker levels over time, is needed.

GAP-43

TBI affects neuronal transport mechanisms (Pettus et al., 1994), membrane functioning and microvascular function (Dietrich et al., 1994) each of which are likely associated with behavioral deficits common after TBI. These pathological mechanisms often occur in the absence of histopathological or ultrastructural changes. The genes involved in facilitating neuroplasticity and neural remodeling are upregulated following neural injury, and this phenomenon may be associated with at least some elements of behavioral recovery observed post-TBI (Stroemer et al., 1993, 1995). Growth-associated protein (GAP-43) is a membrane-bound protein that is thought to be involved in membrane remodeling and neurite growth, and it has been localized in the growth cones of sprouting CNS axons (Benowitz et al., 1988, 1990; Christman et al., 1997; Coggeshall et al., 1991; Gorgels et al., 1989; Masliah et al., 1991; Oestreicher et al., 1997; Strittmatter et al., 1990). GAP-43 may represent a valuable biomarker in the study of TBI allowing researchers to detect membrane disruption that is undetectable by ultrastructural analysis alone. It was previously demonstrated that GAP-43 is upregulated following central Fluid Percussion Injury (FPI) (Hulsebosch et al., 1998), and this upregulation was correlated with behavioral test performance. The anatomical pattern of upregulated expression was consistent with the distribution of mechanical shear forces associated with diffuse axonal injury that follows the curvature of the skull. The anatomical location of the upregulated GAP-43 likely represents areas of membrane disruption, and therefore impairment of membrane function, where there are no otherwise overt pathological changes such as cell death. Interestingly, in the CCI model, a contusion based experimental model of TBI, regional GAP-43 expression has been reported to be reduced at chronic time-points, but is manipulable by clinically relevant treatments such as anti-epileptic drugs, suggesting depressed GAP-43 expression is a function of the extent of regional injury associated with the insult. Areas of altered GAP-43 may be essential for many aspects of neuroplasticity which is crucial for human functional recovery following a TBI (Donnelly et al., 2013). Thus, there may be potential to develop imaging or genomic biomarkers that reflect a role for GAP-43 in plasticity, recovery and prognosis after TBI.

Sex Hormone Physiology

Overview

Hormones are one of the primary chemical messengers within the human body, and their hormone physiology is significantly disrupted following head trauma, contributing to complications that patients experience after TBI. Like proteomics and genetics, post-TBI endocrinology biomarkers have a lot of potential to improve our understanding and management of TBI in the acute, sub-acute and chronic settings. Despite the rapidly growing body of literature demonstrating the importance of improving our understanding of post-TBI endocrinology, and the recent interest in evaluating hormones as treatments for TBI (i.e., estradiol, progesterone, androgel, glucocorticoids), systematic data collection on hormone profiles remains limited.

Hormones, Neuroprotection and Prognosis

There is growing evidence suggesting that supplementation with female sex hormones (i.e., progesterone and estradiol) can provide neuroprotection and facilitate neurorestoration (Stein and Cekic, 2011; Stein and Wright, 2010; Herson et al., 2009; Wright et al., 2007). The evolution of physiological hormone profiles over time following a TBI is complicated, and although interpretation of temporal hormone profiles is also complex, they can be informative in outcome prognostication (Wagner and Zitelli, 2013; Wagner et al., 2012a). Though differences in acute hormone profiles are associated with variability in outcomes, sex differences with regard to prognostication are minimal (Wagner et al., 2011a). This conclusion is supported by the fact that the hypothalamic-pituitary-gonadal axis effectively shuts down following TBI, resulting in hormone profiles for men and women with severe TBI having several similarities (Wagner et al., 2011a). Damage to the hypothalamic-pituitary system can be secondary to traumatic damage in the hypophysial portal veins of the pituitary stalk (Klein, 2009), compression from edema, skull fracture, hypoxic insult or direct trauma (Agha and Thompson, 2005). The stress response that occurs after TBI also leads to HPG axis deficits as well as peripheral aromatase activation and elevated serum estradiol (E2) levels, which are associated with poor long term prognosis and mortality (Wagner et al., 2011a).

Hypogonadism

In men following TBI, hypogonadotropic (secondary) hypogonadism is the most common presentation for both early and chronic post-traumatic hypopituitarism and is defined as low testosterone attributable to low pituitary hormone production (Wagner et al., 2011a). In men hypogonadism can lead to fatigue, mood dysfunction, sleep cycle dysregulation, osteoporosis, and sexual dysfunction. Decreased cognitive function has also been shown to be associated with low testosterone levels in the elderly population (Wagner et al., 2011a). In a small cohort of men with severe

TBI, chronic persistent hypogonadotropic hypogonadism occurred for up to 1 year post injury and was associated with lower E2 levels and worse outcome (Wagner et al., 2012a). While deemed a marker of poor prognosis and recovery in the acute phase post-injury, this study implicates a role for E2 levels chronically as a marker that may influence plasticity and recovery over the long term. Further, this study showed that patients whose hypogonadism resolved over time began to show signs of recovery in hormone levels by approximately 6 weeks, suggesting that neuroendocrine screening for persistent hypogonadism might be useful around this time window and suggesting that further studies are needed to inform clinical screening and treatment guidelines for this complication.

Aromatase, Estradiol and Testosterone

Estradiol is a sex steroid classically characterized as having both neuroplasticity and neuroprotection abilities (Azcoitia et al., 2011; Arevalo et al., 2010). It is also a potent transcription factor for many genes. Estradiol's effects in the brain derive from both local production and from the serum based bioavailable hormone fraction from peripheral sources such as the ovaries and adipose tissue that cross the BBB (Hojo et al., 2008; Simpson, 2004).

Estradiol influences brain development and plasticity via receptors located in several brain regions (Azcoitia et al., 2011). After CNS damage, estradiol acts on oligodendrocytes, T-cells, microglia, and astroglia to promote remyelination and proliferation of progenitor cells, via growth factor expression, to decrease acute edema and inflammation (Arevalo et al., 2010), and other neuroprotective mechanisms have been proposed. Animal models have demonstrated these neuroprotective qualities, however elevated systemic levels early after severe TBI in humans are associated with increased mortality and poor outcomes (Wagner et al., 2011a). It is also important to note that persistent elevations in serum testosterone over the initial days post-TBI are linked to poor outcome (Wagner et al., 2011a) and testosterone pre-treatment in models of brain injury can increase histological damage (Herson et al., 2009). In a recent study, our group showed that TBI results in significantly decreased CSF estradiol levels and increased CSF testosterone (Garringer et al., 2013), compared to CSF estradiol and testosterone levels in healthy controls (Wagner et al., 2011a). Higher CSF estradiol/testosterone (E2:T) ratios were associated with lower acute mortality rates and better outcomes (GOS-6 scores) (Garringer et al., 2013), data which are consistent with the traditional concept that estradiol is neuroprotective after injury. Also, genetic variation in the aromatase gene has been associated with CSF and serum hormone levels after severe TBI as well as GOS based outcomes, outside of its genetic associations with (E2:T) ratios (Garringer et al., 2013).

The enzyme aromatase produces estrogens by converting androgens (Azcoitia et al., 2011), and this enzyme is located within several brain regions, primarily in neuronal cell bodies and in pre-synaptic terminals (Azcoitia et al., 2011; Hojo et al., 2008). The overall role of aromatase in CNS hormone physiology and neuroplasticity after TBI is not well known. Androgens have an influence on neuronal growth, and under some conditions, testosterone administration can modulate plasticity through neural growth and plasticity within the hippocampus CA1 neurons in animal studies (MacLusky et al., 2006). However, testosterone may exert its effects on plasticity both through aromatization to estradiol and through aromatase-independent pathways. Our clinical research supports the potential for indirect effects of testosterone on CNS neuroplasticity, by showing that both higher chronic endogenous serum testosterone and estrogen levels over the first year post-injury are associated with better post-TBI outcomes (Wagner et al., 2012a).

Conclusions

We have overviewed the importance of biomarker research with informing care and outcomes of individuals with TBI. The body of literature is still relatively small, and much work is required to bring biomarkers effectively into clinical care to support diagnosis, prognosis, clinical decision making, treatment and management. However, several types of biomarkers hold significant promise with regard to clinical translation, particularly when evaluated from a multidimensional outcome and recovery perspective. A Rehabilomics framework is a natural fit for understanding how the heterogeneity inherent to TBI pathophysiology may contribute to the wide array and severity of complications, impairments and recovery patterns observed in these populations. Importantly, a Rehabilomics framework provides the necessary footing from which to design comparative effectiveness studies aimed at individualizing care to optimize outcomes.

References

Agha, A. and C.J. Thompson. 2005. High risk of hypogonadism after traumatic brain injury: Clinical implications. *Pituitary*. 8(3-4): 245–249.

Alexander, G.E. and M.D. Crutcher. 1990. Functional architecture of basal ganglia circuits: Neural substrates of parallel processing. *Trends Neurosci*. 13: 266–271.

Alexander, S., M.E. Kerr, Y. Kim, M.I. Kamboh, S.R. Beers and Y.P. Conley. 2006. Apolipoprotein E4 allele presence and functional outcome after severe traumatic brain injury. *J Neurotrauma*. 24(5): 790–797.

Altar, C.A., C.B. Boylan, C. Jackson, S. Hershenson, J. Miller, S.J. Wiegand, R.M. Lindsay and C. Hyman. 1992. Brain-derived neurotrophic factor augments rotational behavior and nigrostriatal dopamine turnover *in vivo*. *Proc Natl Acad Sci USA.* 89(23): 11347–11351.

Anderson, A.E., L.O. Hansson, O. Nilsson, R. Dijlai-Merzoug and G. Settergren. 2001. High serum S100b levels for patients without head injuries. *Neurosurgery.* 48: 1255–1260.

Anderson, G.D., N.R. Temkin, S.S. Dikmen, R. Diaz-Arrastia, J.E. Machamer, C. Farhrenbruch, J.W. Miller and S.M. Sadrzadeh. 2009. Haptoglobin phenotype and apolipoprotein E polymorphism: Relationship to posttraumatic seizures and neuropsychological functioning after traumatic brain injury. *Epilepsy Behav.* 16(3): 501–506.

Andiné, P., K.A. Rudolphi, B.B. Fredholm and H. Hagberg. 1990. Effect of propentofylline (HWA 285) on extracellular purines and excitatory amino acids in CA1 of rat hippocampus during transient ischaemia. *Br J Pharmacol.* 100: 814–818.

Antonsson, B., F. Conti, A. Ciavatta, S. Montessuit, S. Lewis, I. Martinou, L. Bernasconi, A. Bernard, J.J. Mermod, G. Mazzei, K. Maundrell, F. Gambale, R. Sadoul and J.C. Martinou. 1997. Inhibition of Bax channel-forming activity by Bcl-2. *Science.* 277: 370–372.

Arabi, Y.M., S. Haddad, H.M Tamim, A. Al-Dawood, S. Al-Qahtani, A. Ferayan, I. Al-Abdulmughni, J. Al-Oweis and A. Rugaan. 2010. Mortality reduction after implementing a clinical practice guidelines-based management protocol for severe traumatic brain injury. *J Crit Care.* 25(2): 190–195.

Arevalo, M.A., M. Santos-Galindo, M.J. Bellini, I. Azcoitia and L.M. Garcia-Segura. 2010. Actions of estrogens on glial cells: Implications for neuroprotection. *Biochim Biophys Acta.* 1800: 1106–1112.

Azcoitia, I., J.G. Yague and L.M. Garcia-Segura. 2011. Estradiol synthesis within the human brain. *Neuroscience.* 191: 139–347.

Badjatia, N., N. Carney, T.J. Crocco, M.E. Fallat, H.M. Hennes, A.S. Jagoda, S. Jernigan, P.B. Letarte, E.B. Lerner, T.M. Moriarty, P.T. Pons, S. Sasser, T. Scalea, C.L. Schleien, D.W. Wright, Brain Trauma Foundation and BTF Center for Guidelines Management. 2008. Guidelines for prehospital management of traumatic brain injury 2nd edition. *Prehop Emerg Care.* 12 Suppl 1: S1–S2.

Bales, J., A.K. Wagner, A.E. Kline and C.E. Dixon. 2009. Persistent cognitive dysfunction during rehabilitation of traumatic brain injury: Towards a dopamine hypothesis. *Neurosci Biobehav Rev.* 33(7): 981–1003.

Bannon, M.J. and R.H. Roth. 1983. Pharmacology of mesocortical dopamine neurons. *Pharmacol Rev.* 35(1): 53–68.

Bao, Y.H., H.M. Bramlett, C.M. Atkins, J.S. Truettner, G. Lotocki, O.F. Alonso and W.D. Dietrich. 2011. Post-traumatic seizures exacerbate histopathological damage after fluid-percussion brain injury. *J Neurotrauma.* 28(1): 35–42.

Bell, M.J., C.S. Robertson, P.M. Kochanek, J.C. Goodman, S.P. Gopinath, J.A. Carcillo, R.S. Clark, D.W. Marion, Z. Mi and E.K. Jackson. 2001. Interstitial brain adenosine and xanthine increase during jugular venous oxygen desaturations in humans after traumatic brain injury. *Crit Care Med.* 29(2): 399–404.

Benowitz, L., W. Rodriguez and R. Neve. 1990. The pattern of GAP-43 immunostaining changes in the rat hippocampal formation during reactive synaptogenesis. *Mol Brain Res.* 8: 17–23.

Benowitz, L.I., P.J. Apostolides, N. Perrone-Bizzozero, S.P. Finklestein and H. Zwiers. 1988. Anatomical distribution of the growth-associated protein GAP-43/B-50 in the adult rat brain. *J Neurosci.* 8: 339–352.

Berger, R. 2006. The use of serum biomarkers to predict outcome after traumatic brain injury in adults and children. *J Head Trauma Rehabil.* 21(4): 315–333.

Berger, R.P., B. Bazaco, A.K. Wagner, P.M. Kochanek and A. Fabio. 2010. Trajectory analysis of serum biomarker concentrations facilitates outcome prediction after pediatric traumatic and hypoxemic brain injury. *Dev Neurosci.* 32(5-6): 396–405.

Berger, R.P., S.R. Beers, R. Richichi, D. Wiesman and P.D. Adelson. 2007. Serum biomarker concentrations and outcome after pediatric traumatic brain injury. *J Neurotrauma.* 24: 1793–1801.

Berman, B.P., D.J. Weisenberger and P.W. Laird. 2009. Locking in on the human methylome. *Nat Biotechnol.* 27: 341–342.

Binder, D.K. and H.E. Scharfman. 2004. Brain-derived neurotrophic factor. Growth Factors 22(3): 123–1231.

Boehning, D., R.L. Patterson, L. Sedaghat, N.O. Glebova, T. Kurosaki and S.H. Snyder. 2003. Cytochrome c binds to inositol (1,4,5) trisphosphate receptors, amplifying calcium-dependent apoptosis. *Nat Cell Biol.* 5: 1051–1061.

Bonneh-Barkay, D., P. Zagadailov, H. Zou, C. Niyonkuru, M. Figley, A. Starkey, G. Wang, S. Bissel, C.A. Wiley and A.K. Wagner. 2010. YKL-40 expression in traumatic brain injury-An initial analysis. *J Neurotrauma.* 27(7): 1215–1223.

Bottiglieri, T., K. Hyland, M. Laundy, P. Godfrey, M.W. Carney, B.K. Toone and E.H. Reynolds. 1992. Folate deficiency, biopterin and monoamine metabolism in depression. *Psychol Med.* 22: 871–876.

Brain Trauma Foundation, American Association of Neurological Surgeons and Congress of Neurological Surgeons. 2007. Guidelines for the management of severe traumatic brain injury. *J Neurotrauma.* 24 Suppl. 1: S1–106.

Brophy, G.M., S. Mondello, L. Papa, S.A. Robicsek, A. Gabrielli, J. Tepas 3rd, A. Buki, C. Robertson, F.C. Tortella, R.L. Hayes and K.K. Wang. 2011. Biokinetic Analysis of Ubiquitin C-Terminal Hydrolase-L1 (UCH-L1) in Severe Traumatic Brain Injury Patient Biofluids. *J Neurotrauma.* 28(6): 861–870.

Brozoski, T.J., R.M. Brown, H.E. Rosvold and P.S. Goldman. 1979. Cognitive deficit caused by regional depletion of dopamine in prefrontal cortex of rhesus monkey. *Science.* 205(4409): 929–932.

Bu, D.F., M.G. Erlander, B.C. Hitz, N.J. Tillakaratne, D.L. Kaufman, C.B. Wagner-McPherson, G.A. Evans and A.J. Tobin. 1992. Two human glutamate decarboxylases, 65-kDa GAD and 67-kDA GAD, are each encoded by a single gene. *Proc Natl Acad Sci USA.* 89(6): 2115–2119.

Bueller, J.A., M. Aftab, S. Sen, D. Gomez-Hassan, M. Burmeister and J.K. Zubieta. 2006. BDNF Val66Met allele is associated with reduced hippocampal volume in healthy subjects. *Biol Psychiatry.* 59: 812–815.

Burgess, J.D., O. Pedraza, N.R. Graff-Radford, M. Hirpa, F. Zou, R. Miles, T. Nguyen, M. Li , J.A. Lucas, R.J. Ivnik, J. Crook, V.S. Pankratz, D.W. Dickson, R.C. Petersen, S.G. Younkin and N. Ertekin-Taner. 2011. Association of common KIBRA variants with episodic memory and AD risk. *Neurobiol Aging.* 32: 557.e1–557.e9.

Bushnik, T. and W. Gordon. 2012. Updates from the Third Federal Interagency Conference on traumatic brain injury. *J Head Trauma Rehabil.* 27(3): 222–223.

Büther, K., C. Plaas, A. Barnekow and J. Kremerskothen. 2004. KIBRA is a novel substrate for protein kinase C. *Biochem Biophys Res Commun.* 317: 703–707.

Caspi, A., K. Sugden, T.E. Moffitt, A. Taylor, I.W. Craig, H. Harrington, J. McClay, J. Mill, J. Martin, A. Braithwaite and R. Poulton. 2003. Influence of life stress on depression: Moderation by a polymorphism in the 5-htt gene. *Science.* 301: 386–389.

Castren, E., V. Voikar and T. Rantamaki. 2007. Role of neurotrophic factors in depression. *Curr Opin Pharmacol.* 7: 18–21.

Cavalleri, G.L., N.M. Walley, N. Soranzo, J. Mulley, C.P. Doherty, A. Kapoor, C. Depondt, J.M. Lynch, I.E. Scheffer, A. Heils, A. Gehrmann, P. Kinirons, S. Gandhi, P. Satishchandra, N.W. Wood, A. Anand, T. Sander, S.F. Berkovic, N. Delanty, D.B. Goldstein and S.M. Sisodiya. 2007. A multicenter study of BRD2 as a risk factor for juvenile myoclonic epilepsy. *Epilepsia.* 48(4): 706–712.

Chan, F., K.L. Lanctôt, A. Feinstein, N. Herrmann, J. Strauss, T. Sicard, J.L. Kennedy, S. McCullagh and M.J. Rapoport. 2008. *Brain Inj.* 22(6): 471–479.

Chatfield, D.A., F.P. Zemlan, D.J. Day and D.K. Menon. 2002. Discordant temporal patterns of S100 beta and cleaved tau protein elevation after head injury: A pilot study. *Br Neurosurg.* 16: 471–476.

Chen, J., B.K. Lipska, N. Halim, Q.D. Ma, M. Matsumoto, S. Melhem, B.S. Kolachana, T.M. Hyde, M.M. Herman, J. Apud, M.F. Egan, J.E. Kleinman and D.R. Weinberger. 2004. Functional analysis of genetic variation in catechol-O-methyltransferase (COMT): Effects on mRNA, protein, and enzyme activity in postmortem human brain. *Am J Hum Genet.* 75(5): 807–821.

Chen, J.W.Y. and C.G. Westerlain. 2006. Status epilepticus: Pathophysiology and management in adults. *Lancet.* 5(3): 246–256.

Christman, C.W., J.B. Salvant, S.A. Walker and J.T. Povlishock. 1997. Characterization of a prolonged regenerative attempt by diffusely injured axons following traumatic brain injury in adult cat: A light and electron microscopic immunocytochemical study. *Acta Neuropathol.* 94: 329–337.

Ciruela, F., V. Casadó, R.J. Rodrigues, R. Luján, J. Burgueño, M. Canals, J. Borycz, N. Rebola, S.R. Goldberg, J. Mallol, A. Cortés, E.I. Canela, J.F. López-Giménez, G. Milligan, C. Lluis, R.A. Cunha, S. Ferré and R. Franco. 2006. Presynaptic control of striatal glutamatergic neurotransmission by adenosine A1-A2A receptor heteromers. *J Neurosci.* 26(7): 2080–2087.

Clark, R.S., H. Bayir and L.W. Jenkins. 2005. Posttranslational protein modifications. *Crit Care Med.* 33: S407–S409.

Clark, R.S., J. Chen, S.C. Watkins, P.M. Kochanek, M. Chen, R.A. Stetler, J.E. Loeffert and S.H. Graham. 1997. Apoptosis-suppressor gene bcl-2 expression after traumatic brain injury in rats. *J Neurosci.* 17: 9172–9182.

Clark, R.S., P.M. Kochanek, M. Chen, S.C. Watkins, D.W. Marion, J. Chen, R.L. Hamilton, J.E. Loeffert and S.H. Graham. 1999. Increases in Bcl-2 and cleavage of caspase-1 and caspase-3 in human brain after head injury. *FASEB J.* 13: 813–821.

Clark, R.S., P.M. Kochanek, P.D. Adelson, M.J. Bell, J.A. Carcillo, M. Chen, S.R. Wisniewski, K. Janesko, M.J. Whalen and S.H. Graham. 2000. Increases in bcl-2 protein in cerebrospinal fluid and evidence for programmed cell death in infants and children after severe traumatic brain injury. *J Pediatr.* 137: 197–204.

Coggeshall, R.E., M.L. Reynolds and C.J. Woolf. 1991. Distribution of the growth associated protein GAP-43 in the central processes of axotomized primary afferents in the adult rat spinal cord; presence of growth cone-like structures. *Neurosci Lett.* 131: 37–41.

Cohen, J.D. and D. Servan-Schreiber. 1992. Context, cortex, and dopamine: A connectionist approach to behavior and biology in schizophrenia. *Psychol Rev.* 99: 45–77.

Conley, Y.P. and S. Alexander. 2011. Genomic, transcriptomic, and epigenomic approaches to recovery after acquired brain injury. *PM R* 3(6 Suppl. 1): S52–58.

Cook, E.H. Jr., M.A. Stein, M.D. Krasowski, N.J. Cox, D.M. Olkon, J.E. Kieffer and B.L. Leventhal. 1995. Association of attention-deficit disorder and the dopamine transporter gene. *Am J Hum Genet.* 56(4): 993–998. PMID: 7717410.

Corneveaux, J.J., W.S. Liang, E.M. Reiman, J.A. Webster, A.J. Myers, V.L. Zismann, K.D. Joshipura, J.V. Pearson, D. Hu-Lince, D.W. Craig, K.D. Coon, T. Dunckley, D. Bandy, W. Lee, K. Chen, T.G. Beach, D. Mastroeni, A. Grover, R. Ravid, S.B. Sando, J.O. Aasly, R. Heun, F. Jessen, H. Kölsch, J. Rogers, M.L. Hutton, S. Melquist, R.C. Petersen, G.E. Alexander, R.J. Caselli, A. Papassotiropoulos, D.A. Stephan and M.J. Huentelman. 2010. Evidence for an association between KIBRA and late-onset Alzeimer's disease. *Neurobiol Aging.* 31: 901–909.

Crack, P.J., J. Gould, N. Bye, S. Ross, U. Ali, M.D. Habgood, C. Morganti-Kossman, N.R. Saunders, P.J. Hertzog and Victorian Neurotrauma Research Group. 2009. The genomic profile of the cerebral cortex after closed head injury in mice: Effects of minocycline. *J Neural Transm.* 116: 1–12.

Darrah, S.D., M.A. Miller, D. Ren, N.Z. Hoh, J.M. Scanlon, Y.P. Conley and A.K. Wagner. 2013. Genetic variability in glutamic acid decarboxylase genes: Associations with post-traumatic seizures after severe TBI. *Epilepsy Res.* 103(2-3): 180–194.

Dash, P.K., J. Zhao, G. Hergenroeder and A.N. Moore. 2010. Biomarkers for the diagnosis, prognosis, and evaluation of treatment efficacy for traumatic brain injury. *Neurotherapeutics.* 7: 100–114.

Daws, L.C., J.L. Munn, M.F. Valdez, T. Frosto-Burke and J.G. Hensler. 2007. Serotonin transporter function, but not expression, is dependent on brain-derived neurotrophic factor (BDNF): *in vivo* studies in BDNF-deficient mice. *J Neurochem.* 101: 641–651. PubMed: 17254018.

Daws, L.C., P.D. Callaghan, J.A. Moron, K.M. Kahlig, T.S. Shippenberg, J.A. Javitch and A. Galli. 2002. Cocaine increases dopamine uptake and cell surface expression of dopamine transporters. *Bioch Biophys Res Commun.* 290(5): 1545–1550.

De Kruijk, J.R., P. Leffers, P.P.C.A. Menheere, S. Meerhoff, J. Rutten and A. Twijnstra. 2002. Prediction of post-traumatic complaints after mild traumatic brain injury: Early symptoms and biochemical markers. *J Neurol Neurosurg Psychiatry.* 73: 727–732.

Deckert, J. and M.B. Jorgensen. 1988. Evidence for pre- and postsynaptic localization of adenosine A1 receptors in the CA1 region of rat hippocampus: A quantitative autoradiographic study. *Brain Res.* 446: 161–164.

Diaz-Arrastia, R., Y. Gong, S. Fair, K.D. Scott, M.C. Garcia, M.C. Carlile, M.A. Agostini and P.C. Van Ness. 2003. Increased risk of late posttraumatic seizures associated with inheritance of APOE epsilon4 allele. *Arch Neurol.* 60(6): 818–822.

Dietrich, W.D., O. Alonso and M. Halley. 1994. Early microvascular and neuronal consequences of traumatic brain injury: A light and electron microscopic study in rats. *J Neurotrauma.* 11: 289–301.

Dimopoulou, I., S. Korfias, U. Dafni, A. Anthi, C. Psachoulia, G. Jullien, D.E. Sakas and C. Roussos. 2003. Protein S-100b serum levels in trauma-induced brain death. *Neurology.* 60: 947–951.

Dixon, C.E., M.F. Kraus, A.E. Kline, X. Ma, H.Q. Yan, R.G. Griffith, B.M. Wolfson and D.W. Marion. 1999. Amantadine improves water maze performance without affecting motor behavior following traumatic brain injury in rats. *Restor Neurol Neurosci.* 14(4): 285–294. PMID: 12671249.

Donato, R. 2003. Intracellular and extracellular roles of S100 proteins. *Microsc Res Tech.* 60: 540–551.

Donnelly. C.J., M. Park, M. Spillane, S. Yoo, A. Pacheco, C. Gomes, D. Vuppalanchi, M. McDonald, H.H. Kim, T.T. Merianda, G. Gallo and J.L. Twiss. 2013. Axonally synthesized β-actin and GAP-43 proteins support distinct modes of axonal growth. *J Neurosci.* 33(8): 3311–3322.

Dresel, S., J. Krause, K.H. Krause, C. LaFougere, K. Brinkbaumer, H.F. Kung, K. Hahn and K. Tatsch. 2000. Attention deficit hyperactivity disorder: Binding of [99mTc]TRODAT-1 to the dopamine transporter before and after methylphenidate treatment. *Eur J Nucl Med.* 27(10): 1518–1524.

Duman, R.S. and L.M. Monteggia. 2006. A neurotrophic model for stress-related mood disorders. *Biol Psychiatry.* 59: 1116–1127.

Dziedzic, T. 2006. Systemic inflammatory markers and risk of dementia. *Am J Alzheimers Dis Othr Demen.* 21: 258–262.

Egan, M.F., M. Kojima, J.H. Callicott, T.E. Goldberg, B.S. Kolachana, A. Bertolino, E. Zaitsev, B. Gold, D. Goldman, M. Dean, B. Lu and D.R. Weinberger. 2003. The BDNF val66met polymorphism affects activity-dependent secretion of BDNF and human memory and hippocampal function. *Cell.* 112: 257–269.

Eley, T.C., K. Sugden, A. Corsico, A.M. Gregory, P. Sham, P. McGuffin, R. Plomin and I.W. Craig. 2004. Gene-environment interaction analysis of serotonin system markers with adolescent depression. *Mol Psychiatry.* 9: 908–915.

Elting, J.W., A.E. de Jager, A.W. Teelken, M.J. Schaaf, N.M. Maurits, J. van der Naalt, C.T. Sibinga, G.A. Sulter and J. De Keyser. 2000. Comparison of serum S-100 protein levels following stroke and traumatic brain injury. *J Neurol Sci.* 181: 104–110.

Eng, L.F., J.J. Vanderhaeghen, A. Bignami and B. Gerstl. 1971. An acidic protein isolated from fibrous astrocytes. *Brain Res.* 28: 351–354.

Espinosa-Aguilar, A., H. Reyes-Morales, C.E. Huerta-Posada, I.L. de León, F. López-López, M. Mejía-Hernández, M.A. Mondragón-Martínez, L.M. Calderón-Téllez, R.L. Amezcua-Cuevas and J.A. Rebollar-González. 2008. Design and validation of a critical pathway for hospital management of patients with severe traumatic brain injury. *J Trauma.* 64(5): 1327–1341.

Failla, M.D., J.N. Burkhardt, M.A. Miller, J.M. Scanlon, Y.P. Conley, R.E. Ferrell and A.K. Wagner. 2013. Variants of the SLC6A4 gene in depression risk following severe TBI. *Brain Inj.* 27: 696–706.

Fedele, D.E., N. Gouder, M. Güttinger, L. Gabernet, L. Scheurer, T. Rülicke, F. Crestani and D. Boison. 2005. Astrogliosis in epilepsy leads to overexpression of adenosine kinase, resulting in seizure aggravation. *Brain.* 128: 2383–2395.

Fedoroff, J.P., S.E. Starkstein, A.W. Forrester, F.H. Geisler, R.E. Jorge, S.V. Arndt and R.G. Robinson. 1992. Depression in patients with acute traumatic brain injury. *Am J Psychiatry.* 149: 918–923.

Floresco, S.B. and O. Magyar. 2006. Mesocortical dopamine modulation of executive functions: Beyond working memory. *Psychopharmacology (Berl).* 188: 567–585.

Franco, R., S. Ferré, L. Agnati, M. Torvinen, S. Ginés, J. Hillion, V. Casadó, P. Lledó, M. Zoli, C. Lluis and K. Fuxe. 2000. Evidence for adenosine/dopamine receptor interactions: Indications for heteromerization. *Neuropsychopharmacology.* 23(4 Suppl.): S50–59.

Frieden, T.R., R. Ikeda, R. Hunt, M.M. Faul, L. Xu, M. Wald and V.G. Coronado. 2006. Traumatic Brain injury In the United States: Emergency Department Visits, Hospitalizations and Deaths 2002–2006. *National Center for Injury Prevention and Control, Centers for Disease Control and Prevention.*

Frisen, J., V.M. Verge, S. Cullheim, H. Persson, K. Fried, D.S. Middlemas, T. Hunter, T. Hökfelt and M. Risling. 1992. Increased levels of trkB mRNA and trkB protein-like immunoreactivity in the injured rat and cat spinal cord. *Proc Natl Acad Sci USA.* 89: 11282–11286.

Garringer, J., C. Niyonkuru, E. McCullough, T.L. Loucks, C.E. Dixon, Y.P. Conley, S. Berga and A.K. Wagner. 2013. Impact of Aromatase Genetic Variation on Hormone Levels and Global Outcome after Severe TBI. *J Neurotrauma* (Epub ahead of print).

Giacino, J.T., J. Whyte, E. Bagiella, K. Kalmar, N. Childs, A. Khademi, B. Eifert, D. Long, D.I. Katz, S. Cho, S.A. Yablon, M. Luther, F.M. Hammond, A. Nordenbo, P. Novak, W. Mercer, P. Maurer-Karattup and M. Sherer. 2012. Placebo-controlled trial of amantadine for severe traumatic brain injury. *N Engl J Med.* 366(9): 819–826.

Gilbody, S., S. Lewis and T. Lightfoot. 2007. Methylenetetrahydrofolate reductase (MTHFR) genetic polymorphisms and psychiatric disorders: A HuGE review. *Am J Epidemiol.* 165: 1–13.

Gillespie, N.A., J.B. Whitfield, B. Williams, A.C. Heath and N.G. Martin. 2005. The relationship between stressful life events, the serotonin transporter (5-httlpr) genotype and major depression. *Psychol Med.* 35: 101–111.

Goeders, N.E. and J.E. Smith. 1993. Intracranial cocaine self-administration into the medial prefrontal cortex increases dopamine turnover in the nucleus accumbens. *J Pharmacol Exp Therap.* 265(2): 592–600.

Goldstein, L.B. 2003. Neuropharmacology of TBI-induced plasticity. *Brain Inj.* 17: 685–694.

Gomez-Hernandez, R., J.E. Max, T. Kosier, S. Paradiso and R.G. Robinson. 1997. Social impairment and depression after traumatic brain injury. *Arch Phys Med Rehabil.* 78: 1321–1326.

Gong, B. and E. Leznik. 2007. The role of ubiquitin C-terminal hydrolase L1 in neurodegenerative disorders. *Drug News Perspect.* 20: 365–370.

Gorgels, T.G., M. Ven Lookeren-Compagne, A.B. Oestreicher, A.A. Gribnau and W.H. Gispen. 1989. B-50/Gap43 is localized at the cytoplasmic side of the plasma membrane in developing and adult rat pyramidal tract. *J Neurosci.* 9: 3861–3869.

Goyal, A., M. Carter, C. Niyonkuru, A. Fabio, K. Amin, R.P. Berger and A.K. Wagner. 2012. S100b as a prognostic Biomarker in Outcome Prediction for Patients with severe TBI. *J Neurotrauma.* 2013 Jun 1; 30(11): 946–57.

Hall, D., A. Dhilla, A. Charalambous, J.A. Gogos and M. Karayiorgou. 2003. Sequence variants of the brain-derived neurotrophic factor (BDNF) gene are strongly associated with obsessive-compulsive disorder. *Am J Hum Genet.* 73: 370–376.

Hariri, A.R., T.E. Goldberg, V.S. Mattay, B.S. Kolachana, J.H. Callicott, M.F. Egan and D.R. Weinberger. 2003. Brain-derived neurotrophic factor val66met polymorphism affects human memory-related hippocampal activity and predicts memory performance. *J Neurosci.* 23: 6690–6694.

Hashimoto, T., A. Serrano-Pozo, Y. Hori, K.W. Adams, S. Takeda, A.O. Banerji, A. Mitani, D. Joyner, D.H. Thyssen, B.J. Bacskai, M.P. Frosch, T.L. Spires-Jones, M.B. Finn, D.M. Holtzman and B.T. Hyman. 2012. Apolipoprotein E, especially apolipoprotein E4, increases the oligomerization of amyloid β peptide. *J Neurosci.* 32(43): 15181–15192.

Hayakata, T., T. Shiozaki, O. Tasaki, H. Ikegawa, Y. Inoue, F. Toshiyuki, H. Hosotubo, F. Kieko, T. Yamashita, H. Tanaka, T. Shimazu and H. Sugimoto. 2004. Changes in Csf S100b and Cytokine Concentrations in Early-Phase Severe Traumatic Brain Injury. *Shock.* 22(2): 102–107.

Heils, A., A. Teufel, S. Petri, G. Stober, P. Riederer, D. Bengel and K.P. Lesch. 1996. Allelic variation of human serotonin transporter gene expression. *J Neurochem.* 66: 2621–2624.

Heldt, S.A., L. Stanek, J.P. Chhatwal and K.J. Ressler. 2007. Hippocampus-specific deletion of BDNF in adult mice impairs spatial memory and extinction of aversive memories. *Mol Psychiatry.* 12: 656–670.

Helisalmi, S., M. Hiltunen, S. Vepsäläinen, S. Iivonen, A. Mannermaa, M. Lehtovirta, A.M. Koivisto, I. Alafuzoff and H. Soininen. 2004. Polymorphisms in neprilysin gene affect the risk of Alzheimer's disease in Finnish patients. *J Neurol Neurosurg Psychiatry.* 75(12): 1746–1748.

Herrmann, N., M.J. Rapoport, R.D. Rajaram, F. Chan, A. Kiss, A.K. Ma, A. Feinstein, S. McCullagh and K.L. Lanctot. 2009. Factor analysis of the Rivermead post-concussion symptoms questionnaire in mild-tomoderate traumatic brain injury patients. *J Neuropsychiatry Clin Neurosci.* 21: 181–188.

Herson, P.S., I.P. Koerner and P.D. Hurn. 2009. Sex, sex steroids, and brain injury. *Semin Reprod Med.* 27: 229–239.

Hicks, R.R., S. Numan, H.S. Dhillon, M.R. Prasad and K.B. Seroogy. 1997. Alterations in BDNF and NT-3 mRNAs in rat hippocampus after experimental brain trauma. *Brain Res Mol Brain Res.* 48: 401–406.

Hoh, N.Z., A.K. Wagner, S.A. Alexander, R.B. Clark, S.R. Beers, D.O. Okonkwo, D. Ren and Y.P. Conley. 2010. BCL2 genotypes: Functional and neurobehavioral outcomes after severe traumatic brain injury. *J Neurotrauma.* 27(8): 1413–1427.

Hojo, Y., G. Murakami, H. Mukai, S. Higo, Y. Hatanaka, M. Ogiue-Ikeda, H. Ishii, T. Kimoto and S. Kawato. 2008. Estrogen synthesis in the brain—role in synaptic plasticity and memory. *Mol Cell Endocrinol.* 290: 31–43.

Horn, A.S. 1990. Dopamine uptake: A review of progress in the last decade. *Prog Neurobiol.* 34(5): 387–400.

Huang, E.J. and L.F. Reichardt. 2001. Neurotrophins: Roles in neuronal development and function. *Annu Rev Neurosci.* 24: 677–736.

Huang, E.J. and L.F. Reichardt. 2003. Trk receptors: Roles in neuronal signal transduction. *Annu Rev Biochem.* 72: 609–642.

Huang, R., J. Huang, H. Cathcart, S. Smith and S.E. Poduslo. 2007. Genetic variants in brain-derived neurotrophic factor associated with Alzheimer's disease. *J Med Genet.* 44: e66.

Hwang, J.P., S.J. Tsai, C.J. Hong, C.H. Yang, J.F. Lirng and Y.M. Yang. 2006. The Val66Met polymorphism of the brain-derived neurotrophic-factor gene is associated with geriatric depression. *Neurobiol Aging.* 27: 1834–1837.

Ingebrigtsen, T. and B. Romner. 2002. Biochemical serum markers of traumatic brain injury. *J Trauma.* 52: 798–808.

Jacobs, N., G. Kenis, F. Peeters, C. Derom, R. Vlietinck and J. van Os. 2006. Stress-related negative affectivity and genetically altered serotonin transporter function: Evidence of synergism in shaping risk of depression. *Arch Gen Psychiatry.* 63: 989–996.

Jacobsen, L.K., M.R. Picciotto, C.J. Heath, W.E. Mencl and J. Gelernter. 2009. Allelic variation of calsyntenin 2 (CLSTN2) modulates the impact of developemental tobacco smoke exposure on mnemonic processing in adolescents. *Biol Psychiatry.* 65: 671–679.

Jamali, S., A. Salzmann, N. Perroud, M. Ponsole-Lenfant, J. Cillario, P. Roll, N. Roeckel-Trevisiol, A. Crespel, J. Balzar, K. Schlachter, U. Gruber-Sedlmayr, E. Pataraia, C. Baumgartner, A. Zimprich, F. Zimprich, A. Malafosse and P. Szepetowski. 2010. Functional variant in complement C3 gene promoter and genetic susceptibility to temporal lobe epilepsy and febrile seizures. *PLoS One.* 5(9): pii:e12740.

Johannsen, S., K. Duning, H. Pravendstadt, J. Kremerskothen and T.M. Boeckers. 2008. Temporal-spatial expression and novel biochemical properties of the memory-related protein KIBRA. *Neuroscience.* 155: 1165–1173.

Johnson, V.E., W. Stewart and D.H. Smith. 2010. Traumatic brain injury and amyloid-B pathology: A link to Alzheimer's disease? *Nat Rev Neurosci.* 11(5): 361–370.

Johnson, V.E., W. Stewart, J.E. Stewart, D.I. Graham, A.H. Praestgaad, D.H. Smith and A. Neprilysin. 2009. Polymorphism and Amyloid-beta Plaques Following Traumatic Brain Injury. *J Neurotrauma.* 26: 1197–1202.

Jones, S.R., R.R. Gainetdinov, M. Jaber, B. Giros, R.M. Wightman and M.G. Caron. 1998. Profound neuronal plasticity in response to inactivation of the dopamine transporter. *Proc Natl Acad Sci USA.* 95(7): 4029–4034. PMID: 9520487.

Jönsson, E.G., M.M. Nöthen, F. Grünhage, L. Farde, Y. Nakashima, P. Propping and G.C. Sedvall. 1999. Polymorphisms in the dopamine D2 receptor gene and their relationships to striatal dopamine receptor density of healthy volunteers. *Mol Psychiatry.* 4(3): 290–296.

Jorge, R.E., R.G. Robinson, S.V. Arndt, S.E. Starkstein, A.W. Forrester and F. Geisler. 1993. Depression following traumatic brain injury: A 1 year longitudinal study. *J Affect Disord.* 27: 233–243.

Juengst, S.B., R. Kumar, M.D. Failla, A. Goyal, A.K. Wagner. 2014. Acute Inflammatory Associations with Depression after severe TBI. J. Head Trauma Rehabilitation. *J Head Trauma Rehabil.* 2014, Feb 28 [Epub ahead of print].

Kendler, K.S., J.W. Kuhn, J. Vittum, C.A. Prescott and B. Riley. 2005. The interaction of stressful life events and a serotonin transporter polymorphism in the prediction of episodes of major depression: A replication. *Arch Gen Psychiatry.* 62: 529–535.

Klein, M.J. 2009. Post head injury endocrine complications. *In:* C.T. Lorenzo [ed.]. WebMD LL, New York, NY.

Kleindienst, A. and M.R. Bullock. 2006. A critical analysis of the role of the neurotrophic protein S100B in acute brain injury. *J Neurotrauma.* 23: 1185–1200.

Kleindienst, A., C. Schmidt, H. Parsch, I. Emtmann, Y. Xu and M. Buchfelder. 2010. The passage of S100b from brain to blood is not specifically related to the blood-brain barrier integrity. *Cardiovasc Psychiatry Neurol.* 1–8.

Kline, A.E., H.Q. Yan, J. Bao, D.W. Marion and C.E. Dixon. 2000. Chronic methylphenidate treatment enhances water maze performance following traumatic brain injury in rats. *Neurosci Lett.* 280: 163–166.

Kochanek, K., T. Novack, R. Nakase-Richardson, M. Sherer, A.B. Frol, W. Gordon, R.A. Hanks, J.T. Giacino and J.H. Ricker. 2008. Feasibility of a brief neuropsychological test battery during acute inpatient rehabilitation after TBI. *Arch Phys Med Rehabil.* 89(5): 242–249.

Kochanek, P.M., V.A. Vagni, K.L. Janesko, C.B. Washington, P.K. Crumrine, R.H. Garman, L.W. Jenkins, R.S. Clark, G.E. Homanics, C.E. Dixon, J. Schnermann and E.K. Jackson. 2006. Adenosine A1 receptor knockout mice develop lethal status epilepticus after experimental traumatic brain injury. *J Cereb Blood Flow Metab.* 26: 565–575.

Korfias, S., G. Stranjalis, A. Papadimitriou, C. Psachoulia, G. Daskalakis, A. Antsaklis and D.E. Sakas. 2006. Serum S-100B protein as a biochemical marker of brain injury: A review of current concepts. *Curr Med Chem.* 13: 3719–3731.

Krause, K.H., S.H. Dresel, J. Krause, H.F. Kung and K. Tatsch. 2000. Increased striatal dopamine transporter in adult patients with attention deficit hyperactivity disorder: Effects of methylphenidate as measured by single photon emission computed tomography. *Neurosci Lett.* 285(2): 107–110.

Kremerskothen, J., C. Plaas, K. Büther, I. Finger, S. Veltel, T. Matanis, T. Liedtke and A. Barnekow. 2002. Characterization of KIBRA, a novel WW domain-containing protein. *Biochem Biophys Res Commun.* 300: 862–867.

Krueger, F., M. Pardini, E.D. Huey, V. Raymont, J. Solomon, R.H. Lipsky, C.A. Hodgkinson, D. Goldman and J. Grafman. 2011. The role of the Met66 brain-derived neurotrophic factor allele in the recovery of executive functioning after combat-related traumatic brain injury. *J. Neurosci.* 31: 598–606.

Kumar, R., J.A. Boles, M.D. §Failla and A.K. Wagner. 2014. Chronic Inflammation Characterization after severe TBI and associations with outcome. *J Head Trauma Rehabil.* 2014, Jun 4 [Epub ahead of print].

Kuo, P.H., G. Kalsi, C.A. Prescott, C.A. Hodgkinson, D. Goldman, J. Alexander, E.J. van den Oord, X. Chen, P.F. Sullivan, D.G. Patterson, D. Walsh, K.S. Kendler and B.P. Riley. 2009. Associations of glutamate decarboxylase genes with initial sensitivity and age-at-onset of alcohol dependence in the Irish affected Sib Pair Study of Alcohol Dependence. *Drug Alcohol Depend.* 101: 80–87.

Kwan, P., L. Baum and V. Wong. 2007. Association between ABCB1 C3435T polymorphism and drug-resistant epilepsy in Han Chinese. *Epilepsy Behav.* 11(1): 112–117.

Kwan, P., V. Wong, P.W. Ng, C.H. Lui, N.C. Sin, W.S. Poon, H.K. Ng, K.S. Wong and L. Baum. 2009. Gene-wide tagging study of association between ABCB1 polymorphisms and multidrug resistance in epilepsy in Han Chinese. *Pharmacogenomics.* 10(5): 723–732. doi: 10.2217/pgs.09.32.

Lanctôt, K.L., M.J. Rapoport, F. Chan, R.D. Rajaram, J. Strauss, T. Sicard, S. McCullagh, A. Feinstein, A. Kiss, J.L. Kennedy, A.S. Bassett and N. Herrmann. 2010. Genetic predictors of response to treatment with citalopram in depression secondary to traumatic brain injury. *Brain Inj.* 24(7-8): 959–969.

Leblanc, J., E. de Guise, N. Gosselin and M. Feyz. 2006. Comparison of functional outcome following acute care in young, middle aged, and elderly patients with traumatic brain injury. *Brain Inj.* 20(8): 779–790.

Leibrock, J., F. Lottspeich, A. Hohn, M. Hofer, B. Hengerer, P. Masiakowski, H. Thoenen and Y.A. Barde. 1989. Molecular cloning and expression of brain-derived neurotrophic factor. *Nature.* 341: 149–152.

Liou, A.K., R.S. Clark, D.C. Henshall, X.M. Yin and J. Chen. 2003. To die or not to die for neurons in ischemia, traumatic brain injury and epilepsy: A review on the stress-activated signaling pathways and apoptotic pathways. *Prog Neurobiol.* 69: 103–142.

Lipsky, R.H., M.B. Sparling, L.M. Ryan, K. Xu, A.M. Salazar, D. Goldman and D.L. Warden. 2005. Association of COMT Val158Met genotype with executive functioning following traumatic brain injury. *J Neuropsychiatry Clin Neurosci.* 17(4): 465–471. PMID: 16387984.

Liu, Y., H. Zou, A.C. Michael and A.K. Wagner. 2009. Gender comparisons in neurochemical effects of methylphenidate on striatal neurotransmission of rats with experimental traumatic brain injury. Program No. 618.27-Tu 2009 Neuroscience Meeting Planner. Chicago, IL: Society for Neuroscience. Online.

Lu, B. 2003. Pro-region of neurotrophins: Role in synaptic modulation. *Neuron.* 39: 735–738.

Luikart, B.W. and L.F. Parada. 2006. Receptor tyrosine kinase B-mediated excitatory synaptogenesis. *Prog Brain Res.* 157: 15–24.

Lumpkins, K.M., G.V. Bochicchio, K. Keledjian, J.M. Simard, M. McCunn and T. Scalea. 2008. Glial fibrillary acidic protein is highly correlated with brain injury. *J Trauma.* 65: 778–782.

MacLusky, N.J., T. Hajszan, J. Prange-Kiel and C. Leranth. 2006. Androgen modulation of hippocampal synaptic plasticity. *Neuroscience.* 138: 957–965.

Maher, B.S., M.L. Marazita, R.E. Ferrell and M.M. Vanyukov. 2002. Dopamine system genes and attention deficit hyperactivity disorder: A meta-analysis. *Psychiatr Genet.* 12(4): 207–215.

Manuck, S.B., J.D. Flory, R.E. Ferrell, J.J. Mann and M.F. Muldoon. 2000. A regulatory polymorphism of the monoamine oxidase-A gene may be associated with variability in aggression, impulsivity, and central nervous system serotonergic responsivity. *Psychiatry Res.* 95(1): 9–23.

Masel, B.E. and D.S. DeWitt. 2010. Traumatic brain injury: A disease process, not an event. *J Neurotrauma.* 27: 1529–1540.

Masliah, E., A.M. Fagan, R.D. Terry, R. DeTeresa, M. Mallory and F.H. Gage. 1991. Reactive synaptogenesis assessed by synaptophysin immunoreactivity is associated with GAP-43 in the dentate gyrus of the adult rat. *Exp Neurol.* 113: 131–142.

Mauri, M., E. Sinforiani, G. Bono, R. Cittadella, A. Quattrone, F. Boller and G. Nappi. 2006. Interaction between apolipoprotein epsilon 4 and traumatic brain injury in patients with Alzeimer's disease and mild cognitive impairment. *Funct Neurol.* 21(4): 223–228.

McAllister, A.K., D.C. Lo and L.C. Katz. 1995. Neurotrophins regulate dendritic growth in developing visual cortex. *Neuron.* 15(4): 791–803.

McAllister, T.W. 2010. Genetic factors modulating outcome after neurotrauma. *PMR.* 2: S241–S252.

McAllister, T.W., L.A. Flashman, C. Harker Rhodes, A.L. Tyler, J.H. Moore, A.J. Saykin, B.C. McDonald, T.D. Tosteson and G.J. Tsongalis. 2008. Single nucleotide polymorphisms in ANKK1 and the dopamine D2 receptor gene affect cognitive outcome shortly after traumatic brain injury: A replication and extension study. *Brain Inj.* 22: 705–714.

Metting, Z., N. Wilczak, L.A. Rodiger, J.M. Schaaf and J. van der Naalt. 2012. GFAP and S100B in the acute phase of mild traumatic brain injury. *Neurology.* 78(18): 1428–1433.

Michelhaugh, S.K., C. Fiskerstrand, E. Lovejoy, M.J. Bannon and J.P. Quinn. 2001. The dopamine transporter gene (SLC6A3) variable number of tandem repeats domain enhances transcription in dopamine neurons. *J Neurochem.* 79(5): 1033–1038.

Miller, M.A., Y. Conley, J.M. Scanlon, D. Ren, M. Ilyas Kamboh, C. Niyonkuru and A.K. Wagner. 2010. APOE genetic associations with seizure development after severe traumatic brain injury. *Brain Inj.* 24(12): 1468–1477.

Missler, U., M. Wiesmann, G. Wittmann, O. Magerkurth and H. Hagenström. 1999. Measurement of glial firbrillary acidic protein in human blood: Analytical method and preliminary clinical results. *Clin Chem.* 45: 138–141.

Momose, Y., M. Murata, K. Kobayashi, M. Tachikawa, Y. Nakabayashi, I. Kanazawa and T. Toda. 2002. Association studies of multiple candidate genes for Parkinson's disease using single nucleotide polymorphisms. *Ann Neurol.* 51(1): 133–136. erratum in *Ann Neurol.* 51(4): 534.

Mondello, S., L. Papa, A. Buki, M.R. Bullock, E. Czeiter, F.C. Tortella, K.K. Wang and R.L. Hayes. 2011. Neuronal and glial markers are differently associated with computed tomography findings and outcome in patients with severe traumatic brain injury: A case control study. *Crit Care.* 15(3): R156.

Muir, J.K., D. Lobner, H. Monyer and D.W. Choi. 1996. GABAA receptor activation attenuates excitotoxicity but exacerbates oxygen-glucose deprivation-induced neuronal injury *in vitro.* *J Cereb Blood Flow Metab.* 16(6): 1211–1218.

Nacmias, B., V. Bessi, S. Bagnoli, A. Tedde, E. Cellini, C. Piccini, S. Sorbi and L. Bracco. 2008. KIBRA gene variants are associated with episodic memory performance in subjective memory complaints. *Neurosci Lett.* 436: 145–147.

Naeimi, Z.S., A. Weinhofer, K. Sarahrudi, T. Heinz and V. Vécsei. 2006. Predictive value of S-100B protein and neuron specific-enolase as markers of traumatic brain damage in clinical use. *Brain Inj.* 20: 463–468.

Nakamura, M., R. Raghupathi, D.E. Merry, U. Scherbel, K.E. Saatman and T.K. McIntosh. 1999. Overexpression of Bcl-2 is neuroprotective after experimental brain injury in transgenic mice. *J Comp Neurol.* 412: 681–692.

Need, A.C., D.K. Attix, J.M. McEvoy, E.T. Cirulli, K.N. Linney, A.P. Wagoner, C.E. Gumbs, I. Giegling, H.J. Möller, P. Francks, P. Muglia, A. Roses, G. Gibson, M.E. Weale, D. Rujescu and D.B. Goldstein. 2008. Failure to replicate effect of Kibra on human memory in two large cohorts of European origin. *Am J Med Genet B Neuropsychiatr Genet.* 147B: 667–668.

Neese, S.L., L.K. Sherill, A.A. Tan, R.W. Roosevelt, R.A. Browning, D.C. Smith, A. Duke and R.W. Clough. 2007. Vagus nerve stimulation may protect GABAergic neurons following traumatic brain injury in rats: An immunocytochemical study. *Brain Res.* 1128(1): 157–163.

Neurobehavioral Guidelines Working Group, D.L. Warden, B. Gordon, T.W. McAllister, J.M. Silver, J.T. Barth, J. Bruns, A. Drake, T. Gentry, A. Jagoda, D.I. Katz, J. Kraus, L.A. Labbate, L.M. Ryan, M.B. Sparling, B. Walters, J. Whyte, A. Zapata and G. Zitnay. 2006. Guidelines for the pharmacologic treatment of neurobehavioral sequelae of traumatic brain injury. *J Neurotrauma.* 23(10): 1468–1501.

Neves-Pereira, M., E. Mundo, P. Muglia, N. King, F. Macciardi and J.L. Kennedy. 2002. The brain-derived neurotrophic factor gene confers susceptibility to bipolar disorder: Evidence from a family-based association study. *Am J Hum Genet.* 71: 651–655.

Newcorn, J.H. 2001. New treatments and approaches for attention deficit hyperactivity disorder. *Curr Psychiatry Rep.* 3(2): 87–91.

Nilsson, P., E. Ronne-Engstrom, R. Flink, U. Ungerstedt, H. Carlson and L. Hillered. 1994. Epileptic seizure activity in the acute phase following cortical impact trauma in rat. *Brain Res.* 637: 227–232.

Niyonkuru, C., A.K. Wagner, H. Ozawa, E.H. McCullough, A. Goyal, K. Amin and A. Fabio. 2013. Biomarker applications for group based trajectory analysis and prognostic model development in severe TBI: A practical example. *J Neurotrauma.* 2013 Jun 1; 30(11): 938–45.

Noble, E.P. 2000. Addiction and its reward process through polymorphisms of the D2 dopamine receptor gene: a review. *Eur Psychiatry.* 15(2): 79–89.

Noble, E.P., L.A. Gottschalk, J.H. Fallon, T.L. Ritchie and J.C. Wu. 1997. D2 dopamine receptor polymorphism and brain regional glucose metabolism. *Am J Med Genet.* 74(2): 162–166. PMID: 9129716.

Nylén, K., M. Ost, L.Z. Csajbok, I. Nilsson, C. Hall, K. Blennow, B. Nellgård and L. Rosengren. 2008. Serum levels of S100B, S100A1B and S100BB are all related to outcome after severe traumatic brain injury. *Acta Neurochir.* 150: 221–227.

Nylén, K., M. Ost, L.Z. Csajbok, I. Nilsson, K. Blennow, B. Nellgård and L. Rosengren. 2006. Increased serum-GFAP in patients with severe traumatic brain injury is related to outcome. *J Neurol Sci.* 240: 85–91.

O'Dell, D.M., C.J. Gibson, M.S. Wilson, S.M. DeFord and R.J. Hamm. 2000. Positive and negative modulation of the GABA(A) receptor and outcome after traumatic brain injury in rats. *Brain Res.* 861(2): 325–332.

Oechsner, M., C. Buhmann, J. Strauss and H.J. Stuerenberg. 2002. COMT-inhibition increases serum levels of dihydroxyphenylacetic acid (DOPAC) in patients with advanced Parkinson's disease. *J Neural Transm.* 109(1): 69–75.

Oestreicher, A.B., N.E. De Graan, W.H. Gispen, J. Verhaagen and L.H. Schrama. 1997. B-50, the growth associated protein-43: Modulation of cell morphology and communication in the nervous system. *Prog Neurobiol.* 53: 627–686.

Okada, T., R. Hasimoto, T. Numakawa, Y. Iijima, A. Kosuga, M. Tatsumi, K. Kamijima, T. Kato and H. Kunugi. 2006. A complex polymorphic region in the brain-derived neurotrophic factor (BDNF) gene confers susceptibility to bipolar disorder and affects transcriptional activity. *Mol Psychiatry.* 11: 695–703.

Oswald, P., J. Del-Favero, I. Massat, D. Souery, S. Claes, C. Van Broeckhoven and J. Mendlewicz. 2005. No implication of brain-derived neurotrophic factor (BDNF) gene in unipolar affective disorder: Evidence from Belgian first and replication patient-control studies. *Eur Neuropsychopharmacol.* 15: 491–495.

Palmer, A.M., D.W. Marion, M.L. Botscheller, P.E. Swedlow, S.D. Styren and S.T. DeKosky. 1993. Traumatic brain injury-induced excitotoxicity assessed in a controlled cortical impact model. *J Neurochem.* 61(6): 2015–2024.

Papa, L., L. Akinyi, M.C. Liu, J.A. Pineda, J.J. Tepas, M.W. Oli, W. Zheng, G. Robinson, S.A. Robicsek, A. Gabrielli, S.C. Heaton, H.J. Hannay, J.A. Demery, G.M. Brophy, J. Layon, C.S. Robertson, R.L. Hayes and K.K. Wang. 2010. Ubiquitin C terminal hydrolase is a novel biomarker in humans for severe traumatic brain injury. *Crit Care Med.* 38: 138–144.

Papa, L., L.M. Lewis, J.L. Falk, Z. Zhang, S. Silvestri, P. Giordano, G.M. Brophy, J.A. Demery, N.K. Dixit, I. Ferguson, M.C. Liu, J. Mo, L. Akinyi, K. Schmid, S. Mondello, C.S. Robertson, F.C. Tortella, R.L. Hayes and K.K. Wang. 2012. Elevated levels of serum glial fibrillary acidic protein breakdown products in mild and moderate traumatic brain injury are associated with intracranial lesions and neurosurgical intervention. *Ann Emerg Med.* 59(6): 471–483.

Papassotiropoulos, A., D.A. Stephen, M.J. Huentelman, F.J. Hoerndli, D.W. Craig, J.V. Pearson, K.D. Huynh, F. Brunner, J. Corneveaux, D. Osborne, M.A. Wollmer, A. Aerni, D. Coluccia, J. Hänggi, C.R. Mondadori, A. Buchmann, E.M. Reiman, R.J. Caselli, K. Henke and D.J. de Quervain. 2006. Common Kibra alleles are associated with human memory performance. *Science.* 314: 475–478.

Pazos, A., A. Probst and J.M. Palacios. 1987. Serotonin receptors in the human brain–iv. Autoradiographic mapping of serotonin-2 receptors. *Neuroscience.* 21: 123–139.

Pelinka, L.E., A. Kroepfl, M. Leixnering, W. Buchinger, A. Raabe and H. Redl. 2004a. GFAP versus S100B in serum after traumatic brain injury: Relationship to brain damage and outcome. *J Neurotrauma.* 21: 1553–1561.

Pelinka, L.E., A. Kroepfl, R. Schmidhammer, M. Krenn, W. Buchinger, H. Redl and A. Raabe. 2004b. Glial fibrillary acidic protein in serum after traumatic brain injury and multiple trauma. *J Trauma.* 57: 1006–1012.

Pelinka, L.E., L. Szalay, M. Jafarmadar, R. Schmidhammer, H. Redl and S. Bahrami. 2003. Circulating S100B is increased after bilateral femur fracture without brain injury in the rat. *Br J Anaesth.* 91: 595–597.

Pettus, E.H., C.W. Chritman, M.L. Giebel and J.T. Povlishock. 1994. Traumatically induced altered membrane permeability: Its relationship to traumatically induced reactive axonal change. *J Neurotrauma.* 11: 507–522.

Petzold, A., G. Keir, D. Lim, M. Smith and E.J. Thompson. 2003. Cerebrospinal fluid (CSF) and serum S100B: Release and wash-out pattern. *Brain Res Bull.* 61: 281–285.

Pezawas, L., B.A. Verchinski, V.S. Mattay, J.H. Callicott, B.S. Kolachana, R.E. Straub, M.F. Egan, A. Meyer-Lindenberg and D.R. Weinberger. 2004. The brain-derived neurotrophic factor val66met polymorphism and variation in human cortical morphology. *J Neurosci.* 24: 10099–10102.

Pohjalainen, T., J.O. Rinne, K. Någren, P. Lehikoinen, K. Anttila, E.K. Syvälahti and J. Hietala. 1998. The A1 allele of the human D2 dopamine receptor gene predicts low D2 receptor availability in healthy volunteers. *Mol Psychiatry.* 3(3): 256–260.

Poo, M.M. 2001. Neurotrophins as synaptic modulators. *Nat Rev Neurosci.* 2: 24–32.

Raabe, A., C. Grolms, O. Sorge, M. Zimmermann and V. Seifert. 1999. Serum S-100B protein in severe head injury. *Neurosurgery.* 45: 477–483.

Raghupathi, R., S.C. Fernandez, H. Murai, S.P. Trusko, R.W. Scott, W.K. Nishioka and T.K. McIntosh. 1998. BCL-2 overexpression attenuates cortical cell loss after traumatic brain injury in transgenic mice. *J Cereb Blood Flow Metab.* 18: 1259–1269.

Rainey, T., M. Lesko, R. Sacho, F. Lecky and C. Childs. 2009. Predicting outcome after severe traumatic brain injury using the serum S100B biomarker: Results using a single (24 h) time-point. *Resuscitation.* 80: 341–345.

Rapoport, M.J., N. Herrmann, P. Shammi, A. Kiss, A. Phillips and A. Feinstein. 2006. Outcome after traumatic brain injury sustained in older adulthood: A one-year longitudinal study. *Am J Geriatr Psychiatry.* 14: 456–465.

Rapoport, M.J., S. McCullagh, D. Streiner and A. Feinstein. 2003. The clinical significance of major depression following mild traumatic brain injury. *Psychosomatics.* 44: 31–37.

Rapoport, M.J., S. McCullagh, P. Shammi and A. Feinstein. 2005. Cognitive impairment associated with major depression following mild and moderate traumatic brain injury. *J Neuropsychiatry Clin Neurosci.* 17: 61–65.

Raymont, V., A. Greathouse, K. Reding, R. Lipsky, A. Salazar and J. Grafman. 2008. Demographic, structural and genetic predictors of late cognitive decline after penetrating head injury. *Brain.* 131: 543–558.

Ribases, M., M. Gratacos, L. Armengol, R. de Cid, A. Badia, L. Jiménez, R. Solano, J. Vallejo, F. Fernández and X. Estivill. 2003. Met66 in the brain-derived neurotrophic factor (BDNF) precursor is associated with anorexia nervosa restrictive type. *Mol Psychiatry*. 8: 745–751.

Robertson, C.L., M.J. Bell, P.M. Kochanek, P.D. Adelson, R.A. Ruppel, J.A. Carcillo, S.R. Wisniewski, Z. Mi, K.L. Janesko, R.S. Clark, D.W. Marion, S.H. Graham and E.K. Jackson. 2001. Increased adenosine in cerebrospinal fluid after severe traumatic brain injury in infants and children: Association with severity of injury and excitotoxicity. *Crit Care Med*. 29: 2287–2293.

Robinson, R.G. and B. Szetela. 1981. Mood change following left hemispheric brain injury. *Ann Neurol*. 9: 447–453.

Rodriguez-Rodriguez, E., J. Infante, J. Llorca, I. Mateo, C. Sánchez-Quintana, I. García-Gorostiaga, P. Sánchez-Juan, J. Berciano and O. Combarros. 2009. Age-dependent association of KIBRA genetic and Alzheimer's disease risk. *Neurobiol Aging*. 30: 322–324.

Rostami, E., F. Krueger, S. Zoubak, O. Dal Monte, V. Raymont, M. Pardini, C.A. Hodgkinson, D. Goldman, M. Risling and J. Grafman. 2011. BDNF polymorphism predicts general intelligence after penetrating traumatic brain injury. *PLoS One*. 6(11): e27389.

Rowe, D.C., E.J. den Oord, C. Stever, L.N. Giedinghagen, J.M. Gard, H.H. Cleveland, M. Gilson, S.T. Terris, J.H. Mohr, S. Sherman, A. Abramowitz and I.D. Waldman. 1999. The DRD2 TaqI polymorphism and symptoms of attention deficit hyperactivity disorder. *Mol Psychiatry*. 4(6): 580–586.

Rudolphi, K.A., P. Schubert, F.E. Parkinson and B.B. Fredholm. 1992. Adenosine and brain ischemia. *Cerebrovasc Brain Metab Rev*. 4: 346–369. Review.

Rujescu, D., I. Giegling, A. Gietl, A.M. Hartmann and H.J. Möller. 2003. A functional single nucleotide polymorphism (V158M) in the COMT gene is associated with aggressive personality traits. *Biol Psychiatry*. 54(1): 34–39.

Ruppel, R.A., P.M. Kochanek, P.D. Adelson, M.E. Rose, S.R. Wisniewski, M.J. Bell, R.S. Clark, D.W. Marion and S.H. Graham. 2001. Excitatory amino acid concentrations in ventricular cerebrospinal fluid after severe traumatic brain injury in infants and children: The role of child abuse. *J Pediatr*. 138(1): 18–25.

Salonia, R., P.E. Empey, S.M. Poloyac, S.R. Wisniewski, A. Klamerus, H. Ozawa, A.K. Wagner, R. Ruppel, M.J. Bell, K. Feldman, P.D. Adelson, R.S.B. Clark and M. Kochanek. 2010. Endothelin-1 is increased in cerebrospinal fluid and associated with unfavorable outcomes in children after severe TBI. *J Neurotrauma*. 27(10): 1819–1825.

Santarsieri, M., C. Niyonkuru, E.H. McCullough, T. Loucks, J. §Dobos, C.E. Dixon, S. Berga and A.K. Wagner. 2014. CNS Cortisol and Progesterone Profiles and Outcomes after Severe TBI. *J Neurotrauma* 2014, Feb 6 [Epub ahead of print].

Satchell, M.A., Y. Lai, P.M. Kochanek, S.R. Wisniewski, E.L. Fink, N.A. Siedberg, R.P. Berger, S.T. DeKosky, P.D. Adelson and R.S. Clark. 2005. Cytochrome c, a biomarker of apoptosis, is increased in cerebrospinal fluid from infants with inflicted brain injury from child abuse. *J Cereb Blood Flow Metab*. 25: 919–927.

Scanlon, J.M., Y.P. Conley, J. Ricker, C. Niyonkuru, C.E. Dixon and A.K. Wagner. 2009. Genetic variability for the dopamine transporter influences TBI outcome. *J Neurotrauma*. 26(8): A-90-Online.

Scher, A.I., H. Wu, J.W. Tsao, H.J. Blom, P. Feit, R.L. Nevin and K.A. Schwab. 2011. MTHFR C677T genotype as a risk factor for epilepsy including post-traumatic epilepsy in a representative military cohort. *J Neurotrauma*. 28(9): 1739–1745.

Schulze, T.G., D.J. Müller, H. Krauss, H. Scherk, S. Ohlraun, Y.V. Syagailo, C. Windemuth, H. Neidt, M. Grässle, A. Papassotiropoulos, R. Heun, M.M. Nöthen, W. Maier, K.P. Lesch and M. Rietschel. 2000. Association between a functional polymorphism in the monoamine oxidase: A gene promoter and major depressive disorder. *Am J Med Genet*. 96(6): 801–803.

Schumacher, J., R.A. Jamra, T. Becker, T. Ohlraun, N. Klopp, E.B. Binder, T.G. Schulze, M. Deschner, C. Schmäl, S. Höfels, A. Zobel, T. Illig, P. Propping, F. Holsboer, M. Rietschel, M.M. Nöthen and S. Cichon. 2005. Evidence for a relationship between genetic variants at the brain-derived neurotrophic factor (BDNF) locus and major depression. *Biol Psychiatry.* 58: 307–314.

Sesack, S.R., V.A. Hawrylak, C. Matus, M.A. Guido and A.I. Levey. 1998. Dopamine axon varicosities in the prelimbic division of the rat prefrontal cortex exhibit sparse immunoreactivity for the dopamine transporter. *J Neurosci.* 18: 2697–2708.

Silver, J.M., T.W. McAllister and D.B. Arciniegas. 2009. Depression and cognitive complaints following mild traumatic brain injury. *Am J Psychiatry.* 166: 653–661.

Simpson, E.R. 2004. Aromatase: Biologic relevance of tissue-specific expression. *Semin Reprod Med.* 22: 11–23.

Sivanandam, T. and M.K. Thakur. 2012. Traumatic brain injury: A risk factor for Alzheimer's disease. *Neurosci Biobehav Rev.* 36: 1376–1381.

Sklar, P., S.B. Gabriel, M.G. McInnis, P. Bennett, Y.M. Lim, G. Tsan, S. Schaffner, G. Kirov, I. Jones, M. Owen, N. Craddock, J.R. DePaulo and E.S. Lander. 2002. Family-based association study of 76 candidate genes in bipolar disorder: BDNF is a potential risk locus. Brain-derived neutrophic factor. *Mol Psychiatry.* 7: 579–593.

Stein, D.G. and D.W. Wright. 2010. Progesterone in the clinical treatment of acute traumatic brain injury. *Expert Opin Investig Drugs.* 19: 847–857.

Stein, D.G. and M.M. Cekic. 2011. Progesterone and vitamin d hormone as a biologic treatment of traumatic brain injury in the aged. *PMR.* 3: S100–S110.

Strauss, J., C.L. Barr, C.J. George, B. Devlin, A. Vetro, E. Kiss, I. Baji, N. King, S. Shaikh, M. Lanktree, M. Kovacs and J.L. Kennedy. 2005. Brain-derived neurotrophic factor variants are associated with childhood-onset mood disorder: Confirmation in a Hungarian sample. *Mol Psychiatry.* 10: 861–867.

Strittmatter, S.M., D. Valenzuela, T.E. Kennedy, E.J. Neer and M.C. Fishman. 1990. GO is a major growth cone protein subject to regulation by GAP-43. *Nature.* 344: 836–841.

Stroemer, R.P., T.A. Kent and C.E. Hulsebosch. 1993. Acute increases in GAP-43 expression following cortical ischemia. *Neurosci Lett.* 162: 51–54.

Stroemer, R.P., T.A. Kent and C.E. Hulsebosch. 1995. Neocortical neural sprouting, synaptogenesis and behavioral recovery following neocortical infarction in rats. *Stroke.* 26: 2135–2144.

Surtees, P.G., N.W. Wainwright, S.A. Willis-Owen, R. Luben, N.E. Day and J. Flint. 2006. Social adversity, the serotonin transporter (5-httlpr) polymorphism and major depressive disorder. *Biol Psychiatry.* 59: 224–229.

Swanson, J.M. 2000. Dopamine-transporter density in patients with ADHD. [comment]. *Lancet.* 355(9213): 1461–1462.

Szeszko, P.R., R. Lipsky, C. Mentschel, D. Robinson, H. Gunduz-Bruce, S. Sevy, M. Ashtari, B. Napolitano, R.M. Bilder, J.M. Kane, D. Goldman and A.K. Malhotra. 2005. Brain-derived neurotrophic factor val66met polymorphism and volume of the hippocampal formation. *Mol Psychiatry.* 10: 631–636.

Tassin, J.P., J. Bockaert, G. Blanc, L. Stinus, A.M. Thierry, S. Lavielle, J. Prémont and J. Glowinski. 1978. Topographical distribution of dopaminergic innervation and dopaminergic receptors of the anterior cerebral cortex of the rat. *Brain Res.* 154(2): 241–251.

Teasdale, G.M., G.D. Murray and J.A. Nicoll. 2005. The association between APOE epsilon4, age, and outcome after head injury: A prospective cohort study. *Brain.* 128: 2556–2561.

Tehranian, R., M.E. Rose, V. Vagni, R.P. Griffith, S. Wu, S. Maits, X. Zhang, R.S. Clark, C.E. Dixon, P.M. Kochanek, O. Bernard and S.H. Graham. 2006. Transgenic mice that overexpress the anti-apoptotic Bcl-2 protein have improved histological outcome but unchanged behavioral outcome after traumatic brain injury. *Brain Res.* 1101: 126–135.

Testa, J.A., J.F. Malec, A.M. Moessner and A.W. Brown. 2005. Outcome after traumatic brain injury: Effects of aging on recovery. *Arch Phys Med Rehabil.* 86(9): 1815–1823.

Thompson, J., N. Thomas, A. Singleton, M. Piggott, S. Lloyd, E.K. Perry, C.M. Morris, R.H. Perry, I.N. Ferrier and J.A. Court. 1997. D2 dopamine receptor gene (DRD2) Taq1 A polymorphism: Reduced dopamine D2 receptor binding in the human striatum associated with the A1 allele. *Pharmacogenetics.* 7(6): 479–484.

Thurman, D.J., C. Alverson, K.A. Dunn, J. Guerrero and J.E. Sniezek. 1999. Traumatic Brain Injury in the United States: A public health perspective. *J Head Trauma Rehabil.* 14(6): 602–615.

Timmons, S.D. and S.A. Toms. 2012. Comparative effectiveness research in neurotrauma. *Neurosurg Focus.* 33(1): E3.

Tongaonkar, P., L. Chen, D. Lambertson, B. Ko and K. Madura. 2000. Evidence for an interaction between ubiquitin-conjugating enzymes and the 26S proteasome. *Mol Cell Biol.* 20: 4691–4698.

Townend, W. and T. Ingebrigtsen. 2006. Head injury outcome prediction: a role for protein S-100B? *Injury.* 37: 1098–1108.

Townend, W.J., M.J. Guy, M.A. Pani, B. Martin and D.W. Yates. 2002. Head injury outcome prediction in the emergency department: A role for protein S-100B? *J Neurol Neurosurg Psychiatry.* 73: 542–546.

Trivedi, M.H., A.J. Rush, S.R. Wisniewski, A.A. Nierenberg, D. Warden, L. Ritz, G. Norquist, R.H. Howland, B. Lebowitz, P.J. McGrath, K. Shores-Wilson, M.M. Biggs, G.K. Balasubramani, M. Fava and STAR*D Study Team. 2006. Evaluation of outcomes with citalopram for depression using measurement-based care in STAR*D: Implications for clinical practice. *Am J Psychiatry.* 163: 28–40.

Tsai, S.J., C.J. Hong, Y.W. Yu and T.J. Chen. 2004. Association study of a brain-derived neurotrophic factor (BDNF) Val66Met polymorphism and personality trait and intelligence in healthy young females. *Neuropsychobiology.* 49: 13–16.

Ucar, T., A. Baykal, M. Akyuz, L. Dosemeci and B. Toptas. 2004. Comparison of serum and cerebrospinal fluid protein S-100b levels after severe head injury and their prognostic importance. *J Trauma.* 57: 95–98.

Uzan, M., H. Erman, T. Tanriverdi, G.Z. Sanus, A. Kafadar and H. Uzun. 2006. Evaluation of apoptosis in cerebrospinal fluid of patients with severe head injury. *Acta Neurochir (Wien).* 148: 1157–1164.

Virgos, C., L. Martorell, J. Valero, L. Figuera, F. Civeira, J. Joven, A. Labad and E. Vilella. 2001. Association study of schizophrenia with polymorphisms at six candidate genes. *Schizophr Research.* 49(1-2): 65–71.

Vos, P.E., B. Jacobs, T.M.J.C. Andriessen, K.J. Lamers, G.F. Borm, T. Beems, M. Edwards, C.F. Rosmalen and J.L. Vissers. 2010. GFAP and S100B are biomarkers of traumatic brain injury. *Neurology.* 75: 1786–1793.

Vos, P.E., K.J. Lamers, J.C. Hendriks, M. van Haaren, T. Beems, C. Zimmerman, W. van Geel, H. de Reus, J. Biert and M.M. Verbeek. 2004. Glial and neuronal proteins in serum predict outcome after severe traumatic brain injury. *Neurology.* 62: 1303–1310.

Wagner, A.K. 2010. Translational TBI rehabilitation research in the 21st century: Exploring a Rehabilomics research model. *Eur J Phys Rehabil Med.* 46(4): 549–556.

Wagner, A.K. 2011. Rehabilomics: A conceptual framework to drive biologics research. *PMR.* 3(6 suppl. 1): S28–30.

Wagner, A.K. and K.T. Zitelli. 2013. A Rehabilomics focused perspective on molecular mechanisms underlying neurological injury, complications, and recovery after severe TBI. *Pathophysiology.* 20(1): 39–48.

Wagner, A.K., A. Fabio, A.M. Puccio, R. Hirchberg, W. Li, R.D. Zafonte and D.W. Marion. 2005a. Gender associations with cerebrospinal fluid glutamate and lactate/pyruvate levels after severe traumatic brain injury. *Crit Care Med.* 33(2): 407–413.

Wagner, A.K., J.E. Sokoloski, D. Ren, X. Chen, A.S. Khan, R.D. Zafonte, A.C. Michael and C.E. Dixon. 2005b. Controlled cortical impact injury affects dopaminergic transmission in the rat striatum. *J Neurochem.* 95(2): 457–465.

Wagner, A.K., X. Chen, A.E. Kline, R.D. Zafonte and C.E. Dixon. 2005c. Gender and environmental enrichment impact dopamine transporter expression after experimental traumatic brain injury. *Exp Neurol.* 195(2): 475–483.

Wagner, A.K., D. Ren, Y.P. Conley, X. Ma, M.E. Kerr, R.D. Zafonte, A.M. Puccio, D.W. Marion and C.E. Dixon. 2007. Sex and Genetic associations with cerebral fluid dopamine and metabolite production after severe traumatic brain injury. *J Neurosurg.* 106: 538–547.

Wagner, A.K., E.H. McCullough, C. Niyonkuru, H. Ozawa, T.L. Loucks, J.A. Dobos, C.A. Brett, M. Santarsieri, C.E. Dixon, S.L. Berga and A. Fabio. 2011a. Acute serum hormone levels: Characterization and prognosis after severe traumatic brain injury. *J Neurotrauma.* 28(6): 871–888.

Wagner, A.K., K. Amin, C. Niyonkuru, B.A. Postal, E.H. McCullough, H. Ozawa, H. Bayir, R.S. Clark, P.M. Kochanek and A. Fabio. 2011b. CSF Bcl-2 and Cytochrome C temporal profiles in outcome prediction for adults with severe TBI. *J Cereb Blood Flow Metab.* 31(9): 1886–1896. doi: 10.1038/jcbfm.2011.31.

Wagner, A.K., H. Bayir, D. Ren, A. Puccio, R.D. Zafonte and P. Kochanek. 2004. Relationships between cerebrospinal fluid markers of excitotoxicity, ischemia, and oxidative damage after severe TBI: The impact of gender, age, and hypothermia. *J Neurotrauma.* 21(2): 125–136. PMID: 15000754.

Wagner, A.K., J.E. Sokoloski, X. Chen, R. Harun, D.P. Clossin, A.S. Khan, M. Andes-Koback, A.C. Michael and C.E. Dixon. 2009a. Controlled cortical impact injury influences methylphenidate-induced changes in striatal dopamine neurotransmission. *J Neurochem.* 110: 801–810. PMID: 19457094.

Wagner, A.K., L. Drewencki, X. Chen, F.R. Santos, A.S. Khan, R. Harun, G. Torres, A.C. Michael and C.E. Dixon. 2009b. Chronic methylphenidate treatment enhances striatal dopamine neurotransmission after experimental traumatic brain injury. *J Neurochem.* 108(4): 986–997.

Wagner, A.K., C.A. Brett, E.H. McCullough, C. Niyonkuru, T.L. Loucks, C.E. Dixon, J. Ricker, P. Arenth and S.L. Berga. 2012a. Persistent hypogonadism influences estradiol synthesis, cognition and outcome in males after severe TBI. *Brain Inj.* 26(10): 1226–1242.

Wagner, A.K., L.E. Hatz, J.M. Scanlon, C. Niyonkuru, M.A. Miller, J.H. Ricker, Y.P. Conley and R.E. Ferrell. 2012b. Association of KIBRA rs17070145 polymorphism and episodic memory in individuals with severe TBI. *Brain Inj.* 26(13-14): 1658–1669.

Wagner, A.K., M.A. Miller, J. Scanlon, D. Ren, P.M. Kochanek and Y.P. Conley. 2010. Adenosine A1 receptor gene variants associated with post-traumatic seizures after severe TBI. *Epilepsy Res.* 90(3): 259–272. PMID: 20609566.

Wagner, A.K. and S.A. Sowa. 2014. Rehabilomics Research: a Model for Translational Rehabilitation and Comparative Effectiveness Rehabilitation Research. *AJPMR.* 2014, Jun 4 [Epub ahead of print].

Wagner, A.K., J.M. Scanlon, C. Niyonkuru, C.E. Dixon, Y.P. Conley, C. Becker and J. Price. 2014. Genetic variation in the Dopamine Transporter gene and the D2 receptor gene influences Striatal DAT Binding in Adults with Severe TBI. *JCBFM.* 2014, May 21. doi: 10.1038/jcbfm.2014.87 [Epub ahead of Print].

Webster, M.J., M.B. Knable, N. Johnston-Wilson, K. Nagata, M. Inagaki and R.H. Yolken. 2001. Immunohistochemical localization of phosphorylated glial fibrillary acidic protein in the prefrontal cortex and hippocampus from patients with schizophrenia, bipolar disorder, and depression. *Brain Behav Immun.* 15: 388–400.

Whyte, J., T. Hart, M. Vaccaro, P. Grieb-Neff, A. Risser, M. Polansky and H.B. Coslett. 2004. Effects of methylphenidate on attention deficits after traumatic brain injury: A multidimensional, randomized, controlled trial. *Am J Phys Med Rehabil.* 83(6): 401–420.

Wilhelm, K., P.B. Mitchell, H. Niven, A. Finch, L. Wedgwood, A. Scimone, I.P. Blair, G. Parker and P.R. Schofield. 2006. Life events, first depression onset and the serotonin transporter gene. *Br J Psychiatry.* 188: 210–215.

Wishart, H.A., R.M. Roth, A.J. Saykin, C.H. Rhodes, G.J. Tsongalis, K.A. Pattin, J.H. Moore and T.W. McAllister. 2011. Comt val158met genotype and individual differences in executive function in healthy adults. *J Int Neuropsychol Soc.* 17(1): 174–180.

Woertgen, C., R.D. Rothoerl, C. Metz and A. Brawanski. 1999. Comparison of clinical, radiologic, and serum marker as prognostic factors after severe head injury. *J Trauma.* 47: 1126–1130.

World Health Organization. 2001. International Classification of Functioning, Disability and Health. Geneva, Switzerland: World Health Organization.

Wright, D.W., A.L. Kellermann, V.S. Hertzberg, P.L. Clark, M. Frankel, F.C. Goldstein, J.P. Salomone, L.L. Dent, O.A. Harris, D.S. Ander, D.W. Lowery, M.M. Patel, D.D. Denson, A.B. Gordon, M.M. Wald, S. Gupta, S.W. Hoffman and D.G. Stein. 2007. ProTECT: A randomized clinical trial of progesterone for acute traumatic brain injury. *Ann Emerg Med.* 49: 391–402.

Yablon, S.A. and V.G. Dostrow. 2001. Pot-traumatic seizures and epilepsy. *Phys Med Rehabil: State Art Rev.* 15: 301–326.

Youmans, K.L., L.M. Tai, E. Nwabuisi-Heath, L. Jungbauer, T. Kanekiyo, M. Gan, J. Kim, W.A. Eimer, S. Estus, G.W. Rebeck, E.J. Weeber, G. Bu, C. Yu and M.J. Ladu. 2012. APOE4-specific changes in Aβ accumulation in a new transgenic mouse model of Alzheimer disease. *J Biol Chem.* 287(50): 41774–41786. doi: 10.1074/jbc.M112.407957.

Young, G.B., K.G. Jordan and G.S. Doig. 1996. An assessment of non-convulsive seizures in the intensive care unit using continuous EEG monitoring: An investigation of variables associated with mortality. *Neurology.* 47(1): 83–89.

Zalsman, G., Y.Y. Huang, M.A. Oquendo, A.K. Burke, X.Z. Hu, D.A. Brent, S.P. Ellis, D. Goldman and J.J. Mann. 2006. Association of a triallelic serotonin transporter gene promoter region (5-httlpr) polymorphism with stressful life events and severity of depression. *Am J Psychiatry.* 163: 1588–1593.

Zhang, X., Y. Chen, L.W. Jenkins, P.M. Kochanek and R.S. Clark. 2005. Bench-to-bedside review: Apoptosis/programmed cell death triggered by traumatic brain injury. *Crit Care.* 9: 66–75.

Zintzaras, E. 2006. C677T and A1298C methylenetetrahydrofolate reductase gene polymorphisms in schizophrenia, bipolar disorder and depression: A meta-analysis of genetic association studies. *Psychiatr Genet.* 16: 105–115.

9

Clinical Biomarkers: Neurotherapeutics and Recovery from Traumatic Brain Injury

Jacob W. Van Landingham,[1], Arielle Schreck[1] and Vedrana Marin[2],**

INTRODUCTION

Biomarkers have a vast and promising potential to impact the treatment of various traumatic brain injury-related complications and the subsequent recovery of TBI patients both acutely and in the long-term. These biomarkers, obtained from the cerebrospinal fluid (CSF) and blood serum, will provide clinicians with more tools to treat patients. Physicians, nurses and other members of the healthcare team will be able to utilize biomarkers to: 1) *predict* which therapies will be effective on a patient-by-patient basis; and 2) *monitor* the effectiveness of these said therapies (Manne et al., 2005). Researchers are working to identify molecular targets for novel and existing therapies; however, therapeutic research for brain injury presents unique challenges compared to therapeutic research for other diseases. Unlike a

[1] Department of Biomedical Sciences, College of Medicine, 1115 W. Call Street, Florida State University, Tallahassee, FL 32306-4300.
[2] EnCor Biotechnology, Inc. 4949 SW 41st Blvd, Suites 40&50, Gainesville, FL 32608.
* Corresponding authors: jacob.vanlandingham@med.fsu.edu; Vedrana@encorbio.com

breast tumor or intestinal inflammation, brain tissue cannot be biopsied easily and safely in living patients—biomarker analyses must generally be restricted to blood, urine, saliva and CSF.

These unique challenges to therapeutic research and clinical usage of biomarkers for the treatment of TBI will be discussed. We will then explain which biomarkers have been identified as therapeutic targets for the treatment of TBI. Some of these therapeutic targets include factors of inflammation such as interleukin-1 (IL-1); factors of cell death such as spectrin breakdown products (SBDPs); factors indicative of metabolic disturbances such as N-acetylaspartate; and factors of oxidative stress such as glutathione. We will further predict why some of these protein expressions may operate better than others in a clinical setting. We will comment on how different biomarkers may be used depending on the treatment setting— the neuro-Intensive Care Unit (neuro ICU), the neurosurgery setting and the rehabilitation center. We will explain how biomarkers may be used to monitor the therapeutic efficacy of current therapies, therapies in clinical trials, and therapies that have not even been translated yet from the bench to the bedside.

Unique Challenges

The utilization of biomarkers in a clinical setting to optimize the treatment of TBI has unique challenges—especially considering that the development of neurotherapeutics for the treatment of TBI also has been and still is a complex process. Fortunately, the process of developing therapies through research and discovering biomarkers that may be translated into a clinical setting are not mutually exclusive processes: when researchers determine which biomarkers indicate treatment efficacy in the laboratory, these same biomarkers may be used to choose that particular treatment or monitor that treatment when administered to patients. These unique challenges to determining biomarker protocols (by which clinicians may follow) stem from the system-specific complexity of the Central Nervous System (CNS). The CNS has many pathways that can lead to seemingly similar injury responses (Gallen, 2004); for example, caspase-3, a protein mediator and factor of cell death in the brain, may be activated via an extrinsic receptor-mediated caspase-8 pathway or through an intrinsic mitochondrial caspase-9 pathway (Budihardjo et al., 1999). The post-injury cascade can differ from patient to patient, affecting their physical and mental health in different ways and to different degrees of severities (Gallen, 2004). Symptoms of mild TBI are variable, including fatigue, headaches, visual disturbances, memory loss, poor attention, sleep disturbances, dizziness, emotional disturbances and seizures; symptoms of moderate to severe TBI are likewise variable, including difficulties with memory, concentration, attention, processing, vision, sleep, appetite, regulation of body temperature, and various social-emotional problems, to name a few. Due to the variability within the CNS, physiologically, and between patients, regarding symptomology, spinal cord and brain injury expert Kathryn E. Saatman has dubbed the "heterogeneity of TBI" the most significant barrier to finding effective therapeutic interventions (Saatman et al., 2008).

In order to define this heterogeneity, researchers have created different classification systems by which to categorize TBI patients and aid clinicians in prescribing appropriate treatment plans. These classification systems include: systems based on severity, such as the Glascow Coma Scale; the pathoanatomical system that defines the location of a penetrating head injury or blunt trauma; or, the physical mechanism of injury (Saatman et al., 2008)—is the head trauma caused by explosive blast waves, a sports concussion, a motor vehicle accident or a fall? These classification systems and the questions that follow aid clinicians in accessing primary post-injury damage, such as edema and inflammation; however, secondary post-injury effects may affect patients' physiologies so individualistically that currently no reliable protocols exist by which to follow in order to predict crucial outcomes such as stroke, hypertension and seizures, nor debilitating long-term effects, ranging from problems with cognition and coordination to the development of neurodegenerative diseases. The heterogeneity of TBI is all the more reason why we need biomarkers in order to determine when to prescribe which therapies and treatment plans.

One challenge to using biomarkers as a tool for the clinical setting is that differences in injury severity may lead to different physiological responses, or varying degrees of a particular physiological response. Higher incidences of ischemia have been associated with the severity of brain injury (Bouma et al., 1991): the more severe the case of TBI, the greater the chance that the patient will have insufficient blood flow to the brain, leading to decreased oxygen, greater metabolic demand than oxygen supply and eventually cell death. Higher degrees of brain injury severity have also been correlated with greater Purkinje neuron cell death (Park et al., 2006), which may lead to many of the symptoms that can be observed with severe TBI such as memory loss and seizures. With biomarkers developed for diagnostic use, it should be no surprise that they may correlate with injury severity; for example, undetectable serum levels of S-100B indicate a normal finding on a CT scan (Romner et al., 2000).

The problem with undetectable serum levels is that they do not always indicate the absence of injury—in the previous example, normal CT findings just indicate the absence of moderate-severe injury. Added challenges to employing biomarkers as a therapeutic tool is finding biomarkers that are sensitive enough to detect physiological abnormalities for mild injury. With other types of diseases, such as cancer, physicians can actually biopsy the disease site—the tumor. One type of biomarker for inflammatory breast cancer has determined with 90% efficacy if tumors will respond to a certain epidermal growth factor receptor inhibitor (Johnston et al., 2008). Unlike a tumor, however, brain tissue cannot be biopsied easily and safely in living patients—biomarker analyses must generally be restricted to blood, saliva, urine and CSF. Scientists hypothesize that certain biomarkers that may be found in the blood such as NSE, S-100B and GFAP pass directly from blood circulation to systemic circulation when the Blood-Brain Barrier (BBB) is disturbed by the brain injury. With mild injuries, the disruption may or may not be enough to influence serum protein levels.

There are many other factors that influence the biomarker values that may appear in serum or CSF. One of these factors is the type of injury. While biomarkers typically appear in lower levels with the severity of injury, the type of injury is also important: serum levels of NSE and S-100 are significantly higher in patients with a diffuse injury rather than a focal injury (Antonelli et al., 2007). Diffuse injuries are a trademark component of the mild TBI, also known as concussion when used in sports terminology— so, while the severity of injury is important, certain types of mild injuries may be more prone to exhibiting higher levels of certain biomarkers than others. If the biomarker cannot be detected by noninvasive means, it cannot have any clinical usage. For focal injuries, the distance between the area in which the brain is injured and the CSF compartment also influences the biomarker level in the CSF (Lamers et al., 1995).

Certain types of injuries typically have unique complications as well—the example already noted was that of mild diffuse injury with two opposing factors indicating the biomarker level that appears in serum. Another example is blast injury endured by military personnel. Blast injury typically manifests as mild and diffuse, just like a sports concussion; as, the waves from Improvised Explosive Devices (IEDs) cause shearing of the neurons throughout the brain. Military personnel also have concomitant disorders such as Post Traumatic Stress Disorder (PTSD) or Substance Use Disorder (SUD) that often complicate and exacerbate the already variable symptoms—particularly psychiatric—that TBI patients already experience, such as anxiety, depression, sleep disturbances, mental fogginess, confusion and irritability.

While concomitant diseases do complicate the treatment of TBI, the use of biomarkers may actually be able to aid clinicians in treating these patients. In one laboratory, gene expression markers have enabled scientists to determine the efficacy of antidepressant, antipsychotic and opioid drug action with between 83.3 and 88.9% accuracy (Gunther et al., 2003). Like the military population, another population that both suffers from TBI and frequently suffers from concomitant mental illness are the elderly.

The elderly frequently endure falls; further, falls are more common in demented patients than controls, even after correcting for cognitive deficits (Horikawa et al., 2005). The demented population could benefit from the use of biomarkers in pinpointing which types of therapies may be effective for their concomitant TBI and cognitive illness. Age is also known to be a factor of TBI outcomes: elderly individuals have greater mortality rates and decreased functional outcome when discharged from a treatment setting (Susman et al., 2002); while, the youngest age group (18–25) in the military has the strongest correlation between Post Concussion Syndrome (PCS) and PTSD, which is a stronger indicator of PCS than is age (Schneiderman et al., 2008).

Clearly, there are many complications surrounding the treatment of TBI—whether the complications are due to the age of the patient, the type of injury, concomitant disorders or just the physiology of the patient. Some of these issues may make the utilization of biomarkers more difficult—such as the restriction on types of biomarkers that may be used. Regardless of the challenges to finding effective biomarkers for therapeutic use; however, the heterogeneity of TBI symptomology and the differing factors that affect the physiology of the disease call for the employment of more accurate strategies toward prescribing and determining the efficacy of therapies in treatment settings. While challenging, biomarkers have the potential to provide clinicians with measurable techniques to monitor TBI patients. The most promising therapeutic targets that researchers have identified and how they might be used in a clinical setting are described below.

Therapeutic Targets

Different classes of therapeutic targets for biomarkers have been identified based on their type of physiological responses to the TBI. However, some of these biomarkers are not mutually exclusive based on category; for example, certain factors of inflammation like IL-1 may also activate factors of cell death, some of which result in glutamate mediated excitotoxicity (Rothwell, 2003). When targeting these biomarkers, it is also important to keep in mind that certain biomarkers may be upregulated in the serum due to peripheral injuries in addition to brain injury. For this reason and also just for accuracy, these biomarkers are best when used in combination—and often markers from different classes are also best used rather than two or three markers from the same class. For prognosis, one group of researchers advocates using the S100b marker with either L-selectin or IL-6 to determine unfavorable outcomes with 96 and 100% accuracy, respectively (Lo et al., 2009). For the purposes of therapeutics, though, it may be important to monitor groups of biomarkers from all the categories in order to ensure that the patient is recovering from injury in all the aspects.

Targets of Inflammation

Inflammation is one of the hallmark characteristics of TBI. The rationale behind its manifestation involves the fact that astrocytes and microglia, cells of the CNS, are capable of synthesize cytokines and chemokines and their receptors in low levels (Benveniste, 1998). Under normal physiological conditions, the brain cells do not express their immunological function; however, disruption of the BBB during and after injury causes leukocytes from the systemic circulation to enter the cerebral circulation and release pro-inflammatory cytokines—factors that in turn activate immune functions of the cerebral cells, one function of which is inflammation (Morganti-Kossmann et al., 2005). The inflammatory factors that are released following injury often in turn activate more and more cytokines and chemokines; the factors of inflammation that are upregulated or downregulated following TBI practically form a repository of potential biomarkers. The most common factors of inflammation reviewed in the literature are covered here.

IL-1 has been identified as one of the most promising biomarkers for determining therapeutic efficacy due to the cascade of multiple functions it has on cerebral physiology, as mentioned before. However, the most common function is inflammation: the secretion of IL-1 leads to the release of more and more cytokines and chemokines as the cerebral cells mount an immune response similar to that which is found in the systemic circulation (Rothwell, 2003). The realm of its therapeutic usages is vast; however, we will mention only a few examples. One study has demonstrated that seizures in rats are induced by IL-1 expression, enhanced by IL-1 and inhibited by IL-1ra (inhibitor of IL-1) (De Simoni et al., 2000). Post traumatic seizures occur in up to 10% of TBI cases; when seizures are chronic they often last two years post-injury before tapering off (Yablon, 1993). Antiepileptic drugs are often prescribed to these patients in order to prevent the onset of seizures. It is possible that the monitoring of IL-1 could be used in order to manage the efficacy of these drugs and predict the dosages that need to be prescribed to patients in order to minimize side effects.

Another interesting correlation between IL-1 and one of the less-commonly publicized effects of brain injury is the role it plays in the development of Alzheimer's Disease (AD). TBI with the development of Chronic Traumatic Encephalopathy (CTE) leads to the same abnormal TAR DNA-binding protein 43 (TDP-43) expression levels that are found in AD patients in about 83% of cases (McKee et al., 2010). Further, the brains of AD patients have many of the same pathophysiological features of TBI-related inflammation such as the expression of inflammatory molecules—including complement, eicosanoids and nitric oxide (Marx et al., 1998). Anti-inflammatory agents have actually been suggested as therapies to protect against the development of AD (McGeer and McGeer, 1996). While prescribing broad-range anti-inflammatory drugs is not recommended in the acute period following brain injury (due to peripheral side effects), it is possible that an IL-1 inhibitor such as IL-1ra could be a potential candidate. For patients with CTE from repetitive injury, monitoring IL-1 in the long-term could also be beneficial in order to diagnose and subsequently reduce cases of long-term neurodegenerative disorders that result from brain injury. IL-1 may play a larger role in injury severity in the long term depending on the individual: approximately 50% of people have one copy of a polymorphism in this gene, which results in an increased production of the cytokine (McDowell et al., 1995). Measuring this *genetic* biomarker may actually determine for which individuals an IL-1ra therapy may be appropriate.

Tumor Necrosis Factor-α (TNF-α) is another type of pro-inflammatory cytokine cited in the literature as often as IL-1. Unlike IL-1, TNF-α does not have as broad a usage in a clinical setting because its expression varies based on the type of injury: in focal injury models, TNF-α is found in brain tissue and indicative of brain swelling (Fan et al., 1996); in diffuse injury models, TNF-α is found in the serum but not the brain tissue (Kamm et al., 2006). It has been suggested that in a clinical setting TNF-α be used in conjunction with IL-8. Like TNF-α, IL-8 is more relevant to severely brain injured patients; IL-8 recruits peripheral neutrophils that mediate secondary brain injury (Morganti-Kossmann et al., 2001; Sherwood and Prough, 2000). Together TNF-α and IL-8 can be used to diagnose the intracranial hypertension (ICH) that results from cerebral inflammation.

ICH is a very serious medical problem that can result from TBI. Increased pressure often damages the brain and spinal cord by restricting blood flow to the brain. Surgical procedures often need to be performed in order to relieve the ICH. Neuro-surgery is also a treatment that many people regard as risky, stressful to the patient and their family and invasive. Biomarkers such as the combination TNF-α/IL-8 have potential to be used in order to determine the degree of invasiveness necessary in order to relieve ICH that results from inflammation. For patients that elect to endure less invasive procedures such as drainage of the CSF rather than undergoing surgery on the brain, biomarkers could be used to measure the efficacy less "risky" procedures before proceeding to less desirable ones.

Some many researchers regard IL-6 the most promising biomarker in terms of prognosis. IL-6 is the inflammatory cytokine that is present in the highest concentration in the CSF after TBI, and early increases in CSF IL-6 levels are associated with improved neurological outcome in children with severe TBI (Hergenroeder et al., 2008). Elevated levels of serum IL-6 are in contrast associated with multi-organ failure, death and poor neurological outcome following hemorrhagic stroke (Chiaretti et al., 2008; Harris et al., 1999). Measuring IL-6 is a powerful indicator that drastic measures need to be taken to prevent stroke and other serious side-effects of the most severe types of brain injuries. However, from a therapeutic point of view, IL-6 may not be the best indicator of therapeutic efficacy: One reason is because it is most relevant to only the most severe types of patients, as stated above; another reason involves the fact that the physiological properties of IL-6 are very complex, even when limiting the scope to only inflammation.

IL-6 acts in two opposing mechanisms within the CNS following TBI. While it induces IL-1 and is ultimately correlated with chemotaxis and upregulated chemokine production, it inhibits TNF in a seemingly opposing fashion (Morganti-Kossmann et al., 2007). Too much of a therapy targeting IL-6 could cause an upregulation of TNF, which is not desirable; the two biomarkers would have to be carefully monitored in coordination while administering this type of therapy. Another inflammatory factor that has paradoxical properties is IL-10, which inhibits pro-inflammatory cytokines and correlated with neurological recovery (Knoblach and Faden, 1998). Administering a therapy that upregulates IL-10 could also have effects on the entire body when it enters the systemic circulation—the effects of such a therapy could be catastrophic for the polytrauma patient. IL-10 is in fact correlated with bodily septic infection in polytrauma patients due to this immunosuppression (Neidhardt et al., 1997).

The fact that so many patients present with complex cases does introduce added challenges to determining which biomarkers may be appropriate for use in the clinical setting. While biomarkers such as IL-6 may be appropriate for prognosis, other biomarkers, especially when used in combination, may be more appropriate for choosing and monitoring therapies. Several of the biomarkers described here that are typically most associated with inflammation also have relevance on cell death, below.

Targets of Cell Death

Two types of cell death may occur in the injured brain following TBI: necrosis and apoptosis. Necrosis occurs from external factors outside the cell, such as ischemia or reactive oxygen species—both of which relate to the following sections on metabolic disturbances and oxidative stress, respectively. Inflammation is also both a cause of and characteristic of necrosis, which leads to the swelling of the brain. When dividing biomarkers into therapeutic targets, the interrelatedness of their function and implications further complicates their usage in a clinical setting: clinicians cannot just prescribe a therapy to prevent one type of symptom. Apoptosis—which in contrast to necrosis, may occur due to either intrinsic or extrinsic factors— is similarly controlled by a broad range of factors including cytokines. Apoptosis can result in glial scar formation, which is the accumulation of an extracellular matrix around the neuronal cell that blocks cell to cell communication following injury. Death of cells in the CNS typically occurs in an early necrotic phase and a long-term apoptotic phase following TBI (Colicos and Dash, 1996; Hausmann et al., 2004; Marciano et al., 2004).

Several of the inflammatory factors mentioned earlier also have implications for cell death and cell growth. IL-1 has an opposing duality of function: the projection of IL-1 leads to glutamate-mediated excitotoxicity (which will be discussed later) as well as the induction of Nerve Growth Factor (NGF), a factor of cell growth in astrocytes (DeKosky et al., 1996). IL-6 has a similar duality of function, but not opposing; as, it both induces NGF production while defending against glutamate mediated excitotoxicity (Kossmann et al., 1996; Penkowa et al., 2000). In addition to enhancing inflammation, the recruitment of leukocytes by TNF also inhibits neuronal regeneration (Morganti-Kossmann et al., 2007). In terms of a therapy target, TNF does not have a specific opposing duality of function with respect to cell death; so, as also mentioned earlier on inflammation, it may be a good biomarker indicative of therapeutic efficacy when used in combination with other biomarkers. The suggestion earlier was to use a combination of TNF and IL-8, which—by the way—also enhances NFG production. Once again, this combination may be effective when analyzed from a new therapeutic target of cell death: TNF should clearly be downregulated, while IL-8 should be upregulated. TNF has also been shown to have an inverse relationship with the anti-inflammatory IL-10 (Schmidt et al., 2005); so, a triplit analysis of inflammatory factors including TNF, IL-8 and IL-10 (that are also indicative of cell death), may ensure that therapies targeting TNF do not have an implicit, undesirable consequences.

Three of the most widely studied biomarkers of TBI in relation to diagnosis and prognosis are also inherently related to mechanisms of cell death (Hergenroeder et al., 2008): Glial Fibrillary Acidic Protein (GFAP) and S 100 calcium-binding protein B (S100B), which are expressed by astrocytes and significantly upregulated following TBI; and Myelin Basic Protein (MBP) that when upregulated enhances remyelination following TBI and may be thought of as a factor of cell "growth". While these biomarkers may be excellent indicators for TBI diagnosis in terms of sensitivity (related to true positives) and specificity (related to true negatives), the sensitivity and specificity of these factors need to be even more accurate and able to detect all types of severities in order to employ their usage for therapeutics.

While S100B may be a strong indicator of whether or not someone has a TBI, the S100B levels must exist on a continuum related to injury severity in order to steadily demonstrate improvement following a therapeutic intervention. One of the promising aspects of GFAP following TBI is that the levels are strictly related to brain-injury and independent of other factors related to polytrauma (Pelinka et al., 2004); so, GFAP may be a better tool to monitor patients rather than factors of inflammation, which may be influenced by other injuries, infections or medications to reduce the risk of infection when operating on the brain or other parts of the body. MBP may be the most promising out of these three popularly-studied biomarkers in terms of therapeutic usage, because it remains elevated for up to two weeks post-injury (Berger et al., 2007). When biomarkers are only elevated for the first 24 to 72 hours after injury, they are not overly useful for monitoring therapeutic efficacy in a rehabilitation setting—MBP, in contrast, has promise for this type of use.

The clinical utility of biomarkers may drastically differ for prognosis and therapeutics. With most of the TBI research thus far geared toward diagnosis and prognosis, it is important that a shift in scientific thinking toward recognizing therapeutic utilization as a separate category with distinct needs occurs. It is possible that some of the more newly-recognized biomarkers for TBI actually have more promise than the more widely-studied ones. One of the newer groups of biomarkers indicative of cell death are called spectrin breakdown products (SBDPs)—particularly related to the α-II-spectrin protein that is a structural component of the cortical membrane cytoskeleton (Goodman et al., 1995; Riederer et al., 1987). Interestingly, different subsets of the α-II-SBDP remain elevated in the CSF for different periods of time following injury: the SBDP 150 remains elevated for 24 hours; the SBDP 145 remains elevated for 72 hours; and the SBDP 120 has a sustained elevation for at least five days (Pineda et al., 2007). Depending on the injury severity, the different elevations in the SBDP protein fragments could potentially mirror the physical transition of a patient from the emergency department to the neuro-ICU to the rehabilitation center. Another possibility is that therapies could reduce the levels of SBDP proteins faster so that peaks may reach lower maximum levels or transition from peak to peak faster that would controls. Using these proteins may even enable clinicians to determine when it is safe to move a patient from the neuro-ICU to the rehabilitation center.

The reason that SBDPs appear in the CSF following brain injury is that the enzymes of the calpain and caspase family cleave the α-II-spectrin protein, thus raising these levels above normal as they fall out of their structure (Pineda et al., 2007). Similar to how there are many biomarkers in the interleukin family that indicate inflammatory trends following TBI—there are also many biomarkers in the calpain and caspase families that indicate trends of apoptotic and necrotic cell death (Budihardjo et al., 1999; Newcomb-Fernandez et al., 2001; Pike et al., 1998; Zhao et al., 1999). As will interleukins, these indicators are also interrelated with other processes, mainly metabolic, and may have duality of function. For example, caspase-8 may directly activate caspase-3 or indirectly activate caspase-3 via a BID mechanism that also activates caspase-9 (Budihardjo et al., 1999).

One study compared many types of biomarkers in this family to look at trends following TBI: caspase-3, caspase-9, caspase-8, BH3 interacting domain death agonist (BID), a pro-apoptotic protein, calpain-1, calpain-2 and calpastatin, an inhibitor of calpain (Ringger et al., 2004); this study actually allowed clearer trends to emerge that enables us to make predictions on which molecules have the most potential for clinical use. In the Ringger 2004 study, Caspase-8, caspase-3 and BID all increased concurrently following TBI; so, therapeutic interventions should ideally cause all three molecules to decrease following treatment. By measuring multiple factors within the same family, clinicians may be able to obtain greater accuracy in their evaluations of treatment efficacy. Another benefit to using BID as a biomarker for therapeutic usage is that it is increased up to 7 days following injury, which makes it ideal for both the acute and the rehabilitation setting (Franz et al., 2002).

Just as targets of inflammation are very interrelated with targets of cell death, targets of cell death are very interrelated with targets of metabolic disturbances. In fact, the loss of integrity of the mitochondrial membrane is actually what is associated with casaspe-9 and caspase-3 production (Budihardjo et al., 1999; Verweij et al., 1997). Biomarkers that indicate more than one type of therapeutic target may actually have more potential because medications that target these biomarkers do not treat one aspect of the TBI disease at the expense of another. Targets of metabolic disturbances, such as the glutamate-mediated excitotoxicity already mentioned earlier, will be further elaborated on.

Targets of Metabolic Disturbances

Metabolic disturbances following TBI occur via several mechanisms, and are most often correlated with increased factors of cell death. One type of metabolic disturbance occurs when the cerebral circulation is disrupted: the brain does not receive enough oxygen and nutrients in order for aerobic cell respiration to occur which leads to decreased mitochondrial functioning, not enough ATP formed for cell survival and eventually cell death. One of the most promising groups of biomarkers to detect Cerebral Hypoxia (CH) is the IL-8/TNF pair described in the inflammation section as also a promising indicator for ICH (Stein, 2011). CH is an acute emergency and must be treated quickly. Following treatment biomarkers indicative of ICH could represent a decreased risk that the ICH will occur a second time.

Another indicator of CH is the biomarker Neuron Specific Enolase (NSE), an enzyme that aids with glycolysis within the cytoplasm of the neuron. Researchers have indicated that it is not the exact level of the biomarker in the serum or CSF that has therapeutic relevance; it is changes in this particular biomarker that may indicate impending metabolically-related cell death. When the NSE concentration increases between the onset of injury and 12 hours after injury, it may indicate impending CH and poor outcomes following brain injury (Schoerkhuber et al., 1999). Glycolytic factors like NSE increase due to poor cell oxygenation and nutrition and result in an upregulation of factors that will promote a greater formation of ATP. Instead of measuring NSE as an indicator of prognosis, NSE could be measured every few hours in severely brain-injured patients in order to identify trends before, during and following treatments.

Glutamate is another important factor of cell metabolism in addition to being responsible for postsynaptic excitation of neural cells: glutamate helps to regulate cell-to-cell communication, memory, learning and regulation. Metabolic disturbances can cause an increase in glutamate concentrations—which, due to its dual neurotransmitter function, leads to a calcium influx, swelling of the mitochondria, and release of reactive oxygen species and other factors of cell death. ATP function may be further disrupted in a cycle that is self-promoting. While it may seem like glutamate levels would be an optimal biomarker indicative of metabolic disturbances, glutamate in the gray matter of patients is only elevated in those with the most severe injury and outcomes (Shutter et al., 2004). So, one issue with using glutamate as a therapeutic target is: it does not translate into clinical practice across a broad range of injury severities.

Another issue is that studies identifying therapeutic targets of metabolic disturbances like glutamate use techniques to measure the protein levels directly from the brain tissue—which is not as logistically easy to put into clinical practice as the traditional techniques of obtaining serum or CSF samples. In order to measure protein levels in living tissues that cannot be biopsied, proton Magnetic Resonance Spectroscopy (MRS) may be used to evaluate individuals. MRS has a lower availability to clinicians than does traditional Magnetic Resonance Imaging (MRI) and is costly; simply stated, it may only be made available for the most severely brain-injured patients. In an animal model, treatment with the common psychiatric drug, dextrorphan, N-Methyl-D-Aspartate (NMDA) antagonist, was able to lower levels of the extracellular excitatory molecules glutamate and aspartate at the injury site (Faden et al., 1989). While glutamate and/or aspartate antagonists may improve clinical outcomes in some patients, MRS must be used to measure these levels in tissues in order to determine if that type of therapy would be appropriate. Further, glutamate levels should be monitored during and after therapy as well in order to determine when cessation of the therapy would be appropriate.

For patients with more mild TBIs, like concussions, the levels of another metabolite, N-acetylasparate (NAA), may be used to determine metabolic recovery. MRS techniques that measure the ratio of NAA to creatine-containing compounds have shown a steady increase in NAA levels between 0 and 30 days post-injury (Vagnozzi et al., 2010). While the biological function of NAA is not fully understood, the compound has been linked to metabolic indicators such as excitotoxicity, ATP levels and acetyl-CoA availability. The raising of NAA levels following TBI indicate that the neuronal cells are recovering energy and that the mitochondria are functioning better. One study found that NAA values are actually not a good predictor of outcome following brain injury; as, measurements are more effective for prognosis when taken a few weeks, rather than two or three days, after injury (Shutter et al., 2004). The case of NAA is a good example of how the best biomarkers for prognosis may not be the best biomarkers for therapeutic usage and vice versa. When the measurements are taken daily, there is a clear trend toward recovery that mirrors the gradual 30 day recovery of a typical concussion patient; however, when taken singularly—values may vary based on things like injury severity and the time the image was captured.

Other than NSE, we have observed the following trend regarding the most useful biomarkers that target recovery from metabolic disturbances: these biomarkers tend to result from analyses of cerebral tissues, rather than serum or CSF. The other trend that we have observed is that all of these therapeutic targets are very much interrelated; and the section that follows is no exception—metabolic disturbances initiate oxidative stress just as they do cell death. When the mitochondrial barrier is disrupted, a translocation of cytochrome c to the cytosol initiates both the apoptotic cell-death pathway, as well as the necrotic cell-death pathway in severe brain injury (Lewen et al., 2001). The release of cytochrome c into the cytosol leads to the overexpression of certain reactive oxygen species (ROS) (Cai and Jones, 1998; Fujimura et al., 1999; Fujimura et al., 2000). Antioxidants are needed to scavenge these ROS before they cause toxicity and cell death. Oxidative stress as a therapeutic target will be further elaborated later.

Targets of Oxidative Stress

With metabolic factors being the main cause of oxidative stress, many of the metabolic factors can indirectly measure oxidative stress. The metabolic factor, NAA, mentioned in the earlier, has been correlated with factors of lipid peroxidation such as malondialdehyde (MDA) to show how it can indirectly measure oxidative stress (Tavazzi et al., 2005). MDA, along with other thiobarbituric acid reactive species (TBARS) is a compound formed when reactive oxygen species reduce lipids in the brain. Other metabolic factors indicative of oxidative stress include nonproteinthiols (NPSH), Glutathione Peroxidase (GPx), and Glutathione Reductase (GR); these factors have been significantly associated with oxidative stress up to 14 days following injury in an animal model (Schwarzbold et al., 2010).

Many of the other therapeutic targets that we have mentioned, with the exception of NAA, have been more appropriate for measuring patient status within one week at the most after injury. Some of these oxidative stress targets, like NAA, may be appropriate for measuring more long-term recovery and response to therapy. Unfortunately, just like metabolic factors, most of the research on factors of oxidative stress has been done in tissues. Oxidative stress has a high potential for treatment with compounds that have antioxidant activity; however, in order to measure a patient's response during treatment, the biomarkers must be able to be sensitive enough to be detected in blood or CSF.

When taking multiple samples within a 30 day period, for example, serum would the best, least invasive, method of detection. Both direct and indirect measures of oxidative stress have been measured in serum oxidation: MDA and pyruvate dehydrogenase (PDH), respectively. PDH contributes to the hyperglycemia and lactic acidosis due to reduced metabolism in the brain, as measured by increases in lactate and lactate to pyruvate rations, following brain injury (Cady et al., 1973; Marmarou, 1992). Treatment with pyruvate has increased serum MDA levels by five fold at 72 hours post-TBI (Sharma et al., 2009). This research has been done in an animal model, but could translate to humans; as, it is shown how biomarkers may be used to measure a certain response to treatment based on therapeutic target category. Being able to measure a certain therapeutic target would also prevent clinicians from overdosing patients with certain therapeutics. Further research is needed to ensure that serum biomarkers can detect response to therapy as well as can tissue biomarkers.

Utilizing Biomarkers in Treatment Settings

The Neuro-ICU

While in the acute setting, the primary role of the clinicians is to detect and reverse secondary neuronal injury (Menon, 1999). When patients arrive in the neuro-ICU, primary injury has already occurred; however, secondary injury is highly correlated with outcome (Jones et al., 1994). The use of biomarkers could greatly enhance the detection of factors of secondary injury more rapidly than current methods. Some of the current methods used in the neuro-ICU involve the detection of things like hypoxia and intracranial hypertension (ICH) (Menon, 1999). Unfortunately, the clinical signs of hypertension are late and inconsistent; patients with normal CT scans can later, sporadically, develop ICH (O'Sullivan et al., 1994). Once discovered, ICH may be often treated by hyperventilation in order to quickly reduce the cerebral pressure; however, prolonged treatment of hyperventilation can actually cause poorer outcomes in the severe patient— patients who are the most susceptible to ICH in the first place (Muizelaar et al., 1991).

Transcranial Doppler (TCD) ultrasonography, which measures the velocity of blood cells flowing through the base of the brain, can be used to estimate problems with ICP and predict the onset of stroke. Other techniques used to predict hypoxia involve brain oximetry measurements; however, this involves implanting a microsensor in the brain to measure the pressure of oxygen, carbon dioxide and hydrogen (Dings et al., 1998; Valadka et al., 1998). Newer techniques have and are being developed in order to less-invasively measure the partial pressure of oxygen in the brain. One of these techniques is called frequency-wave spectroscopy, which involves placing sensors on the head to measure wavelengths of hemoglobin concentration (indirect measurement of oxygen pressure) (Fantini et al., 1999). Regardless of whether or not newer techniques are less invasive, though, is it realistic to perform these measurements every day—or rather, multiple times per day—in neuro-ICU patients?

Biomarkers—especially serum biomarkers—but also those observed from CSF samples are a much more efficient and direct method of detecting abnormalities in patients. One of the examples discussed above is the TNF/ IL-8 combination that may be used to diagnose ICH and CH that results from cerebral inflammation. Another example that we discussed above was NSE measured in the CSF. Physicians may induce coma using high doses of barbiturates in order to reduce ICH; however, side effects like increased infection rate (Stover and Stocker, 1998) and cognitive problems following the procedure (Schalen et al., 1992) are deleterious to the recovery of patients from brain injury. If biomarkers were able to detect the onset of ICH earlier—rather than after the onset—perhaps earlier treatment would decrease side effects. Further reduction in side effects may occur if drastic treatments did not have to be performed in high doses.

Other medications that may be prescribed to patients in the neuro-ICU include anti-epileptics to reduce the risk of seizures and excitatory amino acid antagonists to reduce problems related to metabolic disturbances. Currently there is no great way to determine the exact risk of seizure or the exact type of metabolic disturbance that a patient may have. A biomarker would be a good way to measure the problem in order to prescribe the correct doses of a therapy. As mentioned above, IL-1 has been identified a biomarker correlated with risk of seizure; and, there has been research with NAA antagonists that show how treatment can be monitored and measured as it is being performed. Utilizing biomarkers in a clinical setting would enable physicians to detect the physiological details behind the problem: ICH and CH are more or less the result of the inflammation, metabolic disturbances and associated oxidative stress discussed as therapeutic targets above. Treating the problem more efficiently, rapidly and precisely would be a better strategy than blindly administrating treatment based on a less quantifiable injury severity measure or in response to rapid symptom onset.

The Operating Room

Earlier, we described methods of monitoring patients and how the use of biomarkers may aide in preventing the need for drastic therapies and improve the administration of treatment in the pre-operational neuro-ICU. The most pressing concern for the intensivist is the management of ICH— the treatment of which may require drastic and rather risky measures. The strategy of the acute management team is to generally begin treatment with less invasive procedures and move up to more invasive procedures pending failed or insufficient results. For the treatment of ICH, patients may receive sedation or osmotherapy with diuretics like mannitol, hypothermia or induction of hyperventilation; if these strategies do not produce the desired results, or if the patient's condition worsens, CSF drainage or the most risky procedures like barbiturate-induced coma or decompressivecraniectomy (DECRA) may be chosen (Winter et al., 2005).

There have been conflicting results on whether or not DECRA effectively treats severe TBI. It is difficult to prove whether or not the short-term benefits outweigh the long-term results. In the short-term, DECRA lowers ICP and improves therapeutic intensity levels with shorter time required on mechanical ventilation. However, Extended Glasgow Outcome scores are significantly worse at six months for patients who receive a DECRA (Chi, 2011). There are really no studies to date that can prove whether or not DECRA is effective in patients suffering from TBI (Citerio and Andrews, 2007): if patients are unresponsive to therapies in the short-term, how can scientists prove that the long-term effects of DECRA are better or worse than the effect of not receiving effective therapy in the neuro-ICU? DECRA improves the mean ICP in patients by over 100%, and does slightly improve the brain tissue oxygenation (Stiefel et al., 2004). The best way to improve brain tissue oxygenation, though, would be to prevent the sustained increase in ICP before it happens in the first place. The use of biomarkers could enable clinicians to better predict if less invasive surgical techniques are effective before moving toward more invasive techniques like DECRA.

One of these less invasive surgical techniques is the CSF shunt surgery. Shunt surgery is recommended when ICP pressure is consistently high—but not necessarily continuously increasing. Shunt surgery may also be recommended when CSF is high, even if ICP pressure is in the normal range. It can be difficult to determine these measurements, however, so certain criteria have been developed in order to decide whether or not to perform a shunt surgery: onset of symptoms within 6 months after head trauma; dilation of ventricles in the absence of brain atrophy, according to CT scans; CSF flow dynamics measurements (Marmarou, 1992; Missori et al., 2006). SPECT and CT have also been used as predictive measures (Mazzini et al., 2003; Yang et al., 2003); however, these tests are expensive and can be invasive, as mentioned earlier. The need for finding less invasive and more accurate biomarkers to measure factors associated with ICP and CSF pressure are heightened when considering whether or not to perform surgery, though. There are more risks associated with surgery and more side effects than medication. Biomarkers are needed to monitor patient outcome following less invasive techniques and moving up to more invasive techniques—whether the patient is in the neuro-ICU or a rehabilitation setting.

The Rehabilitation Center

Rehabilitation from TBI occurs in two phases: the inpatient phase, which lasts between one and three months; and the outpatient phase, which involves community reintegration and can last between one and two years (Khan, 2003). Chua et al. (2007) define the role of the rehabilitation physician as follows: 1) Identifying physical impairments, including those due to secondary injury; 2) two assessing injury severity in terms of prognostics; 3) managing medical complications such as thromboembolism, epilepsy, amnesia and agitation; 4) timing transfer to the rehabilitation unit and discharge; 5) coordinating rehabilitation programs and preventing a second TBI; 6) assessing permanent disability and end of life issues for those in a permanent vegetative state.

The utilization of biomarkers has the potential to improve the physician's role in every aspect of his or her responsibilities to the patient, as outlined by Chua et al. (2007). Physicians could potentially identify all the aspects of secondary injury, pinpointing metabolic disturbances and inflammation in addition to the resultant ICH and CH. Physicians can more accurately assess the prognosis of a patient, especially the most severe ones in order to help caregivers gauge their long-term responsibilities to the most severe patients. Physicians can more accurately assess the risk of thrombosis and epilepsy, as well as determine the patient's response to those therapies and resultant reduced risk. Certain target biomarker levels could be set to more accurately determine when it is appropriate to transition between the neuro-ICU and the rehabilitation center, as evidenced by reduced risk for ICH and CH. Further, for the most severe patients, the use of biomarkers could determine the risk of patients waking from a coma or death.

For rehabilitation programs such as cognitive therapy and physical therapy, the use of biomarkers actually has potential to determine the appropriateness of such therapies and the extent to which these therapies may benefit or overburden patients in a rehabilitation setting. The two main factors that can affect patient recovery in a rehabilitation setting include mood disorders and fatigue (Ziino and Ponsford, 2006). The NAA curve, specific to the patient, may be a good indicator of patient recovery status: perhaps they should receive two days of physical therapy per week when they have improved by 50%; of three days when they have improved by 75%. More direct measures of oxidative stress could be used to determine to what percentage of maximum oxygen capacity to work the patients during physical therapy sessions. Biomarkers can be used to determine which neuropharmacological interventions or nutritional programs (discussed later) may be of most use to patients in order to enhance their recovery during therapy. Genetic biomarker testing can be used to pinpoint optimal neuropsychological management by enabling psychiatrists to prescribe better drugs to enhance sessions with neuropsychologists and community reintegration.

One question that needs further research is: Can biomarkers be used to measure response to rehabilitation therapies like cognitive therapy, physical therapy and neuropsychological therapy in the same way they can be used to measure response to neurotherapeutics or surgical interventions? The cognitive deficits in attention and memory following TBI have been noted to mimic those seen in Alzheimer's Disease (AD); so, researchers have proposed using cholinergic agents that are used to treat AD in order to alleviate some of the cognitive deficits of TBI patients (Griffin et al., 2003). NMDA and AMPA antagonists have been proposed for brain injury recovery (Nilsson et al., 1990) due to the glutamate and asparate activities involved as metabolic mediators of TBI (See section on Therapeutic Targets of Metabolic Disturbances). Biomarkers (NAA, NSE) could be used as indirect indicators of response to treatment with cholinergic agents. They may also indicate response to cognitive therapy and overall recovery during the rehabilitation phase. The most promising therapeutics and nutritional programs that may be administered during the acute or rehabilitative recovery phases will be discussed later.

Determining and Monitoring Therapeutic Efficacy

Nutritional Programs

For the nutritional treatment of TBI, two important factors to consider are monitoring diet—in intensive care, the rehabilitation center and at home, depending on TBI severity—and supplementing the diet with vitamins and minerals intended to promote recovery. Recommendations for monitoring diet may be simple; for example, patients should avoid High-Fat Sucrose (HFS) diets because they can reduce levels of Brain-Derived Neurotrophic Factor (BDNF), leading to reductions in neuronal and behavioral plasticity and increasing susceptibility of brain to injury (Hoane et al., 2011). BDNF can be measured in serum as a function of antidepressant medications (Sen et al., 2008), so could potentially be used to assess the progress of nutritional programs of TBI patients—in addition to the progress of mental health programs.

Inpatient nutritional programs may be complex, however, with stringent daily monitoring of caloric requirements, glucose and protein levels, as well as timing of feedings from onset of injury. Determining the proper caloric requirements of brain-injured patients is challenging, and over-feeding can result in hyperglycemia, electrolyte imbalances, as well as damage to the liver and lungs (Cook et al., 2008). While it may seem like determining exact caloric and glucose requirements may not be important, nutrition is actually a significant predictor of death due to TBI: for every 10 kcal/kg decrease in caloric intake from normal within the first 5 to 7 days after TBI, patients have a 30–40% increase in mortality rates (Hartl et al., 2008). Further, just as glucose administration must be measured and monitored, sometimes protein administration must also be monitored and administered as a dietary supplement. Inflammatory factors often cause protein catabolism—breakdown of protein correlates with injury severity. While protein administration cannot balance nitrogen excretion or necessarily maintain muscle mass, in order to promote long-term recovery, clinicians supplement patients' diets with between 1.5 to 2 g/kg/day for acute TBI patients (Clifton et al., 1986; Hatton et al., 1993; Young et al., 1987). Biomarkers of proteolytic damage can be used to monitor protein catabolism and the effectiveness of protein nutritional supplementation. These biomarkers include those described as therapeutic targets of cell death, such as SBDPs and those in the calpain and caspase family.

In addition to dietary supplementation with protein, vitamin and mineral supplements have been proven effective when added to patients' diets. The most promising dietary supplement given to patients is vitamin D. The rationale behind administering vitamin D dietary supplementation involves the fact that the same inflammatory cytokines that cause brain damage in TBI patients, also result in the frailty of older adults with vitamin D deficiency (Boxer et al., 2008; Morley et al., 2006). Vitamin D deficiency has been found to exacerbate inflammation and reduce the benefits of Progesterone treatment (Prog, See Neurosteroids) after TBI in an animal model (Cekic and Stein, 2010). Targets of inflammation such as those in the interleukin family and TNF may be used to monitor response to Vitamin D supplementation. Dietary omega-3 supplementation has also been shown to reduce oxidative damage and normalize BDNF levels in a rodent model (Wu et al., 2004). Omega-3 supplementation is recommended for the brain-injured population because it is neuroprotective and enhances neuron and glial cell survival; it also reduces lipid peroxidation (Michael-Titus, 2007). Markers of oxidative stress (MDA, PDH) and upregulation of MBP may be appropriate markers of response to omega-3 fatty acid supplementation.

Other effective supplements include zinc, which has reduced anxiety and depression-like behaviors in an animal model (Cope et al., 2012), and magnesium. Zinc supplementation has improved protein metabolism and neurological outcome one month after TBI, as measured by serum prealbumin and retinol binding protein levels (Young et al., 1996). These protein levels in the serum may be good markers as response to zinc treatments, in addition to markers of cell death. Magnesium is thought to be protective due to its activation of the NMDA receptor, modulation of cellular energy production, and calcium influx (McKee et al., 2010). Markers of metabolic activity and oxidative stress would be appropriate to measure response to magnesium therapy. Another supplement that works to both modulate cellular injury by lowering blood glucose (like magnesium) and improve protein conservation (like zinc) is exogenous IGF-1 (Hatton et al., 1993). Unfortunately, excessive administration of IGF-1 is not recommended due to long-term side effects (Herndon et al., 1999). Biomarkers could be used to administer the lowest dosages possible of supplements like IGF-1 that may have harmful long-term side effects.

In addition to using biomarkers to monitor responses to nutritional treatment programs, they may also be used to tailor treatment programs to patients' specific needs. Overall, nutritional changes have been considered the most promising and thus far the only approved treatment plans for TBI patients. However, these nutritional plans must be closely monitored—especially when administering supplements like IGF-1, and high levels of protein and glucose. Due to side-effects, small levels of different supplements in combination may be more effective than high levels of any single supplement alone. Biomarkers can aid clinicians in determining patients' responses to these treatments, as different patients may require different programs due to differences in ages and injury severity.

Neurosteroids

Although still in the clinical trial phase and not part of approved programs like nutrition, the most exciting treatment programs on the verge of becoming the new standard of care for patients with TBI involve the administration of neurosteroids. The neurosteroid progesterone (Prog) has been considered the best candidate thus far for the treatment of TBI (Stein, 2011) because it crosses the blood-brain barrier (Duvdevani et al., 1995; Schumacher et al., 2007), reduces edema (Roof et al., 1996; Roof et al., 1992) and factors of inflammation (VanLandingham et al., 2007) and cell death (Djebaili et al., 2005), and enhances neuronal remyelination in an animal model (Ibanez et al., 2003). Currently, Prog is being administered in 31 centers across the US in an NIH-sponsored, Phase III clinical trial for the treatment of TBI (Stein, 2011). Prog is also being administered in international clinical trials (BHR-Pharma-SyNAPSe®). Patients treated with Prog are more than twice as likely to survive TBI as un-treated patients (Wright et al., 2007). Prog treatment has also been correlated with better cognitive status and motor-coordination after acutely (Cutler et al., 2006) after injury.

The efficacy of the neurosteroid Prog in the treatment of TBI lies in that it targets several physiological mechanisms, both primary and secondary, that result from brain injury, including the cascade of events that affect other organs such as the gut, spleen, thymus and liver (Stein, 2011). In the evaluation of Prog as an effective neurosteroid, scientists have measured biomarkers in tissue and serum to gauge the spectrum and level of efficacy. For example, Prog reduces levels of the pro-apoptotic precursor protein, brain-derived neurotrophic factor (BDNF) in an animal model—until recent psychiatric studies, these measurements have not been taken in a clinical setting because the brain must be biopsied. Other markers of cell death and inflammation could be used in a clinical setting to measure Prog efficacy, though; as, the therapeutic targets so many of the different treatment areas.

Another option is to use neurosteroids like Prog in order to broadly reduce poor physiological outcomes and then use nutritional programs to target any specific physiological issues that remain following neurosteroid treatment. Biomarkers are essential in order to determine which areas need improvement following brain injury. Some new biomarkers that we have not yet mentioned could also measure response to Prog treatment: tissue biomarkers of the expression of inflammatory factors such as Tumor Necrosis Factor-alpha (TNF-α), IL-1β, and Nuclear Factor-kappa β (NF-$\kappa\beta$) in the intestinal mucosa (Chen et al., 2007). If serum levels are not sufficient enough to measure nuances in response to therapy, other bodily tissues may be biopsied to measure response to inflammation and cell death. In the serum, the level of circulating Endothelial Progenitor Cells (EPCs) has been identified as a marker of vascular regeneration and remodeling in the injured brain (Li et al., 2011). We have already identified the most promising biomarkers to be used in the therapeutic setting, but this does not mean that new biomarkers and novel ways to obtain them cannot be identified. With a therapy that targets so many of the different treatment areas, using a neurosteroid like Prog could be used in research to identify new biomarkers associated with brain injury recovery.

Like Prog, its derivative, Allopregnanolone (Allo), has also been identified as a potential alternative treatment for TBI patients: Both Prog and Allo reduce biomarkers of apoptosis, namely caspase-3 and Bax, in the injured brain and also reduce the size of the GFAP astrocyte (Djebaili et al., 2004). Both protein treatments decrease factors of inflammation via the production of the cell-surface protein, CD55, an inhibitor of complement convertases that are activators of the inflammatory cascade (VanLandingham et al., 2007). Inflammatory and cell death biomarkers have identified that Prog and Allo operate via similar mechanisms, and thus may be comparable treatments for patients. However, the pharmacological properties of the two potential treatments are different: their coagulation properties are not the same. Allo has a decreased tendency for coagulation as evidenced by lower thrombin levels in the injured brain (VanLandingham et al., 2008). Prog is considered a better all-inclusive drug for the treatment of TBI; however, by measuring these subtle differences in side-effects via biomarker analyses, researchers have suggested that Allo may be a better treatment for patients at a high risk of thrombotic stroke, as evidenced by biomarkers indicative of CH.

Other Prog-related compounds that have also been identified as treatment options for TBI include the enantiomer of Prog (*ent*-Prog) and medroxyprogesterone (MPA), a synthetic form of Prog. *Ent*-Prog has been identified as a viable alternative to Prog for the following reasons: it reduces edema, reactive gliosis, inflammatory cytokines and factors of cell death as well as does Prog, and increases anti-oxidant activity in the brain better than Prog does (VanLandingham et al., 2006). As with comparing Prog and Allo, however, the physiological complexities of the two compounds differ from each other—*ent*-Prog does not activate the traditional Prog receptor and actually inhibits Prog binding to the Prog receptor (VanLandingham et al., 2006). While *ent*-Prog may be a good alternative to Prog to be used repetitively or in high doses in males, it may prevent pregnancy or induce parturition in females by inhibiting the PR (Condon et al., 2003). These are important therapeutic considerations when choosing the best therapies for TBI patients.

For patients for which neither Prog nor *ent*-Prog may be a good therapy, MPA (available in a variety of combinations) may offer a second alternative to natural Prog. While MPA reduces cerebral edema at 48 hours post-TBI with equal efficacy as does Prog, it does not enhance behavioral recovery, meaning improve cognition, motor coordination or anxiety-like behaviors (Wright et al., 2008). Currently, for Prog and Prog-related compounds and derivatives, only natural Prog has reached clinical trials and is expected to become the new standard of care. Biomarkers have enabled research scientists to distinguish between the nuances of these different compounds. As further options reach clinical trials (and ultimately the drug-approval stages), biomarkers will further aid clinicians in determining which nuances will work best for individual patients.

Dehydroepiandrosterone (DHEA) and related compounds are another class of drugs that has shown potential in the treatment of TBI. Dehydroepiandrosterone sulfate (DHEAS) has improved behavioral recovery on both motor and cognitive tasks in an animal model (Hoffman et al., 2003) and is a potent inhibitor of gliosis (Garcia-Estrada et al., 1999). The rationale behind DHEAS research with TBI is that it plays a role in brain development and aging—further research with DHEAS may find that it is an appropriate treatment specifically for the young or elderly. There really are a plethora of neurosteroids for different therapeutic usages in different populations with different considerations. Another example is methylprednisolone, which leads to increased expression of bcl-2 (another biomarker, not as common in research) in the cardiac tissue of severe TBI patients and thus protects the heart from mitochondrial-related cell death (Emir et al., 2005); administration of methylprednisonone could also, however, lead to increased factors of cell death (Chen et al., 2011) and decreased neuronal plasticity (Zhang et al., 2011), which are not ideal physiological conditions for TBI patients. With TBI being so heterogeneous a disease, all patients may not respond the same way—and markers of cell death and myelin protection can be used to monitor potential side effects of different neurotherapeutics.

With the wide spectrum of conditions to consider when evaluating the treatment needs of a TBI patient—acutely and in the long term; primarily in the brain and secondarily in other organs—many drugs have been researched, particularly neurosteroids, and some have reached clinical trials, mainly progesterone but also crossover drugs for the treatment of other things like immunosuppression, anemia and cranial hypertension (Stein, 2011). While biomarkers have enabled researchers to pinpoint which factors are activated by different drugs and thus evaluate drug efficacy in a variety of different realms, they have further applications in various clinical settings.

Conclusions

We have made suggestions regarding how biomarkers may be employed in clinical settings in order to optimize treatment for TBI. While there are trends regarding TBI and the types of therapeutic targets that are elevated during the disease, due to individual differences between patients—gender, age, injury severity and mechanism of injury—just to name a few, not all these therapeutic targets are elevated in the same ratios. It is important to monitor therapies in order to control therapy side effects. Further, as we enter an era of personalized medicine, the expectation is that clinicians will start to tailor programs more and more specifically to individual needs, especially with a disease as heterogeneous as TBI. In the future, genetic biomarkers, measuring a patient's individual genes may even be able to optimize treatment strategies in addition to measuring physiological levels of biomarkers.

Acknowledgement

We would like to thank Ms. Helen Phipps, M.S. for her valuable contribution in editing this chapter.

References

Antonelli, M., M. Levy, P.J. Andrews, J. Chastre, L.D. Hudson, C. Manthous, G.U. Meduri, R.P. Moreno, C. Putensen, T. Stewart and A. Torres. 2007. Hemodynamic monitoring in shock and implications for management. International Consensus Conference, Paris, France, 27–28 April 2006. *Intensive Care Med.* 33: 575–90.

Benveniste, E.N. 1998. Cytokine actions in the central nervous system. *Cytokine Growth Factor Rev.* 9: 259–75.

Berger, R.P., S.R. Beers, R. Richichi, D. Wiesman and P.D. Adelson. 2007. Serum biomarker concentrations and outcome after pediatric traumatic brain injury. *J Neurotrauma.* 24: 1793–801.

Bouma, G.J., J.P. Muizelaar, S.C. Choi, P.G. Newlon and H.F. Young. 1991. Cerebral circulation and metabolism after severe traumatic brain injury: the elusive role of ischemia. *J Neurosurg.* 75: 685–93.

Boxer, R.S., D.A. Dauser, S.J. Walsh, W.D. Hager and A.M. Kenny. 2008. The association between vitamin D and inflammation with the 6-minute walk and frailty in patients with heart failure. *J Am Geriatr Soc.* 56: 454–61.

Budihardjo, I., H. Oliver, M. Lutter, X. Luo and X. Wang. 1999. Biochemical pathways of caspase activation during apoptosis. *Annu Rev Cell Dev Biol.* 15: 269–90.

Cady, L.D., Jr., M.H. Weil, A.A. Afifi, S.F. Michaels, V.Y. Liu and H. Shubin. 1973. Quantitation of severity of critical illness with special reference to blood lactate. *Crit Care Med.* 1: 75–80.

Cai, J. and D.P. Jones. 1998. Superoxide in apoptosis. Mitochondrial generation triggered by cytochrome c loss. *J Biol Chem.* 273: 11401–4.

Cekic, M. and D.G. Stein. 2010. Traumatic brain injury and aging: is a combination of progesterone and vitamin D hormone a simple solution to a complex problem? *Neurotherapeutics.* 7: 81–90.

Chen, G., J. Shi, Y. Ding, H. Yin and C. Hang. 2007. Progesterone prevents traumatic brain injury-induced intestinal nuclear factor kappa B activation and proinflammatory cytokines expression in male rats. *Mediators Inflamm.* 93431.

Chen, X., B. Zhang, Y. Chai, B. Dong, P. Lei, R. Jiang and J. Zhang. 2011. Methylprednisolone exacerbates acute critical illness-related corticosteroid insufficiency associated with traumatic brain injury in rats. *Brain Res.* 1382: 298–307.

Chi, J.H. 2011. Craniectomy for traumatic brain injury: results from the DECRA trial. *Neurosurgery.* 68: N19–20.

Chiaretti, A., A. Antonelli, A. Mastrangelo, P. Pezzotti, L. Tortorolo, F. Tosi and O. Genovese. 2008. Interleukin-6 and nerve growth factor upregulation correlates with improved outcome in children with severe traumatic brain injury. *J Neurotrauma.* 25: 225–34.

Chua, K.S., Y.S. Ng, S.G. Yap and C.W. Bok. 2007. A brief review of traumatic brain injury rehabilitation. *Ann Acad Med Singapore.* 36: 31–42.

Citerio, G. and P.J. Andrews. 2007. Refractory elevated intracranial pressure: intensivist's role in solving the dilemma of decompressive craniectomy. *Intensive Care Med.* 33: 45–8.

Clifton, G.L., C.S. Robertson and S.C. Choi. 1986. Assessment of nutritional requirements of head-injured patients. *J Neurosurg.* 64: 895–901.

Colicos, M.A. and P.K. Dash. 1996. Apoptotic morphology of dentate gyrus granule cells following experimental cortical impact injury in rats: possible role in spatial memory deficits. *Brain Res.* 739: 120–31.

Condon, J.C., P. Jeyasuria, J.M. Faust, J.W. Wilson and C.R. Mendelson. 2003. A decline in the levels of progesterone receptor coactivators in the pregnant uterus at term may antagonize progesterone receptor function and contribute to the initiation of parturition. *Proc Natl Acad Sci USA.* 100: 9518–23.

Cook, A.M., A. Peppard and B. Magnuson. 2008. Nutrition considerations in traumatic brain injury. *Nutr Clin Pract.* 23: 608–20.

Cope, E.C., D.R. Morris, A.G. Scrimgeour and C.W. Levenson. 2012. Use of Zinc as a Treatment for Traumatic Brain Injury in the Rat: Effects on Cognitive and Behavioral Outcomes. *Neurorehabil Neural Repair.*

Cutler, S.M., J.W. VanLandingham, A.Z. Murphy and D.G. Stein. 2006. Slow-release and injected progesterone treatments enhance acute recovery after traumatic brain injury. *Pharmacol Biochem Behav.* 84: 420–8.

De Simoni, M.G., C. Perego, T. Ravizza, D. Moneta, M. Conti, F. Marchesi, A. De Luigi, S. Garattini and A. Vezzani. 2000. Inflammatory cytokines and related genes are induced in the rat hippocampus by limbic status epilepticus. *Eur J Neurosci.* 12: 2623–33.

DeKosky, S.T., S.D. Styren, M.E. O'Malley, J.R. Goss, P. Kochanek, D. Marion, C.H. Evans and P.D. Robbins. 1996. Interleukin-1 receptor antagonist suppresses neurotrophin response in injured rat brain. *Ann Neurol.* 39: 123–7.

Dings, J., A. Jager, J. Meixensberger and K. Roosen. 1998. Brain tissue pO_2 and outcome after severe head injury. *Neurol Res.* 20 Suppl 1: S71–5.

Djebaili, M., Q. Guo, E.H. Pettus, S.W. Hoffman and D.G. Stein. 2005. The neurosteroids progesterone and allopregnanolone reduce cell death, gliosis, and functional deficits after traumatic brain injury in rats. *J Neurotrauma.* 22: 106–18.

Djebaili, M., S.W. Hoffman and D.G. Stein. 2004. Allopregnanolone and progesterone decrease cell death and cognitive deficits after a contusion of the rat pre-frontal cortex. *Neuroscience.* 123: 349–59.

Duvdevani, R., R.L. Roof, Z. Fulop, S.W. Hoffman and D.G. Stein. 1995. Blood-brain barrier breakdown and edema formation following frontal cortical contusion: does hormonal status play a role? *J Neurotrauma.* 12: 65–75.

Emir, M., K. Ozisik, K. Cagli, P. Ozisik, S. Tuncer, V. Bakuy, E. Yildirim, K. Kilinc and K. Gol. 2005. Beneficial effect of methylprednisolone on cardiac myocytes in a rat model of severe brain injury. *Tohoku J Exp Med.* 207: 119–24.

Faden, A.I., P. Demediuk, S.S. Panter and R. Vink. 1989. The role of excitatory amino acids and NMDA receptors in traumatic brain injury. *Science.* 244: 798–800.

Fan, L., P.R. Young, F.C. Barone, G.Z. Feuerstein, D.H. Smith and T.K. McIntosh. 1996. Experimental brain injury induces differential expression of tumor necrosis factor-alpha mRNA in the CNS. *Brain Res Mol Brain Res.* 36: 287–91.

Fantini, S., M. Franceschini, E. Gratton, D. Hueber, W. Rosenfeld, D. Maulik, P. Stubblefield and M. Stankovic. 1999. Non-invasive optical mapping of the piglet brain in real time. *Opt Express.* 4: 308–14.

Franz, G., R. Beer, D. Intemann, S. Krajewski, J.C. Reed, K. Engelhardt, B.R. Pike, R.L. Hayes, K.K. Wang, E. Schmutzhard and A. Kampfl. 2002. Temporal and spatial profile of Bid cleavage after experimental traumatic brain injury. *J Cereb Blood Flow Metab.* 22: 951–8.

Fujimura, H., K. Ohsawa, M. Funaba, T. Murata, E. Murata, M. Takahashi, M. Abe and K. Torii. 1999. Immunological localization and ontogenetic development of inhibin alpha subunit in rat brain. *J Neuroendocrinol.* 11: 157–63.

Fujimura, M., Y. Morita-Fujimura, N. Noshita, T. Sugawara, M. Kawase and P.H. Chan. 2000. The cytosolic antioxidant copper/zinc-superoxide dismutase prevents the early release of mitochondrial cytochrome c in ischemic brain after transient focal cerebral ischemia in mice. *J Neurosci.* 20: 2817–24.

Gallen, C.C. 2004. Strategic challenges in neurotherapeutic pharmaceutical development. *NeuroRx.* 1: 165–80.

Garcia-Estrada, J., S. Luquin, A.M. Fernandez and L.M. Garcia-Segura. 1999. Dehydroepiandrosterone, pregnenolone and sex steroids down-regulate reactive astroglia in the male rat brain after a penetrating brain injury. *Int J Dev Neurosci.* 17: 145–51.

Goodman, S.R., W.E. Zimmer, M.B. Clark, I.S. Zagon, J.E. Barker and M.L. Bloom. 1995. Brain spectrin: of mice and men. *Brain Res Bull.* 36: 593–606.

Griffin, S.L., R. van Reekum and C. Masanic. 2003. A review of cholinergic agents in the treatment of neurobehavioral deficits following traumatic brain injury. *J Neuropsychiatry Clin Neurosci.* 15: 17–26.

Gunther, E.C., D.J. Stone, R.W. Gerwien, P. Bento and M.P. Heyes. 2003. Prediction of clinical drug efficacy by classification of drug-induced genomic expression profiles *in vitro*. *Proc Natl Acad Sci USA.* 100: 9608–13.

Harris, T.B., L. Ferrucci, R.P. Tracy, M.C. Corti, S. Wacholder, W.H. Ettinger, Jr., H. Heimovitz, H.J. Cohen and R. Wallace. 1999. Associations of elevated interleukin-6 and C-reactive protein levels with mortality in the elderly. *Am J Med.* 106: 506–12.

Hartl, R., L.M. Gerber, Q. Ni and J. Ghajar. 2008. Effect of early nutrition on deaths due to severe traumatic brain injury. *J Neurosurg.* 109: 50–6.

Hatton, J., M.S. Luer and R.P. Rapp. 1993. Growth factors in nutritional support. *Pharmacotherapy.* 13: 17–27.

Hausmann, R., T. Biermann, I. Wiest, J. Tubel and P. Betz. 2004. Neuronal apoptosis following human brain injury. *Int J Legal Med.* 118: 32–6.

Hergenroeder, G.W., J.B. Redell, A.N. Moore and P.K. Dash. 2008. Biomarkers in the clinical diagnosis and management of traumatic brain injury. *Mol Diagn Ther.* 12: 345–58.

Herndon, D.N., P.I. Ramzy, M.A. DebRoy, M. Zheng, A.A. Ferrando, D.L. Chinkes, J.P. Barret, R.R. Wolfe and S.E. Wolf. 1999. Muscle protein catabolism after severe burn: effects of IGF-1/IGFBP-3 treatment. *Ann Surg.* 229: 713–20; discussion 720–2.

Hoane, M.R., A.A. Swan and S.E. Heck. 2011. The effects of a high-fat sucrose diet on functional outcome following cortical contusion injury in the rat. *Behav Brain Res.* 223: 119–24.

Hoffman, S.W., S. Virmani, R.M. Simkins and D.G. Stein. 2003. The delayed administration of dehydroepiandrosterone sulfate improves recovery of function after traumatic brain injury in rats. *J Neurotrauma.* 20: 859–70.

Horikawa, E., T. Matsui, H. Arai, T. Seki, K. Iwasaki and H. Sasaki. 2005. Risk of falls in Alzheimer's disease: a prospective study. *Intern Med.* 44: 717–21.

Ibanez, C., S.A. Shields, M. El-Etr, E. Leonelli, V. Magnaghi, W.W. Li, F.J. Sim, E.E. Baulieu, R.C. Melcangi, M. Schumacher and R.J. Franklin. 2003. Steroids and the reversal of age-associated changes in myelination and remyelination. *Prog Neurobiol.* 71: 49–56.

Johnston, S., M. Trudeau, B. Kaufman, H. Boussen, K. Blackwell, P. LoRusso, D.P. Lombardi, S. Ben Ahmed, D.L. Citrin, M.L. DeSilvio, J. Harris, R.E. Westlund, V. Salazar, T.Z. Zaks and N.L. Spector. 2008. Phase II study of predictive biomarker profiles for response targeting human epidermal growth factor receptor 2 (HER-2) in advanced inflammatory breast cancer with lapatinib monotherapy. *J Clin Oncol.* 26: 1066–72.

Jones, P.A., P.J. Andrews, S. Midgley, S.I. Anderson, I.R. Piper, J.L. Tocher, A.M. Housley, J.A. Corrie, J. Slattery, N.M. Dearden, N. Mark and J. Miller, Douglas. 1994. Measuring the burden of secondary insults in head-injured patients during intensive care. *J Neurosurg Anesthesiol.* 6: 4–14.

Kamm, K., W. Vanderkolk, C. Lawrence, M. Jonker and A.T. Davis. 2006. The effect of traumatic brain injury upon the concentration and expression of interleukin-1beta and interleukin-10 in the rat. *J Trauma.* 60: 152–7.

Knoblach, S.M. and A.I. Faden. 1998. Interleukin-10 improves outcome and alters proinflammatory cytokine expression after experimental traumatic brain injury. *Exp Neurol.* 153: 143–51.

Kossmann, T., V. Hans, H.G. Imhof, O. Trentz and M.C. Morganti-Kossmann. 1996. Interleukin-6 released in human cerebrospinal fluid following traumatic brain injury may trigger nerve growth factor production in astrocytes. *Brain Res.* 713: 143–52.

Lamers, K.J., B.G. van Engelen, F.J. Gabreels, O.R. Hommes, G.F. Borm and R.A. Wevers. 1995. Cerebrospinal neuron-specific enolase, S-100 and myelin basic protein in neurological disorders. *Acta Neurol Scand.* 92: 247–51.

Lewen, A., T. Sugawara, Y. Gasche, M. Fujimura and P.H. Chan. 2001. Oxidative cellular damage and the reduction of APE/Ref-1 expression after experimental traumatic brain injury. *Neurobiol Dis.* 8: 380–90.

Li, N., Y.J. Je, M. Yang, X.H. Jiang and J.H. Ma. 2011. Pharmacokinetics of baicalin-phospholipid complex in rat plasma and brain tissues after intranasal and intravenous administration. *Pharmazie.* 66: 374–7.

Lo, T.Y., P.A. Jones and R.A. Minns. 2009. Pediatric brain trauma outcome prediction using paired serum levels of inflammatory mediators and brain-specific proteins. *J Neurotrauma.* 26: 1479–87.

Manne, U., R.G. Srivastava and S. Srivastava. 2005. Recent advances in biomarkers for cancer diagnosis and treatment. *Drug Discov Today.* 10: 965–76.

Marciano, P.G., J. Brettschneider, E. Manduchi, J.E. Davis, S. Eastman, R. Raghupathi, K.E. Saatman, T.P. Speed, C.J. Stoeckert, Jr., J.H. Eberwine and T.K. McIntosh. 2004. Neuron-specific mRNA complexity responses during hippocampal apoptosis after traumatic brain injury. *J Neurosci.* 24: 2866–76.

Marmarou, A. 1992. Intracellular acidosis in human and experimental brain injury. *J Neurotrauma.* 9 Suppl. 2: S551–62.

Marx, F., I. Blasko, M. Pavelka and B. Grubeck-Loebenstein. 1998. The possible role of the immune system in Alzheimer's disease. *Exp Gerontol.* 33: 871–81.

Mazzini, L., R. Campini, E. Angelino, F. Rognone, I. Pastore and G. Oliveri. 2003. Posttraumatic hydrocephalus: a clinical, neuroradiologic, and neuropsychologic assessment of long-term outcome. *Arch Phys Med Rehabil.* 84: 1637–41.

McDowell, T.L., J.A. Symons, R. Ploski, O. Forre and G.W. Duff. 1995. A genetic association between juvenile rheumatoid arthritis and a novel interleukin-1 alpha polymorphism. *Arthritis Rheum.* 38: 221–8.

McGeer, P.L. and E.G. McGeer. 1996. Anti-inflammatory drugs in the fight against Alzheimer's disease. *Ann N Y Acad Sci.* 777: 213–20.

McKee, A.C., B.E. Gavett, R.A. Stern, C.J. Nowinski, R.C. Cantu, N.W. Kowall, D.P. Perl, E.T. Hedley-Whyte, B. Price, C. Sullivan, P. Morin, H.S. Lee, C.A. Kubilus, D.H. Daneshvar, M. Wulff and A.E. Budson. 2010. TDP-43 proteinopathy and motor neuron disease in chronic traumatic encephalopathy. *J Neuropathol Exp Neurol.* 69: 918–29.

Menon, D.K. 1999. Cerebral protection in severe brain injury: physiological determinants of outcome and their optimisation. *Br Med Bull.* 55: 226–58.

Michael-Titus, A.T. 2007. Omega-3 fatty acids and neurological injury. *Prostaglandins Leukot Essent Fatty Acids.* 77: 295–300.

Missori, P., M. Miscusi, R. Formisano, S. Peschillo, F.M. Polli, A. Melone, S. Martini, S. Paolini and R. Delfini. 2006. Magnetic resonance imaging flow void changes after cerebrospinal fluid shunt in post-traumatic hydrocephalus: clinical correlations and outcome. *Neurosurg Rev.* 29: 224–8.

Morganti-Kossmann, M.C., N. Bye, P. Nguyen and T. Kossmann. 2005. Influence of brain trauma on blood-brain barrier properties. pp. 457–79. *In*: D.V. E and P. A [eds.]. The blood brain barrier and its microenvironment: basic physiology to neurological disease. Taylor and Francis Group, New York.

Morganti-Kossmann, M.C., M. Rancan, V.I. Otto, P.F. Stahel and T. Kossmann. 2001. Role of cerebral inflammation after traumatic brain injury: a revisited concept. *Shock.* 16: 165–77.

Morganti-Kossmann, M.C., L. Satgunaseelan, N. Bye and T. Kossmann. 2007. Modulation of immune response by head injury. *Injury.* 38: 1392–400.

Morley, J.E., M.T. Haren, Y. Rolland and M.J. Kim. 2006. Frailty. *Med Clin North Am.* 90: 837–47.

Muizelaar, J.P., A. Marmarou, J.D. Ward, H.A. Kontos, S.C. Choi, D.P. Becker, H. Gruemer and H.F. Young. 1991. Adverse effects of prolonged hyperventilation in patients with severe head injury: a randomized clinical trial. *J Neurosurg.* 75: 731–9.

Neidhardt, R., M. Keel, U. Steckholzer, A. Safret, U. Ungethuem, O. Trentz and W. Ertel. 1997. Relationship of interleukin-10 plasma levels to severity of injury and clinical outcome in injured patients. *J Trauma.* 42: 863–70; discussion 870–1.

Newcomb-Fernandez, J.K., X. Zhao, B.R. Pike, K.K. Wang, A. Kampfl, R. Beer, S.M. DeFord and R.L. Hayes. 2001. Concurrent assessment of calpain and caspase-3 activation after oxygen-glucose deprivation in primary septo-hippocampal cultures. *J Cereb Blood Flow Metab.* 21: 1281–94.

Nilsson, O.G., P. Kalen, E. Rosengren and A. Bjorklund. 1990. Acetylcholine release from intrahippocampal septal grafts is under control by the host brain: a microdialysis study. *Prog Brain Res.* 82: 321–8.

O'Sullivan, M.G., P.F. Statham, P.A. Jones, J.D. Miller, N.M. Dearden, I.R. Piper, S.I. Anderson, A. Housley, P.J. Andrews, S. Midgley, J. Corrie, J.I. Tocher and R. Sellar. 1994. Role of intracranial pressure monitoring in severely head-injured patients without signs of intracranial hypertension on initial computerized tomography. *J Neurosurg.* 80: 46–50.

Park, E.M., S. Cho, K.A. Frys, S.B. Glickstein, P. Zhou, J. Anrather, M.E. Ross and C. Iadecola. 2006. Inducible nitric oxide synthase contributes to gender differences in ischemic brain injury. *J Cereb Blood Flow Metab.* 26: 392–401.

Pelinka, L.E., A. Kroepfl, R. Schmidhammer, M. Krenn, W. Buchinger, H. Redl and A. Raabe. 2004. Glial fibrillary acidic protein in serum after traumatic brain injury and multiple trauma. *J Trauma.* 57: 1006–12.

Penkowa, M., M. Giralt, J. Carrasco, H. Hadberg and J. Hidalgo. 2000. Impaired inflammatory response and increased oxidative stress and neurodegeneration after brain injury in interleukin-6-deficient mice. *Glia.* 32: 271–85.

Pike, B.R., X. Zhao, J.K. Newcomb, R.M. Posmantur, K.K. Wang and R.L. Hayes. 1998. Regional calpain and caspase-3 proteolysis of alpha-spectrin after traumatic brain injury. *Neuroreport.* 9: 2437–42.

Pineda, J.A., S.B. Lewis, A.B. Valadka, L. Papa, H.J. Hannay, S.C. Heaton, J.A. Demery, M.C. Liu, J.M. Aikman, V. Akle, G.M. Brophy, J.J. Tepas, K.K. Wang, C.S. Robertson and R.L. Hayes. 2007. Clinical significance of alphaII-spectrin breakdown products in cerebrospinal fluid after severe traumatic brain injury. *J Neurotrauma.* 24: 354–66.

Riederer, B.M., I.S. Zagon and S.R. Goodman. 1987. Brain spectrin (240/235) and brain spectrin (240/235E): differential expression during mouse brain development. *J Neurosci.* 7: 864–74.

Ringger, N.C., B.E. O'Steen, J.G. Brabham, X. Silver, J. Pineda, K.K. Wang, R.L. Hayes and L. Papa. 2004. A novel marker for traumatic brain injury: CSF alphaII-spectrin breakdown product levels. *J Neurotrauma.* 21: 1443–56.

Romner, B., T. Ingebrigtsen, P. Kongstad and S.E. Borgesen. 2000. Traumatic brain damage: serum S-100 protein measurements related to neuroradiological findings. *J Neurotrauma.* 17: 641–7.

Roof, R.L., R. Duvdevani, J.W. Heyburn and D.G. Stein. 1996. Progesterone rapidly decreases brain edema: treatment delayed up to 24 hours is still effective. *Exp Neurol.* 138: 246–51.

Roof, R.L., R. Duvdevani and D.G. Stein. 1992. Progesterone treatment attenuates brain edema following contusion injury in male and female rats. *Restor Neurol Neurosci.* 4: 425–7.

Rothwell, N. 2003. Interleukin-1 and neuronal injury: mechanisms, modification, and therapeutic potential. *Brain Behav Immun.* 17: 152–7.

Saatman, K.E., A.C. Duhaime, R. Bullock, A.I. Maas, A. Valadka and G.T. Manley. 2008. Classification of traumatic brain injury for targeted therapies. *J Neurotrauma.* 25: 719–38.

Schalen, W., B. Sonesson, K. Messeter, G. Nordstrom and C.H. Nordstrom. 1992. Clinical outcome and cognitive impairment in patients with severe head injuries treated with barbiturate coma. *Acta Neurochir (Wien).* 117: 153–9.

Schmidt, O.I., C.E. Heyde, W. Ertel and P.F. Stahel. 2005. Closed head injury—an inflammatory disease? *Brain Res Brain Res Rev.* 48: 388–99.

Schneiderman, A.I., E.R. Braver and H.K. Kang. 2008. Understanding sequelae of injury mechanisms and mild traumatic brain injury incurred during the conflicts in Iraq and Afghanistan: persistent postconcussive symptoms and posttraumatic stress disorder. *Am J Epidemiol.* 167: 1446–52.

Schoerkhuber, W., H. Kittler, F. Sterz, W. Behringer, M. Holzer, M. Frossard, S. Spitzauer and A.N. Laggner. 1999. Time course of serum neuron-specific enolase. A predictor of neurological outcome in patients resuscitated from cardiac arrest. *Stroke.* 30: 1598–603.

Schumacher, M., R. Guennoun, D.G. Stein and A.F. De Nicola. 2007. Progesterone: therapeutic opportunities for neuroprotection and myelin repair. *Pharmacol Ther.* 116: 77–106.

Schwarzbold, M.L., D. Rial, T. De Bem, D.G. Machado, M.P. Cunha, A.A. dos Santos, D.B. dos Santos, C.P. Figueiredo, M. Farina, E.M. Goldfeder, A.L. Rodrigues, R.D. Prediger and R. Walz. 2010. Effects of traumatic brain injury of different severities on emotional, cognitive, and oxidative stress-related parameters in mice. *J Neurotrauma.* 27: 1883–93.

Sen, S., R. Duman and G. Sanacora. 2008. Serum brain-derived neurotrophic factor, depression, and antidepressant medications: meta-analyses and implications. *Biol Psychiatry.* 64: 527–32.

Sharma, P., B. Benford, Z.Z. Li and G.S. Ling. 2009. Role of pyruvate dehydrogenase complex in traumatic brain injury and Measurement of pyruvate dehydrogenase enzyme by dipstick test. *J Emerg Trauma Shock.* 2: 67–72.

Sherwood, E.R. and D.S. Prough. 2000. Interleukin-8, neuroinflammation, and secondary brain injury. *Crit Care Med.* 28: 1221–3.

Shutter, L., K.A. Tong and B.A. Holshouser. 2004. Proton MRS in acute traumatic brain injury: role for glutamate/glutamine and choline for outcome prediction. *J Neurotrauma.* 21: 1693–705.

Stein, D.G. 2011. Is progesterone a worthy candidate as a novel therapy for traumatic brain injury? *Dialogues Clin Neurosci.* 13: 352–9.

Stiefel, P., M.L. Miranda, M.J. Rodriguez-Puras, S. Garcia-Morillo, J. Carneado, E. Pamies and J. Villar. 2004. Glucose effectiveness is strongly related to left ventricular mass in subjects with stage I hypertension or high-normal blood pressure. *Am J Hypertens.* 17: 146–53.

Stover, J.F. and R. Stocker. 1998. Barbiturate coma may promote reversible bone marrow suppression in patients with severe isolated traumatic brain injury. *Eur J Clin Pharmacol.* 54: 529–34.

Susman, M., S.M. DiRusso, T. Sullivan, D. Risucci, P. Nealon, S. Cuff, A. Haider and D. Benzil. 2002. Traumatic brain injury in the elderly: increased mortality and worse functional outcome at discharge despite lower injury severity. *J Trauma.* 53: 219–23; discussion 223–4.

Tavazzi, B., S. Signoretti, G. Lazzarino, A.M. Amorini, R. Delfini, M. Cimatti, A. Marmarou and R. Vagnozzi. 2005. Cerebral oxidative stress and depression of energy metabolism correlate with severity of diffuse brain injury in rats. *Neurosurgery.* 56: 582–9; discussion 582–9.

Vagnozzi, R., S. Signoretti, L. Cristofori, F. Alessandrini, R. Floris, E. Isgro, A. Ria, S. Marziale, G. Zoccatelli, B. Tavazzi, F. Del Bolgia, R. Sorge, S.P. Broglio, T.K. McIntosh and G. Lazzarino. 2010. Assessment of metabolic brain damage and recovery following mild traumatic brain injury: a multicentre, proton magnetic resonance spectroscopic study in concussed patients. *Brain.* 133: 3232–42.

Valadka, A.B., S.P. Gopinath, C.F. Contant, M. Uzura and C.S. Robertson. 1998. Relationship of brain tissue PO2 to outcome after severe head injury. *Crit Care Med.* 26: 1576–81.

VanLandingham, J.W., M. Cekic, S. Cutler, S.W. Hoffman and D.G. Stein. 2007. Neurosteroids reduce inflammation after TBI through CD55 induction. *Neurosci Lett.* 425: 94–8.

VanLandingham, J.W., M. Cekic, S.M. Cutler, S.W. Hoffman, E.R. Washington, S.J. Johnson, D. Miller and D.G. Stein. 2008. Progesterone and its metabolite allopregnanolone differentially regulate hemostatic proteins after traumatic brain injury. *J Cereb Blood Flow Metab.* 28: 1786–94.

VanLandingham, J.W., S.M. Cutler, S. Virmani, S.W. Hoffman, D.F. Covey, K. Krishnan, S.R. Hammes, M. Jamnongjit and D.G. Stein. 2006. The enantiomer of progesterone acts as a molecular neuroprotectant after traumatic brain injury. *Neuropharmacology.* 51: 1078–85.

Verweij, B.H., J.P. Muizelaar, F.C. Vinas, P.L. Peterson, Y. Xiong and C.P. Lee. 1997. Mitochondrial dysfunction after experimental and human brain injury and its possible reversal with a selective N-type calcium channel antagonist (SNX-111). *Neurol Res.* 19: 334–9.

Winter, C.D., A. Adamides and J.V. Rosenfeld. 2005. The role of decompressive craniectomy in the management of traumatic brain injury: a critical review. *J Clin Neurosci.* 12: 619–23.

Wright, D.W., S.W. Hoffman, S. Virmani and D.G. Stein. 2008. Effects of medroxyprogesterone acetate on cerebral oedema and spatial learning performance after traumatic brain injury in rats. *Brain Inj.* 22: 107–13.

Wright, D.W., A.L. Kellermann, V.S. Hertzberg, P.L. Clark, M. Frankel, F.C. Goldstein, J.P. Salomone, L.L. Dent, O.A. Harris, D.S. Ander, D.W. Lowery, M.M. Patel, D.D. Denson, A.B. Gordon, M.M. Wald, S. Gupta, S.W. Hoffman and D.G. Stein. 2007. ProTECT: a randomized clinical trial of progesterone for acute traumatic brain injury. *Ann Emerg Med.* 49: 391–402, 402 e1–2.

Wu, A., Z. Ying and F. Gomez-Pinilla. 2004. Dietary omega-3 fatty acids normalize BDNF levels, reduce oxidative damage, and counteract learning disability after traumatic brain injury in rats. *J Neurotrauma.* 21: 1457–67.

Yablon, S.A. 1993. Posttraumatic seizures. *Arch Phys Med Rehabil.* 74: 983–1001.

Yang, X.J., G.L. Hong, S.B. Su and S.Y. Yang. 2003. Complications induced by decompressive craniectomies after traumatic brain injury. *Chin J Traumatol.* 6: 99–103.

Young, B., L. Ott, E. Kasarskis, R. Rapp, K. Moles, R.J. Dempsey, P.A. Tibbs, R. Kryscio and C. McClain. 1996. Zinc supplementation is associated with improved neurologic recovery rate and visceral protein levels of patients with severe closed head injury. *J Neurotrauma.* 13: 25–34.

Young, B., L. Ott, D. Twyman, J. Norton, R. Rapp, P. Tibbs, D. Haack, B. Brivins and R. Dempsey. 1987. The effect of nutritional support on outcome from severe head injury. *J Neurosurg.* 67: 668–76.

Zhang, B., X. Chen, Y. Lin, T. Tan, Z. Yang, C. Dayao, L. Liu, R. Jiang and J. Zhang. 2011. Impairment of synaptic plasticity in hippocampus is exacerbated by methylprednisolone in a rat model of traumatic brain injury. *Brain Res.* 1382: 165–72.

Zhao, X., B.R. Pike, J.K. Newcomb, K.K. Wang, R.M. Posmantur and R.L. Hayes. 1999. Maitotoxin induces calpain but not caspase-3 activation and necrotic cell death in primary septo-hippocampal cultures. *Neurochem Res.* 24: 371–82.

Ziino, C. and J. Ponsford. 2006. Vigilance and fatigue following traumatic brain injury. *J Int Neuropsychol Soc.* 12: 100–10.

10

Inflammatory Biomarkers of Brain Injury and Disease

Erik A. Johnson,[1], Walid Yassin[2] and Michelle Guignet[1]*

INTRODUCTION

The brain is a complex organization of specialized cells responsible for controlling the function of an entire organism. As such, this highly regulated system must be able to quickly respond to a number of potential pathological states including ischemia, injury or infection to mediate damage and promote repair (Dheen et al., 2007). This response involves the upregulation of numerous proteins associated with neuroinflammation, referred to as "sterile inflammation" when it occurs in the absence of pathogen (Chen and Nunez, 2010). Neuroinflammation following injury begins with the release of Damage-Associated Molecular Patterns (DAMPs) from dead cells. Recognition of these molecular patterns by pattern recognition receptors on Central Nervous System (CNS) glial cells begins the inflammatory cascade (Berda-Haddad et al., 2011). Pattern recognition receptors are primarily found on astrocytes and reactive microglia; their activation initiates the expression of inflammation-related proteins to recruit and activate more glial cells and peripheral immune cells to the site of injury (Suffredini et al., 1999). In addition, endothelial cells, neurons and infiltrating leukocytes can also produce inflammation-related proteins, generally described

[1] US Army Medical Research Institute of Chemical Defense, Research Division/ Pharmacology Branch, 3100 Ricketts Point, Aberdeen Proving Ground, MD 21010.
[2] Department of Neuropsychiatry, Graduate School of Medicine, Kyoto University, Japan.
* Corresponding author: erik.a.johnson1.civ@mail.mil

as cytokines, which are immunomodulating proteins, and chemokines, which are cytokines that are specifically chemotactic. Together, these factors propagate and amplify the inflammatory cascade and moderate the eventual resolution of the event. Increased glial activation, increased pro-inflammatory cytokine concentrations, loss of Blood Brain Barrier (BBB) integrity, and leukocyte infiltration occur following brain injury and during neurodegenerative disease progression. The response of resident inflammatory cells, microglia in particular, can be quite extensive as rapid activation at the injury site can progress to microglial activation throughout the brain (Price et al., 2006). The rapid upregulation of pro-inflammatory factors in response to injury helps to recruit and activate immune cells to aid in debris removal (Suffredini et al., 1999). Prolonged expression of other inflammatory proteins likely helps in the reparative process through chemotaxis of glia and stem cells and through trophic support (Bye et al., 2012; Ransohoff and Brown, 2012). These properties make inflammatory factors candidates for biomarker development to gauge injury progression and potentially predict varying outcomes associated with brain injury.

Neuroinflammatory proteins meet several of the criteria for potential biomarker development. First, most cytokines and chemokines have low basal expression in healthy CNS tissue and are undetectable in serum and cerebrospinal fluid (CSF), reducing the potential for false positive detection and potentially increasing assay sensitivity. Second, many cytokines are rapidly upregulated in response to injury and are therefore specific for tissue damage. Third, though cytokines are produced locally at the site of brain injury, they can be released into both CSF and blood serum, which are more easily accessed clinically. Lastly, many cytokines appear to be effective biomarkers in certain animal models and are currently being validated in the clinic. Numerous studies have shown that inflammatory protein concentration can directly correlate to the severity of damage and that changes in these concentrations can predict long-term outcome and mortality in various conditions (Bartha et al., 2004; Girard et al., 2010; Sotgiu et al., 2006; Stein et al., 2012; Yuen et al., 2007). This chapter will explore the utility and limitations of select cytokines (interleukin-1 [IL-1] family members, IL-6, Tumor Necrosis Factor-α [TNF-α]), chemokines (C-X-C chemokine family members) and related inflammatory proteins (leukocyte binding proteins) as biomarkers for brain injury.

The IL-1 Superfamily of Cytokines

The interleukin-1 (IL-1) cytokine family consists of IL-1 receptor ligands that are structurally homologous and share the presence of beta trefoil motifs. All IL-1 family members, however, have profound effects on the progression of the inflammatory state. The most prominent and widely studied members of this family are the acute phase cytokines IL-1α and IL-1β, the natural IL-1 receptor antagonist (IL-1Ra) and IL-18.

IL-1 α and β

IL-1α and β are pro-inflammatory cytokines that are rapidly upregulated at the onset of injury and play a critical role in the initiation and propagation of the acute phase response (Allan et al., 2005; Giulian and Lachman, 1985; Woodroofe et al., 1991). They are functionally equivalent in many models because of their almost identical 3-dimensional structure (Boutin et al., 2001). The importance of IL-1 expression to neuroinflammatory progression and the exacerbation of pathology have been demonstrated, most dramatically by introduction of exogenous IL-1 into the CNS (Allan and Rothwell, 2000; Giulian et al., 1988; Loddick and Rothwell, 1996; Quagliarello et al., 1991; Stroemer and Rothwell, 1998; Yamasaki et al., 1995). Conversely, knocking out the genes in mice for both IL-1α and β, but not individually, significantly reduces infarct volume and edema in an ischemic brain damage model (Boutin et al., 2001). The prominent role of IL-1α and β in neuroinflammation and neuronal cell death has been well established in multiple CNS injury models including cerebral ischemia (Hillhouse et al., 1998; Wang et al., 1994; Zhang et al., 1998), severe seizure activity (Johnson et al., 2011; Johnson and Kan, 2010), stroke (Pinteaux et al., 2009), blast injury (Svetlov et al., 2012) and Traumatic Brain Injury (TBI) (Lu et al., 2005). Rapid and prolonged increases in the level of these cytokines and their activating enzymes also occur in humans following brain injury of various causes (Clark et al., 1999; Emsley et al., 2005; Hutchinson et al., 2007).

In cases where IL-1 expression is upregulated as a result of pathogens, the main activity of IL-1 is to activate and stimulate lymphocytes to fight infection among other physiological attributes (Kannan et al., 1996; Niijima et al., 1991; Rothwell, 1994; Takahashi et al., 1992). During sterile inflammation, IL-1 can upregulate the expression or release of many pro-inflammatory factors in the CNS (Chen et al., 2007). For example, IL-1 can induce the expression of TNFα and IL-8 (Kasahara et al., 1991), which in turn can cause direct and indirect recruitment and activation of neutrophil granulocytes. IL-1 can also act as a chemoattractant for leukocytes and promotes the adhesion of leukocytes and lymphocytes by enhancing the expression of adhesion molecules (Wang et al., 1995). Though IL-1 does not appear directly toxic to neurons, IL-1 and IL-1-induced factors can cause neuronal cell death through secondary actions like increasing astrocyte/microglial activation with subsequent release of neurotoxic factors (Herx and Yong, 2001; Parker et al., 2002; Rothwell, 2003; Tsakiri et al., 2008) and endothelial-mediated changes enhancing leukocyte infiltration and BBB breakdown, further magnifying brain injury (Dinkel et al., 2004; Nguyen et al., 2007).

IL-1 as a Biomarker

Platelets have been shown to express significantly higher concentrations of IL-1α, but not IL-1β, in rodent models of focal cerebral ischemia, and this expression induced cerebrovascular endothelial cells to produce chemokine (C-X-C motif) ligand 1 (CXCL1), a potent neutrophil chemoattractant (Thornton et al., 2010). This expression pattern suggests that IL-1α concentrations in blood may correlate with ischemic injury, at least in rodents, though a formal correlation between injury severity and IL-1α concentration in this particular experiment was not pursued. In humans with CNS injury, circulating IL-1 levels appear to be a better predictor of poor outcome than do levels in the CSF or the brain. For example, following brain trauma, recovered concentrations of both IL-1α and β from brain dialysates did not correlate with markers of tissue hypoxia (lactate/pyruvate ratio, glutamate, glucose, lactate, pyruvate), intracranial pressure, cerebral perfusion pressure, Glasgow Coma Scale (GCS) score or 6-month outcome (Hutchinson et al., 2007). In CSF, IL-1 levels do not significantly change following seizure activity (Peltola et al., 2000); however, they do increase in the blood (Lehtimaki et al., 2008; Lehtimaki et al., 2007). In neonates, high serum levels of many pro-inflammatory cytokines, including IL-1β, were predictive of later cerebral palsy development (Foster-Barber et al., 2001; Nelson et al., 1998). Similarly, in children with neonatal encephalopathy, blood levels of IL-1β were significantly associated with a high lactate/choline ratio in the deep gray nuclei, indicative of low oxidative metabolism, and with impaired development at 30 months (Bartha et al., 2004).

IL-1 Receptor Antagonist (IL-1Ra)

Equally important to IL-1 family signaling is the expression of IL-1 family inhibitors. Because IL-1 is integral to the inflammatory response, many endogenous negative regulators can attenuate the inflammatory cascade by decreasing IL-1 signaling. IL-1 receptor antagonist (IL-1Ra) is probably the most well known endogenous IL-1 inhibitor. IL-1Ra is a highly selective and competitive antagonist for the type I IL-1 receptor (IL-1RI) and has a higher receptor binding affinity than both IL-lα and IL-lβ. Binding prevents IL-1 signaling, thus mitigating the damaging effects of IL-1 (Dinarello, 1996). Though IL-1Ra is normally expressed as part of an inflammatory negative feedback loop, endogenous production of IL-1Ra in the brain is typically not sufficient to quickly reduce this response during injury (Ban et al., 1991) likely because of the delayed expression of the antagonist compared to expression of IL-1 (Eriksson et al., 1999) and the lower amounts of IL-1Ra being expressed overall (Dripps et al., 1991; Opp and Krueger, 1991). Additionally, the smaller increases in IL-1β compared to IL-1Ra appear to be endemic in the response of the CNS to damage and set up a temporal negative feedback loop for IL-1 activity (Beamer et al., 1995; Dinarello and Thompson, 1991).

IL-1Ra therapy has proved remarkably effective at reducing many of the deleterious effects of CNS injury, including neuronal loss, lesion size, edema, glial activation, peripheral immune cell infiltration and BBB permeability, while simultaneously improving various behavioral outcomes in multiple models (i.e., ischemia, brain trauma, excitotoxicity and heat stroke) (Mulcahy et al., 2003; Relton and Rothwell, 1992; Touzani et al., 1999). Specifically in a model of Status Epilepticus (SE), systemic IL-1Ra pretreatment significantly reduces the overall incidence of SE by 50%, and in those animals that do develop SE, IL-1Ra increases onset latency while decreasing the intensity of seizures (Marchi et al., 2009). IL-1Ra neuroprotection can be achieved as late as 3 hours after focal ischemia (Mulcahy et al., 2003) and 8+ hours after both global ischemia and traumatic brain injury (Touzani et al., 1999). Inhibiting the IL-1 receptor with IL-1Ra can reduce lesion size, edema, glial activation and immune cell infiltration while increasing neuronal survival and positive behavioral outcomes in Middle Cerebral Artery Occlusion (MCAO) models of ischemia (Mulcahy et al., 2003; Relton and Rothwell, 1992; Rothwell, 2003). In humans, IL-1Ra administration can reduce leukocyte and lymphocyte numbers in circulation as well as the common inflammatory markers C-reactive protein and IL-6, suggesting that IL-1 inhibition may attenuate the systemic inflammatory response (Emsley et al., 2005) in addition to the central inflammatory response (Betz et al., 1995; Garcia et al., 1995; Lin et al., 1995; Loddick and Rothwell, 1996; Relton and Rothwell, 1992; Stroemer and Rothwell, 1998; Yamasaki et al., 1995; Yang et al., 1998). Inhibition of IL-1 signaling by recombinant IL-1Ra reduces brain damage in many CNS injury rodent models to include focal cerebral ischemia, permanent and transient MCAO, and traumatic and excitotoxic injury (Relton and Rothwell, 1992) and can significantly reduce the intensity of SE in rodents (Vezzani et al., 2000).

IL-1Ra as a Biomarker

IL-1Ra is expressed at low to undetectable ranges in the CNS tissue, blood and CSF in a non-injury state, making the detection of the protein a potentially useful biomarker. Relatively few studies have investigated the clinical correlation between IL-1Ra and tissue damage severity, and the results have been highly dependent on what tissue was sampled and when the samples were taken. For example, expression of IL-1Ra and IL-1 receptor mRNA in rodent brain tissue following focal stroke was significantly increased and paralleled mRNA increases in IL-1β. However, IL-1Ra mRNA expression continued much longer than for IL-1β (Liu et al., 1993; Wang et al., 1997). In brain tissue following white matter damage in humans, IL-1β, IL-1 receptors and IL-1Ra are overexpressed in injured tissues and absent in both controls and non-damaged tissues. However, upregulation was lower for IL-1Ra compared to IL-1β. In brain dialysates of human TBI patients, IL-1Ra inversely correlated with intracranial pressure but was not correlated with cerebral perfusion pressure or admittance GCS score. Additionally, there was a significant positive correlation between IL-1Ra concentrations and the 6-month Glasgow Outcome Scale (GOS) score in that those patients attaining a favorable outcome had over twice the amount of IL-1Ra as those with an unfavorable outcome (Hutchinson et al., 2007). These data, while limited, seem to indicate that IL-1Ra does increase in response to CNS injury.

While tissue expression is not a viable medium for biomarker detection, limited studies detecting IL-1Ra in CSF and plasma were less definitive regarding the injury state of CNS tissue. In the CSF of humans with Alzheimer's Disease (AD), levels of IL-1Ra were significantly lower than in healthy age-matched controls (Tarkowski et al., 2003). However, IL-1Ra was superior to other potential biomarkers (IL-6 and IL-1β) at identifying increased levels of cell loss in AD (Tarkowski et al., 2001). In the CSF of humans with recurrent tonic-clonic or prolonged partial seizures, IL-1Ra levels were significantly increased, whereas IL-1β levels were significantly lower than in the controls (Lehtimaki et al., 2010). In the blood serum of human neonates with seizure activity, IL-1Ra levels decreased more quickly than in normal controls and in general were much lower in the seizure group (Youn et al., 2012). While these data suggest that these subjects would be more vulnerable to the effects of IL-1 induced following seizure, they are in opposition to the data found in affected brain tissue. To complicate the issue, one case had highly elevated cytokine concentrations, including IL-1Ra, within the first 24 hours, which precluded a diagnosis of quadriplegic cerebral palsy caused by severe clinical asphyxia. These studies show that IL-1Ra expression varies greatly with respect to the location of the protein (i.e., tissue, CSF or plasma), CNS injury mechanism and time of collection. Alternately, the actual molecular interaction between IL-1 and IL-1Ra may factor into pathology progression and has been shown to be a better predictor of injury than have individual concentrations of inflammatory factors (Bass et al., 2008).

The IL-1/IL-1Ra Ratio

The molar imbalance between IL-1 and IL-1Ra expression may be more critical to the progression of pathology than their individual concentrations because of the need for a several thousand-fold excess of IL-1Ra to IL-1 (most prominently IL-1β) for complete receptor inhibition (Dripps et al., 1991; Opp and Krueger, 1991). This ratio has been used to determine the inflammatory state of a tissue and can help guide or determine treatment effectiveness (Arend, 2002). In human TBI patients, the ratio of IL-1Ra/IL-1β, but not IL-1Ra/IL-1α, was significantly higher for the favorable outcome group compared to the unfavorable outcome group at 6 months post-injury. However, there was no correlation between admittance GCS scores for either IL-1Ra/IL-1β or IL-1Ra/IL-1α (Hutchinson et al., 2007). Similarly, the decreasing levels of intracerebrally produced IL-1Ra in the brain of AD patients suggest an analogous imbalance as the pathology continues to progress in this disease (Tarkowski et al., 2003).

IL-18

IL-18 is a pleiotropic inflammatory cytokine that plays a pivotal role in regulating innate and acquired immune responses in the periphery, though its actions in the CNS are still not fully understood. It is clear, however, that IL-18 can activate microglia, and its expression pattern suggests a regulatory role for infiltrating peripheral immune cells (Jander et al., 2002). Knockout of the IL-18 gene in mice exacerbates hippocampal neurodegeneration in the kainic acid excitotoxicity model (Zhang et al., 2007) and following 1-methyl-4-phenyl-1,2,3,6-tetrahydropyridine (MPTP) treatment (Sugama et al., 2004), whereas inhibition by IL-18-binding protein can attenuate hyperoxic brain injury (Felderhoff-Mueser et al., 2005). IL-18 is expressed following various brain insults such as demyelinating brain lesions in multiple sclerosis (MS) (Furlan et al., 1999; Huang et al., 2004; Nicoletti et al., 2001), stroke (Zaremba and Losy, 2003), and AD (Ojala et al., 2009) as well as in animal models of pediatric brain trauma (Sifringer et al., 2007), hypoxic-ischemia (Hedtjarn et al., 2002) and focal ischemia (Jander et al., 2002).

IL-18 as a Biomarker

A few studies have investigated the use of IL-18 as a biomarker for CNS injury. Serum levels of IL-18 were significantly higher in patients with MS (Huang et al., 2004) and following acute ischemic stroke at 48 (Yuen et al., 2007) and 72 hours after injury (Ormstad et al., 2011b). In addition, serum levels of IL-18 positively correlated with the interference measure in the modified Stroop test and with white matter lesions in patients with executive dysfunction caused by cerebral small-vessel disease (Hoshi et al., 2010). Plasma levels of IL-18 were also a positive predictor of stroke recurrence or death 90 days following initial stroke (Yuen et al., 2007), though a subsequent study did not find this correlation (Welsh et al., 2008). In the CSF, elevated levels of IL-18 are seen in MS patients (Huang et al., 2004) and in men with AD where they positively correlated with tau levels, the putative marker for AD progression (Ojala et al., 2009). It should be noted, though, that IL-18, like many other inflammatory proteins, is not specific to CNS injury. IL-18 is also highly elevated in the plasma of patients with cardiovascular disease (Blankenberg et al., 2002; Yamaoka-Tojo et al., 2002) and myocardial infarction (Youssef et al., 2007), so care must be taken when using IL-18 as a biomarker for CNS injury/disease.

The pro-inflammatory Cytokine TNFα

Members of the Tumor Necrosis Factor (TNF) family of cytokines are released in response to stress on the CNS and are produced by resident astrocytes and microglia during the acute phase response (Moro et al., 2004). While two major forms of the TNF family exist, α and β, TNFα plays the central role in mediating the inflammatory response (Fiers, 1992). TNFα expression has been linked to both neurodegeneration and neuroprotection following brain injury. As a potent pro-inflammatory factor, TNFα expression occurs following brain injury (Frugier et al., 2010; Silveira and Procianoy, 2003). Additionally, specific inhibition of TNFα can reduce neuropathology and improve behavioral outcomes in multiple *in vivo* models including Parkinson's Disease (PD) (Zhou et al., 2011), intraventricular hemorrhage (Vinukonda et al., 2010) and AD (Ryu and McLarnon, 2008). Following CNS insult, stimulated glial cells produce TNFα to induce the proliferation of astrocytes and microglia (Mehta et al., 1994), which can interfere with normal signal transmission (Gadient et al., 1990). TNFα also exerts detrimental effects on the CNS by mediating apoptosis in oligodendrocytes (Mehta et al., 1994), which can further impede nervous signal transmission and remyelination after brain injury (Dheen et al., 2007). In contrast, TNFα can also be neuroprotective in certain *in vitro* (Carlson et al., 1998; Dolga et al., 2008) and *in vivo* models (Nijboer et al., 2009). Non-acute TNFα expression may be neuroprotective by stimulating neurite outgrowth (Schmitt et al., 2010).

TNFα exerts these pleiotropic biological functions through two cell surface receptors, TNF Receptor-1(TNFR1) and -2, with each having contrasting roles on neuronal function associated with the downstream effectors coupled with each. TNFR1 has generally been associated with activation of the intrinsic apoptotic pathway and exacerbation of neuropathology, while TNFR2 activation is neuroprotective because of NF-κB antioxidant activity (Fontaine et al., 2002), though TNFR1 activation may also be neuroprotective (Carlson et al., 1998). The dichotomy of TNFα function is also partially explained by the differential expression ratios in neurons and glia (Marchetti et al., 2004). In addition, the signal duration is critical; prolonged TNFR2 signaling activates continual NF-κB activity which can negate transient TNFR1 induction (Marchetti et al., 2004). Overall, the activation pattern of these TNF receptors significantly impacts the neurodegenerative or neuroprotective role that TNFα plays following brain injury.

TNFα as a Biomarker

Though TNFα has neuroprotective properties, the detection of the cytokine after brain injury is almost exclusively linked to injury severity and poor neurological outcome. In patients with TBI, increases in serum levels of TNFα positively correlate with cerebral hypoperfusion and intracranial hypertension, a severe secondary insult with high morbidity. TNFα expression appears to be a good predictor of secondary injury and can allow for early treatment to reduce the deleterious effects of TBI (Stein et al., 2012; Stein et al., 2011b). TNFα levels are also elevated both in the CNS and in the serum following ischemic stroke (Zhao et al., 2011), and early elevated serum concentrations of TNFα correlate with neurological impairment (low GCS or GOS scores) and poor outcome at 3 months (Sotgiu et al., 2006). Similar results are seen in patients with subarachnoid hemorrhage (Chou et al., 2012). In addition, increases in serum TNFα following lacunar stroke are correlated with early neurological deterioration and poor functional outcome (Castellanos et al., 2002). TNFα also has value as a biomarker for MS. Elevated levels in serum and CSF are indicative of progressive chronic MS and predict a poor 2-year prognosis, whereas lower levels are found in those with stable MS or other neurodegenerative disorders. While TNFα levels are good indicators of disease progression, whether this cytokine has a direct role in neurodegeneration or only reflects other processes that ultimately lead to neurodegeneration (Sharief and Hentges, 1991) is not well understood. Conversely, some studies have found no association with serum TNFα levels and acute injury (Schneider Soares et al., 2012; Silveira and Procianoy, 2003) or disease (Tarkowski et al., 1999).

TNFα is detectable in the CSF as well following various CNS injuries (Fassbender et al., 1997; Otto et al., 2000; Silveira and Procianoy, 2003) and has been investigated as a putative biomarker. In adults, TNFα in the CSF is directly correlated to poor prognosis following ischemic conditions (Vila et al., 2000). In infants, TNFα levels are increased in the CSF after a perinatal hypoxic insult that resulted in neurological impairments at 12 months of age (Oygur et al., 1998). Unlike IL-6 and IL-1β, TNFα levels are elevated in the CSF of patients with AD compared to healthy patients. Therefore, TNFα expression in AD has been thought to be specifically triggered by the pathogenesis of the disease, potentially through the production of the Aβ peptide (Tarkowski et al., 2003). TNFα levels in the CSF are, on average, 25-fold higher in AD patients as compared to the controls (Tarkowski et al., 1999). The high TNFα CSF/serum expression ratio suggests that the production of this cytokine occurs locally within the CNS and is not an influx of cytokines from the periphery via a deficient BBB (Tarkowski et al., 2003). Interestingly, it is thought that TNFα may play a neuroprotective role in response to AD pathogenesis. For example, high levels of TNFα in the CSF also correlated with decreased levels of Fas/APO1, a pro-apoptotic protein responsible for triggering neural apoptosis (Tarkowski et al., 2003). Further evidence indicates that high CSF levels of TNFα are also associated with decreased tau levels and increased levels of the anti-apoptotic protein Bcl2 in the CSF. These relationships suggest a potential neuroprotective role of TNFα in preventing AD-associated neural apoptosis.

TNFα is under tight regulatory control with multiple inhibitors such as the anti-inflammatory soluble TNF receptors (sTNFRs) p55 and p75. Similar to IL-1, the molar ratio of pro-inflammatory TNFα and the anti-inflammatory sTNFRs may determine the inflammatory environment. While this relationship is not as well understood for the TNFα system as it is for IL-1, the concentrations of p55 and p75 are elevated in the serum and CSF of TBI patients. Expression levels in serum and CSF were discordant with respect to time and magnitude, with an early yet sustained elevation observed in the serum and a delayed and attenuated elevation in the CSF. These observations suggest a molar imbalance of pro- and anti-inflammatory TNF system components following brain injury, skewed strongly towards pro-inflammation and neurodegeneration early followed by a delayed increase in anti-inflammatory mediators (Maier et al., 2006). Further studies may yield a better model that would potentially use multiple components of the TNF signaling system for biomarker discovery and outcome prognostication.

Members and Targets of the ELR-positive CXC Family of Chemokines

Chemokines are small (~8 to 14 kD) polypeptides secreted by a number of cells in the body, including those in the CNS, in order to regulate cell movement important to development, homeostasis, angiogenesis and immune function. There are four families of chemokines that are classified based on the sequence motif of the cysteine residues at the N-terminus of the molecule: C (Υ chemokines), CC (β chemokines), CXC (α chemokines), and CX_3C (δ chemokines) where C is cysteine and X is any amino acid. The CXC chemokines are further characterized as either having a glutamic acid-leucine-arginine (ELR-positive) motif prior to the first cysteine or not (ELR-negative). Two well studied members of the ELR-positive CXC family are CXCL1 (Gro-α/KC/CINC-1) and IL-8 (CXCL8). During development, chemokines have a role in regulating cell proliferation and migration, as well as in maintaining neuronal excitability and survival (Bajetto et al., 2002). For example, CXCL1 plays a role in regulating oligodendrocyte precursor cell proliferation in the developing CNS, potentially mediated by constitutive expression in astrocytes (Robinson et al., 1998; Wu et al., 2000). Likewise, IL-8 is an important factor for angiogenesis by enhancing endothelial cell migration (Heidemann et al., 2003; Li et al., 2003). However, following injury, ELR-positive CXC chemokines, like CXCL1 and IL-8, are critical in propagating the inflammatory response (Bazan et al., 1997) by promoting neutrophil infiltration to the site of injury to assist with wound healing (Bajetto et al., 2002), though over expression can further exacerbate secondary tissue loss (Jin et al., 2010; Kenne et al., 2012).

CXCL-1

Following CNS injury, neutrophils are directed to the site of injury by CXCL1, which is released by resident brain cells including astrocytes (Ransohoff et al., 1993), microglia (Janabi et al., 1999), endothelial cells (Johnson et al., 2011) and neurons (Johnson et al., 2011; Xia and Hyman, 2002). Though CXCL1 is not constitutively released in the non-developing brain (Popivanova et al., 2003), its release is increased in the CNS during neuroinflammation induced by injury or disease (Semple et al., 2010) through activation of NF-κB, a key immune response transcription factor (Bonizzi and Karin, 2004). CXCL1 levels increase before neutrophil infiltration, suggesting that its expression precedes and is essential for neutrophil recruitment (Johnson et al., 2011). The CXCL1 signaling system (CXCL1 and the CXCL1 receptor, CXCR2) is upregulated in the brain following seizure activity (Johnson et al., 2011), in active MS lesions (Omari et al., 2006), ischemic stroke (Losy et al., 2005) and AD (Xia and Hyman, 2002).

Upregulation of CXCL1 and CXCR2 is critical for the promotion of leukocyte infiltration and plays an important role in neuroprotection following brain injury. Many of these effects are initially focused at the level of the brain endothelium. Following exposure to IL-1, brain endothelial cells are activated and begin producing adhesion molecules required for leukocyte infiltration, such as intercellular adhesion molecule-1 (ICAM-1) and vascular cell adhesion molecule-1 (VCAM-1), begin breaking down endothelial tight junctions and begin to release chemokines, including CXCL1. The net effect of this activation is a loss of BBB integrity and an increase in neutrophil adhesion/infiltration into the injured brain tissue (Bolton et al., 1998). Excessive upregulation of CXCL1 and/or CXCR2, however, can overly increase neutrophil accumulation and initiate neurodegenerative processes (Valles et al., 2006). These processes can be prevented by neutrophil depletion (Kenne et al., 2012), MMP-9 blockade (McColl et al., 2007; McColl et al., 2008) or loss of CXCR2 (Semple et al., 2010). These dynamics remain true even with chronic CNS diseases such as MS or AD; CXCL1 and CXCR2 are upregulated in brain tissues in both conditions, with a strong positive correlation between CXCL1 or CXCR2 levels and progression of the disease (Omari et al., 2006; Xia and Hyman, 2002).

CXCL1 as a Biomarker

Very few studies, either in animal models or in human patient populations, have investigated CXCL1 as a biomarker for injury. In MS animal models that exhibit excessive demyelination of axons, a direct correlation existed with increased levels of CXCL1 and the number of oligodendrocyte precursor cells in the developing spinal cord. This suggests a role for CXCL1 in early proliferating oligodendrocytes to help with the repair of damaged cells and may be an indicator of MS severity (Wu et al., 2000). In a rodent model of severe seizure activity using the organophosphate soman, CXCL1 is significantly upregulated in regions where brain damage is progressing (Johnson et al., 2011). In humans with ischemic stroke, increased concentrations of CXCL1 in the CSF directly correlated with neutrophil infiltration during the acute phase of stroke and subsequent brain damage (Losy et al., 2005).

IL-8

Like CXCL1, IL-8 is a CXC chemokine that is secreted by many cell types such as macrophages, epithelial cells, neutrophils, microglia, astrocytes and endothelial cells (Croitoru-Lamoury et al., 2003; Ehrlich et al., 1998; Proost et al., 1996; Yamasaki et al., 1997). It is also a potent mediator of the neuroinflammatory response with strong neutrophil recruitment and activating properties (Baggiolini et al., 1994; Lopez-Cortes et al., 1995), though it can also inhibit the adhesion of leukocytes to activated endothelial cells as a protective mechanism (Gimbrone et al., 1989). IL-8 can bind both

CXCR1 and CXCR2 receptors with a higher affinity for CXCR1 (Hoch et al., 1996; Tsai et al., 2002). Multiple studies have shown that IL-8 levels are elevated in the CSF and serum following TBI in rat models and in human patients (Ehrlich et al., 1998; Gopcevic et al., 2007; Ott et al., 1994; Whalen et al., 2000), though upregulation may also be associated with secondary injury (Mussack et al., 2002).

IL-8 as a Biomarker

Like most inflammatory mediators, IL-8 does not appear to be constitutively expressed in the non-developing CNS, thus makings its presence indicative of injury. In fact, IL-8 plays an important role in inflammatory progression and is expressed in varying brain injury or disease states (Billiau et al., 2007; Gopcevic et al., 2007; Kushi et al., 2003; Whalen et al., 2000). Injury-induced expression makes IL-8 an attractive biomarker for CNS injury, and it has been investigated as such in CNS injury conditions. For instance, serum levels of IL-8 significantly increase in humans with neonatal seizures induced by hypoxic-ischemic encephalitis. IL-8 has a long detection window, up to 72 hours following birth, and its expression reflects disease severity (Youn et al., 2012) and can predict poor outcomes and mortality (Ramaswamy et al., 2009). In patients with TBI, increases in serum levels of IL-8 positively correlate with cerebral hypoperfusion, a severe secondary insult with high morbidity. It also has high specificity/low sensitivity in predicting cerebral perfusion pressure (<60 mmHg, 50 mmHg), but does not predict changes in intracranial pressure. IL-8 expression, in this case, preceded the clinical manifestations of cerebral hypoperfusion and may act as a predictor of poor outcome, allowing for earlier, more aggressive treatment to attenuate secondary injury (Stein et al., 2011a; Stein et al., 2012). In patients with contusional TBI, IL-8 positively correlated with mortality, though no correlation between the size of the contusion and the serum levels of IL-8 was found despite the presence of more neutrophils in the contused tissue (Rhodes et al., 2009). Similarly, in patients with severe head injury, IL-8 was significantly higher in both the serum and CSF of those who died of their injuries than in those who survived at 72 hours (Kushi et al., 2003); a similar result was seen in children with severe TBI (Whalen et al., 2000). Conversely, no significant increases in CNS-attributable IL-8 production were detected in the serum of TBI or subarachnoid hemorrhage patients, though a high lowest limit of detection may have masked any changes (McKeating et al., 1997).

Although serum is readily accessible, levels of IL-8 in CSF can be a 1,000 times higher than in the periphery (Kushi et al., 2003). Differential detection of IL-8 in the CSF and serum implies primary CNS production, shown after a variety of CNS insults (Asano et al., 2010; Billiau et al., 2007; Semple et al., 2010), and therefore CSF may provide more sensitivity for determining CNS injury severity than the serum. While some studies detect significant changes in both peripheral and central IL-8 production (Kushi et al., 2003), others have found the CSF to correlate better with various consequences of CNS injury. For example, in severe TBI patients, increases in CSF-derived IL-8 correlate with severe BBB compromise (Kossmann et al., 1997) and increased mortality (Whalen et al., 2000), though no correlation existed between mortality and IL-8 plasma levels (Kossmann et al., 1997). Increased IL-8 levels are also detected in the CSF of MS patients compared with controls (Lund et al., 2004). CSF levels of IL-8 also appear to be protective in newborns with hypoxic brain damage (Hussein et al., 2010). It should be noted, though, that IL-8 serum levels are also commonly used as prognosis predictors of various cancers, including hepatocellular carcinoma (Welling et al., 2012), pancreatic cancer (Chen et al., 2012) and medullary thyroid carcinoma (Broutin et al., 2011), and is consistently high in the CSF of patients with bacterial or viral meningitis (Van Meir et al., 1992).

Neutrophil-related Biomarker Molecules

Leukocytes directed to the injured brain by CXC chemokines require endothelial cells to express Cell Adhesion Molecules (CAMs) to bind and stop the leukocytes in areas of damage. Therefore CAMs, including intracellular (I) CAM-1, vascular (V)CAM-1 and endothelial (E)-selectin, have been used as biomarkers for CNS injury. For instance, serum soluble (s)ICAM-1, integral to neutrophil infiltration and upregulated at the same time as CXC chemokines, is inversely correlated to endothelial progenitor cell mobilization, a marker of favorable stroke outcome, in human patients with stroke (Bogoslovsky et al., 2011). Similarly, after TBI in humans, serum levels of sVCAM-1 and sE-selectin were also elevated (Rhind et al., 2010). In patients with acute ischemic stroke, serum levels of sICAM-1 were positively correlated with early death (Rallidis et al., 2009), though other studies do not find any increase in sICAM-1 or E-selectin with the same injury (Bleecker et al., 1998). CSF levels of sICAM-1 were increased in TBI patients and correlated with brain contusion size and BBB dysfunction (Pleines et al., 1998). Collectively, these data from CXCL1, IL-8 and CAMs do point to a significant role for the CXC/neutrophil system that can be used to access CNS injury and potentially predict mortality.

The Pleiotropic Cytokine IL-6

Interleukin-6 (IL-6) is a cytokine belonging to a family of neuropoietins that contribute to homeostatic processes and modulate the inflammatory response in the CNS. Unlike most other inflammatory cytokines, IL-6 is constitutively expressed, though at low levels, in the mature brain (Pitossi et al., 1997) and is important for moderating normal brain function (Quintana et al., 2007). IL-6 affects a myriad of physiological parameters to include temperature (LeMay et al., 1990), weight and body-mass index (Himmerich et al., 2006), sleep and wakefulness (Weschenfelder et al., 2012), depression (Kubera et al., 2011), memory (Balschun et al., 2004) and emotional regulation (Armario et al., 1998). During development, IL-6 can induce neurons to become cholinergic (Fann and Patterson, 1994; Marz et al., 1998; Oh and O'Malley, 1994) and can promote synaptic sprouting following injury (Murphy et al., 1999; Ramer et al., 1998). IL-6 is also involved in neuronal neurogenesis in the mature brain, though whether it enhances (Islam et al., 2009; Oh et al., 2010) or inhibits (Monje et al., 2003) this process *in vivo* is dependent on the presence of other localized factors. Much clearer is the positive proliferative effects on astrocytes (Nakanishi et al., 2007), microglia (Streit et al., 2000) and oligodendrocytes (Valerio et al., 2002; Zhang et al., 2006). IL-6 is also an integral physiological mediator involved in the acute inflammatory response. In the periphery, IL-6 is important for differentiation, maturation and activation of B- and T-cells in response to injury or infection and induces hepatic expression of acute phase proteins (Juttler et al., 2002). Following brain trauma, IL-6 is considered a key mediator of inflammation as it is one of the first cytokines to be released as part of the acute phase response and plays an important role in evolving biological events post-trauma (Chiaretti et al., 2008a; Ley et al., 2011; Quintana et al., 2007).

The pro- and anti-inflammatory functions of IL-6 emphasize the dichotomy of IL-6 release during brain injury (Bell et al., 1997; Huang et al., 2006; Kossmann et al., 1996; Lenzlinger et al., 2001; Ley et al., 2011), wherein moderate increases in IL-6 activate glial cells and neurotrophic factor release, which may aid in recovery, while greater increases may exacerbate pathology (Spooren et al., 2011). This duality is exemplified in studies with transgenic mice that over-express IL-6 in the brain (GFAP-IL-6). In the absence of injury, over-expression of IL-6 significantly increases inflammatory cytokine production in the mouse brain (Di Santo et al., 1996), which leads to a neurological syndrome whose severity positively correlates with IL-6 expression (Campbell et al., 1993). However, following TBI, GFAP-IL-6 mice also recover and heal more quickly than wild-type mice, potentially because of increased revascularization (Penkowa et al., 2003; Swartz et al., 2001). Similarly, in a permanent focal cerebral ischemia rodent model, an intracerebrovascular injection of IL-6 resulted in a reduced infract volume (Loddick et al., 1998; Marklund et al., 2005), potentially through the biosynthesis of nerve growth factors (Chiaretti et al., 2008b; Hama et al., 1989; Morganti-Kossmann et al., 1992; Rhodes et al., 2002). Furthermore, higher levels of IL-6 can reduce neural apoptosis through a decrease in glutamate and NMDA-induced neural death following ischemia (Ali et al., 2000) or excitotoxin exposure (Carlson et al., 1999). Decreasing levels of IL-6 may lead to neurodegeneration and increased oxidative stress (Hergenroeder et al., 2010; Penkowa et al., 2000), though contrasting data also exist (Pang et al., 2006). To further complicate the picture, the effects of IL-6 expression after CNS injury appear to be time dependent. Studies have shown that acute expression of IL-6 may be beneficial by improving healing and decreasing oxidative stress and apoptosis, but long-term expression promotes neuroinflammation and long-term damage (Quintana et al., 2007; Rhodes et al., 2002). Beyond animal models, IL-6 plays an integral role in human pathophysiology after CNS injury as well. Increased expression of IL-6 is a consequence of CNS injury (Amick et al., 2001; Ormstad et al., 2011a; Stein et al., 2011b), CNS infection (Frei et al., 1988) and chronic neurodegenerative disease (Holmes et al., 2011; Lindqvist et al., 2012; Rubio-Perez and Morillas-Ruiz, 2012), and IL-6 has been investigated as a biomarker for these conditions.

IL-6 as a Biomarker

Though IL-6 is constitutively expressed, significant changes in IL-6 expression have been documented following brain injury in tissue (Suehiro et al., 2004), the cerebral interstitial space (Winter et al., 2004), CSF and plasma (Csuka et al., 1999; McClain et al., 1991). These sizable shifts in IL-6 concentration allow for the use of this cytokine as a biomarker for various CNS injuries. It is possible via micro dialysis to measure cytokine concentrations from the interstitial space fluid in the brain, and this technique can be accomplished in conjunction with other neurointensive care interventions (Hillman et al., 2007). In doing so, one study found an inverse correlation between IL-6 levels and mortality and a positive correlation between peak IL-6 levels and higher GOS scores, suggesting a neuroprotective effect for IL-6 (Winter et al., 2004). A similar study in patients with traumatic brain injury with diffuse lesions found no correlation between IL-6 and intracranial pressure, brain tissue oxygenation or swelling, proving that this procedure may yet hold some practical limitations (Perez-Barcena et al., 2011).

More accessible than interstitial fluid, CSF has proved an invaluable medium for IL-6 detection and has become more accessible by the more routine use of the intraventricular catheter as a part of brain injury treatment.

In addition, the concentration of IL-6 in human CSF is much higher than in the plasma at the same time points (Hans et al., 1999; Kossmann et al., 1995), which can make CSF sampling more sensitive (Singhal et al., 2002; Tarkowski et al., 1995), though patient variability may undermine its usefulness (McClain et al., 1991). As an example of IL-6 as a biomarker, elevated levels of IL-6 in the CSF of hypoxic-ischemic term newborns have been correlated with brain injury severity and poor outcome (Savman et al., 1998). Following single tonic-clonic, prolonged or recurrent seizure activity, of the assayed inflammatory mediators, only levels of IL-6 increased significantly (Peltola et al., 1998; Peltola et al., 2000), and the amount of IL-6 positively correlated with seizure severity (Lehtimaki et al., 2004). Elevated levels of IL-6 in the CSF are correlated with infarct volume after stroke (Tarkowski et al., 1995). In pediatric cases of TBI, IL-6 increases in the CSF within 24 hours of injury were significantly correlated with a low (<4) admission GCS score, low GOS scores, injury severity and mortality (Amick et al., 2001; Bell et al., 1997; Chiaretti et al., 2005; Minambres et al., 2003). In contrast, studies show that this increase of IL-6 in the CSF is associated with favorable outcome as measured by GOS scores (Chiaretti et al., 2008a,b) as far as 3 months after injury (Singhal et al., 2002). Further, no correlation between early (<24 hour) IL-6 increases and head injury severity was found as measured by admission GCS scores (Chiaretti et al., 2008a,b; Hayakata et al., 2004). It should be noted that in at least one of these studies (Chiaretti et al., 2008a), as IL-6 increased, IL-1β also significantly decreased and may have contributed to improved neurological outcome as well.

Unlike CSF, serum is readily accessible and can be used for IL-6 detection following brain injury, though the results are more equivocal. IL-6 is certainly detectable in serum after seizure (Lehtimaki et al., 2008; Lehtimaki et al., 2004; Lehtimaki et al., 2007; Peltola et al., 2000), in term newborns with hypoxic-ischemic encephalopathy (Silveira and Procianoy, 2003) and following TBI (McClain et al., 1991; Woiciechowsky et al., 2002). A decrease in plasma IL-6 has been correlated with GCS improvement in patients with severe TBI (McClain et al., 1991), whereas high serum levels of IL-6 correlated with poor neurological outcome following hemorrhagic stroke (Oto et al., 2008), head injury severity in children (Kalabalikis et al., 1999) and mortality following severe isolated TBI (Arand et al., 2001). In addition, in neonates with hypoxic-ischemic injury, those with elevated concentrations of serum IL-6 had poorer neurological outcomes at 30 months than those who had lower concentrations (Bartha et al., 2004). In patients with severe brain injury, increased IL-6 plasma concentrations of >100 pg/mL within 1 day of injury were predictive of 7-day outcome, serious infection, mortality and injury severity (Woiciechowsky et al., 2002), though other studies have found no association between increased IL-6 concentrations and longer term outcome (Kalabalikis et al., 1999). In patients with isolated TBI, early increases (<17 hours) in IL-6 serum concentrations above 5 pg/ml can be used as an effective biomarker for the prediction of ICP development, and the predominant source of the IL-6 appeared to be from the brain as opposed to other peripheral injuries (Hergenroeder et al., 2010). However, multiple trauma is a significant confounding variable that negates the prognostic ability of serum IL-6 for brain injury (Hergenroeder et al., 2010). Lastly, in one study with acute ischemic stroke patients, IL-6 was inversely correlated with neurological outcome and infarct volume, suggesting a neuroprotective effect (Sotgiu et al., 2006).

Limitations of Inflammatory Biomarkers for CNS Injury

While the inflammatory factors discussed here can be useful biomarkers, contradictory results are still commonplace in both experimental and clinical settings. Some of these discrepancies can be explained by the dynamic and ever-changing inflammatory cascade that is activated following brain injury, underscoring the need for constant or consistent temporal monitoring and an increased accuracy in reporting times. However, other fundamental biological differences can affect cytokine expression patterns. Factors such as obesity (Pistell et al., 2010), sex (Lewis et al., 2010) and age (Mosenthal et al., 2004; Sandhir et al., 2004; Timaru-Kast et al., 2012) can greatly alter basal expression levels and injury-induced expression and must be taken into consideration when analyzing biomarker data. Also, many individual markers have not been properly validated to date, and it is likely that combinations of markers or expression patterns of multiple markers are truly needed for outcome prediction (Ramaswamy et al., 2009). To that end, below are some limitations of inflammatory markers that must be accounted for during proper biomarker development.

Inflammation is not specific to any injury or disease

Inflammation is initiated in response to any tissue injury regardless of mechanism and typically initiates a similar activation and abatement pattern in response to that injury. In the CNS, this means that disease or injuries that have different causes and different neuropathological changes may look very similar when observing proteins expressed during neuroinflammation. For example, IL-8 expression has been found in a wide range of CNS injury and disease states such as epilepsy (Billiau et al., 2007), severe head injury (Kushi et al., 2003; Whalen et al., 2000), multiple sclerosis (Lund et al., 2004), cancer (Rosenkilde and Schwartz, 2004), bacterial and viral meningitis (Bociaga-Jasik et al., 2012) and schizophrenia (Reale et al., 2011). Alternatively, similar mechanisms of injury may have varied inflammatory profiles. For example, seizure activity has been shown to activate the inflammatory response in experimental models using acetylcholinesterase inhibitors (Dhote et al., 2007; Johnson and Kan, 2010) or kainic acid (Lehtimaki et al., 2003; Vezzani et al., 1999), but each shows variations in the regional and temporal progression of the inflammatory response. While perhaps not specific for injury type, certain inflammatory biomarkers in both the CSF and serum have proven to be excellent indicators of severity and can predict long-term outcome and mortality. However, in multi-trauma scenarios, these biomarkers become less useful since there are now multiple contributing sources (injuries) to the overall measurable inflammatory state (Hergenroeder et al., 2010). In fact, this peripheral inflammation can potentially exacerbate damage in the CNS (McColl et al., 2007).

Localization of inflammatory mediators is important

Neuroinflammation is a highly regulated process with numerous competing positive and negative feedback loops. Following CNS insult, mRNA transcription is quickly upregulated followed shortly by translation, with the exact concentrations and times depending on mode of injury and severity (Touzani et al., 1999). Ultimately, the function of certain neuroinflammatory mediators is highly dependent on the receptor to which it has bound, the cell by which that receptor is expressed, the activation state of that cell, the input of other inflammatory mediators on that cell, the micro-environment in which that cell finds itself, the responding immune cell phenotype and the timing of the interaction between the immune cell and the damaged neural cells, all of which are unique to each injury type (Li et al., 2007; Patzer et al., 2008). In short, at a given injury location, the concentration at specific time points in the progression of the inflammatory process can determine whether the environment is overall pro- or anti-inflammatory, which can have a profound effect on the efficacy of a single biomarker to predict outcome, injury severity or mortality. The IL-1 family of proteins exemplifies these issues. IL-1 is a potent pro-inflammatory cytokine, while IL-1Ra is a potent anti-inflammatory receptor antagonist. Based on these functions, poor outcome would be assumed to correlate with increases in IL-1, whereas increases in IL-1Ra would correlate with better outcome, though this was not always the case. Instead, a more accurate predictor of outcome is the molar balance between the two at the time of assay, not the concentrations present, because the relationship between the two defines where on the continuum of inflammation the tissue is currently positioned (Hutchinson et al., 2007; Mikuniya et al., 2000). While micro-environmental precision is not attainable with serum or CSF, it is important to keep in mind when interpreting biomarker and pathology data.

When you assay is as important as what you assay

While the presence of individual inflammatory proteins at the injured tissue determines the inflammatory environment, when they are there is also imperative. Being a tightly regulated process, many potent pro-inflammatory factors have a short window of detection because of their own upregulation of anti-inflammatory factors. Since many studies use only a few data points, the time at which the sample is taken is crucial. To complicate matters, progression of the inflammatory response and individual biology are variable, further increasing the chances of missing the expression of the factor being investigated (Bartha et al., 2004). Limited assaying only provides a temporal snapshot of the constantly evolving inflammatory process and provides no information about the velocity of cytokine expression. Data analysis can be further complicated by inconsistent temporal sampling (i.e., "early" samples taken between 4–24 hours), which can greatly increase variability and severely limit biomarker discovery and is common with most clinical studies (Bartha et al., 2004). True biomarker discovery requires serial sampling to correlate pro- and anti-inflammatory markers to the initiation and progressive pathology of injury and to determine how these markers are interrelated and change over time. Lastly, studying multiple potential biomarkers may help mitigate timing issues associated with limited sampling and emphasizes the importance of studying whole inflammatory systems rather than single inflammatory components for biomarker development.

The target medium for the biomarker assay can make all the difference

It is often found that the timing and concentration of individual cytokines in the blood and CSF following injury are discordant and that one or the other may be more predictive of outcomes. This is in some ways not surprising given the proximity and volumes of the tissues assayed (CSF has less volume and is more proximate to the brain injury) and the half-lives of most cytokines. Therefore, each fluid may have different predictive abilities dependent on the injury mechanism and marker investigated. For instance, IL-1 appears to be a better predictor of poor outcome after brain injury when assayed in the blood serum than in the CSF (Lehtimaki et al., 2008; Lehtimaki et al., 2007; Peltola et al., 2000), while the opposite appears to be true for IL-8 (Kossmann et al., 1997; Whalen et al., 2000). Care must be taken when selecting the medium for biomarker analysis.

Conclusions

Inflammation is a nearly ubiquitous response to injury and the brain is no exception. Most inflammatory mediators are not expressed constitutively; therefore, concentration increases can be more easily detected following tissue damage. Many of the inflammatory factors involved in this process are expressed by injured brain tissues and have been found in increasing concentrations in clinically relevant fluids such as the CSF and blood. The low expression in normal tissue and the large increases in response to injury have made inflammatory mediators potential biomarkers for CNS injury and disease and, in some cases, effective predictors of outcome. Biomarker development for inflammatory factors in many ways is still in its infancy, as with other biomarkers for CNS injury, and may prove to be more complicated than the simple presence or concentration change of any one protein. While the current literature certainly points at several candidates as being reflective of the current injury level in the brain and as having predictive value for long-term outcome based on concentration, such as with TNFα (Sotgiu et al., 2006; Stein et al., 2012; Stein et al., 2011b; Vila et al., 2000), even this does not appear to be applicable for all CNS injuries or, in some cases, even replicable for the same CNS injury. Much of the variability lies in the nature of inflammation itself; it is a constantly changing process that can be either pro- or anti-inflammatory depending on multiple variables. At a minimum, time and concentration of related cytokines may offer better insight into injury states and potential recovery following CNS damage. Some of these issues are already being addressed, most notably by using ratios of pro- and anti-inflammatory factors to predict outcome, with early success coming from the relationships of IL-1/IL-1Ra (Girard et al., 2010; Hutchinson et al., 2007) and TNFα/sTNFR (Maier et al., 2006). Further development of inflammatory biomarkers will probably need to be based on upregulation patterns of multiple inflammatory mediators and will likely include additional non-inflammatory biomarker candidates.

Disclaimer

The views expressed in this chapter are those of the author(s) and do not reflect the official policy of the Department of Army, Department of Defense, or the U.S. Government.

References

Ali, C., O. Nicole, F. Docagne, S. Lesne, E.T. MacKenzie, A. Nouvelot, A. Buisson and D. Vivien. 2000. Ischemia-induced interleukin-6 as a potential endogenous neuroprotective cytokine against NMDA receptor-mediated excitotoxicity in the brain. *J Cereb Blood Flow Metab*. 20: 956–66.

Allan, S.M. and N.J. Rothwell. 2000. Cortical death caused by striatal administration of AMPA and interleukin-1 is mediated by activation of cortical NMDA receptors. *J Cereb Blood Flow Metab*. 20: 1409–13.

Allan, S.M., P.J. Tyrrell and N.J. Rothwell. 2005. Interleukin-1 and neuronal injury. *Nat Rev Immunol*. 5: 629–40.

Amick, J.E., K.A. Yandora, M.J. Bell, S.R. Wisniewski, P.D. Adelson, J.A. Carcillo, K.L. Janesko, S.T. DeKosky, T.M. Carlos, R.S. Clark and P.M. Kochanek. 2001. The Th1 versus Th2 cytokine profile in cerebrospinal fluid after severe traumatic brain injury in infants and children. *Pediatr Crit Care Med*. 2: 260–264.

Arand, M., H. Melzner, L. Kinzl, U.B. Bruckner and F. Gebhard. 2001. Early inflammatory mediator response following isolated traumatic brain injury and other major trauma in humans. *Langenbecks Arch Surg*. 386: 241–8.

Arend, W.P. 2002. The balance between IL-1 and IL-1Ra in disease. *Cytokine Growth Factor Rev*. 13: 323–40.

Armario, A., J. Hernandez, H. Bluethmann and J. Hidalgo. 1998. IL-6 deficiency leads to increased emotionality in mice: evidence in transgenic mice carrying a null mutation for IL-6. *J Neuroimmunol*. 92: 160–9.

Asano, T., K. Ichiki, S. Koizumi, K. Kaizu, T. Hatori, O. Fujino, K. Mashiko, Y. Sakamoto, T. Miyasho and Y. Fukunaga. 2010. IL-8 in Cerebrospinal Fluid from Children with Acute Encephalopathy is Higher than in that from Children with Febrile Seizure. *Scandinavian Journal of Immunology*. 71: 447–451.

Baggiolini, M., B. Dewald and B. Moser. 1994. Interleukin-8 and related chemotactic cytokines—CXC and CC chemokines. *Adv Immunol*. 55: 97–179.

Bajetto, A., R. Bonavia, S. Barbero and G. Schettini. 2002. Characterization of chemokines and their receptors in the central nervous system: physiopathological implications. *J Neurochem*. 82: 1311–29.

Balschun, D., W. Wetzel, A. Del Rey, F. Pitossi, H. Schneider, W. Zuschratter and H.O. Besedovsky. 2004. Interleukin-6: a cytokine to forget. *FASEB J*. 18: 1788–90.

Ban, E., G. Milon, N. Prudhomme, G. Fillion and F. Haour. 1991. Receptors for interleukin-1 (alpha and beta) in mouse brain: mapping and neuronal localization in hippocampus. *Neuroscience*. 43: 21–30.

Bartha, A.I., A. Foster-Barber, S.P. Miller, D.B. Vigneron, D.V. Glidden, A.J. Barkovich and D.M. Ferriero. 2004. Neonatal encephalopathy: association of cytokines with MR spectroscopy and outcome. *Pediatr Res*. 56: 960–6.

Bass, W.T., E.S. Buescher, P.S. Hair, L.E. White, J.C. Welch and B.L. Burke. 2008. Proinflammatory cytokine-receptor interaction model improves the predictability of cerebral white matter injury in preterm infants. *Am J Perinatol*. 25: 211–8.

Bazan, J.F., K.B. Bacon, G. Hardiman, W. Wang, K. Soo, D. Rossi, D.R. Greaves, A. Zlotnik and T.J. Schall. 1997. A new class of membrane-bound chemokine with a CX3C motif. *Nature*. 385: 640–4.

Beamer, N.B., B.M. Coull, W.M. Clark, J.S. Hazel and J.R. Silberger. 1995. Interleukin-6 and interleukin-1 receptor antagonist in acute stroke. *Ann Neurol*. 37: 800–5.

Bell, M.J., P.M. Kochanek, L.A. Doughty, J.A. Carcillo, P.D. Adelson, R.S. Clark, S.R. Wisniewski, M.J. Whalen and S.T. DeKosky. 1997. Interleukin-6 and interleukin-10 in cerebrospinal fluid after severe traumatic brain injury in children. *J Neurotrauma*. 14: 451–7.

Berda-Haddad, Y., S. Robert, P. Salers, L. Zekraoui, C. Farnarier, C.A. Dinarello, F. Dignat-George and G. Kaplanski. 2011. Sterile inflammation of endothelial cell-derived apoptotic bodies is mediated by interleukin-1alpha. *Proc Natl Acad Sci USA.* 108: 20684–9.

Betz, A.L., G.Y. Yang and B.L. Davidson. 1995. Attenuation of stroke size in rats using an adenoviral vector to induce overexpression of interleukin-1 receptor antagonist in brain. *J Cereb Blood Flow Metab.* 15: 547–51.

Billiau, A.D., P. Witters, B. Ceulemans, A. Kasran, C. Wouters and L. Lagae. 2007. Intravenous immunoglobulins in refractory childhood-onset epilepsy: effects on seizure frequency, EEG activity, and cerebrospinal fluid cytokine profile. *Epilepsia.* 48: 1739–49.

Blankenberg, S., L. Tiret, C. Bickel, D. Peetz, F. Cambien, J. Meyer and H.J. Rupprecht. 2002. Interleukin-18 is a strong predictor of cardiovascular death in stable and unstable angina. *Circulation.* 106: 24–30.

Bleecker, J.D., I. Coulier, C. Fleurinck and J.D. Reuck. 1998. Circulating intercellular adhesion molecule-1 and E-selectin in acute ischemic stroke. *J Stroke Cerebrovasc Dis.* 7: 192–5.

Bociaga-Jasik, M., A. Garlicki, A. Ciesla, A. Kalinowska-Nowak, I. Sobczyk-Krupiarz and T. Mach. 2012. The diagnostic value of cytokine and nitric oxide concentrations in cerebrospinal fluid for the differential diagnosis of meningitis. *Adv Med Sci.* 57: 142–7.

Bogoslovsky, T., M. Spatz, A. Chaudhry, D. Maric, M. Luby, J. Frank and S. Warach. 2011. Stromal-derived factor-1[alpha] correlates with circulating endothelial progenitor cells and with acute lesion volume in stroke patients. *Stroke.* 42: 618–25.

Bolton, S.J., D.C. Anthony and V.H. Perry. 1998. Loss of the tight junction proteins occludin and zonula occludens-1 from cerebral vascular endothelium during neutrophil-induced blood-brain barrier breakdown *in vivo. Neuroscience.* 86: 1245–57.

Bonizzi, G. and M. Karin. 2004. The two NF-kappaB activation pathways and their role in innate and adaptive immunity. *Trends Immunol.* 25: 280–8.

Boutin, H., R.A. LeFeuvre, R. Horai, M. Asano, Y. Iwakura and N.J. Rothwell. 2001. Role of IL-1alpha and IL-1beta in ischemic brain damage. *J Neurosci.* 21: 5528–34.

Broutin, S., N. Ameur, L. Lacroix, T. Robert, B. Petit, N. Oumata, M. Talbot, B. Caillou, M. Schlumberger, C. Dupuy and J.M. Bidart. 2011. Identification of soluble candidate biomarkers of therapeutic response to sunitinib in medullary thyroid carcinoma in preclinical models. *Clin Cancer Res.* 17: 2044–54.

Bye, N., A.M. Turnley and M.C. Morganti-Kossmann. 2012. Inflammatory regulators of redirected neural migration in the injured brain. *Neurosignals.* 20: 132–46.

Campbell, I.L., C.R. Abraham, E. Masliah, P. Kemper, J.D. Inglis, M.B. Oldstone and L. Mucke. 1993. Neurologic disease induced in transgenic mice by cerebral overexpression of interleukin 6. *Proc Natl Acad Sci USA.* 90: 10061–5.

Carlson, N.G., A. Bacchi, S.W. Rogers and L.C. Gahring. 1998. Nicotine blocks TNF-alpha-mediated neuroprotection to NMDA by an alpha-bungarotoxin-sensitive pathway. *J Neurobiol.* 35: 29–36.

Carlson, N.G., W.A. Wieggel, J. Chen, A. Bacchi, S.W. Rogers and L.C. Gahring. 1999. Inflammatory cytokines IL-1 alpha, IL-1 beta, IL-6, and TNF-alpha impart neuroprotection to an excitotoxin through distinct pathways. *J Immunol.* 163: 3963–8.

Castellanos, M., J. Castillo, M.M. Garcia, R. Leira, J. Serena, A. Chamorro and A. Davalos. 2002. Inflammation-mediated damage in progressing lacunar infarctions: a potential therapeutic target. *Stroke.* 33: 982–7.

Chen, C.J., H. Kono, D. Golenbock, G. Reed, S. Akira and K.L. Rock. 2007. Identification of a key pathway required for the sterile inflammatory response triggered by dying cells. *Nat Med.* 13: 851–6.

Chen, G.Y. and G. Nunez. 2010. Sterile inflammation: sensing and reacting to damage. *Nat Rev Immunol.* 10: 826–37.

Chen, Y., M. Shi, G.Z. Yu, X.R. Qin, G. Jin, P. Chen and M.H. Zhu. 2012. Interleukin-8, a promising predictor for prognosis of pancreatic cancer. *World J Gastroenterol.* 18: 1123–9.

Chiaretti, A., A. Antonelli, A. Mastrangelo, P. Pezzotti, L. Tortorolo, F. Tosi and O. Genovese. 2008a. Interleukin-6 and nerve growth factor upregulation correlates with improved outcome in children with severe traumatic brain injury. *J Neurotrauma*. 25: 225–34.

Chiaretti, A., A. Antonelli, R. Riccardi, O. Genovese, P. Pezzotti, C. Di Rocco, L. Tortorolo and G. Piedimonte. 2008b. Nerve growth factor expression correlates with severity and outcome of traumatic brain injury in children. *Eur J Paediatr Neurol*. 12: 195–204.

Chiaretti, A., O. Genovese, L. Aloe, A. Antonelli, M. Piastra, G. Polidori and C. Di Rocco. 2005. Interleukin 1beta and interleukin 6 relationship with paediatric head trauma severity and outcome. *Childs Nerv Syst*. 21: 185–93; discussion 194.

Chou, S.H., S.K. Feske, J. Atherton, R.G. Konigsberg, P.L. De Jager, R. Du, C.S. Ogilvy, E.H. Lo and M. Ning. 2012. Early elevation of serum tumor necrosis factor-alpha is associated with poor outcome in subarachnoid hemorrhage. *J Investig Med*. 60: 1054–8.

Clark, R.S., P.M. Kochanek, M. Chen, S.C. Watkins, D.W. Marion, J. Chen, R.L. Hamilton, J.E. Loeffert and S.H. Graham. 1999. Increases in Bcl-2 and cleavage of caspase-1 and caspase-3 in human brain after head injury. *FASEB J*. 13: 813–21.

Croitoru-Lamoury, J., G.J. Guillemin, D. Dormont and B.J. Brew. 2003. Quinolinic acid up-regulates chemokine production and chemokine receptor expression in astrocytes. *Adv Exp Med Biol*. 527: 37–45.

Csuka, E., M.C. Morganti-Kossmann, P.M. Lenzlinger, H. Joller, O. Trentz and T. Kossmann. 1999. IL-10 levels in cerebrospinal fluid and serum of patients with severe traumatic brain injury: relationship to IL-6, TNF-alpha, TGF-beta1 and blood-brain barrier function. *J Neuroimmunol*. 101: 211–21.

Dheen, S.T., C. Kaur and E.A. Ling. 2007. Microglial activation and its implications in the brain diseases. *Curr Med Chem*. 14: 1189–97.

Dhote, F., A. Peinnequin, P. Carpentier, V. Baille, C. Delacour, A. Foquin, G. Lallement and F. Dorandeu. 2007. Prolonged inflammatory gene response following soman-induced seizures in mice. *Toxicology*. 238: 166–76.

Di Santo, E., T. Alonzi, E. Fattori, V. Poli, G. Ciliberto, M. Sironi, P. Gnocchi, P. Ricciardi-Castagnoli and P. Ghezzi. 1996. Overexpression of interleukin-6 in the central nervous system of transgenic mice increases central but not systemic proinflammatory cytokine production. *Brain Res*. 740: 239–44.

Dinarello, C.A. 1996. Biologic basis for interleukin-1 in disease. *Blood*. 87: 2095–147.

Dinarello, C.A. and R.C. Thompson. 1991. Blocking IL-1: interleukin 1 receptor antagonist *in vivo* and *in vitro*. *Immunol Today*. 12: 404–10.

Dinkel, K., F.S. Dhabhar and R.M. Sapolsky. 2004. Neurotoxic effects of polymorphonuclear granulocytes on hippocampal primary cultures. *Proc Natl Acad Sci USA*. 101: 331–6.

Dolga, A.M., I. Granic, T. Blank, H.G. Knaus, J. Spiess, P.G. Luiten, U.L. Eisel and I.M. Nijholt. 2008. TNF-alpha-mediates neuroprotection against glutamate-induced excitotoxicity via NF-kappaB-dependent up-regulation of K2.2 channels. *J Neurochem*. 107: 1158–67.

Dripps, D.J., E. Verderber, R.K. Ng, R.C. Thompson and S.P. Eisenberg. 1991. Interleukin-1 receptor antagonist binds to the type II interleukin-1 receptor on B cells and neutrophils. *J Biol Chem*. 266: 20311–5.

Ehrlich, L.C., S. Hu, W.S. Sheng, R.L. Sutton, G.L. Rockswold, P.K. Peterson and C.C. Chao. 1998. Cytokine regulation of human microglial cell IL-8 production. *J Immunol*. 160: 1944–8.

Emsley, H.C., C.J. Smith, R.F. Georgiou, A. Vail, S.J. Hopkins, N.J. Rothwell and P.J. Tyrrell. 2005. A randomised phase II study of interleukin-1 receptor antagonist in acute stroke patients. *J Neurol Neurosurg Psychiatry*. 76: 1366–72.

Eriksson, C., A.M. Van Dam, P.J. Lucassen, J.G. Bol, B. Winblad and M. Schultzberg. 1999. Immunohistochemical localization of interleukin-1beta, interleukin-1 receptor antagonist and interleukin-1beta converting enzyme/caspase-1 in the rat brain after peripheral administration of kainic acid. *Neuroscience*. 93: 915–30.

Fann, M.J. and P.H. Patterson. 1994. Neuropoietic cytokines and activin A differentially regulate the phenotype of cultured sympathetic neurons. *Proc Natl Acad Sci USA*. 91: 43–7.

Fassbender, K., U. Schminke, S. Ries, A. Ragoschke, U. Kischka, M. Fatar and M. Hennerici. 1997. Endothelial-derived adhesion molecules in bacterial meningitis: association to cytokine release and intrathecal leukocyte-recruitment. *J Neuroimmunol*. 74: 130–4.

Felderhoff-Mueser, U., M. Sifringer, O. Polley, M. Dzietko, B. Leineweber, L. Mahler, M. Baier, P. Bittigau, M. Obladen, C. Ikonomidou and C. Buhrer. 2005. Caspase-1-processed interleukins in hyperoxia-induced cell death in the developing brain. *Ann Neurol*. 57: 50–9.

Fiers, W. 1992. Precursor structures and structure-function analysis of TNF and lymphotoxin. *Immunol Ser*. 56: 79–92.

Fontaine, V., S. Mohand-Said, N. Hanoteau, C. Fuchs, K. Pfizenmaier and U. Eisel. 2002. Neurodegenerative and neuroprotective effects of tumor Necrosis factor (TNF) in retinal ischemia: opposite roles of TNF receptor 1 and TNF receptor 2. *Journal of Neuroscience*. 22: RC216.

Foster-Barber, A., B. Dickens and D.M. Ferriero. 2001. Human perinatal asphyxia: correlation of neonatal cytokines with MRI and outcome. *Dev Neurosci*. 23: 213–8.

Frei, K., T.P. Leist, A. Meager, P. Gallo, D. Leppert, R.M. Zinkernagel and A. Fontana. 1988. Production of B cell stimulatory factor-2 and interferon gamma in the central nervous system during viral meningitis and encephalitis. Evaluation in a murine model infection and in patients. *J Exp Med*. 168: 449–53.

Frugier, T., M.C. Morganti-Kossmann, D. O'Reilly and C.A. McLean. 2010. *In situ* detection of inflammatory mediators in post mortem human brain tissue after traumatic injury. *J Neurotrauma*. 27: 497–507.

Furlan, R., M. Filippi, A. Bergami, M.A. Rocca, V. Martinelli, P.L. Poliani, L.M. Grimaldi, G. Desina, G. Comi and G. Martino. 1999. Peripheral levels of caspase-1 mRNA correlate with disease activity in patients with multiple sclerosis; a preliminary study. *J Neurol Neurosurg Psychiatry*. 67: 785–8.

Gadient, R.A., K.C. Cron and U. Otten. 1990. Interleukin-1 beta and tumor necrosis factor-alpha synergistically stimulate nerve growth factor (NGF) release from cultured rat astrocytes. *Neurosci Lett*. 117: 335–40.

Garcia, J.H., K.F. Liu and J.K. Relton. 1995. Interleukin-1 receptor antagonist decreases the number of necrotic neurons in rats with middle cerebral artery occlusion. *Am J Pathol*. 147: 1477–86.

Gimbrone, M.A., Jr., M.S. Obin, A.F. Brock, E.A. Luis, P.E. Hass, C.A. Hebert, Y.K. Yip, D.W. Leung, D.G. Lowe, W.J. Kohr et al. 1989. Endothelial interleukin-8: a novel inhibitor of leukocyte-endothelial interactions. *Science*. 246: 1601–3.

Girard, S., G. Sebire and H. Kadhim. 2010. Proinflammatory orientation of the interleukin 1 system and downstream induction of matrix metalloproteinase 9 in the pathophysiology of human perinatal white matter damage. *J Neuropathol Exp Neurol*. 69: 1116–29.

Giulian, D. and L.B. Lachman. 1985. Interleukin-1 stimulation of astroglial proliferation after brain injury. *Science*. 228: 497–9.

Giulian, D., J. Woodward, D.G. Young, J.F. Krebs and L.B. Lachman. 1988. Interleukin-1 injected into mammalian brain stimulates astrogliosis and neovascularization. *J Neurosci*. 8: 2485–90.

Gopcevic, A., B. Mazul-Sunko, J. Marout, A. Sekulic, N. Antoljak, M. Siranovic, Z. Ivanec, M. Margaritoni, M. Bekavac-Beslin and N. Zarkovic. 2007. Plasma interleukin-8 as a potential predictor of mortality in adult patients with severe traumatic brain injury. *Tohoku J Exp Med*. 211: 387–93.

Hama, T., M. Miyamoto, H. Tsukui, C. Nishio and H. Hatanaka. 1989. Interleukin-6 as a neurotrophic factor for promoting the survival of cultured basal forebrain cholinergic neurons from postnatal rats. *Neurosci Lett*. 104: 340–4.

Hans, V.H., T. Kossmann, H. Joller, V. Otto and M.C. Morganti-Kossmann. 1999. Interleukin-6 and its soluble receptor in serum and cerebrospinal fluid after cerebral trauma. *Neuroreport.* 10: 409–12.

Hayakata, T., T. Shiozaki, O. Tasaki, H. Ikegawa, Y. Inoue, F. Toshiyuki, H. Hosotubo, F. Kieko, T. Yamashita, H. Tanaka, T. Shimazu and H. Sugimoto. 2004. Changes in CSF S100B and cytokine concentrations in early-phase severe traumatic brain injury. *Shock.* 22: 102–7.

Hedtjarn, M., A.L. Leverin, K. Eriksson, K. Blomgren, C. Mallard and H. Hagberg. 2002. Interleukin-18 involvement in hypoxic-ischemic brain injury. *J Neurosci.* 22: 5910–9.

Heidemann, J., H. Ogawa, M.B. Dwinell, P. Rafiee, C. Maaser, H.R. Gockel, M.F. Otterson, D.M. Ota, N. Lugering, W. Domschke and D.G. Binion. 2003. Angiogenic effects of interleukin 8 (CXCL8) in human intestinal microvascular endothelial cells are mediated by CXCR2. *J Biol Chem.* 278: 8508–15.

Hergenroeder, G.W., A.N. Moore, J.P. McCoy, Jr., L. Samsel, N.H. Ward, 3rd, G.L. Clifton and P.K. Dash. 2010. Serum IL-6: a candidate biomarker for intracranial pressure elevation following isolated traumatic brain injury. *J Neuroinflammation.* 7: 19.

Herx, L.M. and V.W. Yong. 2001. Interleukin-1 beta is required for the early evolution of reactive astrogliosis following CNS lesion. *J Neuropathol Exp Neurol.* 60: 961–71.

Hillhouse, E.W., S. Kida and F. Iannotti. 1998. Middle cerebral artery occlusion in the rat causes a biphasic production of immunoreactive interleukin-1beta in the cerebral cortex. *Neurosci Lett.* 249: 177–9.

Hillman, J., O. Aneman, M. Persson, C. Andersson, C. Dabrosin and P. Mellergard. 2007. Variations in the response of interleukins in neurosurgical intensive care patients monitored using intracerebral microdialysis. *J Neurosurg.* 106: 820–5.

Himmerich, H., S. Fulda, J. Linseisen, H. Seiler, G. Wolfram, S. Himmerich, K. Gedrich and T. Pollmacher. 2006. TNF-alpha, soluble TNF receptor and interleukin-6 plasma levels in the general population. *Eur Cytokine Netw.* 17: 196–201.

Hoch, R.C., I.U. Schraufstatter and C.G. Cochrane. 1996. *In vivo, in vitro,* and molecular aspects of interleukin-8 and the interleukin-8 receptors. *J Lab Clin Med.* 128: 134–45.

Holmes, C., C. Cunningham, E. Zotova, D. Culliford and V.H. Perry. 2011. Proinflammatory cytokines, sickness behavior, and Alzheimer disease. *Neurology.* 77: 212–8.

Hoshi, T., H. Yamagami, S. Furukado, K. Miwa, M. Tanaka, M. Sakaguchi, S. Sakoda and K. Kitagawa. 2010. Serum inflammatory proteins and frontal lobe dysfunction in patients with cardiovascular risk factors. *Eur J Neurol.* 17: 1134–40.

Huang, J., U.M. Upadhyay and R.J. Tamargo. 2006. Inflammation in stroke and focal cerebral ischemia. *Surg Neurol.* 66: 232–45.

Huang, W.X., P. Huang and J. Hillert. 2004. Increased expression of caspase-1 and interleukin-18 in peripheral blood mononuclear cells in patients with multiple sclerosis. *Mult Scler.* 10: 482–7.

Hussein, M.H., G.A. Daoud, H. Kakita, S. Kato, T. Goto, M. Kamei, K. Goto, M. Nobata, Y. Ozaki, T. Ito, S. Fukuda, I. Kato, S. Suzuki, H. Sobajima, F. Hara, T. Hashimoto and H. Togari. 2010. High cerebrospinal fluid antioxidants and interleukin 8 are protective of hypoxic brain damage in newborns. *Free Radic Res.* 44: 422–9.

Hutchinson, P.J., M.T. O'Connell, N.J. Rothwell, S.J. Hopkins, J. Nortje, K.L. Carpenter, I. Timofeev, P.G. Al-Rawi, D.K. Menon and J.D. Pickard. 2007. Inflammation in human brain injury: intracerebral concentrations of IL-1alpha, IL-1beta, and their endogenous inhibitor IL-1ra. *J Neurotrauma.* 24: 1545–57.

Islam, O., X. Gong, S. Rose-John and K. Heese. 2009. Interleukin-6 and neural stem cells: more than gliogenesis. *Mol Biol Cell.* 20: 188–99.

Janabi, N., I. Hau and M. Tardieu. 1999. Negative feedback between prostaglandin and alpha- and beta-chemokine synthesis in human microglial cells and astrocytes. *J Immunol.* 162: 1701–6.

Jander, S., M. Schroeter and G. Stoll. 2002. Interleukin-18 expression after focal ischemia of the rat brain: association with the late-stage inflammatory response. *J Cereb Blood Flow Metab.* 22: 62–70.

Jin, R., G. Yang and G. Li. 2010. Inflammatory mechanisms in ischemic stroke: role of inflammatory cells. *J Leukoc Biol.* 87: 779–89.

Johnson, E.A., T.L. Dao, M.A. Guignet, C.E. Geddes, A.I. Koemeter-Cox and R.K. Kan. 2011. Increased expression of the chemokines CXCL1 and MIP-1alpha by resident brain cells precedes neutrophil infiltration in the brain following prolonged soman-induced status epilepticus in rats. *J Neuroinflammation.* 8: 41.

Johnson, E.A. and R.K. Kan. 2010. The acute phase response and soman-induced status epilepticus: temporal, regional and cellular changes in rat brain cytokine concentrations. *J Neuroinflammation.* 7: 40.

Juttler, E., V. Tarabin and M. Schwaninger. 2002. Interleukin-6 (IL-6): a possible neuromodulator induced by neuronal activity. *Neuroscientist.* 8: 268–75.

Kalabalikis, P., K. Papazoglou, D. Gouriotis, N. Papadopoulos, M. Kardara, F. Papageorgiou and J. Papadatos. 1999. Correlation between serum IL-6 and CRP levels and severity of head injury in children. *Intensive Care Med.* 25: 288–92.

Kannan, H., Y. Tanaka, T. Kunitake, Y. Ueta, Y. Hayashida and H. Yamashita. 1996. Activation of sympathetic outflow by recombinant human interleukin-1 beta in conscious rats. *Am J Physiol.* 270: R479–85.

Kasahara, T., N. Mukaida, K. Yamashita, H. Yagisawa, T. Akahoshi and K. Matsushima. 1991. IL-1 and TNF-alpha induction of IL-8 and monocyte chemotactic and activating factor (MCAF) mRNA expression in a human astrocytoma cell line. *Immunology.* 74: 60–7.

Kenne, E., A. Erlandsson, L. Lindbom, L. Hillered and F. Clausen. 2012. Neutrophil depletion reduces edema formation and tissue loss following traumatic brain injury in mice. *J Neuroinflammation.* 9: 17.

Kossmann, T., V. Hans, H.G. Imhof, O. Trentz and M.C. Morganti-Kossmann. 1996. Interleukin-6 released in human cerebrospinal fluid following traumatic brain injury may trigger nerve growth factor production in astrocytes. *Brain Res.* 713: 143–52.

Kossmann, T., V.H. Hans, H.G. Imhof, R. Stocker, P. Grob, O. Trentz and C. Morganti-Kossmann. 1995. Intrathecal and serum interleukin-6 and the acute-phase response in patients with severe traumatic brain injuries. *Shock.* 4: 311–7.

Kossmann, T., P.F. Stahel, P.M. Lenzlinger, H. Redl, R.W. Dubs, O. Trentz, G. Schlag and M.C. Morganti-Kossmann. 1997. Interleukin-8 released into the cerebrospinal fluid after brain injury is associated with blood-brain barrier dysfunction and nerve growth factor production. *J Cereb Blood Flow Metab.* 17: 280–9.

Kubera, M., E. Obuchowicz, L. Goehler, J. Brzeszcz and M. Maes. 2011. In animal models, psychosocial stress-induced (neuro)inflammation, apoptosis and reduced neurogenesis are associated to the onset of depression. *Prog Neuropsychopharmacol Biol Psychiatry.* 35: 744–59.

Kushi, H., T. Saito, K. Makino and N. Hayashi. 2003. IL-8 is a key mediator of neuroinflammation in severe traumatic brain injuries. *Acta Neurochir Suppl.* 86: 347–50.

Lehtimaki, K., T. Keranen, M. Huuhka, J. Palmio, M. Hurme, E. Leinonen and J. Peltola. 2008. Increase in plasma proinflammatory cytokines after electroconvulsive therapy in patients with depressive disorder. *J ECT.* 24: 88–91.

Lehtimaki, K.A., T. Keranen, H. Huhtala, M. Hurme, J. Ollikainen, J. Honkaniemi, J. Palmio and J. Peltola. 2004. Regulation of IL-6 system in cerebrospinal fluid and serum compartments by seizures: the effect of seizure type and duration. *J Neuroimmunol.* 152: 121–5.

Lehtimaki, K.A., T. Keranen, J. Palmio, R. Makinen, M. Hurme, J. Honkaniemi and J. Peltola. 2007. Increased plasma levels of cytokines after seizures in localization-related epilepsy. *Acta Neurol Scand.* 116: 226–30.

Lehtimaki, K.A., T. Keranen, J. Palmio and J. Peltola. 2010. Levels of IL-1beta and IL-1ra in cerebrospinal fluid of human patients after single and prolonged seizures. *Neuroimmunomodulation.* 17: 19–22.

Lehtimaki, K.A., J. Peltola, E. Koskikallio, T. Keranen and J. Honkaniemi. 2003. Expression of cytokines and cytokine receptors in the rat brain after kainic acid-induced seizures. *Brain Res Mol Brain Res.* 110: 253–60.

LeMay, L.G., A.J. Vander and M.J. Kluger. 1990. Role of interleukin 6 in fever in rats. *Am J Physiol.* 258: R798–803.

Lenzlinger, P.M., M.C. Morganti-Kossmann, H.L. Laurer and T.K. McIntosh. 2001. The duality of the inflammatory response to traumatic brain injury. *Mol Neurobiol.* 24: 169–81.

Lewis, D.K., S. Bake, K. Thomas, M.K. Jezierski and F. Sohrabji. 2010. A high cholesterol diet elevates hippocampal cytokine expression in an age and estrogen-dependent manner in female rats. *J Neuroimmunol.* 223: 31–8.

Ley, E.J., M.A. Clond, M.B. Singer, D. Shouhed and A. Salim. 2011. IL6 deficiency affects function after traumatic brain injury. *J Surg Res.* 170: 253–6.

Li, A., S. Dubey, M.L. Varney, B.J. Dave and R.K. Singh. 2003. IL-8 directly enhanced endothelial cell survival, proliferation, and matrix metalloproteinases production and regulated angiogenesis. *J Immunol.* 170: 3369–76.

Li, L., J. Lu, S.S. Tay, S.M. Moochhala and B.P. He. 2007. The function of microglia, either neuroprotection or neurotoxicity, is determined by the equilibrium among factors released from activated microglia *in vitro*. *Brain Res.* 1159: 8–17.

Lin, M.T., T.Y. Kao, Y.T. Jin and C.F. Chen. 1995. Interleukin-1 receptor antagonist attenuates the heat stroke-induced neuronal damage by reducing the cerebral ischemia in rats. *Brain Res Bull.* 37: 595–8.

Lindqvist, D., E. Kaufman, L. Brundin, S. Hall, Y. Surova and O. Hansson. 2012. Non-motor symptoms in patients with Parkinson's disease—correlations with inflammatory cytokines in serum. *PLoS One.* 7: e47387.

Liu, T., P.C. Mcdonnell, P.R. Young, R.F. White, A.L. Siren, J.M. Hallenbeck, F.C. Barone and G.Z. Feurerstein. 1993. Interleukin-1-Beta Messenger-Rna Expression in Ischemic Rat Cortex. *Stroke.* 24: 1746–1751.

Loddick, S.A. and N.J. Rothwell. 1996. Neuroprotective effects of human recombinant interleukin-1 receptor antagonist in focal cerebral ischaemia in the rat. *J Cereb Blood Flow Metab.* 16: 932–40.

Loddick, S.A., A.V. Turnbull and N.J. Rothwell. 1998. Cerebral interleukin-6 is neuroprotective during permanent focal cerebral ischemia in the rat. *J Cereb Blood Flow Metab.* 18: 176–9.

Lopez-Cortes, L.F., M. Cruz-Ruiz, J. Gomez-Mateos, P. Viciana-Fernandez, F.J. Martinez-Marcos and J. Pachon. 1995. Interleukin-8 in cerebrospinal fluid from patients with meningitis of different etiologies: its possible role as neutrophil chemotactic factor. *J Infect Dis.* 172: 581–4.

Losy, J., J. Zaremba and P. Skrobanski. 2005. CXCL1 (GRO-alpha) chemokine in acute ischaemic stroke patients. *Folia Neuropathol.* 43: 97–102.

Lu, K.T., Y.W. Wang, J.T. Yang, Y.L. Yang and H.I. Chen. 2005. Effect of interleukin-1 on traumatic brain injury-induced damage to hippocampal neurons. *J Neurotrauma.* 22: 885–95.

Lund, B.T., N. Ashikian, H.Q. Ta, Y. Chakryan, K. Manoukian, S. Groshen, W. Gilmore, G.S. Cheema, W. Stohl, M.E. Burnett, D. Ko, N.J. Kachuck and L.P. Weiner. 2004. Increased CXCL8 (IL-8) expression in Multiple Sclerosis. *J Neuroimmunol.* 155: 161–71.

Maier, B., M. Lehnert, H.L. Laurer, A.E. Mautes, W.I. Steudel and I. Marzi. 2006. Delayed elevation of soluble tumor necrosis factor receptors p75 and p55 in cerebrospinal fluid and plasma after traumatic brain injury. *Shock.* 26: 122–7.

Marchetti, L., M. Klein, K. Schlett, K. Pfizenmaier and U.L. Eisel. 2004. Tumor necrosis factor (TNF)-mediated neuroprotection against glutamate-induced excitotoxicity is enhanced by N-methyl-D-aspartate receptor activation. Essential role of a TNF receptor 2-mediated phosphatidylinositol 3-kinase-dependent NF-kappa B pathway. *J Biol Chem.* 279: 32869–81.

Marchi, N., Q. Fan, C. Ghosh, V. Fazio, F. Bertolini, G. Betto, A. Batra, E. Carlton, I. Najm, T. Granata and D. Janigro. 2009. Antagonism of peripheral inflammation reduces the severity of status epilepticus. *Neurobiol Dis.* 33: 171–81.

Marklund, N., C. Keck, R. Hoover, K. Soltesz, M. Millard, D. LeBold, Z. Spangler, A. Banning, J. Benson and T.K. McIntosh. 2005. Administration of monoclonal antibodies neutralizing the inflammatory mediators tumor necrosis factor alpha and interleukin-6 does not attenuate acute behavioral deficits following experimental traumatic brain injury in the rat. *Restor Neurol Neurosci*. 23: 31–42.

Marz, P., J.G. Cheng, R.A. Gadient, P.H. Patterson, T. Stoyan, U. Otten and S. Rose-John. 1998. Sympathetic neurons can produce and respond to interleukin 6. *Proc Natl Acad Sci USA*. 95: 3251–6.

McClain, C., D. Cohen, R. Phillips, L. Ott and B. Young. 1991. Increased plasma and ventricular fluid interleukin-6 levels in patients with head injury. *J Lab Clin Med*. 118: 225–31.

McColl, B.W., N.J. Rothwell and S.M. Allan. 2007. Systemic inflammatory stimulus potentiates the acute phase and CXC chemokine responses to experimental stroke and exacerbates brain damage via interleukin-1- and neutrophil-dependent mechanisms. *J Neurosci*. 27: 4403–12.

McColl, B.W., N.J. Rothwell and S.M. Allan. 2008. Systemic inflammation alters the kinetics of cerebrovascular tight junction disruption after experimental stroke in mice. *J Neurosci*. 28: 9451–62.

McKeating, E.G., P.J. Andrews, D.F. Signorini and L. Mascia. 1997. Transcranial cytokine gradients in patients requiring intensive care after acute brain injury. *Br J Anaesth*. 78: 520–3.

Mehta, K., T. McQueen, S. Tucker, R. Pandita and B.B. Aggarwal. 1994. Inhibition by all-trans-retinoic acid of tumor necrosis factor and nitric oxide production by peritoneal macrophages. *J Leukoc Biol*. 55: 336–42.

Mikuniya, T., S. Nagai, M. Takeuchi, T. Mio, Y. Hoshino, H. Miki, M. Shigematsu, K. Hamada and T. Izumi. 2000. Significance of the interleukin-1 receptor antagonist/interleukin-1 beta ratio as a prognostic factor in patients with pulmonary sarcoidosis. *Respiration*. 67: 389–96.

Minambres, E., A. Cemborain, P. Sanchez-Velasco, M. Gandarillas, G. Diaz-Reganon, U. Sanchez-Gonzalez and F. Leyva-Cobian. 2003. Correlation between transcranial interleukin-6 gradient and outcome in patients with acute brain injury. *Crit Care Med*. 31: 933–8.

Monje, M.L., H. Toda and T.D. Palmer. 2003. Inflammatory blockade restores adult hippocampal neurogenesis. *Science*. 302: 1760–5.

Morganti-Kossmann, M.C., T. Kossmann and S.M. Wahl. 1992. Cytokines and neuropathology. *Trends Pharmacol Sci*. 13: 286–91.

Moro, M.A., A. Cardenas, O. Hurtado, J.C. Leza and I. Lizasoain. 2004. Role of nitric oxide after brain ischaemia. *Cell Calcium*. 36: 265–75.

Mosenthal, A.C., D.H. Livingston, R.F. Lavery, M.M. Knudson, S. Lee, D. Morabito, G.T. Manley, A. Nathens, G. Jurkovich, D.B. Hoyt and R. Coimbra. 2004. The effect of age on functional outcome in mild traumatic brain injury: 6-month report of a prospective multicenter trial. *J Trauma*. 56: 1042–8.

Mulcahy, N.J., J. Ross, N.J. Rothwell and S.A. Loddick. 2003. Delayed administration of interleukin-1 receptor antagonist protects against transient cerebral ischaemia in the rat. *Br J Pharmacol*. 140: 471–6.

Murphy, P.G., M.S. Ramer, L. Borthwick, J. Gauldie, P.M. Richardson and M.A. Bisby. 1999. Endogenous interleukin-6 contributes to hypersensitivity to cutaneous stimuli and changes in neuropeptides associated with chronic nerve constriction in mice. *Eur J Neurosci*. 11: 2243–53.

Mussack, T., P. Biberthaler, K.G. Kanz, E. Wiedemann, C. Gippner-Steppert, W. Mutschler and M. Jochum. 2002. Serum S-100B and interleukin-8 as predictive markers for comparative neurologic outcome analysis of patients after cardiac arrest and severe traumatic brain injury. *Crit Care Med*. 30: 2669–74.

Nakanishi, M., T. Niidome, S. Matsuda, A. Akaike, T. Kihara and H. Sugimoto. 2007. Microglia-derived interleukin-6 and leukaemia inhibitory factor promote astrocytic differentiation of neural stem/progenitor cells. *Eur J Neurosci*. 25: 649–58.

Nelson, K.B., J.M. Dambrosia, J.K. Grether and T.M. Phillips. 1998. Neonatal cytokines and coagulation factors in children with cerebral palsy. *Ann Neurol*. 44: 665–75.

Nguyen, H.X., T.J. O'Barr and A.J. Anderson. 2007. Polymorphonuclear leukocytes promote neurotoxicity through release of matrix metalloproteinases, reactive oxygen species, and TNF-alpha. *J Neurochem.* 102: 900–12.

Nicoletti, F., R. Di Marco, K. Mangano, F. Patti, E. Reggio, A. Nicoletti, K. Bendtzen and A. Reggio. 2001. Increased serum levels of interleukin-18 in patients with multiple sclerosis. *Neurology.* 57: 342–4.

Niijima, A., T. Hori, S. Aou and Y. Oomura. 1991. The effects of interleukin-1 beta on the activity of adrenal, splenic and renal sympathetic nerves in the rat. *J Auton Nerv Syst.* 36: 183–92.

Nijboer, C.H., C.J. Heijnen, F. Groenendaal, F. van Bel and A. Kavelaars. 2009. Alternate pathways preserve tumor necrosis factor-alpha production after nuclear factor-kappaB inhibition in neonatal cerebral hypoxia-ischemia. *Stroke.* 40: 3362–8.

Oh, J., M.A. McCloskey, C.C. Blong, L. Bendickson, M. Nilsen-Hamilton and D.S. Sakaguchi. 2010. Astrocyte-derived interleukin-6 promotes specific neuronal differentiation of neural progenitor cells from adult hippocampus. *J Neurosci Res.* 88: 2798–809.

Oh, Y.J. and K.L. O'Malley. 1994. IL-6 increases choline acetyltransferase but not neuropeptide transcripts in sympathetic neurons. *Neuroreport.* 5: 937–40.

Ojala, J., I. Alafuzoff, S.K. Herukka, T. van Groen, H. Tanila and T. Pirttila. 2009. Expression of interleukin-18 is increased in the brains of Alzheimer's disease patients. *Neurobiol Aging.* 30: 198–209.

Omari, K.M., G. John, R. Lango and C.S. Raine. 2006. Role for CXCR2 and CXCL1 on glia in multiple sclerosis. *Glia.* 53: 24–31.

Opp, M.R. and J.M. Krueger. 1991. Interleukin 1-receptor antagonist blocks interleukin 1-induced sleep and fever. *Am J Physiol.* 260: R453–7.

Ormstad, H., H.C. Aass, N. Lund-Sorensen, K.F. Amthor and L. Sandvik. 2011a. Serum levels of cytokines and C-reactive protein in acute ischemic stroke patients, and their relationship to stroke lateralization, type, and infarct volume. *J Neurol.* 258: 677–85.

Ormstad, H., H.C.D. Aass, N. Lund-Sorensen, K.F. Amthor and L. Sandvik. 2011b. Serum levels of cytokines and C-reactive protein in acute ischemic stroke patients, and their relationship to stroke lateralization, type, and infarct volume. *Journal of Neurology.* 258: 677–685.

Oto, J., A. Suzue, D. Inui, Y. Fukuta, K. Hosotsubo, M. Torii, S. Nagahiro and M. Nishimura. 2008. Plasma proinflammatory and anti-inflammatory cytokine and catecholamine concentrations as predictors of neurological outcome in acute stroke patients. *J Anesth.* 22: 207–12.

Ott, L., C.J. McClain, M. Gillespie and B. Young. 1994. Cytokines and metabolic dysfunction after severe head injury. *J Neurotrauma.* 11: 447–72.

Otto, V.I., U.E. Heinzel-Pleines, S.M. Gloor, O. Trentz, T. Kossmann and M.C. Morganti-Kossmann. 2000. sICAM-1 and TNF-alpha induce MIP-2 with distinct kinetics in astrocytes and brain microvascular endothelial cells. *J Neurosci Res.* 60: 733–42.

Oygur, N., O. Sonmez, O. Saka and O. Yegin. 1998. Predictive value of plasma and cerebrospinal fluid tumour necrosis factor-alpha and interleukin-1 beta concentrations on outcome of full term infants with hypoxic-ischaemic encephalopathy. *Arch Dis Child Fetal Neonatal Ed.* 79: F190–3.

Pang, Y., L.W. Fan, B. Zheng, Z. Cai and P.G. Rhodes. 2006. Role of interleukin-6 in lipopolysaccharide-induced brain injury and behavioral dysfunction in neonatal rats. *Neuroscience.* 141: 745–55.

Parker, L.C., G.N. Luheshi, N.J. Rothwell and E. Pinteaux. 2002. IL-1 beta signalling in glial cells in wildtype and IL-1RI deficient mice. *Br J Pharmacol.* 136: 312–20.

Patzer, A., Y. Zhao, I. Stock, P. Gohlke, T. Herdegen and J. Culman. 2008. Peroxisome proliferator-activated receptorsgamma (PPARgamma) differently modulate the interleukin-6 expression in the peri-infarct cortical tissue in the acute and delayed phases of cerebral ischaemia. *Eur J Neurosci.* 28: 1786–94.

Peltola, J., M. Hurme, A. Miettinen and T. Keranen. 1998. Elevated levels of interleukin-6 may occur in cerebrospinal fluid from patients with recent epileptic seizures. *Epilepsy Res.* 31: 129–33.

Peltola, J., J. Palmio, L. Korhonen, J. Suhonen, A. Miettinen, M. Hurme, D. Lindholm and T. Keranen. 2000. Interleukin-6 and interleukin-1 receptor antagonist in cerebrospinal fluid from patients with recent tonic-clonic seizures. *Epilepsy Res*. 41: 205–11.

Penkowa, M., M. Giralt, J. Carrasco, H. Hadberg and J. Hidalgo. 2000. Impaired inflammatory response and increased oxidative stress and neurodegeneration after brain injury in interleukin-6-deficient mice. *Glia*. 32: 271–85.

Penkowa, M., M. Giralt, N. Lago, J. Camats, J. Carrasco, J. Hernandez, A. Molinero, I.L. Campbell and J. Hidalgo. 2003. Astrocyte-targeted expression of IL-6 protects the CNS against a focal brain injury. *Exp Neurol*. 181: 130–48.

Perez-Barcena, J., J. Ibanez, M. Brell, C. Crespi, G. Frontera, J.A. Llompart-Pou, J. Homar and J.M. Abadal. 2011. Lack of correlation among intracerebral cytokines, intracranial pressure, and brain tissue oxygenation in patients with traumatic brain injury and diffuse lesions. *Crit Care Med*. 39: 533–40.

Pinteaux, E., P. Trotter and A. Simi. 2009. Cell-specific and concentration-dependent actions of interleukin-1 in acute brain inflammation. *Cytokine*. 45: 1–7.

Pistell, P.J., C.D. Morrison, S. Gupta, A.G. Knight, J.N. Keller, D.K. Ingram and A.J. Bruce-Keller. 2010. Cognitive impairment following high fat diet consumption is associated with brain inflammation. *J Neuroimmunol*. 219: 25–32.

Pitossi, F., A. del Rey, A. Kabiersch and H. Besedovsky. 1997. Induction of cytokine transcripts in the central nervous system and pituitary following peripheral administration of endotoxin to mice. *J Neurosci Res*. 48: 287–98.

Pleines, U.E., J.F. Stover, T. Kossmann, O. Trentz and M.C. Morganti-Kossmann. 1998. Soluble ICAM-1 in CSF coincides with the extent of cerebral damage in patients with severe traumatic brain injury. *J Neurotrauma*. 15: 399–409.

Popivanova, B.K., K. Koike, A.B. Tonchev, Y. Ishida, T. Kondo, S. Ogawa, N. Mukaida, M. Inoue and T. Yamashima. 2003. Accumulation of microglial cells expressing ELR motif-positive CXC chemokines and their receptor CXCR2 in monkey hippocampus after ischemia-reperfusion. *Brain Res*. 970: 195–204.

Price, C.J., D. Wang, D.K. Menon, J.V. Guadagno, M. Cleij, T. Fryer, F. Aigbirhio, J.C. Baron and E.A. Warburton. 2006. Intrinsic activated microglia map to the peri-infarct zone in the subacute phase of ischemic stroke. *Stroke*. 37: 1749–53.

Proost, P., A. Wuyts and J. van Damme. 1996. The role of chemokines in inflammation. *Int J Clin Lab Res*. 26: 211–23.

Quagliarello, V.J., B. Wispelwey, W.J. Long, Jr. and W.M. Scheld. 1991. Recombinant human interleukin-1 induces meningitis and blood-brain barrier injury in the rat. Characterization and comparison with tumor necrosis factor. *J Clin Invest*. 87: 1360–6.

Quintana, A., M. Giralt, A. Molinero, I.L. Campbell, M. Penkowa and J. Hidalgo. 2007. Analysis of the cerebral transcriptome in mice subjected to traumatic brain injury: importance of IL-6. *Neuroimmunomodulation*. 14: 139–43.

Rallidis, L.S., M.G. Zolindaki, M. Vikelis, K. Kaliva, C. Papadopoulos and D.T. Kremastinos. 2009. Elevated soluble intercellular adhesion molecule-1 levels are associated with poor short-term prognosis in middle-aged patients with acute ischaemic stroke. *Int J Cardiol*. 132: 216–20.

Ramaswamy, V., J. Horton, B. Vandermeer, N. Buscemi, S. Miller and J. Yager. 2009. Systematic review of biomarkers of brain injury in term neonatal encephalopathy. *Pediatr Neurol*. 40: 215–26.

Ramer, M.S., P.G. Murphy, P.M. Richardson and M.A. Bisby. 1998. Spinal nerve lesion-induced mechanoallodynia and adrenergic sprouting in sensory ganglia are attenuated in interleukin-6 knockout mice. *Pain*. 78: 115–21.

Ransohoff, R.M. and M.A. Brown. 2012. Innate immunity in the central nervous system. *J Clin Invest*. 122: 1164–71.

Ransohoff, R.M., T.A. Hamilton, M. Tani, M.H. Stoler, H.E. Shick, J.A. Major, M.L. Estes, D.M. Thomas and V.K. Tuohy. 1993. Astrocyte expression of mRNA encoding cytokines IP-10 and JE/MCP-1 in experimental autoimmune encephalomyelitis. *FASEB J*. 7: 592–600.

Reale, M., A. Patruno, M.A. De Lutiis, M. Pesce, M. Felaco, M. Di Giannantonio, M. Di Nicola and A. Grilli. 2011. Dysregulation of chemo-cytokine production in schizophrenic patients versus healthy controls. *BMC Neurosci.* 12: 13.

Relton, J.K. and N.J. Rothwell. 1992. Interleukin-1 receptor antagonist inhibits ischaemic and excitotoxic neuronal damage in the rat. *Brain Res Bull.* 29: 243–6.

Rhind, S.G., N.T. Crnko, A.J. Baker, L.J. Morrison, P.N. Shek, S. Scarpelini and S.B. Rizoli. 2010. Prehospital resuscitation with hypertonic saline-dextran modulates inflammatory, coagulation and endothelial activation marker profiles in severe traumatic brain injured patients. *J Neuroinflammation.* 7: 5.

Rhodes, J., J. Sharkey and P. Andrews. 2009. Serum IL-8 and MCP-1 concentration do not identify patients with enlarging contusions after traumatic brain injury. *J Trauma.* 66: 1591–7; discussion 1598.

Rhodes, J.K., P.J. Andrews, M.C. Holmes and J.R. Seckl. 2002. Expression of interleukin-6 messenger RNA in a rat model of diffuse axonal injury. *Neurosci Lett.* 335: 1–4.

Robinson, S., M. Tani, R.M. Strieter, R.M. Ransohoff and R.H. Miller. 1998. The chemokine growth-regulated oncogene-alpha promotes spinal cord oligodendrocyte precursor proliferation. *J Neurosci.* 18: 10457–63.

Rosenkilde, M.M. and T.W. Schwartz. 2004. The chemokine system—a major regulator of angiogenesis in health and disease. *APMIS.* 112: 481–95.

Rothwell, N. 2003. Interleukin-1 and neuronal injury: mechanisms, modification, and therapeutic potential. *Brain Behav Immun.* 17: 152–7.

Rothwell, N.J. 1994. CNS regulation of thermogenesis. *Crit Rev Neurobiol.* 8: 1–10.

Rubio-Perez, J.M. and J.M. Morillas-Ruiz. 2012. A review: inflammatory process in Alzheimer's disease, role of cytokines. *Scientific World Journal.* 756357.

Ryu, J.K. and J.G. McLarnon. 2008. Thalidomide inhibition of perturbed vasculature and glial-derived tumor necrosis factor-alpha in an animal model of inflamed Alzheimer's disease brain. *Neurobiol Dis.* 29: 254–66.

Sandhir, R., V. Puri, R.M. Klein and N.E. Berman. 2004. Differential expression of cytokines and chemokines during secondary neuron death following brain injury in old and young mice. *Neurosci Lett.* 369: 28–32.

Savman, K., M. Blennow, K. Gustafson, E. Tarkowski and H. Hagberg. 1998. Cytokine response in cerebrospinal fluid after birth asphyxia. *Pediatr Res.* 43: 746–51.

Schmitt, K.R., F. Boato, A. Diestel, D. Hechler, A. Kruglov, F. Berger and S. Hendrix. 2010. Hypothermia-induced neurite outgrowth is mediated by tumor necrosis factor-alpha. *Brain Pathol.* 20: 771–9.

Schneider Soares, F.M., N. Menezes de Souza, M. Liborio Schwarzbold, A. Paim Diaz, J. Costa Nunes, A. Hohl, P. Nunes Abreu da Silva, J. Vieira, R. Lisboa de Souza, M. More Bertotti, R.D. Schoder Prediger, M. Neves Linhares, A. Bafica and R. Walz. 2012. Interleukin-10 is an independent biomarker of severe traumatic brain injury prognosis. *Neuroimmunomodulation.* 19: 377–85.

Semple, B.D., N. Bye, J.M. Ziebell and M.C. Morganti-Kossmann. 2010. Deficiency of the chemokine receptor CXCR2 attenuates neutrophil infiltration and cortical damage following closed head injury. *Neurobiol Dis.* 40: 394–403.

Sharief, M.K. and R. Hentges. 1991. Association between tumor necrosis factor-alpha and disease progression in patients with multiple sclerosis. *N Engl J Med.* 325: 467–72.

Sifringer, M., V. Stefovska, S. Endesfelder, P.F. Stahel, K. Genz, M. Dzietko, C. Ikonomidou and U. Felderhoff-Mueser. 2007. Activation of caspase-1 dependent interleukins in developmental brain trauma. *Neurobiol Dis.* 25: 614–22.

Silveira, R.C. and R.S. Procianoy. 2003. Interleukin-6 and tumor necrosis factor-alpha levels in plasma and cerebrospinal fluid of term newborn infants with hypoxic-ischemic encephalopathy. *J Pediatr.* 143: 625–9.

Singhal, A., A.J. Baker, G.M. Hare, F.X. Reinders, L.C. Schlichter and R.J. Moulton. 2002. Association between cerebrospinal fluid interleukin-6 concentrations and outcome after severe human traumatic brain injury. *J Neurotrauma*. 19: 929–37.

Sotgiu, S., B. Zanda, B. Marchetti, M.L. Fois, G. Arru, G.M. Pes, F.S. Salaris, A. Arru, A. Pirisi and G. Rosati. 2006. Inflammatory biomarkers in blood of patients with acute brain ischemia. *Eur J Neurol*. 13: 505–13.

Spooren, A., K. Kolmus, G. Laureys, R. Clinckers, J. De Keyser, G. Haegeman and S. Gerlo. 2011. Interleukin-6, a mental cytokine. *Brain Res Rev*. 67: 157–83.

Stein, D.M., J.A. Kufera, A. Lindell, K.R. Murdock, J. Menaker, G.V. Bochicchio, B. Aarabi and T.M. Scalea. 2011a. Association of CSF biomarkers and secondary insults following severe traumatic brain injury. *Neurocrit Care*. 14: 200–7.

Stein, D.M., A. Lindell, K.R. Murdock, J.A. Kufera, J. Menaker, K. Keledjian, G.V. Bochicchio, B. Aarabi and T.M. Scalea. 2011b. Relationship of serum and cerebrospinal fluid biomarkers with intracranial hypertension and cerebral hypoperfusion after severe traumatic brain injury. *J Trauma*. 70: 1096–103.

Stein, D.M., A.L. Lindel, K.R. Murdock, J.A. Kufera, J. Menaker and T.M. Scalea. 2012. Use of serum biomarkers to predict secondary insults following severe traumatic brain injury. *Shock*. 37: 563–8.

Streit, W.J., S.D. Hurley, T.S. McGraw and S.L. Semple-Rowland. 2000. Comparative evaluation of cytokine profiles and reactive gliosis supports a critical role for interleukin-6 in neuron-glia signaling during regeneration. *J Neurosci Res*. 61: 10–20.

Stroemer, R.P. and N.J. Rothwell. 1998. Exacerbation of ischemic brain damage by localized striatal injection of interleukin-1beta in the rat. *J Cereb Blood Flow Metab*. 18: 833–9.

Suehiro, E., H. Fujisawa, T. Akimura, H. Ishihara, K. Kajiwara, S. Kato, M. Fujii, S. Yamashita, T. Maekawa and M. Suzuki. 2004. Increased matrix metalloproteinase-9 in blood in association with activation of interleukin-6 after traumatic brain injury: influence of hypothermic therapy. *J Neurotrauma*. 21: 1706–11.

Suffredini, A.F., G. Fantuzzi, R. Badolato, J.J. Oppenheim and N.P. O'Grady. 1999. New insights into the biology of the acute phase response. *J Clin Immunol*. 19: 203–14.

Sugama, S., S.A. Wirz, A.M. Barr, B. Conti, T. Bartfai and T. Shibasaki. 2004. Interleukin-18 null mice show diminished microglial activation and reduced dopaminergic neuron loss following acute 1-methyl-4-phenyl-1,2,3,6-tetrahydropyridine treatment. *Neuroscience*. 128: 451–8.

Svetlov, S.I., V. Prima, O. Glushakova, A. Svetlov, D.R. Kirk, H. Gutierrez, V.L. Serebruany, K.C. Curley, K.K. Wang and R.L. Hayes. 2012. Neuro-glial and systemic mechanisms of pathological responses in rat models of primary blast overpressure compared to "composite" blast. *Front Neurol*. 3: 15.

Swartz, K.R., F. Liu, D. Sewell, T. Schochet, I. Campbell, M. Sandor and Z. Fabry. 2001. Interleukin-6 promotes post-traumatic healing in the central nervous system. *Brain Res*. 896: 86–95.

Takahashi, H., M. Nishimura, M. Sakamoto, I. Ikegaki, T. Nakanishi and M. Yoshimura. 1992. Effects of interleukin-1 beta on blood pressure, sympathetic nerve activity, and pituitary endocrine functions in anesthetized rats. *Am J Hypertens*. 5: 224–9.

Tarkowski, E., K. Blennow, A. Wallin and A. Tarkowski. 1999. Intracerebral production of tumor necrosis factor-alpha, a local neuroprotective agent, in Alzheimer disease and vascular dementia. *J Clin Immunol*. 19: 223–30.

Tarkowski, E., A.M. Liljeroth, L. Minthon, A. Tarkowski, A. Wallin and K. Blennow. 2003. Cerebral pattern of pro- and anti-inflammatory cytokines in dementias. *Brain Research Bulletin*. 61: 255–260.

Tarkowski, E., A.M. Liljeroth, A. Nilsson, L. Minthon and K. Blennow. 2001. Decreased levels of intrathecal interleukin 1 receptor antagonist in Alzheimer's disease. *Dementia and Geriatric Cognitive Disorders*. 12: 314–317.

Tarkowski, E., L. Rosengren, C. Blomstrand, C. Wikkelso, C. Jensen, S. Ekholm and A. Tarkowski. 1995. Early intrathecal production of interleukin-6 predicts the size of brain lesion in stroke. *Stroke*. 26: 1393–8.

Thornton, P., B.W. McColl, A. Greenhalgh, A. Denes, S.M. Allan and N.J. Rothwell. 2010. Platelet interleukin-1alpha drives cerebrovascular inflammation. *Blood*. 115: 3632–9.

Timaru-Kast, R., C. Luh, P. Gotthardt, C. Huang, M.K. Schafer, K. Engelhard and S.C. Thal. 2012. Influence of age on brain edema formation, secondary brain damage and inflammatory response after brain trauma in mice. *PLoS One*. 7: e43829.

Touzani, O., H. Boutin, J. Chuquet and N. Rothwell. 1999. Potential mechanisms of interleukin-1 involvement in cerebral ischaemia. *J Neuroimmunol*. 100: 203–15.

Tsai, H.H., E. Frost, V. To, S. Robinson, C. Ffrench-Constant, R. Geertman, R.M. Ransohoff and R.H. Miller. 2002. The chemokine receptor CXCR2 controls positioning of oligodendrocyte precursors in developing spinal cord by arresting their migration. *Cell*. 110: 373–83.

Tsakiri, N., I. Kimber, N.J. Rothwell and E. Pinteaux. 2008. Differential effects of interleukin-1 alpha and beta on interleukin-6 and chemokine synthesis in neurones. *Mol Cell Neurosci*. 38: 259–65.

Valerio, A., M. Ferrario, M. Dreano, G. Garotta, P. Spano and M. Pizzi. 2002. Soluble interleukin-6 (IL-6) receptor/IL-6 fusion protein enhances *in vitro* differentiation of purified rat oligodendroglial lineage cells. *Mol Cell Neurosci*. 21: 602–15.

Valles, A., L. Grijpink-Ongering, F.M. de Bree, T. Tuinstra and E. Ronken. 2006. Differential regulation of the CXCR2 chemokine network in rat brain trauma: implications for neuroimmune interactions and neuronal survival. *Neurobiol Dis*. 22: 312–22.

Van Meir, E., M. Ceska, F. Effenberger, A. Walz, E. Grouzmann, I. Desbaillets, K. Frei, A. Fontana and N. de Tribolet. 1992. Interleukin-8 is produced in neoplastic and infectious diseases of the human central nervous system. *Cancer Res*. 52: 4297–305.

Vezzani, A., M. Conti, A. De Luigi, T. Ravizza, D. Moneta, F. Marchesi and M.G. De Simoni. 1999. Interleukin-1beta immunoreactivity and microglia are enhanced in the rat hippocampus by focal kainate application: functional evidence for enhancement of electrographic seizures. *J Neurosci*. 19: 5054–65.

Vezzani, A., D. Moneta, M. Conti, C. Richichi, T. Ravizza, A. De Luigi, M.G. De Simoni, G. Sperk, S. Andell-Jonsson, J. Lundkvist, K. Iverfeldt and T. Bartfai. 2000. Powerful anticonvulsant action of IL-1 receptor antagonist on intracerebral injection and astrocytic overexpression in mice. *Proceedings of the National Academy of Sciences of the United States of America*. 97: 11534–11539.

Vila, N., J. Castillo, A. Davalos and A. Chamorro. 2000. Proinflammatory cytokines and early neurological worsening in ischemic stroke. *Stroke*. 31: 2325–9.

Vinukonda, G., A. Csiszar, F. Hu, K. Dummula, N.K. Pandey, M.T. Zia, N.R. Ferreri, Z. Ungvari, E.F. LaGamma and P. Ballabh. 2010. Neuroprotection in a rabbit model of intraventricular haemorrhage by cyclooxygenase-2, prostanoid receptor-1 or tumour necrosis factor-alpha inhibition. *Brain*. 133: 2264–80.

Wang, X., F.C. Barone, N.V. Aiyar and G.Z. Feuerstein. 1997. Interleukin-1 receptor and receptor antagonist gene expression after focal stroke in rats. *Stroke*. 28: 155–61; discussion 161–2.

Wang, X., G.Z. Feuerstein, J.L. Gu, P.G. Lysko and T.L. Yue. 1995. Interleukin-1 beta induces expression of adhesion molecules in human vascular smooth muscle cells and enhances adhesion of leukocytes to smooth muscle cells. *Atherosclerosis*. 115: 89–98.

Wang, X., T.L. Yue, F.C. Barone, R.F. White, R.C. Gagnon and G.Z. Feuerstein. 1994. Concomitant cortical expression of TNF-alpha and IL-1 beta mRNAs follows early response gene expression in transient focal ischemia. *Mol Chem Neuropathol*. 23: 103–14.

Welling, T.H., S. Fu, S. Wan, W. Zou and J.A. Marrero. 2012. Elevated serum IL-8 is associated with the presence of hepatocellular carcinoma and independently predicts survival. *Cancer Invest*. 30: 689–97.

Welsh, P., G.D. Lowe, J. Chalmers, D.J. Campbell, A. Rumley, B.C. Neal, S.W. MacMahon and M. Woodward. 2008. Associations of proinflammatory cytokines with the risk of recurrent stroke. *Stroke*. 39: 2226–30.

Weschenfelder, J., C. Sander, M. Kluge, K.C. Kirkby and H. Himmerich. 2012. The influence of cytokines on wakefulness regulation: clinical relevance, mechanisms and methodological problems. *Psychiatr Danub.* 24: 112–26.

Whalen, M.J., T.M. Carlos, P.M. Kochanek, S.R. Wisniewski, M.J. Bell, R.S. Clark, S.T. DeKosky, D.W. Marion and P.D. Adelson. 2000. Interleukin-8 is increased in cerebrospinal fluid of children with severe head injury. *Crit Care Med.* 28: 929–34.

Winter, C.D., A.K. Pringle, G.F. Clough and M.K. Church. 2004. Raised parenchymal interleukin-6 levels correlate with improved outcome after traumatic brain injury. *Brain.* 127: 315–20.

Woiciechowsky, C., B. Schoning, J. Cobanov, W.R. Lanksch, H.D. Volk and W.D. Docke. 2002. Early IL-6 plasma concentrations correlate with severity of brain injury and pneumonia in brain-injured patients. *J Trauma.* 52: 339–45.

Woodroofe, M.N., G.S. Sarna, M. Wadhwa, G.M. Hayes, A.J. Loughlin, A. Tinker and M.L. Cuzner. 1991. Detection of interleukin-1 and interleukin-6 in adult rat brain, following mechanical injury, by *in vivo* microdialysis: evidence of a role for microglia in cytokine production. *J Neuroimmunol.* 33: 227–36.

Wu, Q., R.H. Miller, R.M. Ransohoff, S. Robinson, J. Bu and A. Nishiyama. 2000. Elevated levels of the chemokine GRO-1 correlate with elevated oligodendrocyte progenitor proliferation in the jimpy mutant. *J Neurosci.* 20: 2609–17.

Xia, M. and B.T. Hyman. 2002. GROalpha/KC, a chemokine receptor CXCR2 ligand, can be a potent trigger for neuronal ERK1/2 and PI-3 kinase pathways and for tau hyperphosphorylation-a role in Alzheimer's disease? *J Neuroimmunol.* 122: 55–64.

Yamaoka-Tojo, M., T. Tojo, T. Inomata, Y. Machida, K. Osada and T. Izumi. 2002. Circulating levels of interleukin 18 reflect etiologies of heart failure: Th1/Th2 cytokine imbalance exaggerates the pathophysiology of advanced heart failure. *J Card Fail.* 8: 21–7.

Yamasaki, Y., Y. Matsuo, J. Zagorski, N. Matsuura, H. Onodera, Y. Itoyama and K. Kogure. 1997. New therapeutic possibility of blocking cytokine-induced neutrophil chemoattractant on transient ischemic brain damage in rats. *Brain Res.* 759: 103–11.

Yamasaki, Y., N. Matsuura, H. Shozuhara, H. Onodera, Y. Itoyama and K. Kogure. 1995. Interleukin-1 as a pathogenetic mediator of ischemic brain damage in rats. *Stroke.* 26: 676–80; discussion 681.

Yang, G.Y., X.H. Liu, C. Kadoya, Y.J. Zhao, Y. Mao, B.L. Davidson and A.L. Betz. 1998. Attenuation of ischemic inflammatory response in mouse brain using an adenoviral vector to induce overexpression of interleukin-1 receptor antagonist. *J Cereb Blood Flow Metab.* 18: 840–7.

Youn, Y.A., S.J. Kim, I.K. Sung, S.Y. Chung, Y.H. Kim and I.G. Lee. 2012. Serial examination of serum IL-8, IL-10 and IL-1Ra levels is significant in neonatal seizures induced by hypoxic-ischaemic encephalopathy. *Scand J Immunol.* 76: 286–93.

Youssef, A.A., L.T. Chang, C.L. Hang, C.J. Wu, C.I. Cheng, C.H. Yang, J.J. Sheu, H.T. Chai, S. Chua, K.H. Yeh and H.K. Yip. 2007. Level and value of interleukin-18 in patients with acute myocardial infarction undergoing primary coronary angioplasty. *Circ J.* 71: 703–8.

Yuen, C.M., C.A. Chiu, L.T. Chang, C.W. Liou, C.H. Lu, A.A. Youssef and H.K. Yip. 2007. Level and value of interleukin-18 after acute ischemic stroke. *Circ J.* 71: 1691–6.

Zaremba, J. and J. Losy. 2003. Interleukin-18 in acute ischaemic stroke patients. *Neurol Sci.* 24: 117–24.

Zhang, P.L., M. Izrael, E. Ainbinder, L. Ben-Simchon, J. Chebath and M. Revel. 2006. Increased myelinating capacity of embryonic stem cell derived oligodendrocyte precursors after treatment by interleukin-6/soluble interleukin-6 receptor fusion protein. *Mol Cell Neurosci.* 31: 387–98.

Zhang, X.M., R.S. Duan, Z. Chen, H.C. Quezada, E. Mix, B. Winblad and J. Zhu. 2007. IL-18 deficiency aggravates kainic acid-induced hippocampal neurodegeneration in C57BL/6 mice due to an overcompensation by IL-12. *Exp Neurol.* 205: 64–73.

Zhang, Z., M. Chopp, A. Goussev and C. Powers. 1998. Cerebral vessels express interleukin 1beta after focal cerebral ischemia. *Brain Res.* 784: 210–7.

Zhao, D., N. Hou, M. Cui, Y. Liu, X. Liang, X. Zhuang, Y. Zhang, L. Zhang, D. Yin, L. Gao and C. Ma. 2011. Increased T cell immunoglobulin and mucin domain 3 positively correlate with systemic IL-17 and TNF-alpha level in the acute phase of ischemic stroke. *J Clin Immunol.* 31: 719–27.

Zhou, Q.H., R. Sumbria, E.K. Hui, J.Z. Lu, R.J. Boado and W.M. Pardridge. 2011. Neuroprotection with a brain-penetrating biologic tumor necrosis factor inhibitor. *J Pharmacol Exp Ther.* 339: 618–23.

11

Biomarkers in Spinal Cord Injury

Shoji Yokobori,[1,2] Khadil Hosein,[1] Michael Y. Wang,[1]
Shyam Gajavelli,[1] Ahmed Moghieb,[2] Zhiqun Zhang,[3]
Kevin K.W. Wang,[3,4] M. Ross Bullock[1] and
W. Dalton Dietrich[1,]*

INTRODUCTION

Severe Central Nervous System (CNS) injuries, including Spinal Cord Injury (SCI), are common causes of disability resulting from trauma. In CNS injury, the important factors which determine the prognosis of patients are the severity of the primary injury (at the time of traumatic impact) and the ensuing secondary injury. Successful management of SCI also depends on accurate diagnosis of primary injury and prevention of delayed secondary injury. Recently innovations in the field of biomarkers have been made; as yet there is no consensus on their clinical use for SCI. The aims of this chapter are therefore to evaluate the current status of protein biomarkers in SCI, and to discuss their potential diagnostic and prognostic value for SCI. We will also present novel biomarker data from CSF and plasma in SCI, both in animal models and patients.

[1] Department of Neurosurgery, University of Miami Miller School of Medicine, Miami, Florida, USA.
[2] Department of Emergency and Critical Care Medicine, Nippon Medical School, Tokyo, Japan.
[3] Center for Neuroproteomics & Biomarkers Research, Department of Psychiatry, University of Florida, Gainesville, FL 32611, USA.
[4] Neuroscience and Physiological Science, University of Florida, Gainesville, FL 32611, USA.
* Corresponding author: DDietrich@med.miami.edu

Why do we Need an SCI biomarker?—The Limitations of Imaging

The clinical diagnosis of spinal trauma has been traditionally based mainly on the neurological symptoms. More recently, diagnosis of SCI has been greatly relied on the imaging procedures, including Magnetic Resonance Imaging (MRI). Conventional MRI is currently the best imaging modality for evaluating SCI during the acute phase. However, in many circumstances, MRI is not available. Even when it is accessible, patients with concomitant multiple injuries maybe too unstable for an early MRI due to the delay in medical care for up to an hour. Also, MRI cannot be used in the patients with medical implants and devices, such as Implantable Cardioverter Defibrillator (ICD), pacemakers, deep brain stimulation devices, or bullets or shrapnel. In addition to these limitations in MRI accessibility, there is a well-known phenomenon of Spinal Cord Injury Without Radiographic Abnormality (SCIWORA), which has been well described in previous literature (Lustrin et al., 2003; Pang and Wilberger, 1982). SCIWORA, may present as complete paralysis, but without obvious radiographic abnormality. Spinal ischemia may also account for SCIWORA, either through direct injury of spinal vessels or through transient hypoperfusion of the spinal cord (Choi et al., 1986). These problems in image diagnosis may present challenges in early diagnosis of SCI. Consequently, the concept of a biomarker-based diagnosis would be clearly of a significant benefit in supporting a diagnosis.

Diagnostic and Prognostic Property of SCI Biomarkers

Currently, there still are not many reports focused on diagnostic or prognostic biomarkers in SCI. Guez et al. (2003) first reported the diagnostic property of a biomarker in SCI. CSF concentrations of NFL (neurofilament light chain) and GFAP (Glial Fibrillary Acid Protein) were analyzed in acute SCI patients, and compared to a control group. All six cases with cord injury and pronounced neurological deficits showed significantly increased concentrations of both GFAP and NFL in CSF (Guez et al., 2003), in this landmark study.

In 2009, Kwon et al. (2011) conducted a larger prospective study to assess prognostic aspects of biomarkers in human acute SCI patients. They obtained CSF and blood samples over a period of 72 hours in 27 SCI patients with complete SCI (American Spinal Injury Association or ASIA A) or incomplete SCI (ASIA B or C). Cytokines and proteins were measured and compared among three injury severities (ASIA A, B, and C). The CSF concentrations of three putative biomarkers (S100β, GFAP, IL-8) were able to classify the injury severity of the patients, and predicted motor outcome better at 6 months than the standard clinical serial evaluation (Kwon et al., 2011). Hayakawa et al. (2012) also evaluated plasma NFH (Neurofilament heavy chain) concentrations in 14 acute cervical SCI patients. Plasma NFH values were compared between different ASIA Impairment Scale (AIS) grades. In patients with complete SCI (ASIA A), NFH became detectable at 12 hours after injury and remained elevated at 21 days after injury. They concluded NFH was a potential biomarker to independently distinguish complete from incomplete SCI in patients.

Candidate Biomarkers for SCI Biomarkers
(see also Table 11.1)

Neuron-Specific Enolase (NSE)

There have been numerous animal and human studies describing NSE as a biomarker in SCI. In 2005, Loy et al. (2005) evaluated serum levels of NSE and S-100β protein in 34 rats after weight drop SCI. At 6 hours after injury, the serum concentration of NSE and S-100β were significantly elevated, compared with the control rat group. Furthermore, Cao et al. (2008) used 60 SCI rats and evaluated the protein levels of NSE and S-100β in serum and CSF. Compared with the control group, the protein levels of NSE and S-100β in serum and CSF significantly increased from 2 hours after injury (p <0.05) and reached a maximum at 6 hours after SCI. Also, the protein levels of NSE and S-100β in serum and CSF were closely related to the severity of injury level (p <0.05). They conclude that protein levels of NSE and S-100β in serum and CSF significantly increased after experimental spinal cord injury in a time-dependent manner and thus may be considered relatively specific biomarkers for acute SCI.

In another recent animal study, expression levels of Myelin Basic Protein (MBP), Neuron-Specific Enolase (NSE), and the glial cytoplasmic protein S-100β were investigated in the serum and cerebrospinal fluid, using enzyme-linked immunosorbent assays, in a SCI rat model with Behind Armor Blunt Trauma (BABT) (Zhang et al., 2011). The concentrations of neuron-specific enolase, myelin basic protein, and S-100β were significantly increased in the serum and cerebrospinal fluid 3 hours after trauma (p <0.05) (Zhang et al., 2011).

Thus, there have been several experimental studies describing the usefulness of NSE as a SCI "structural biomarker" protein. However, there are still limited clinical reports on SCI.

Brisby et al. (1999) measured CSF-NSE concentration in 15 patients who underwent surgery due to a lumbar disc herniation and the levels

Table 11.1. Potential spinal cord biomarkers and their attributes.

Candidate Biomarkers for SCI	Marker Origin	Attributes	Use in Animal SCI Model	Use in Human SCI Study	Main References
Neuron-specific enolase (NSE)	Cytoplasm of neurons	Neural damage marker	Serum/CSF	CSF	Loy et al., 2005; Cao et al., 2008; Brisby et al.,1999
S100β	Glia	CNS injury marker benchmark	CSF/Serum	CSF/Serum	Loy et al., 2005; Cao et al., 2008; Ma et al., 2001; Kwon et al., 2010
Neurofilament proteins (NFL, NFM, NFH)	Cytoskeletal component	Axonal injury markers	Tissue/Serum	CSF/Serum	Kang et al., 2006; Ueno et al., 2011; Guez et al., 2003; Hayakawa et al., 2012
Cleaved-Tau (c-Tau)	Assembles axonal microtubule bundles	Axonal injury markers	Tissue	CSF	Shiiya et al., 2004; Kwon et al., 2010
Myelin basic protein (MBP)	Oligodendrocytes/Schwann cells	Demyelination	Serum/CSF	CSF/Serum	Zhang et al., 2011
Microtubule-associated protein 2 (MAP2)	Dendrites	Dendritic injury	Tissue	-	Zhang et al., 2000
Glial fibrillary acidic protein (GFAP)	Glia	Gliosis/Glial cell injury	Tissue*/CSF*	CSF*/Serum*	Guez et al., 2003; Winnerkvist et al., 2007 *From our pilot studies
Ubiquitin C-terminal hydrolase-L1(UCH-L1)	Neuronal cell body	Neuronal cell body injury	CSF*	CSF*/Serum*	*From our pilot studies
SBDP150/SBDP145	Axon (calpain-generated)	Acute necrosis	Tissue	CSF*/Serum*	Schumacher et al., 1999 *From our pilot studies
SBDP120	Axon (caspase-3-generated)	Delayed apoptosis	-	CSF*/Serum*	*From our pilot studies

compared with those seven patients without lumbar disc herniation. NSE concentration in CSF was higher in lumbar disc herniated patients than control. However, as far as we know, there have been no reports in human trauma cases or SCI. In addition, NSE has a problem with specificity because this structural biomarker is also present in red blood cells and platelets. This may cause difficulty with translation from basic to clinical situations, as it would be confounding especially in trauma patients. The use of NSE as a biomarker thus seems to be limited in traumatic SCI patients. Further studies are needed to improve specificity.

S100β

In the field of SCI, there are also several basic and clinical reports relating S100β. Ma et al. (2001) measured serum S100 in 40 SCI injured rats with weight drop injury. After induction of spinal cord injury, the concentration of S100β rapidly increased and within 72 hours had reached a concentration about five times that of the control animals. These results suggested that the concentration of S100 protein in serum might be useful as an early diagnostic tool for detecting neuronal damage caused by spinal cord injury (Ma et al., 2001). Kwon et al. (2010), also reported the potential of serum and CSF S100β concentration as a good candidate biomarker in human SCI.

One problem relating to the use of S100β has been its lower specificity. Although S100β is glial and neuron specific and is expressed primarily by astrocytes and Schwann cells, it is also found in several non–nervous system cells, such as adipocytes (white and brown fat), chondrocytes, skin, glioblastoma and melanoma cells (Zimmer et al., 1995). The presence of multiple extracranial injuries can influence the levels of S100β, yielding inadequate or even confusing results. This might be one of the reason why the use of S100β has not spread widely especially in the context of acute cord trauma or cord injury.

Neurofilaments

Neurofilament (NF) is a major cytoskeletal component in axons. The NF heteropolymer consists of light chain (NFL, 68 kDa), medium chain (NFM, 150 kDa), heavy chain (NFH, 190–210 kDa), and α-internexin polypeptides (Petzold, 2005). One experimental study with a SCI rat model revealed that NFM was upregulated between 6–24 hours after injury (Kang et al., 2006). After spinal axonal injury in rats, significant serum levels of NFH fragments were seen at as early as 6 hours; the levels peaked between 12 and 48 hours and then decreased to baseline levels, by 7 days (Shaw et al., 2005). The two peaks in the serum levels of NFH fragments corresponded to primary and secondary axotomy, respectively. But because the secondary axotomy was more severe than the primary, the second peak was sharper than the first (Shaw et al., 2005). As blood sampling is faster, safer, and more convenient than CSF, the assay of serum NFH has the potential to be a novel, specific, and convenient tool for assessing axonal damage in SCI patients (Anderson et al., 2008).

Cleaved-Tau (c-Tau)

There is still little evidence of c-Tau in SCI patients. In one study, CSF tau levels were evaluated in 28 patients undergoing complex aortic surgery and the values were compared with the incidence and severity of neurological complications. In this study, Tau had excellent sensitivity for detecting a cerebral neurological complication, including cerebral infarction. However, the Tau levels were not significantly elevated in patients with postoperative spinal cord ischemia and paraparesis when compared with the group without any neurologic complications (Shiiya et al., 2004). The authors speculated that this result could be explained by the difference in the relative volume of the gray and white matter between the brain and the spinal cord (Shiiya et al., 2004). Because the white matter, which is rich in Tau proteins, occupies a larger part in the brain than in the spinal cord, brain injury is more likely to be associated with elevated CSF Tau levels than in the spinal cord ischemia cases (Shiiya et al., 2004).

In another human SCI study, Tau protein measurement was performed upon CSF samples, obtained over a period of 72 hours, in 27 patients with complete or incomplete SCI. The CSF Tau concentrations were elevated in a severity-dependent fashion (Kwon et al., 2010). Based on these studies Tau protein clearly merits more research to establish its diagnostic potential.

Myelin Basic Protein (MBP)

At present, only a single report exists relating to the concentrations of myelin basic protein in an SCI animal model (Zhang et al., 2011). The concentrations of MBP was significantly increased in the serum and cerebrospinal fluid 3 hours after trauma in a Behind Armor Blunt Trauma (BABT)—SCI model (Zhang et al., 2011), designed to mimic military blast SCI.

Microtubule-associated Protein 2 (MAP2)

In SCI, there is only minimal basic and clinical research regarding MAP2 as a biomarker. In one animal study, the extent and time course of MAP2 loss were examined following SCI (Zhang et al., 2000). Within 1 to 6 hours following SCI, there is rapid loss of MAP2 at the injury site. This study suggests that MAP2 may have potential as a structural biomarker in TBI.

Glial Fibrillary Acidic Protein (GFAP)

In SCI, some human studies supporting the use of GFAP as a biomarker have been done. Guez et al. (2003) measured CSF GFAP concentration in six traumatic SCI patients, and compared with healthy controls. This result did not show significant elevation of GFAP concentration in SCI group. Winnerkvist et al. (2007) also measured CSF GFAP concentration in 39 patients undergoing elective thoracoabdominal aortic aneurysm surgery and compared the values in patients with or without ischemic complications of the spinal cord during the postoperative period. The patients with spinal cord symptoms had significant increases of CSF biomarkers GFAP (571-fold), NFL (14-fold) and S100β (18-fold) compared to asymptomatic patients, during the postoperative period. This report concluded that GFAP is the most promising biomarker for identifying patients at risk for postoperative delayed paraplegia (Winnerkvist et al., 2007), after aortic aneurysm surgery.

Ubiquitin C-terminal Hydrolase-L1 (UCH-L1)

Ubiquitin C-terminal hydrolase-L1 (UCH-L1), also called neuronal-specific protein gene product (PGP 9.3), is highly abundant in the neuronal soma (Jackson and Thompson, 1981; Papa et al., 2010). Recently, Mondello et al. (2012) reported a case-control study in which 95 severe TBI subjects were enrolled, and they studied the temporal profile of CSF and serum UCH-L1 levels over 7 days for these severe TBI patients. They concluded serum and CSF levels of UCH-L1 appear to have potential clinical utility in diagnosing TBI, including correlating to injury severity and survival outcome (Mondello et al., 2012). In the field of SCI, there are however no published reports relating to the potential utility of UCH-L1 as biomarker.

α-II Spectrin Breakdown Products (SBDPs)-SDBP120, 145 and 150

The α-II spectrin protein forms part of the axolemmal cytoskeleton, anchoring to ankyrin, which links axolemma constituents and the presynaptic terminal (Buki et al., 1999; Reeves et al., 2010), and stabilizes the nodal and paranodal structure of myelinated axons (Reeves et al.,

2010). The latest study showed the efficacy of measurement of SBDPs for diagnosis and outcome in severe traumatic brain injury patients (Mondello et al., 2011; Mondello et al., 2010a; Pike et al., 2001).

In view of the previous data from traumatic brain injury, SBDPs also seem to be good candidates in SCI. However, for the present, there are no published clinical reports proposing SBDPs as biomarkers in SCI patients. Clearly further investigations are needed in animal SCI models and humans.

New Findings in SCI Biomarker Candidates

As we mentioned above, diagnostic or prognostic protein biomarkers remain poorly studied in the field of SCI. Innovation of reliable SCI biomarkers is an intense topic of research for many studies. With the advances in "High-throughput" detection methods, such as mass spectrophotometry combined with immunoblotting, ELISA, proteomics and genomics, the field of biomarker research has very rapid progressed in the last few years. Our group also has recently been working to develop SCI biomarkers. In next paragraph, we report pilot data from our studies.

SBDPs, GFAP, MBP, and UCH-L1 as Potential SCI Biomarkers in a Rat SCI Model

Methodology of SCI

Adult female Fischer rats (220–250 gms) were used for the weight drop SCI rat model. Briefly, the anesthetized and ventilated rat is placed ventrally on sterile gauze, and the T9–10 surgical site is exposed and prepared. A 2 cm longitudinal skin incision is next centered over the T9 spinal process along the midline, perispinal nerves and ligaments are laterally dissected and retracted followed by removal of the bony elements of the posterior spine including the spinus process, and lamina, using a micro rongeurs. The ninth thoracic spinal segment is then exposed without removing the dura by dorsal laminectomy. The exposed cord is then contused by a 10 gm weight dropped at a height of either 12.5 mm (moderate), or 25 mm (severe) by using the quantitative New York University (NYU)-MASCIS impactor (Gruner, 1992). This model and its injury severities have been shown in our laboratory as well as by others to produce a well-described pattern of electrophysiological, behavioral and histopathological consequences (Agrawal et al., 2010) and paraparesis.

Rat Spinal Cord Tissue Collection and Processing

Following serum collection, spinal cord tissue was collected after extending the laminectomy to allow three segments of tissue to be removed: (i) rostral to injury site, (ii) epicenter (injury site), and (iii) caudal to injury site. Each segment was approximately 6–7 mm in length. The fresh tissue was rinsed gently with cold Phosphate-Buffered Saline (PBS) and placed in a microcentrifuge or cryofreeze tube with cap, snap-frozen on dry ice or liquid nitrogen, and stored at –80°C until used. Spinal cord regions (caudal, rostral and epicenter) thus extracted were pulverized in liquid nitrogen. The powder was then homogenized in lysis buffer containing 50 mM Tris (pH 7.4), 5 mM EDTA, 1% (v/v) Triton X-100, 1 mM DTT, 1x protease inhibitor cocktail (Roche). The protein concentration of the supernatant was measured by DC Protein Assay (Bio-Rad, Hercules, CA).

Following SCI, CSF and blood samples were drawn at 4, 24 or 48 hours post trauma.

Biomarker Sandwich ELISA Biomarker and Autoantibody Analysis (CSF and Serum)

UCH-L1, SBDP150 (similar to SBDP145), SBDP120, GFAP sandwich ELISA (swELISA) were performed as described in our recently studies (Liu et al., 2010b; Mondello et al., 2012; Mondello et al., 2010b; Papa et al., 2010; Papa et al., 2011). These assays have a detection limit of at least 0.1 ng/ml and samples with inter-assay and intra-assay CV of 11–13 and 9–8% respectively (Liu et al., 2010a; Mondello et al., 2010b; Papa et al., 2010). Commercial MBP sandwich ELISA (IPOC) were performed as based on manufacturer's instructions.

Results in SCI Experimental Model

In rat spinal cord tissue we found increases in the proteolytic biomarkers of alphaII-spectrin—SBDP150 and SBDP145. An Increase in SBDP120 generated by caspase-3 and increase of GFAP (gliosis) and GFAP-BDP (38 K; glial injury) and MBP fragment (myelin damage) in rat spinal cord tissue (caudal to epicenter) 6 hours after experimental SCI were also observed (Figure 11.1). These patterns were very similar to what we discovered in TBI (Liu et al., 2006; Papa et al., 2011; Pike et al., 2004; Pike et al., 2001). In addition, in rat CSF samples after SCI, we performed immunblotting assays for SBDP150, SBDP120 and UCH-L1 biomarker release into CSF 4 h, 24 hours to 48 hours after injury (Figure 11.2A). These studies demonstrated

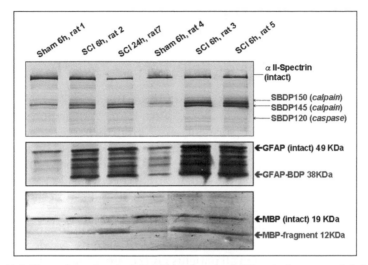

Figure 11.1. Tissue-based proteolytic biomarkers for SCI in rats: Increases of all-Spectrin-breakdown products (BDP): SBDP150 & SBDP145 generated by calpain. SBDP120 generated by caspase-3, increase of intact GFAP (gliosis) and GFAP-BDP (38 K; glial injury) and MBP fragment (myelin damage) in rat spinal cord tissue 6 hours after experimental SCI (caudal to epicenter) by immunoblotting (20 μg protein loading for tissue lysate) were observed.

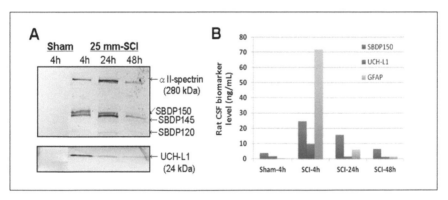

Figure 11.2. all-Spectrin-BDPs, UCH-L1, GFAP biomarker release into rat CSF after experimental spinal cord injury. Biomarkers were detected by (A) immunblotting showing increase release of all-spectrin breakdown products SBDP150, SBDP145, SBDP120 and UCH-L1 at 4, 24 and 48 hours after injury, and (B) ELISA showing increase CSF levels of SBDP150, UCH-L1 and GFAP at 4, 24 and 48 hours after SCI. Results shown are representative of two experiments.

Color image of this figure appears in the color plate section at the end of the book.

that SBDP150/145, UCH-L1 and GFAP levels are elevated in CSF as early as 4 hours after SCI in rats (Figure 11.2B).

SBDPs, GFAP, MBP, and UCH-L1 are Promising Biomarkers in Human SCI

Methods of Biofluid Sample Collection

SCI patients (moderate-severe, AIS Grades A, B & C; n=7) were recruited and classified according to American Spinal Injury Association scale (AIS) of impairment (degree of impairment) and AIS on discharge (improvement). At 30 days and 90 days, these SCI patients were reassessed by the AIS.

Fifteen non-SCI CSF control samples collected intraoperatively from either hydrocephalic patients with Ventriculo-Peritoneal (VP) shunts or patients with unruptured aneurysm. Single CSF and serum samples were collected from this group.

Fifteen normal control serum samples (n=15 individuals) were also obtained from a commercial source (Tissue Solutions, Clydebank, UK). Consent was obtained from all patients and controls, and studies were IRB approved beforehand.

Results in a Human Pilot Study

As for the rat SCI studies, we found robust immunoblotting evidence of SBDP150, UCH-L1 and GFAP and its major 38 K BDP in CSF, even out to day 2–3 after injury in a human patient (Figure 11.3). In the same patient,

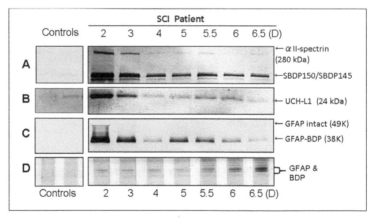

Figure 11.3. Time course of SBDP150, UCH-L1, GFAP biomarker release into human biofluid after spinal cord injury. Biomarkers were detected by immunblotting (A) CSF αll-spectrin and SBDPs, (B) CSF UCH-L1, (C) CSF GFAP and BDP and (D) serum autoantibody to GFAP and BDP in one SCI patient (various time points-2 day to 6.5 days post-injury) in comparison to biofluid samples from two controls.

Color image of this figure appears in the color plate section at the end of the book.

we also observed that this SCI patient had developed an autoantibody against GFAP and its BDP beginning on day 6 after the initial spinal cord injury (Figure 11.3).

swELISA data further showed elevation of UCH-L1 and SBDP150 at day 2 after SCI (both CSF and serum), while MBP was also elevated at Day 2, and was sustained to day 4–5 in CSF; in contrast, serum MBP levels were highest at 4.5–5 days (Figure 11.4). CSF and serum samples from various time points of seven (7) SCI patients were analyzed.

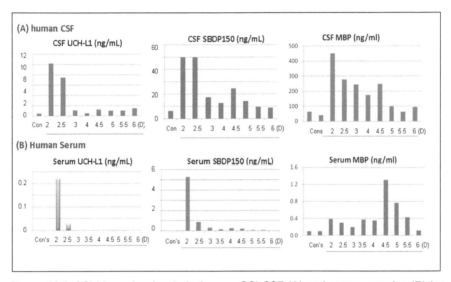

Figure 11.4. SCI biomarker levels in human SCI CSF (A) and serum samples (B) by swELISA. In comparison to CSF control and normal serum control, in one SCI human patient, Both CSF and serum UCH-L1 and SBDP145 levels are elevated in day 2 after SCI. In contrast, CSF MBP elevation peaks in day 2 but also sustained to 4 days, while serum MBP elevation peaks at day 4–5.

Color image of this figure appears in the color plate section at the end of the book.

Conclusions

Over the previous two decades, many types of biomarkers have been researched and applied to clinical practice in TBI. However, none as yet found a useful place in the clinic, as compared to cancer or cardiology. As we mentioned earlier, there is little evidence to support biomarkers in SCI.

Most SCI studies in the biomarker field have focused upon NSE and S100β. However, with their lower specificity in multiple trauma patients, these markers seem to be inadequate in accuracy or specificity for diagnosis and outcome prediction. Our preliminary data, and that of others, suggests that structural proteins, such as SBDPs, GFAP, MBP, and UCH-L1 have potential as promising biomarkers in an animal SCI model and in human SCI patients. This pilot data will support a larger clinical, multi-center trial to further examine the efficacy of these biomarker proteins as a diagnostic or prognostic tool in SCI patients.

References

Agrawal, G., C. Kerr, N.V. Thakor and A.H. All. 2010. Characterization of graded multicenter animal spinal cord injury study contusion spinal cord injury using somatosensory-evoked potentials. *Spine (Phila Pa 1976)*. 35: 1122–7.

Anderson, K.J., S.W. Scheff, K.M. Miller, K.N. Roberts, L.K. Gilmer, C. Yang and G. Shaw. 2008. The phosphorylated axonal form of the neurofilament subunit NF-H (pNF-H) as a blood biomarker of traumatic brain injury. *J Neurotrauma*. 25: 1079–85.

Brisby, H., K. Olmarker, L. Rosengren, C.G. Cederlund and B. Rydevik. 1999. Markers of nerve tissue injury in the cerebrospinal fluid in patients with lumbar disc herniation and sciatica. *Spine (Phila Pa 1976)*. 24: 742–6.

Buki, A., R. Siman, J.Q. Trojanowski and J.T. Povlishock. 1999. The role of calpain-mediated spectrin proteolysis in traumatically induced axonal injury. *J Neuropathol Exp Neurol*. 58: 365–75.

Cao, F., X.F. Yang, W.G. Liu, W.W. Hu, G. Li, X.J. Zheng, F. Shen, X.Q. Zhao and S.T. Lv. 2008. Elevation of neuron-specific enolase and S-100beta protein level in experimental acute spinal cord injury. *J Clin Neurosci*. 15: 541–4.

Choi, J.U., H.J. Hoffman, E.B. Hendrick, R.P. Humphreys and W.S. Keith. 1986. Traumatic infarction of the spinal cord in children. *J Neurosurg*. 65: 608–10.

Gruner, J.A. 1992. A monitored contusion model of spinal cord injury in the rat. *Journal of Neurotrauma*. 9: 123–6; discussion 126–8.

Guez, M., C. Hildingsson, L. Rosengren, K. Karlsson and G. Toolanen. 2003. Nervous tissue damage markers in cerebrospinal fluid after cervical spine injuries and whiplash trauma. *J Neurotrauma*. 20: 853–8.

Hayakawa, K., R. Okazaki, K. Ishii, T. Ueno, N. Izawa, Y. Tanaka, S. Toyooka, N. Matsuoka, K. Morioka, Y. Ohori, K. Nakamura, M. Akai, Y. Tobimatsu, Y. Hamabe and T. Ogata. 2012. Phosphorylated neurofilament subunit NF-H as a biomarker for evaluating the severity of spinal cord injury patients, a pilot study. *Spinal Cord*. 2012 Jul; 50(7): 493–6.

Jackson, P. and R.J. Thompson. 1981. The demonstration of new human brain-specific proteins by high-resolution two-dimensional polyacrylamide gel electrophoresis. *J Neurol Sci*. 49: 429–38.

Kang, S.K., H.H. So, Y.S. Moon and C.H. Kim. 2006. Proteomic analysis of injured spinal cord tissue proteins using 2-DE and MALDI-TOF MS. *Proteomics*. 6: 2797–812.

Kwon, B.K., S. Casha, R.J. Hurlbert and V.W. Yong. 2011. Inflammatory and structural biomarkers in acute traumatic spinal cord injury. *Clin Chem Lab Med*. 49: 425–33.

Kwon, B.K., A.M. Stammers, L.M. Belanger, A. Bernardo, D. Chan, C.M. Bishop, G.P. Slobogean, H. Zhang, H. Umedaly, M. Giffin, J. Street, M.C. Boyd, S.J. Paquette, C.G. Fisher and M.F. Dvorak. 2010. Cerebrospinal fluid inflammatory cytokines and biomarkers of injury severity in acute human spinal cord injury. *J Neurotrauma*. 27: 669–82.

Liu, M.C., L. Akinyi, D. Scharf, J. Mo, S.F. Larner, U. Muller, M.W. Oli, W. Zheng, F. Kobeissy, L. Papa, X.C. Lu, J.R. Dave, F.C. Tortella, R.L. Hayes and K.K. Wang. 2010a. Ubiquitin C-terminal hydrolase-L1 as a biomarker for ischemic and traumatic brain injury in rats. *Eur J Neurosci*. 31: 722–32.

Liu, M.C., L. Akinyi, D. Scharf, J. Mo, S.F. Larner, U. Muller, M.W. Oli, W. Zheng, F. Kobeissy, L. Papa, X.C. Lu, J.R. Dave, F.C. Tortella, R.L. Hayes and K.K. Wang. 2010b. Ubiquitin

C-terminal hydrolase-L1 as a biomarker for ischemic and traumatic brain injury in rats. *Eur J Neurosci.* 31: 722–32.

Liu, M.C., V. Akle, W. Zheng, J. Kitlen, B. O'Steen, S.F. Larner, J.R. Dave, F.C. Tortella, R.L. Hayes and K.K. Wang. 2006. Extensive degradation of myelin basic protein isoforms by calpain following traumatic brain injury. *J Neurochem.* 98: 700–12.

Loy, D.N., A.E. Sroufe, J.L. Pelt, D.A. Burke, Q.L. Cao, J.F. Talbott and S.R. Whittemore. 2005. Serum biomarkers for experimental acute spinal cord injury: rapid elevation of neuron-specific enolase and S-100beta. *Neurosurgery.* 56: 391–7; discussion 391–7.

Lustrin, E.S., S.P. Karakas, A.O. Ortiz, J. Cinnamon, M. Castillo, K. Vaheesan, J.H. Brown, A.S. Diamond, K. Black and S. Singh. 2003. Pediatric cervical spine: normal anatomy, variants, and trauma. *Radiographics.* 23: 539–60.

Ma, J., L.N. Novikov, K. Karlsson, J.O. Kellerth and M. Wiberg. 2001. Plexus avulsion and spinal cord injury increase the serum concentration of S-100 protein: an experimental study in rats. *Scandinavian J Plast Reconstr Surg Hand Surg.* 35: 355–9.

Mondello, S., A. Linnet, A. Buki, S. Robicsek, A. Gabrielli, J. Tepas, L. Papa, G.M. Brophy, F. Tortella, R.L. Hayes and K.K. Wang. 2012. Clinical utility of serum levels of ubiquitin C-terminal hydrolase as a biomarker for severe traumatic brain injury. *Neurosurgery.* 70: 666–75.

Mondello, S., U. Muller, A. Jeromin, J. Streeter, R.L. Hayes and K.K. Wang. 2011. Blood-based diagnostics of traumatic brain injuries. *Expert Rev Mol Diagn.* 11: 65–78.

Mondello, S., S.A. Robicsek, A. Gabrielli, G.M. Brophy, L. Papa, J. Tepas, C. Robertson, A. Buki, D. Scharf, M. Jixiang, L. Akinyi, U. Muller, K.K. Wang and R.L. Hayes. 2010a. alphaII-spectrin breakdown products (SBDPs): diagnosis and outcome in severe traumatic brain injury patients. *J Neurotrauma.* 27: 1203–13.

Mondello, S., S.A. Robicsek, A. Gabrielli, G.M. Brophy, L. Papa, J. Tepas, C. Robertson, A. Buki, D. Scharf, M. Jixiang, L. Akinyi, U. Muller, K.K. Wang and R.L. Hayes. 2010b. alphaII-spectrin breakdown products (SBDPs): diagnosis and outcome in severe traumatic brain injury patients. *J Neurotrauma.* 27: 1203–13.

Pang, D. and J.E. Wilberger, Jr. 1982. Spinal cord injury without radiographic abnormalities in children. *J Neurosurg.* 57: 114–29.

Papa, L., L. Akinyi, M.C. Liu, J.A. Pineda, J.J. Tepas, 3rd, M.W. Oli, W. Zheng, G. Robinson, S.A. Robicsek, A. Gabrielli, S.C. Heaton, H.J. Hannay, J.A. Demery, G.M. Brophy, J. Layon, C.S. Robertson, R.L. Hayes and K.K. Wang. 2010. Ubiquitin C-terminal hydrolase is a novel biomarker in humans for severe traumatic brain injury. *Crit Care Med.* 38: 138–44.

Papa, L., L.M. Lewis, J.L. Falk, Z. Zhang, S. Silvestri, P. Giordano, G.M. Brophy, J.A. Demery, N.K. Dixit, I. Ferguson, M.C. Liu, J. Mo, L. Akinyi, K. Schmid, S. Mondello, C.S. Robertson, F.C. Tortella, R.L. Hayes and K.K. Wang. 2011. Elevated Levels of Serum Glial Fibrillary Acidic Protein Breakdown Products in Mild and Moderate Traumatic Brain Injury Are Associated With Intracranial Lesions and Neurosurgical Intervention. *Ann Emerg Med.* 2012 Jun; 59(6): 471–83.

Petzold, A. 2005. Neurofilament phosphoforms: surrogate markers for axonal injury, degeneration and loss. *J Neurol Sci.* 233: 183–98.

Pike, B.R., J. Flint, J.R. Dave, X.C. Lu, K.K. Wang, F.C. Tortella and R.L. Hayes. 2004. Accumulation of calpain and caspase-3 proteolytic fragments of brain-derived alphaII-spectrin in cerebral spinal fluid after middle cerebral artery occlusion in rats. *Journal of Cerebral Blood Flow and Metabolism.* 24: 98–106.

Pike, B.R., J. Flint, S. Dutta, E. Johnson, K.K. Wang and R.L. Hayes. 2001. Accumulation of non-erythroid alpha II-spectrin and calpain-cleaved alpha II-spectrin breakdown products in cerebrospinal fluid after traumatic brain injury in rats. *J Neurochem.* 78: 1297–306.

Reeves, T.M., J.E. Greer, A.S. Vanderveer and L.L. Phillips. 2010. Proteolysis of submembrane cytoskeletal proteins ankyrin-G and alphaII-spectrin following diffuse brain injury: a role in white matter vulnerability at Nodes of Ranvier. *Brain Pathol.* 20: 1055–68.

Shaw, G., C. Yang, R. Ellis, K. Anderson, J. Parker Mickle, S. Scheff, B. Pike, D.K. Anderson and D.R. Howland. 2005. Hyperphosphorylated neurofilament NF-H is a serum biomarker of axonal injury. *Biochem Biophys Res Commun.* 336: 1268–77.

Shiiya, N., T. Kunihara, T. Miyatake, K. Matsuzaki and K. Yasuda. 2004. Tau protein in the cerebrospinal fluid is a marker of brain injury after aortic surgery. *Ann Thorac Surg.* 77: 2034–8.

Winnerkvist, A., R.E. Anderson, L.O. Hansson, L. Rosengren, A.E. Estrera, T.T. Huynh, E.E. Porat and H.J. Safi. 2007. Multilevel somatosensory evoked potentials and cerebrospinal proteins: indicators of spinal cord injury in thoracoabdominal aortic aneurysm surgery. *Eur J Cardiothorac Surg.* 31: 637–42.

Zhang, B., Y. Huang, Z. Su, S. Wang, J. Wang, A. Wang and X. Lai. 2011. Neurological, functional, and biomechanical characteristics after high-velocity behind armor blunt trauma of the spine. *J Trauma.* 71: 1680–8.

Zhang, S.X., M. Underwood, A. Landfield, F.F. Huang, S. Gison and J.W. Geddes. 2000. Cytoskeletal disruption following contusion injury to the rat spinal cord. *J Neuropathol Exp Neurol.* 59: 287–96.

Zimmer, D.B., E.H. Cornwall, A. Landar and W. Song. 1995. The S100 protein family: history, function, and expression. *Brain Res Bull.* 37: 417–29.

Section C

OTHER CNS DISORDER BIOMARKERS

12

miRNA as Biomarkers for Multiple Sclerosis: Quest for Identification and Treatment

Deepak Kumar[1],* and *Roopali Gandhi*[2]

INTRODUCTION

Multiple Sclerosis

Multiple Sclerosis (MS) is a chronic inflammatory demyelinating autoimmune disease of the central nervous system whose precise etiology is still obscure. The disease primarily affects young adults and is quite heterogeneous with respect to clinical manifestations, disease course, brain Magnetic Resonance Imaging (MRI) findings, composition of lesion pathology and response to treatment (Noseworthy et al., 2000). The main characteristic features of MS include recurrent episodes of MS plaques which include destruction of myelin sheath, oligodendrocyte damage, axonal damage, glial scar formation and presence of inflammatory cells such as T cells, macrophages, microglial cells, astrocytes, and mast cells, which in turn incite a pro-inflammatory reaction leading to local tissue injury (Bar-Or, 2005; Morales et al., 2006; Pittock and Lucchinetti, 2007). Based upon the clinical course of the disease, MS is divided into three

[1] Biological Chemistry and Molecular Pharmacology, Harvard Medical School, Boston, MA 02115.

[2] Center for Neurologic Diseases, Brigham and Women's Hospital, Boston, MA 02115.

* Corresponding author: Deepakkjaiswal@gmail.com

major subtypes: 1) Relapsing Remitting; 2) Secondary Progressive; and 3) Primary Progressive. Relapsing Remitting MS (RRMS) affects about 80% of individuals and is characterized by unpredictable relapse followed by a period of remission. Secondary Progressive MS (SPMS) affects about 65% of individuals who initially exhibit RRMS but eventually show progression of disability. In Primary Progressive MS (PPMS), there is no remission after the initial symptoms. This type of MS affects about 10–20% of individuals with signs of progression of disability from onset. Based upon the clinical course, it has been proposed that while both adaptive and innate immune responses contribute to RRMS, sustained activation of CNS innate immunity drives SPMS (Compston and Coles, 2008). Owing to recent advances in understanding the immune mechanism of MS, we now have about 10 FDA approved immune modulating therapies. Nonetheless, the major challenge in MS is to develop biomarkers that could help in understanding individual MS patient, whether they are a responder or non-responder to therapy, which medicine is more effective, and the degree to which they may be entering the progressive phase of disease. In the last few years, a lot of efforts have been drawn toward identification of diagnostic, prognostic, process-specific and treatment-related biomarkers for MS. In this chapter, we will focus on the micro RNAs (miRNAs) as a potential candidate for MS biomarker.

miRNA as Potential Biomarkers

miRNAs are a class of endogenous, small, on an average 22 nt long, non-coding RNAs that regulate gene expression at the post-transcriptional level by binding to complementary sequences in the 3' UTR of the target mRNA. This usually results in translational repression or mRNA degradation leading to decrease in encoded protein (Bartel, 2004). miRNAs play a critical role in multiple phenomenon including development, organogenesis and homeostasis (Ebert and Sharp, 2012). The dysregulated expression of miRNAs is associated with onset of diseases such as immune disease, neurological disease, cardiovascular disease and cancer (Mendell and Olson, 2012). Until now, about 1900 mature human miRNAs have been identified (Kozomara and Griffiths-Jones, 2011). Besides their existence in the cells, the miRNAs have been detected in several body fluids, including plasma or serum, cerebrospinal fluid (CSF), breast milk, urine, tears, semen and saliva (Cortez et al., 2011). Interestingly, researchers have observed significant differences in serum miRNA levels during disease or certain physiological state such as pregnancy (Chim et al., 2008). However our knowledge about the origin and functions of circulating miRNAs is very limited. It is not clear whether the presence of miRNAs in circulation is simply a byproduct of cellular degradation or they are actively secreted into the body fluids to regulate intercellular gene expression (Kosaka et al., 2013). Yet, miRNAs are highly stable in the blood even at severe conditions such as extended storage, freeze-thaw, extreme pH and circulating ribonucleases due to their packaging in lipid vesicles such as exosomes, binding to RNA-binding protein and their association with high-density lipoprotein (Chen et al., 2008; Mitchell et al., 2008). Their stability along with development of sensitive methods of detection and quantification such as miRNA microarray, Taqman-based array, Bead-based assay, NanoString technique and deep sequencing (Guerau-de-Arellano et al., 2012) makes them a potential candidate for a novel class of non-invasive and sensitive biomarkers.

miRNA Expression in MS Patients

miRNA Expression in MS Lesions, Whole Blood or PBMC

A number of studies have highlighted the importance of miRNAs as biomarkers in the field of cancer. However, in case of MS the use of miRNAs as biomarkers is still evolving. A few studies that are published till date have used either whole blood, Peripheral Blood Mononuclear Cells (PBMCs), plasma, cerebrospinal fluid (CSF), T cells or the multiple sclerosis lesion for miRNA expression analysis.

In the first report of comprehensive miRNA expression analysis in MS lesions, Junker et al. (2009) used both active and inactive MS lesions to establish a miRNA expression profiles. Using cells isolated by laser capture microdissection from MS lesions, they analyzed the expression level of 365 different mature miRNAs by qPCR in 16 active and five inactive white matter MS brain lesions and in nine control white matter specimens. They found that in active lesions, 20 miRNAs were upregulated and 8 miRNAs were downregulated as compared to normal white matter. In inactive lesions, 22 miRNAs were upregulated and 13 miRNAs were downregulated. Further analysis of 10 most strongly upregulated miRNAs in active MS lesions showed that astrocytes expressed all 10 miRNAs. Among the upregulated miRNAs, miR-34a, miR-155 and miR-326 could putatively target CD47 and may play a role in MS pathology by releasing macrophages from inhibitory control. On similar lines, Du et al. (2009) also reported that miR-326 is increased in PBMCs and CD4$^+$ T cells isolated from relapsing MS patients. They identified that miR-326 may contribute to disease by increasing pathogenic Th17 cell development via targeting the transcription factor ets-1. miR-155 is another miRNA that is involved in Th17 cell development and has been linked to MS and rheumatoid arthritis previously (Junker et al., 2009; Stanczyk et al., 2008).

The first study on miRNA expression analysis in PBMCs was conducted by Otaegui et al. (2009). They obtained PBMCs from MS patients in relapse status (n = 4), in remission status (n = 9) and healthy controls (HCs; n = 8) and analyzed the expression patterns of 364 miRNAs using RT-qPCR. Their results showed that expression of miR-18b, miR-493 and miR-599 is increased in relapsing group when compared to the control group. However, they found differential expression of miR-148a, miR-184, miR-193a and miR-96 in remission group when compared to control group using co-expression networks. The validation of these miRNAs in two separate MS and HCs group (Group B: 14 remission, 13 relapse, 15 HCs; and Group C: 4 relapse, three remission, 7HCs) revealed that only two miRNAs (miR-18b and miR-599) may be relevant at the time of relapse and that another miRNA (miR-96) may be involved in remission. The miRNA miR-96 has been suggested to play a role in interleukin signaling.

In another study involving blood cells, Keller et al. (2009) used miRNA array to analyze the miRNA expression patterns in blood cells obtained from RRMS (n = 20) and HCs (n = 19). They found 165 significantly dysregulated miRNAs in blood cells of patients with RRMS compared to HCs. Most significantly, dysregulated miRNA, miR-145, has differentiated between MS patients and HCs with an accuracy, specificity, and sensitivity of 89.7, 89.5 and 90.0%, respectively. However, some of the MS patients used in this study were treated with either Glatiramer Acetate or Interferon-β. In a more recent study, Keller et al. (2013) used Next Generation Sequencing (NGS), microarray analysis, and RT-qPCR in whole blood samples from treatment-naïve patients with CIS (n = 25) or RRMS (n = 25) and 50 HCs to study miRNA expression pattern. They found that in patients with CIS/RRMS, NGS and microarray analysis identified 38 and eight significantly dysregulated miRNAs, respectively. Among the 8 miRNAs identified by microarray, 5 miRNAs (miR-146b-5p, miR-7-1-3p, miR-20a-5p, miR-3653 and miR-20b) were downregulated and three were upregulated (miR-16-2-3p, miR-574-5p and miR-1202). Three of these miRNAs were found to be significantly downregulated (miR-20a-5p, miR-7-1-3p) or upregulated (miR-16-2-3p) by both methods. Downregulation of miR-20a-5p was the reconfirmation of their previous finding (Keller et al., 2009). In another study, Martinelli-Boneschi et al. (2012) analyzed the expression profile of 1145 miRNAs in PBMCs of 19 MS patients including RRMS (n = 7), SPMS (n = 6), PPMS (n = 6) and 14 controls using miRNA array. Some of the patients involved in this study had been given either immunosuppressive drugs or Interferon-β or both. They found a total of 104 deregulated miRNAs in MS patients. However, during validation in a separate group of individuals only two miRNA let-7g and miR-150 were found to be different. Altered expression of let-7g was also noticed by Cox et al. (2010) who performed miRNA analysis in 18PPMS, 17SPMS and 24RRMS patients. In addition to let-7g, they also found that the expression of miR-20a and miR-17 are significantly downregulated in MS. These miRNAs are involved in T cell activation and their T cell targets were upregulated in MS whole blood RNA. Down regulation of miR-20a was also reported in two previous studies (Keller et al., 2009; Keller et al., 2013).

In a more recent study, Sondergaard et al. (2013) analyzed miRNA expression profiling of PBMCs isolated from 20 treatment-naïve patients with RRMS and 21 HCs using miRNA array. They identified 46 differentially expressed miRNAs in MS patients, of which 17 miRNAs were upregulated and 29 miRNAs were downregulated. For validation, selected miRNAs were analyzed again in 12 treatment-naïve RRMS patients and 20 HCs. However, in the validation cohort, only two miRNAs, let-7d and miR-744 showed significant upregulation in MS patients as compared to the control. They also found that miR-145 is also significantly upregulated in MS patients.

miRNA Expression in T Cells of MS Patients

MS is an inflammatory disease which in part is driven by IFN-γ producing Th1 and IL-17 producing Th17 cells (Lock et al., 2002; Olsson et al., 1990; Voskuhl et al., 1993). The development of T cells in thymus and their differentiation in periphery is controlled by both extracellular signaling and miRNAs. While a lot of work has been done to understand the role of miRNAs in T cell differentiation, there are only a few reports that emphasized their role as a T cell biomarker in MS. In a first study to analyze the miRNA expression in T cells, De Santis et al. (2010) used miRNA microarray for CD4$^+$CD25^{high} T cells from peripheral blood of 12 RRMS patients (treatment free for at least 6 months) and 14 HCs. They found that 23 miRNAs are differentially expressed between CD4$^+$CD25^{high} Treg cells from MS patients and HCs. Validation study using real time-PCR also confirmed the upregulation of miR-106b, miR-19a, miR-19b and miR-25 in CD4$^+$CD25^{high} of MS subjects as compared to healthy subjects. miR-106b and miR-25 modulates TGF-β signaling pathway and dysregulated expression of these miRNAs may alter regulatory T cell biology in MS. In another study, Lindberg et al. (2010) used CD4$^+$, CD8$^+$ T cells and B cells from 8 RRMS and 10 HCs to analyze the miRNA expression profile by Taqman array. They found that miR-17-5p was upregulated in CD4$^+$ T cells isolated from MS patients. The miR-17-5p belongs to miR-17-92 cluster and plays an important role in development of autoimmunity and lymphoproliferative disease in mice (Xiao et al., 2008).

Another study by Guerau-de-Arellano et al. (2011) suggested increased expression of miR-128 and miR-27b in naïve T cells and miR-340 in memory T cells isolated from MS patients. These miRNA have a combined effect on suppression of Th2 differentiation and thus could enhance Th1 mediated autoimmune responses. A similar study by Smith et al. (2012) showed altered expression of miR-29b in MS and EAE. miR-29b is involved in controlling expression of IFN-γ and Th1 transcription factor, T-bet. A more recent study on miRNA analysis in T cells by miRNA microarray, Jernas et al. (2013) used 11 RRMS patients (treated with Interferon β) and 9 HCs. They identified 21 miRNAs whose expression was decreased in MS patients as compared to controls. Of these 21 differentially expressed miRNAs, 20 were shown to affect the expression of immune system related target genes. They confirmed the expression of two selected miRNAs (miR-494 and miR-197) by RT-qPCR.

miRNA Expression in Serum or Plasma of MS Patients

Stable expression of miRNA in serum and plasma makes them an important candidate for biomarker study. However, in case of MS there are only two reported studies that have analyzed circulating miRNA. In the first study on circulating miRNA in MS, Siegel et al. (2012) conducted a miRNA microarray analysis of over 900 known miRNA transcripts from plasma samples collected from four MS patients and 4 HCs. They identified six plasma miRNA (miR-614, miR-572, miR-648, miR-1826, miR-422a and miR-22) that were significantly upregulated and one plasma miRNA (miR-1979) that was significantly downregulated in MS patients.

In a more recent study by our group, we analyzed a total of 368 miRNAs in plasma samples of 10 RRMS, 9 SPMS, and 9 HCs using miRNA RT-qPCR. We found that 29 miRNAs distinguished RRMS from HCs, 35 miRNAs distinguished SPMS from HCs, and 16 miRNAs distinguished RRMS from SPMS. The 20 selected miRNAs were further validated on a larger cohort of 100 MS patients, 15 amyotrophic lateral sclerosis (ALS) and 32 controls using RT-qPCR. Among the validated miRNAs, we identified miR-92a-1* in the largest number of comparisons. It was different in RRMS versus SPMS, and RRMS versus HCs, and showed an association with EDSS and disease duration. The miRNA miR-92 has target genes involved in cell cycle regulation and cell signaling (Tsuchida et al., 2011). The let-7a differentiated SPMS from HCs and RRMS from SPMS. The let-7 family of miRNAs are known to regulate stem cell differentiation and T cell activation, activate Toll-like receptor 7, and are linked to neurodegeneration (Grigoryev et al., 2011; Kumar et al., 2011; Lehmann et al., 2012; Roush and Slack, 2008; Swaminathan et al., 2012). The miR-454 differentiated RRMS from SPMS, and miR-145 differentiated RRMS from HCs and RRMS from SPMS. miR-145 findings were in consensus with two previous studies in MS (Keller et al., 2009; Sondergaard et al., 2013). Interestingly, the same circulating miRNAs (let-7 and miR-92) that were differentially expressed in RRMS versus SPMS also differentiated ALS from RRMS subjects, but were not different between SPMS and ALS, suggesting that similar processes may occur in SPMS and ALS (Gandhi et al., 2013).

miRNA Expression in CSF of MS Patient

In the only published study involving miRNA analysis in CSF of MS patients, Haghikia et al. (2012) tested global miRNA profile to screen for reported miRNAs, followed by RT-qPCR to validate candidate miRNAs. They used CSF of 53 patients with MS and 39 patients with Other Neurologic Diseases (OND) for global miRNA profiling. Their results showed that miR-922, miR-181c, and miR-633 are differentially regulated in patients with MS as compared with OND. Importantly, miR-181c and miR-633 differentiated RRMS from SPMS with specificity of 82% and a sensitivity of 69%.

miRNAs as Treatment Response Biomarkers of MS

For MS, there are about 10 FDA approved disease modifying therapies available. However, Injection of recombinant interferon-β (IFN-β) and Glatiramer acetate (Copaxone®) are considered as first-line option in the treatment of RRMS. These treatments reduce the number of relapse and suppress the accumulation of new inflammatory lesions in the brain. Several studies have evaluated the effect of these therapies on the regulation of mRNAs (reviewed in Minagar, 2013), however, studies involving their effect on the regulation of miRNAs are lacking. Such studies are not only important to identify a marker to analyze treatment response but also to understand the molecular mechanism of treatment. Till date, there are only three published reports on the expression of miRNAs upon treatment in MS patients. In the first published study, Waschbisch et al. (2011) obtained PBMCs from patients with RRMS, and analyzed the relative expression of five selected miRNAs (miR-20b, miR-142-3p, miR-146a, miR-155 and miR-326) by real-time PCR. These miRNAs were selected based upon their previous reported function in Th17 cell differentiation (miR-326 and miR-155, discussed earlier), regulation of immune tolerance (miR-142-3p and miR-146a) or innate immunity (miR-146a). They compared the miRNA levels between treatment-naive patients ($n = 36$), IFN-β-treated patients ($n = 18$; Avonex® n = 5, Rebif® n = 8, Betaferon® n = 5), and patients treated with Glatiramer acetate (Copaxone®; $n = 20$). They excluded the patients that were treated with glucocorticoids during the last four weeks before study entry. They observed the expression of miR-142-3p, miR-146a, miR-155 and miR-326 was remarkably increased in untreated RRMS patients compared to HCs. Further, they showed that a combination of miR-155, miR-146a and miR-142-3p yielded the best results in predicting disease with a sensitivity of 77.8% and a specificity of 88.0% (AUC 0.77). While the expression of selected five miRNAs did not differ between treatment-naïve and IFN-β treated RRMS patients, the expression of miR-142-3p and miR-146a was significantly decreased in Glatiramer acetate treated RRMS patients.

In another study, Sievers et al. (2012) tested 1059 miRNAs in B cells isolated from 10 untreated patients, 10 Natalizumab treated patients and 10 HCs. Natalizumab (Tysabri®) is a FDA approved monoclonal antibody that binds to integrins on T cells and prevents their migration across blood brain barrier and thus decreases inflammation in the brain. The treatment effects are associated with reduced relapse rate, disability progression and reduced number of lesion in the brain as measured by MRI. Their results showed that 10 miRNAs were differentially expressed in B cells from Natalizumab-treated patients compared to untreated patients. Expression of miR-106b-25 and miR-17-92 cluster was dysregulated both in untreated and Natalizumab-treated patients suggesting a role of these miRNA in MS pathogenesis. Functional cooperation among these two miRNA clusters, characterized by reduction in pre-B cells and increased apoptosis, was suggested previously in mice (Ventura et al., 2008).

In a recent study, Hecker et al. (2013) used microarrays to investigate the expression of miRNAs and mRNAs in PBMC of patients with CIS or RRMS in response to IFN-β therapy. The blood samples were obtained longitudinally from six patients at four time points in the early phase of therapy, namely before the first (baseline), after second, and third IFN-β injection as well as after one month of treatment. Upon comparing the expression levels at the three time points during therapy to the expression levels at baseline, 20 different miRNAs were filtered at various time points. Of these, seven miRNAs appeared as upregulated and 13 miRNAs appeared as downregulated in response to the therapy. Two of the 20 miRNAs, miR-149-5p and miR-708-5p were filtered at two different time points. For the remaining miRNAs, the expression changes were not very stable in the course of therapy. Most of the miRNAs ($n = 14$) were filtered as upregulated or downregulated one month after IFN-β treatment initiation, which corresponded with gene expression profiling. They used Affymetrix miRNA microarrays to replicate the miRNA measurements of the PBMC samples from three patients before the start of IFN-β therapy as well as after one month. These microarrays had a lower measurement range than the TaqMan miRNA arrays and 13 of the 20 filtered miRNAs showed the same trend of upregulation (let-7a-5p, let-7b-5p, miR-16-5p, miR-342-5p, miR-346, miR-518b) or downregulation (miR-29a-3p, miR-29b-1-5p, miR-29c-3p, miR-95, miR-149-5p, miR-181c-3p, miR-708-5p) with significance threshold alpha = 0.10, four miRNAs (miR-29a-3p, miR-29c-3p, miR-193a-3p, and miR-532-5p) were confirmed to be downregulated during treatment. For further validation, they selected five of the 20 filtered miRNAs to quantify their expression in PBMC of an independent cohort of 12 patients using TaqMan single-tube assays. These 12 patients (8 RRMS/4 CIS; 7 females/5 males) also started a therapy with IFN-β-1b. The PBMC were obtained again in a longitudinal manner before the first drug injection and after one month of treatment. In this data set, miR-29a-3p and miR-29c-3p could be confirmed as differentially expressed in response to IFN-β therapy. miR-29c-3p was expressed at lower levels during therapy in comparison to pretreatment levels in all 12 patients. Additionally, miR-532-5p was confirmed to be downregulated. miR-16-5p and miR-149-5p showed the same trend of expression change as in the other data sets, but this was not statistically significant.

Conclusion

There is considerable excitement regarding the use of miRNAs in the biomarker field, which is well founded due to their stability and relative ease of detection. In this chapter we have reviewed the miRNA based studies carried out in MS patients in order to evaluate their potential as a biomarker. We observed a wide variety of results obtained by different groups (summarized in Table 12.1), which not only raised questions but also, showed the limitations associated with such studies. In general, we find that there is a paucity of comprehensive studies to evaluate miRNA as biomarker in MS.

In order to develop a miRNA-based biomarker for MS or any disease, a most important prerequisite is the ability to quantify miRNAs from a variety of samples with sufficient sensitivity and precision. A number of variables such as specimen collection, RNA extraction, RT-qPCR, data analysis and normalization are some of the major challenges associated with miRNA-based studies. There is also considerable inter-sample variability in both the protein and lipid content of each individual's serum and plasma, which could affect the efficiency of RNA extraction and can introduce potential inhibitors to RT-qPCR. Another major issue associated with miRNA based study is normalization, amplification and contamination. Due to low concentration of miRNA in body fluids, their quantitation is not only difficult but also requires occasional amplification, which in turn creates additional variables. Most importantly, there is no consensus on suitable small RNA reference genes that could be used as internal controls for biological variability. A common practice is to process the identical input volume of samples, which is later corrected for technical variability using spiked-in synthetic non-human miRNA as a normalizing control (Kroh et

Table 12.1. miRNA-based biomarker studies in Multiple Sclerosis.

Author	MS type	Tissue	Method	Treatment	Key miRNAs
Junker et al., 2009	SPMS, RRMS, PPMS	MS brain lesions	TLDA	Unspecified	miR-650, miR-155, miR-326, miR-142-3p, miR-146a, miR-34a, miR-21, miR-23a, miR-199a and miR-27a
Du et al., 2009	RRMS	PBMC, CD4$^+$ T cells	Real-time PCR	Unspecified	miR-326
Otaegui et al., 2009	RRMS	PBMC	TLDA	Unspecified	miR-18b, miR-493, miR-96 and miR-599
Keller et al., 2009	RRMS	Whole blood	Microarray	Glatiramer acetate, IFNβ or none	miR-145, miR-186, miR-664, miR-20b, miR-422a, miR-142-3p, miR-584, miR-223, miR-1275 and miR-491-5p
Keller et al., 2013	CIS, RRMS	Whole blood	NGS, Microarray, Real-time PCR	Treatment naïve	miR-146b-5p, miR-7-1-3p, miR-20a-5p, miR-3653, miR-20b, miR-16-2-3p, miR-574-5p and miR-1202
Martinelli-Boneschi et al., 2012	RRMS, SPMS, PPMS	PBMC	BeadArray	Immunosuppressive, IFNβ or both	let-7g, miR-150
Cox et al., 2010	RRMS, SPMS, PPMS	Whole blood	Microarray	No treatment for 3 months	miR-17 and miR-20a
Sondergaard et al, 2013	RRMS	PBMC, Plasma	LNA Microarray, Real-time PCR	Treatment naïve	let-7d, miR-744 and miR-145
De Santis et al., 2010	RRMS	CD4$^+$ CD25$^+$ T cells	Microarray	Treatment naïve and steroid free for 6 months	miR-106b, miR-93, miR-19a, miR-19b and miR-25
Lindberg et al., 2010	RRMS	CD4$^+$ T cells	TLDA	No treatment for 6 months	miR-485-3p, miR-376a, miR-497, miR-193a, miR-126, miR-17-5p and miR-34a

Reference	MS type	Sample	Method	Treatment	miRNAs
Guerau-de-Arellano et al., 2011	RRMS, SPMS, PPMS	Naïve CD4+ T cells	TLDA	Treatment naïve	miR-128, miR-27b and miR-340
Smith et al., 2012	RRMS, SPMS, PPMS	CD4+CD45RO+ T cells	Nano Stringn Counter	Treatment naïve	miR-29b
Jernas et al., 2013	RRMS	T cells	Microarray, Real-time PCR	IFNβ	miR-494, miR-15b, miR-30c, miR-23a, miR-197, miR-1260b, miR-125a-5p, miR-361-5p, miR-320d, miR-423-3p, miR-1280, miR-663, miR-423-5p, miR-99b, miR-339-5p, let-7a, miR-1979, miR-3178, miR-625, miR-150 and miR-3153
Siegel et al., 2012	Unspecified	Plasma	Microarray	Unspecified	miR-614, miR-572, miR-648, miR-1826, miR-422a, miR-22 and miR-1979
Gandhi et al., 2013	RRMS, SPMS, PPMS	Plasma	TLDA	Untreated	miR-30e, miR-92a-1*, Let-7a, miR-454 and miR-145
Haghikia et al., 2012	RRMS, SPMS, PPMS	CSF	PCR array, RT-qPCR	Unspecified	miR-922, miR-181c and miR-633
Waschbisch et al., 2011	RRMS	PBMC	Real-time PCR	Treatment naïve, IFNβ, Glatiramer acetate	miR-142-3p, miR-146a, miR-155 and miR-326
Sievers et al., 2012	RRMS	B cells	Microarray	Nata izumab	miR-19b, miR-551a, miR-106b and miR-191
Hecker et al., 2013	CIS, RRMS	PBMC	TLDA	IFNβ	miR-29a-3p, miR-29c-3p, miR-532-5p

al., 2010; Mitchell et al., 2008). In addition, quantification of miRNAs in the serum/plasma can be altered by contamination by miRNAs leaked from cellular components of blood, either through hemolysis during sampling or processing or by carry-over of whole cells in the serum/plasma.

However, the studies published till date point towards miRNAs having a role in the future of MS biomarkers and to design proof of concept studies. It will require considerable effort from researchers in different fields to develop consensus protocols. We believe it will not be long before the current challenges are overcome and a miRNA-based biomarker is established for MS.

References

Bar-Or, A. 2005. Immunology of multiple sclerosis. *Neurol Clin*. 23: 149–75, vii.

Bartel, D.P. 2004. MicroRNAs: genomics, biogenesis, mechanism, and function. *Cell*. 116: 281–97.

Chen, X., Y. Ba, L. Ma, X. Cai, Y. Yin, K. Wang, J. Guo, Y. Zhang, J. Chen, X. Guo, Q. Li, X. Li, W. Wang, Y. Zhang, J. Wang, X. Jiang, Y. Xiang, C. Xu, P. Zheng, J. Zhang, R. Li, H. Zhang, X. Shang, T. Gong, G. Ning, J. Wang, K. Zen, J. Zhang and C.Y. Zhang. 2008. Characterization of microRNAs in serum: a novel class of biomarkers for diagnosis of cancer and other diseases. *Cell Res*. 18: 997–1006.

Chim, S.S., T.K. Shing, E.C. Hung, T.Y. Leung, T.K. Lau, R.W. Chiu and Y.M. Lo. 2008. Detection and characterization of placental microRNAs in maternal plasma. *Clin Chem*. 54: 482–90.

Compston, A. and A. Coles. 2008. Multiple sclerosis. *Lancet*. 372: 1502–17.

Cortez, M.A., C. Bueso-Ramos, J. Ferdin, G. Lopez-Berestein, A.K. Sood and G.A. Calin. 2011. MicroRNAs in body fluids—the mix of hormones and biomarkers. *Nat Rev Clin Oncol*. 8: 467–77.

Cox, M.B., M.J. Cairns, K.S. Gandhi, A.P. Carroll, S. Moscovis, G.J. Stewart, S. Broadley, R.J. Scott, D.R. Booth and J. Lechner-Scott. 2010. MicroRNAs miR-17 and miR-20a inhibit T cell activation genes and are under-expressed in MS whole blood. *PLoS One*. 5: e12132.

De Santis, G., M. Ferracin, A. Biondani, L. Caniatti, M. Rosaria Tola, M. Castellazzi, B. Zagatti, L. Battistini, G. Borsellino, E. Fainardi, R. Gavioli, M. Negrini, R. Furlan and E. Granieri. 2010. Altered miRNA expression in T regulatory cells in course of multiple sclerosis. *J Neuroimmunol*. 226: 165–71.

Du, C., C. Liu, J. Kang, G. Zhao, Z. Ye, S. Huang, Z. Li, Z. Wu and G. Pei. 2009. MicroRNA miR-326 regulates TH-17 differentiation and is associated with the pathogenesis of multiple sclerosis. *Nat Immunol*. 10: 1252–9.

Ebert, M.S. and P.A. Sharp. 2012. Roles for microRNAs in conferring robustness to biological processes. *Cell*. 149: 515–24.

Gandhi, R., B. Healy, T. Gholipour, S. Egorova, A. Musallam, M.S. Hussain, P. Nejad, B. Patel, H. Hei, S. Khoury, F. Quintana, P. Kivisakk, T. Chitnis and H.L. Weiner. 2013. Circulating microRNAs as biomarkers for disease staging in multiple sclerosis. *Ann Neurol*. 73: 729–40.

Grigoryev, Y.A., S.M. Kurian, T. Hart, A.A. Nakorchevsky, C. Chen, D. Campbell, S.R. Head, J.R. Yates, 3rd and D.R. Salomon. 2011. MicroRNA regulation of molecular networks mapped by global microRNA, mRNA, and protein expression in activated T lymphocytes. *J Immunol*. 187: 2233–43.

Guerau-de-Arellano, M., H. Alder, H.G. Ozer, A. Lovett-Racke and M.K. Racke. 2012. miRNA profiling for biomarker discovery in multiple sclerosis: from microarray to deep sequencing. *J Neuroimmunol*. 248: 32–9.

Guerau-de-Arellano, M., K.M. Smith, J. Godlewski, Y. Liu, R. Winger, S.E. Lawler, C.C. Whitacre, M.K. Racke and A.E. Lovett-Racke. 2011. Micro-RNA dysregulation in multiple sclerosis favours pro-inflammatory T-cell-mediated autoimmunity. *Brain*. 134: 3578–89.

Haghikia, A., A. Haghikia, K. Hellwig, A. Baraniskin, A. Holzmann, B.F. Decard, T. Thum and R. Gold. 2012. Regulated microRNAs in the CSF of patients with multiple sclerosis: a case-control study. *Neurology*. 79: 2166–70.

Hecker, M., M. Thamilarasan, D. Koczan, I. Schroder, K. Flechtner, S. Freiesleben, G. Fullen, H.J. Thiesen and U.K. Zettl. 2013. MicroRNA expression changes during interferon-beta treatment in the peripheral blood of multiple sclerosis patients. *Int J Mol Sci*. 14: 16087–110.

Jernas, M., C. Malmestrom, M. Axelsson, I. Nookaew, H. Wadenvik, J. Lycke and B. Olsson. 2013. MicroRNA regulate immune pathways in T-cells in multiple sclerosis (MS). *BMC Immunol*. 14: 32.

Junker, A., M. Krumbholz, S. Eisele, H. Mohan, F. Augstein, R. Bittner, H. Lassmann, H. Wekerle, R. Hohlfeld and E. Meinl. 2009. MicroRNA profiling of multiple sclerosis lesions identifies modulators of the regulatory protein CD47. *Brain*. 132: 3342–52.

Keller, A., P. Leidinger, J. Lange, A. Borries, H. Schroers, M. Scheffler, H.P. Lenhof, K. Ruprecht and E. Meese. 2009. Multiple sclerosis: microRNA expression profiles accurately differentiate patients with relapsing-remitting disease from healthy controls. *PLoS One*. 4: e7440.

Keller, A., P. Leidinger, F. Steinmeyer, C. Stahler, A. Franke, G. Hemmrich-Stanisak, A. Kappel, I. Wright, J. Dorr, F. Paul, R. Diem, B. Tocariu-Krick, B. Meder, C. Backes, E. Meese and K. Ruprecht. 2013. Comprehensive analysis of microRNA profiles in multiple sclerosis including next-generation sequencing. *Mult Scler*.

Kosaka, N., Y. Yoshioka, K. Hagiwara, N. Tominaga, T. Katsuda and T. Ochiya. 2013. Trash or Treasure: extracellular microRNAs and cell-to-cell communication. *Front Genet*. 4: 173.

Kozomara, A. and S. Griffiths-Jones. 2011. miRBase: integrating microRNA annotation and deep-sequencing data. *Nucleic Acids Res*. 39: D152–7.

Kroh, E.M., R.K. Parkin, P.S. Mitchell and M. Tewari. 2010. Analysis of circulating microRNA biomarkers in plasma and serum using quantitative reverse transcription-PCR (qRT-PCR). *Methods*. 50: 298–301.

Kumar, M., T. Ahmad, A. Sharma, U. Mabalirajan, A. Kulshreshtha, A. Agrawal and B. Ghosh. 2011. Let-7 microRNA-mediated regulation of IL-13 and allergic airway inflammation. *J Allergy Clin Immunol*. 128: 1077–85 e1–10.

Lehmann, S.M., C. Kruger, B. Park, K. Derkow, K. Rosenberger, J. Baumgart, T. Trimbuch, G. Eom, M. Hinz, D. Kaul, P. Habbel, R. Kalin, E. Franzoni, A. Rybak, D. Nguyen, R. Veh, O. Ninnemann, O. Peters, R. Nitsch, F.L. Heppner, D. Golenbock, E. Schott, H.L. Ploegh, F.G. Wulczyn and S. Lehnardt. 2012. An unconventional role for miRNA: let-7 activates Toll-like receptor 7 and causes neurodegeneration. *Nat Neurosci*. 15: 827–35.

Lindberg, R.L., F. Hoffmann, M. Mehling, J. Kuhle and L. Kappos. 2010. Altered expression of miR-17-5p in CD4+ lymphocytes of relapsing-remitting multiple sclerosis patients. *Eur J Immunol*. 40: 888–98.

Lock, C., G. Hermans, R. Pedotti, A. Brendolan, E. Schadt, H. Garren, A. Langer-Gould, S. Strober, B. Cannella, J. Allard, P. Klonowski, A. Austin, N. Lad, N. Kaminski, S.J. Galli, J.R. Oksenberg, C.S. Raine, R. Heller and L. Steinman. 2002. Gene-microarray analysis of multiple sclerosis lesions yields new targets validated in autoimmune encephalomyelitis. *Nat Med*. 8: 500–8.

Martinelli-Boneschi, F., C. Fenoglio, P. Brambilla, M. Sorosina, G. Giacalone, F. Esposito, M. Serpente, C. Cantoni, E. Ridolfi, M. Rodegher, L. Moiola, B. Colombo, M. De Riz, V. Martinelli, E. Scarpini, G. Comi and D. Galimberti. 2012. MicroRNA and mRNA expression profile screening in multiple sclerosis patients to unravel novel pathogenic steps and identify potential biomarkers. *Neurosci Lett*. 508: 4–8.

Mendell, J.T. and E.N. Olson. 2012. MicroRNAs in stress signaling and human disease. *Cell*. 148: 1172–87.

Minagar, A. 2013. Current and Future Therapies for Multiple Sclerosis. *Scientifica (Cairo)*. 249101.

Mitchell, P.S., R.K. Parkin, E.M. Kroh, B.R. Fritz, S.K. Wyman, E.L. Pogosova-Agadjanyan, A. Peterson, J. Noteboom, K.C. O'Briant, A. Allen, D.W. Lin, N. Urban, C.W. Drescher, B.S. Knudsen, D.L. Stirewalt, R. Gentleman, R.L. Vessella, P.S. Nelson, D.B. Martin and M. Tewari. 2008. Circulating microRNAs as stable blood-based markers for cancer detection. *Proc Natl Acad Sci USA.* 105: 10513–8.

Morales, Y., J.E. Parisi and C.F. Lucchinetti. 2006. The pathology of multiple sclerosis: evidence for heterogeneity. *Adv Neurol.* 98: 27–45.

Noseworthy, J.H., C. Lucchinetti, M. Rodriguez and B.G. Weinshenker. 2000. Multiple sclerosis. *N Engl J Med.* 343: 938–52.

Olsson, T., W.W. Zhi, B. Hojeberg, V. Kostulas, Y.P. Jiang, G. Anderson, H.P. Ekre and H. Link. 1990. Autoreactive T lymphocytes in multiple sclerosis determined by antigen-induced secretion of interferon-gamma. *J Clin Invest.* 86: 981–5.

Otaegui, D., S.E. Baranzini, R. Armananzas, B. Calvo, M. Munoz-Culla, P. Khankhanian, I. Inza, J.A. Lozano, T. Castillo-Trivino, A. Asensio, J. Olaskoaga and A. Lopez de Munain. 2009. Differential micro RNA expression in PBMC from multiple sclerosis patients. *PLoS One.* 4: e6309.

Pittock, S.J. and C.F. Lucchinetti. 2007. The pathology of MS: new insights and potential clinical applications. *Neurologist.* 13: 45–56.

Roush, S. and F.J. Slack. 2008. The let-7 family of microRNAs. *Trends Cell Biol.* 18: 505–16.

Siegel, S.R., J. Mackenzie, G. Chaplin, N.G. Jablonski and L. Griffiths. 2012. Circulating microRNAs involved in multiple sclerosis. *Mol Biol Rep.* 39: 6219–25.

Sievers, C., M. Meira, F. Hoffmann, P. Fontoura, L. Kappos and R.L. Lindberg. 2012. Altered microRNA expression in B lymphocytes in multiple sclerosis: towards a better understanding of treatment effects. *Clin Immunol.* 144: 70–9.

Smith, K.M., M. Guerau-de-Arellano, S. Costinean, J.L. Williams, A. Bottoni, G. Mavrikis Cox, A.R. Satoskar, C.M. Croce, M.K. Racke, A.E. Lovett-Racke and C.C. Whitacre. 2012. miR-29ab1 deficiency identifies a negative feedback loop controlling Th1 bias that is dysregulated in multiple sclerosis. *J Immunol.* 189: 1567–76.

Sondergaard, H.B., D. Hesse, M. Krakauer, P.S. Sorensen and F. Sellebjerg. 2013. Differential microRNA expression in blood in multiple sclerosis. *Mult Scler.* 19: 1849–57.

Stanczyk, J., D.M. Pedrioli, F. Brentano, O. Sanchez-Pernaute, C. Kolling, R.E. Gay, M. Detmar, S. Gay and D. Kyburz. 2008. Altered expression of MicroRNA in synovial fibroblasts and synovial tissue in rheumatoid arthritis. *Arthritis Rheum.* 58: 1001–9.

Swaminathan, S., K. Suzuki, N. Seddiki, W. Kaplan, M.J. Cowley, C.L. Hood, J.L. Clancy, D.D. Murray, C. Mendez, L. Gelgor, B. Anderson, N. Roth, D.A. Cooper and A.D. Kelleher. 2012. Differential regulation of the Let-7 family of microRNAs in CD4[+] T cells alters IL-10 expression. *J Immunol.* 188: 6238–46.

Tsuchida, A., S. Ohno, W. Wu, N. Borjigin, K. Fujita, T. Aoki, S. Ueda, M. Takanashi and M. Kuroda. 2011. miR-92 is a key oncogenic component of the miR-17-92 cluster in colon cancer. *Cancer Sci.* 102: 2264–71.

Ventura, A., A.G. Young, M.M. Winslow, L. Lintault, A. Meissner, S.J. Erkeland, J. Newman, R.T. Bronson, D. Crowley, J.R. Stone, R. Jaenisch, P.A. Sharp and T. Jacks. 2008. Targeted deletion reveals essential and overlapping functions of the miR-17 through 92 family of miRNA clusters. *Cell.* 132: 875–86.

Voskuhl, R.R., R. Martin, C. Bergman, M. Dalal, N.H. Ruddle and H.F. McFarland. 1993. T helper 1 (Th1) functional phenotype of human myelin basic protein-specific T lymphocytes. *Autoimmunity.* 15: 137–43.

Waschbisch, A., M. Atiya, R.A. Linker, S. Potapov, S. Schwab and T. Derfuss. 2011. Glatiramer acetate treatment normalizes deregulated microRNA expression in relapsing remitting multiple sclerosis. *PLoS One.* 6: e24604.

Xiao, C., L. Srinivasan, D.P. Calado, H.C. Patterson, B. Zhang, J. Wang, J.M. Henderson, J.L. Kutok and K. Rajewsky. 2008. Lymphoproliferative disease and autoimmunity in mice with increased miR-17-92 expression in lymphocytes. *Nat Immunol.* 9: 405–14.

13

Putative Protein Biomarkers of Multiple Sclerosis

Swetha Mahesula,[1,3,6-9,#] Itay Raphael,[2,#]
David Black,[3,6,7] Sean Leonard,[2] Madeleine Zaehringer,[2]
Anjali B. Purkar,[1-2,6-9] Jonathan A.L. Gelfond,[11,12]
Thomas G. Forsthuber[2,8,*] and William E. Haskins[1-3, 6-10,12,*]

INTRODUCTION

Multiple Sclerosis (MS) is a debilitating neurological disease that affects approximately 2.5 million people globally (Mateen et al., 2012). Clinical diagnosis of MS, versus other similar neurological diseases, and

[1] Pediatric Biochemistry Laboratory, University of Texas at San Antonio, San Antonio, TX, 78249.
[2] Department of Biology, University of Texas at San Antonio, San Antonio, TX, 78249.
[3] Department of Chemistry, University of Texas at San Antonio, San Antonio, TX, 78249.
[4] Department of Computer Science, University of Texas at San Antonio, San Antonio, TX, 78249.
[5] High Performance Computing, University of Texas at San Antonio, San Antonio, TX, 78249.
[6] RCMI Proteomics Core, University of Texas at San Antonio, San Antonio, TX, 78249.
[7] RCMI Protein Biomarkers Core, University of Texas at San Antonio, San Antonio, TX, 78249.
[8] Center for Interdisciplinary Health Research, University of Texas at San Antonio, San Antonio, TX, 78249.
[9] Center for Research & Training in the Sciences; University of Texas at San Antonio, San Antonio, TX, 78249.
[10] Department of Medicine, Division of Hematology & Medical Oncology, University of Texas Health Science Center at San Antonio, San Antonio, TX, 78229.
[11] Department of Epidemiology & Biostatistics, University of Texas Health Science Center at San Antonio, San Antonio, TX, 78229.
[12] Cancer Therapy & Research Center, University of Texas Health Science Center at San Antonio, San Antonio, TX, 78229.
[#] These authors contributed equally.
[*] Corresponding authors: WEH.scholar@gmail.com

classification into the following consensus definitions of the four major subtypes of MS, is based on a wide variety of tests (Confavreux et al., 2000; Lublin and Reingold, 1996; Noseworthy et al., 2000; Sospedra and Martin, 2005; Steinman, 1996; Thompson et al., 1997): (A) Relapsing–Remitting (RRMS), (B) Secondary Progressive (SPMS), (C) Primary Progressive (PPMS) and (D) Progressive Relapsing (PRMS). Approximately 80% of the patients initially develop the RRMS form of the disease-characterized by clinical attacks (relapses) with diverse neurological dysfunctions, followed by a full recovery or partial recovery with a residual disability. Clinical symptoms include: difficulty walking, muscle spasms, double vision, loss of balance and incontinence. More than half of RRMS patients will eventually develop SPMS-characterized by a steady worsening of clinical symptoms, with or without attacks during the progressive phase. Less than 15% of all patients have PPMS-characterized either by steady, or by periods of slower or faster, progression of clinical worsening with no attacks. The subgroup of MS with the lowest incidence (less than 5%) is PRMS-characterized by progressive disease course with occasional relapses, with either recovery or no recovery between these attacks. A graphical overview of each of these subtypes of MS is shown in Figure 13.1A-D, adapted from (Thomson, 2006).

Current immunomodulatory treatments that ameliorate these symptoms of MS patients but do not cure the disease include: beta-interferons, therapeutic antibodies, glucocorticoids, and glatiramer acetate. Individual responses to treatment are typically evaluated by clinical measures of disease progression such as the Expanded Disability Status Scale (EDSS) (Poonawalla et al., 2010), oligoclonal bands (OCBs) produced by plasma cells in MS patients and detected with electrophoretic methods (Lourenco et al., 2012), and Magnetic Resonance Imaging (MRI) of brain

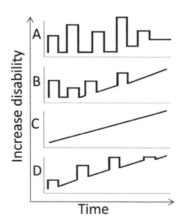

Figure 13.1. Subtypes of MS at-a-glance: (A) RRMS; (B) SPMS; (C) PPMS; (D) PRMS.

lesion volume, cerebral spinal fluid flow, etc. (Bakshi et al., 2008; Filippi and Agosta, 2010; Neema et al., 2007; Tourdias and Dousset, 2012). None of these clinical measures show a strong correlation to disease progression in individual MS patients.

The hallmark of MS biology involves episodes of inflammation and demyelination, believed to be mediated by an autoimmune attack directed against components of the central nervous system (CNS) including proteins such as Myelin Basic Protein (MBP), proteolipid protein (PLP), and Myelin Oligodendrocyte Glycoprotein (MOG). This autoimmune attack generates inflammation in the brain acting as a biological distress signal, and recruits innate- and adaptive immune-cells to the site of inflammation. Then, immune system cells initiate damage to the protective myelin sheath on axons and a breakdown of nerve-to-nerve signal transmission. Once activated, autoimmune T cells migrate and infiltrate the CNS, crossing the Blood-Brain-Barrier (BBB) in a multistep process (Steinman, 1996). Both CD8+ and CD4+ T cells infiltrate the brain of MS patients and have active roles in disease pathogenesis (Goverman, 2009). Infiltrating T cells release pro- and anti-inflammatory cytokines that modulate the activation of microglia and infiltrating macrophages and dendritic cells to release neurotoxic mediators, including nitric oxide and Reactive Oxygen Species (ROS) (Dhib-Jalbut et al., 2006; Rovaris et al., 2006). Macrophages, microglia cells and dendritic cells are also actively involved in the inflammatory response (Benveniste, 1997; Hartung et al., 1992). B cells might also be involved in MS, conceivably by secreting myelin-specific antibodies to further exacerbate demyelination (von Budingen et al., 2011). Finally, microbes or viruses might contribute to the disease process by initially activating autoreactive T cells due to structure similarity shared between pathogenic-antigens and neuro-self antigens (also termed molecular mimicry) (Azoulay-Cayla, 2000). In addition, bacteria and viruses may promote disease via shared pathogen-associated molecular patterns that trigger inflammatory events through Toll-like receptor signaling. While MS research has historically focused on inflammatory events in the CNS, for example on the pathological role of cytokines (Olsson, 1992; Veldhoen, 2009), a more detailed molecular understanding of the biology of other proteins, particularly those released across the BBB into CSF or blood, will be critical to further our understanding of progressive disease and MS and develop and test new treatments to reduce the burden of disease borne by MS patients and their families. Thus, development of biomarkers for MS will be critical to move forward.

The most sensitive and selective molecular biomarkers of disease progression of MS are expected to be proteins. More than 24,000 genes are translated into an estimated 2 million protein isoforms in humans, encoding far more molecular diversity than the relatively static genome or transcriptome. Paradoxically, less than 100 proteins are routinely quantified in blood today (Anderson and Anderson, 2002; Rifai et al., 2006). Proteins must be measured directly due to the poor correlation between the transcriptome and proteome due to alternative splicing, Post-Translational Modifications (PTMs), single nucleotide polymorphisms, limiting ribosomes available for translation, mRNA and protein stability and unknown actors (e.g., microRNA).

In this chapter, we describe progress in the quest for novel predictive/prognostic biomarkers that accurately reflect disease progression in individual MS patients. Caution should be taken when interpreting any discussion of putative protein biomarkers of MS because of the limited and heterogeneous nature of published studies. Indeed, none of the proteins discussed below have both high sensitivity and specificity for a large population of MS patients.[1] Therefore, we focus on a relatively small number of putative protein biomarkers of MS that exemplify selected biological function categories (Bielekova and Martin, 2004), known isoforms and PTMs thereof (Table 13.1). Lastly, putative protein biomarkers revealed by studies of animal models of MS, and their potential translation to studies of MS patients, are also presented.

Putative Protein Biomarkers of MS

(i) Biomarkers reflecting alteration of the immune system:

a. Interleukins such as interleukin 17 A-F (IL-17 A-F) produced by CD4[+] T cells, also known as Th17 cells, might be the dominant mediators of MS pathogenesis and autoimmunity inflammation (Cua et al., 2003; Harrington et al., 2005; Park et al., 2005). IL-17 is associated with non-responsiveness to interferon-beta (IFN-β) therapy in RRMS patients (Axtell et al., 2010). IL-10, a potent anti-inflammatory cytokine (Saraiva and O'Garra, 2010), might also be an important predictive biomarker for IFN-β treatment. IL-12 and IL-23, important for the differentiation of naïve T cells into Th1 and Th17 cells respectively (Kobayashi et al., 1989; Park et al., 2005; Trinchieri et al., 1992), were observed at higher levels in CSF from MS patients compared to healthy controls (Bielekova et al., 2012). Numerous studies have shown the differential expression of other interleukins (e.g., IL-1, IL-2, IL-4, IL-6, IL-10, IL-13, IL-18) in

---

[1] The advantages and disadvantages of: individual study designs (number of MS patient specimens, clinical correlates, statistical analysis of fold-changes, etc.); sampling (post-mortem brain tissue, CSF, blood, serum, peripheral blood mononuclear cells, etc.); and protein measurement methods (antibody-based, two-dimensional electrophoresis, mass spectrometry, tandem mass spectrometry, etc.) are beyond the scope of this work.

Table 13.1. Chapter at-a-glance: Selected putative protein biomarkers of MS and their biological function categories.

Putative protein biomarker of MS	Biological function category
IL-17 A-F	Biomarkers reflecting alteration of the immune system
IL-10	
IFN-β	
IL-12 and IL-23	
IL-1, IL-2, IL-4, IL-6, IL-10, IL-13, IL-18	
MIF	
TNF-α and TNF-β	
CXCL12 , CXCL13 and CXCR5	
HLA	
C3, C4, C4a, C1 inhibitor, factor H and C9	
VLA-4	
OPN	
ICAM-1	
NCAM-1	
MMP-3 and MMP-9	Biomarkers of BBB disruption
MMP-2/TIMP-2	
THBD	
MBP	Biomarkers of demyelination
PLP1	
PRX1-6	Biomarkers of oxidative stress and excitotoxicity
Tubulins and actins	Biomarkers of axonal/neuronal damage
NF-L and NF-H	
S100β	
Tau	
GFAP	Biomarkers of gliosis
14-3-3ε	Biomarkers of remyelination and repair
CNTF	
MAP-2+-13	
B-CPK	
NSE	
PAM	
DBP	Other

MS patients compared to healthy controls (Chiaramonte et al., 2003; Gironi et al., 2013; Martins et al., 2011; Oehninger-Gatti et al., 2000).

b. Macrophage Migration Inhibitory Factor (MIF) is a proinflammatory cytokine involved in autoimmune disease that may counter the effects of IL-2 (Denkinger et al., 2004). MIF has also been linked to various proliferative, anti-apoptotic, inflammatory and angiogenic survival pathways of p53, CD74, NFκB, Bcl2, IL-8 and VEGF (Binsky et al., 2007; Fingerle-Rowson et al., 2003). Interestingly, MIF was found to interact with peroxiredoxin 1 (PRX1), where the complex decreased the D-dopachrometautomerase activity to inhibit chemotaxis (Chen and Moy, 2000; Jung et al., 2001). Recently, elevated MIF levels were correlated to worsening clinical measures of disease progression in MS patients (Hagman et al., 2011).

c. Tumor Necrosis Factors (TNF) is a family of cytokines that includes pro-inflammatory TNF-alpha (TNF-α) and anti-inflammatory TNF-beta (TNF-β). These cytokines play important roles in various physiological process, inflammation and apoptosis (Chu, 2013). TNF-α is secreted by macrophages and T-cells present in the active brain lesions of MS patients. Studies in the CSF have shown significantly higher levels of TNF-α in patients with chronic progressive MS when compared to patients with stable MS (Hauser et al., 1986; Sharief and Hentges, 1991).

d. CXCL13 is a selective chemoattractant for B lymphocytes and B helper T cells via its specific receptor CXCR5 (Legler et al., 1998; Ziemann et al., 2011). Elevated levels of CXCL12 and CXCL13 have been also been reported in the CSF of MS patients (Krumbholz et al., 2006; Romme Christensen et al., 2012). Significantly, IFNγ-producing and Myelin-Basic Protein (MBP)-reactive CCR2+CCR5+CCR6+T cells were enriched in the CSF of MS patients compared to patients with other neurological diseases (Sato et al., 2012).

e. Human Leukocyte Antigen (HLA) variants have been associated with genetic susceptibility for MS for some time (Ebers and Paty, 1979; Francis et al., 1991). Recent studies suggest that soluble HLA measurements in saliva from RRMS patients might be comparable to those found in CSF (Adamashvili et al., 2005).

f. The complement system is part of the innate immune response and involves circulating latent precursor proteins in the bloodstream. In the presence of an infection, the complement system produces a cascade of enzymatic cleavages that activate precursor proteins involved in the opsonization of myelin. Complement mediated attack complex proteins have been identified in the plasma and CSF of MS patients at various stages of disease severity (Ingram et al., 2012; Ingram et al., 2011; Li et al., 2011; Pieragostino et al., 2010; Sawai et al., 2010; Teunissen et al., 2011). For example, factor H levels were significantly increased in progressive MS patients (Ingram et al., 2010). Likewise, increased plasma levels of C3, C4, C4a, C1 inhibitor, and factor H, and reduced levels of C9 were observed in MS patients compared to controls (Ingram et al., 2012).

g. Very Late Antigen-4 (VLA-4) is an integrin expressed in a wide variety of immune cells that facilitate cell adhesion and T-cell activation (Bornsen et al., 2012). VLA-4 is essential for leukocytes to cross the BBB and suggested to be a marker for long term interferon therapy in MS patients (Ohkuma et al., 2007; Soilu-Hanninen et al., 2005). Natalizumab, the first approved therapeutic antibody for treating MS patients, blocks VLA-4 and is highly effective in improving disease in RRMS patients (Wipfler et al., 2011).

h. Osteopontin (OPN) has been shown to play important roles in mediating inflammation, immunity and autoimmune demyelinating disease (Ashkar et al., 2000; Chabas et al., 2001; O'Regan et al., 1999). High levels of OPN have been observed in the CSF of MS patients (Romme Christensen et al., 2012).

i. Intercellular Adhesion Molecule-1 (ICAM-1) is an early marker of immune activation and is bound to the cell surface. Previous studies have shown that ICAM-1 is differentially expressed in the serum and CSF of MS patients compared to healthy controls (Sharief et al., 1993).

j. Neural Cell Adhesion Molecule (NCAM) is a glycoprotein found within the plasma membrane of neurons and glial cells. It contains extracellular fibronectin type III and Ig-like domains that are associated with cell-cell adhesion and the initiation of neuronal growth, and a variety of isoforms are found on the cell surface of neurons (Ditlevsen and Kolkova, 2010; Niethammer et al., 2002). Polysialyated NCAM is a marker of neuronal development and plasticity (Quartu et al., 2008), and when NCAM is expressed on the axonal surface it has an inhibitory effect on remyelination that believed to be involved in MS progression (Charles et al., 2002). CSF from MS patients exhibited lower levels of soluble NCAM than healthy controls (Jia et al., 2012).

(ii) Biomarkers of BBB disruption

a. Matrix metalloproteinases (MMPs) are endopeptidases that are moderated by their corresponding tissue inhibitors of MMPs (TIMPs) (Romi et al., 2012; Yong et al., 2007). A strong correlation between an increase in MMP-3 and MMP-9 levels in serum for RRMS patients in remission has been observed (Kanesaka et al., 2006). MMP-9, in particular, has being implicated in enhancing BBB permeability during inflammation (Gijbels et al., 1994; Sato et al., 2012). MMP-2/TIMP-2 levels in CSF and serum were significantly higher for patients with RRMS than for those patients with other inflammatory neurological disorders or non-inflammatory neurological disorders, and a possible correlation between MMP-2/TIMP-2 levels in serum and the duration of disease progression has been proposed (Fainardi et al., 2009).

b. Thrombomodulin (THBD) is a membrane protein found in the brain capillary endothelial cells (BCECs) that make up the BBB. It is known to play a role in the activation of protein C, which has various roles, including, as an inhibitor to inflammatory cell movement across the BBB (Festoff et al., 2005). Previous studies have shown that THBD in blood and serum are elevated in patients with MS, consistent with the paradigm of increased release of THBD though a damaged BBB (Tsukada et al., 1995). THBD in the plasma of RRMS patients in remission was observed at higher levels than healthy controls (Festoff et al., 2012).

(iii) Biomarkers of demyelination

a. MBP is the second most abundant protein in CNS myelin, comprising ~30% of the total protein in the myelin sheath (Boggs, 2006). MBP is a positively charged protein that binds to negatively charged lipids, present at the cytosolic surface of myelin. Alternative splicing and PTMs generate numerous MBP isoforms (Akiyama et al., 2002; Boggs, 2006; Boggs et al., 1997; Givogri et al., 2000; Kimura et al., 1998; Nakajima et al., 1993; Ottens et al., 2008). One of the charged isoforms, known as C8, is overexpressed in MS and may be involved in its pathogenesis (Kim et al., 2003). Most RRMS patients have significantly elevated MBP or MBP fragment (i.e., breakdown product) levels in the CSF during relapse compared to healthy controls or non-relapsing MS patients, providing an index for active demyelination (Martin et al., 2006; Sellebjerg et al., 1998; Ziemann et al., 2011).

b. Proteolipid protein **(PLP1)** is the most abundant membrane protein of CNS myelin, playing an important role in maintaining axonal integrity. Consequently, PLP1 is a well-studied protein for MS and other neurological disorders (Tuohy, 1994). The full length PLP1 gene encodes an integral membrane protein of 276 amino acid residues with four hydrophobic membrane-spanning domains (Lazzarini, 2004). Alternative splicing generates a functionally distinct isoform known as DM20 that lacks a 35 residue intracellular region. DM20 is selectively expressed in early development, thymus, and is the most abundant isoform in immature oligodendrocytes. In contrast, PLP1 is restricted to myelinating Schwann cells (Ikenaka et al., 1992). Autoimmunity against PLP1 has been extensively studied in MS. B-cells secreting autoantibodies against PLP1 and PLP1-reactive T-cells have been identified in the peripheral blood and CSF of MS patients (Kaushansky et al., 2012; Sun et al., 1991).

(iv) Biomarkers of oxidative stress and excitotoxicity

a. Peroxiredoxins **(PRXs)** are a family of enzymes comprised of six isoforms that protect against oxidation in various parts of the cell by reactive oxygen species (ROS): PRX 1,2 and 6 are cytosolic; PRX 3 and 5 are mitochondrial/peroxisomal; and PRX 4 is extracellular (Rhee et al., 2005). **PRXs** reduce and eliminate peroxides and other ROS that lead to cell apoptosis (Kang et al., 1998). PRX1 also inhibits ASK1-mediated apoptosis, particularly in high peroxide environments, and long-term exposure of BCECs to ROS or activation of microglia by lipopolysaccharide stimulation leads to up-regulation of PRX1 (Kim et al., 2008b). Therefore, PRX1 may be an indicator for activated microglia, where it functions as a free radical scavenger to protect cells from ROS-induced apoptosis, particularly in the inflammatory lesions (Kim et al., 2008a; Schreibelt et al., 2008a,b).

(v) Biomarkers of axonal/neuronal damage

a. Microfilaments and microtubules are major components of the neuronal cytoskeleton. They are primarily composed of cytoskeletal proteins such as actins and tubulins, with numerous subunits and isoforms. Tubulins comprises 20% of the brain cellular protein and functions in axonal growth, migration and transport (Downing, 2000). Actins are present in microfilaments to provide stability for membrane integrity and act as substrates for transport (Letourneau, 2009). Studies in CSF have shown upregulation of actin and tubulin subunits for MS patients vs. healthy controls (Teunissen et al., 2005). Tubulin levels in particular were observed to be upregulated in progressive MS when compared to RRMS patients or healthy controls (Semra et al., 2002).

b. Neurofilament (NF) proteins are major cytoskeletal structural proteins of neurons and are found heavily concentrated in axons (Khalil et al., 2012; Mateen et al., 2012; Teunissen and Khalil, 2012). NF-Light (NF-L) shows some promise as an indicator of acute axonal damage because higher levels have been observed during relapses compared to remission in MS patients. Significantly, elevated NF-L levels measured in patients with clinically definite MS have been shown to be predictive of more rapid disease progression and increased risk of conversion to progressive disease, albeit NF-L levels of MS patients in remission can be comparable to healthy controls (Gunnarsson et al., 2011). Unfortunately, NF-L is degraded relatively rapidly in the CSF or blood, thus potentially confounding NF-L measurements (Koel-Simmelink et al., 2011). In contrast, NF-heavy (NF-H) is usually heavily phosphorylated and particularly resistant to proteolysis when released into the CSF or blood by axonal degeneration. Lastly, recent studies found higher levels of NF-H in RRMS or SPMS patients with higher EDSS scores (Gresle et al., 2011).

c. S100β belongs to the S100 family of calcium binding proteins. It is one of the highly expressed proteins in the CNS and found primarily expressed in astrocytes (Zimmer et al., 1995). S100B elicits an increase in Ca^{+2} concentration in both glial and neuronal cells and plays a role in neurite outgrowth, glial cell proliferation and neuronal maturation (Sorci et al., 2010). Neurotrophic and gliotrophic actions of S100β suggest it could also function in normal CNS development and recovery after injury (Yardan et al., 2011). S100β levels are elevated in adult MS patients compared to healthy controls, and differences in S100β expression were observed among PPMS, SPMS and RRMS patients (Petzold et al., 2002).

d. Tau, including cleaved and phosphorylated isoforms, functions in the assembly and stabilization of microtubules and is released into the CSF during axonal damage (Frederiksen et al., 2012). A common feature in many neurological disorders is the formation of insoluble neurotoxic Tau protein aggregates due to phosphorylation (Ballatore et al., 2007; Dujmovic, 2011). Tau levels increased with disease progression in adult SPMS and PPMS patients (Brettschneider et al., 2005), and Tau levels were also higher for childhood MS patients compared to controls (Rostasy et al., 2005).

(vi) Biomarkers of gliosis

a. Glial fibrillary protein (GFAP) is an intermediate filament protein present in astrocytes and other CNS cells (Avsar et al., 2012). Where neuronal injury occurs, gliosis leads to an influx of astrocytes and elevated GFAP levels. This glial scar might serve as a protective barrier around the damaged axon. GFAP is also released into the CSF or blood, with significantly higher levels in MS patients compared to healthy controls (Axelsson et al., 2011; Festoff et al., 2005).

(vii) Biomarkers of remyelination and repair

a. The 14-3-3 proteins are an evolutionarily conserved family consisting of seven isoforms that are abundantly expressed in the nervous system and involved in a variety of cellular functions, including: cell cycle; transcriptional control; regulation of ion channels; apoptosis; and neurodegeneration (Berg et al., 2003; Fu, 2000; Satoh et al., 2006). 14-3-3ε plays an important role in nerve apoptosis by interaction with downstream molecules (Kimura et al., 2001). For example, 14-3-3ε interacts with and blocks pro-apoptotic proteins that can be released by caspase-3 cleavage during apoptosis (Won et al., 2003). 14-3-3ε has been observed in the CSF of MS patients presenting with severe myelitis and in reactive astrocytes in demyelinating lesions (Satoh et al., 2004; Satoh et al., 2003). Other studies have shown that 14-3-3ε is differentially expressed in human BCECs when treated with sera from MS patients. Lastly, 14-3-3ε interacts with GFAP in cultured human astrocytes, suggesting a role in the activation of astrocytes during CNS inflammation. Taken together, this suggests that 14-3-3ε serves as a neurotrophic protein that may induce or prevent neuronal cell death.

b. Cytokine ciliaryneurotrophic factor (CNTF) is found to have augmenting effect on oligodendrocyte apoptosis (Linker et al., 2002). CNTF is a member of the IL-6 cytokine family expressed in glial cells that promotes survival of neurons and oligodendrocytes (Vergara and Ramirez, 2004). Circulating CNTF binds with the CNTF receptor alpha complex thereby initiating gene expression via the JAK-STAT signaling cascade (Krady et al., 2008). Peripheral blood mononuclear cells from MS patients contained increased levels of CNTF after corticosteroid treatment (Lindquist et al., 2011).

c. Microtubule-associated protein-2 exon 13 (MAP-2+-13) is expressed in oligodendrycytes during human fetal development and declines as cells reach maturation. Surprisingly, MAP-2+13 is also expressed in remyelinating oligodendrocytes found in adult patients with MS lesions (Shafit-Zagardo et al., 1999).

d. Brain creatine phosphokinase (B-CPK) is a cytosolic enzyme that catalyzes the conversion of creatine to phosphocreatine. Reduced levels were observed in the serum and white matter of MS patients compared to healthy controls (Steen et al., 2010).

e. Neuron-specific enolase (NSE) is a glycolytic enzyme localized in the neuronal cytoplasm and responsible for a critical step of glycolysis, the conversion of 2-phospho-D-glycerate into phosphoenolpyruvate (Royds et al., 1983). Plasma levels of NSE have been inversely correlated to MS disease progression using EDSS scores (Hein Nee Maier et al., 2008; Koch et al., 2007). Anti-NSE antibodies have been isolated from CSF in MS patients, suggesting release of NSE across the BBB (Forooghian et al., 2007; Nakamura et al., 2007).

f. Peptidylglycine a-amidating monooxygenase (PAM), is an enzyme expressed by Schwann cells, glia, and neurons that is responsible for activation of peptides in the brain via c-terminal amidation (Bolkenius and Ganzhorn, 1998). High levels of PAM were identified in the CSF of MS patients potentially released from dying neurons and glial cells (Tsukamoto et al., 1995).

(viii) Other

a. Vitamin D binding protein **(DBP)** is a glycoprotein found in blood and CSF that transports vitamin D and related its metabolites. A possible link between vitamin D deficiency and the risk of acquiring MS has been reported (Ascherio et al., 2010). Therefore, DBP might be a good candidate biomarker for MS, but conflicting results have been reported (Disanto et al., 2011). For example, significantly lower levels of DBP were observed in the CSF of RRMS compared to patients with other neurological disorders (Qin et al., 2009). However, higher levels of DBP protein were observed for RRMS patients compared to healthy controls (Rithidech et al., 2009; Yang et al., 2013).

Putative Protein Biomarkers of EAE

The most commonly studied animal model of MS is murine Experimental Autoimmune Encephalomyelitis (EAE). While there are other animal models for MS and various demyelinating diseases (Denic et al., 2011), T cell-mediated EAE closely mimics disease progression in human RRMS patients. EAE can be induced by injection of myelin-derived antigens in adjuvant, adoptive-transfer of antigen specific auto-reactive T cells, or develop spontaneously in mice transgenic for T cell receptors of neuroantigen-reactive T cells (Goverman, 2009; Lafaille et al., 1994; Zamvil et al., 1985). Significantly, the generation of knockout mice for genes thought to be relevant for the disease process enabled dissection of their function *in vivo* and *in vitro*, and will be useful for identifying the biological role of putative protein biomarkers for EAE/MS. Several notable discoveries have been made by EAE studies of proteins for some of the selected biological function categories discussed above (Rosenling et al., 2011).

Biomarkers reflecting alteration of the immune system: IL-17A was found to induce activation of NADPH-oxidase in monocytes and ROS production to disrupt the BBB (Huppert et al., 2010). Anti-MIF antibodies and Edaravone, a ROS scavenger, have been shown to ameliorate EAE (Denkinger et al., 2003; Moriya et al., 2008). Similarly, MIF knockout mice have been shown to respond better to glucocorticoid treatment compared to wild-type mice, with a delay in disease onset (Ji et al., 2010). Complement factor H has been shown to be critical for myelin opsonization (Griffiths et al., 2009).

Demyelination: Calcium-activated neutral proteinase (calpain), a cytosolic cysteine endopeptidase (expressed by astrocytes, T-cells, activated microglia and macrophages (Kim et al., 2003)), was shown to cleave MBP and other proteins (Shields and Banik, 1999; Shields et al., 1998). Calpain inhibitors have been shown to suppress disease progression (Hassen et al., 2006; Hassen et al., 2008).

Biomarkers of axonal and neuronal damage: Serum levels of phosphorylated NF-H were observed to peak with disease severity (Gresle et al., 2008).

Gliosis: Bioluminescent GFAP expression has been strongly correlated with disease progression (Luo et al., 2008).

Markers of remyelination and repair: The actions of IL-17A and other pro-inflammatory cytokines might be abrogated by CNTF (Linker et al., 2009; Lu et al., 2009), and CNTF knockout mice have been shown to recover more slowly from injury (Yao et al., 1999). Differential expression of B-CPK and other proteins have also been observed (Jastorff et al., 2009).

Other: High levels of DBP have been shown to adversely affect recovery, suggesting DBP as a promising biomarker for progressive disease (Yang et al., 2013). Our laboratory has shown that glucose-6-phosphate isomerase (GPI) and other proteins (e.g., 14-3-3ε, PLP1; PRX1, etc.) closely mirror disease progression (Raphael et al., 2012). We also precisely quantified MBP, MIF, and CD47 (Mahesula et al., 2012). GPI is a dimeric enzyme that catalyzes the reversible isomerization of glucose-6-phosphate and fructose-6-phosphate. In the cytoplasm, GPI is involved in glycolysis and gluconeogenesis, while outside the cell it functions as a neurotrophic factor for spinal and sensory neurons. Major metabolic pathways converging from G6P include the pentose phosphate pathway to produce NADPH and ribose (nucleotides) and the glycolytic pathway/TCA cycle to produce ATP. Lastly, 14-3-3γ depletion causes an increase of GPI expression in the brain, suggestion a mechanism by which GPI compensates for the lack of 14-3-3 to enhance cell survival (Steinacker et al., 2005). CD47, is a "marker of self" overexpressed by myelin that prevents phagocytosis, or "cellular devouring", by inappropriately activated, signal regulatory protein alpha (SIRPα)-expressing, microglia and CNS-infiltrating macrophages (Oldenborg et al., 2000; Tsai and Discher, 2008; Tsai et al., 2010; Van et al., 2006). Based on our own results and previous work, we proposed that regulation of inhibitory CD47-SIRPα interactions and/or ROS signaling may promote removal of neurotoxic myelin debris and CNS repair. The mechanism, by which CD47+ intact myelin and CD47+ myelin debris are distinguished by phagocytically active microglia/macrophages remains an intriguing biological question (Han et al., 2012). Nonetheless, discovering protein biomarkers of EAE is expected to provide critical information that might not be revealed by other MS studies, including biomarkers that are predictive of disease onset, severity, chronic progression and response to treatment.

Conclusions

A combination of putative protein biomarkers might enable a more precise molecular stratification of different subtypes of MS patients: beyond RRMS,

PPMS, SPMS and PRMS. However, as delineated in other chapters in this book, the majority of putative protein biomarkers of MS are not exclusive to a single disorder. Rather, they may also be found in other diseases and neurological disorders. Poorly understood variations of socioeconomic, environmental and genetic factors in the MS patient population are profound challenges for protein biomarker research. Studying the correlation of protein biomarker expression to disease progression within individual MS patients is expected to mitigate population variability to a certain degree and account for potential patient-specific factors. Time-course analyses is also expected to distinguish leading and lagging indicators of disease states that are clinically important for improved diagnosis and treatment of different phases of disease progression. For these reasons and others, we anticipate that longitudinal measurements of a combination of protein biomarkers with high sensitivity and selectivity, for which at least some are expected to be observable independent of any clinical measures of disease progression (EDSS, OCBs or MRI), are urgently needed to personalize MS diagnostics and treatment. Proof-of-principle studies in animal models of human MS that assess the longitudinal biomarker fingerprint of individual mice at specific time points will allow us to identify associations of biomarkers with disease trajectories and treatment conditions. Ideally, taking multiple measurements from individual mice from blood or CSF will provide powerful proof-of-principle for the feasibility of developing homologous biomarkers for clinical monitoring. Finally, we anticipate that novel protein biomarkers will lead to novel biological insights on MS that might be leveraged towards the discovery of novel therapeutic targets, particularly for the treatment of progressive MS.

Acknowledgements

This project was supported by a RCMI grant from the National Institute on Minority Health and Health Disparities (G12MD007591) (WEH, TGF), NS52177 (TGF), and NIH5U54RR022762-05 (WEH) from the National Institute of Health, and grant RG3701 from the National Multiple Sclerosis Society (TGF). We thank the RCMI program and facilities at UTSA for assistance. The authors also acknowledge the support of the Cancer Therapy and Research Center at the University of Texas Health Science Center San Antonio, a National Cancer Institute -designated Cancer Center (NIHP30CA054174). Lastly, the authors acknowledge the following persons for their contributions: Karan Kalsaria, Venkat Kotagiri, Anusha Manne, Monisha Kandala, Soumita Bose and Okame Sanders.

References

Adamashvili, I., A. Minagar, E. Gonzalez-Toledo, L. Featherston and R.E. Kelley. 2005. Soluble HLA measurement in saliva and cerebrospinal fluid in Caucasian patients with multiple sclerosis: a preliminary study. *J Neuroinflammation*. 2: 13.

Akiyama, K., S. Ichinose, A. Omori, Y. Sakurai and H. Asou. 2002. Study of expression of myelin basic proteins (MBPs) in developing rat brain using a novel antibody reacting with four major isoforms of MBP. *J Neurosci Res*. 68: 19–28.

Anderson, N.L. and N.G. Anderson. 2002. The human plasma proteome: history, character, and diagnostic prospects. *Mol Cell Proteomics*. 1: 845–867.

Ascherio, A., K.L. Munger and K.C. Simon. 2010. Vitamin D and multiple sclerosis. *Lancet Neurol*. 9: 599–612.

Ashkar, S., G.F. Weber, V. Panoutsakopoulou, M.E. Sanchirico, M. Jansson, S. Zawaideh, S.R. Rittling, D.T. Denhardt, M.J. Glimcher and H. Cantor. 2000. Eta-1 (osteopontin): an early component of type-1 (cell-mediated) immunity. *Science*. 287: 860–864.

Avsar, T., D. Korkmaz, M. Tutuncu, N.O. Demirci, S. Saip, M. Kamasak, A. Siva and E.T. Turanli. 2012. Protein biomarkers for multiple sclerosis: semi-quantitative analysis of cerebrospinal fluid candidate protein biomarkers in different forms of multiple sclerosis. *Mult Scler J*. 18: 1081–1091.

Axelsson, M., C. Malmestrom, S. Nilsson, S. Haghighi, L. Rosengren and J. Lycke. 2011. Glial fibrillary acidic protein: a potential biomarker for progression in multiple sclerosis. *J Neurol*. 258: 882–888.

Axtell, R.C., B.A. de Jong, K. Boniface, L.F. van der Voort, R. Bhat, P. De Sarno, R. Naves, M. Han, F. Zhong, J.G. Castellanos, R. Mair, A. Christakos, I. Kolkowitz, L. Katz, J. Killestein, C.H. Polman, R. de Waal Malefyt, L. Steinman and C. Raman. 2010. T helper type 1 and 17 cells determine efficacy of interferon-beta in multiple sclerosis and experimental encephalomyelitis. *Nature Medicine*. 16: 406–412.

Azoulay-Cayla, A. 2000. Is multiple sclerosis a disease of viral origin? *Pathol Biol (Paris)*. 48: 4–14.

Bakshi, R., A.J. Thompson, M.A. Rocca, D. Pelletier, V. Dousset, F. Barkhof, M. Inglese, C.R. Guttmann, M.A. Horsfield and M. Filippi. 2008. MRI in multiple sclerosis: current status and future prospects. *Lancet Neurol*. 7: 615–625.

Ballatore, C., V.M. Lee and J.Q. Trojanowski. 2007. Tau-mediated neurodegeneration in Alzheimer's disease and related disorders. *Nature Reviews Neuroscience*. 8: 663–672.

Benveniste, E.N. 1997. Role of macrophages/microglia in multiple sclerosis and experimental allergic encephalomyelitis. *J Mol Med (Berl)*. 75: 165–173.

Berg, D., C. Holzmann and O. Riess. 2003. 14-3-3 proteins in the nervous system. *Nat Rev Neurosci*. 4: 752–762.

Bielekova, B., M. Komori, Q. Xu, D.S. Reich and T. Wu. 2012. Cerebrospinal Fluid IL-12p40, CXCL13 and IL-8 as a Combinatorial Biomarker of Active Intrathecal Inflammation. *PloS One*. 7: e48370.

Bielekova, B. and R. Martin. 2004. Development of biomarkers in multiple sclerosis. *Brain*. 127: 1463–1478.

Binsky, I., M. Haran, D. Starlets, Y. Gore, F. Lantner, N. Harpaz, L. Leng, D.M. Goldenberg, L. Shvidel, A. Berrebi, R. Bucala and I. Shachar. 2007. IL-8 secreted in a macrophage migration-inhibitory factor- and CD74-dependent manner regulates B cell chronic lymphocytic leukemia survival. *Proc Natl Acad Sci USA*. 104: 13408–13413.

Boggs, J.M. 2006. Myelin basic protein: a multifunctional protein. *Cell Mol Life Sci*. 63: 1945–1961.

Boggs, J.M., P.M. Yip, G. Rangaraj and E. Jo. 1997. Effect of posttranslational modifications to myelin basic protein on its ability to aggregate acidic lipid vesicles. *Biochemistry*. 36: 5065–5071.

Bolkenius, F.N. and A.J. Ganzhorn. 1998. Peptidylglycine α-Amidating Mono-Oxygenase: Neuropeptide Amidation as a Target for Drug Design. *General Pharmacology: The Vascular System*. 31: 655–659.

Bornsen, L., J.R. Christensen, R. Ratzer, A.B. Oturai, P.S. Sorensen, H.B. Sondergaard and F. Sellebjerg. 2012. Effect of natalizumab on circulating CD4+ T-cells in multiple sclerosis. *PloS One.* 7: e47578.

Brettschneider, J., M. Maier, S. Arda, A. Claus, S.D. Sussmuth, J. Kassubek and H. Tumani. 2005. Tau protein level in cerebrospinal fluid is increased in patients with early multiple sclerosis. *Mult Scler.* 11: 261–265.

Chabas, D., S.E. Baranzini, D. Mitchell, C.C. Bernard, S.R. Rittling, D.T. Denhardt, R.A. Sobel, C. Lock, M. Karpuj, R. Pedotti, R. Heller, J.R. Oksenberg and L. Steinman. 2001. The influence of the proinflammatory cytokine, osteopontin, on autoimmune demyelinating disease. *Science.* 294: 1731–1735.

Charles, P., R. Reynolds, D. Seilhean, G. Rougon, M.S. Aigrot, A. Niezgoda, B. Zalc and C. Lubetzki. 2002. Re-expression of PSA-NCAM by demyelinated axons: an inhibitor of remyelination in multiple sclerosis? *Brain: A Journal of Neurology.* 125: 1972–1979.

Chen, A. and V.T. Moy. 2000. Cross-linking of cell surface receptors enhances cooperativity of molecular adhesion. *Biophys J.* 78: 2814–2820.

Chiaramonte, M.G., M. Mentink-Kane, B.A. Jacobson, A.W. Cheever, M.J. Whitters, M.E. Goad, A. Wong, M. Collins, D.D. Donaldson, M.J. Grusby and T.A. Wynn. 2003. Regulation and function of the interleukin 13 receptor alpha 2 during a T helper cell type 2-dominant immune response. *The Journal of Experimental Medicine.* 197: 687–701.

Chu, W.M. 2013. Tumor necrosis factor. *Cancer Letters.* 328: 222–225.

Confavreux, C., S. Vukusic, T. Moreau and P. Adeleine. 2000. Relapses and progression of disability in multiple sclerosis. *The New England Journal of Medicine.* 343: 1430–1438.

Cua, D.J., J. Sherlock, Y. Chen, C.A. Murphy, B. Joyce, B. Seymour, L. Lucian, W. To, S. Kwan, T. Churakova, S. Zurawski, M. Wiekowski, S.A. Lira, D. Gorman, R.A. Kastelein and J.D. Sedgwick. 2003. Interleukin-23 rather than interleukin-12 is the critical cytokine for autoimmune inflammation of the brain. *Nature.* 421: 744–748.

Denic, A., A.J. Johnson, A.J. Bieber, A.E. Warrington, M. Rodriguez and I. Pirko. 2011. The relevance of animal models in multiple sclerosis research. *Pathophysiology.* 18: 21–29.

Denkinger, C.M., M. Denkinger, J.J. Kort, C. Metz and T.G. Forsthuber. 2003. *In vivo* blockade of macrophage migration inhibitory factor ameliorates acute experimental autoimmune encephalomyelitis by impairing the homing of encephalitogenic T cells to the central nervous system. *J Immunol.* 170: 1274–1282.

Denkinger, C.M., C. Metz, G. Fingerle-Rowson, M.D. Denkinger and T. Forsthuber. 2004. Macrophage migration inhibitory factor and its role in autoimmune diseases. *Arch Immunol Ther Exp (Warsz).* 52: 389–400.

Dhib-Jalbut, S., D.L. Arnold, D.W. Cleveland, M. Fisher, R.M. Friedlander, M.M. Mouradian, S. Przedborski, B.D. Trapp, T. Wyss-Coray and V.W. Yong. 2006. Neurodegeneration and neuroprotection in multiple sclerosis and other neurodegenerative diseases. *J Neuroimmunol.* 176: 198–215.

Disanto, G., S.V. Ramagopalan, A.E. Para and L. Handunnethi. 2011. The emerging role of vitamin D binding protein in multiple sclerosis. *J Neurol.* 258: 353–358.

Ditlevsen, D.K. and K. Kolkova. 2010. Signaling pathways involved in NCAM-induced neurite outgrowth. *Adv Exp Med Biol.* 663: 151–168.

Downing, K.H. 2000. Structural basis for the interaction of tubulin with proteins and drugs that affect microtubule dynamics. *Annual Review of Cell and Developmental Biology.* 16: 89–111.

Dujmovic, I. 2011. Cerebrospinal fluid and blood biomarkers of neuroaxonal damage in multiple sclerosis. *Multiple Sclerosis International.* 767083.

Ebers, G. and D.W. Paty. 1979. The major histocompatibility complex, the immune system and multiple sclerosis. *Clinical Neurology and Neurosurgery.* 81: 69–86.

Fainardi, E., M. Castellazzi, C. Tamborino, A. Trentini, M.C. Manfrinato, E. Baldi, M.R. Tola, F. Dallocchio, E. Granieri and T. Bellini. 2009. Potential relevance of cerebrospinal fluid and serum levels and intrathecal synthesis of active matrix metalloproteinase-2 (MMP-2) as markers of disease remission in patients with multiple sclerosis. *Mult Scler.* 15: 547–554.

Festoff, B.W., C. Li and S.G. Lynch. 2005. Soluble thrombomodulin, a biomarker of blood brain barrier damage and/or neuroprotection in multiple sclerosis. *Mult Scler*. 11: S139–S139.

Festoff, B.W., C.Y. Li, B. Woodhams and S. Lynch. 2012. Soluble thrombomodulin levels in plasma of multiple sclerosis patients and their implication. *J Neurol Sci*. 323: 61–65.

Filippi, M. and F. Agosta. 2010. Imaging biomarkers in multiple sclerosis. *J Magn Reson Imaging*. 31: 770–788.

Fingerle-Rowson, G., O. Petrenko, C. Metz, T. Forsthuber, R. Mitchell, R. Huss, U. Moll, W. Müller and R. Bucala. 2003. The p53-dependent effects of macrophage migration inhibitory factor revealed by gene targeting. *Proc Natl Acad Sci USA*. 100: 9354–9359.

Forooghian, F., R.K. Cheung, W.C. Smith, P. O'Connor and H.M. Dosch. 2007. Enolase and arrestin are novel nonmyelin autoantigens in multiple sclerosis. *Journal of Clinical Immunology*. 27: 388–396.

Francis, D.A., A.J. Thompson, P. Brookes, N. Davey, R.I. Lechler, W.I. McDonald and J.R. Batchelor. 1991. Multiple sclerosis and HLA: is the susceptibility gene really HLA-DR or -DQ? *Human Immunology*. 32: 119–124.

Frederiksen, J., K. Kristensen, J.M. Bahl and M. Christiansen. 2012. Tau protein: a possible prognostic factor in optic neuritis and multiple sclerosis. *Mult Scler*. 18: 592–599.

Fu, H. 2000. 14-3-3 Proteins: Structure, Function, and Regulation. *Annual Review of Pharmacology and Toxicology*. 40: 617–647.

Gijbels, K., R.E. Galardy and L. Steinman. 1994. Reversal of experimental autoimmune encephalomyelitis with a hydroxamate inhibitor of matrix metalloproteases. *The Journal of Clinical Investigation*. 94: 2177–2182.

Gironi, M., M. Saresella, M. Rovaris, M. Vaghi, R. Nemni, M. Clerici and E. Grossi. 2013. A novel data mining system points out hidden relationships between immunological markers in multiple sclerosis. *Immunity & Ageing: I & A*. 10: 1.

Givogri, M.I., E.R. Bongarzone and A.T. Campagnoni. 2000. New insights on the biology of myelin basic protein gene: the neural-immune connection. *J Neurosci Res*. 59: 153–159.

Goverman, J. 2009. Autoimmune T cell responses in the central nervous system. *Nat Rev Immunol*. 9: 393–407.

Gresle, M.M., H. Butzkueven and G. Shaw. 2011. Neurofilament proteins as body fluid biomarkers of neurodegeneration in multiple sclerosis. *Mult Scler Int*. 315406.

Gresle, M.M., G. Shaw, B. Jarrott, E.N. Alexandrou, A. Friedhuber, T.J. Kilpatrick and H. Butzkueven. 2008. Validation of a novel biomarker for acute axonal injury in experimental autoimmune encephalomyelitis. *J Neurosci Res*. 86: 3548–3555.

Griffiths, M.R., J.W. Neal, M. Fontaine, T. Das and P. Gasque. 2009. Complement factor H, a marker of self protects against experimental autoimmune encephalomyelitis. *J Immunol*. 182: 4368–4377.

Gunnarsson, M., C. Malmestrom, M. Axelsson, P. Sundstrom, C. Dahle, M. Vrethem, T. Olsson, F. Piehl, N. Norgren, L. Rosengren, A. Svenningsson and J. Lycke. 2011. Axonal damage in relapsing multiple sclerosis is markedly reduced by natalizumab. *Ann Neurol*. 69: 83–89.

Hagman, S., M. Raunio, M. Rossi, P. Dastidar and I. Elovaara. 2011. Disease-associated inflammatory biomarker profiles in blood in different subtypes of multiple sclerosis: prospective clinical and MRI follow-up study. *Journal of Neuroimmunology*. 234: 141–147.

Han, M.H., D.H. Lundgren, S. Jaiswal, M. Chao, K.L. Graham, C.S. Garris, R.C. Axtell, P.P. Ho, C.B. Lock, J.I. Woodard, S.E. Brownell, M. Zoudilova, J.F. Hunt, S.E. Baranzini, E.C. Butcher, C.S. Raine, R.A. Sobel, D.K. Han, I. Weissman and L. Steinman. 2012. Janus-like opposing roles of CD47 in autoimmune brain inflammation in humans and mice. *J Exp Med*. 209: 1325–1334.

Harrington, L.E., R.D. Hatton, P.R. Mangan, H. Turner, T.L. Murphy, K.M. Murphy and C.T. Weaver. 2005. Interleukin 17-producing CD4[+] effector T cells develop via a lineage distinct from the T helper type 1 and 2 lineages. *Nature Immunology*. 6: 1123–1132.

Hartung, H.P., S. Jung, G. Stoll, J. Zielasek, B. Schmidt, J.J. Archelos and K.V. Toyka. 1992. Inflammatory mediators in demyelinating disorders of the CNS and PNS. *J Neuroimmunol*. 40: 197–210.

Hassen, G.W., J. Feliberti, L. Kesner, A. Stracher and F. Mokhtarian. 2006. A novel calpain inhibitor for the treatment of acute experimental autoimmune encephalomyelitis. *J Neuroimmunol*. 180: 135–146.

Hassen, G.W., J. Feliberti, L. Kesner, A. Stracher and F. Mokhtarian. 2008. Prevention of axonal injury using calpain inhibitor in chronic progressive experimental autoimmune encephalomyelitis. *Brain Res*. 1236: 206–215.

Hauser, S.L., A.K. Bhan, F. Gilles, M. Kemp, C. Kerr and H.L. Weiner. 1986. Immunohistochemical analysis of the cellular infiltrate in multiple sclerosis lesions. *Annals of Neurology*. 19: 578–587.

Hein Nee Maier, K., A. Kohler, R. Diem, M.B. Sattler, I. Demmer, P. Lange, M. Bahr and M. Otto. 2008. Biological markers for axonal degeneration in CSF and blood of patients with the first event indicative for multiple sclerosis. *Neuroscience Letters*. 436: 72–76.

Huppert, J., D. Closhen, A. Croxford, R. White, P. Kulig, E. Pietrowski, I. Bechmann, B. Becher, H.J. Luhmann, A. Waisman and C.R. Kuhlmann. 2010. Cellular mechanisms of IL-17-induced blood-brain barrier disruption. *FASEB J*. 24: 1023–1034.

Ikenaka, K., T. Kagawa and K. Mikoshiba. 1992. Selective expression of DM-20, an alternatively spliced myelin proteolipid protein gene product, in developing nervous system and in nonglial cells. *J Neurochem*. 58: 2248–2253.

Ingram, G., S. Hakobyan, C.L. Hirst, C.L. Harris, S. Loveless, J.P. Mitchell, T.P. Pickersgill, N.P. Robertson and B.P. Morgan. 2012. Systemic complement profiling in multiple sclerosis as a biomarker of disease state. *Multiple Sclerosis*. 18: 1401–1411.

Ingram, G., S. Hakobyan, C.L. Hirst, C.L. Harris, T.P. Pickersgill, M.D. Cossburn, S. Loveless, N.P. Robertson and B.P. Morgan. 2010. Complement regulator factor H as a serum biomarker of multiple sclerosis disease state. *Brain: A Journal of Neurology*. 133: 1602–1611.

Ingram, G., S. Hakobyan, S. Loveless, N. Robertson and B.P. Morgan. 2011. Complement regulator factor H in multiple sclerosis. *Journal of Cellular Biochemistry*. 112: 2653–2654.

Jastorff, A.M., K. Haegler, G. Maccarrone, F. Holsboer, F. Weber, T. Ziemssen and C.W. Turck. 2009. Regulation of proteins mediating neurodegeneration in experimental autoimmune encephalomyelitis and multiple sclerosis. *Proteomics Clinical Applications*. 3: 1273–1287.

Ji, N.E., M. Talley and T.G. Forsthuber. 2010. Macrophage migration inhibitory factor promotes resistance to glucocorticoids in experimental autoimmune encephalomyelitis. *Mult Scler J*. 16: 1009–1010.

Jia, Y., T. Wu, C.A. Jelinek, B. Bielekova, L. Chang, S. Newsome, S. Gnanapavan, G. Giovannoni, D. Chen, P.A. Calabresi, A. Nath and R.J. Cotter. 2012. Development of protein biomarkers in cerebrospinal fluid for secondary progressive multiple sclerosis using selected reaction monitoring mass spectrometry (SRM-MS). *Clin Proteomics*. 9: 9.

Jung, H., T. Kim, H.Z. Chae, K.T. Kim and H. Ha. 2001. Regulation of macrophage migration inhibitory factor and thiol-specific antioxidant protein PAG by direct interaction. *J Biol Chem*. 276: 15504–15510.

Kanesaka, T., M. Mori, T. Hattori, T. Oki and S. Kuwabara. 2006. Serum matrix metalloproteinase-3 levels correlate with disease activity in relapsing-remitting multiple sclerosis. *J Neurol Neurosur Ps*. 77: 185–188.

Kang, S.W., H.Z. Chae, M.S. Seo, K. Kim, I.C. Baines and S.G. Rhee. 1998. Mammalian peroxiredoxin isoforms can reduce hydrogen peroxide generated in response to growth factors and tumor necrosis factor-alpha. *J Biol Chem*. 273: 6297–6302.

Kaushansky, N., D.M. Altmann, C.S. David, H. Lassmann and A. Ben-Nun. 2012. DQB1*0602 rather than DRB1*1501 confers susceptibility to multiple sclerosis-like disease induced by proteolipid protein (PLP). *J Neuroinflammation*. 9: 29.

Khalil, M., C. Enzinger, C. Langkammer, S. Ropele, A. Mader, A. Trentini, M. Vane, M. Wallner-Blazek, G. Bachmaier, J.J. Archelos, M. Koel-Simmelink, M. Blankenstein, S. Fuchs, F. Fazekas and C. Teunissen. 2012. CSF neurofilament and N-acetylaspartate related brain changes in clinically isolated syndrome. *Mult Scler*.

Kim, J.K., F.G. Mastronardi, D.D. Wood, D.M. Lubman, R. Zand and M.A. Moscarello. 2003. Multiple sclerosis: an important role for post-translational modifications of myelin basic protein in pathogenesis. *Mol Cell Proteomics*. 2: 453–462.

Kim, S.U., C.N. Hwang, H.N. Sun, M.H. Jin, Y.H. Han, H. Lee, J.M. Kim, S.K. Kim, D.Y. Yu, D.S. Lee and S.H. Lee. 2008a. Peroxiredoxin I is an indicator of microglia activation and protects against hydrogen peroxide-mediated microglial death. *Biol Pharm Bull*. 31: 820–825.

Kim, S.Y., T.J. Kim and K.Y. Lee. 2008b. A novel function of peroxiredoxin 1 (Prx-1) in apoptosis signal-regulating kinase 1 (ASK1)-mediated signaling pathway. *FEBS Lett*. 582: 1913–1918.

Kimura, M., M. Sato, A. Akatsuka, S. Saito, K. Ando, M. Yokoyama and M. Katsuki. 1998. Overexpression of a minor component of myelin basic protein isoform (17.2 kDa) can restore myelinogenesis in transgenic shiverer mice. *Brain Res*. 785: 245–252.

Kimura, M.T., S. Irie, S. Shoji-Hoshino, J. Mukai, D. Nadano, M. Oshimura and T.A. Sato. 2001. 14-3-3 is involved in p75 neurotrophin receptor-mediated signal transduction. *J Biol Chem*. 276: 17291–17300.

Kobayashi, M., L. Fitz, M. Ryan, R.M. Hewick, S.C. Clark, S. Chan, R. Loudon, F. Sherman, B. Perussia and G. Trinchieri. 1989. Identification and purification of natural killer cell stimulatory factor (NKSF), a cytokine with multiple biologic effects on human lymphocytes. *The Journal of Experimental Medicine*. 170: 827–845.

Koch, M., J. Mostert, D. Heersema, A. Teelken and J. De Keyser. 2007. Plasma S100beta and NSE levels and progression in multiple sclerosis. *Journal of the Neurological Sciences*. 252: 154–158.

Koel-Simmelink, M.J., C.E. Teunissen, P. Behradkia, M.A. Blankenstein and A. Petzold. 2011. The neurofilament light chain is not stable *in vitro*. *Ann Neurol*. 69: 1065–1066; author reply 1066–1067.

Krady, J.K., H.W. Lin, C.M. Liberto, A. Basu, S.G. Kremlev and S.W. Levison. 2008. Ciliary neurotrophic factor and interleukin-6 differentially activate microglia. *J Neurosci Res*. 86: 1538–1547.

Krumbholz, M., D. Theil, S. Cepok, B. Hemmer, P. Kivisakk, R.M. Ransohoff, M. Hofbauer, C. Farina, T. Derfuss, C. Hartle, J. Newcombe, R. Hohlfeld and E. Meinl. 2006. Chemokines in multiple sclerosis: CXCL12 and CXCL13 up-regulation is differentially linked to CNS immune cell recruitment. *Brain: A Journal of Neurology*. 129: 200–211.

Lafaille, J.J., K. Nagashima, M. Katsuki and S. Tonegawa. 1994. High incidence of spontaneous autoimmune encephalomyelitis in immunodeficient anti-myelin basic protein T cell receptor transgenic mice. *Cell*. 78: 399–408.

Lazzarini, R.A. 2004. Myelin biology and disorders. Elsevier Academic Press.

Legler, D.F., M. Loetscher, R.S. Roos, I. Clark-Lewis, M. Baggiolini and B. Moser. 1998. B cell-attracting chemokine 1, a human CXC chemokine expressed in lymphoid tissues, selectively attracts B lymphocytes via BLR1/CXCR5. *The Journal of Experimental Medicine*. 187: 655–660.

Letourneau, P.C. 2009. Actin in axons: stable scaffolds and dynamic filaments. *Results and Problems in Cell Differentiation*. 48: 65–90.

Li, Y., Z. Qin, M. Yang, Y. Qin, C. Lin and S. Liu. 2011. Differential expression of complement proteins in cerebrospinal fluid from active multiple sclerosis patients. *Journal of Cellular Biochemistry*. 112: 1930–1937.

Lindquist, S., S. Hassinger, J.A. Lindquist and M. Sailer. 2011. The balance of pro-inflammatory and trophic factors in multiple sclerosis patients: effects of acute relapse and immunomodulatory treatment. *Mult Scler*. 17: 851–866.

Linker, R.A., P. Brechlin, S. Jesse, P. Steinacker, D.H. Lee, A.R. Asif, O. Jahn, H. Tumani, R. Gold and M. Otto. 2009. Proteome profiling in murine models of multiple sclerosis: identification of stage specific markers and culprits for tissue damage. *PLoS One*. 4: e7624.

Linker, R.A., M. Maurer, S. Gaupp, R. Martini, B. Holtmann, R. Giess, P. Rieckmann, H. Lassmann, K.V. Toyka, M. Sendtner and R. Gold. 2002. CNTF is a major protective factor in demyelinating CNS disease: a neurotrophic cytokine as modulator in neuroinflammation. *Nature Medicine*. 8: 620–624.

Lourenco, P., A. Shirani, J. Saeedi, J. Oger, W.E. Schreiber and H. Tremlett. 2012. Oligoclonal bands and cerebrospinal fluid markers in multiple sclerosis: associations with disease course and progression. *Mult Scler.*

Lu, Z., X. Hu, C. Zhu, D. Wang, X. Zheng and Q. Liu. 2009. Overexpression of CNTF in Mesenchymal Stem Cells reduces demyelination and induces clinical recovery in experimental autoimmune encephalomyelitis mice. *J Neuroimmunol.* 206: 58–69.

Lublin, F.D. and S.C. Reingold. 1996. Defining the clinical course of multiple sclerosis: results of an international survey. National Multiple Sclerosis Society (USA) Advisory Committee on Clinical Trials of New Agents in Multiple Sclerosis. *Neurology.* 46: 907–911.

Luo, J., P. Ho, L. Steinman and T. Wyss-Coray. 2008. Bioluminescence *in vivo* imaging of autoimmune encephalomyelitis predicts disease. *J Neuroinflammation.* 5: 6.

Mahesula, S., I. Raphael, R. Raghunathan, K. Kalsaria, V. Kotagiri, A.B. Purkar, M. Anjanappa, D. Shah, V. Pericherla, Y.L. Jadhav, J.A. Gelfond, T.G. Forsthuber and W.E. Haskins. 2012. Immunoenrichment microwave and magnetic proteomics for quantifying CD47 in the experimental autoimmune encephalomyelitis model of multiple sclerosis. *Electrophoresis.* 33: 3820–3829.

Martin, R., B. Bielekova, R. Hohlfeld and U. Utz. 2006. Biomarkers in multiple sclerosis. *Disease Markers.* 22: 183–185.

Martins, T.B., J.W. Rose, T.D. Jaskowski, A.R. Wilson, D. Husebye, H.S. Seraj and H.R. Hill. 2011. Analysis of proinflammatory and anti-inflammatory cytokine serum concentrations in patients with multiple sclerosis by using a multiplexed immunoassay. *American Journal of Clinical Pathology.* 136: 696–704.

Mateen, F.J., T. Dua, G.C. Shen, G.M. Reed, R. Shakir and S. Saxena. 2012. Neurological disorders in the 11th revision of the International Classification of Diseases: now open to public feedback. *Lancet Neurol.* 11: 484–485.

Moriya, M., Y. Nakatsuji, K. Miyamoto, T. Okuno, M. Kinoshita, A. Kumanogoh, S. Kusunoki and S. Sakoda. 2008. Edaravone, a free radical scavenger, ameliorates experimental autoimmune encephalomyelitis. *Neurosci Lett.* 440: 323–326.

Nakajima, K., K. Ikenaka, T. Kagawa, J. Aruga, J. Nakao, K. Nakahira, C. Shiota, S.U. Kim and K. Mikoshiba. 1993. Novel isoforms of mouse myelin basic protein predominantly expressed in embryonic stage. *J Neurochem.* 60: 1554–1563.

Nakamura, M., A. Kuramasu, I. Nakashima, K. Fujihara and Y. Itoyama. 2007. Candidate antigens specifically detected by cerebrospinal fluid-IgG in oligoclonal IgG bands-positive multiple sclerosis patients. *Proteomics Clinical Applications.* 1: 681–687.

Neema, M., J. Stankiewicz, A. Arora, Z.D. Guss and R. Bakshi. 2007. MRI in multiple sclerosis: what's inside the toolbox? *Neurotherapeutics.* 4: 602–617.

Niethammer, P., M. Delling, V. Sytnyk, A. Dityatev, K. Fukami and M. Schachner. 2002. Cosignaling of NCAM via lipid rafts and the FGF receptor is required for neuritogenesis. *The Journal of Cell Biology.* 157: 521–532.

Noseworthy, J.H., C. Lucchinetti, M. Rodriguez and B.G. Weinshenker. 2000. Multiple sclerosis. *N Engl J Med.* 343: 938–952.

O'Regan, A.W., G.L. Chupp, J.A. Lowry, M. Goetschkes, N. Mulligan and J.S. Berman. 1999. Osteopontin is associated with T cells in sarcoid granulomas and has T cell adhesive and cytokine-like properties *in vitro. Journal of Immunology.* 162: 1024–1031.

Oehninger-Gatti, C., R. Buzo, J.C. Alcantara, C. Chouza, A. Gomez, D. Cibils and S. Gordon-Firing. 2000. The use of biological markers in the diagnosis and follow-up of patients with multiple sclerosis. Test of five fluids. *Rev Neurol.* 30: 977–979.

Ohkuma, K., T. Sasaki, S. Kamei, S. Okuda, H. Nakano, T. Hamamoto, K. Fujihara, I. Nakashima, T. Misu and Y. Itoyama. 2007. Modulation of dendritic cell development by immunoglobulin G in control subjects and multiple sclerosis patients. *Clinical and Experimental Immunology.* 150: 397–406.

Oldenborg, P.A., A. Zheleznyak, Y.F. Fang, C.F. Lagenaur, H.D. Gresham and F.P. Lindberg. 2000. Role of CD47 as a marker of self on red blood cells. *Science.* 288: 2051–2054.

Olsson, T. 1992. Cytokines in neuroinflammatory disease: role of myelin autoreactive T cell production of interferon-gamma. *Journal of Neuroimmunology.* 40: 211–218.

Ottens, A.K., E.C. Golden, L. Bustamante, R.L. Hayes, N.D. Denslow and K.K. Wang. 2008. Proteolysis of multiple myelin basic protein isoforms after neurotrauma: characterization by mass spectrometry. *J Neurochem.* 104: 1404–1414.

Park, H., Z. Li, X.O. Yang, S.H. Chang, R. Nurieva, Y.H. Wang, Y. Wang, L. Hood, Z. Zhu, Q. Tian and C. Dong. 2005. A distinct lineage of CD4 T cells regulates tissue inflammation by producing interleukin 17. *Nature Immunology.* 6: 1133–1141.

Petzold, A., M.J. Eikelenboom, D. Gveric, G. Keir, M. Chapman, R.H. Lazeron, M.L. Cuzner, C.H. Polman, B.M. Uitdehaag, E.J. Thompson and G. Giovannoni. 2002. Markers for different glial cell responses in multiple sclerosis: clinical and pathological correlations. *Brain: A Journal of Neurology.* 125: 1462–1473.

Pieragostino, D., F. Petrucci, P. Del Boccio, D. Mantini, A. Lugaresi, S. Tiberio, M. Onofrj, D. Gambi, P. Sacchetta, C. Di Ilio, G. Federici and A. Urbani. 2010. Pre-analytical factors in clinical proteomics investigations: impact of *ex vivo* protein modifications for multiple sclerosis biomarker discovery. *Journal of Proteomics.* 73: 579–592.

Poonawalla, A.H., S. Datta, V. Juneja, F. Nelson, J.S. Wolinsky, G. Cutter and P.A. Narayana. 2010. Composite MRI scores improve correlation with EDSS in multiple sclerosis. *Mult Scler.* 16: 1117–1125.

Qin, Z.Y., Y.J. Qin and S.L. Liu. 2009. Alteration of DBP Levels in CSF of Patients with MS by Proteomics Analysis. *Cell Mol Neurobiol.* 29: 203–210.

Quartu, M., M.P. Serra, M. Boi, V. Ibba, T. Melis and M. Del Fiacco. 2008. Polysialylated-neural cell adhesion molecule (PSA-NCAM) in the human trigeminal ganglion and brainstem at prenatal and adult ages. *BMC Neurosci.* 9: 108.

Raphael, I., S. Mahesula, K. Kalsaria, V. Kotagiri, A.B. Purkar, M. Anjanappa, D. Shah, V. Pericherla, Y.L. Jadhav, R. Raghunathan, M. Vaynberg, D. Noriega, N.H. Grimaldo, C. Wenk, J.A. Gelfond, T.G. Forsthuber and W.E. Haskins. 2012. Microwave and magnetic (M(2)) proteomics of the experimental autoimmune encephalomyelitis animal model of multiple sclerosis. *Electrophoresis.* 33: 3810–3819.

Rhee, S.G., H.Z. Chae and K. Kim. 2005. Peroxiredoxins: a historical overview and speculative preview of novel mechanisms and emerging concepts in cell signaling. *Free Radic Biol Med.* 38: 1543–1552.

Rifai, N., M.A. Gillette and S.A. Carr. 2006. Protein biomarker discovery and validation: the long and uncertain path to clinical utility. *Nat Biotechnol.* 24: 971–983.

Rithidech, K.N., L. Honikel, M. Milazzo, D. Madigan, R. Troxell and L.B. Krupp. 2009. Protein expression profiles in pediatric multiple sclerosis: potential biomarkers. *Mult Scler.* 15: 455–464.

Romi, F., G. Helgeland and N.E. Gilhus. 2012. Serum Levels of Matrix Metalloproteinases: Implications in Clinical Neurology. *Eur Neurol.* 67: 121–128.

Romme Christensen, J., L. Bornsen, M. Khademi, T. Olsson, P.E. Jensen, P.S. Sorensen and F. Sellebjerg. 2012. CSF inflammation and axonal damage are increased and correlate in progressive multiple sclerosis. *Multiple Sclerosis.*

Rosenling, T., A. Attali, T.M. Luider and R. Bischoff. 2011. The experimental autoimmune encephalomyelitis model for proteomic biomarker studies: from rat to human. *Clin Chim Acta.* 412: 812–822.

Rostasy, K., E. Withut, D. Pohl, P. Lange, B. Ciesielcyk, R. Diem, J. Gartner and M. Otto. 2005. Tau, phospho-tau, and S-100B in the cerebrospinal fluid of children with multiple sclerosis. *Journal of Child Neurology.* 20: 822–825.

Rovaris, M., C. Confavreux, R. Furlan, L. Kappos, G. Comi and M. Filippi. 2006. Secondary progressive multiple sclerosis: current knowledge and future challenges. *Lancet Neurol.* 5: 343–354.

Royds, J.A., G.A. Davies-Jones, N.A. Lewtas, W.R. Timperley and C.B. Taylor. 1983. Enolase isoenzymes in the cerebrospinal fluid of patients with diseases of the nervous system. *Journal of Neurology, Neurosurgery, and Psychiatry.* 46: 1031–1036.

Saraiva, M. and A. O'Garra. 2010. The regulation of IL-10 production by immune cells. *Nature Reviews Immunology.* 10: 170–181.

Sato, W., A. Tomita, D. Ichikawa, Y. Lin, H. Kishida, S. Miyake, M. Ogawa, T. Okamoto, M. Murata, Y. Kuroiwa, T. Aranami and T. Yamamura. 2012. CCR2(+)CCR5(+) T cells produce matrix metalloproteinase-9 and osteopontin in the pathogenesis of multiple sclerosis. *Journal of Immunology.* 189: 5057–5065.

Satoh, J., H. Tabunoki, Y. Nanri, K. Arima and T. Yamamura. 2006. Human astrocytes express 14-3-3 sigma in response to oxidative and DNA-damaging stresses. *Neurosci Res.* 56: 61–72.

Satoh, J., T. Yamamura and K. Arima. 2004. The 14-3-3 protein epsilon isoform expressed in reactive astrocytes in demyelinating lesions of multiple sclerosis binds to vimentin and glial fibrillary acidic protein in cultured human astrocytes. *Am J Pathol.* 165: 577–592.

Satoh, J., M. Yukitake, K. Kurohara, H. Takashima and Y. Kuroda. 2003. Detection of the 14-3-3 protein in the cerebrospinal fluid of Japanese multiple sclerosis patients presenting with severe myelitis. *J Neurol Sci.* 212: 11–20.

Sawai, S., H. Umemura, M. Mori, M. Satoh, S. Hayakawa, Y. Kodera, T. Tomonaga, S. Kuwabara and F. Nomura. 2010. Serum levels of complement C4 fragments correlate with disease activity in multiple sclerosis: proteomic analysis. *Journal of Neuroimmunology.* 218: 112–115.

Schreibelt, G., J. van Horssen, R.F. Haseloff, A. Reijerkerk, S.M. van der Pol, O. Nieuwenhuizen, E. Krause, I.E. Blasig, C.D. Dijkstra, E. Ronken and H.E. de Vries. 2008a. Protective effects of peroxiredoxin-1 at the injured blood-brain barrier. *Free Radic Biol Med.* 45: 256–264.

Schreibelt, G., J. van Horssen, R.F. Haseloff, A. Reijerkerk, S.M. van der Pol, O. Nieuwenhuizen, E. Krause, I.E. Blasig, C.D. Dijkstra, E. Ronken and H.E. de Vries. 2008b. Protective effects of peroxiredoxin-1 at the injured blood-brain barrier. *Free Radical Biology & Medicine.* 45: 256–264.

Sellebjerg, F., M. Christiansen and P. Garred. 1998. MBP, anti-MBP and anti-PLP antibodies, and intrathecal complement activation in multiple sclerosis. *Mult Scler.* 4: 127–131.

Semra, Y.K., O.A. Seidi and M.K. Sharief. 2002. Heightened intrathecal release of axonal cytoskeletal proteins in multiple sclerosis is associated with progressive disease and clinical disability. *Journal of Neuroimmunology.* 122: 132–139.

Shafit-Zagardo, B., Y. Kress, M.L. Zhao and S.C. Lee. 1999. A novel microtubule-associated protein-2 expressed in oligodendrocytes in multiple sclerosis lesions. *J Neurochem.* 73: 2531–2537.

Sharief, M.K. and R. Hentges. 1991. Association between tumor necrosis factor-alpha and disease progression in patients with multiple sclerosis. *The New England Journal of Medicine.* 325: 467–472.

Sharief, M.K., M.A. Noori, M. Ciardi, A. Cirelli and E.J. Thompson. 1993. Increased levels of circulating ICAM-1 in serum and cerebrospinal fluid of patients with active multiple sclerosis. Correlation with TNF-alpha and blood-brain barrier damage. *J Neuroimmunol.* 43: 15–21.

Shields, D.C. and N.L. Banik. 1999. Pathophysiological role of calpain in experimental demyelination. *Journal of Neuroscience Research.* 55: 533–541.

Shields, D.C., W.R. Tyor, G.E. Deibler, E.L. Hogan and N.L. Banik. 1998. Increased calpain expression in activated glial and inflammatory cells in experimental allergic encephalomyelitis. *Proceedings of the National Academy of Sciences of the United States of America.* 95: 5768–5772.

Soilu-Hanninen, M., M. Laaksonen, A. Hanninen, J.P. Eralinna and M. Panelius. 2005. Downregulation of VLA-4 on T cells as a marker of long term treatment response to interferon beta-1a in MS. *Journal of Neuroimmunology.* 167: 175–182.

Sorci, G., R. Bianchi, F. Riuzzi, C. Tubaro, C. Arcuri, I. Giambanco and R. Donato. 2010. S100B Protein, A Damage-Associated Molecular Pattern Protein in the Brain and Heart, and Beyond. *Cardiovascular Psychiatry and Neurology.*

Sospedra, M. and R. Martin. 2005. Immunology of multiple sclerosis. *Annu Rev Immunol.* 23: 683–747.

Steen, C., N. Wilczak, J.M. Hoogduin, M. Koch and J. De Keyser. 2010. Reduced creatine kinase B activity in multiple sclerosis normal appearing white matter. *PLoS One*. 5: e10811.

Steinacker, P., P. Schwarz, K. Reim, P. Brechlin, O. Jahn, H. Kratzin, A. Aitken, J. Wiltfang, A. Aguzzi, E. Bahn, H.C. Baxter, N. Brose and M. Otto. 2005. Unchanged survival rates of 14-3-3gamma knockout mice after inoculation with pathological prion protein. *Mol Cell Biol*. 25: 1339–1346.

Steinman, L. 1996. Multiple sclerosis: a coordinated immunological attack against myelin in the central nervous system. *Cell*. 85: 299–302.

Sun, J.B., T. Olsson, W.Z. Wang, B.G. Xiao, V. Kostulas, S. Fredrikson, H.P. Ekre and H. Link. 1991. Autoreactive T and B cells responding to myelin proteolipid protein in multiple sclerosis and controls. *Eur J Immunol*. 21: 1461–1468.

Teunissen, C.E., C. Dijkstra and C. Polman. 2005. Biological markers in CSF and blood for axonal degeneration in multiple sclerosis. *Lancet Neurol*. 4: 32–41.

Teunissen, C.E. and M. Khalil. 2012. Neurofilaments as biomarkers in multiple sclerosis. *Mult Scler*. 18: 552–556.

Teunissen, C.E., M.J. Koel-Simmelink, T.V. Pham, J.C. Knol, M. Khalil, A. Trentini, J. Killestein, J. Nielsen, H. Vrenken, V. Popescu, C.D. Dijkstra and C.R. Jimenez. 2011. Identification of biomarkers for diagnosis and progression of MS by MALDI-TOF mass spectrometry. *Multiple Sclerosis*. 17: 838–850.

Thompson, A.J., C.H. Polman, D.H. Miller, W.I. McDonald, B. Brochet, M.M.X. Filippi and J. De Sa. 1997. Primary progressive multiple sclerosis. *Brain: A Journal of Neurology*. 120(Pt 6): 1085–1096.

Thomson, A. 2006. FTY720 in multiple sclerosis: the emerging evidence of its therapeutic value. *Core Evidence*. 1: 157–167.

Tourdias, T. and V. Dousset. 2012. Neuroinflammatory Imaging Biomarkers: Relevance to Multiple Sclerosis and its Therapy. *Neurotherapeutics*.

Trinchieri, G., M. Wysocka, A. D'Andrea, M. Rengaraju, M. Aste-Amezaga, M. Kubin, N.M. Valiante and J. Chehimi. 1992. Natural killer cell stimulatory factor (NKSF) or interleukin-12 is a key regulator of immune response and inflammation. *Progress in Growth Factor Research*. 4: 355–368.

Tsai, R.K. and D.E. Discher. 2008. Inhibition of "self" engulfment through deactivation of myosin-II at the phagocytic synapse between human cells. *J Cell Biol*. 180: 989–1003.

Tsai, R.K., P.L. Rodriguez and D.E. Discher. 2010. Self inhibition of phagocytosis: the affinity of 'marker of self' CD47 for SIRPalpha dictates potency of inhibition but only at low expression levels. *Blood Cells Mol Dis*. 45: 67–74.

Tsukada, N., M. Matsuda, K. Miyagi and N. Yanagisawa. 1995. Thrombomodulin in the Sera of Patients with Multiple-Sclerosis and Human Lymphotropic Virus Type-1-Associated Myelopathy. *J Neuroimmunol*. 56: 113–116.

Tsukamoto, T., M. Noguchi, H. Kayama, T. Watanabe, T. Asoh and T. Yamamoto. 1995. Increased peptidylglycine alpha-amidating monooxygenase activity in cerebrospinal fluid of patients with multiple sclerosis. *Internal Medicine*. 34: 229–232.

Tuohy, V.K. 1994. Peptide determinants of myelin proteolipid protein (PLP) in autoimmune demyelinating disease: a review. *Neurochem Res*. 19: 935–944.

Van, V.Q., S. Lesage, S. Bouguermouh, P. Gautier, M. Rubio, M. Levesque, S. Nguyen, L. Galibert and M. Sarfati. 2006. Expression of the self-marker CD47 on dendritic cells governs their trafficking to secondary lymphoid organs. *EMBO J*. 25: 5560–5568.

Veldhoen, M. 2009. The role of T helper subsets in autoimmunity and allergy. *Current Opinion in Immunology*. 21: 606–611.

Vergara, C. and B. Ramirez. 2004. CNTF, a pleiotropic cytokine: emphasis on its myotrophic role. *Brain Res Brain Res Rev*. 47: 161–173.

von Budingen, H.C., A. Bar-Or and S.S. Zamvil. 2011. B cells in multiple sclerosis: connecting the dots. *Curr Opin Immunol*. 23: 713–720.

Wipfler, P., K. Oppermann, G. Pilz, S. Afazel, E. Haschke-Becher, A. Harrer, M. Huemer, A. Kunz, S. Golaszewski, W. Staffen, G. Ladurner and J. Kraus. 2011. Adhesion molecules are promising candidates to establish surrogate markers for natalizumab treatment. *Multiple Sclerosis.* 17: 16–23.

Won, J., D.Y. Kim, M. La, D. Kim, G.G. Meadows and C.O. Joe. 2003. Cleavage of 14-3-3 protein by caspase-3 facilitates bad interaction with Bcl-x(L) during apoptosis. *J Biol Chem.* 278: 19347–19351.

Yang, M., Z. Qin, Y. Zhu, Y. Li, Y. Qin, Y. Jing and S. Liu. 2013. Vitamin D-binding Protein in Cerebrospinal Fluid is Associated with Multiple Sclerosis Progression. *Mol Neurobiol.*

Yao, M., M.S. Moir, M.Z. Wang, M.P. To and D.J. Terris. 1999. Peripheral nerve regeneration in CNTF knockout mice. *Laryngoscope.* 109: 1263–1268.

Yardan, T., A.K. Erenler, A. Baydin, K. Aydin and C. Cokluk. 2011. Usefulness of S100B protein in neurological disorders. *JPMA. The Journal of the Pakistan Medical Association.* 61: 276–281.

Yong, V.W., R.K. Zabad, S. Agrawal, A.G. DaSilva and L.M. Metz. 2007. Elevation of matrix metalloproteinases (MMPs) in multiple sclerosis and impact of immunomodulators. *J Neurol Sci.* 259: 79–84.

Zamvil, S., P. Nelson, J. Trotter, D. Mitchell, R. Knobler, R. Fritz and L. Steinman. 1985. T-cell clones specific for myelin basic protein induce chronic relapsing paralysis and demyelination. *Nature.* 317: 355–358.

Ziemann, U., M. Wahl, E. Hattingen and H. Tumani. 2011. Development of biomarkers for multiple sclerosis as a neurodegenerative disorder. *Progress in Neurobiology.* 95: 670–685.

Zimmer, D.B., E.H. Cornwall, A. Landar and W. Song. 1995. The S100 protein family: history, function, and expression. *Brain Research Bulletin.* 37: 417–429.

14

Various Types of Multiple Sclerosis Biomarkers

Kenkichi Nozaki[1],* and Naren L. Banik[2],*

INTRODUCTION

Multiple Sclerosis (MS) is an inflammatory demyelinating disease of Central Nervous System (CNS) (Poser and Brinar, 2004). Dysregulation of Th1, Th2, and Th17 cytokines is a key concept for the pathogenesis of MS (McFarland and Martin, 2007). It also shows axonal transection as a consequence of demyelination (Trapp et al., 1998), suggesting neurodegenerative component in its pathology. Typical presenting symptoms of MS include sensory disturbances, unilateral Optic Neuritis (ON), diplopia (internuclear ophthalmoplegia), Lhermitte's sign (trunk and limb paresthesi as evoked by neck flexion), limb weakness, clumsiness, gait ataxia, and neurogenic bladder and bowel symptoms (Noseworthy et al., 2000). Currently, there is no specific and definite test to diagnose MS, and thus MS is a diagnosis with constellation of clinical features and test results, as well as with exclusion of other autoimmune CNS demyelinating diseases, including acute disseminating encephalomyelitis (ADEM), transverse myelitis (idiopathic or secondary to other autoimmune disease such as systemic lupus erythematosus and Sjogren's disease), neuromyelitis optica (NMO), and neurosarcoidosis.

[1] Department of Neurology, University of Alabama at Birmingham, 1720 7th Avenue South, suite 350 Sparks Center, Birmingham, AL 35233
[2] Division of Neurology, Department of Neurology and Neurosurgery, Medical University of South Carolina, 96 Jonathan Lucas Street, suite 301 CSB, Charleston, SC 29425.
* Corresponding authors: kenkichinozaki@gmail.com; baniknl@musc.edu

Based on its clinical course, MS is subtyped in 1) Relapsing-Remitting MS (RRMS), 2) Primary Progressing MS (PPMS), 3) Secondary Progressing MS (SPMS), and 4) Progressive-Relapsing MS (PRMS) (Lublin and Reingold, 1996). The majority (about 85%) of MS patients have a biphasic disease course, starting with relapse-remitting neurological symptoms. Within 25 years from onset, 90% of them move into the secondary progressing phase. In general, immune/inflammatory process corresponds to the relapse-remitting phase, while neurodegenerative process develops in the secondary progressing phase (Dutta and Trapp, 2010). At the point of initial onset, however, it is difficult to identify whether it is only a one-time event without further relapse-remitting and/or progression (Clinically Isolated Syndrome (CIS)) or the first manifestation of MS. In the latter case, it is also difficult to predict which clinical course a patient will take in future.

Currently, most of the Disease Modifying Agents (DMAs) approved by United States Food and Drug Administration (FDA) are indicated only for RRMS or CIS. Although mitoxantrone is approved for SPMS, PRMS and worsening RRMS, its use has steadily declined in most of the MS centers in the U.S. due to its severe side effects including cardiotoxicity. Since multiple new agents have been developed and increasing number of DMAs will be available, we can expect those agents that are indicated for the other MS subtypes. In near future, MS treatment may be "tailor-made" for each patient. Therefore, accurate assessment of clinical subtype and disease stage is essential in order to select the best matching agent for each patient. When a treatment does not work or stops working, investigation of its underlying cause is important, such as neutralizing antibody to interferon-beta (Ross et al., 2000).

Based on this background, ideal biomarkers for MS are those possessing the following values including (1) diagnostics and stratification of subcategories for MS and of disease stages, (2) prediction of disease course, (3) treatment selection and improved prognosis for treatment success, and (4) the evaluation of novel therapeutics (http://www.ninds.nih.gov/news_and_events/proceedings/BiomarkerWorkshopSummary.htm, 2004).

In this chapter, we will first summarize biomarkers used as a diagnostic test in the patient care of MS, including Magnetic Resonance Imaging (MRI), and cerebrospinal fluid (CSF) studies. We will then review biomarkers related to the pathogenesis of MS and its animal model, Experimental Allergic Encephalomyelitis (EAE), especially regarding its neuroimmunological and neurodegenerative aspects. Since each pathogenetical aspect represents a different phase of the disease, it may give some clues to the biomarker research targeting to the disease stage and sub-classification. Finally, we will discuss biomarkers for treatment response.

Biomarkers as a Diagnostic Test

Magnetic Resonance Imaging (MRI)

MRI plays critical roles in the diagnosis of MS, as well as monitoring of the disease progression and evaluation of treatment efficacy (Ceccarelli et al., 2012). Currently, MRI is considered to be the most sensitive diagnostic imaging modality for revealing demyelinating plaques. Typical MS lesions are in oval shape and located in periventricular and callosal areas. In the spinal cord, it is located intramedullary. Usually spinal cord lesions are small and typically over two spine segments or less.

T2-weighted images are a classic method for visualizing MS demyelinating plaque in hyperintensity and are highly sensitive for disease activity (Paty and Li, 1993). However, T2 hyperintense lesion is not specific for MS and can be positive in other disease processes such ischemic changes, neoplasms and infection (Pirko, 2010). In T2-weighted scans, early active lesions form a border of decreased intensity compared with the lesion center and the perifocal edema. The histological correlate of this pattern is activated macrophages in a zone of myelin destruction at the plaque border (Bruck et al., 1997). Fluid-attenuated inversion recovery (FLAIR) is an MRI technique that shows areas of tissue T2 prolongation as bright while suppressing (darkening) CSF signal, thus clearly revealing lesions in proximity to CSF (De Coene et al., 1992). FLAIR has increased sensitivity in hemispheric lesions (Stevenson et al., 1997), especially those located in the cortical and juxtacortical areas (Bakshi et al., 2001). However, it has lower sensitivity in the posterior fossa and spinal cord (Stevenson et al., 1997). For assessment of MS plaque in the spinal cord, fast short-inversion-time inversion recovery (STIR) MR sequence is the best choice (Hittmair et al., 1996; Rocca et al., 1999). STIR sequences also provide fat suppression, which results in increased sensitivity in imaging of optic nerves (Pirko, 2010).

Gadolinium-diethylene-triaminepenta-acetic acid (Gd-DTPA) provides positive contrast, as the hyperintensity area on T1-weighted images. Gd enhancement detects blood-brain barrier breakdown and inflammation in new and reactivated chronic lesions (Katz et al., 1993; Kermode et al., 1990). The enhancement is usually considered to last for 4 to 6 weeks or less (Pirko, 2010). In a weekly interval Gd enhancement MRI study for RRMS patients, however, the average duration of Gd-DTPA enhancement in individual new lesions was 3.07 weeks (median, 2 weeks) (Cotton et al., 2003). Recommended standard dosage is 0.1 mmol/kg with a scanning delay of 5 to 10 minutes. Using a triple dose (0.3 mmol/kg) with a long delay (40–60 min), 126% more enhancing lesions were detected compared to the standard single-dose administration (Silver et al., 1997).

Approximately 65–80% of contrast enhancing lesions appear acutely hypointense on corresponding unenhanced T1-weighted images (Bagnato et al., 2003; van Waesberghe et al., 1998), becoming isointense with normal-appearing white matter once contrast-enhancement ends (van Waesberghe et al., 1998). A part of T2 hyperintensity lesions are identified as hypointensity on T1 weighted images with low signal intensity relative to the surrounding white matter, especially in late active lesions (Bruck et al., 1997). The degree of hypointensity on T1-weighted images does not correlate with that of demyelination or number of reactive astrocytes, but is associated with axonal density (van Waesberghe et al., 1999; van Walderveen et al., 1998). A "black hole" is usually defined as a persistent or chronic hypointense lesion on a T1 pre-constrast image that coincides with a region of high signal intensity on the T2-weighted images (Sahraian et al., 2010). In a monthly MRI study for consecutive 48 months, 55.7% of T1 hypointensitiy lesions with positive enhancement evolved to persistent black holes (Bagnato et al., 2003). Correlation between black holes and clinical disability, as well as that between T1 hypointensity lesions and relapse rates is inconsistent between studies (Sahraian et al., 2010).

Cerebrospinal Fluid (CSF) Studies (Table 14.1)

Increased intrathecal immunoglobulin production is seen in MS, implying B cell related immune processes. However, it is not specific to MS and can be seen associated with various other pathologies such as CNS inflammatory or infectious diseases. Qualitative and quantitative tests are available for its evaluation.

Table 14.1. CSF studies in MS.

Biomarkers	Clinical/Diagnostic values in MS
OCB	• Positive: ≥3 bands • ≥10 bands: highly suggest MS rather than other neurological disorders
IgG index	• Positive (>0.7): implies immune/inflammatory process in CNS/CSF • Less specific for MS compared to OCB • May help for the prediction of CIS to CDMS
MBP	• Elevated levels correlate with clinical severity in MS • Not specific for MS
Cell count	• Mild elevation seen in MS • Small diagnostic value for MS

CDMS, clinically definite multiple sclerosis; CIS, clinically isolated syndrome; CNS, central nervous system; CSF, cerebrospinal fluid; MBP, myelin basic protein; MS, multiple sclerosis; OCB, oligoclonal band

Oligoclonal Bands (OCBs)

Detection of oligoclonal bands (OCBs) is a qualitative test and has shown high sensitivities in clinically definite MS patients (>95%) (Andersson et al., 1994; Fredrikson, 2010; Miller et al., 1983). OCBs are immunoglobulins separated by isoelectric focusing. The same amounts of immunoglobulin G (IgG) in parallel CSF and serum samples are used and OCBs are revealed with IgG specific antibody staining (Andersson et al., 1994). Identical band pattern in CSF and serum does not suggest intrathecal synthesis. In a large cohort of MS and other neurological diseases, three or more OCBs showed a sensitivity of 85% and a specificity of 92% for the diagnosis of MS. The positive and negative predictive values were 86.5 and 90%, respectively. In addition, 10 or more bands were more frequently found in MS than in other neurological diseases (Bourahoui et al., 2004). The presence of OCBs is highly specific and sensitive for early prediction of conversion from CIS to MS (Masjuan et al., 2006). It also doubles the risk for having a second attack, independent from MRI findings, but it does not seem to influence the development of disability (Tintore et al., 2008). In a large retrospective case control study, OCB negative MS patient group exhibited clinical features atypical for MS more frequently and better prognosis compared to the OCB positive one (Joseph et al., 2009). The OCB pattern in CSF remains in individual MS patients during the course of their disease and OCBs do not disappear by treatment with glucocorticosteroids or disease modifying agents (Fredrikson, 2010).

IgG Studies

Quantitative measurements of IgG production in the CNS are less sensitive than isoelectric focusing (Andersson et al., 1994). Several different formula and quotients have been presented as quantitative methods for intrathecal IgG synthesis. One simple and widely used quantitative method is the IgG index (CSF IgG/serum IgG: CSF albumin/serum albumin), which compensates for the influence of serum levels of IgG and albumin. Increased IgG index (>0.7) indicates IgG synthesis within CNS/CSF compartment (Fredrikson, 2010).

The diagnostic value of these CSF studies has changed over the last decade. The original McDonald criteria (McDonald et al., 2001) mandated positive OCBs or an elevated IgG index for the diagnosis of MS. In 2005 revisions (Polman et al., 2005), the positive CSF result is not mandatory, though it is used as additional data when clinical and MRI features do not fill the criteria. In 2010 revisions (Polman et al., 2011), it is used as an additional data only for PPMS. However, the Panel reaffirmed that positive

CSF findings can be important to support the inflammatory demyelinating nature of the underlying condition, to evaluate alternative diagnoses, and to predict clinically definite MS (CDMS).

Myelin basic protein (MBP)

MBP is a major protein component of myelin sheath and comprises 30% of myelin proteins. MS activation is accompanied with myelin breakdown and MBP release in CSF (Ohta and Ohta, 2002). MBP level in CSF correlates closely with clinical activity of MS (Cohen et al., 1980). It is higher in active demyelinating disease and acute MS exacerbation compared to slowly progressing MS and MS in remission (Cohen et al., 1976). It also shows significant correlation with clinical severity in MS (Sellebjerg et al., 1998). However, its elevation is seen in other neurological disorders, and thus is not specific for MS (Biber et al., 1981).

Other studies

Albumin is a major CSF protein. Since it is only synthesized by hepatocytes but not within CNS, albumin detected in CSF derives from serum and its increase in CSF indicates altered blood-CSF barrier function. The preferred method for detection of blood-CSF barrier dysfunction is the albumin quotient (CSF albumin/serum albumin). Most MS patients have values for the albumin quotient below the upper reference limit. Higher values suggest a different neurological disorder (Andersson et al., 1994).

Cytological examination should be done but is considered to be a complementary test. The normal value is no more than 4 cells/microliters. About 50% of patients with CDMS show a normal cell count and only 1% of patients with MS have cell counts of more than 35/µl. Cell counts >35/µl make MS unlikely (Andersson et al., 1994).

Blood Test

Aquaporin-4 antibody

During the process of establishing the diagnosis of MS, it is important to differentiate it from other CNS immune/inflammatory diseases. One of such diseases is NMO (as known as Devic's disease). NMO is a severe idiopathic inflammatory demyelinating disease that selectively affects optic nerves and the spinal cord, but typically spares the brain (Mandler et al., 1993). Features of NMO distinct from typical MS includes >50 cells/mm^3 in CSF (often polymorphonuclear), normal initial brain MRI, and lesions extending over three or more vertebral segments on the spinal cord MRI (Wingerchuk et al., 1999). In 2004, Lennon et al. (2004) identified IgG autoantibodies in serum of NMO patients by indirect immunofluorescence assay, with 73% sensitivity and 91% specificity. None of the patients with classic MS or miscellaneous autoimmune or paraneoplastic neurological disorders showed positive results. Later, the IgG autoantibody was identified to bind selectively to the aquaporin-4 water channel located in astrocytic foot processes at the blood-brain barrier (Lennon et al., 2005). Thus, the IgG autoantibody is currently known as anti-aquqporin-4 or -NMO antibody and is used as a clinical test, not only for the diagnosis of NMO but also for the differentiation from classic MS or other CNS immune/inflammatory demyelinating diseases.

Evoked Potentials

Evoked potentials have long been used as a diagnostic tool in MS. In contrast to MRI, each pathophysiological finding represents an alteration of function due to at least one lesion. Therefore, an alteration in evoked potentials correlates with an alteration in the load of functionally relevant lesions (Fuhr et al., 2001). The other advantage is that evoked potentials can detect subclinical or asymptomatic abnormalities of the tested systems (Sanders et al., 1985). When different methods are used in combination, the diagnostic sensitivity reaches 100% for definite MS patients (Khoshbin and Hallett, 1981). However, the study is non-specific and prolonged latencies can be seen in any type of diseases affecting each tract.

The following three tests are most frequently used to measure nerve conduction latencies in the corresponding pathway: Visual Evoked Potentials (VEPs) in prechiasmatic visual pathways; brain stem auditory evoked potentials in brain stem auditory pathways; and somatosensory evoked potentials in central somatosensory pathway. Among them, VEP is most reliable. If retinal abnormalities are excluded, delays in P100 latencies or significant side-to-side differences in P100 latencies would indicate abnormality in one or both optic nerves (Birnbaum, 2008). Improvement of VEP latency is associated with contrast and central visual field sensitivities (Brusa et al., 2001).

Biomarkers Related to the Pathogenesis (Table 14.2)

The immune response in MS patients differs from that in non-affected individuals in terms of the regulation and activation of the immune reaction. Identification of the disrupted critical pathways that lead to

Table 14.2. Key biomarkers related the pathogenesis of MS.

Biomarkers	Characteristics
Th1 cytokines (IL-2, -12, IFN-γ, TNF-α)	• Pro-inflammatory • Upregulated during relapses
Th2 cytokines (IL-4, -5, -10, -13, TGF-β)	• Anti-inflammatory • Associated with remissions and stable clinical course
VCAM-1	• Adhesion molecule expressed on endothelial cells • Upregulated during treatment • Increased levels correlate with favorable MRI findings and clinical outcome
VLA-4	• Integrin dimer expressed on leukocyte plasma membranes • Downregulated during treatment • Decreased levels correlate with favorable clinical outcome • Antigenic target of natalizumab
NF-L	• Axonal cytoskeletal protein • CSF levels correlate with clinical severity
Tau protein	• Microtubule-associated protein • CSF levels correlate with clinical disability and gadolinium enhancing lesions in MRI

IFN-γ , interferon-gamma; IL-2, interleukin-2; MRI, magnetic resonance imaging; MS, multiple sclerosis; NF-L, neurofilament-L; TGF-β, transforming growth factor-beta TNF-α, tumor necrosis factor-alpha; VCAM-1, vascular cell adhesion molecule-1; VLA-4, very late antigen-4

the development of MS has shown that the progression of this disease is multifaceted (Kasper and Shoemaker, 2010).

Cytokine Dysregulation

Cytokine dysregulation is well established in MS. Human CD4$^+$ T helper cells, the key immune cells in MS, are separated into different subtypes based on their cytokine profiles (Kasper and Shoemaker, 2010). The decision for a T-cell to which subprofile depends on cellular environment, antigen concentration, types of Antigen-Presenting Cells (APCs), and co-stimulatory factors, as well as exposure to specific interleukins (ILs). An uncommitted CD4$^+$T cell on exposure to IL-12 will be polarized to Th1 cells which are pro-inflammatory and produce interleukin-2 (IL-2), IL-12, interferon-gamma (IFN-γ), and Tumor Necrosis Factor-alpha (TNF-α) (Zamvil and Steinman, 1990)l. In contrast, the same uncommitted CD4$^+$T cell exposed to IL-4 becomes polarized to Th2 cells which are anti-inflammatory and produce IL-4, IL-5, IL-10, IL-13 and Transforming Growth Factor-beta (TGF-β) (Rocken et al., 1996). A recently identified Th subset, IL-17-expressing Th17 cells, becomes differentiated in the presence of transforming growth factor (TGF)-β, IL-1, IL-6, IL-21, and IL-23 (Chen and O'Shea, 2008).

Cytokines are one of the most intensively studied biomarkers in EAE and MS. In RRMS, Th1 cytokines are upregulated during relapses, while Th2 cytokines are associated with clinical remissions and stable clinical course (Hollifield et al., 2003; Link, 1998). Indeed, several established MS therapies, including interferon-beta and glatiramer acetate, are thought to mediate part of their benefit by shifting immune responses in patients from Th1 to Th2 responses (Th2-immune deviation), and/or by promoting function of regulatory immune cells (Bar-Or, 2008). In SPMS patients, IL-12 production is higher compared to those with RRMS (Filion et al., 2003; Soldan et al., 2004). IL-12 may have potential to be a biomarker to differentiate between RRMS and SPMS (Bielekova and Martin, 2004).

Recently, Th17 related cytokines are considered to play major roles in the pathogenesis of EAE and MS. Passive transfer of IL-23 dependent CD4⁺ T-cells is sufficient to induce EAE (Langrish et al., 2005). In MS patients, IL-17 mRNA expression is increased in blood and CSF mononuclear cells (Matusevicius et al., 1999), and IL-17 and IL-23 levels are increased in serum (Chen et al., 2012). There is correlation between production of IL-17 and IL-5 by MBP induced CD4⁺ T cells and the number of active plaques on MRI (Hedegaard et al., 2008).

T-cell Activation

TNF-α and lymphotoxin mRNA levels in Peripheral Blood Mononuclear Cells (PBMCs) are increased prior to relapse in RRMS patients (Rieckmann et al., 1995). In addition, EAE can be induced by peripheral immunization with myelin components including MBP and myelin oligodendrocyte protein (MOG) (Kabat et al., 1947; Rivers and Schwentker, 1935), as well as adoptive transfer of activated myelin reactive CD4⁺ T cells from animals with established disease (Paterson, 1960). These findings suggest that peripheral activation of CNS autoreactive T-cells results in CNS inflammatory disease. Currently, no single CNS antigen has been established as a major target in MS, though multiple CD4⁺ T-cell targets have been implicated in MS including myelin antigens (Bar-Or, 2008).

Stimulation of CD4⁺ helper T cells requires two separate signals, which lead to production of Th1 and Th2 cytokines, depending on local environment. The primary signal is given by the interaction of the T-cell receptor (TCR) on the surface of the T cell with the major histocompatibility class II protein (MHC II) upon surface of Antigen Presenting Cell (APC) (Thome and Acuto, 1995). One major secondary or co-stimulatory signal is given by the interaction of CD 28 on the T cell surface with B7-1 (CD80) or B7-2 (CD86) upon the surface of the APC (Rudd et al., 1994). CD28 has been shown to participate in T cell proliferation (Turka et al., 1990) and induction of cytokine and chemokine production (Herold et al., 1997). Co-stimulation through CD28 is responsible for the activation of a number of signal transduction pathways that eventually lead to the synthesis of cytokines through activation of transcription factors including Nuclear Factor of Activated T-cells (NFAT) (Timmerman et al., 1996) and Nuclear Factor kappa B (NFκB) (Christman et al., 1998). NFAT is a Ca^{2+} dependent transcription factor that is involved in IL-2 gene expression.

Activation of NFκB after recognition of antigen by the T cell receptors is critical for regulation of the genes involved in pivotal mechanisms such as proliferation of activated lymphocytes, IL2 production, and differentiation of Th cells to Th1, Th2 or Th17 phenotypes (Yan and Greer, 2008). Although multiple studies have been performed to evaluate the activation of NFκB in Peripheral Blood Mononuclear Cells (PBMCs), results are conflicting (Yan and Greer, 2008). Thus, further studies are required to conclude its role as a MS biomarker.

T-cell Migration into the Central Nervous System

Infiltration of immune cells into the CNS and demyelination are hall marks of MS pathology (Noseworthy et al., 2000). One of the most important cytokines for chemotaxis and migration of immune cells is IL-8. IL-8 is produced by many immune cells, including T cells, neutrophils, and monocytes (Mukaida, 2000), and local injection of IL-8 in endothelial venules results direct migration of T-cells to the injection sites (Larsen et al., 1989). IL-8 levels in serum and its secretion from PBMCs are higher in untreated MS patients compared to controls, and they are significantly reduced in MS patients treated with interferon-beta-1a (Lund et al., 2004). In addition, the expression of its receptors for IL-8 is elevated in MS lesions (Muller-Ladner et al., 1996). Th17 related cytokines also play a major role in T-cell migration. In MS lesions, IL-17 and IL-22 receptors are expressed on Blood-Brain Barrier Endothelial Cells (BBB-ECs). In addition, Th17 lymphocytes transmigrate efficiently across BBB-ECs, highly express granzyme B, kill human neurons, and promote CNS inflammation through CD4+ lymphocyte recruitment (Kebir et al., 2007).

Adhesion molecules are abnormally expressed on the endothelial cells in MS lesions, including intercellular adhesion molecule (ICAM)-1 and vascular cell adhesion molecule (VCAM)-1. Their ligands, including Lymphocyte Function-associated Antigen (LFA)-1 and Very Late Antigen (VLA)-4 have been identified on the perivascular inflammatory cells within MS lesions (Bar-Or, 2008). Levels of VCAM were significantly increased during treatment with interferon-beta-1b, associated with a decrease in the number of contrast-enhancing lesions on MRI (Calabresi et al., 1997b). VCAM-1 was upregulated in serum and VLA-4 was downregulated on T cells during treatment with interferon-beta-1a associated with favorable long term outcome (Soilu-Hanninen et al., 2005). VLA-4 fluorescence in lymphocytes of MS patients was significantly decreased during interferon-beta treatment (Calabresi et al., 1997a). Monoclonal antibody against VLA-4 (natalizumab) is currently approved for aggressive RRMS by FDA. VLA-4 expression levels on circulating plasmacytoid and myeloid dentritic cells decreased significantly after one year treatment with natalizumab (de Andres et al., 2012).

Demyelination

Demyelination and the subsequent impairment of conduction block are thought to be responsible for disabilities associated with MS and EAE (McFarlin and McFarland, 1982). As described above, MRI is currently the most sensitive study to detect demyelinating plaque in MS, though it is not specific. Clinical significance of autoantibodies against myelin component has been studied. Among them, those targeted to MBP and MOG have been especially focused. Berger et al. (2003) reported their predicting roles of early conversion from CIS to CDMS. However, subsequent studies produced controversial results, with correlation ranging from highly significant to not significant at all (Jelcic and Martin, 2010).

Axonal Damage

MS has been recognized as an immune/inflammatory demyelinating disease of CNS for a long time. Since late 1990s, various studies have identified axonal damage in demyelinating lesions. In recent years, it has become more evident that axonal damage is the major morphological substrate of permanent clinical disability in MS. The highest number of axonal damage is observed in patients with a disease duration of less than one year, suggesting that acute axonal damage is an early event during the development of multiple sclerosis lesions (Kuhlmann et al., 2002). Amyloid Precursor Protein (APP) positive axons are a marker of acute axonal damage. APP is expressed in damaged axons within acute MS lesions (Ferguson et al., 1997; Kuhlmann et al., 2002). Axonal loss is more prominent in PPMS compared to SPMS (Tallantyre et al., 2009). There is a significant correlation between the extent of axonal damage and the degree of inflammation, including T cell and B-cell infiltration and macrophages/microglia (Frischer et al., 2009; Kuhlmann et al., 2002; Trapp et al., 1998).

Neurofilaments are one of the major axonal cytoskeleton proteins. Trapp et al. (1998) demonstrated dramatically increased neurofilament (SMI-32) expression in demyelinated axons in MS lesions. CSF neurofilament light chain (NF-L) levels peak during acute relapses (Malmestrom et al., 2003), and correlate with clinical severity and increased risk for severe MS (Salzer et al., 2010) and disability, exacerbation rate, and time from the start of the previous exacerbation to the time of the lumbar puncture (Lycke et al., 1998). Tau protein is a microtubule-associated protein and is mostly found in CNS neurons. Tau protein is abnormally phosphorylated in EAE and SPMS, and the accumulation of insoluble tau correlates with progression from relapse-remitting to chronic stages in EAE (Anderson et al., 2008). CSF tau protein levels are elevated in all forms of MS (Terzi et al., 2007), without significant difference among different subtypes (Brettschneider et al., 2005). They correlate with disability (Frederiksen et al., 2012) and Gd-enhancing brain lesions in MRI (Brettschneider et al., 2005). They are elevated in patients with RRMS and those with ON who eventually progressed to MS, but not in those with monosymptomatic ON. Hence measurement of CSF tau protein may provide information for predicting progression from ON to MS (Frederiksen et al., 2012). Overall, NF-L and tau protein might be the most likely candidate for biomarker of axonal damage (Bielekova and Martin, 2004).

MR finding representing axonal damage includes brain atrophy and hypointense lesions in T1-weighted images ("black holes"). These findings correlate with disease duration and clinical disability (Paolillo et al., 2000). Significant brain atrophy, affecting both gray and white matter, occurs early in the clinical course of MS (Chard et al., 2002). There is a significant correlation between callosal atrophy and functional impairment of interhemispheric transfer (Pelletier et al., 2001). Diffusion Tensor Imaging (DTI) has shown promise in specifically detecting axonal damage. There is a significant correlation among DTI parameters, histological metrics, and EAE scores. In addition, axial diffusivity is the primary correlate of quantitative staining for neurofilaments (SMI-31), markers of axonal integrity (Budde et al., 2009). MR spectroscopy studies have also shown findings to suggest axonal changes including correlation of disability and reduced levels of N-acetylaspartate, a biomarker found exclusively in neurons and their process (Matthews et al., 1998), and axonal dysfunction remote from cerebral demyelinating lesions (De Stefano et al., 1999).

Calpain

Our group has studied the pathogenesis of MS and EAE, especially focusing on the role of calpain, a calcium-activated protease. We have identified that calpain activation is correlated to demyelination (Shields and Banik, 1998; Shields et al., 1999), inflammation and glial activation (Shields et al., 1998), encephalitogenecity (Guyton et al., 2009), axonal degeneration (Guyton et al., 2010), neuronal cell death with apoptotic changes (Guyton et al., 2010) and sublesional muscle atrophy and associated molecular changes (Park et al., 2012). In the study of PBMCs obtained from MS patients, we have shown increased calpain activity and expression, associated with higher Th1 cytokines and lower Th2 cytokines during relapse compared to controls. Treatment of PBMCs obtained from MS patients with calpeptin (CP), a calpain inhibitor decreased Th1 cytokine level (Imam et al., 2007). We have also identified that CP downregulates Th1/Th17 cytokine and mRNA levels in PBMCs obtained from MS patients (Smith et al., 2011).

Biomarkers for Treatment Response

Neutralizing Antibodies to Interferon-beta

Interferon-beta 1a and 1b are approved for RRMS (and some for CIS) by FDA. They are biological agents and show immunogenic properties, and thus treated patients may develop neutralizing antibodies (NAbs) to them. NAbs generally develop between 6 and 24 months after therapy starts (Goodin et al., 2007). Patients who remain NAb-negative during the first 24 months of therapy rarely develop NAbs. On the contrary, the majority of patients, who were NAb-positive from 12 through 30 months after starting therapy, remain NAb-positive (Sorensen et al., 2005b). Weekly intramuscular interferon-beta-1a is less immunogenic than subcutaneous interferon-beta-1a and -1b preparations given administration frequency (1996). Multiple studies provide evidence for the reduced therapeutic benefits of interferon-beta by NAbs, both clinically (1996; Boz et al., 2007; Francis et al., 2005; Kappos et al., 2005; Malucchi et al., 2008; Malucchi et al., 2004; Sorensen et al., 2003) and radiologically (1996; Francis et al., 2005; Kappos et al., 2005). In spite of such evidence, there are a couple of concerns for the role and significance of NAbs. First, it is still unclear whether NAbs eliminate or merely attenuate the effect of interferon-1b. There are cases with excellent response to interferon-1b in spite of very high NAb titer. In addition, NAb status fluctuates in a minor portion of MS patients with interferon-1b therapy, and the antibody can disappear in spite of continuous administration (Sorensen et al., 2005b).

Thus, there have been debates regarding the use of NAbs in the management of MS patients. American Academy of Neurology concluded that there was insufficient information on the utilization of NAb testing to provide specific recommendations regarding when to test, which test to use, how many tests are necessary, and which cutoff titer to apply (Goodin et al., 2007). Meanwhile, European groups are more aggressive for NAb measurement and its application for treatment plan (Polman et al., 2010; Sorensen et al., 2005a).

Conclusions

Currently, only a limited number of biomarkers are accepted as effective clinical tests in MS. Further studies are required to confirm the efficacy of other biomarkers in MS.

References

Anderson, J.M., D.W. Hampton, R. Patani, G. Pryce, R.A. Crowther, R. Reynolds, R.J. Franklin, G. Giovannoni, D.A. Compston, D. Baker, M.G. Spillantini and S. Chandran. 2008. Abnormally phosphorylated tau is associated with neuronal and axonal loss in experimental autoimmune encephalomyelitis and multiple sclerosis. *Brain.* 131: 1736–48.

Andersson, M., J. Alvarez-Cermeno, G. Bernardi, I. Cogato, P. Fredman, J. Frederiksen, S. Fredrikson, P. Gallo, L.M. Grimaldi, M. Gronning, G. Keir, K. Lamers, H. Link, A. Magalhaes, A.R. Massaro, S. Ohman, H. Reiber, L. Ronnback, M. Schluep, E. Schuller, C.J.M. Sindic, E.J. Thompson, M. Trojano and U. Wurster. 1994. Cerebrospinal fluid in the diagnosis of multiple sclerosis: a consensus report. *J Neurol Neurosurg Psychiatry.* 57: 897–902.

Bagnato, F., N. Jeffries, N.D. Richert, R.D. Stone, J.M. Ohayon, H.F. McFarland and J.A. Frank. 2003. Evolution of T1 black holes in patients with multiple sclerosis imaged monthly for 4 years. *Brain.* 126: 1782–9.

Bakshi, R., S. Ariyaratana, R.H. Benedict and L. Jacobs. 2001. Fluid-attenuated inversion recovery magnetic resonance imaging detects cortical and juxtacortical multiple sclerosis lesions. *Archives of Neurology.* 58: 742–8.

Bar-Or, A. 2008. The immunology of multiple sclerosis. *Semin Neurol.* 28: 29–45.

Berger, T., P. Rubner, F. Schautzer, R. Egg, H. Ulmer, I. Mayringer, E. Dilitz, F. Deisenhammer and M. Reindl. 2003. Antimyelin antibodies as a predictor of clinically definite multiple sclerosis after a first demyelinating event. *N Engl J Med.* 349: 139–45.

Biber, A., D. Englert, D. Dommasch and K. Hempel. 1981. Myelin basic protein in cerebrospinal fluid of patients with multiple sclerosis and other neurological diseases. *J Neurol.* 225: 231–6.

Bielekova, B. and R. Martin. 2004. Development of biomarkers in multiple sclerosis. *Brain.* 127: 1463–78.

Birnbaum, G. 2008. Multiple Sclerosis. Clinician's Guide to Diagnosis and Treatment. Oxford University Press, New York.

Bourahoui, A., J. De Seze, R. Guttierez, B. Onraed, B. Hennache, D. Ferriby, T. Stojkovic and P. Vermersch. 2004. CSF isoelectrofocusing in a large cohort of MS and other neurological diseases. *Eur J Neurol.* 11: 525–9.

Boz, C., J. Oger, E. Gibbs and S.E. Grossberg. 2007. Reduced effectiveness of long-term interferon-beta treatment on relapses in neutralizing antibody-positive multiple sclerosis patients: a Canadian multiple sclerosis clinic-based study. *Mult Scler.* 13: 1127–37.

Brettschneider, J., M. Maier, S. Arda, A. Claus, S.D. Sussmuth, J. Kassubek and H. Tumani. 2005. Tau protein level in cerebrospinal fluid is increased in patients with early multiple sclerosis. *Mult Scler.* 11: 261–5.

Bruck, W., A. Bitsch, H. Kolenda, Y. Bruck, M. Stiefel and H. Lassmann. 1997. Inflammatory central nervous system demyelination: correlation of magnetic resonance imaging findings with lesion pathology. *Ann Neurol.* 42: 783–93.

Brusa, A., S.J. Jones and G.T. Plant. 2001. Long-term remyelination after optic neuritis: A 2-year visual evoked potential and psychophysical serial study. *Brain.* 124: 468–79.

Budde, M.D., M. Xie, A.H. Cross and S.K. Song. 2009. Axial diffusivity is the primary correlate of axonal injury in the experimental autoimmune encephalomyelitis spinal cord: a quantitative pixelwise analysis. *J Neurosci.* 29: 2805–13.

Calabresi, P.A., C.M. Pelfrey, L.R. Tranquill, H. Maloni and H.F. McFarland. 1997a. VLA-4 expression on peripheral blood lymphocytes is downregulated after treatment of multiple sclerosis with interferon beta. *Neurology.* 49: 1111–6.

Calabresi, P.A., L.R. Tranquill, J.M. Dambrosia, L.A. Stone, H. Maloni, C.N. Bash, J.A. Frank and H.F. McFarland. 1997b. Increases in soluble VCAM-1 correlate with a decrease in MRI lesions in multiple sclerosis treated with interferon beta-1b. *Ann Neurol.* 41: 669–74.

Ceccarelli, A., R. Bakshi and M. Neema. 2012. MRI in multiple sclerosis: a review of the current literature. *Curr Opin Neurol.* 25: 402–9.

Chard, D.T., C.M. Griffin, G.J. Parker, R. Kapoor, A.J. Thompson and D.H. Miller. 2002. Brain atrophy in clinically early relapsing-remitting multiple sclerosis. *Brain.* 125: 327–37.

Chen, Y.C., S.D. Chen, L. Miao, Z.G. Liu, W. Li, Z.X. Zhao, X.J. Sun, G.X. Jiang and Q. Cheng. 2012. Serum levels of interleukin (IL)-18, IL-23 and IL-17 in Chinese patients with multiple sclerosis. *J Neuroimmunol.* 243: 56–60.

Chen, Z. and J.J. O'Shea. 2008. Th17 cells: a new fate for differentiating helper T cells. *Immunol Res.* 41: 87–102.

Christman, J.W., L.H. Lancaster and T.S. Blackwell. 1998. Nuclear factor kappa B: a pivotal role in the systemic inflammatory response syndrome and new target for therapy. *Intensive Care Med.* 24: 1131–8.

Cohen, S.R., B.R. Brooks, R.M. Herndon and G.M. McKhann. 1980. A diagnostic index of active demyelination: myelin basic protein in cerebrospinal fluid. *Ann Neurol.* 8: 25–31.

Cohen, S.R., R.M. Herndon and G.M. McKhann. 1976. Radioimmunoassay of myelin basic protein in spinal fluid. An index of active demyelination. *N Engl J Med.* 295: 1455–7.

Cotton, F., H.L. Weiner, F.A. Jolesz and C.R. Guttmann. 2003. MRI contrast uptake in new lesions in relapsing-remitting MS followed at weekly intervals. *Neurology.* 60: 640–6.

de Andres, C., R. Teijeiro, B. Alonso, F. Sanchez-Madrid, M.L. Martinez, J. Guzman de Villoria, E. Fernandez-Cruz and S. Sanchez-Ramon. 2012. Long-term decrease in VLA-4 expression and functional impairment of dendritic cells during natalizumab therapy in patients with multiple sclerosis. *PLoS One.* 7: e34103.

De Coene, B., J.V. Hajnal, P. Gatehouse, D.B. Longmore, S.J. White, A. Oatridge, J.M. Pennock, I.R. Young and G.M. Bydder. 1992. MR of the brain using fluid-attenuated inversion recovery (FLAIR) pulse sequences. *AJNR Am J Neuroradiol.* 13: 1555–64.

De Stefano, N., S. Narayanan, P.M. Matthews, G.S. Francis, J.P. Antel and D.L. Arnold. 1999. *In vivo* evidence for axonal dysfunction remote from focal cerebral demyelination of the type seen in multiple sclerosis. *Brain.* 122(Pt 10): 1933–9.

Dutta, R. and R.D. Trapp. 2010. Multiple Sclerosis 3. Saunders, Philadelphia, PA, USA.

Ferguson, B., M.K. Matyszak, M.M. Esiri and V.H. Perry. 1997. Axonal damage in acute multiple sclerosis lesions. *Brain.* 120(Pt 3): 393–9.

Filion, L.G., D. Matusevicius, G.M. Graziani-Bowering, A. Kumar and M.S. Freedman. 2003. Monocyte-derived IL12, CD86 (B7-2) and CD40L expression in relapsing and progressive multiple sclerosis. *Clin Immunol.* 106: 127–38.

Francis, G.S., G.P. Rice and J.C. Alsop. 2005. Interferon beta-1a in MS: results following development of neutralizing antibodies in PRISMS. *Neurology.* 65: 48–55.

Frederiksen, J., K. Kristensen, J.M. Bahl and M. Christiansen. 2012. Tau protein: a possible prognostic factor in optic neuritis and multiple sclerosis. *Mult Scler.* 18: 592–9.

Fredrikson, S. 2010. Clinical Usefulness of Cerebrospinal Fluid Evaluation. *The International MS Journal.* 17: 24–27.

Frischer, J.M., S. Bramow, A. Dal-Bianco, C.F. Lucchinetti, H. Rauschka, M. Schmidbauer, H. Laursen, P.S. Sorensen and H. Lassmann. 2009. The relation between inflammation and neurodegeneration in multiple sclerosis brains. *Brain*. 132: 1175–89.

Fuhr, P., A. Borggrefe-Chappuis, C. Schindler and L. Kappos. 2001. Visual and motor evoked potentials in the course of multiple sclerosis. *Brain*. 124: 2162–8.

Goodin, D.S., E.M. Frohman, B. Hurwitz, P.W. O'Connor, J.J. Oger, A.T. Reder and J.C. Stevens. 2007. Neutralizing antibodies to interferon beta: assessment of their clinical and radiographic impact: an evidence report: report of the Therapeutics and Technology Assessment Subcommittee of the American Academy of Neurology. *Neurology*. 68: 977–84.

Guyton, M.K., S. Brahmachari, A. Das, S. Samantaray, J. Inoue, M. Azuma, S.K. Ray and N.L. Banik. 2009. Inhibition of calpain attenuates encephalitogenicity of MBP-specific T cells. *J Neurochem*. 110: 1895–907.

Guyton, M.K., A. Das, S. Samantaray, G.C.t. Wallace, J.T. Butler, S.K. Ray and N.L. Banik. 2010. Calpeptin attenuated inflammation, cell death, and axonal damage in animal model of multiple sclerosis. *J Neurosci Res*. 88: 2398–408.

Hedegaard, C.J., M. Krakauer, K. Bendtzen, H. Lund, F. Sellebjerg and C.H. Nielsen. 2008. T helper cell type 1 (Th1), Th2 and Th17 responses to myelin basic protein and disease activity in multiple sclerosis. *Immunology*. 125: 161–9.

Herold, K.C., J. Lu, I. Rulifson, V. Vezys, D. Taub, M.J. Grusby and J.A. Bluestone. 1997. Regulation of C-C chemokine production by murine T cells by CD28/B7 costimulation. *J Immunol*. 159: 4150–3.

Hittmair, K., R. Mallek, D. Prayer, E.G. Schindler and H. Kollegger. 1996. Spinal cord lesions in patients with multiple sclerosis: comparison of MR pulse sequences. *AJNR Am J Neuroradiol*. 17: 1555–65.

Hollifield, R.D., L.S. Harbige, D. Pham-Dinh and M.K. Sharief. 2003. Evidence for cytokine dysregulation in multiple sclerosis: peripheral blood mononuclear cell production of pro-inflammatory and anti-inflammatory cytokines during relapse and remission. *Autoimmunity*. 36: 133–41.

http://www.ninds.nih.gov/news_and_events/proceedings/BiomarkerWorkshopSummary. htm. 2004. Biomarkers in Multiple Sclerosis. Workshop Summary, Washington, D.C., USA.

Imam, S.A., M.K. Guyton, A. Haque, A. Vandenbark, W.R. Tyor, S.K. Ray and N.L. Banik. 2007. Increased calpain correlates with Th1 cytokine profile in PBMCs from MS patients. *J Neuroimmunol*. 190: 139–45.

Jelcic, I. and R. Martin. 2010. Multiple Sclerosis 3. Saunders, Philadelphia, PA, USA.

Joseph, F.G., C.L. Hirst, T.P. Pickersgill, Y. Ben-Shlomo, N.P. Robertson and N.J. Scolding. 2009. CSF oligoclonal band status informs prognosis in multiple sclerosis: a case control study of 100 patients. *J Neurol Neurosurg Psychiatry*. 80: 292–6.

Kabat, E.A., A. Wolf and A.E. Bezer. 1947. The Rapid Production of Acute Disseminated Encephalomyelitis in Rhesus Monkeys by Injection of Heterologous and Homologous Brain Tissue with Adjuvants. *J Exp Med*. 85: 117–30.

Kappos, L., M. Clanet, M. Sandberg-Wollheim, E.W. Radue, H.P. Hartung, R. Hohlfeld, J. Xu, D. Bennett, A. Sandrock and S. Goelz. 2005. Neutralizing antibodies and efficacy of interferon beta-1a: a 4-year controlled study. *Neurology*. 65: 40–7.

Kasper, L.H. and J. Shoemaker. 2010. Multiple sclerosis immunology: The healthy immune system vs. the MS immune system. *Neurology*. 74 Suppl 1: S2–8.

Katz, D., J.K. Taubenberger, B. Cannella, D.E. McFarlin, C.S. Raine and H.F. McFarland. 1993. Correlation between magnetic resonance imaging findings and lesion development in chronic, active multiple sclerosis. *Ann Neurol*. 34: 661–9.

Kebir, H., K. Kreymborg, I. Ifergan, A. Dodelet-Devillers, R. Cayrol, M. Bernard, F. Giuliani, N. Arbour, B. Becher and A. Prat. 2007. Human TH17 lymphocytes promote blood-brain barrier disruption and central nervous system inflammation. *Nat Med*. 13: 1173–5.

Kermode, A.G., A.J. Thompson, P. Tofts, D.G. MacManus, B.E. Kendall, D.P. Kingsley, I.F. Moseley, P. Rudge and W.I. McDonald. 1990. Breakdown of the blood-brain barrier precedes symptoms and other MRI signs of new lesions in multiple sclerosis. Pathogenetic and clinical implications. *Brain*. 113(Pt 5): 1477–89.

Khoshbin, S. and M. Hallett. 1981. Multimodality evoked potentials and blink reflex in multiple sclerosis. *Neurology*. 31: 138–44.

Kuhlmann, T., G. Lingfeld, A. Bitsch, J. Schuchardt and W. Bruck. 2002. Acute axonal damage in multiple sclerosis is most extensive in early disease stages and decreases over time. *Brain*. 125: 2202–12.

Langrish, C.L., Y. Chen, W.M. Blumenschein, J. Mattson, B. Basham, J.D. Sedgwick, T. McClanahan, R.A. Kastelein and D.J. Cua. 2005. IL-23 drives a pathogenic T cell population that induces autoimmune inflammation. *J Exp Med*. 201: 233–40.

Larsen, C.G., A.O. Anderson, E. Appella, J.J. Oppenheim and K. Matsushima. 1989. The neutrophil-activating protein (NAP-1) is also chemotactic for T lymphocytes. *Science*. 243: 1464–6.

Lennon, V.A., T.J. Kryzer, S.J. Pittock, A.S. Verkman and S.R. Hinson. 2005. IgG marker of optic-spinal multiple sclerosis binds to the aquaporin-4 water channel. *J Exp Med*. 202: 473–7.

Lennon, V.A., D.M. Wingerchuk, T.J. Kryzer, S.J. Pittock, C.F. Lucchinetti, K. Fujihara, I. Nakashima and B.G. Weinshenker. 2004. A serum autoantibody marker of neuromyelitis optica: distinction from multiple sclerosis. *Lancet*. 364: 2106–12.

Link, H. 1998. The cytokine storm in multiple sclerosis. *Mult Scler*. 4: 12–5.

Lublin, F.D. and S.C. Reingold. 1996. Defining the clinical course of multiple sclerosis: results of an international survey. National Multiple Sclerosis Society (USA) Advisory Committee on Clinical Trials of New Agents in Multiple Sclerosis. *Neurology*. 46: 907–11.

Lund, B.T., N. Ashikian, H.Q. Ta, Y. Chakryan, K. Manoukian, S. Groshen, W. Gilmore, G.S. Cheema, W. Stohl, M.E. Burnett, D. Ko, N.J. Kachuck and L.P. Weiner. 2004. Increased CXCL8 (IL-8) expression in Multiple Sclerosis. *J Neuroimmunol*. 155: 161–71.

Lycke, J.N., J.E. Karlsson, O. Andersen and L.E. Rosengren. 1998. Neurofilament protein in cerebrospinal fluid: a potential marker of activity in multiple sclerosis. *J Neurol Neurosurg Psychiatry*. 64: 402–4.

Malmestrom, C., S. Haghighi, L. Rosengren, O. Andersen and J. Lycke. 2003. Neurofilament light protein and glial fibrillary acidic protein as biological markers in MS. *Neurology*. 61: 1720–5.

Malucchi, S., F. Gilli, M. Caldano, F. Marnetto, P. Valentino, L. Granieri, A. Sala, M. Capobianco and A. Bertolotto. 2008. Predictive markers for response to interferon therapy in patients with multiple sclerosis. *Neurology*. 70: 1119–27.

Malucchi, S., A. Sala, F. Gilli, R. Bottero, A. Di Sapio, M. Capobianco and A. Bertolotto. 2004. Neutralizing antibodies reduce the efficacy of betaIFN during treatment of multiple sclerosis. *Neurology*. 62: 2031–7.

Mandler, R.N., L.E. Davis, D.R. Jeffery and M. Kornfeld. 1993. Devic's neuromyelitis optica: a clinicopathological study of 8 patients. *Ann Neurol*. 34: 162–8.

Masjuan, J., J.C. Alvarez-Cermeno, N. Garcia-Barragan, M. Diaz-Sanchez, M. Espino, M.C. Sadaba, P. Gonzalez-Porque, J. Martinez San Millan and L.M. Villar. 2006. Clinically isolated syndromes: a new oligoclonal band test accurately predicts conversion to MS. *Neurology*. 66: 576–8.

Matthews, P.M., N. De Stefano, S. Narayanan, G.S. Francis, J.S. Wolinsky, J.P. Antel and D.L. Arnold. 1998. Putting magnetic resonance spectroscopy studies in context: axonal damage and disability in multiple sclerosis. *Semin Neurol*. 18: 327–36.

Matusevicius, D., P. Kivisakk, B. He, N. Kostulas, V. Ozenci, S. Fredrikson and H. Link. 1999. Interleukin-17 mRNA expression in blood and CSF mononuclear cells is augmented in multiple sclerosis. *Mult Scler*. 5: 101–4.

McDonald, W.I., A. Compston, G. Edan, D. Goodkin, H.P. Hartung, F.D. Lublin, H.F. McFarland, D.W. Paty, C.H. Polman, S.C. Reingold, M. Sandberg-Wollheim, W. Sibley, A. Thompson, S. van den Noort, B.Y. Weinshenker and J.S. Wolinsky. 2001. Recommended diagnostic criteria for multiple sclerosis: guidelines from the International Panel on the diagnosis of multiple sclerosis. *Ann Neurol*. 50: 121–7.

McFarland, H.F. and R. Martin. 2007. Multiple sclerosis: a complicated picture of autoimmunity. *Nat Immunol*. 8: 913–9.

McFarlin, D.E. and H.F. McFarland. 1982. Multiple sclerosis (second of two parts). *N Engl J Med*. 307: 1246–51.

Miller, J.R., A.M. Burke and C.T. Bever. 1983. Occurrence of oligoclonal bands in multiple sclerosis and other CNS diseases. *Ann Neurol*. 13: 53–8.

Mukaida, N. 2000. Interleukin-8: an expanding universe beyond neutrophil chemotaxis and activation. *Int J Hematol*. 72: 391–8.

Muller-Ladner, U., J.L. Jones, R.A. Wetsel, S. Gay, C.S. Raine and S.R. Barnum. 1996. Enhanced expression of chemotactic receptors in multiple sclerosis lesions. *J Neurol Sci*. 144: 135–41.

Noseworthy, J.H., C. Lucchinetti, M. Rodriguez and B.G. Weinshenker. 2000. Multiple sclerosis. *N Engl J Med*. 343: 938–52.

Ohta, M. and K. Ohta. 2002. Detection of myelin basic protein in cerebrospinal fluid. *Expert Rev Mol Diagn*. 2: 627–33.

Paolillo, A., C. Pozzilli, C. Gasperini, E. Giugni, C. Mainero, S. Giuliani, V. Tomassini, E. Millefiorini and S. Bastianello. 2000. Brain atrophy in relapsing-remitting multiple sclerosis: relationship with 'black holes', disease duration and clinical disability. *J Neurol Sci*. 174: 85–91.

Park, S., K. Nozaki, M.K. Guyton, J.A. Smith, S.K. Ray and N.L. Banik. 2012. Calpain inhibition attenuated morphological and molecular changes in skeletal muscle of experimental allergic encephalomyelitis rats. *J Neurosci Res*. 90: 2134–45.

Paterson, P.Y. 1960. Transfer of allergic encephalomyelitis in rats by means of lymph node cells. *J Exp Med*. 111: 119–36.

Paty, D.W. and D.K. Li. 1993. Interferon beta-1b is effective in relapsing-remitting multiple sclerosis. II. MRI analysis results of a multicenter, randomized, double-blind, placebo-controlled trial. UBC MS/MRI Study Group and the IFNB Multiple Sclerosis Study Group. *Neurology*. 43: 662–7.

Pelletier, J., L. Suchet, T. Witjas, M. Habib, C.R. Guttmann, G. Salamon, O. Lyon-Caen and A.A. Cherif. 2001. A longitudinal study of callosal atrophy and interhemispheric dysfunction in relapsing-remitting multiple sclerosis. *Archives of Neurology*. 58: 105–11.

Pirko, I. 2010. Multiple Sclerosis 3. Saunders, Philadelphia, PA, USA.

Polman, C.H., A. Bertolotto, F. Deisenhammer, G. Giovannoni, H.P. Hartung, B. Hemmer, J. Killestein, H.F. McFarland, J. Oger, A.R. Pachner, J. Petkau, A.T. Reder, S.C. Reingold, H. Schellekens and P.S. Sorensen. 2010. Recommendations for clinical use of data on neutralising antibodies to interferon-beta therapy in multiple sclerosis. *Lancet Neurol*. 9: 740–50.

Polman, C.H., S.C. Reingold, B. Banwell, M. Clanet, J.A. Cohen, M. Filippi, K. Fujihara, E. Havrdova, M. Hutchinson, L. Kappos, F.D. Lublin, X. Montalban, P. O'Connor, M. Sandberg-Wollheim, A.J. Thompson, E. Waubant, B. Weinshenker and J.S. Wolinsky. 2011. Diagnostic criteria for multiple sclerosis: 2010 revisions to the McDonald criteria. *Ann Neurol*. 69: 292–302.

Polman, C.H., S.C. Reingold, G. Edan, M. Filippi, H.P. Hartung, L. Kappos, F.D. Lublin, L.M. Metz, H.F. McFarland, P.W. O'Connor, M. Sandberg-Wollheim, A.J. Thompson, B.G. Weinshenker and J.S. Wolinsky. 2005. Diagnostic criteria for multiple sclerosis: 2005 revisions to the "McDonald Criteria". *Ann Neurol*. 58: 840–6.

Poser, C.M. and V.V. Brinar. 2004. The nature of multiple sclerosis. *Clin Neurol Neurosurg*. 106: 159–71.

Rieckmann, P., M. Albrecht, B. Kitze, T. Weber, H. Tumani, A. Broocks, W. Luer, A. Helwig and S. Poser. 1995. Tumor necrosis factor-alpha messenger RNA expression in patients with relapsing-remitting multiple sclerosis is associated with disease activity. *Ann Neurol*. 37: 82–8.

Rivers, T.M. and F.F. Schwentker. 1935. Encephalomyelitis Accompanied by Myelin Destruction Experimentally Produced in Monkeys. *J Exp Med*. 61: 689–702.

Rocca, M.A., G. Mastronardo, M.A. Horsfield, C. Pereira, G. Iannucci, B. Colombo, L. Moiola, G. Comi and M. Filippi. 1999. Comparison of three MR sequences for the detection of cervical cord lesions in patients with multiple sclerosis. *AJNR Am J Neuroradiol.* 20: 1710–6.

Rocken, M., M. Racke and E.M. Shevach. 1996. IL-4-induced immune deviation as antigen-specific therapy for inflammatory autoimmune disease. *Immunol Today.* 17: 225–31.

Ross, C., K.M. Clemmesen, M. Svenson, P.S. Sorensen, N. Koch-Henriksen, G.L. Skovgaard and K. Bendtzen. 2000. Immunogenicity of interferon-beta in multiple sclerosis patients: influence of preparation, dosage, dose frequency, and route of administration. Danish Multiple Sclerosis Study Group. *Ann Neurol.* 48: 706–12.

Rudd, C.E., O. Janssen, Y.C. Cai, A.J. da Silva, M. Raab and K.V. Prasad. 1994. Two-step TCR zeta/CD3-CD4 and CD28 signaling in T cells: SH2/SH3 domains, protein-tyrosine and lipid kinases. *Immunol Today.* 15: 225–34.

Sahraian, M.A., E.W. Radue, S. Haller and L. Kappos. 2010. Black holes in multiple sclerosis: definition, evolution, and clinical correlations. *Acta Neurol Scand.* 122: 1–8.

Salzer, J., A. Svenningsson and P. Sundstrom. 2010. Neurofilament light as a prognostic marker in multiple sclerosis. *Mult Scler.* 16: 287–92.

Sanders, E.A., J.P. Reulen, L.A. Hogenhuis and E.A. van der Velde. 1985. Electrophysiological disorders in multiple sclerosis and optic neuritis. *Can J Neurol Sci.* 12: 308–13.

Sellebjerg, F., M. Christiansen, P.M. Nielsen and J.L. Frederiksen. 1998. Cerebrospinal fluid measures of disease activity in patients with multiple sclerosis. *Mult Scler.* 4: 475–9.

Shields, D.C. and N.L. Banik. 1998. Upregulation of calpain activity and expression in experimental allergic encephalomyelitis: a putative role for calpain in demyelination. *Brain Res.* 794: 68–74.

Shields, D.C., K.E. Schaecher, T.C. Saido and N.L. Banik. 1999. A putative mechanism of demyelination in multiple sclerosis by a proteolytic enzyme, calpain. *Proc Natl Acad Sci USA.* 96: 11486–91.

Shields, D.C., W.R. Tyor, G.E. Deibler, E.L. Hogan and N.L. Banik. 1998. Increased calpain expression in activated glial and inflammatory cells in experimental allergic encephalomyelitis. *Proc Natl Acad Sci USA.* 95: 5768–72.

Silver, N.C., C.D. Good, G.J. Barker, D.G. MacManus, A.J. Thompson, I.F. Moseley, W.I. McDonald and D.H. Miller. 1997. Sensitivity of contrast enhanced MRI in multiple sclerosis. Effects of gadolinium dose, magnetization transfer contrast and delayed imaging. *Brain.* 120(Pt 7): 1149–61.

Smith, A.W., B.P. Doonan, W.R. Tyor, N. Abou-Fayssal, A. Haque and N.L. Banik. 2011. Regulation of Th1/Th17 cytokines and IDO gene expression by inhibition of calpain in PBMCs from MS patients. *J Neuroimmunol.* 232: 179–85.

Soilu-Hanninen, M., M. Laaksonen, A. Hanninen, J.P. Eralinna and M. Panelius. 2005. Downregulation of VLA-4 on T cells as a marker of long term treatment response to interferon beta-1a in MS. *J Neuroimmunol.* 167: 175–82.

Soldan, S.S., A.I. Alvarez Retuerto, N.L. Sicotte and R.R. Voskuhl. 2004. Dysregulation of IL-10 and IL-12p40 in secondary progressive multiple sclerosis. *J Neuroimmunol.* 146: 209–15.

Sorensen, P.S., F. Deisenhammer, P. Duda, R. Hohlfeld, K.M. Myhr, J. Palace, C. Polman, C. Pozzilli and C. Ross. 2005a. Guidelines on use of anti-IFN-beta antibody measurements in multiple sclerosis: report of an EFNS Task Force on IFN-beta antibodies in multiple sclerosis. *Eur J Neurol.* 12: 817–27.

Sorensen, P.S., N. Koch-Henriksen, C. Ross, K.M. Clemmesen and K. Bendtzen. 2005b. Appearance and disappearance of neutralizing antibodies during interferon-beta therapy. *Neurology.* 65: 33–9.

Sorensen, P.S., C. Ross, K.M. Clemmesen, K. Bendtzen, J.L. Frederiksen, K. Jensen, O. Kristensen, T. Petersen, S. Rasmussen, M. Ravnborg, E. Stenager and N. Koch-Henriksen. 2003. Clinical importance of neutralising antibodies against interferon beta in patients with relapsing-remitting multiple sclerosis. *Lancet.* 362: 1184–91.

Stevenson, V.L., M.L. Gawne-Cain, G.J. Barker, A.J. Thompson and D.H. Miller. 1997. Imaging of the spinal cord and brain in multiple sclerosis: a comparative study between fast FLAIR and fast spin echo. *J Neurol.* 244: 119–24.

Tallantyre, E.C., L. Bo, O. Al-Rawashdeh, T. Owens, C.H. Polman, J. Lowe and N. Evangelou. 2009. Greater loss of axons in primary progressive multiple sclerosis plaques compared to secondary progressive disease. *Brain.* 132: 1190–9.

Terzi, M., A. Birinci, E. Cetinkaya and M.K. Onar. 2007. Cerebrospinal fluid total tau protein levels in patients with multiple sclerosis. *Acta Neurol Scand.* 115: 325–30.

The IFNB Multiple Sclerosis Study Group and the University of British Columbia MS/MRI Analysis Group. 1996. Neutralizing antibodies during treatment of multiple sclerosis with interferon beta-1b: experience during the first three years. *Neurology.* 47: 889–94.

Thome, M. and O. Acuto. 1995. Molecular mechanism of T-cell activation: role of protein tyrosine kinases in antigen receptor-mediated signal transduction. *Res Immunol.* 146: 291–307.

Timmerman, L.A., N.A. Clipstone, S.N. Ho, J.P. Northrop and G.R. Crabtree. 1996. Rapid shuttling of NF-AT in discrimination of Ca^{2+} signals and immunosuppression. *Nature.* 383: 837–40.

Tintore, M., A. Rovira, J. Rio, C. Tur, R. Pelayo, C. Nos, N. Tellez, H. Perkal, M. Comabella, J. Sastre-Garriga and X. Montalban. 2008. Do oligoclonal bands add information to MRI in first attacks of multiple sclerosis? *Neurology.* 70: 1079–83.

Trapp, B.D., J. Peterson, R.M. Ransohoff, R. Rudick, S. Mork and L. Bo. 1998. Axonal transection in the lesions of multiple sclerosis. *N Engl J Med.* 338: 278–85.

Turka, L.A., J.A. Ledbetter, K. Lee, C.H. June and C.B. Thompson. 1990. CD28 is an inducible T cell surface antigen that transduces a proliferative signal in CD^{3+} mature thymocytes. *J Immunol.* 144: 1646–53.

van Waesberghe, J.H., W. Kamphorst, C.J. De Groot, M.A. van Walderveen, J.A. Castelijns, R. Ravid, G.J. Lycklama a Nijeholt, P. van der Valk, C.H. Polman, A.J. Thompson and F. Barkhof. 1999. Axonal loss in multiple sclerosis lesions: magnetic resonance imaging insights into substrates of disability. *Ann Neurol.* 46: 747–54.

van Waesberghe, J.H., M.A. van Walderveen, J.A. Castelijns, P. Scheltens, G.J. Lycklama a Nijeholt, C.H. Polman and F. Barkhof. 1998. Patterns of lesion development in multiple sclerosis: longitudinal observations with T1-weighted spin-echo and magnetization transfer MR. *AJNR Am J Neuroradiol.* 19: 675–83.

van Walderveen, M.A., W. Kamphorst, P. Scheltens, J.H. van Waesberghe, R. Ravid, J. Valk, C.H. Polman and F. Barkhof. 1998. Histopathologic correlate of hypointense lesions on T1-weighted spin-echo MRI in multiple sclerosis. *Neurology.* 50: 1282–8.

Wingerchuk, D.M., W.F. Hogancamp, P.C. O'Brien and B.G. Weinshenker. 1999. The clinical course of neuromyelitis optica (Devic's syndrome). *Neurology.* 53: 1107–14.

Yan, J. and J.M. Greer. 2008. NFκB, a Potential Therapeutic Target for the Treatment of Multiple Sclerosis. *CNS & Neurological Disorders-Drug Targets.* 7: 536–557.

Zamvil, S.S. and L. Steinman. 1990. The T lymphocyte in experimental allergic encephalomyelitis. *Annu Rev Immunol.* 8: 579–621.

15

Biomarkers and Neurodegenerative Diseases: Promising Inroads Toward a Distant Goal

Richard Rubenstein

"What we know is a drop, what we don't know is an ocean"
Sir Isaac Newton

INTRODUCTION

A major emphasis in the field of neurodegenerative diseases has been the identification of reliable and disease-specific biomarkers (neuroimaging, biochemical, molecular). The immediate need of defining and assessing biomarkers is to provide clinicians and research scientists with information on the nature of the disorders. The long-term goals include biomarker quantitation to: define disease progression, provide insight for the pathogenic mechanisms, establish and monitor the effects of therapeutic interventions. Neurodegenerative diseases can be divided into those that are associated with dementia such as Alzheimer's Disease (AD), Prion Diseases (PD), dementia with Lewy Bodies, frontotemporal dementia, and most

SUNY Downstate Medical Center, Departments of Neurology and Physiology/Pharmacology, Box #1213, 450 Clarkson Avenue, Brooklyn, NY, 11203, USA.
Email: richard.rubenstein@downstate.edu

recently, Chronic Traumatic Encephalopathy and those that are not which include the motor neuron diseases, Parkinson's Disease and Amyotropic Lateral Sclerosis. Although there has been increasing studies on biomarkers for all these diseases, the ones that are in the mainstream, and the focus of this discussion, are AD because of its prevalence and PD because of their transmissibility and fatal outcome.

Prion Diseases (Transmissible Spongiform Encephalopathies)

PD, or Transmissible Spongiform Encephalopathies (TSEs), are a unique group of diseases that affect both animals and humans (Table 15.1). PD include scrapie in sheep and goats, bovine spongiform encephalopathy (BSE or "mad cow disease") in cattle and chronic wasting disease in deer and elk. Human prion diseases include Kuru, which is isolated to the Fore tribe in New Guinea, Gerstmann Sträussler Scheinker syndrome, fatal familial insomnia and Creutzfeldt–Jakob Disease (CJD). CJD has been classified into four distinct forms: classic or sporadic (sCJD), genetic or familial (fCJD), iatrogenic (iCJD) and the variant form (vCJD) which is associated with ingestion of BSE-infected tissue. PD are associated with an abnormally-folded and β-sheet rich, protease-resistant isoform (PrP^{Sc}) of the host-coded cellular prion protein (PrP^C).

The initiation of the infectious process is associated with the PrP^{Sc} serving as a template for the conversion of PrP^C into PrP^{Sc}. It has been proposed that PrP^{Sc} molecules generated *de novo* (sporadic or genetic PD) or introduced through infection binds to PrP^C and induces a refolding process, which converts the cellular protein into the pathological and β-sheet rich PrP^{Sc} isoform. The incubation period for PD can vary from months to decades during which time PrP^{Sc} is accumulating (presumably as a result of the ongoing PrP^C-PrP^{Sc} conversion process) with the absence of clinical symptoms. Following the onset of symptoms, however, the disease progresses rapidly, leading to brain damage and, ultimately, death. The absence of effective treatments for PD stresses the need for developing diagnostic screening assays to prevent a potential spread of the disease from animal to human or from human to human.

Clinical Evaluation and Neuroimaging

A definite diagnosis for CJD may be formulated only by postmortem brain autopsy or by immunohistochemistry for the identification of abnormal PrP^{Sc} on biopsy in the neuronal parenchyma. In sCJD the manifestation of

Table 15.1. Prion Diseases.

Animal Prion Diseases	Host	Etiology
Scrapie	Sheep, Goats	Natural infection with PD agent of unknown origin
Chronic Wasting Disease	Cervids	Infection with PD agent of unknown origin
Transmissible Mink Encephalopathy	Mink	Infection with PD agent from sheep and/or cattle proposed
Bovine Spongiform Encephalopathy (BSE, Mad Cow Disease)	Cattle	Infection with PD agent from sheep proposed
Encephalopathy of Exotic Ungulates	Nyala, Kudo	Infection with BSE-tainted meat and bone meal
Feline Spongiform Encephalopathy	Cats	Infection with BSE-tainted meat and bone meal
Non-Human Primates	Lemurs	Infection with BSE-tainted meat and bone meal
Human Prion Diseases		
Kuru	Human	Ingestion of prion-infected material through ritualistic cannibalism: Fore tribe, New Guinea
Creutzfeldt-Jakob Disease sporadic	Human	Unknown: spontaneous conversion of PrP^C to PrP^{Sc} or somatic mutation proposed
familial/genetic		Mutations in human PrP gene
iatrogenic		Infection with PD agent of human origin (corneal transplant, dura mater grafts, growth hormone)
variant		Infection with PD agent of BSE origin
Gerstmann-Straussler Scheinker	Human	Mutations in human PrP gene
Fatal Familial Insomnia	Human	Mutations in human PrP gene
Sporadic Fatal Insomnia	Human	Unknown: spontaneous conversion of PrP^C to PrP^{Sc} or somatic mutation proposed
Variably protease-sensitive prionopathy	Human	Unknown: spontaneous conversion of PrP^C to PrP^{Sc} or somatic mutation proposed

non-specific clinical symptoms and signs, such as rapid cognitive decline, myoclonus and ataxia, may complicate the diagnostic interpretation and thus require differential diagnosis against other dementia disorders, such as AD. An electroencephalogram, depending on the stage of the disease, exhibits characteristic changes in sCJD ranging from nonspecific findings in early stages, to typical disease characteristics in middle and late stages. Careful evaluation of the motor disturbances may be useful in the clinical differentiation between CJD and AD. Indeed, the presence of ataxia and dysmetria, along with the absence of hypokinesia, is suggestive for the diagnosis of CJD rather than AD (Edler et al., 2009). Magnetic Resonance Imaging (MRI) has become increasingly important in the clinical diagnosis of sCJD. The use of sensitive fluid attenuated inversion recovery and diffusion-weight imaging sequences allows the detection of basal ganglia hyperintensity and signal increase in at least two cortical regions (temporal, parietal or occipital) or both caudate nucleus and putamen (Zerr et al., 2009). Characteristic MRI lesion patterns corresponding to individual CJD subtypes have also been reported (Meissner et al., 2009).

Prion Agent Infection: PrP^{Sc}, PrP^{Sc}-Associated Proteins, and Protein Dysregulation

PrP^{Sc} is associated with the infectious agent making it a highly specific biomarker for PD. However, there are instances where the PrP^{Sc}: infectivity ratio can vary in different *in vivo* models such that infectivity can be measured (by bioassays) with low or undetectable PrP^{Sc} (detected by immunoassays) and, conversely, detectable PrP^{Sc} in the absence of infectivity (Lasmezas et al., 1997; Shaked et al.,1999; Race et al., 2001; Baron et al., 2007; Piccardo et al., 2007). It is therefore possible that in addition to PrP^{Sc}, infectivity is associated with accessory molecules such as other endogenous proteins which might then also serve as surrogate biomarkers. Their physiological roles, if any, regarding association with PrP^{C}, the PrP^{C}-PrP^{Sc} conversion, and infectivity or their ability to serve as biomarkers are, as yet, unknown. In spite of this, currently, all commercial diagnostic tests for PD are carried out by the direct detection of PrP^{Sc} in brain tissue. Conventionally, histological and immunohistological methods have served as reference for clinical diagnosis by detection of characteristic vacuolar or spongiform changes in specific areas of the brain (Fraser, 1976). Antibody-based assays provide rapid and large-scale analysis which is now a routine PD diagnostic assay for use in animals (Oesch et al., 2000; Schaller et al., 1999; Grassi et al., 2000). Due to the fact that the concentration of PrP^{Sc} in peripheral tissue is substantially lower than that in neural tissues, the sensitivity of conventional immunoassays is not high enough for the detection of PrP^{Sc} in easily accessible body fluids for preclinical diagnosis. The capability of a Mass Spectrometry (MS)-based method for more sensitive detection and quantification of PrP^{Sc} has been reported (Onisko et al., 2007; Silva et al., 2011).

An assay, termed Quaking-Induced Conversion (QuIC), has been developed using hamster-adapted sheep scrapie brain tissue and Western blotting for PrP^{Sc} detection (Atarashi et al., 2007, 2008). It has shown to be sufficiently sensitive to detect PrP^{Sc} in cerebrospinal fluid (CSF) from scrapie-infected hamsters, vCJD brain and CSF from scrapie-infected sheep

(Atarashi et al., 2008; Orru et al., 2009). More recently, the QuIC assay has incorporated aspects of the amyloid seeding assay employing thioflavin T (ThT) as a reporter of recombinant PrP conversion in real time, producing a realtime QuIC (RT-QuIC) assay for hamster scrapie prions and CJD (Atarashi et al., 2011; Peden et al., 2012).

In addition, a highly sensitive laser-based immunoassay, termed surround optical fiber immunoassay (SOFIA) incorporating Protein Misfolding Cyclic Amplification (PMCA) technology (Saborio et al., 2001) with ultrasensitive PrP^{Sc} biomarker detection has been described for antemortem PD diagnosis (Chang et al., 2009; Rubenstein et al., 2010, 2011, 2012). PMCA is based on the ability of a small amount of PrP^{Sc} to convert an excess amount of normal PrP^{C} to a proteinase K-resistant form to the level that can be detected by immunoassays. In spite of PMCA, the use of PrP^{Sc} as a preclinical biomarker for surveillance will be challenging due to extremely low PrP^{Sc} levels in easily accessible tissue or body fluids. To circumvent these problems and develop an effective premortem diagnostic method, an alternative strategy is to identify proteins (other than PrP^{Sc}) that are indicative of disease and can be used as surrogate biomarkers for early detection of prion infection.

Recent efforts have been described to identify potential diagnostic surrogate biomarkers by proteomic technologies involving MS (Table 15.2). Both the advantages and disadvantages of a proteomic approach is that massive volumes of data are collected. The conundrum here is that while strict sets of guidelines for parameters evaluated need to be established to perform differential proteomics, those guidelines may be too strict so that important, yet subtle, differences are overlooked. MS-based biomarker discovery for PD exists in different formats but all involve several techniques in combination. Two dimensional gel electrophoresis MS (2D-GE MS) is the most widely used method in studies of proteomic profiling of prion-infected samples. Variations on this technique includes: 2D-DIGE MS, SELDI-TOF MS and shotgun proteomic analysis by Liquid Chromatography (LC)-MS/MS. 2D-DIGE adds a quantitative dimension to this technique enabling multiple protein extracts to be separated on the same 2D gel, thus providing a promising approach for comparative analysis of proteomes in complex samples. SELDI-TOF MS selectively captures proteins of interest on a surface modified with chemical functionality prior to co-crystallization with an energy absorbing matrix followed by ionization with a laser for TOF MS analysis. Liquid-based shotgun proteomic separation techniques have become increasingly popular in proteomic research because they are reproducible, highly automated and have a greater likelihood of detecting low-abundance proteins. In this case, the protein mixture is digested and the resulting peptides are separated by LC prior to tandem MS detection, which allows for the identification of different proteins by peptide sequencing.

Table 15.2. PrP$_{Sc}$-Associated Proteins as Possible Surrogate Biomarkers Identified Using Proteomics.

Category
Cell Structure

	Ferritin	Tubulin
	Collagen	Colifin-1
	Keratins	Desmoglein 1
	Actin	Versican V3
	Myosin	Claudin-11

Protein Folding and Degradation

	Heat shock protein 90α	Ubiquitin
	Heat shock protein 60	Guanine nucleotide binding protein
	Heat shock protein 70kDa	Proteasome subunit a type-1
	T complex protein -1-α	

Translation

	Histones	DNA methyltransferase
	60S ribosomal proteins	Tuftelin-interacting protein 11
	40S ribosomal protein S6	

Myelin

Myelin proteolipid protein
Myelin-oligodendrocyte glycoprotein
Myelin-associated oligodendrocyte basic protein
2', 3'-cyclic-nucleotide 3'-phosphodiesterase

Cell-to-Cell Communication

	Na⁺-K⁺ ATPase	Exocyst complex component 6
	Ras-related protein Rab-3D	Reticulon-3
	Vesicle-fusing ATPase	TBC1 domain family, member 10b
	Gap junction a-1 protein	Synapsin
	Syntaxin-binding protein 1	Dynamin 1

Cellular Functions

Apo E
Alpha enolase
ADP/ATP translocase 1 or 2
Creatine kinase B-type
Macrophage-expressed gene 1 protein
Glyceraldehyde-3-phosphate dehydrogenase
Calcium/calmodulin dependent protein kinase subunit α or β

With all these methods thousands of proteins can be separated on a single gel according to isoelectric point and molecular weight. Individual protein spots can be digested into peptides and analyzed by MALDI-TOF MS or LC-MS/MS for protein identification.

Several *in vivo* PD model systems and different proteomic approaches have been used to identify candidate proteins that are modified during prion agent infection and thus may serve as useful surrogate biomarkers (Krapfenbaner et al., 2002; Steinacker et al., 2005; Chich et al., 2007; Jang et al., 2008, 2010; Barr et al., 2009; Provansal et al., 2010; Wagner et al., 2010).

A difficulty in the interpretation of this data is inconsistency in that not all of the candidate proteins are affected across the different model systems. The majority of the dysregulated proteins are involved in protein folding in the Endoplasmic Reticulum (ER), the mitochondria and the cytoplasm. For example, the chaperones Binding Immunoglobulin Protein (BiP) (also known as Hspa5 or Grp78), heat shock protein (Hsp) 90 (Grp 94/Tra1), Hsp70 (Hspa4, T-complex protein 1 and Hsp8/Hsc71), Cpn10, protein disulfide isomerase A6, and Grp75 are all upregulated presumably as a function of protein misfolding in only some *in vitro* and *in vivo* PD models (Chich et al., 2007; Provansal et al., 2010; Wagner et al., 2010; Tang et al., 2010). Other dysregulated proteins during the infection are tubulin and laminin receptors, vinculin, lamin B2, actinin and gelsolin (Chich et al., 2007; Provansal et al., 2010) while vimentin, glial fibrillary acidic protein, actin and statin are all covalently modified (Jang et al., 2008, 2010; Wagner et al., 2010).

Additional studies have analyzed the proteins that are associated, or co-purify with PrP^{Sc}. These proteins also have the potential to serve as surrogate biomarkers. It has been reported that PrP^{Sc} is associated with more than 50 proteins belonging to different families and functional groups (Giorgi et al., 2009; Petrakis et al., 2009; Moore et al., 2011; Graham et al., 2011). Significant amounts of several proteins, including ferritin, collagen and keratins are associated with PrP^{Sc} due to similar biochemical properties and appear to be contaminants and thus will not be useful disease-specific biomarkers. However, other PrP^{Sc}-associated proteins, such as actin and tubulin likely have a specific function since they are involved in the active transport, localization and microtubule polymerization of PrP^{C} in the cell (Hachiya et al., 2004; Mange et al., 2004; Nieznansk et al., 2005, 2006; Dong et al., 2008). Other PrP^{Sc}-associated proteins (and thus potential biomarkers) include: factors involved in the maintenance of cell and tissue architecture such as extracellular matrix (ECM) and cell-adhesion components, ubiquitin, histones, ribosomal proteins, and calcium-calmodulin kinase II α. Another relevant biomarker candidate is ApoE. ApoE colocalizes with PrP^{Sc} deposits and is upregulated during PD pathogenesis (Baumann et al., 2000; Armstrong et al., 2008; Giorgi et al., 2009).

PD Candidate Biomarkers in CSF

Being in close anatomical contact with brain interstitial fluid, CSF has been the most studied sample for diagnostic biomarker identification for neurological diseases due to its obvious association with the CNS. Biochemical changes in CSF caused by PD and associated with progressive neuropathogenesis can be reflected by altered expression or post-translational modification of certain proteins. A comparison of proteomic data between uninfected and CJD-infected individuals has led to useful biomarkers. However, since most of this data was collected from tissues and body fluids from terminally ill patients, their relevance for early or preclinical diagnosis in humans is uncertain. While biomarkers for preclinical disease confirmation has been addressed using experimentally infected laboratory animal models, the data must be interpreted with caution since there is no guarantee that the animal fully recapitulates the human response. Several studies have used CSF for the detection of PrP^{Sc} and involved protein amplification by PMCA followed by immunodetection (Arashi et al., 2011; McGuire et al., 2012; Rubenstein et al., 2013). Investigations of the CSF has led to the identification of numerous proteins, which may play a direct or complementary role in the diagnosis of PD (Table 15.3). It has been shown that the 14-3-3 protein in the CSF of CJD patients can serve as a useful biomarker (Harrington et al., 1986; Hsich et al., 1996; Kerr et al.,2000; Sanchez-Juan, 2006). Several other candidate biomarkers identified using a proteomic approach includes the heart isoform of the fatty acid binding protein, apolipoproteins, cystatin and α-1 anti-chymotrypsin. Additional CSF proteins that have been reported as potential CJD biomarkers include total tau (t-tau), S100β and neuron-specific enolase (Otto et al., 2002; Jesse et al., 2009). Only the detection of 14-3-3 protein is included in the diagnostic criteria approved by the WHO for the premortem diagnosis of clinically suspected cases of sCJD. However, the diagnostic accuracy of this test is debatable because of its high false-positive rate where elevated 14-3-3 protein levels were also

Table 15.3. Candidate Surrogate Biomarkers in CJD CSF.*

Proteins with Increased Expression Levels in CJD
14-3-3
α1-antichymotrypsin
Apolipoprotein A1
Apolipoprotein A4
Apoliprotein E
Apolipoprotein J
Cystatin C
Fatty acid-binding protein
Hp2-α-haptoglobin
Neuron-specific enolase
S 100β
Tau
Ubiquitin

Proteins with Decreased Expression Levels in CJD
Aβ(1–42)
Apolipoprotein J
Gelsolin

*Qualtieri et al., 2010

reported in other conditions associated with acute neuronal damage (Saiz et al., 1999; Bartosik-Psujek and Archelos, 2004). Due to the high rate of false positives, differential diagnosis between CJD and other neurodegenerative diseases can be challenging, and the use of a combination of multiple biomarkers is often necessary to improve the sensitivity and specificity of CSF 14-3-3 protein analysis (Sanchez-Juan et al., 2006; Huzarewich et al., 2010; Qualtieri et al., 2010). CJD specific biomarkers could include proteins that are directly involved in the pathological process. One such potential protein is extracellular signal regulating kinase ERK1/2 as MAP kinases are activated during pathogenesis in prion diseases (Lee et al., 2005). The levels of total ERK1/2 and phosphorylated ERK1/2 are significantly elevated in the CSF of CJD patients (Steinacker et al., 2010). The increase of ERK1/2 was also observed in a CJD case that was negative for 14-3-3 protein or had low levels of tau protein, suggesting that ERK1/2 can be used as an alternative CSF biomarker for CJD (Steinacker et al., 2010). In previous studies, transthyretin has been shown as a potential CSF biomarker with altered levels in sCJD patients (Brechlin et al., 2008).

PD Candidate Biomarkers in Blood and Urine

Blood is a useful source for biomarker analysis since it is relatively non-invasive, safe and simple to access. The exchange of material between the CSF and the blood, which is increased following damage to the blood-brain barrier in neurodegenerative diseases, makes blood a viable alternative for evaluation and detection of biomarkers. The importance of blood is strengthened by the studies reporting that vCJD can be transmitted to humans through blood transfusions (Hunter et al., 2002; Llewelyn et al., 2004). This stresses the need to develop and validate an effective diagnosis of PD during the preclinical phase of disease using blood samples. However, a major challenge in using blood has been detecting low levels of disease-specific biomarkers in samples containing high amounts of normal blood proteins. Approaches to address this issue have included immunodepletion of high abundance contaminating plasma proteins (Bjorhall et al., 2005), or enrichment of disease-associated protein subpopulations followed by proteomic analysis (Wei et al., 2011). An alternative approach that has been used successfully focuses on PrP^{Sc} detection in blood using PMCA for PrP^{Sc} amplification followed by immunodetection by either Western blotting or SOFIA (Castilla et al., 2005; Chang et al., 2007; Thorne and Terry, 2008; Rubenstein et al., 2010).

PrP^{Sc} has been identified by PMCA in the urine of scrapie-infected sheep, hamsters and mice and infrequently in both Chronic Wasting Disease (CWD)-infected deer and transgenic mice (Kariv-Inbal et al., 2006; Murayama et al., 2007; Andrievskaia et al., 2008; Gonzalez-Romero et al., 2008; Gregori et al., 2008; Rubenstein et al., 2011). Using proteomic approaches the urine of BSE-infected cattle was found to contain the immunoglobulin γ-2 chain C region and clusterin were significantly increased in abundance (Simon et al., 2008). Increased abundance of immunoglobulins has also been reported in the urine of scrapie-infected hamsters and sheep (Serban et al., 2004; Rubenstein et al., 2011).

Given the infectious and fatal nature of prion diseases, early diagnosis is a major goal. Although current antibody-based assays allow specific and sensitive detection of biomarker PrP^{Sc} in brain tissue, readily detection for premortem screening in body fluids is not routinely available due to its low concentrations. MS-based proteomics is being used to identify surrogate protein biomarkers that are associated with PD in complex biological samples such as plasma/serum, CSF and urine and may eventually be able to substitute for, or complement, PrP^{Sc} detection.

Alzheimer's Disease

AD is a severe neurodegenerative disorder of the brain that is characterized by loss of memory and cognitive decline. The majority of AD cases are sporadic (risk age >60 years), and <2.5% have a genetic disposition. The estimated current prevalence of AD in the United States is 4.5 million individuals. The incidence of disease doubles every 5 years after the age of 60 resulting in an estimated prevalence by the year 2050 of 14 million individuals in the US alone. AD is the most common illness leading to nursing home placement. Care for affected individuals currently costs up to US$ 140 billion per year in health care, nursing home and lost productivity. Thus, it is a great challenge to establish reliable surrogate markers to diagnose and monitor disease progression.

Definitive diagnosis requires both clinical assessment of the disease and post-mortem verification of the AD pathology. A probable diagnosis of AD can be established with a confidence of >90%, based on clinical criteria, including medical history, physical examination, laboratory tests, neuroimaging and neuropsychological evaluation. Accurate, early diagnosis of AD is still difficult because early symptoms of the disease are shared by a variety of disorders, which reflects common neuropathological features. An ideal biomarker would distinguish AD from other types of dementia, such as Mild Cognitive Impairment (MCI), CJD, vasculardementia, frontotemporal lobe dementia, or Lewy body dementia. This is important because treatment for these diseases might differ.

The major pathologic features that characterize AD are amyloid or Senile Plaques (SP), neurofibrillary tangles (NFT), decreased synaptic density, neuron loss, and cerebral atrophy. Intense debate has centered on which of two primary neuropathological protein deposits are causative-SP vs. NFT. The anatomic distribution of NFT is much more closely associated with the pattern of neuron loss and therefore with the clinical symptoms in AD. However, the available genetic evidence strongly implicates a derangement in amyloid metabolism as the primary instigating factor in AD. Based on data using transgenic mouse models, the most favorable hypothesis is that the molecular event(s) in the cascade leading to AD are related to amyloid dysmetabolism.

The consensus (Growdon et al., 1998) is that an ideal biomarker for AD should fulfill the following criteria: a) detect a fundamental feature of the neuropathology, b) be validated in neuropathologically confirmed cases, c) have a sensitivity higher than 80% for detecting AD and have a specificity of more than 75% for distinguishing AD from other causes of dementia, preferably established by two independent studies appropriately powered, d) be precise, reliable and inexpensive, and e) be convenient to use and not harmful to the patient. Currently, AD biomarkers can be divided into two categories: (1) imaging, and (2) abnormal levels of certain proteins in tissues and/or body fluids.

Imaging Biomarkers

The most promising imaging biomarkers are Positron Emission Tomography (PET) and MRI based techniques. The most widely studied PET based technique is fluorodeoxyglucose (FDG), which reveals decreased glucose metabolism in a characteristic temporal parietal association cortex distribution. Fouquet et al. (2009) assessed longitudinal changes in FDG regional brain uptake in MCI, reporting that progression from MCI to AD was associated with a faster decline of FDG uptake in two medial brain regions (left anterior cingulate and subgenual region) that have been implicated in early AD. However, FDG-PET, although validated, does not measure a specific disease mechanism or treatment target. Advances in AD imaging includes the development of PET-based amyloid binding radiolabeled ligands. The most commonly utilized amyloid tracer is the "Pittsburgh Compound B" (PiB) (Herholz et al., 2011; Prvulovic and Hampel, 2011). PET using PiB (PiB-PET) can be employed as a tool for monitoring changes in the Aβ plaque burden (Klunk et al., 2003). Studies have demonstrated a correlation between PiB binding and histological Aβ plaque load (Bacskai et al., 2007; Ikonomovic et al., 2008). In one study, Jack et al. (2008) found areas of both concordance and discordance between the

^{11}C-PiB uptake and gray matter volume loss in AD, confirming pathological findings that plaque deposition and neuronal loss proceed at different rates in different regions of the AD brain. Gray matter volume loss correlated more strongly with cognitive deficits than PiB uptake suggesting that PiB uptake occurs early in AD and does not track disease severity closely at later stages. A major drawback with all PiB studies is that it is radiolabeled using the ^{11}C isotope, which has a short half-life time (about 20 minutes), making PiB-PET difficult to manage and expensive. As a result, the use of this neuroimaging in routine screening of individuals at risk of AD currently appears to be impractical (Klunk and Mathis, 2008; Ho et al., 2010).

The most widely studied MRI technique in AD and therefore a likely biomarker candidate is structural MRI (Fennema-Notestine et al., 2009). Structural MRI demonstrates atrophy, which begins in medial temporal limbic areas and then spreads to the neocortex. These are associated with decreased N-acetyl-aspartate and increased myo-Inositol. Choline levels have been less consistent with some studies reporting increase and others decrease in this metabolite. Diffusion imaging reveals increased apparent diffusion coefficients and decreased fractional anisotropy. Arterial spin labeling reveals decreased perfusion. Functional activation studies reveal decreased activation in more severely AD-impaired subjects while increased activation has been reported in very mildly impaired subjects early in the course of the disease. Finally, resting state studies indicated decreased functional connectivity in the default mode network.

As an additional candidate biomarker, a recent study (Frost et al., 2013) demonstrated relationships between retinal vascular abnormalities, neocortical brain amyloid plaque burden and AD. This suggests that retinal photography might provide a sensitive method (or adjunct to blood or other tests) for detecting preclinical AD, allowing the possibility of population screening.

AD Biomarkers in CSF

CSF biomarkers that have been well-established and validated to diagnose AD are: β-amyloid [Aβ(1-42)], total (t)-tau, phosphorylated tau at position 181 (phospho-tau181) and phospho-tau231. The combination of Aβ(1–42), total tau (t-tau) and phospho-tau as CSF biomarkers significantly increases the diagnostic validity for sporadic AD (Blennow, 2004, 2005; Marksteiner et al., 2007; Blennow et al., 2010). The principal component of the SP is the hydrophobic Aβ(1–42), whereas hyperphosphorylated tau, a fraction of the concentration of t-tau, is a characteristic component of NFT (Masters et al., 1985; Vandermeeren et al., 1993; Vanmechelen et al., 2000).

Aβ is generated from the amyloid precursor protein (APP) through proteolytic action of the enzymes β-secretase and γ-secretase (Haass, 2004) (Figure 15.1). APP is a membrane protein with the amino terminus oriented extracellularly and the carboxyl terminus in the cytosol. In the amyloidogenic pathway, β-site APP cleaving enzyme-1 (BACE1) cleaves APP in its ectodomain, releasing a soluble fragment of APP (APPsβ). The remainder of APP is then cleaved in its intra-membranous region by γ-secretase, which generates Aβ. The two main Aβ species, Aβ(1–40) and Aβ(1–42), are generated depending on the γ-secretase cleavage site. APP may also be processed in an nonamyloidogenic pathway. It is then cleaved by α-secretase, releasing APPsα and preventing the generation of Aβ. These pathways may be monitored by CSF biomarkers and several studies have evaluated their potential as biomarkers for AD neuropathology. For example, increases of BACE1 in the CSF have been found in patients with AD suggesting that upregulation of BACE1 might be an early event in AD (Zhong et al., 2007).

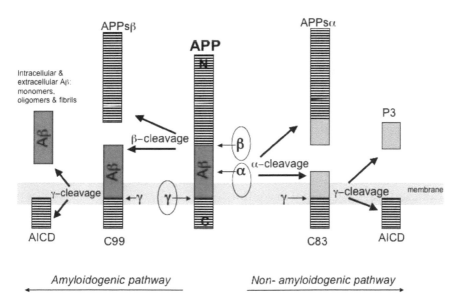

Amyloidogenic pathway   *Non- amyloidogenic pathway*

Figure 15.1. Amyloid precursor protein (APP) processing pathways. The transmembrane protein APP can be processed by two pathways, the non-amyloidogenic α-secretase pathway and the amyloidogenic β-secretase pathway. In the non-amyloidogenic pathway, α-secretase cleaves in the middle of the β-amyloid (Aβ) region to release the soluble APP-fragment APPsα. The APP C-terminal fragment 83 (C83) is then cleaved by γ-secretase to release the APP intracellular domain (AICD) and P3 fragment. In the amyloidogenic pathway, β-secretase (BACE1) cleaves APP to produce the soluble fragment APPsβ. The remaining C-terminal fragment (C99) is then cleaved by γ-secretase to produce the various Aβ species (Aβ40, Aβ42, Aβ oligomers, etc.) and AICD (Cole and Vassar, 2007).

CSF concentrations of APPsα and APPsβ are highly correlated and similar in both AD patients and controls (Zetterberg et al., 2008). Some studies have found unaltered concentrations of APPsα and APPsβ in AD patients compared with controls (Zetterberg et al., 2008; Johansson et al., 2011; Rosen et al., 2012). Other studies found increased concentrations of CSF and blood APPsβ in probable AD and MCI patients who later progressed to AD (Perneczky et al., 2011, 2013). Compared to age-matched controls, the levels of Aβ(1–42) in the CSF of AD patients are significantly reduced to approximately 50% (Blennow, 2004). These reduced levels of Aβ(1–42) in the CSF are presumed to be due to an inability of brain Aβ to be cleared via the blood/CSF, as well as enhanced aggregation and plaque deposition in the brain. In contrast, CSF levels of Aβ(1–40) are unchanged or increased in AD suggesting that the Aβ(1–42)/Aβ(1–40) ratio can improve AD diagnosis, although this is not universally accepted (Sunderland et al., 2004; Schoonenboom et al., 2005).

The aggregation of soluble Aβ peptides to form insoluble fibrillar Aβ in plaques has been regarded to be the central pathogenic event in AD. It has further been suggested (Walsh and Selkoe, 2007) that soluble Aβ oligomers inhibit long-term potentiation and thereby have a role in AD pathogenesis. However, although CSF Aβ oligomers might be an additional AD biomarker, the levels are low in the CSF and do not seem to be consistent with the presence or severity of the disease in AD patients. Large scale clinical studies using MS or other forms of analysis are necessary to increase detection sensitivity and to also determine the specificity of Aβ oligomers in AD pathogenesis.

The intranuclear inclusion levels of the microtubule-associated protein, tau increase with age in the CSF of healthy individuals. Compared to age-matched controls, the levels of t-tau in AD patients are significantly higher (for example, <400 pg/ml in controls vs. >600 pg/ml in AD at 70 years of age) (Humpel, 2011) and can approach an increase of up to 300% (Blennow, 2004). Tau is present in the axons of neurons and its normal function is to promote their stability through the binding to microtubules. In AD, this ability is disrupted due to its phosphorylated state, which leads to dissociation from microtubules, polymerization into paired helical filaments and the formation of NFT causing impaired axonal transport and neuronal degeneration. Of the 39 possible target phosphorylation sites, position 181 and 231 have been associated with AD. The detection of tau phosphorylated at position 181 or 231 are significantly enhanced in AD compared to controls (Hampel, 2010, 2011). In addition to position 181 and 231 the analysis of other phosphorylated forms of tau (phospho-tau-199, -235, -396 and -404) will help to differentiate AD from non-AD forms of dementia (Blennow, 2005). Taken together, the studies have found that patients with AD have a marked increase in CSF levels of t-tau and phospho-tau and a substantial reduction in Aβ(1–42) levels. Each of these biomarkers has been reported to differentiate patients with AD from healthy elderly individuals with an 80–90% sensitivity and specificity (Blennow, 2004). A combined analysis of at least two of these biomarkers improves the sensitivity of a diagnosis for AD (Maddalena et al., 2003). Total tau levels, but not phospho-tau, are dramatically enhanced in CJD (>3000 pg/ml) when compared to controls (200–300 pg/ml) (Humpel, 2011) allowing tau to aid in the differential diagnosis between CJD and AD.

Studies (Lee et al., 2008) have reported that changes in some neuronal-related proteins might be promising AD biomarkers. Elevated levels of visinin-like protein (VLP-1), a neuronal calcium sensor protein, was found in the CSF of AD patients who carried the apolipoprotein E ε4 allele (a defined genetic risk factor for AD). The ability of changes in VLP-1 to diagnose AD was similar to the sensitivity and specificity of t-tau, phospho-tau and Aβ(1–42). Another potential neuronal-related biomarker is the growth-associated protein, GAP-43. The levels of GAP-43 was found to increase in the CSF of AD patients and had a positive correlation with the levels of t-tau suggesting that both proteins might be useful to monitor axonal degeneration (Sjogren et al., 2001).

In addition to the presence of SP and NFT, the AD brain exhibits oxidative stress in the form of oxygen radical-mediated damage. This damage occurs early in the pathogenesis of the disease making it a potential biomarker candidate. Free radical damage of proteins and polyunsaturated fatty acids results in modified forms that can be measured in fluids as markers of oxidation state. In particular, F2-isoprostanes, one group of lipid peroxidation products derived from arachidonic acid, may serve as uesful biomarkers since they were elevated in AD brain and CSF but not in control samples (Pratico et al., 1998; Pratico and Delanty, 2000).

Although additional candidate biomarkers in the CSF from AD patients have been reported (Blasko et al., 2006; Olson et al., 2010), they have not been validated and useful for disease diagnosis due to the heterogeneous nature of AD both clinically and neuropathologically. CSF transforming growth factor (TGF)-β was increased in the AD groups compared with control groups (Swardfager et al., 2010). High levels of the monocyte chemoattractant protein, CCL2, which is important for monocyte recruitment in the CNS, was found in the CSF of AD patients in some studies but not others (Galimberti et al., 2006; Mattsson et al., 2011). Candidate inflammatory biomarkers that need further validation include the cytokine osteopontin, which has been found in higher levels in CSF in AD patients (Comi et al., 2010), and the TNF-α-induced proinflammatory agent lipocalin 2, which has been found in lower levels in CSF of MCI and AD patients compared with controls (Naude et al., 2012).

AD Biomarkers in Blood

To date, biomarkers obtained from both CSF examination and neuroimaging approaches have not achieved extensive clinical application. As a result, research efforts have focused on the development and validation of biomarkers in blood, plasma or serum (Schneider et al., 2009). In contrast to CSF, blood plasma levels of Aβ are inconsistent in sporadic AD (Cedazo-Minguez et al., 2010). Levels of Aβ(1–42) and Aβ(1–40) can be elevated, reduced or even unchanged in AD versus control patients (Cedazo-Minguez et al., 2010; Zetterberg et al., 2010). Longitudinal studies (Borroni et al., 2006) have shown that high plasma Aβ(1–42) levels are not sensitive and specific for early diagnosis, though it might be used in selected cases for predicting AD risk.

As reviewed by Humpel (2011), there are several reasons why plasma levels of Aβ(1–42) are unstable: (i) Aβ expression is influenced by medications, (ii) Aβ binds to other proteins and thus becomes trapped, (iii) Aβ levels in blood fluctuate over time and among individuals, and might differ in mild, early and late AD, and (iv) blood platelets contain high amounts of Aβ, which directly affects plasma levels. Platelets express APP and the secretase machinery, and mainly process Aβ(1–40), which plays a role in platelet aggregation. In platelets of AD patients, three subtypes of APP (106, 110 and 130 kDa) have been found and the AD patients show enhanced processing of the 130-kDa APP subtype (Tang et al., 2006; Borroni et al., 2010).

Protein processing involves enzymatic activity which may prove to be useful as biomarkers. Tau is a brain-specific protein, and altered function of protein kinases and phosphatases has been implicated in tau pathology (Hampel et al., 2010). A number of kinases contribute to tau hyperphosphorylation, including glycogen synthase kinase-3 (GSK-3), cyclin-dependent kinase 5, and microtubule affinity-regulating kinase; conversely, protein phosphatase 2A dephosphorylates tau. At present, tau-related biomarkers in the periphery do not seem to be useful for AD diagnosis.

There are relatively few plasma/serum proteomics analyses of AD. Proteome-based plasma/serum biomarkers need to be validated independently by several groups to increase confidence in the reproducibility of the data. The multiplicity of the experimental strategies and the wide heterogeneity of the technologies employed account for the degree of the variability in the findings between groups. For example, proteomics technologies have been used to search for biomarkers in human plasma and serum with varying results. Two-dimensional gel electrophoresis of plasma and MS analysis has been reported by Hye et al. (2006) to identify a number of proteins previously implicated in the disease pathology

(Table 15.4), including complement factor H precursor and α-2-macroglobulin precursor which correlated with AD severity.

In another study, Ray et al. (2007) found 18 signaling proteins (Table 15.5) in blood plasma that can be used to distinguish between AD and control patients with close to 90% accuracy and to identify patients who had MCI that progressed to AD 2–6 years later. The biological significance of the signaling proteins points to AD-related systemic dysregulation of hematopoiesis, immune responses, apoptosis and neuronal functions.

O'Bryant et al. (2010) used multiplex fluorescent immunoassay to identify a panel of 25 proteins that were either differentially overexpressed (n = 15) or underexpressed (n = 10) in AD patients relative to controls and established a serum protein-based algorithm to discriminate AD individuals from human controls. The sensitivity and specificity values for biomarkers were equal to 80 and 91%, respectively. Several of the total

Table 15.4. Proteomic Identification of Proteins/Biomarkers in AD Plasma.

Plasma Proteins Exhibiting Increased Expression in AD
Desmoplakin
Ig kappa chain C region
Ig kappa chain V-II region
Serum amyloid P-component precursor
Serum albumin precursor
Galectin-7
Complement factor H precursor
α-2-macroglobulin precursor
Ceruloplasmin precursor
Ig lambda chain C regions
Ig lambda chain V-III region
Complement factor H related protein 2 precursor
Ig lambda chain V-II region
Ig kappa chain V-I region
Ig kappa chain V-IV region
Ig alpha-1 chain C region
Plasma Proteins Exhibiting Decreased Expression in AD
Inter-alpha-trypsin inhibitor heavy chain H4 precursor
Ceruloplasmin precursor
Complement C4 precursor
Ig gamma-1 chain C region
AD Serum albumin precursor
Histone H2B.a/g/h/k/l
AD CD5 antigen-like precursor
Serum albumin precursor
Ig mu chain C region

Table 15.5. Plasma signaling proteins as predictors of AD.

Plasma signaling proteins whose expression increases in AD
Angiopoietin-2 (ANG-2)
Chemokine (C-C motif) ligand 18 (CCL18)
Chemokine(C-X-C motif) ligand 8 (CXCL8)
Intercellular adhesion molecule-1 (ICAM-1)
Insulin-like growth factor–binding protein-6 (IGFBP-6)
Interleukin-11 (IL-11)
TNF-related apoptosis-inducing ligand receptor-4 (TRAIL-R4)

Plasma signaling proteins whose expression decreases in AD
Chemokine (C-C motif) ligand 5 (CCL5)
Chemokine (C-C motif) ligand 7 (CCL7)
Chemokine (C-C motif) ligand 15 (CCL15)
Epidermal growth factor (EGF)
Glial-derived neurotrophic factor (GDNF)
Granulocyte-colony stimulating factor (G-CSF)
Interleukin-1α (IL-1α)
Interleukin-3 (IL-3)
Macrophage colony stimulating factor(M-CSF)
Platelet-derived growth factor-β polypeptide (PDGF-BB)
Tumor necrosis factor-α (TNF-α)

markers detected were primarily represented by inflammatory and vascular factors as macrophage inflammatory protein-1α, eotaxin-3, tumor necrosis factor-α, C-reactive protein, von Willebrand factor and some interleukins. The inflammatory nature of the upregulated or downregulated proteins may offer targeted therapeutic strategies for future AD treatment. Using a different proteomic approach, Zhang et al. (2004) reported that a series of molecules including some acute response proteins such as haptoglobin and hemoglobin were elevated in AD serum. In contrast, levels of α-1 acid glycoprotein were lower. The concentrations of other proteins such as apolipoprotein B-100, fragment of factor H, histidine rich glycoprotein, vitronectin and α-2 macroglobulin were demonstrated to be elevated in AD patients.

Conclusions

Advances in technologies for biomarker assay development are very dynamic. In the case of PD, a proteomic approach has resulted in transcriptional profiles from homogeneous cell populations at different stages of disease and identification of individual proteins from a single sample. However, current obstacles include small numbers of clinical cases for validation studies, the long incubation period, and the variability of pathogenesis between strains of infectious agent and routes of infection. It is this heterogeneity among prion disease phenotypes that requires careful choice of tissues and time points to use as starting materials for biomarker discovery. Given these factors it is unlikely that a single biomarker will be useful to the diagnosis of all PD. Similarly, biomarker development for AD is needed to: evaluate responses to treatment, improve understanding of the effect of drugs that target disease mechanisms, and identify AD in its preclinical stage. The combination of body fluid biomarkers and structural and/or functional neuroimaging increases the diagnostic accuracy compared with CSF/blood biomarkers or imaging alone (Vemuri et al., 2009).

Studies to define biomarkers for human neurodegenerative diseases are typically dependent on the availability of appropriate animal models for the particular disease although final validation requires human patients. For example, using animal models for PD, PrP^{Sc} has been identified as the cause of neurodegeneration and neuronal death (Aguzzi and Haass, 2003). Moreover, the PD field has an advantage in that its animal models have a defined and clear endpoint: death. Recent advances with organotypic brain slices even permit rapid *ex vivo* analysis of cell types involved and their own unique set of potential biomarkers. In other diseases, such as AD, the toxic protein or species is not as evident. Researchers investigating AD lack such models and, at best, have models that can recapitulate MCI and the pathology of AD but not true neurodegeneration. Although the amyloid cascade hypothesis (Hardy and Selkoe, 2002) and the generation of oligomeric Aβ(1–42) is the forerunner as the cause of neurodegeneration, and thus serves as a biomarker, definitive proof of what initiates and causes neurodegeneration is unclear. The pathologic characteristics of AD, amyloid plaques and NFT, have confirmed that both Aβ and phospho-tau are neurotoxic and thus potential biomarkers. Additional pathways (and sources of potential biomarkers) are also involved in neuropathogenesis (Table 15.6) since plaques are found in cognitively normal individuals, and plaque burden does not necessarily correlate with memory decline (Iacono, 2009).

Here, we described the ongoing research in biomarker discovery for PD and AD. It is apparent that, although progress continues to be made, there are still many unanswered questions regarding the discovery, verification and clinical validation of biomarkers for neuropathogenesis-associated human neurodegenerative diseases.

Table 15.6. AD-Associated Neuropathogenesis and Verified or Candidate CSF Biomarkers.

Pathology	Biomarkers
Amyloidogenesis	Aβ(1–42); Aβ(1–42): Aβ(1–40) ratio; Aβ(1–42): Aβ(1–38) ratio; carboxy-terminal truncated Aβ peptides <1–2; Aβ oligomers; BACE1 activity; sAPPα and sAPPβ.
Axonal/neuronal degeneration	t-tau.
NFT	phospho-tau 181, 231; Protein kinases (glycogen synthase kinase 3, cyclin-dependent kinase 5, and microtubule-affinity regulating kinase); Protein phosphatase 2A.
Neuronal/Synaptic dysfunction	GAP-43; SNAP-25; Synaptotagmin; Rab-3a; Neurogranin; VLP-1.
Inflammation	Cytokines (TGF-β); F2-isoprostanes; Cytochrome c; Proinflammatory agents (lipocalin).
Gliosis	CCL2; YKL-40.

References

Aguzzi, A. and C. Haass. 2003. Games Played by Rogue Proteins in Prion Disorders and Alzheimer's Disease. *Science.* 302: 814–818.

Andrievskaia, O., J. Algire, A. Balachandran and K. Nielsen.2008. Prion protein in sheep urine. *J Vet Diagn Invest.* 20: 141–146.

Armstrong, R.A., P.L. Lantos and N.J. Cairns. 2008. What determines the molecular composition of abnormal protein aggregates in neurodegenerative disease? *Neuropathology.* 28: 351–365.

Atarashi, R., J.M. Wilham, L. Christensen, A.G. Hughson, R.A. Moore, L.M. Johnson, H.A. Onwubiko, S.A. Priola and B. Caughey. 2008. Simplified ultrasensitive prion detection by recombinant PrP conversion with shaking. *Nat Methods.* 5: 211–212.

Atarashi, R., K. Satoh, K. Sano, T. Fuse, N. Yamaguchi, D. Ishibashi, T. Matsubara, T. Nakagaki, H. Yamanaka, S. Shirabe, M. Yamada, H. Mizusawa, T. Kitamoto, G. Klug, A. McGlade and N. Nishida. 2011. Ultrasensitive human prion detection in cerebrospinal fluid by realtime quaking-induced conversion. *Nat Med.* 17: 175–178.

Atarashi, R., R.A. Moore, V.L. Sim, A.G. Hughson, D.W. Dorward, H.A. Onwubiko, S.A. Priola and B. Caughey. 2007. Ultrasensitive detection of scrapie prion protein using seeded conversion of recombinant prion protein. *Nat Methods.* 4: 645–650.

Bacskai, B.J., M.P. Frosch, S.H. Freeman, S.B. Raymond, J.C. Augustinack, K.A. Johnson, M.C. Irizarry, W.E. Klunk, C.A. Mathis, S.T. Dekosky, S.M. Greenberg, B.T. Hyman and J.H. Growdon. 2007. Molecular imaging with Pittsburgh Compound B confirmed at autopsy: a case report. *Arch Neurol.* 64: 431–434.

Barr, J.B., M. Watson, M.W. Head, J.W. Ironside, N. Harris, C. Hogarth, J.R. Fraser and R. Barron. 2009. Differential protein profiling as a potential multi-marker approach for TSE diagnosis. *BMC Infect Dis.* 9: 188.

Barron, R.M., S.L. Campbell, D. King, A. Bellon, K.E. Chapman, R.A. Williamson and J.C. Manson. 2007. High Titers of transmissible spongiform encephalopathy infectivity associated with extremely low levels of PrPSc *in vivo. J Biol Chem.* 282: 35878–35886.

Bartosik-Psujek, H. and J.J. Archelos. 2004. Tau protein and 14-3-3 are elevated in the cerebrospinal fluid of patients with multiple sclerosis and correlate with intrathecal synthesis of IgG. *J Neurol.* 251: 414–420.

Baumann, M.H., J. Kallijarvi, H. Lankinen, C. Soto and M. Haltia. 2000. Apolipoprotein E includes a binding site which is recognized by several amyloidogenic polypeptides. *Biochem J.* 349: 77–84.

Bjorhall, K., T. Miliotis and P. Davidsson. 2005. Comparison of different depletion strategies for improved resolution in proteomic analysis of human serum samples. *Proteomics.* 5: 307–317.

Blasko, I., W. Lederer, H. Oberbauer, T. Walch, G. Kemmler, H. Hinterhuber, J. Marksteiner and C. Humpel. 2006. Measurement of thirteen biological markers in CSF of patients with Alzheimer's disease and other dementias. *Dement Geriatr Cogn Disord.* 21: 9–15.

Blennow, K. 2004a. Cerebrospinal fluid protein biomarkers for Alzheimer's disease. *NeuroRx.* 1: 213–225.

Blennow, K. 2004b. CSF biomarkers for mild cognitive impairment. *J Intern Med.* 256: 224–234.

Blennow, K. 2005. CSF biomarkers for Alzheimer's disease: use inearly diagnosis and evaluation of drug treatment. *Expert Rev Mol Diagn.* 5: 661–672.

Blennow, K., H. Hampel, M. Weiner and H. Zetterberg. 2010. Cerebrospinal fluid and plasma biomarkers in Alzheimer disease. *Nat Rev Neurol.* 6: 131–144.

Borroni, B., C. Agosti, E. Marcello, M. Di Luca and A. Padovani. 2010. Blood cell markers in Alzheimer's disease: amyloid precursor protein form ratio in platelets. *Exp Gerontol.* 45: 53–56.

Borroni, B., M. Di Luca and A. Padovani. 2006. Predicting Alzheimer dementia in mild cognitive impairment patients. Are biomarkers useful? *Eur J Pharmacol.* 545: 73–80.

Brechlin, P., O. Jahn, P. Steinacker, L. Cepek, H. Kratzin, S. Lehnert, S. Jesse, B. Mollenhauer, H.A. Kretzschmar, J. Wiltfang and M. Otto. 2008. Cerebrospinal fluid-optimized two-dimensional difference gel electrophoresis (2-D DIGE) facilitates the differential diagnosis of Creutzfeldt-Jakob disease. *Proteomics.* 8: 4357–4366.

Budka, J.L. Laplanche, R.G. Will and S. Poser. 2000. Analysis of EEG and CSF 14-3-3 proteins as aids to the diagnosis of Creutzfeldt-Jakob disease. *Neurology.* 55: 811–815.

Castilla, J., P. Saa and C. Soto. 2005. Detection of prions in blood. *Nat Med.* 11: 982–985.

Cedazo-Minguez, A. and B. Winblad. 2010. Biomarkers of Alzheimer's disease and other forms of dementia: clinical needs, limitations and future aspects. *Exp Gerontol.* 45: 5–14.

Chang, B., P. Gray, M. Piltch, M.S. Bulgin, S. Sorensen-Melson, M.W. Miller, P. Davies, D.R. Brown, D.R. Coughlin and R. Rubenstein. 2009. Surround optical fiber immunoassay (SOFIA): More than an ultra-sensitive assay for PrP detection. *J Virol Methods.* 159: 15–22.

Chang, B., X. Cheng, S. Yin, T. Pan, H. Zhang, P. Wong, S.-C. Kang, F. Xiao, H. Yan, C. Li, L.L. Wolfe, M.W. Miller, T. Wisniewski, M.I. Greene and M.-S. Sy. 2007. Test for detection of disease-associated prion aggregate in the blood of infected but asymptomatic animals. *Clin Vaccine Immunol.* 14: 36–43.

Chich, J.F., B. Schaeffer, A.P. Bouin, F. Mouthon, V. Labas, C. Larramendy, J.-P. Deslys and J. Grosclaude. 2007. Prion infection-impaired functional blocks identified by proteomics enlighten the targets and the curing pathways of an anti-prion drug. *Biochim Biophys Acta.* 1774: 154–167.

Cole, S.L. and R. Vassar. 2007. The Alzheimer's disease β-secretase enzyme, BACE1. *Mol Neurodegen.* 2: 22. doi:10.1186/1750-1326-2-22.

Comi, C., M. Carecchio, A. Chiocchetti, S. Nicola, D. Galimberti, C. Fenoglio, G. Cappellano, F. Monaco, E. Scarpini and U. Dianzani. 2010. Osteopontin is increased in the cerebrospinal fluid of patients with Alzheimer's disease and its levels correlate with cognitive decline. *J Alzheimers Dis.* 19: 1143–1148.

Dong, C.F., Shi, S., Wang, X.F.R. An, P. Li, J.-M. Chen, X. Wang, G.-R. Wang, B. Shan, B.-Y. Zhang, J. Han and X.-P. Dong. 2008. The N-terminus of PrP is responsible for interacting with tubulin and fCJD related PrP mutants possess stronger inhibitive effect on microtubule assembly *in vitro*. *Arch Biochem Biophys.* 470: 83–92.

Edler, J., B. Mollenhauer, U. Heinemann, D. Varges, I. Zerr and W.J. Schulz-Schaeffer. 2009. Movement disturbances in the differential diagnosis of Creutzfeldt-Jakob disease. *Mov Disord.* 24: 350–356.

Fennema-Notestine, C., D.J. Hagler Jr., L.K. McEvoy, A.S. Fleisher, E.H. Wu, D.S. Karow and A.M. Dale. 2009. Structural MRI biomarkers for preclinical and mild Alzheimer's disease. *Human Brain Mapping.* 30: 3238–3253.

Fouquet, M., B. Desgranges, B. Landeau, E. Duchesnay, F. Mezenge, F. Viader, J.-C. Baron, F. Eustache and G. Chételat. 2009. Longitudinal brain metabolic changes from amnestic mild cognitive impairment to Alzheimer's disease. *Brain.* 132: 2058–2067.

Fraser, H. 1976. The pathology of a natural and experimental scrapie. *Front Biol.* 44: 267–305.

Frost, S., Y. Kanagasingam, H. Sohrabi, J. Vignarajan, P. Bourgeat, O. Salvado, V. Villemagne, C.C. Rowe, S. Lance Macaulay, C. Szoeke, K.A. Ellis, D. Ames, C.L. Masters, S. Rainey-Smith, R.N. Martins and the AIBL Research Group. 2013. Retinal vascular biomarkers for early detection and monitoring of Alzheimer's disease. *Transl Psychiatry.* 3: e233. doi:10.1038/tp.2012.150.

Galimberti, D., N. Schoonenboom, P. Scheltens, C. Fenoglio, E. Venturelli, Y.A.L. Pijnenburg, N. Bresolin and E. Scarpini. 2006. Intrathecal chemokine levels in Alzheimer disease and frontotemporal lobar degeneration. *Neurology.* 66: 146–147.

Giorgi, A., L. Di Francesco, S. Principe, G. Mignogna, L. Sennels, C. Mancone, T. Alonzi, M. Sbriccoli, A. De Pascalis, J. Rappsilber, F. Cardone, M. Pocchiari, B. Maras and M. Eugenia Schinina. 2009. Proteomic profiling of PrP27-30-enriched preparations extracted from the brain of hamsters with experimental scrapie. *Proteomics.* 9: 3802–3814.

Gonzalez-Romero, D., M.A. Barria, P. Leon, R. Morales and C. Soto. 2008. Detection of infectious prions in urine. *FEBS Lett.*582: 3161–3166.

Graham, J.F., D. Kurian, S. Agarwal, L. Toovey, L. Hunt, L. Kirby, T.J.T. Pinheiro, S.J. Banner and A.C. Gill. 2011. Na+/K+-ATPase is present in scrapie-associated fibrils, modulates PrP misfolding *in vitro* and links PrP function and dysfunction. *PLoS One.* 6: e26813.

Grassi, J., C. Créminon, Y. Frobert, P. Frétier, I. Turbica and H. Rezaei. 2000. Specific determination of the proteinase K-resistant form of the prion protein using two-site immunometric assays. Application to the post-mortem diagnosis of BSE. *Arch Virol Suppl.* 16: 197–205.

Gregori, L., G.G. Kovacs, I. Alexeeva, H. Budka and R.G. Rohwer. 2008. Spongiform encephalopathy infectivity in urine. *Emerg Infect Dis.* 14: 1406–1412.

Growdon, J.H., D.J. Selkoe and A. Roses. 1998. Consensus report of the working group on biological markers of Alzheimer's disease. *Neurobiol Aging.* 19: 109–116.

Haass, C. 2004. Take five–BACE and the gamma-secretase quartet conduct Alzheimer's amyloid beta-peptide generation. *EMBO J.* 23: 483–488.

Hachiya, N.S., K. Watanabe, M. Yamada, Y. Sakasegawa and K. Kaneko. 2004. Anterograde and retrograde intracellular trafficking of fluorescent cellular prion protein. *Biochem Biophys Res Commun.* 315: 802–807.

Hampel, H., K. Blennow, L.M. Shaw, Y.C. Hoessler, H. Zetterberg and J.Q. Trojanowski. 2010. Total and phosphorylated tau protein as biological markers of Alzheimer's disease. *Exp Gerontol.* 45: 30–40.

Hardy, J. and D.J. Selkoe. 2002. The Amyloid Hypothesis of Alzheimer's Disease: Progress and Problems on the Road to Therapeutics. *Science.* 297: 353–356.

Harrington, M.G., C.R. Merril, D.M. Asher and D.C. Gajdusek. 1986. Abnormal proteins in the cerebrospinal fluid of patients with Creutzfeldt-Jakob disease. *N Engl J Med.* 315: 279–283.

Herholz, K. and K. Ebmeier. 2011. Clinical amyloid imaging in Alzheimer's disease. *Lancet Neurology.* 10: 667–670.

Ho, L., H. Fivecoat, J. Wang and G.M. Pasinetti. 2010. Alzheimer's disease biomarker discovery in symptomatic and asymptomatic patients: experimental approaches and future clinical applications. *Exp Gerontol.* 45: 15–22.

Hsich, G., K. Kenney, C.J. Gibbs, K.H. Lee and M.G. Harrington. 1996. The 14-3-3 brain protein in cerebrospinal fluid as a marker for transmissible spongiform encephalopathies. *N Engl J Med.* 335: 924–930.

Humpel, C. 2011. Identifying and validating biomarkers for Alzheimer's disease. *Trends in Biotechnology.* 29: 26–32.

Hunter, N., J. Foster, A. Chong, S. McCutcheon, D. Parnham, S. Eaton, C. MacKenzie and F. Houston. 2002. Transmission of prion diseases by blood transfusion. *J Gen Virol.* 83: 2897–2905.

Huzarewich, R.L., C.G. Siemens and S.A. Booth. 2010. Application of "omics" to prion biomarker discovery. *J Biomed Biotechnol.* 613504.

Hye, A., S. Lynham, M. Thambisetty, M. Causevic, J. Campbell, H.L. Byers, C. Hooper, F. Rijsdijk, S.J. Tabrizi, S. Banner, C.E. Shaw, C. Foy, M. Poppe, N. Archer, G. Hamilton, J. Powell, R.G. Brown, P. Sham, M. Ward and S. Lovestone. 2006. Proteome-based plasma biomarkers for Alzheimer's disease. *Brain.* 129: 3042–3050.

Iacono, D., W.R. Markesbery, M. Gross, O. Pletnikova, G. Rudow, P. Zandi and J.C. Troncoso. 2009. The Nun Study. Clinically silent AD, neuronal hypertrophy, and linguistic skills in early life. *Neurology.* 73: 665–673.

Ikonomovic, M.D., W.E. Klunk, E.E. Abrahamson, C.A. Mathis, J.C. Price, N.D. Tsopelas, B.J. Lopresti, S. Ziolko, W. Bi, W.R. Paljug, M.L. Debnath, C.E. Hope and B.A. Isanski, 2008. Post-mortem correlates of in vivo PiB-PET amyloid imaging in a typical case of Alzheimer's disease. *Brain.*131: 1630–1645.

Jack, C.R. Jr, V.J. Lowe, M.L. Senjem, S.D. Weigand, B.J. Kemp, M.M. Shiung, D.S. Knopman, B.F. Boeve, W.E. Klunk, C.A. Mathis and R.C. Petersen. 2008. ^{11}C-PiB and structural MRI provide complementary information in imaging of Alzheimer's disease and amnestic mild cognitive impairment. *Brain.* 131: 665–680.

Jang, B., E. Kim, J.K. Choi, J.-K. Jin, J.-L. Kim, A. Ishigami, N. Maruyama, R.I. Carp, Y.-S. Kim and E.-K. Choi. 2008. Accumulation of citrullinated proteins by up-regulated peptidylarginine deiminase 2 in brains of scrapie-infected mice: a possible role in pathogenesis. *Am J Pathol.* 173: 1129–1142.

Jang, B., J.K. Jin, Y.C. Jeon, H.J. Cho, A. Ishigami, K.-C. Choi, R.I. Carp, N. Maruyama, Y.-S. Kim and E.-K. Choi. 2010. Involvement of peptidyl arginine deiminase-mediated post-translational citrullination in pathognesis of sporadic Creutzfeldt-Jakob disease. *Acta Neuropathol.* 119: 199–210.

Jesse, S., P. Steinacker, L. Cepek, C.V. Arnim, H. Tumani, S. Lehnert, H.A. Kretzschmar, M. Baier and M. Otto. 2009. Glial fibrillary acidic protein and protein S100B: different concentration pattern of glial proteins in cerebrospinal fluid of patients with Alzheimer's disease and Creutzfeldt-Jakob disease. *J Alzheimers Dis.* 17: 541–551.

Johansson, P., N. Mattsson, O. Hansson, A. Wallin, J. Johansson, U. Andreasson, H. Zetterberg, K. Blennow and J. Svensson. 2011. Cerebrospinal fluid biomarkersfor Alzheimer's disease: diagnostic performance in a homogeneous monocenterpopulation. *J Alzheimers Dis.* 24: 537–546.

Kariv-Inbal, Z., T. Ben-Hur, N.C. Grigoriadis, R. Engelstein and R. Gabizon. 2006. Urine from scrapie- infected hamsters comprises low levels of prion infectivity. *Neurodegener Dis.*3: 123–128.

Klunk, W.E. and C.A. Mathis. 2008. The future of amyloid-beta imaging: a tale of radionuclides and tracer proliferation. *Current Opinion in Neurology.* 21: 683–687.

Klunk, W.E., H. Engler, A. Nordberg, B.J. Bacskai, Y. Wang, J.C. Price, M. Bergstrom, B.T. Hyman, B. Langstrom and C.A. Mathis. 2003. Imaging the pathology of Alzheimer's disease: amyloid-imaging with positron emission tomography. *Neuroimaging Clinics of North America.* 13: 781–789.

Krapfenbaner, K., B.C. Yoo, M. Fountoulakis, E. Mitrova and G. Lubec. 2002. Expression patterns of antioxident proteins in brains of patients with sporadic Creutzfeldt-Jakob disease. *Electrophoresis.* 23: 2541–2547.

Lasmézas, C.I., J.P. Deslys, O. Robain, A. Jaegly, V. Beringue, J.-M. Peyrin, J.-G. Fournier, J.-J. Hauw, J. Rossier and D. Dormont. 1997. Transmission of the BSE agent to mice in the absence of detectable abnormal prion protein. *Science.* 275: 402–405.

Lee, H.P., Y.C. Jun, J.K. Choi, J.-L. Kim, R.I. Carp and Y.-S. Kim. 2005. Activation of mitogen-activated protein kinases in hamster brains infected with 263K scrapie agent. *J Neurochem.* 95: 584–593.

Lee, J.M., K. Blennow, N. Andreasen, O. Laterza, V. Modur, J. Olander, F. Gao, M. Ohlendorf and J.H. Ladenson. 2008. The brain injury biomarker VLP-1 is increased in the cerebrospinal fluid of Alzheimer disease patients. *Clin Chem.* 54: 1617–1623.

Llewelyn, C.A., P.E. Hewitt, R.S. Knight, K. Amar, S. Cousens, J. Mackenzie and R.G. Will. 2004. Possible transmission of variant Creutzfeldt-Jakob disease by blood transfusion. *Lancet.* 363: 417–421.

Maddalena, A., A. Papassotiropoulos, B. Müller-Tillmanns, H.H. Jung, T. Hegi, R.M. Nitsch and C. Hock. 2003. Biochemical diagnosis of Alzheimer disease by measuring the cerebrospinal fluid ratio of phosphorylated tau protein to β-amyloid peptide. *Arch Neurol.* 59: 1729–1734.

Mange, A., C. Crozet, S. Lehmann and F. Beranger. 2004. Scrapie-like prion protein is translocated to the nuclei of infected cells independently of proteasome inhibition and interacts with chromatin. *J Cell Sci.* 117: 2411–2416.

Marksteiner, J., H. Hinterhuber and C. Humpel. 2007. Cerebrospinal fluid biomarkers fordiagnosis of Alzheimer's disease: Beta-amyloid(1–42), tau, phosphotau-181 and total protein. *Drugs Today.* 43: 423431.

Masters, C.L., G. Simms, N.A. Weinman, G. Multhaup, B.L. McDonald and K. Beyreuther. 1985. Amyloid plaque core protein in Alzheimer disease and Down syndrome. *Proc Natl Acad Sci USA.* 82: 4245–4249.

Mattsson, N., S. Tabatabaei, P. Johansson, O. Hansson, U. Andreasson, J.E. Månsson, J.O. Johansson, B. Olsson, A. Wallin, J. Svensson, K. Blennow and H. Zetterberg. 2011. Cerebrospinal fluid microglial markers in Alzheimer's disease: elevated chitotriosidase activity but lack of diagnostic utility. *Neuromolecular Med.* 13: 151–159.

Meissner, B., K. Kallenberg, P. Sanchez-Juan, D. Collie, D.M. Summers, S. Almonti, S.J. Collins, P. Smith, P. Cras, G.H. Jansen, J.P. Brandel, M.B. Coulthart, H. Roberts, B. van Everbroeck, D. Galanaud, V. Mellina, R.G. Will and I. Zerr. 2009. MRI lesion profiles in sporadic Creutzfeldt-Jakob disease. *Neurology.* 72: 1994–2001.

McGuire, L.I., A.H. Peden, C.D. Orrú, J.M. Wilham, N.E. Appleford, G. Mallinson, M. Andrews, M.W. Head, B. Caughey, R.G. Will, R.S.G. Knight and A.J.E. Green. 2012. Real time quaking-induced conversion analysis of cerebrospinal fluid in sporadic Creutzfeldt–Jakob disease. *Annals Neurol.* 72: 278–285.

Moore, R.A., A.G. Timmes, P.A. Wilmarth, D. Safronetz and S.A. Priola. 2011. Identification and removal of proteins that co-purify with infectious prion protein improves the analysis of its secondary structure. *Proteomics.* 11: 3853–3865.

Murayama, Y., M. Yoshioka, H. Okada, M. Takata, T. Yokoyama and S. Mohri. 2007. Urinary excretion and blood level of prions in scrapie-infected hamsters. *J Gen Virol.* 88: 2890–2898.

Naudé, P.J.W., C. Nyakas, L.E. Eiden, D. Ait-Ali, R. van der Heide, S. Engelborghs, P.G.M. Luiten, P.P. De Deyn, J.A. den Boer and U.L.M. Eisel. 2012. Lipocalin 2: novel component of proinflammatory signaling in Alzheimer's disease. *FASEB J.* 26: 2811–2823.

Nieznansk, K., H. Nieznanska, K.J. Skowronek, K.M. Osjecka and D. Stepkowski. 2005. Direct interaction between prion protein and tubulin. *Biochem Biophys Res Commun.* 334: 403–411.

Nieznanski, K., Z.A. Podlubnaya and H. Nieznanska. 2006. Prion protein inhibits microtubule assembly by inducing tubulin oligomerization. *Biochem Biophys Res Commun.* 349: 391–399.

O'Bryant, S.E., G. Xiao, R. Barber, J. Reisch, R. Doody, T. Fairchild, P. Adams, S. Waring and R. Diaz- Arrastia. 2010. Texas Alzheimer's Research Consortium. A serum protein-based algorithm for the detection of Alzheimer disease. *Arch Neurology.* 67: 1077–1081.

Oesch, B., M. Doherr, D. Heim, K. Fischer, S. Egli, S. Bolliger, K. Biffiger, O. Schaller, M. Vandevelde and M. Moser. 2000. Application of Prionics Western blotting procedure to screen for BSE in cattle regularly slaughtered at Swiss abattoirs. *Arch Virol Suppl.* 16: 189–195.

Olson, L. and C. Humpel. 2010. Growth factors and cytokines/chemokines as surrogate biomarkers in cerebrospinal fluid and blood for diagnosing Alzheimer's disease and mild cognitive impairment. *Exp Gerontol.* 45: 41–46.

Onisko, B., I. Dynin, J.R. Requena, C.J. Silva, M. Erickson and J.M. Carter. 2007. Mass spectrometric detection of attomole amounts of the prion protein by nanoLC/MS/MS. *J Am Soc Mass Spectrom.* 18: 1070–1079.

Orru, C.D., J.M. Wilham, A.G. Hughson, L.D. Raymond, K.L. McNally, A. Bossers, C. Ligios and B. Caughey. 2009. Human variant Creutzfeldt–Jakob disease and sheep scrapie PrPres detection using seeded conversion of recombinant prion protein. *Protein Eng Des Sel.* 22: 515–521.

Otto, M., J. Wiltfang, L. Cepek, M. Neumann, B. Mollenhauer, P. Steinacker, B. Ciesielczyk, W. Schulz–Schaeffer, H.A. Kretzschmar and S. Poser. 2002. Tau protein and 14-3-3 protein in the differential diagnosis of Creutzfeldt-Jakob disease. *Neurology.* 58: 192–197.

Peden, A.H., L.I. McGuire, N.E.J. Appleford, G. Mallinson, J.M. Wilham, C.D. Orru, B. Caughey, J.W. Ironside, R.S. Knight, R.G. Will, A.J.E. Green and M.W. Head. 2012. Sensitive and specific detection of sporadic Creutzfeldt–Jakob disease brain prion protein using real-time quaking-induced conversion. *J Gen Virol.* 93: 438–449.

Perneczky, R., A. Tsolakidou, A. Arnold, J. Diehl-Schmid, T. Grimmer, H. Förstl, A. Kurz and P. Alexopoulos. 2011. CSF soluble amyloid precursor proteins in the diagnosis of incipient Alzheimer disease. *Neurology.* 77: 35–38.

Perneczky, R., L.-H. Guo, S.M. Kagerbauer, L. Werle, A. Kurz, J. Martin and P. Alexopoulos. 2013. Soluble amyloid precursor protein β as blood-based biomarker of Alzheimer's disease. *Transl Psychiatry.* 3: e227.

Petrakis, S., A. Malinowska, M. Dadlez and T. Sklaviadis. 2009. Identification of proteins co-purifying with scrapie infectivity. *J Proteomics.* 72: 690–694.

Piccardo, P., J.C. Manson, D. King, B. Ghetti and R.M. Barron. 2007. Accumulation of prion protein in the brain that is not associated with transmissible disease. *Proc Natl Acad Sci USA.* 104: 4712–4717.

Pratico, D. and N. Delanty. 2000. Oxidative injury in diseases of the central nervous system: focus on Alzheimer's disease. *Am J Med.* 109: 577–585.

Pratico, D., V.M.-Y. Lee, J.Q. Trojanowski, J. Rokach and G.A. Fitzgerald. 1998. Increased F2-isoprostanes in Alzheimer's disease: evidence for enhanced lipid peroxidation *in vivo.* *FASEB J.* 12: 1777–1783.

Provansal, M., S. Roche, M. Pastore, D. Casanova, M. Belondrade, S. Alais, P. Leblanc, O. Windl and S. Lehmann. 2010. Proteomic consequences of expression and pathological conversion of the prion protein in inducible neuroblastoma N2a cells. *Prion.* 4: 292–301.

Prvulovic, D. and H. Hampel. 2011. Amyloid β (Aβ) and phospho-tau (p-tau) as diagnostic biomarkers in Alzheimer's disease. *Clin Chem and Lab Med.* 49: 367–374.

Qualtieri, A., E. Urso, M. Le Pera, T. Sprovieri, S. Bossio, A. Gambardella and A. Quattrone. 2010. Proteomic profiling of cerebrospinal fluid in Creutzfeldt-Jakob disease. *Expert Rev Proteomics.* 7: 907–917.

Race, R., A. Raines, G.I. Raymond, B. Caughey and B. Chesebro. 2001. Long-term subclinical carrier state precedes scrapie replication and adaptation in a resistant species: analogies to bovine spongiform encephalopathy and variant Creutzfeldt-Jakob Disease in humans. *J Virol.* 75: 10106–10112.

Ray, S., M. Britschgi, C. Herbert, Y. Takeda-Uchimura, A. Boxer, K. Blennow, L.F. Friedman, D.R. Galasko, M. Jutel, A. Karydas, J.A. Kaye, J. Leszek, B.L. Miller, L. Minthon, J.F. Quinn, G.D. Rabinovici, W.H. Robinson, M.N. Sabbagh, Y.T. So, D.L. Sparks, M. Tabaton, J. Tinklenberg, J.A. Yesavage, R. Tibshirani and T. Wyss-Coray. 2007. Classification and prediction of clinical Alzheimer's diagnosis based on plasma signaling proteins. *Nature Med.* 13: 1359–1362.

Rosen, C., U. Andreasson, N. Mattsson, J. Marcusson, L. Minthon, N. Andreasen, K. Blennow and H. Zetterberg. 2012. Cerebrospinal fluid profiles ofAβ-related biomarkers in Alzheimer's disease. *Neuromolecular Med.*14: 65–73.

Rubenstein, R., B. Chang, P. Gray, M. Piltch, M.S. Bulgin, S. Sorensen-Melson and M.W. Miller. 2010. A novel method for preclinical detection of PrP^{Sc} in blood. *J Gen Virol.* 91: 1883–1892.

Rubenstein, R. and B. Chang. 2013. Re-Assessment of PrP^{Sc} Distribution in Sporadic and Variant CJD. *PLoS One.* 8: e66352. doi:10.1371/journal.pone.0066352.

Rubenstein, R., B. Chang, P. Gray, M. Piltch, M. Bulgin, S. Sorensen-Melson and M.W. Miller. 2011. Prion Disease detection, PMCA kinetics, and IgG in urine from naturally/experimentally infected scrapie sheep and preclinical/clinical CWD deer. *J Virol.* 85: 9031–9038.

Rubenstein, R., M.S. Bulgin, B. Chang, S. Sorensen-Melson, R.B. Petersen and G. LaFauci. 2012. PrP^{Sc} detection and infectivity in semen from scrapie-infected sheep. *J Gen Virol.* 93: 1375–1383.

Saborio, G.P., B. Permanne and C. Soto. 2001. Sensitive detection of pathological prion protein by cyclic amplification of protein misfolding. *Nature.* 411: 810–813.

Saiz, A., F. Graus, J. Dalmau, A. Pifarre, C. Marin and E. Tolosa. 1999. Detection of 14-3-3 brain protein in the cerebrospinal fluid of patients with paraneoplastic neurological disorders. *Ann Neurol.* 46: 774–777.

Sanchez-Juan, P., A. Green, A. Ladogana, N. Cuadrado-Corrales, R. Sáanchez-Valle, E. Mitrováa, K. Stoeck, T. Sklaviadis, J. Kulczycki, K. Hess, M. Bodemer, D. Slivarichová, A. Saiz, M. Calero, L. Ingrosso, R. Knight, A.C.J.W. Janssens, C. M. van Duijn and I. Zerr. 2006. CSF tests in the differential diagnosis of Creutzfeldt-Jakob disease. *Neurology.* 67: 637–643.

Schaller, O. R. Fatzer, M. Stack, J. Clark, W. Cooley, K. Biffiger, S. Egli, M. Doherr, M. Vandevelde, D. Heim, O. Oesch and M. Moser. 1999. Validation of a Western immunoblotting procedure for bovine PrP(Sc) detection and its use as a rapid surveillance method for the diagnosis of bovine spongiform encephalopathy (BSE). *Acta Neuropathol.* 98: 437–443.

Schneider, P., H. Hampel and K. Buerger. 2009. Biological marker candidates of Alzheimer's disease in blood, plasma, and serum. *CNS Neurosci and Therapeutics.* 15: 358–374.

Schoonenboom, N.S., C. Mulder, G.J. Van Kamp, S.P. Mehta, P. Scheltens, M.A. Blankenstein and P.D. Mehta. 2005. Amyloid beta 38, 40 and 42 species in cerebrospinal fluid: More of the same? *Ann Neurol.* 58: 139–142.

Serban, A., G. Legname, K. Hansen, N. Kovaleva and S.B. Prusiner. 2004. Immunoglobulins in urine of hamsters with scrapie. *J Biol Chem.* 279: 48817–48820.

Shaked, G.M., G. Fridlander, Z. Meiner, A. Taraboulos and R. Gabizon. 1999. Protease-resistant and detergent-insoluble prion protein is not necessarily associated with prion infectivity. *J Biol Chem.* 274: 17981–17986.

Silva, C.J., B.C. Onisko, I. Dynin, M.L. Erickson, J.R. Requena and J.M. Carter. 2011. Utility of mass spectrometry in the diagnosis of prion diseases. *Anal Chem.* 83: 1609–1615.

Simon, S.L.R., L. Lamoureux, M. Plews, M. Stobart, J. LeMaistre, U. Ziegler, C. Graham, S. Czub, M. Groschup and J.D. Knox. 2008. The identification of disease-induced biomarkers in the urine of BSE infected cattle. *Proteome Sci.* 6: 23.

Sjögren, M., P. Davidsson, J. Gottfries, H. Vanderstichele, A. Edman, E. Vanmechelen, A. Wallin and K. Blennow. 2001. The cerebrospinal fluid levels of tau, growth-associated protein-43 and soluble amyloid precursor protein correlate in Alzheimer's disease, reflecting a common pathophysiological process. *Dement Geriatr Cogn Disord.* 12: 257–264.

Steinacker, P., H. Klafki, S. Lehnert, S. Jesse, C.A.F.v. Arnim, H. Tumani, A. Pabst, H.A. Kretzschmar, J. Wiltfang and M. Otto. 2010. ERK2 is increased in cerebrospinal fluid of Creutzfeldt-Jakob disease patients. *J Alzheimers Dis.* 22: 119–128.

Steinacker, P., P. Schwarz, K. Reim, K.P. Brechlin, O. Jahn, H. Kratzin, A. Aitken, J. Wiltfang, A. Aguzzi, E. Bahn, H.C. Baxter, N. Brose and M. Otto. 2005. Unchanged survival rates of 14-3-3γ knockout mice after inoculation with pathological prion protein. *Mol Cell Biol.* 25: 1339–1346.

Sunderland, T., N. Mirza, K.T. Putnam, G. Linker, D. Bhupali, R. Durham, H. Soares, L. Kimmel, D. Friedman, J. Bergeson, G. Csako, J.A. Levy, J.J. Bartko and R.M. Cohen. 2004. Cerebrospinal fluid beta-amyloid 1–42 and tau in control subjects at risk for Alzheimer's disease: the effect of APOE e4 allele. *Biol Psychiatry.* 56: 670–676.

Swardfager, W., K. Lanctot, L. Rothenburg, A. Wong, J. Cappell and N. Hermann. 2010. A meta-analysis of cytokines in Alzheimer's disease. *Biol Psychiatry.* 68: 930–941.

Tang, K., L.S. Hynan, F. Baskin and R.N. Rosenberg. 2006. Platelet amyloid precursor protein processing: a bio-marker for Alzheimer's disease. *J Neurol Sci.* 240: 53–58.

Tang, Y., W. Xiang, L. Terry, H.A. Kretzschmar and O. Windl. 2010. Transcriptional analysis implicates endoplasmic reticulum stress in bovine spongiform encephalopathy. *PLos One.* 5: e14207.

Thorne, L. and L.A. Terry. 2008. *In vitro* amplification of PrPSc derived from the brain and blood of sheep infected with scrapie. *J Gen Virol.* 89: 3177–3184.

Vandermeeren, M., M. Mercken, E. Vanmechelen, J. Six, A. van de Voorde, J.J. Martin and P. Cras. 1993. Detection of tau proteins in normal and Alzheimer's disease cerebrospinal fluid with a sensitive sandwich enzyme-linked immunosorbent assay. *J Neurochem.* 61: 1828–1834.

Vanmechelen, E., H. Vanderstichele, P. Davidsson, E. Van Kerschaver, B. Van Der Perre, M. Sjogren, N. Andreasen and K. Blennow. 2000. Quantification of tau phosphorylated at threonine 181 in human cerebrospinal fluid: a sandwich ELISA with a synthetic phosphopeptide for standardization. *Neurosci Lett.* 285: 49–52.

Vemuri, P., H.J. Wiste, S.D. Weigand, L.M. Shaw, J.Q. Trojanowski, M.W. Weiner, D.S. Knopman, R.C. Petersen, C.R. Jack, Jr. and the Alzheimer's Disease Neuroimaging Initiative. 2009. MRI and CSF biomarkers in normal, MCI, and AD subjects: diagnostic discrimination and cognitive correlations. *Neurology.* 73: 287–293.

Wagner, W., P. Ajuh, J. Lower and S. Wessler. 2010. Quantitative phosphoproteomic analysis of prion-infected neuronal cells. *Cell Commun Signal.* 8: 28.

Walsh, D.M. and D.J. Selkoe. 2007. AB oligomers-a decade of discovery. *J Neurochem.* 101: 1172–1184.

Wei, X., A. Herbst, D. Ma, J. Aiken and I. Li. 2011. A quantitative proteomic approach to prion disease biomarker research; delving into the glycoproteome. *Proteome Res.* 10: 2687–702.

Zerr, I., K. Kallenberg, D.M. Summers, C. Romero, A. Taratuto, U. Heinemann, M. Breithaupt, D. Varges, B. Meissner, A. Ladogana, M. Schuur, S. Haik, S.J. Collins, G.H. Jansen, G.B. Stokin, J. Pimentel, E. Hewer, D. Collie, P. Smith, H. Roberts, J.P. Brandel, C. van Duijn, M. Pocchiari, C. Begue, P. Cras, R.G. Will and P. Sanchez-Juan. 2009. Updated clinical diagnostic criteria for sporadic Creutzfeldt-Jakob disease. *Brain.* 132: 2659–2668.

Zerr, I., M. Pocchiari, S. Collins, J.P. Brandel, J. de Pedro Cuesta, R.S.G. Knight, H. Bernheimer, F. Cardone, N. Delasnerie-Lauprêtre, N. Cuadrado Corrales, A. Ladogana, M. Bodemer, A. Fletcher, T. Awan, A. Ruiz Bremón, H. Budka, J.L. Laplanche, R.G. Will and S. Poser. 2000. Analysis of EEG and CSF 14-3- 3 proteins as aids to the diagnosis of Creutzfeldt-Jakob disease. *Neurology.* 55: 811–815.

Zetterberg, H., K. Blennow and E. Hanse. 2010. Amyloid β and APP as biomarkers for Alzheimer's disease. *Exp Gerontol.* 45: 23–29.

Zetterberg, H., U. Andreasson, O. Hansson, G. Wu, S. Sankaranarayanan, M.E. Andersson, P. Buchhave, E. Londos, R.M. Umek, L. Minthon, A.J. Simon and K. Blennow. 2008. Elevated cerebrospinal fluid BACE1 activity in incipient Alzheimer disease. *Arch Neurol.* 65: 1102–1107.

Zhang, R., L. Barker, D. Pinchev, J. Marshall, M. Rasamoelisolo, C. Smith, P. Kupchak, I. Kireeva, L. Ingratta and G. Jackowski. 2004. Mining biomarkers in human sera using proteomic tools. *Proteomics.* 4: 244–256.

Zhong, Z., M. Ewers, S. Teipel, K. Burger, A. Wallin, K. Blennow, P. He, C. McAllister, H. Hampel and Y. Shen. 2007. Levels of β-secretase (BACE1) in cerebrospinal fluid as a predictor of risk in mild cognitive impairment. *Arch Gen Psychiatry.* 64: 718–726.

16

Biomarkers of Apoptosis and Inflammation in Neurodegenerative Disorders

Hayat Harati, [1,]* *Jihane Soueid* [2] *and Rose-Mary Boustany* [2]

INTRODUCTION

Neurodegenerative disease refers to neuronal loss as the principal pathology associated with disorders such as amyotrophic lateral sclerosis, Alzheimer's disease, Huntington disease, Batten disease and Parkinson disease. Neurodegeneration often results from cell death and inflammatory damage in brain areas affected in these diseases. The predominant form of cell death in neurodegenerative diseases is by apoptosis (Smale et al., 1995) but different mechanisms contribute to neuronal loss and disease progression. Accumulation of autophagosomes or disrupted axonal transport are also common features of many neurodegenerative diseases (Anglade et al., 1997; Sapp et al., 1997; Sasaki, 2011; Yu et al., 2005). Autophagic/lysosomal pathways are quality control systems of proteins and organelles in neurons and are essential in both physiological and pathological conditions (Klionsky and Emr, 2000). Dysfunctional autophagy machinery can initiate self-digestion of dying cells (Komatsu et al., 2006) with an inadequate autophagic response increasing vulnerability to stress conditions and induction of cell death (Komatsu et al., 2007; Ravikumar et al., 2005).

---
[1] Medical school, Lebanese University, Hadath, Lebanon.
[2] Neurogenetics Program and Division of Pediatric Neurology, Departments of Pediatrics, Adolescent Medicine and Biochemistry, American University of Beirut.
* Corresponding author: hayat.harati@gmail.com

Apoptosis is a programmed form of cell death that plays a vital physiological and pathological role in organisms. It is crucial for the correct sculpting of organs and tissues during development and for maintenance of tissue homeostasis through elimination of excessive or injured cells (Ellis et al., 1991). During our lifetime, over 99.9% of cells will undergo cell death and be eliminated by the body, thus apoptosis provides an important mechanism for maintaining homeostasis and a process for fighting disease (Vaux and Korsmeyer, 1999). Disruption of homeostasis by excessive or deficient cell death is a hallmark of many pathological conditions including cancer, cardiovascular and autoimmune diseases (Thompson, 1995; van Heerde et al., 2000). Accelerated apoptosis in the brain causes neurodegenerative disease. Two apoptosis pathways, the death receptor pathway (extrinsic) and the mitochondrial (intrinsic) pathway, are recognized (Kroemer et al., 2007). In the extrinsic pathway, apoptosis is initiated by binding of TNF family cytokines (TNFα, Fas and TNF-related apoptosis-inducing ligand, TRAIL) to their cognate death receptors at the cytoplasmic cell membrane. Ligand-induced activation of death receptors activates caspase-8 and -10, which in turn activate downstream executor caspases (caspase-3, -6, and -7) (Wajant et al., 2003). In the intrinsic pathway, apoptosis is triggered by intracellular stress signals, mainly associated with outer mitochondrial membrane permeabilization. The intrinsic pathway involves an increase in Reactive Oxygen Species (ROS) production, or the functional incapacitation of mitochondria by the pro-apoptotic BCL-2 family members Bax, Bak, Bid and Bik. Both pathways cause loss of potential across the mitochondrial membrane and release of cytochrome c which activates initiator caspase-9. This in turn activates executor caspases-3/-7 with destruction of cellular proteins (Wang, 2001). Caspases are a family of intracellular cysteine proteases that exist as zymogens within the cell and are converted by pro-apoptotic signals into active forms (Li and Yuan, 2008). Initiator caspases-8/-9 is activated by homodimerization, whereas effector caspases are activated by cleavage of a catalytic domain. Most morphological changes of apoptotic cells are caused by caspases activated by both pathways. Caspases cleave a restricted group of target substrate peptides that immediately follow an aspartate residue (Debatin and Krammer, 2004; Thornberry, 1998). Apoptosis is characterized by cell morphology changes such as cytoplasm shrinkage, cell detachment, chromatin condensation, nuclear fragmentation and the formation of apoptotic bodies. Biochemically, apoptotic cell death includes activation of caspases, mitochondrial outer membrane permeabilization, DNA fragmentation, generation of reactive oxygen species (ROS), ER stress, Lysosomal Membrane Permeabilization (LMP) and exposure of molecular biomarkers such as phosphatidylserine (PS) on the outer leaflet of the plasma membrane.

Apoptosis is a prominent feature of Central Nervous System (CNS) development resulting in Naturally Occurring Cell Death (NOCD), a necessary and desirable process. NOCD effectively eliminates neurons that have made faulty synapses or have not reached appropriate targets (Purves, 1985). The CNS in the adult, however, has very limited regenerative capacity making it important to limit cell death in that region (Rossi and Cattaneo, 2002). Undesirable and inappropriate neuronal death occurring in the adult brain is the phenotypic expression of neurodegenerative disease (Nishimoto et al., 1997; Smale et al., 1995). In addition, neuroinflammation is a major contributor to neuronal degeneration. Early reports defined the brain as immune privileged, being separated from the peripheral blood system by the blood-brain barrier. It is now known that many neurodegenerative diseases are characterized by microglial activation and infiltration of leukocytes from the periphery (Block and Hong, 2005). Microglial cells, which can be activated to a phagocytic state, are the main cellular components of innate immunity in the brain (Campbell, 2004). Astrocytic cells form processes that surround endothelial cells of the cerebral microvasculature protecting it against foreign invasion. Neuroinflammation is characterized by the reactivity of microglia and astrocytes, activation of inducible nitric ioxide (NO)-synthase (iNOS) and increased expression and release of pro-inflammatory cytokines and chemokines such as Tumor Necrosis Factor α (TNF-α) (John et al., 2003). Uncontrolled inflammation can result in production of neurotoxic factors that exacerbate neurodegenerative pathologic processes. Many neurodegenerative diseases are caused by genetic defects in unrelated genes having a direct or indirect effect on apoptosis and neuroinflammation. In this chapter the focus is on biomarkers of apoptosis and neuroinflammation in Batten disease and Alzheimer's disease.

Alzheimer's disease

Overview

Alzheimer's disease (AD), also called presenile dementia, is one of the most common neurodegenerative diseases in the elderly. There are three known types of Alzheimer's disease. They include:

- Early-onset Alzheimer's which is rare, with diagnosis occurring before age 65.
- Late-onset Alzheimer's is the most common form accounting for ~90% of cases and the age of onset is after 65 years.

- Familial or hereditary Alzheimer's disease is extremely rare, accounting for less than 1% of all cases. It has an earlier onset with onset during the fifth decade.

Changes in brain chemistry affecting the way neurons communicate are more likely to be responsible for memory problems associated with normal aging. Memory loss and other cognitive changes in AD are due to profound neuronal loss in parts of the brain critical for memory. These areas are the first affected by AD, although destruction of cells occurs in other areas of the brain. AD progresses through three stages: 1) A preclinical period where AD biomarkers are present, but symptoms have not yet appeared. 2) Mild Cognitive Impairment (MCI) where patients have cognitive deficits but no functional impairment. 3) Dementia which is described as decline in two or more cognitive domains that gradually progresses to the point that functioning at work becomes impossible and that activities of daily living are affected (Hall and Roberson, 2012).

The main pathological features of AD include Senile Plaques (SPs) that are comprised of aggregates of amyloid-β (Aβ) protein surrounded by damaged neuronal processes and reactive glia. There are also intraneuronal aggregates of the microtubule-associated protein, Tau, forming paired helical filaments that become evident in dystrophic neurites and neurofibrillary tangles (NFTs) and in areas of neuron loss. Aβ and phosphorylated Tau have been widely used as biomarkers in the clinical diagnosis of AD patients, but pathologic changes in AD comprise many other biological alterations.

The ideal model for AD should manifest the full range of clinical and pathological features of cognitive and behavioral deficits, amyloid plaques and neurofibrillary tangles, gliosis, synapse loss, axonopathy, neuronal loss and neurodegeneration. Different mouse lines develop these phenotypes to varying degrees, but no single existing mouse model exhibits all features of AD (for review of AD mouse models see (Hall and Roberson, 2012)).

Brain weight loss in AD of ~20% is due to substantial degeneration of neurons in regions involved in memory and learning processes (e.g., temporal, entorhinal and frontal cortex and hippocampus). Synapse loss is detected early in AD (Masliah et al., 2001; Sze et al., 1997) and precedes accumulation of plaques and tangles in a transgenic AD mouse model (Oddo et al., 2003).

The mechanisms of cell death involved in AD have not been fully elucidated, but there are several reports of neurons dying by apoptosis, suggesting the latter is responsible for the extensive neuronal cell death seen in AD (Ankarcrona and Winblad, 2005; Lassmann et al., 1995; Li et al., 1997). Glial cell death, and Aβ protein deposits and NFTs are closely associated with apoptosis in AD (Kobayashi et al., 2002; Sugaya et al., 1997). Neuronal apoptosis in neurodegenerative diseases often goes hand in hand with an inflammatory response (Abdi et al., 2011). Deposition of plaques and tangles produces inflammatory stress in AD (Potter et al., 2001; Wyss-Coray, 2006). This notion is strengthened by studies showing that non steroidal anti-inflammatory drugs can play a protective role (Choi et al., 2009). Additionally, intra-neuronal accumulation of Aβ may cause cell death and apoptosis, triggering neuronal signals to activate microglia and astrocytes (Pereira et al., 2005), independent of extracellular Aβ deposition. Taken together, inflammation and apoptotic stress devastate AD brain (Naoi and Maruyama, 2010). Rapid advances in AD biology have heightened the urgency to develop reliable biological markers for apoptosis and inflammation to diagnose and monitor AD activity (Bailey, 2007). We will be focusing on such markers.

Markers of Apoptosis in Alzheimer's Disease

General Markers of Apoptosis

Several studies of apoptosis in AD have been performed on post-mortem tissues and the results only reflect the end-stage of AD. Early diagnosis of AD is essential. This is where a reliable marker for cell death could be of help. Apoptosis has been shown to play a crucial role in pathological studies of AD mainly associated with gray matter lesions (Li et al., 1997; Sheng et al., 1998). Other studies (Brown et al., 2000) investigate the relationship between glial apoptosis and white matter lesions in AD. Kobayashi et al. (2002) reported that in the temporal white matter, the density of apoptotic glial cells, mainly oligodendroglia, is high. Additionally, the rate of white matter shrinkage correlates with density of apoptotic glial cells.

Terminal dUDP nick end-labeling or TUNEL-positive neurons and glia in post-mortem AD hippocampus and cortex support the occurrence of DNA fragmentation and apoptosis (Dragunow et al., 1995; Lassmann et al., 1995; Li et al., 1997; Smale et al., 1995; Su et al., 1994; Sugaya et al., 1997). Increased expression of BCL-2 family members (Drache et al., 1997; Giannakopoulos et al., 1999; Kitamura et al., 1998; MacGibbon et al., 1997; Nagy and Esiri, 1997) and elevated caspase activities and cleavage of caspase substrates in AD brain (Ahmed and Gilani, 2011; Chan et al., 1999; Chiu et al., 1996; LeBlanc et al., 1999; Pompl et al., 2003; Rohn et al., 2001; Stadelmann et al., 1999; Uetsuki et al., 1999) have also been reported. More recently, activation of the caspase machinery has been shown to precede tangle formation (Avila, 2010). Caspase-6 and caspase-3 (D'Amelio et al., 2011) appear to be involved in AD pathology (Agostini et al., 2011). Importantly, the neuritic beading induced by amyloid precursor protein (APP) is caspase-6 dependent and is inhibited by the specific caspase-6 inhibitor, z-VEID-fmk (Sivananthan et al., 2010). Anti-apoptotic BCL-2 (an inhibitor of apoptosis) expression decreases (Pan et al., 2008; Yao et al., 2005). Overexpression of BCL-2 provides protection against development of AD pathology (Rohn et al., 2008).

In a recent study, β-amyloid peptide was injected into the rat cortex for induction of neuro-inflammation in hippocampus (Abdi et al., 2011) following which elevation in the inflammatory cytokine TNF-α, as well as an increase in caspase-3 activity and TUNEL-positive cells were observed. The latter two facts support the occurrence of apoptosis triggered by Aβ in the hippocampus. Interestingly, neurons treated with Aβ in culture reveal typical apoptosis, including an increase in JNK (c Jun N terminal Kinase), caspase-2 and caspase-3 activation (Troy et al., 2001), condensation of cell bodies and DNA fragmentation (Xiao et al., 2002). In another study upregulation of p53, Bax, mitochondrial membrane permeability transition, release of cytochrome c and activation of apoptosome formation resulting in caspase-3 activation are also documented (Mattson, 2004).

Some studies support the notion that the effect of Aβ is indirect via activation of microglia causing inflammation (Manelli et al., 2004).

Researchers have previously reported many molecules linked to cell death in AD detected in cerebrospinal fluid (CSF), plasma and urine [for review, see (Ankarcrona and Winblad, 2005)].

The neuronal thread protein AD7c-NTP increases in both urine and CSF in patients suffering from AD (de la Monte and Wands, 2002). Overexpression of this protein leads to neuronal cell death mediated by impaired mitochondrial function and apoptosis. *Tissue transglutaminase (tTG)* is a protein and potential marker of cell death that increases in the CSF in AD (Bonelli et al., 2002b). TTG cross-links the substrate Tau into insoluble aggregates resistant to proteases. Inhibition of tTG prevents apoptosis (Bonelli et al., 2002a). CD95 ligand, or Fas receptor, a member of the TNF/NGF receptor family, are potential markers for apoptosis in AD (Ankarcrona and Winblad, 2005). Its level increases in serum from AD patients (Richartz et al., 2002). Fas receptor expression is increased in the brains of AD patients (Ankarcrona and Winblad, 2005; de la Monte et al., 1997). Apoptosis is activated when the CD95 ligand binds to the Fas receptor (Nagata and Golstein, 1995).

Evidence for ceramide-assisted neuronal death in AD

Sphingolipid metabolism is a dynamic process that modulates the formation of a number of bioactive metabolites or second messengers critical in cellular signaling and apoptosis. In the brain, the regulated levels of sphingolipids are essential for normal neuronal function, as evidenced by a number of severe brain disorders that are the result of deficiencies in enzymes that control sphingolipid metabolism. Lipid rafts in plasma membrane are microdomains that function as platforms for receptor binding and initiation of signaling cascades in a temporally and spatially orchestrated manner. Lipid rafts consist of sphingolipids and cholesterol and are derived from Golgi. They are involved in protein trafficking/endocytotic pathways (Simons and Ikonen, 1997) and harbor ceramide, a proapoptotic lipid second messenger. Ceramide is the building block for sphingomyelin/ galactosylceramide/sulfatide and other glycosphingolipids. Lipid rafts also house caspase-8, an initiator caspase activated in apoptosis. Laboratory and animal studies implicate direct and indirect mechanisms by which sphingolipids contribute to Aβ production and Alzheimer pathogenesis. Later studies have recapitulated these findings in humans. Plasma sphingolipids are shown to predict cognitive impairment and hippocampal volume loss in patients with AD (Mielke et al., 2011).

The majority of post-mortem studies report elevated ceramide levels in gray and white matter brain regions in AD, even at the earliest clinically recognizable stages of the disease (Bandaru et al., 2009; He et al., 2010; Katsel et al., 2007). The most compelling findings were that enzymes controlling ceramide synthesis, particularly the long-chain ceramides (C22:0 and C24:0), were upregulated early in the disease process (Mielke et al., 2012). These results suggest a shift in sphingolipid metabolism towards the accumulation of ceramides at the earliest stages of AD (Han, 2005; He et al., 2010; Mielke and Lyketsos, 2010). Shotgun lipidomics uncovered significant disruptions in the sphingolipidome in plasma obtained from AD patients compared to normal controls. These findings provide new insights into the relationship between lipid biochemistry and neuronal dysfunction in early AD (Han et al., 2010).

Different results of ceramide have been reported depending on the multiple stages of the AD disease process. In one study AD patients had higher levels of ceramide in the middle frontal cortex (Cutler et al., 2004) and white matter (Han et al., 2002). One CSF study reports higher ceramide levels in moderate versus mild or severe AD (Satoi et al., 2005), while another examining brain tissue reported that the gene expression patterns of enzymes participating in the sphingolipid metabolism pathway varied by AD severity (Katsel et al., 2007). Ceramide levels were highest in patients with more than one neuropathologic abnormality (Filippov et al., 2012) and high levels of dihydroceramides (DHCer) and ceramide were associated with severity and a faster progression (Mielke et al., 2011) indicating that these lipids may be sensitive blood-based biomarkers for clinical progression.

Multiple studies suggest a connection between ceramide and the formation of Aβ (Adams, 2008; Cutler et al., 2004; Grimm et al., 2005; Jana and Pahan, 2004; Mattson et al., 2005). Elevated sphingomyelin (SM) levels occur in the inferior parietal lobe of AD patients, and have a strong positive correlation with the number of amyloid-beta plaques (Pettegrew et al., 2001). Consistent with elevations in ceramide levels = acid sphingomyelinase (ASM), which metabolizes SM into ceramides, had increased activities in the frontotemporal gray matter of AD cases compared to controls (He et al., 2010). Importantly, a positive correlation was found between ASM activity and Aβ or phosphorylated tau in this same region, again suggesting elevated ceramide levels are associated with AD pathology (He et al., 2010). A recent study determined that serum SM and ceramides vary according to onset of memory impairment. Also, higher ceramide levels are associated with greater hippocampal atrophy in AD (Mielke et al., 2010).

Ceramides and SM metabolites are second messengers that regulate cellular differentiation, proliferation and apoptosis by activating signaling cascades and promoting free radical generation (Andrieu-Abadie et al., 2001). Aβ induces ceramide production leading to neuronal apoptosis (Jana and Pahan, 2004). Ceramides facilitate the regulation of β-site APP cleaving enzyme 1 (BACE-1) and γ-secretase activity and amyloid precursor protein (APP) processing and trafficking. Evidence also suggests that glycosphingolipids bind Aβ at the cell surface and form domains that facilitate the oligomerization and fibril formation of Aβ (Chi et al., 2008; Ikeda et al., 2011). Once activated, ceramide-associated protein kinases and phosphatases induce pro-apoptotic signaling pathways (Detre et al., 2006; Oh et al., 2006) leading to neurodegeneration.

Exposure of cultured neurons to Aβ directly increases ceramide levels by activating neutral sphingomyelinase (Grimm et al., 2005). Inhibiting ceramide production protects neurons from Aβ-induced cell death (Cutler et al., 2004). In addition, Aβ indirectly increases ceramide production through an oxidative stress-mediated mechanism (Cutler et al., 2004). Ceramide then increases inflammatory stress related neuronal apoptosis (Han et al., 2011) and reactive oxygen species, further exacerbating the pathology. Finally, increased levels of ceramide accelerate the formation of pathogenic forms of amyloid by increasing β- and γ-cleavage of APP (Cordy et al., 2003).

There is also considerable evidence that pro-inflammatory cytokines such as TNFα, IL-1, and Fas/FasL are potent inducers of ceramide production, and increased concentrations of ceramide can stimulate the production of interleukins IL-2 and IL-6 (Ballou et al., 1996). Fas/FasL interaction activates a caspase-8-dependent increase in SMase activity that increases ceramide which promotes formation of large lipid platforms and the assembly of cell death signaling protein complexes (Bourbon et al., 2001). There are also several ceramide-regulated protein kinases (CAPK) and phosphatases (CAPP) that when activated lead to signaling that triggers apoptosis: pro-apoptotic CAPK signaling involves recruitment of

MAPK/ERK kinase kinase (MEKK1), activation of SAPK-kinase (SEK1), Jun N-terminal kinases (JNK 1 and JNK2) and inhibition of the survival factor extracellular signal-regulated kinase-1 and 2 (ERK1 and ERK2) (Huwiler et al., 2004; Shen et al., 2003).

Markers of Inflammation in Alzheimer's Disease

Elevated levels of chemokines and cytokines and their receptors, including IL-1α, CXCR2, CCR3, CCR5, and TGF-β, are reported in post-mortem AD brains (Cartier et al., 2005).

Chemokines

Several chemokines and chemokine receptors are upregulated in AD brain (Xia and Hyman, 1999). Chemokines play an important role for recruiting microglia and astroglia to the site of Aβ deposition. Aβ stimulated human monocytes generate chemokines such as interleukin IL-8 *in vitro*. Monocyte chemoattractant protein-1 (MCP)-1 induces the chemotaxis of astrocytes and contributes to recruitment of astrocytes around plaques as well as macrophage inflammatory protein- (MIP)-1α and MIP-1β. Microglia cultured from rapid autopsies of AD, indicate an increased expression of IL-8, MCP-1, and MIP-1α (Rubio-Perez and Morillas-Ruiz, 2012).

Cytokines

Cytokines play a key role in inflammatory processes in AD. Levels of IL-1α, IL-1β, IL-6, TNF-α, granulocyte-macrophage colony-stimulating factor (GMSF), IFN-α, type B of IL-8 receptor (IL-8RB), and the receptor for CSF-1 are significantly increased in AD brain tissue (McGeer and McGeer, 1997). The Aβ protein potentiates the secretion of IL-6/IL-8 by IL-1β-activated astrocytoma cells and IL-6/TNF-α by lipopolysaccharide (LPS) stimulated astrocytes (Forloni et al., 1997), as well as IL-8 by monocytes (Rubio-Perez and Morillas-Ruiz, 2012). IL-6 is found in high concentration in AD but not in normal aging brain. It also infiltrates early plaques but disappears as plaques age and may play a role in neuritic transformation. Additionally, levels of sIL-6R and sgp130, the receptors for IL6, are decreased in AD brain (Bailey, 2007).

Inflammation can be secondary to protein accumulation in neurodegenerative diseases, including AD [for review, see (Khandelwal et al., 2011)]. Several inflammatory markers have been assessed as potential biomarkers for AD progression (Bailey, 2007). Inflammatory components include microglia and astrocytes, the complement system, as well as cytokines and chemokines [for review see (Rubio-Perez and Morillas-Ruiz, 2012)].

The hypothesis is Aβ plaques and tangles induces a chronic inflammatory reaction to clear debris (Town et al., 2005). These plaques contain dystrophic neurites, activated microglia and reactive astrocytes (Akiyama et al., 2000). Aggregated amyloid fibrils and inflammatory mediators including cytokines, chemokines and complement molecules, secreted by activated microglial and astrocytic cells contribute to neuronal dystrophy (Findeis, 2007). Chronically activated glia can kill adjacent neurons by releasing highly toxic products such as reactive oxygen intermediates, nitric oxide (NO), and proteolytic enzymes (Halliday et al., 2000). In addition, inflammatory mediators and increased cellular stress enhance APP production and the amyloidogenic processing of APP to induce Aβ peptide production. This inhibits the formation of a soluble APP fraction that has a neuronal protective effect (Atwood et al., 2003; Friedlander, 2003).

Glia-mediated inflammatory immune responses are also important components of the pathogenesis of AD and are associated with early activation (McRae et al., 2007). CSF levels of Glial Fibrillary Acidic Protein (GFAP) are increased in AD compared with controls (Jesse et al., 2009). S100B, another glial and inflammatory marker is also increased in response to inflammation, particularly in the vicinity of neuritic plaques (Peskind et al., 2001).

Other activated markers and proinflammatory mediators are found in microglia surrounding plaques and include major immuno-histocompatibility complex (MHC)-II, cyclooxygenase 2 (COX-2). COX-2 is located in neurons and is implicated in synaptic plasticity (Yang and Chen, 2008). Different reports suggest that COX-2 expression varies according to the stage of the disease with upregulation of COX-2 in early AD (Combrinck et al., 2006). Anti-microglial antibodies are also found in the cerebrospinal fluid of the majority of patients with AD (McRae et al., 2007).

Many other markers of neural inflammation are found in AD. These include C1q, a potent facilitator of Aβ aggregation (Webster and Rogers, 1996) which binds Tau and Aß and is increased in AD brain compared to controls (Brachova et al., 1993). Also, melanotransferrin (p97), a protein with a role in iron transport in the brain, that is found in neuroglia surrounding AD plaques, is elevated in cerebrospinal fluid and serum in AD (Kennard et al., 1996). Alpha 1 antichymotrypsin, a serine protease inhibitor, is also elevated in AD and is present in senile plaques. It increases with the severity of dementia (Galasko and Montine, 2010; Licastro et al., 1995).

Batten Disease

Overview

Neuronal Ceroid Lipofuscinoses (NCLs), commonly known as Batten disease, represent a group of the most prevalent (1 in 12,500 births) neurodegenerative storage disorders affecting both children and adults (Cooper et al., 2006; Goebel and Wisniewski, 2004; Haltia, 2006; Hobert and Dawson, 2006; Jalanko et al., 2006).

NCLs are characterized by profound and progressive neurodegeneration which results in clinical symptoms that include seizures, blindness, mental retardation, ataxia and premature death. Storage of autofluorescent and ultrastructurally distinctive material within lysosomes is characteristic for the neuronal ceroid lipofuscinoses (NCLs) or Batten disease (Goebel and Wisniewski, 2004). The pathological hallmark of NCL is neuronal loss and death of photoreceptors in the retina with resultant blindness. Pathophysiological theories include increased lipid peroxidation, altered dolichol turnover, increased inflammation, neuronal apoptosis and accumulation of subunit c of mitochondrial ATP synthase (Boustany, 1996). The latter is the predominant storage material in all NCLs except the infantile CLN1 type (INCL). In INCL, the accumulated material is composed of saposins (SAPs) A and D (Santavuori, 1999).

Fourteen distinct genetic NCL variants are presently described (Haltia and Goebel, 2012). The first gene discovered in the NCLs was the endosomal/Golgi membrane protein CLN3 causing CLN3 disease or JNCL in 1995. Defects in soluble lysosomal proteins such as protein palmitoyl thioesterase 1 or PPT1 and tripeptidyl peptidase 1 or TPP1 cause infantile (INCL) and late infantile (LINCL) forms of the disease, respectively. CLN5p, a soluble glycoprotein is deficient in another LINCL variant or Finnish variant LINCL (fLINCL). The gene for CLN9 variant is now reclassified as CLN5 disease (El Haddad, 2012). Defects in ER-resident proteins CLN6 and CLN8 cause variant late infantile NCLs and defects in CLN6 cause adult NCL or Kuf disease, type A. Cathepsins D and F account for congenital NCL and Kufs disease, type B, respectively.

While neurodegeneration is central to the pathology of all NCLs, the precise cellular mechanisms underlying neuronal death remain unknown. Apoptosis, necrosis and autophagy have been implicated in neuronal death (Mitchison et al., 2004) but other mechanisms may also be involved (Kim et al., 2009). The role played by apoptotic pathways in neuronal death in the NCLs is of fundamental importance as pharmacological intervention in these pathways has been proposed as a therapeutics strategy for such diseases (Dhar et al., 2002; Lane et al., 1996).

Increased apoptosis and dysregulated sphingolipid metabolism are documented in CLN2 disease, CLN3 disease, variant late infantile CLN6-/ Epilepsy with Mental Retardation CLN8-deficient variants, and CLN9 variant now reclassified as CLN5 disease. Accelerated apoptosis which leads to neuronal loss, a hallmark of neurodegenerative disease, is documented in CLN3-/CLN2-/CLN6-/CLN5 (formerly CLN9)-deficient neurons and brain and in CLN1-/CLN2-/CLN3/CLN5-/CLN6- and CLN8-deficient cells. In addition to histological studies showing that cell death in Batten disease is caused by apoptosis (Lane et al., 1996), other studies have shown upregulation of the anti-apoptotic molecule, BCL-2 and elevation of ceramide levels, implicating disrupted regulation of apoptosis (Lane et al., 1996; Puranam et al., 1997).

Ceramide is a pro-apoptotic lipid second messenger and the building block for the phospholipid sphingomyelin, galactosylceramide, sulfatide and other glycosphingolipids. Ceramide is a precursor for galactosylceramide (GalCer) which is also implicated in signal transduction and cell death (Hannun and Obeid, 1995; Obeid et al., 1993). Cer is generated via salvage pathways or *de novo* synthesis (Birbes et al., 2001) and activates caspase-dependent/caspase-independent cell death pathways (Zhao et al., 2004). It limits activation of protective caspase-9b and Bcl-x in favor of apoptotic variants initiating signaling such as caspase-9 and caspase-8 and death receptors from Lipid Rafts (LRs) (Chalfant et al., 2002; Scheel-Toellner et al., 2002). There is also evidence for selective glial activation becoming more widespread with NCL progression (Mitchison et al., 2004). The focus here is on apoptotic mechanisms in the most prevalent and studied forms of NCL disease, and presence of inflammation.

CLN1 disease

CLN1 disease is due to mutations in the CLN1 gene (palmitoyl protein thioesterase or PPT1 protein). Dramatic loss of cortical neurons is documented (Goebel et al., 1988). Apoptosis is a suggested cause of neurodegeneration in CLN1 disease, also called Infantile Neuronal Ceroid Lipofuscinosis (INCL) and was obvious in cortex and retina (Gupta et al., 2001; Riikonen et al., 2000), although the precise mechanism(s) of apoptosis remains unclear (Kim and Klionsky, 2000). The *CLN1* gene codes for the protein PPT1. PPT1 is an enzyme that is responsible for S-acylation of proteins, a process important in anchoring of proteins to membranes (Salonen et al., 2000). The lack of PPT1 results in irreversible thio-ester bond formation that is damaging to cell survival. Overexpression of PPT1 protects neuronal cells from apoptosis induced by ceramide, and transient transfection with antisense-*CLN1* increases their susceptibility to death induced by C2-ceramide (Cho and Dawson, 2000; Cho et al., 2000). Phosphocysteamine and the anti-apoptotic agent N-acetyl-cysteine, cleave thio-ester bonds, similarly to PPT1. This leads to reduction of the number of TUNEL-positive apoptotic *CLN1*-deficient lymphoblasts (Zhang et al., 2001). In line with the link to apoptotic pathways, overexpression of PPT1 protects human neuroblastoma cells from apoptosis induced by short-chain length ceramides or a phosphatidylinositol 3-kinase inhibitor (Cho and Dawson, 2000).

In a PPT1 knockout mouse model, Qiao et al. (2007) profiled brain gene expression and showed an upregulation of a negative regulator of neuronal apoptosis (DAP kinase-1) in these mice compared to wild-type mice. Chemokine ligands that are markers for an intense inflammatory response are also upregulated, and may prove to be good serum markers for disease activity.

It has been reported that brain cells undergo endoplasmic reticulum stress (ER stress) in this same mouse model. This activates both the Unfolded Protein Response (UPR) and the ER-resident cysteine protease, caspase-12, leading to caspase-3 activation and apoptosis (Zhang et al., 2006). There is generation of reactive oxygen species secondary to ER stress and disruption of calcium homeostasis that may induce the mitochondrial caspase-9 pathway (Kim et al., 2006b). In another study, Kim et al. (2006a) reported the same findings in human CLN1 brain, where ER stress-induced activation of the unfolded protein response (UPR) mediates caspase-4 and caspase-3 activation and apoptosis. In addition to caspase activation, CLN1-defective cells displaypoly(ADP)-ribose polymerase cleavage, a compelling sign for apoptosis. Interestingly, saposins A and D accumulate in CLN1 disease. Although the significance of this is unclear, saposins regulate glycosphingolipid flux in the cell and, hence, indirectly, ceramide metabolism as well.

The *Ppt1*$^{\Delta ex4}$ mouse model shows inflammation-associated death of interneurons (Jalanko et al., 2005). In addition, an upregulation of inflammatory genes concomitant with an associated loss of interneurons was observed in PPT1 deficient mouse brain, in addition to upregulation of a number of chemokines, particularly chemokine ligand 21 (Ccl21a/b/c) (Qiao et al., 2007). The onset and progression of glial activation and increased GFAP expression is followed by selective neuronal loss and microglial responses, particularly in the thalamus, with successive death of different classes of interneurons and granule neurons in the cortex of a Ppt1$^{-/-}$ mouse model (Kielar et al., 2007; Macauley et al., 2011; Macauley et al., 2009).

CLN2 disease or LINCL

The gene defect in late infantile neuronal ceroid lipofuscinosis or LINCL decreases tripeptidyl peptidase or TPP1 activity. TPP1 may be implicated in the apoptotic pathway (Dhar et al., 2002). Also CLN2-deficient brain shows elevations in pro-apoptotic ceramide (Puranam et al., 1999). Studies report apoptotic cells identified by TUNEL staining (Lane et al., 1996), alterations of BCL-2 levels and an increase in ceramide levels in the brains of LINCL patients (Puranam et al., 1999). Moreover, flupirtine maleate, a neuroprotective molecule, abrogated cell death in CLN2-deficient cells and neurons (Dhar et al., 2002).

Dhar et al. (2002) speculated that the mechanism by which TPP1 impacts anti-apoptotic pathways may be that TPP1 deficiency is accompanied by an increase in the activity of cathepsin B (Sleat et al., 1998). The latter activates the intrinsic apoptotic pathway via truncated Bid that translocates to the mitochondria inducing cytochrome c release and apoptosis (Stoka et al., 2001).

Other studies speculate that neurodegeneration in CLN2 disease or LINCL may occur via non-apoptotic pathways such as autophagy or necrosis that are independent of p53 or BCL= -2 (Kim et al., 2009; Kurata et al., 1999). They, however, did not exclude a contribution of pro-apoptotic BH3 proteins such as Bim, Noxa and Puma towards cell death in CLN2 disease. In a *Tpp1*$^{-/-}$ mouse model, Kim and collaborators (2009) did not observe a deviation in disease course when modifying activation of caspase-3, p53 or Bcl-2 expression.

In another study, terminal dUDP nick end-labeling TUNEL-immunoreactive nuclei of neurons (Hachiya et al., 2006), and coexistence of the latter with cytoplasmic deposition of 4-hydroxy-2-nonenal-modified protein were seen in CLN2 brain. 4-HNE is a major product of membrane lipid oxidation and is a toxic product in oxidative damage. There was no altered expression of cell death-related proteins including BCL-2 family proteins. 4-HNE expression can induce apoptosis with activation of caspases leading to DNA fragmentation (Liu et al., 2000) and has been demonstrated to increase in the brain in adult-onset neurodegenerative disorders such as Alzheimer and Parkinson disease (Jenner, 2003; Zarkovic, 2003).

CLN3 disease or JNCL

Findings linking apoptosis with juvenile neuronal ceroid lipofuscinosis or JNCL were advanced by discovery of the responsible gene, *CLN3*, in 1995. The CLN3 protein (CLN3p) is a 438-amino acid protein operative in a novel anti-apoptotic pathway. The human disease is associated with deletions/

mutations of the *CLN3* gene, which result in loss of function (Puranam et al., 1999). Defects in the *CLN3* gene coexist with massive neuronal loss and couple this disease most intimately with apoptosis (Goebel and Wisniewski, 2004; Mitchison et al., 2004; Puranam et al., 1997; Puranam et al., 1999).

Apoptotic cells identified by TUNEL staining have been reported in the brain of juvenile or JNCL patients and an ovine model for CLN6 disease (Lane et al., 1996). In this study, apoptosis was established as the mechanism of neuronal and photoreceptor cell death in humans and animal models, and was confirmed by flow cytometry, electron microscopy, and DNA laddering. This study is in contrast with two other studies where apoptotic cells were not observed in a mouse model of JNCL (Cooper et al., 2006; Pontikis et al., 2004). Yet, it was shown that caspase activation occurs in CLN3-deficient. Also, JNCl or CLN3-deficient cells have an increased rate of apoptotic cell death. This begins with caspase-8 activation, strongly supporting the direct link between CLN3p and apoptotic pathways (Persaud-Sawin and Boustany, 2005).

Elevation of pro-apoptotic ceramide occurs in the brain of patients with JNCL. The fact that CLN3-overexpressing cells have low ceramide, whereas absence of CLN3 increases cellular ceramide suggest a link between CLN3p and ceramide production (Puranam et al., 1997). Molecular modeling studies establish that CLN3p harbors a structural galactosylceramide (GalCer) lipid raft binding domain. Also, it has been demonstrated that CLN3p aids in intracellular transport of GalCer from Golgi to lipid rafts in the plasma membrane (Rusyn et al., 2008).

A GalCer deficit in lipid rafts alters physicochemical properties/ structure, impairing raft function. This leads to a secondary increase in ceramide levels as the cell attempts to correct the GalCer deficit. In turn, destabilization of lipid rafts and increased cellular ceramide enhance apoptosis by caspase-8 activation (Rusyn et al., 2008). CLN3p modulates endogenous and vincrisitine stimulated levels of ceramide suggesting intact or wild-type CLN3p mediates its anti-apoptotic effect by attenuating ceramide levels (Puranam et al., 1999). Thus, CLN3p impacts apoptosis by acting upstream of ceramide (Rusyn et al., 2008). Wang et al. (2009) indicate that defects in CLN3p result in ceramide accumulation upstream of mitochondrial membrane permeabilization which induces caspase-dependent and caspase-independent cell death pathways (Wang et al., 2009). Preliminary data suggests an elevation in serum ceramide from CLN3-deficient patients in comparison to healthy age-matched controls (Boustany, pers. comm.). This may be developed as a reliable peripheral and easily measurable biochemical marker for CLN3 disease.

The molecular mechanisms by which ceramide promotes apoptosis is not completely understood, although ceramide is a modulator of protein phosphorylation (Hannun and Obeid, 1995). Proteases related to the mammalian interleukin-1-converting enzyme (ICE2 or caspase-1) are activated in a cascade constituting the effector events of the cell death process (Alnemri et al., 1996). The protein poly(ADP-ribose) polymerase or PARP (Kaufmann et al., 1993) is cleaved during this process and is the key downstream event in the execution of apoptotic cell death. In addition, ceramide activates the protease prICE (protein resembling ICE) and this effect can be blocked by BC-2 (Perry et al., 1997).

Expression of genes involved in immune and inflammatory responses is also altered in CLN3-disease or JNCL (Brooks et al., 2003; Chattopadhyay et al., 2002). A study of the *Cln3$^{-/-}$* mouse knockout model implies astrocytic and glial activation occur before neurodegeneration (Pontikis et al., 2004). In a knock-in *Cln3$^{\Delta ex7/8}$* mouse model, which demonstrates neurological deficits similar to human CLN3 disease, an increase of GFAP and F4/80 is reported in the thalamus, without the occurrence of significant regional atrophy (Pontikis et al., 2005). Preliminary data from this knock-in mouse, however, confirms an elevation of ceramide in brains of affected mice compared to controls (Boustany and Cotman, pers. comm.).

CLN5 disease

CLN5 disease, a variant form of NCL was initially described in Finland (Vesa et al., 1995). Recently, cases have been described from other parts of the world (El Haddad, 2012; Xin et al., 2010). CLN5-deficient cells derived from patients previously classified as having CLN9 disease grow rapidly (Schulz et al., 2004), have increased apoptotic rates and lowered ceramide and dihydroceramide levels. All of this partially corrects following transfection with either the intact CLN5 cDNA or CLN8, a member of the Lag-1 family of proteins with dihydroceramide synthase properties (Schulz et al., 2006; Winter and Ponting, 2002). *CLN5*-deficient, previously identified as CLN9 cells, have significantly increased growth and proliferation rates because of increased DNA synthesis (Schulz et al., 2004). Gene expression studies also support a role for dysregulation of apoptotic and cell adhesion pathways. Activity of serine palmitoyl transferase, a key enzyme in *de novo* ceramide synthesis, was increased, suggesting this could be a compensatory response to low ceramide levels. Glycosphingolipid levels are also low and these play key roles in cell adhesion and apoptosis (Schulz et al., 2004).

CLN6 disease or vLNCL/Costa-Rican

The CLN6 vLINCL is caused by molecular defects in the CLN6 gene which codes for an ER resident transmembrane protein with unknown function. Teixeira et al. (2006) indicated profiling alterations in signal transduction pathways involving apoptosis and immune and inflammatory responses. They also speculate that dysfunctional endosomal and lysosomal vesicles may act as a trigger for apoptosis and for a secondary protective inflammatory response. Additionally, brain and retina from an ovine model for CLN6 disease implicated apoptosis in brain and photoreceptor cell death. This was supported by evidence for TUNEL positive neurons and photoreceptors, flow cytometric studies, chromatin condensation by electron microscopy and DNA laddering (Lane et al., 1996). Another study claimed inability to find apoptotic cells in a CLN6-deficient mouse (Cooper et al., 2006; Pontikis et al., 2004).

It has been reported that neuroinflammation and astrocytic activation and progressive transformation of microglia to brain macrophages precede neurodegeneration (Thelen et al., 2012) with involvement of subcortical nuclei, including the basal ganglia and thalamus, particularly the lateral geniculate nucleus (Oswald et al., 2005). Activated astrocytes were also found early in disease progression in CLN6-deficient New Zealand South Hampshire sheep with localized GFAP immunoreactivity (Oswald et al., 2005).

CLN8 disease or Epilepsy with Mental Retardation (EPMR)

Late-infantile onset forms include those with mutations in the *CLN8* gene, encoding a transmembrane protein that localizes to the Endoplasmic Reticulum (ER). This protein belongs to the Lag-1 family of proteins responsible for dihydroceramide synthase activity (Winter and Ponting, 2002). Significantly, CLN8-deficient *mnd* mouse-derived fibroblasts grow fast, have accelerated rates of apoptosis and lowered levels of a number of ceramide species, including C16/C18/C24 and C24:1 species (El Haddad, 2012; Schulz et al., 2004; Schulz et al., 2006). Guarneri et al. (2004) investigated apoptotic cell death in the mutant motor neuron degeneration (*mnd*) mouse, a model for CLN8 disease (Guarneri et al., 2004). *In situ* DNA fragmentation and TUNEL staining were confirmed in the retina at a young age. This became more pronounced at an older age when severe retinal cell loss set in. Caspase-3 activation also confirmed apoptotic cell death in the mnd mouse model. These results demonstrate the involvement of apoptotic processes in *mnd* mouse retinopathy. Kuronen et al. (2012) show caspase-12-dependent pathway activation in the retina, and caspase-3 cleavage occurring presymptomatically in the cerebellum, hippocampus and retina, and symptomatically in the cerebral cortex and spinal cord of CLN8-deficient or *mnd* mice (Kuronen et al., 2012). Pronounced gliosis was also evident in CLN8 mouse models (Cooper et al., 1999). It is important to note that CLN3, CLN6 and CLN8 cross correct each other for apoptotic defects, suggesting they may act along a common apoptotic pathway (Persaud-Sawin et al., 2007).

Conclusion

This chapter has focused on apoptosis and inflammation in neurodegenerative diseases. Although instigating factors may differ, a common main theme for these diseases consists of dysregulation of apoptosis resulting in increased neuronal death. Apoptosis is supplemented by neuroinflammation which contributes to disease progression. Despite many studies, some of them with conflicting results, there is a lack of established biomarkers of apoptosis and inflammation in neurodegenerative diseases to date. General markers of apoptosis are well studied in neurodegenerative diseases but they are also implicated in many other diseases and are key features of the aging process. Studies have failed to establish a highly sensitive and specific biomarker that can be useful in early diagnosis and for tracking disease progression. Thus, researchers are exploring a combination of markers to complement tools such as brain imaging to increase the reliability of diagnosis.

In view of its direct link to pathogenesis, sphingolipid metabolism is considered to be a target of therapy and biomarker development for several neurodegenerative diseases. Accumulating evidence indicates that ceramide, in particular, is closely involved in neuronal cell death documented in AD and Batten disease. Future studies are needed to investigate the ceramide pathway contribution to other forms of neurodegeneration. Evidence for alteration in ceramide levels has been exposed in brains as well as blood of patients affected by neurodegenerative disorders. Further studies are needed to correlate peripheral and brain ceramide levels, and to identify other mechanisms by which ceramides can increase the risk of neurodegeneration and affect disease progression.

Furthermore, other enzymes and metabolites within the ceramide pathway have the potential of becoming novel and effective targets for intervening in these diseases. Blood sphingolipids/glycosphingolipids may be useful blood-based biomarkers of disease progression. Inhibitors of enzymes involved in ceramide production, such as serine palmitoyltransferase and sphingomyelinase, might offer protection against neurodegeneration and are currently being investigated as potential treatments.

References

Abdi, A., H. Sadraie, L. Dargahi, L. Khalaj and A. Ahmadiani. 2011. Apoptosis inhibition can be threatening in Abeta-induced neuroinflammation, through promoting cell proliferation. *Neurochem Res*. 36: 39–48.

Adams, J.D., Jr. 2008. Alzheimer's disease, ceramide, visfatin and NAD. *CNS Neurol Disord Drug Targets*. 7: 492–8.

Agostini, M., P. Tucci and G. Melino. 2011. Cell death pathology: perspective for human diseases. *Biochem Biophys Res Commun*. 414: 451–5.

Ahmed, T. and A.H. Gilani. 2011. A comparative study of curcuminoids to measure their effect on inflammatory and apoptotic gene expression in an Abeta plus ibotenic acid-infused rat model of Alzheimer's disease. *Brain Res*. 1400: 1–18.

Akiyama, H., S. Barger, S. Barnum, B. Bradt, J. Bauer, G.M. Cole, N.R. Cooper, P. Eikelenboom, M. Emmerling, B.L. Fiebich, C.E. Finch, S. Frautschy, W.S. Griffin, H. Hampel, M. Hull, G. Landreth, L. Lue, R. Mrak, I.R. Mackenzie, P.L. McGeer, M.K. O'Banion, J. Pachter, G. Pasinetti, C. Plata-Salaman, J. Rogers, R. Rydel, Y. Shen, W. Streit, R. Strohmeyer, I. Tooyoma, F.L. Van Muiswinkel, R. Veerhuis, D. Walker, S. Webster, B. Wegrzyniak, G. Wenk and T. Wyss-Coray. 2000. Inflammation and Alzheimer's disease. *Neurobiol Aging*. 21: 383–421.

Alnemri, E.S., D.J. Livingston, D.W. Nicholson, G. Salvesen, N.A. Thornberry, W.W. Wong and J. Yuan. 1996. Human ICE/CED-3 protease nomenclature. *Cell*. 87: 171.

Andrieu-Abadie, N., V. Gouaze, R. Salvayre and T. Levade. 2001. Ceramide in apoptosis signaling: relationship with oxidative stress. *Free Radic Biol Med*. 31: 717–28.

Anglade, P., S. Vyas, F. Javoy-Agid, M.T. Herrero, P.P. Michel, J. Marquez, A. Mouatt-Prigent, M. Ruberg, E.C. Hirsch and Y. Agid. 1997. Apoptosis and autophagy in nigral neurons of patients with Parkinson's disease. *Histol Histopathol*. 12: 25–31.

Ankarcrona, M. and B. Winblad. 2005. Biomarkers for apoptosis in Alzheimer's disease. *Int J Geriatr Psychiatry*. 20: 101–5.

Atwood, C.S., M.E. Obrenovich, T. Liu, H. Chan, G. Perry, M.A. Smith and R.N. Martins. 2003. Amyloid-beta: a chameleon walking in two worlds: a review of the trophic and toxic properties of amyloid-beta. *Brain Res Brain Res Rev*. 43: 1–16.

Avila, J. 2010. Alzheimer disease: caspases first. *Nat Rev Neurol*. 6: 587–8.

Bailey, P. 2007. Biological markers in Alzheimer's disease. *Can J Neurol Sci*. 34 Suppl. 1: S72–6.

Ballou, L.R., S.J. Laulederkind, E.F. Rosloniec and R. Raghow. 1996. Ceramide signalling and the immune response. *Biochim Biophys Acta*. 1301: 273–87.

Bandaru, V.V., J. Troncoso, D. Wheeler, O. Pletnikova, J. Wang, K. Conant and N.J. Haughey. 2009. ApoE4 disrupts sterol and sphingolipid metabolism in Alzheimer's but not normal brain. *Neurobiol Aging*. 30: 591–9.

Birbes, H., S. El Bawab, Y.A. Hannun and L.M. Obeid. 2001. Selective hydrolysis of a mitochondrial pool of sphingomyelin induces apoptosis. *FASEB J*. 15: 2669–79.

Block, M.L. and J.S. Hong. 2005. Microglia and inflammation-mediated neurodegeneration: multiple triggers with a common mechanism. *Prog Neurobiol*. 76: 77–98.

Bonelli, R.M., A. Aschoff and G. Jirikowski. 2002a. Cerebrospinal fluid tissue transglutaminase in vascular dementia. *J Neurol Sci*. 203-204: 207–9.

Bonelli, R.M., A. Aschoff, G. Niederwieser, C. Heuberger and G. Jirikowski. 2002b. Cerebrospinal fluid tissue transglutaminase as a biochemical marker for Alzheimer's disease. *Neurobiol Dis*. 11: 106–10.

Bourbon, N.A., J. Yun, D. Berkey, Y. Wang and M. Kester. 2001. Inhibitory actions of ceramide upon PKC-epsilon/ERK interactions. *Am J Physiol Cell Physiol*. 280: C1403–11.

Boustany. 1996. Batten disease or neuronal ceroid lipofuscinosis. pp. 671–900. *In*: Hugo Moser (ed.). Handbook of Clinical Neurology, Neurodystrophies and Neurolipidoses. Vol. 66. Elsevier, New York.

Brachova, L., L.F. Lue, J. Schultz, T. el Rashidy and J. Rogers. 1993. Association cortex, cerebellum, and serum concentrations of C1q and factor B in Alzheimer's disease. *Brain Res Mol Brain Res*. 18: 329–34.

Brooks, A.I., S. Chattopadhyay, H.M. Mitchison, R.L. Nussbaum and D.A. Pearce. 2003. Functional categorization of gene expression changes in the cerebellum of a Cln3-knockout mouse model for Batten disease. *Mol Genet Metab*. 78: 17–30.

Brown, W.R., D.M. Moody, C.R. Thore and V.R. Challa. 2000. Cerebrovascular pathology in Alzheimer's disease and leukoaraiosis. *Ann N Y Acad Sci*. 903: 39–45.

Campbell, A. 2004. Inflammation, neurodegenerative diseases, and environmental exposures. *Ann N Y Acad Sci*. 1035: 117–32.

Cartier, L., O. Hartley, M. Dubois-Dauphin and K.H. Krause. 2005. Chemokine receptors in the central nervous system: role in brain inflammation and neurodegenerative diseases. *Brain Res Brain Res Rev*. 48: 16–42.

Chalfant, C.E., K. Rathman, R.L. Pinkerman, R.E. Wood, L.M. Obeid, B. Ogretmen and Y.A. Hannun. 2002. *De novo* ceramide regulates the alternative splicing of caspase 9 and Bcl-x in A549 lung adenocarcinoma cells. Dependence on protein phosphatase-1. *J Biol Chem*. 277: 12587–95.

Chan, S.L., W.S. Griffin and M.P. Mattson. 1999. Evidence for caspase-mediated cleavage of AMPA receptor subunits in neuronal apoptosis and Alzheimer's disease. *J Neurosci Res*. 57: 315–23.

Chattopadhyay, S., M. Ito, J.D. Cooper, A.I. Brooks, T.M. Curran, J.M. Powers and D.A. Pearce. 2002. An autoantibody inhibitory to glutamic acid decarboxylase in the neurodegenerative disorder Batten disease. *Hum Mol Genet*. 11: 1421–31.

Chi, E.Y., C. Ege, A. Winans, J. Majewski, G. Wu, K. Kjaer and K.Y. Lee. 2008. Lipid membrane templates the ordering and induces the fibrillogenesis of Alzheimer's disease amyloid-beta peptide. *Proteins*. 72: 1–24.

Chiu, N.C., W.H. Qian, A.L. Shanske, S.S. Brooks and R.M. Boustany. 1996. A common mutation site in the beta-galactosidase gene originates in Puerto Rico. *Pediatr Neurol*. 14: 53–6.

Cho, S. and G. Dawson. 2000. Palmitoyl protein thioesterase 1 protects against apoptosis mediated by Ras-Akt-caspase pathway in neuroblastoma cells. *J Neurochem*. 74: 1478–88.

Cho, S., P.E. Dawson and G. Dawson. 2000. Antisense palmitoyl protein thioesterase 1 (PPT1) treatment inhibits PPT1 activity and increases cell death in LA-N-5 neuroblastoma cells. *J Neurosci Res*. 62: 234–40.

Choi, S.H., S. Aid and F. Bosetti. 2009. The distinct roles of cyclooxygenase-1 and -2 in neuroinflammation: implications for translational research. *Trends Pharmacol Sci*. 30: 174–81.

Combrinck, M., J. Williams, M.A. De Berardinis, D. Warden, M. Puopolo, A.D. Smith and L. Minghetti. 2006. Levels of CSF prostaglandin E2, cognitive decline, and survival in Alzheimer's disease. *J Neurol Neurosurg Psychiatry*. 77: 85–8.

Cooper, J.D., A. Messer, A.K. Feng, J. Chua-Couzens and W.C. Mobley. 1999. Apparent loss and hypertrophy of interneurons in a mouse model of neuronal ceroid lipofuscinosis: evidence for partial response to insulin-like growth factor-1 treatment. *J Neurosci*. 19: 2556–67.

Cooper, J.D., C. Russell and H.M. Mitchison. 2006. Progress towards understanding disease mechanisms in small vertebrate models of neuronal ceroid lipofuscinosis. *Biochim Biophys Acta*. 1762: 873–89.

Cordy, J.M., I. Hussain, C. Dingwall, N.M. Hooper and A.J. Turner. 2003. Exclusively targeting beta-secretase to lipid rafts by GPI-anchor addition up-regulates beta-site processing of the amyloid precursor protein. *Proc Natl Acad Sci USA*. 100: 11735–40.

Cutler, R.G., J. Kelly, K. Storie, W.A. Pedersen, A. Tammara, K. Hatanpaa, J.C. Troncoso and M.P. Mattson. 2004. Involvement of oxidative stress-induced abnormalities in ceramide and cholesterol metabolism in brain aging and Alzheimer's disease. *Proc Natl Acad Sci USA*. 101: 2070–5.

D'Amelio, M., V. Cavallucci, S. Middei, C. Marchetti, S. Pacioni, A. Ferri, A. Diamantini, D. De Zio, P. Carrara, L. Battistini, S. Moreno, A. Bacci, M. Ammassari-Teule, H. Marie and

F. Cecconi. 2011. Caspase-3 triggers early synaptic dysfunction in a mouse model of Alzheimer's disease. *Nat Neurosci.* 14: 69–76.

de la Monte, S.M., Y.K. Sohn and J.R. Wands. 1997. Correlates of p53- and Fas (CD95)-mediated apoptosis in Alzheimer's disease. *J Neurol Sci.* 152: 73–83.

de la Monte, S.M. and J.R. Wands. 2002. The AD7c-ntp neuronal thread protein biomarker for detecting Alzheimer's disease. *Front Biosci.* 7: d989–96.

Debatin, K.M. and P.H. Krammer. 2004. Death receptors in chemotherapy and cancer. *Oncogene.* 23: 2950–66.

Detre, C., E. Kiss, Z. Varga, K. Ludanyi, K. Paszty, A. Enyedi, D. Kovesdi, G. Panyi, E. Rajnavolgyi and J. Matko. 2006. Death or survival: membrane ceramide controls the fate and activation of antigen-specific T-cells depending on signal strength and duration. *Cell Signal.* 18: 294–306.

Dhar, S., R.L. Bitting, S.N. Rylova, P.J. Jansen, E. Lockhart, D.D. Koeberl, A. Amalfitano and R.M. Boustany. 2002. Flupirtine blocks apoptosis in batten patient lymphoblasts and in human postmitotic CLN3- and CLN2-deficient neurons. *Ann Neurol.* 51: 448–66.

Drache, B., G.E. Diehl, K. Beyreuther, L.S. Perlmutter and G. Konig. 1997. Bcl-xl-specific antibody labels activated microglia associated with Alzheimer's disease and other pathological states. *J Neurosci Res.* 47: 98–108.

Dragunow, M., R.L. Faull, P. Lawlor, E.J. Beilharz, K. Singleton, E.B. Walker and E. Mee. 1995. *In situ* evidence for DNA fragmentation in Huntington's disease striatum and Alzheimer's disease temporal lobes. *Neuroreport.* 6: 1053–7.

El Haddad, S., K.M., M. Daoud, R. Kantar, H. Harati, T. Mousallem, O. Alzate, B. Meyer and R.M. Boustany. 2012. CLN5 and CLN8 protein association with Ceramide Synthase: Biochemical and Proteomic Approaches. *Electrophoresis.*

Ellis, R.E., J.Y. Yuan and H.R. Horvitz. 1991. Mechanisms and functions of cell death. *Annu Rev Cell Biol.* 7: 663–98.

Filippov, V., M.A. Song, K. Zhang, H.V. Vinters, S. Tung, W.M. Kirsch, J. Yang and P.J. Duerksen-Hughes. 2012. Increased ceramide in brains with Alzheimer's and other neurodegenerative diseases. *J Alzheimers Dis.* 29: 537–47.

Findeis, M.A. 2007. The role of amyloid beta peptide 42 in Alzheimer's disease. *Pharmacol Ther.* 116: 266–86.

Forloni, G., F. Mangiarotti, N. Angeretti, E. Lucca and M.G. De Simoni. 1997. Beta-amyloid fragment potentiates IL-6 and TNF-alpha secretion by LPS in astrocytes but not in microglia. *Cytokine.* 9: 759–62.

Friedlander, R.M. 2003. Apoptosis and caspases in neurodegenerative diseases. *N Engl J Med.* 348: 1365–75.

Galasko, D. and T.J. Montine. 2010. Biomarkers of oxidative damage and inflammation in Alzheimer's disease. *Biomark Med.* 4: 27–36.

Giannakopoulos, P., E. Kovari, A. Savioz, F. de Bilbao, M. Dubois-Dauphin, P.R. Hof and C. Bouras. 1999. Differential distribution of presenilin-1, Bax, and Bcl-X(L) in Alzheimer's disease and frontotemporal dementia. *Acta Neuropathol.* 98: 141–9.

Goebel, H.H., H. Klein, P. Santavuori and K. Sainio. 1988. Ultrastructural studies of the retina in infantile neuronal ceroid-lipofuscinosis. *Retina.* 8: 59–66.

Goebel, H.H. and K.E. Wisniewski. 2004. Current state of clinical and morphological features in human NCL. *Brain Pathol.* 14: 61–9.

Grimm, M.O., H.S. Grimm, A.J. Patzold, E.G. Zinser, R. Halonen, M. Duering, J.A. Tschape, B. De Strooper, U. Muller, J. Shen and T. Hartmann. 2005. Regulation of cholesterol and sphingomyelin metabolism by amyloid-beta and presenilin. *Nat Cell Biol.* 7: 1118–23.

Guarneri, R., D. Russo, C. Cascio, S. D'Agostino, G. Galizzi, P. Bigini, T. Mennini and P. Guarneri. 2004. Retinal oxidation, apoptosis and age- and sex-differences in the mnd mutant mouse, a model of neuronal ceroid lipofuscinosis. *Brain Res.* 1014: 209–20.

Gupta, P., A.A. Soyombo, A. Atashband, K.E. Wisniewski, J.M. Shelton, J.A. Richardson, R.E. Hammer and S.L. Hofmann. 2001. Disruption of PPT1 or PPT2 causes neuronal ceroid lipofuscinosis in knockout mice. *Proc Natl Acad Sci USA.* 98: 13566–71.

Hachiya, Y., M. Hayashi, S. Kumada, A. Uchiyama, K. Tsuchiya and K. Kurata. 2006. Mechanisms of neurodegeneration in neuronal ceroid-lipofuscinoses. *Acta Neuropathol.* 111: 168–77.

Hall, A.M. and E.D. Roberson. 2012. Mouse models of Alzheimer's disease. *Brain Res Bull.* 88: 3–12.

Halliday, G., S.R. Robinson, C. Shepherd and J. Kril. 2000. Alzheimer's disease and inflammation: a review of cellular and therapeutic mechanisms. *Clin Exp Pharmacol Physiol.* 27: 1–8.

Haltia, M. 2006. The neuronal ceroid-lipofuscinoses: from past to present. *Biochim Biophys Acta.* 1762: 850–6.

Haltia, M., and H.H. Goebel. 2012. The neuronal ceroid-lipofuscinoses: A historical introduction. *Biochim Biophys Acta.*

Han, B.H., M.L. Zhou, A.K. Vellimana, E. Milner, D.H. Kim, J.K. Greenberg, W. Chu, R.H. Mach and G.J. Zipfel. 2011. Resorufin analogs preferentially bind cerebrovascular amyloid: potential use as imaging ligands for cerebral amyloid angiopathy. *Mol Neurodegener.* 6: 86.

Han, X. 2005. Lipid alterations in the earliest clinically recognizable stage of Alzheimer's disease: implication of the role of lipids in the pathogenesis of Alzheimer's disease. *Curr Alzheimer Res.* 2: 65–77.

Han, X., M.H.D., D.W. McKeel, Jr., J. Kelley and J.C. Morris. 2002. Substantial sulfatide deficiency and ceramide elevation in very early Alzheimer's disease: potential role in disease pathogenesis. *J Neurochem.* 82: 809–18.

Han, X., S. Rozen, S.H. Boyle, C. Hellegers, H. Cheng, J.R. Burke, K.A. Welsh-Bohmer, P.M. Doraiswamy and R. Kaddurah-Daouk. 2010. Metabolomics in early Alzheimer's disease: identification of altered plasma sphingolipidome using shotgun lipidomics. *PLoS One.* 6: e21643.

Hannun, Y.A. and L.M. Obeid. 1995. Ceramide: an intracellular signal for apoptosis. *Trends Biochem Sci.* 20: 73–7.

He, X., Y. Huang, B. Li, C.X. Gong and E.H. Schuchman. 2010. Deregulation of sphingolipid metabolism in Alzheimer's disease. *Neurobiol Aging.* 31: 398–408.

Hobert, J.A. and G. Dawson. 2006. Neuronal ceroid lipofuscinoses therapeutic strategies: past, present and future. *Biochim Biophys Acta.* 1762: 945–53.

Huwiler, A., C. Xin, A.K. Brust, V.A. Briner and J. Pfeilschifter. 2004. Differential binding of ceramide to MEKK1 in glomerular endothelial and mesangial cells. *Biochim Biophys Acta.* 1636: 159–68.

Ikeda, K., T. Yamaguchi, S. Fukunaga, M. Hoshino and K. Matsuzaki. 2011. Mechanism of amyloid beta-protein aggregation mediated by GM1 ganglioside clusters. *Biochemistry.* 50: 6433–40.

Jalanko, A., J. Tyynela and L. Peltonen. 2006. From genes to systems: new global strategies for the characterization of NCL biology. *Biochim Biophys Acta.* 1762: 934–44.

Jana, A. and K. Pahan. 2004. Fibrillar amyloid-beta peptides kill human primary neurons via NADPH oxidase-mediated activation of neutral sphingomyelinase. Implications for Alzheimer's disease. *J Biol Chem.* 279: 51451–9.

Jenner, P. 2003. Oxidative stress in Parkinson's disease. *Ann Neurol.* 53 Suppl. 3: S26–36; discussion S36–8.

Jesse, S., P. Steinacker, L. Cepek, C.A. von Arnim, H. Tumani, S. Lehnert, H.A. Kretzschmar, M. Baier and M. Otto. 2009. Glial fibrillary acidic protein and protein S-100B: different concentration pattern of glial proteins in cerebrospinal fluid of patients with Alzheimer's disease and Creutzfeldt-Jakob disease. *J Alzheimers Dis.* 17: 541–51.

John, G.R., S.C. Lee and C.F. Brosnan. 2003. Cytokines: powerful regulators of glial cell activation. *Neuroscientist.* 9: 10–22.

Katsel, P., C. Li and V. Haroutunian. 2007. Gene expression alterations in the sphingolipid metabolism pathways during progression of dementia and Alzheimer's disease: a shift toward ceramide accumulation at the earliest recognizable stages of Alzheimer's disease? *Neurochem Res.* 32: 845–56.

Kaufmann, S.H., S. Desnoyers, Y. Ottaviano, N.E. Davidson and G.G. Poirier. 1993. Specific proteolytic cleavage of poly(ADP-ribose) polymerase: an early marker of chemotherapy-induced apoptosis. *Cancer Res.* 53: 3976–85.

Kennard, M.L., H. Feldman, T. Yamada and W.A. Jefferies. 1996. Serum levels of the iron binding protein p97 are elevated in Alzheimer's disease. *Nat Med.* 2: 1230–5.

Khandelwal, P.J., A.M. Herman and C.E. Moussa. 2011. Inflammation in the early stages of neurodegenerative pathology. *J Neuroimmunol.* 238: 1–11.

Kielar, C., L. Maddox, E. Bible, C.C. Pontikis, S.L. Macauley, M.A. Griffey, M. Wong, M.S. Sands and J.D. Cooper. 2007. Successive neuron loss in the thalamus and cortex in a mouse model of infantile neuronal ceroid lipofuscinosis. *Neurobiol Dis.* 25: 150–62.

Kim, J. and D.J. Klionsky. 2000. Autophagy, cytoplasm-to-vacuole targeting pathway, and pexophagy in yeast and mammalian cells. *Annu Rev Biochem.* 69: 303–42.

Kim, K.H., D.E. Sleat, O. Bernard and P. Lobel. 2009. Genetic modulation of apoptotic pathways fails to alter disease course in tripeptidyl-peptidase 1 deficient mice. *Neurosci Lett.* 453: 27–30.

Kim, S.J., Z. Zhang, E. Hitomi, Y.C. Lee and A.B. Mukherjee. 2006a. Endoplasmic reticulum stress-induced caspase-4 activation mediates apoptosis and neurodegeneration in INCL. *Hum Mol Genet.* 15: 1826–34.

Kim, S.J., Z. Zhang, Y.C. Lee and A.B. Mukherjee. 2006b. Palmitoyl-protein thioesterase-1 deficiency leads to the activation of caspase-9 and contributes to rapid neurodegeneration in INCL. *Hum Mol Genet.* 15: 1580–6.

Kitamura, Y., S. Shimohama, W. Kamoshima, T. Ota, Y. Matsuoka, Y. Nomura, M.A. Smith, G. Perry, P.J. Whitehouse and T. Taniguchi. 1998. Alteration of proteins regulating apoptosis, Bcl-2, Bcl-x, Bax, Bak, Bad, ICH-1 and CPP32, in Alzheimer's disease. *Brain Res.* 780: 260–9.

Klionsky, D.J. and S.D. Emr. 2000. Autophagy as a regulated pathway of cellular degradation. *Science.* 290: 1717–21.

Kobayashi, K., M. Hayashi, H. Nakano, Y. Fukutani, K. Sasaki, M. Shimazaki and Y. Koshino. 2002. Apoptosis of astrocytes with enhanced lysosomal activity and oligodendrocytes in white matter lesions in Alzheimer's disease. *Neuropathol Appl Neurobiol.* 28: 238–51.

Komatsu, M., S. Waguri, T. Chiba, S. Murata, J. Iwata, I. Tanida, T. Ueno, M. Koike, Y. Uchiyama, E. Kominami and K. Tanaka. 2006. Loss of autophagy in the central nervous system causes neurodegeneration in mice. *Nature.* 441: 880–4.

Komatsu, M., Q.J. Wang, G.R. Holstein, V.L. Friedrich, Jr., J. Iwata, E. Kominami, B.T. Chait, K. Tanaka and Z. Yue. 2007. Essential role for autophagy protein Atg7 in the maintenance of axonal homeostasis and the prevention of axonal degeneration. *Proc Natl Acad Sci USA.* 104: 14489–94.

Kroemer, G., L. Galluzzi and C. Brenner. 2007. Mitochondrial membrane permeabilization in cell death. *Physiol Rev.* 87: 99–163.

Kurata, K., M. Hayashi, J. Satoh, H. Kojima, J. Nagata, K. Tamagawa, T. Shinohara, Y. Morimatsu and E. Kominami. 1999. Pathological study on sibling autopsy cases of the late infantile form of neuronal ceroid lipofuscinosis. *Brain Dev.* 21: 63–7.

Kuronen, M., A.E. Lehesjoki, A. Jalanko, J.D. Cooper and O. Kopra. 2012. Selective spatiotemporal patterns of glial activation and neuron loss in the sensory thalamocortical pathways of neuronal ceroid lipofuscinosis 8 mice. *Neurobiol Dis.* 47: 444–57.

Lane, S.C., R.D. Jolly, D.E. Schmechel, J. Alroy and R.M. Boustany. 1996. Apoptosis as the mechanism of neurodegeneration in Batten's disease. *J Neurochem.* 67: 677–83.

Lassmann, H., C. Bancher, H. Breitschopf, J. Wegiel, M. Bobinski, K. Jellinger and H.M. Wisniewski. 1995. Cell death in Alzheimer's disease evaluated by DNA fragmentation *in situ.* *Acta Neuropathol.* 89: 35–41.

LeBlanc, A., H. Liu, C. Goodyer, C. Bergeron and J. Hammond. 1999. Caspase-6 role in apoptosis of human neurons, amyloidogenesis, and Alzheimer's disease. *J Biol Chem.* 274: 23426–36.

Li, J. and J. Yuan. 2008. Caspases in apoptosis and beyond. *Oncogene.* 27: 6194–206.

Li, W.P., W.Y. Chan, H.W. Lai and D.T. Yew. 1997. Terminal dUTP nick end labeling (TUNEL) positive cells in the different regions of the brain in normal aging and Alzheimer patients. *J Mol Neurosci.* 8: 75–82.

Licastro, F., M.C. Morini, E. Polazzi and L.J. Davis. 1995. Increased serum alpha 1-antichymotrypsin in patients with probable Alzheimer's disease: an acute phase reactant without the peripheral acute phase response. *J Neuroimmunol.* 57: 71–5.

Liu, W., M. Kato, A.A. Akhand, A. Hayakawa, H. Suzuki, T. Miyata, K. Kurokawa, Y. Hotta, N. Ishikawa and I. Nakashima. 2000. 4-hydroxynonenal induces a cellular redox status-related activation of the caspase cascade for apoptotic cell death. *J Cell Sci.* 113(Pt 4): 635–41.

Macauley, S.L., M. Pekny and M.S. Sands. 2011. The role of attenuated astrocyte activation in infantile neuronal ceroid lipofuscinosis. *J Neurosci.* 31: 15575–85.

Macauley, S.L., D.F. Wozniak, C. Kielar, Y. Tan, J.D. Cooper and M.S. Sands. 2009. Cerebellar pathology and motor deficits in the palmitoyl protein thioesterase 1-deficient mouse. *Exp Neurol.* 217: 124–35.

MacGibbon, G.A., P.A. Lawlor, E.S. Sirimanne, M.R. Walton, B. Connor, D. Young, C. Williams, P. Gluckman, R.L. Faull, P. Hughes and M. Dragunow. 1997. Bax expression in mammalian neurons undergoing apoptosis, and in Alzheimer's disease hippocampus. *Brain Res.* 750: 223–34.

Manelli, A.M., W.B. Stine, L.J. Van Eldik and M.J. LaDu. 2004. ApoE and Abeta1-42 interactions: effects of isoform and conformation on structure and function. *J Mol Neurosci.* 23: 235–46.

Masliah, E., M. Mallory, M. Alford, R. DeTeresa, L.A. Hansen, D.W. McKeel, Jr. and J.C. Morris. 2001. Altered expression of synaptic proteins occurs early during progression of Alzheimer's disease. *Neurology.* 56: 127–9.

Mattson, M.P. 2004. Pathways towards and away from Alzheimer's disease. *Nature.* 430: 631–9.

Mattson, M.P., R.G. Cutler and D.G. Jo. 2005. Alzheimer peptides perturb lipid-regulating enzymes. *Nat Cell Biol.* 7: 1045–7.

McGeer, E.G. and P.L. McGeer. 1997. The role of the immune system in neurodegenerative disorders. *Mov Disord.* 12: 855–8.

McRae, A., R.N. Martins, J. Fonte, R. Kraftsik, L. Hirt and J. Miklossy. 2007. Cerebrospinal fluid antimicroglial antibodies in Alzheimer disease: a putative marker of an ongoing inflammatory process. *Exp Gerontol.* 42: 355–63.

Mielke, M.M., V.V. Bandaru, N.J. Haughey, J. Xia, L.P. Fried, S. Yasar, M. Albert, V. Varma, G. Harris, E.B. Schneider, P.V. Rabins, K. Bandeen-Roche, C.G. Lyketsos and M.C. Carlson. 2012. Serum ceramides increase the risk of Alzheimer disease: the Women's Health and Aging Study II. *Neurology.* 79: 633-41.

Mielke, M.M., N.J. Haughey, V.V. Bandaru, D.D. Weinberg, E. Darby, N. Zaidi, V. Pavlik, R.S. Doody and C.G. Lyketsos. 2011. Plasma sphingomyelins are associated with cognitive progression in Alzheimer's disease. *J Alzheimers Dis.* 27: 259–69.

Mielke, M.M., N.J. Haughey, V.V. Ratnam Bandaru, S. Schech, R. Carrick, M.C. Carlson, S. Mori, M.I. Miller, C. Ceritoglu, T. Brown, M. Albert and C.G. Lyketsos. 2010. Plasma ceramides are altered in mild cognitive impairment and predict cognitive decline and hippocampal volume loss. *Alzheimers Dement.* 6: 378–85.

Mielke, M.M. and C.G. Lyketsos. 2010. Alterations of the sphingolipid pathway in Alzheimer's disease: new biomarkers and treatment targets? *Neuromolecular Med.* 12: 331–40.

Mitchison, H.M., M.J. Lim and J.D. Cooper. 2004. Selectivity and types of cell death in the neuronal ceroid lipofuscinoses. *Brain Pathol.* 14: 86–96.

Nagata, S., and P. Golstein. 1995. The Fas death factor. *Science.* 267: 1449–56.

Nagy, Z.S. and M.M. Esiri. 1997. Apoptosis-related protein expression in the hippocampus in Alzheimer's disease. *Neurobiol Aging.* 18: 565–71.

Naoi, M. and W. Maruyama. 2010. Monoamine oxidase inhibitors as neuroprotective agents in age-dependent neurodegenerative disorders. *Curr Pharm Des.* 16: 2799–817.

Nishimoto, I., T. Okamoto, U. Giambarella and T. Iwatsubo. 1997. Apoptosis in neurodegenerative diseases. *Adv Pharmacol.* 41: 337–68.

Obeid, L.M., C.M. Linardic, L.A. Karolak and Y.A. Hannun. 1993. Programmed cell death induced by ceramide. *Science.* 259: 1769–71.

Oddo, S., A. Caccamo, J.D. Shepherd, M.P. Murphy, T.E. Golde, R. Kayed, R. Metherate, M.P. Mattson, Y. Akbari and F.M. LaFerla. 2003. Triple-transgenic model of Alzheimer's disease with plaques and tangles: intracellular Abeta and synaptic dysfunction. *Neuron.* 39: 409–21.

Oh, H.L., J.Y. Seok, C.H. Kwon, S.K. Kang and Y.K. Kim. 2006. Role of MAPK in ceramide-induced cell death in primary cultured astrocytes from mouse embryonic brain. *Neurotoxicology.* 27: 31–8.

Oswald, M.J., D.N. Palmer, G.W. Kay, S.J. Shemilt, P. Rezaie and J.D. Cooper. 2005. Glial activation spreads from specific cerebral foci and precedes neurodegeneration in presymptomatic ovine neuronal ceroid lipofuscinosis (CLN6). *Neurobiol Dis.* 20: 49–63.

Pan, R., S. Qiu, D.X. Lu and J. Dong. 2008. Curcumin improves learning and memory ability and its neuroprotective mechanism in mice. *Chin Med J (Engl).* 121: 832–9.

Pereira, C., P. Agostinho, P.I. Moreira, S.M. Cardoso and C.R. Oliveira. 2005. Alzheimer's disease-associated neurotoxic mechanisms and neuroprotective strategies. *Curr Drug Targets CNS Neurol Disord.* 4: 383–403.

Perry, D.K., M.J. Smyth, H.G. Wang, J.C. Reed, P. Duriez, G.G. Poirier, L.M. Obeid and Y.A. Hannun. 1997. Bcl-2 acts upstream of the PARP protease and prevents its activation. *Cell Death Differ.* 4: 29–33.

Persaud-Sawin, D.A. and R.M. Boustany. 2005. Cell death pathways in juvenile Batten disease. *Apoptosis.* 10: 973–85.

Peskind, E.R., W.S. Griffin, K.T. Akama, M.A. Raskind and L.J. Van Eldik. 2001. Cerebrospinal fluid S100B is elevated in the earlier stages of Alzheimer's disease. *Neurochem Int.* 39: 409–13.

Pettegrew, J.W., K. Panchalingam, R.L. Hamilton and R.J. McClure. 2001. Brain membrane phospholipid alterations in Alzheimer's disease. *Neurochem Res.* 26: 771–82.

Pompl, P.N., S. Yemul, Z. Xiang, L. Ho, V. Haroutunian, D. Purohit, R. Mohs and G.M. Pasinetti. 2003. Caspase gene expression in the brain as a function of the clinical progression of Alzheimer disease. *Arch Neurol.* 60: 369–76.

Pontikis, C.C., C.V. Cella, N. Parihar, M.J. Lim, S. Chakrabarti, H.M. Mitchison, W.C. Mobley, P. Rezaie, D.A. Pearce and J.D. Cooper. 2004. Late onset neurodegeneration in the Cln3-/- mouse model of juvenile neuronal ceroid lipofuscinosis is preceded by low level glial activation. *Brain Res.* 1023: 231–42.

Pontikis, C.C., S.L. Cotman, M.E. MacDonald and J.D. Cooper. 2005. Thalamocortical neuron loss and localized astrocytosis in the Cln3Deltaex7/8 knock-in mouse model of Batten disease. *Neurobiol Dis.* 20: 823–36.

Potter, H., I.M. Wefes and L.N. Nilsson. 2001. The inflammation-induced pathological chaperones ACT and apo-E are necessary catalysts of Alzheimer amyloid formation. *Neurobiol Aging.* 22: 923–30.

Puranam, K., W.H. Qian, K. Nikbakht, M. Venable, L. Obeid, Y. Hannun and R.M. Boustany. 1997. Upregulation of Bcl-2 and elevation of ceramide in Batten disease. *Neuropediatrics.* 28: 37–41.

Puranam, K.L., W.X. Guo, W.H. Qian, K. Nikbakht and R.M. Boustany. 1999. CLN3 defines a novel antiapoptotic pathway operative in neurodegeneration and mediated by ceramide. *Mol Genet Metab.* 66: 294–308.

Purves D. and L.J.W. 1985. Principles of Neural Development. S. Associates, editor, Sunderland.

Qiao, X., J.Y. Lu and S.L. Hofmann. 2007. Gene expression profiling in a mouse model of infantile neuronal ceroid lipofuscinosis reveals upregulation of immediate early genes and mediators of the inflammatory response. *BMC Neurosci.* 8: 95.

Ravikumar, B., A. Acevedo-Arozena, S. Imarisio, Z. Berger, C. Vacher, C.J. O'Kane, S.D. Brown and D.C. Rubinsztein. 2005. Dynein mutations impair autophagic clearance of aggregate-prone proteins. *Nat Genet.* 37: 771–6.

Richartz, E., S. Noda, K. Schott, A. Gunthner, P. Lewczuk and M. Bartels. 2002. Increased serum levels of CD95 in Alzheimer's disease. *Dement Geriatr Cogn Disord.* 13: 178–82.

Riikonen, R., S.L. Vanhanen, J. Tyynela, P. Santavuori and U. Turpeinen. 2000. CSF insulin-like growth factor-1 in infantile neuronal ceroid lipofuscinosis. *Neurology.* 54: 1828–32.

Rohn, T.T., E. Head, W.H. Nesse, C.W. Cotman and D.H. Cribbs. 2001. Activation of caspase-8 in the Alzheimer's disease brain. *Neurobiol Dis.* 8: 1006–16.

Rohn, T.T., V. Vyas, T. Hernandez-Estrada, K.E. Nichol, L.A. Christie and E. Head. 2008. Lack of pathology in a triple transgenic mouse model of Alzheimer's disease after overexpression of the anti-apoptotic protein Bcl-2. *J Neurosci.* 28: 3051–9.

Rossi, F. and E. Cattaneo. 2002. Opinion: neural stem cell therapy for neurological diseases: dreams and reality. *Nat Rev Neurosci.* 3: 401–9.

Rubio-Perez, J.M. and J.M. Morillas-Ruiz. 2012. A review: inflammatory process in Alzheimer's disease, role of cytokines. *Scientific World Journal.* 756357.

Rusyn, E., T. Mousallem, D.A. Persaud-Sawin, S. Miller and R.M. Boustany. 2008. CLN3p impacts galactosylceramide transport, raft morphology, and lipid content. *Pediatr Res.* 63: 625–31.

Salonen, T., I. Jarvela, L. Peltonen and A. Jalanko. 2000. Detection of eight novel palmitoyl protein thioesterase (PPT) mutations underlying infantile neuronal ceroid lipofuscinosis (INCL; CLN1). *Hum Mutat.* 15: 273–9.

Santavuori, P., R.J., M. Haltia, J. Tyynela, L. Peltonen and S.E. Mole. 1999. CLN5 Finnish variant late infantile NCL. In The Neuronal Ceroid Lipofuscinoses (Batten Disease). M.S. Goebel HH, Lake BD, Eds, editor. Oxford: IOS Press. 91–101.

Sapp, E., C. Schwarz, K. Chase, P.G. Bhide, A.B. Young, J. Penney, J.P. Vonsattel, N. Aronin and M. DiFiglia. 1997. Huntingtin localization in brains of normal and Huntington's disease patients. *Ann Neurol.* 42: 604–12.

Sasaki, S. 2011. Autophagy in spinal cord motor neurons in sporadic amyotrophic lateral sclerosis. *J Neuropathol Exp Neurol.* 70: 349–59.

Satoi, H., H. Tomimoto, R. Ohtani, T. Kitano, T. Kondo, M. Watanabe, N. Oka, I. Akiguchi, S. Furuya, Y. Hirabayashi and T. Okazaki. 2005. Astroglial expression of ceramide in Alzheimer's disease brains: a role during neuronal apoptosis. *Neuroscience.* 130: 657–66.

Scheel-Toellner, D., K. Wang, R. Singh, S. Majeed, K. Raza, S.J. Curnow, M. Salmon and J.M. Lord. 2002. The death-inducing signalling complex is recruited to lipid rafts in Fas-induced apoptosis. *Biochem Biophys Res Commun.* 297: 876–9.

Schulz, A., S. Dhar, S. Rylova, G. Dbaibo, J. Alroy, C. Hagel, I. Artacho, A. Kohlschutter, S. Lin and R.M. Boustany. 2004. Impaired cell adhesion and apoptosis in a novel CLN9 Batten disease variant. *Ann Neurol.* 56: 342–50.

Schulz, A., T. Mousallem, M. Venkataramani, D.A. Persaud-Sawin, A. Zucker, C. Luberto, A. Bielawska, J. Bielawski, J.C. Holthuis, S.M. Jazwinski, L. Kozhaya, G.S. Dbaibo and R.M. Boustany. 2006. The CLN9 protein, a regulator of dihydroceramide synthase. *J Biol Chem.* 281: 2784–94.

Shen, Y.H., J. Godlewski, J. Zhu, P. Sathyanarayana, V. Leaner, M.J. Birrer, A. Rana and G. Tzivion. 2003. Cross-talk between JNK/SAPK and ERK/MAPK pathways: sustained activation of JNK blocks ERK activation by mitogenic factors. *J Biol Chem.* 278: 26715–21.

Sheng, J.G., X.Q. Zhou, R.E. Mrak and W.S. Griffin. 1998. Progressive neuronal injury associated with amyloid plaque formation in Alzheimer disease. *J Neuropathol Exp Neurol.* 57: 714–7.

Simons, K. and E. Ikonen. 1997. Functional rafts in cell membranes. *Nature.* 387: 569–72.

Sivananthan, S.N., A.W. Lee, C.G. Goodyer and A.C. LeBlanc. 2010. Familial amyloid precursor protein mutants cause caspase-6-dependent but amyloid beta-peptide-independent neuronal degeneration in primary human neuron cultures. *Cell Death Dis.* 1: e100.

Sleat, D.E., I. Sohar, P.S. Pullarkat, P. Lobel and R.K. Pullarkat. 1998. Specific alterations in levels of mannose 6-phosphorylated glycoproteins in different neuronal ceroid lipofuscinoses. *Biochem J.* 334(Pt 3): 547–51.

Smale, G., N.R. Nichols, D.R. Brady, C.E. Finch and W.E. Horton, Jr. 1995. Evidence for apoptotic cell death in Alzheimer's disease. *Exp Neurol.* 133: 225–30.

Stadelmann, C., T.L. Deckwerth, A. Srinivasan, C. Bancher, W. Bruck, K. Jellinger and H. Lassmann. 1999. Activation of caspase-3 in single neurons and autophagic granules of granulovacuolar degeneration in Alzheimer's disease. Evidence for apoptotic cell death. *Am J Pathol.* 155: 1459–66.

Stoka, V., B. Turk, S.L. Schendel, T.H. Kim, T. Cirman, S.J. Snipas, L.M. Ellerby, D. Bredesen, H. Freeze, M. Abrahamson, D. Bromme, S. Krajewski, J.C. Reed, X.M. Yin, V. Turk and G.S. Salvesen. 2001. Lysosomal protease pathways to apoptosis. Cleavage of bid, not pro-caspases, is the most likely route. *J Biol Chem.* 276: 3149–57.

Su, J.H., A.J. Anderson, B.J. Cummings and C.W. Cotman. 1994. Immunohistochemical evidence for apoptosis in Alzheimer's disease. *Neuroreport.* 5: 2529–33.

Sugaya, K., M. Reeves and M. McKinney. 1997. Topographic associations between DNA fragmentation and Alzheimer's disease neuropathology in the hippocampus. *Neurochem Int.* 31: 275–81.

Sze, C.I., J.C. Troncoso, C. Kawas, P. Mouton, D.L. Price and L.J. Martin. 1997. Loss of the presynaptic vesicle protein synaptophysin in hippocampus correlates with cognitive decline in Alzheimer disease. *J Neuropathol Exp Neurol.* 56: 933–44.

Teixeira, C.A., S. Lin, M. Mangas, R. Quinta, C.J. Bessa, C. Ferreira, M.C. Sa Miranda, R.M. Boustany and M.G. Ribeiro. 2006. Gene expression profiling in vLINCL CLN6-deficient fibroblasts: Insights into pathobiology. *Biochim Biophys Acta.* 1762: 637–46.

Thelen, M., M. Damme, M. Schweizer, C. Hagel, A.M. Wong, J.D. Cooper, T. Braulke and G. Galliciotti. 2012. Disruption of the autophagy-lysosome pathway is involved in neuropathology of the nclf mouse model of neuronal ceroid lipofuscinosis. *PLoS One.* 7: e35493.

Thompson, C.B. 1995. Apoptosis in the pathogenesis and treatment of disease. *Science.* 267: 1456–62.

Thornberry, N.A. 1998. Caspases: key mediators of apoptosis. *Chem Biol.* 5: R97–103.

Town, T., V. Nikolic and J. Tan. 2005. The microglial "activation" continuum: from innate to adaptive responses. *J Neuroinflammation.* 2: 24.

Troy, C.M., S.A. Rabacchi, Z. Xu, A.C. Maroney, T.J. Connors, M.L. Shelanski and L.A. Greene. 2001. beta-Amyloid-induced neuronal apoptosis requires c-Jun N-terminal kinase activation. *J Neurochem.* 77: 157–64.

Uetsuki, T., K. Takemoto, I. Nishimura, M. Okamoto, M. Niinobe, T. Momoi, M. Miura and K. Yoshikawa. 1999. Activation of neuronal caspase-3 by intracellular accumulation of wild-type Alzheimer amyloid precursor protein. *J Neurosci.* 19: 6955–64.

van Heerde, W.L., S. Robert-Offerman, E. Dumont, L. Hofstra, P.A. Doevendans, J.F. Smits, M.J. Daemen and C.P. Reutelingsperger. 2000. Markers of apoptosis in cardiovascular tissues: focus on Annexin V. *Cardiovasc Res.* 45: 549–59.

Vaux, D.L. and S.J. Korsmeyer. 1999. Cell death in development. *Cell.* 96: 245–54.

Vesa, J., E. Hellsten, L.A. Verkruyse, L.A. Camp, J. Rapola, P. Santavuori, S.L. Hofmann and L. Peltonen. 1995. Mutations in the palmitoyl protein thioesterase gene causing infantile neuronal ceroid lipofuscinosis. *Nature.* 376: 584–7.

Wajant, H., K. Pfizenmaier and P. Scheurich. 2003. Tumor necrosis factor signaling. *Cell Death Differ.* 10: 45–65.

Wang, S.Y., W.N. Jin and D. Wu. 2009. Mechanisms of juvenile neuronal ceroid lipofuscinosis (JNCL). *Yi Chuan.* 31: 779–84.

Wang, X. 2001. The expanding role of mitochondria in apoptosis. *Genes Dev.* 15: 2922–33.

Webster, S. and J. Rogers. 1996. Relative efficacies of amyloid beta peptide (A beta) binding proteins in A beta aggregation. *J Neurosci Res.* 46: 58–66.

Winter, E. and C.P. Ponting. 2002. TRAM, LAG1 and CLN8: members of a novel family of lipid-sensing domains? *Trends Biochem Sci.* 27: 381–3.

Wyss-Coray, T. 2006. Inflammation in Alzheimer disease: driving force, bystander or beneficial response? *Nat Med.* 12: 1005–15.

Xia, M.Q. and B.T. Hyman. 1999. Chemokines/chemokine receptors in the central nervous system and Alzheimer's disease. *J Neurovirol.* 5: 32–41.

Xiao, A.Y., X.Q. Wang, A. Yang and S.P. Yu. 2002. Slight impairment of Na+,K+-ATPase synergistically aggravates ceramide- and beta-amyloid-induced apoptosis in cortical neurons. *Brain Res*. 955: 253–9.

Xin, W., T.E. Mullen, R. Kiely, J. Min, X. Feng, Y. Cao, L. O'Malley, Y. Shen, C. Chu-Shore, S.E. Mole, H.H. Goebel and K. Sims. 2010. CLN5 mutations are frequent in juvenile and late-onset non-Finnish patients with NCL. *Neurology*. 74: 565–71.

Yang, H. and C. Chen. 2008. Cyclooxygenase-2 in synaptic signaling. *Curr Pharm Des*. 14: 1443–51.

Yao, M., T.V. Nguyen and C.J. Pike. 2005. Beta-amyloid-induced neuronal apoptosis involves c-Jun N-terminal kinase-dependent downregulation of Bcl-w. *J Neurosci*. 25: 1149–58.

Yu, W.H., A.M. Cuervo, A. Kumar, C.M. Peterhoff, S.D. Schmidt, J.H. Lee, P.S. Mohan, M. Mercken, M.R. Farmery, L.O. Tjernberg, Y. Jiang, K. Duff, Y. Uchiyama, J. Naslund, P.M. Mathews, A.M. Cataldo and R.A. Nixon. 2005. Macroautophagy—a novel Beta-amyloid peptide-generating pathway activated in Alzheimer's disease. *J Cell Biol*. 171: 87–98.

Zarkovic, K. 2003. 4-hydroxynonenal and neurodegenerative diseases. *Mol Aspects Med*. 24: 293–303.

Zhang, Z., J.D. Butler, S.W. Levin, K.E. Wisniewski, S.S. Brooks and A.B. Mukherjee. 2001. Lysosomal ceroid depletion by drugs: therapeutic implications for a hereditary neurodegenerative disease of childhood. *Nat Med*. 7: 478–84.

Zhang, Z., Y.C. Lee, S.J. Kim, M.S. Choi, P.C. Tsai, Y. Xu, Y.J. Xiao, P. Zhang, A. Heffer and A.B. Mukherjee. 2006. Palmitoyl-protein thioesterase-1 deficiency mediates the activation of the unfolded protein response and neuronal apoptosis in INCL. *Hum Mol Genet*. 15: 337–46.

Zhao, S., Y.N. Yang and J.G. Song. 2004. Ceramide induces caspase-dependent and -independent apoptosis in A-431 cells. *J Cell Physiol*. 199: 47–56.

17

Neurological Context of Charcot-Marie-Tooth Disease: Implication of Molecular Mechanisms and Therapeutical Approaches

Joelle Makoukji

INTRODUCTION

Hereditary neuropathies represent a genetically heterogeneous group of diseases that affect the Peripheral Nervous System (PNS). The most common form is the Charcot-Marie-Tooth disease (CMT), also called Hereditary Motor and Sensory Neuropathy (HMSN), with a reported prevalence of one in 2,500 people worldwide. Almost 125 years have elapsed since the first contemporary description of the same familial neurological syndrome, "peroneal muscular atrophy", by two French doctors (Charcot and Marie) and an English doctor (Tooth) (Banchs et al., 2009). In general, this syndrome has an infantile or juvenile onset, with motor and sensory polyneuropathic semiology and pes cavus (Harding and Thomas, 1980).

Neurogenetics Program and Division of Pediatric Neurology, Departments of Pediatrics, Adolescent Medicine and Biochemistry, American University of Beirut.
Email: joelle.makoukji@gmail.com

Classification

The mode of transmission, the phenotype and the severity of the different types of CMT are variable. Indeed, the various forms are ranked according to the nature of the nerve defect or the mode of inheritance, as well as the genetic abnormalities involved and the protein deficit:

- **The nature of the nerve defect:**
  1. Demyelinating CMT characterized by Nerve Conduction Velocities (NCVs) less than 38 m/s and demyelinating traits on nerve biopsies.
  2. Axonal CMT showing normal or slightly reduced NCV (NCV >40 m/s) and loss of myelinated axons.
  3. Intermediate CMT in which NCV lies between 30–40 m/s and nerve pathology shows axonal and demyelinating features (Boerkoel et al., 2002; Reilly, 2005).

- **The mode of inheritance and genetic abnormalities involved and protein loss:**
  1. For Autosomal Dominant (AD) type, Charcot-Marie-Tooth disease type one (CMT1) is the demyelinating disease and Charcot-Marie-Tooth disease type two (CMT2) is the axonal form. Then, each genetic form is associated with a letter which corresponds to the gene or locus identified. For example, CMT1A is the demyelinating autosomal dominant form linked to *Peripheral Myelin Protein 22 kDa (PMP22)* gene, while CMT1B is the demyelinating autosomal dominant form related to *Myelin Protein Zero (MPZ)* gene.
  2. For Autosomal Recessive (AR) type, Charcot-Marie-Tooth disease type four (CMT4) corresponds to the demyelinating form and demyelinating autosomal recessive Charcot-Marie-Tooth (AR-CMT) disease belongs to the axonal form.
  3. The dominant forms linked to the X chromosome (CMTX) and intermediate forms (DI-CMT).

In this chapter, we will focus on the most common phenotype which is the demyelinating autosomal dominant neuropathy called Charcot-Marie-Tooth disease type one (CMT1 or HSMN1).

Charcot-Marie-Tooth Type One

The demyelinating neuropathy (CMT1) is a peripheral nerve defect, caused by abnormalities in the myelin sheath surrounding the nerve. In fact, a nerve cell communicates information to distant targets by sending electrical signals down the axon. In order to increase the speed at which these electrical signals travel, the axon is insulated by myelin, which is produced by Schwann cells, in the PNS. CMT is caused by mutations in genes that produce proteins involved in the structure and function of either the peripheral nerve axon or the myelin sheath. Although different proteins are abnormal in different forms of CMT disease, all of the mutations affect the normal function of the peripheral nerves. Consequently, these nerves slowly degenerate and lose the ability to communicate with their distant targets. The degeneration of motor nerves results in muscle weakness and atrophy in the extremities (arms, legs, hands or feet), and in some cases the degeneration of sensory nerves results in a reduced ability to feel heat, cold and pain.

So, CMT1 can be classified into six forms:

- CMT1A, the most common form (about 70% of cases), is associated with a 1.5 Mb duplication of chromosome 17p11.2–12 harboring the *PMP22* gene, leading to the presence of an extra copy of the gene (Lupski et al., 1991). PMP22 constitutes 2–5% of myelin proteins in the PNS, with proposed roles in myelin formation and maintenance (Suter and Scherer, 2003). MPZ is the major myelin protein expressed by Schwann cells, comprising approximately 50% of all PNS myelin proteins (Eylar et al., 1979; Greenfield et al., 1973). MPZ and PMP22 proteins are necessary for both normal myelin function and structure (Giese et al., 1992) (Figure 17.1).

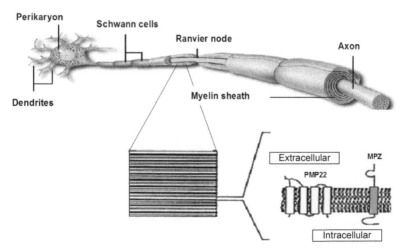

Figure 17.1. Schwann cells myelinating an axon in the PNS, and showing the importance of MPZ and PMP22 as structural proteins.

Peripheral myelin gene expression (*MPZ* and *PMP22*) is tightly regulated in myelinating Schwann cells (Niemann et al., 2006). Slight changes of *MPZ* or *PMP22* gene dosage have a serious impact on the development and preservation of nerves. Thus, the overexpression of *PMP22* gene causes the structure and function of the myelin sheath to be abnormal. *PMP22* duplication causes a peripheral neuropathy that typically occurs before the age of 20 and produces distal muscle weakness and atrophy. Histo-pathologically, the neuropathy is characterized by overmyelination (Gabreels-Festen et al., 1995), followed by demyelination, loss of myelinated nerve fibers and formation of concentric layers of Schwann cells (SC). This SC hypertrophy results in the formation of "onion bulbs"—a sign of repeated cycles of demyelination and remyelination processes (van der Kogel, 1977).

Interestingly, a different neuropathy distinct from CMT1A called hereditary neuropathy with liability to pressure palsy (HNPP) is due to a lack of synthesis of the protein PMP22. This lack of synthesis itself is due, in almost 90% of the cases, to a 1.5 Mb deletion in region 17p11.2 12 including the *PMP22* gene (Chance, 2006; Tyson et al., 1996). The deleted region is the same as the one duplicated in Charcot-Marie-Tooth disease type 1A. In this case, abnormally low levels of the *PMP22* gene result in episodic recurrent demyelinating neuropathy, resulting from minor trauma or compression (Bird, 1993). Peripheral nerve biopsies reveal a characteristic focal sausage-shaped thickening of the myelin sheath in about 25% or less of the internodes, which are designated as "tomacula" (Madrid et al., 1975).

- CMT1B represents less than 10% of CMT1 and is caused by mutations in the *MPZ* gene which is located on chromosome 1q22 and carries the instructions for manufacturing the MPZ protein, which is another critical component of the myelin sheath. As a result of abnormalities in *MPZ*, CMT1B produces symptoms similar to those found in CMT1A.
- The less common CMT1C, CMT1D, CMT1E and CMT1F, which also have symptoms similar to those found in CMT1A, are caused by mutations in the *lipopolysaccharide-induced TNF factor (LITAF), early growth response two (EGR2), MPZ* and *human neurofilament (NEFL)* genes, respectively.

Symptoms

CMT is characterized by a slowly progressive weakness, beginning in the distal limb muscles, and typically in the legs before the arms. Symptoms usually appear in the first two decades of life. Onset of symptoms is most often in adolescence or early adulthood, but some individuals develop symptoms in mid-adulthood. The severity of symptoms varies greatly among individuals and even among family members with the disease.

Progression of symptoms is gradual. Patients may first complain of difficulty walking due to foot and distal leg weakness. Ankle sprains are common at this stage. In its subtlest manifestation, patients may only note clumsiness with running or simply being not very athletic. As weakness becomes more severe, foot drop commonly occurs, resulting in a "steppage" gait in which the patient must lift the leg in an exaggerated fashion in order to clear the foot off the ground. Eventually, atrophy may develop, resulting in the characteristic "stork" leg appearance. Intrinsic foot muscle weakness commonly results in the foot deformity known as pes cavus (Figure 17.2).

Weakness in the upper extremities usually develops later than the lower extremities. Hand weakness results in complaints of poor finger control, poor hand writing, difficulty using zippers and buttons, and clumsiness with manipulating small objects. Despite the fact that the sensory nerves are affected as much as the motor nerve, patients do not complain of numbness. This is thought to be due to the fact that CMT patients never had normal sensation and therefore simply do not perceive their lack of sensation. Muscle cramping is also a common complaint.

On neurological examination, Deep Tendon Reflexes (DTR's) are markedly diminished or absent. Very early in the disease, the DTR's will disappear in the ankles and knees first, and then in the arms. Weakness and atrophy in lower and upper extremities is typical. Despite the lack of complaints of numbness, vibration and proprioception are markedly decreased, and Romberg testing is usually positive. Spinothalamic (pain

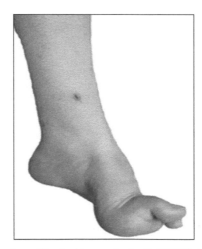

Figure 17.2. Deformation of the foot in a patient with CMT1A (Banchs et al., 2009).

and temperature) sensation is usually intact. Essential tremor is present in 30–50% of CMT patients, neuronal hearing loss in 5%, and scoliosis in 20%. Phrenic nerve involvement with diaphragmatic weakness is rare but has been described.

Therapeutic Approaches

There is no cure for CMT, but physical therapy, occupational therapy, braces and other orthopedic devices, and even orthopedic surgery can help individuals cope with the disabling symptoms of the disease. In addition, pain-killing drugs can be prescribed for individuals who have severe pain. Thus, the research on CMT1A is extremely important in order to learn how to treat, prevent, and even cure this disorder. Ongoing research includes efforts to identify more of the mutant genes and proteins that cause the various disease subtypes, efforts to discover the mechanisms of nerve degeneration and muscle atrophy with the hope of developing interventions to stop or slow down these debilitating processes, efforts to find therapies to reverse nerve degeneration and muscle atrophy, and efforts to reduce the expression of the *PMP22* gene in order to limit the toxic accumulation of abnormal PMP22 protein in Schwann cells.

Experimental studies in mice showed improved motor function and lifespan of mice following the absorption of ascorbic acid (Vitamin C). This molecule would decrease the expression of Pmp22 protein and thus repair the myelin sheath that surrounds the nerve (Lee et al., 2008). Similarly, *in vivo* studies in a rat model of CMT1A show that onapristone, an antagonist of the progesterone receptor, reduces the overexpression of *Pmp22*, while improving muscle mass and strength in rats. This improvement in motor function can be explained by a decrease in axonal loss (Meyer zu Horste et al., 2007).

In addition, research teams are looking into the possible beneficial effects of a natural substance, curcumin, on rare genetic diseases. Curcumin is a component of an Indian curry and is already widely used in traditional medicine in India and China. They have demonstrated that curcumin supplements administered to mice with hereditary neurological diseases decrease the nerve damage, as seen under the microscope. From a functional point of view, this treatment resulted in improved motor functions in these animals. In the absence of toxicity associated with curcumin, tests can be initiated in humans in the coming years (Nakagawa, 2008).

The alteration of PMP22 protein causes degeneration of peripheral axons and a decrease in SC number. Nerve regeneration involves complex interactions between axons, SC and the extracellular matrix. Prolonged absence of contact between axons and SC leads to a gradual decrease in

SC number and loss of ability to regenerate. It has been shown that SC, however, can survive in the absence of axon, through the secretion of various growth factors including the Neurotrophin-three (NT-3) (Lee et al., 2008). Researchers have tested the efficacy of NT-3 treatment, in two mouse models of CMT1A (Sahenk et al., 2005). After a period of eight to 12 weeks, this treatment has significantly improved axonal regeneration and myelination processes in both mouse models. Following these encouraging results, the authors conducted a randomized, double-blind clinical trial NT-3/placebo in order to assess the efficacy of NT-3 subcutaneous treatment in eight patients with CMT1A. After six months of NT-3 administration, significant improvements have been observed in patients treated with NT-3, regarding the regeneration of myelinated fibers in the sural nerve (Pareyson et al., 2006).

Prognostic Biomarkers in CMT1A

Prognostic biomarkers of disease severity could add powerful tools to monitor therapeutic effects in clinical trials (Pareyson et al., 2011; Reilly et al., 2010), could provide patients with important information regarding future life, and influence the decision to start a therapy at certain time points. In CMT1A, the level of *PMP22* appeared variable in some publications and was hypothesized to be secondary to demyelination (Gabriel et al., 1997; Hanemann et al., 1994; Vallat et al., 1996; Yoshikawa et al., 1994). Moreover, Katona et al. (2009) state that dysregulation of *PMP22* expression causes neuropathy in CMT1A. In fact, CMT rats with moderate disease progression showed a significant upregulation of *PMP22* when compared with wild-type controls, in skin biopsies (Makoukji et al., 2012). So, the expression of *PMP22* could potentially provide a sensitive molecular biomarker of disease severity.

Downregulation of PMP22

In peripheral nerve injuries, demyelination occurs as a result of Schwann cell injury and is accompanied or followed by axonal degeneration, and thus by a downregulation of the expression of peripheral myelin genes, *PMP22* and *MPZ*. To date, few cellular signals are known to directly regulate the expression of myelin genes. Tawk et al. (2011) have shown that the Wnt/β-catenin signaling pathway is an essential and direct driver of myelin genes expression and myelin compaction. In fact, the inactivation of Wnt components in zebrafish embryos led to a severe dysmyelination and the inhibition of myelin gene expression. Furthermore, although endogenous Wnt is sufficient for myelin gene expression, the stimulation of canonical Wnt pathway by means of Wnt ligands enhances the expression of peripheral myelin genes, *Mpz* and *Pmp22* (Tawk et al. 2011).

In fact, Wnt ligands bind to the frizzled membrane receptors (Fzl), which interact with co-receptors named the arrow/low-density lipoprotein receptor-related proteins (LRP) (Lai et al., 2009). Wnt binding to Fzl activates different intracellular signaling pathways, including an evolutionarily conserved branch, the so-called canonical Wnt signaling pathway or Wnt/β-catenin pathway. In the absence of Wnt signal, β-catenin is phosphorylated by glycogen synthase kinase three β (GSK3β) and is then targeted for proteasomal degradation. Wnt signaling through Fzl and LRP activates dishevelled (Dvl), which prevents β-catenin degradation. The stabilized β-catenin enters the nucleus, where it regulates gene transcription by interacting with transcription factors called the T-cell factor/lymphoid-enhancer factors (TCF/LEF). Four different forms of TCFs are expressed in mouse tissues: Tcf1, Lef1, Tcf3 and Tcf4. In the absence of nuclear β-catenin, the TCFs interact with transcriptional repressors such as Groucho and prevent Wnt target genes transcription (Bienz, 1998; Chen et al., 1999). However, when Wnt signaling is activated, the interaction with β-catenin converts the TCF into transcriptional activators (Nusse, 1999) (Figure 17.3).

Lithium (LiCl) is an enzymatic inhibitor of GSK3β, and thus stimulates Wnt/β-catenin signaling pathway (Figure 17.6). LiCl enhances myelin gene expression *in vitro* and *in vivo* via the binding of β-catenin to TCF/LEF

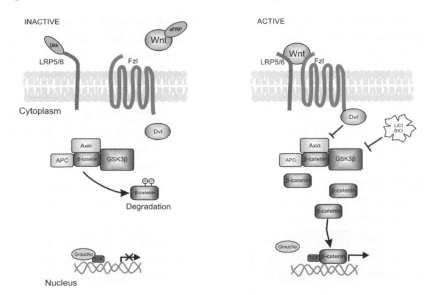

Figure 17.3. Canonical or Wnt/β-catenin signaling pathway. In a cell without a Wnt signal, excess β-catenin is degraded. When an extracellular Wnt binds to its receptor, β-catenin accumulates, binds to Tcf/Lef factors, and alters gene transcription.

transcription factors present in the vicinity of *PMP22* and *MPZ* promoters, as shown by Chromatin Immunoprecipitation (ChIP) assays (Tawk et al., 2011).

In vitro experiments performed by Makoukji et al. (2011), in a mouse Schwann cell line (MSC80), showed that after 24 hours, LiCl stimulated *MPZ* and *PMP22* promoters, transcripts and proteins. Interestingly, in MSC80 cells, LiCl stimulation increased after 48 and 72 hours of treatment (Tawk et al., 2011). LiCl also triggered modification in cell shape and elicited nuclear localization of β-catenin. Atomic force, confocal and light microscopy have shown a significant 2-fold increase in the length of cellular extensions that was dependent on the duration of LiCl treatment. This increase in Schwann cell processes could be a prerequisite to myelination (Schwann cell ensheathment of an axon) (Figure 17.4).

In adult animals, myelin gene expression is low and is only reactivated after nerve injury to stimulate remyelination (Syroid et al., 1996). Therefore, Makoukji et al., tested, *in vivo,* the effects of LiCl on myelin injuries. They performed in mice a nerve crush of the facial nerve which provokes paralysis of the whiskers. They showed that LiCl administration (by i.p. injections) to those mice elicited a rapid functional recovery. This recovery is concomitant with the stimulation of myelin gene expression, the decrease of the g-ratio which is the ratio of the inner axonal diameter to the total outer diameter, and the restoration of the myelin sheath structure around the axons, thus suggesting nerve remyelination (Tawk et al., 2011) (Figure 17.5).

To demonstrate that the beneficial effects of LiCl on remyelinatiom are not limited to a specific type of nerve, Makoukji et al., performed a demyelinating nerve crush of the sciatic nerve this time, and treated the animals with LiCl (added to drinking water). They showed that *Mpz* and *Pmp22* expression are increased, myelin structure was greatly ameliorated (thicker myelin sheath) and g-ratio was decreased, without any alteration in the number of myelinated axons (Tawk et al., 2011) (Figure 17.6).

In conclusion, lithium is known to exert several beneficial effects on axonal regeneration and also has promyelinating effects. Therefore, this molecule could constitute a potential treatment for nerve injuries in which axonal lesions and demyelination coexist. So, GSK3β could be considered as an important player in the expression of peripheral myelin genes, and using modulators like lithium could open novel perspectives in the treatment of demyelinating diseases or nerve injuries, by reinitiating or accelerating the myelination or remyelination processes.

Upregulation of PMP22

As previously stated, in CMT1A, a slight increase in *PMP22* expression has a deep impact on the development and preservation of nerves and their

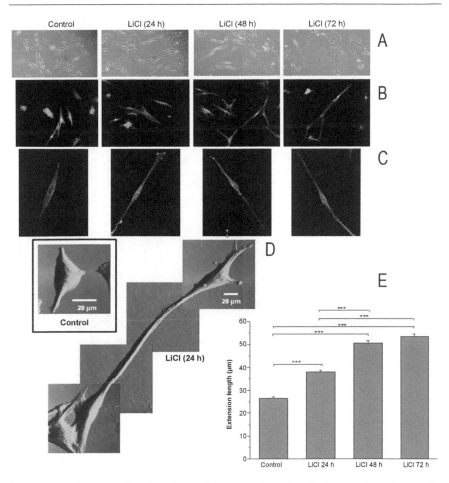

Figure 17.4. Effect of LiCl on the shape of MSC80 cells and on the localization of β-catenin in MSC80 cells (Tawk et al., 2011). MSC80 cells were incubated in the absence (control) or in the presence of LiCl (10 mM) during 24, 48 and/or 72 hours and then monitored by light microscopy (A), confocal microscopy to localize b-catenin (B;C) and atomic force microscopy (D). Scale bar = 20 μm. (B;C) Antibodies directed against b-catenin were added to the cells overnight. (E) Extension lengths of MSC80 cells. Data are presented as mean ± S.E.M. of three independent experiments and a minimum of 100 cells per experiment. Kruskal-Wallis one-way ANOVA with Bonferroni, Scheffe, Tukey and Sidak comparison tests were used to determine the statistical significance between data, ***$P < 0.001$.

Color image of this figure appears in the color plate section at the end of the book.

myelin sheath. Désarnaud et al. (2000) have shown that steroid hormones stimulate the expression of peripheral myelin genes, *Mpz* and *Pmp22*, and thus could enhance the myelination process in the PNS. Oxysterols are oxidized derivatives of cholesterol. There are different oxysterols, in

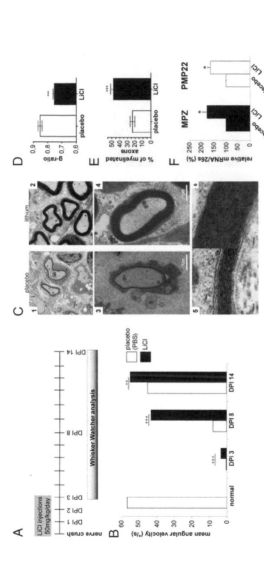

Figure 17.5. Effect of LiCl on functional recovery of whiskers and phenotype of facial nerve after nerve crush (Tawk et al., 2011). (A) Nerve crush of the left facial nerve was performed in eight-week old mice. One group received four i.p. injections of LiCl (50 mg/kg/d) directly after the nerve crush during four days. The other group received PBS solution injection as a placebo (PI). Normal animals (control) did not undergo nerve crush. (B) The mean angular velocity of the whisker movements of all animals was analyzed by means of Whisker Watcher software at DPI3, DPI8, and DPI14. Results are expressed in degrees per second and represent the mean of six animals per group; the angular velocity of the whisker of each animal was analyzed three times (***p<0.005, Kruskal-Wallis test). (C) At DPI8, the facial nerve was isolated. Ultrathin cross-sections were prepared and stained for myelin. 1 and 2, Low-magnification EM images of a section of facial nerve (Scale bar = 2μm). 3 and 4, High-magnification EM images of a representative axon (Scale bar = 1μm). 5 and 6, High-magnification EM images of myelin layers (Scale bar = 100 nm). (D) Myelin thickness was estimated by g-ratio analysis of facial nerves (12 images per animal, three animals per condition). Data are means ± SEM. (E) The percentage of myelinated axons per square micrometer was calculated by using EM images. Three animals per condition were used. Data are means ± SEM. (F) Facial nerves were dissected at DPI3. Quantitative real-time PCR experiments were performed by using primers recognizing Mpz and Pmp22. The RT-PCR was normalized using 26S RNA and Gapdh (*p<0.05 and **p<0.01 by Mann-Whitney's test when compared to control).

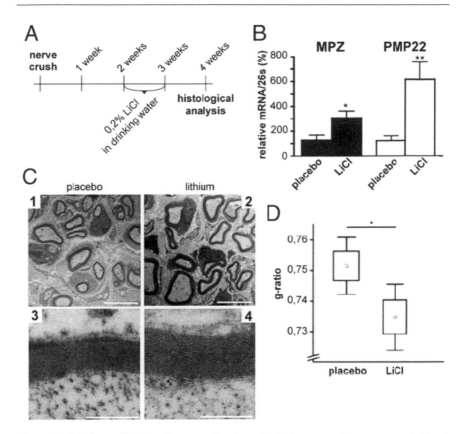

Figure 17.6. Effect of LiCl on the remyelination of sciatic nerve after nerve crush (Tawk et al., 2011). (A) Nerve crush of the right sciatic nerve was performed in wild type mice at the age of eight weeks via a forceps applying a standardized pressure for 40 seconds. One group received, 2 weeks after nerve crush, 0.2% lithium chloride (LiCl) for seven days. All animals were sacrificed 4 weeks post crush, and their sciatic nerves were analyzed. (B) Sciatic nerves distal to the crush were dissected. Quantitative real-time PCR experiments were performed by using primers recognizing Mpz and Pmp22. The RT-PCR was normalized using 26S RNA. **p<0.01 by Mann-Whitney's test when compared to control. (C) Electron micrograph images of sciatic nerves 4 weeks post crush from placebo- (Pl) and LiCl-treated mice. 1 and 2, Low-magnification EM images of section of sciatic nerve (Scale bar = 5 μm). 3 and 4, High-magnification EM images of myelin layers (Scale bar = 0.5 μm). (D) G-ratio analysis of sciatic nerves from placebo-treated animals (n = 8) and LiCl-treated animals (n = 8) 4 weeks after crush. At least 100 axons per animal were quantified on EM level.

particular, 24(S)-hydroxycholesterol (24(S)-OH), 25-hydroxycholesterol (25-OH), and 27-hydroxycholesterol (27-OH). Oxysterols exert their effects by binding to liver X receptors (LXRs), which have two isoforms: LXRα and LXRβ.

In the nucleus, LXR forms a heterodimer with Retinoid X Receptor (RXR) that binds to LXR Response Elements (LXRE), present in the promoter region of their target genes. In the absence of ligands, LXRs bind to corepressors (Chen and Evans, 1995; Horlein et al., 1995). In response to oxysterols binding, the LXR-RXR heterodimer changes conformation, which leads to the release of corepressors and the recruitment of coactivators such as p160 (Huuskonen et al., 2004) or peroxisome-proliferator-activated receptor gamma co-activator one α (PGC-1α) (Oberkofler et al., 2003), to activate the expression of target genes. These coactivators have histone acetyltransferase (HAT) activities that will allow decondensation of chromatin and thus will lead to transcriptional activation (Figure 17.7).

Oxysterols are important signaling molecules involved in the regulation of cholesterol homeostasis. They are also known as important players in the progression of neurodegenerative diseases of the central nervous system such as Alzheimer's disease (Lutjohann et al., 2000) and multiple sclerosis (Teunissen et al., 2003). Makoukji et al. reported, for the first time, a major role of oxysterols in the PNS. They showed, by *in vitro* studies, that oxysterols are produced by the mouse Schwann cell line, MSC80, as well as by sciatic nerves and on the contrary, they downregulate the expression of peripheral myelin genes. In fact, they inhibit *PMP22* and *MPZ* promoter activities, transcripts and proteins. The inhibitory effects of oxysterols are mediated by mechanisms involving LXRs, as demonstrated by ChIP experiments, revealing an increased recruitment of LXR at the levels of *MPZ* and *PMP22* promoters (Tawk et al., 2011).

Furthermore, *in vivo* studies demonstrated that LXRα and LXRβ are crucial for the correct myelination of the peripheral nerve axons. Indeed, LXRα/β$^{-/-}$ knock-out (KO) mice exhibit thinner myelin sheaths around their axons as shown in Figure 17.8 and a significant increase in the expression of

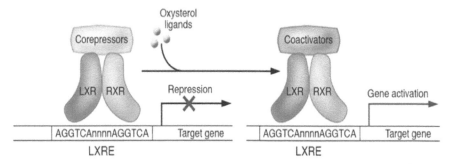

Figure 17.7. Mechanism of action of LXRs (Zelcer and Tontonoz, 2006). Within the nucleus, LXR/RXR heterodimers are bound to LXREs in the promoters of target genes and in complex with corepressors (e.g., SMRT, N-CoR). In response to the binding of oxysterol ligands, the corepressor complexes are exchanged for coactivator complexes, and target gene expression is induced.

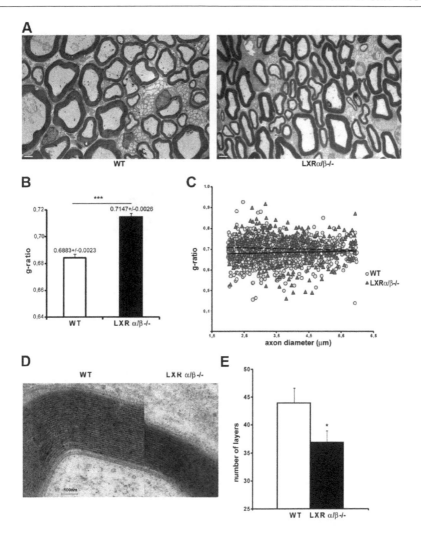

Figure 17.8. Incidence of LXR KO on myelin structure *in vivo* (Tawk et al., 2011). (A) Ultrathin (50–90 nm) cross-sections were prepared from Epon-embedded adult sciatic nerves and stained for myelin. Heterogeneity and reduced thickness of myelin sheaths is apparent in LXRα/β$^{-/-}$ mice when compared with WT. Scale bar = 2 μm. (B) Myelin thickness was estimated by g-ratio determination using electronic microscopy pictures obtained from adult sciatic nerves. Myelin thickness is altered in LXRα/β$^{-/-}$ mice. Three animals per genotype were used. Data are given as means ± SEM. (C) g-ratios were plotted against axonal diameters. The reduction in myelin thickness in LXRα/β$^{-/-}$ mice when compared with WT was observed for nearly all types of axon calibers. Dotted line, linear regression curve for WT animals; solid line, linear regression curve for LXRα/β$^{-/-}$ mice. (D) High-magnification electron micrographs of myelin layers from LXRα/β$^{-/-}$ compared with WT animals. Scale bar = 100 nm. The two axons shown are of approximately equal diameters. (E) Quantification of the number of layers around the axons of WT or LXRα/β$^{-/-}$ mice. *p<0.05 and ***p<0.001 by Student's t test when compared with control (WT).

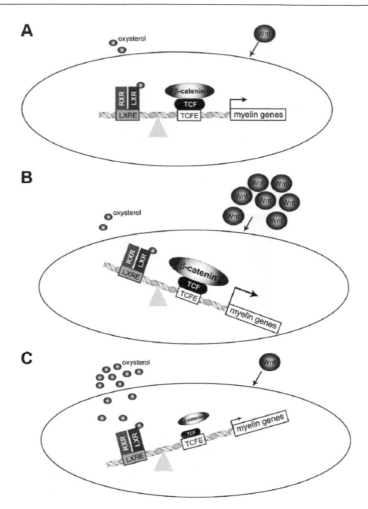

Figure 17.9. Schematic model of crosstalk between oxysterol and Wnt pathways in the expression of myelin genes MPZ and PMP22 (Tawk et al., 2011). (A) At basal levels, the interplay between LXR and Wnt pathways induces the fine-tuning of myelin gene expression. Wnt ligands activate myelin genes by means of the canonical Wnt pathway. We have identified active TCF/LEF response elements (TCFE) on the level of MPZ and PMP22 promoters. These TCFEs bind TCF/LEF transcription factors that interact with β-catenin. Oxysterols are present in Schwann cells. They repress myelin gene expression by means of LXRs that bind to LXREs located near the TCFEs. Furthermore, oxysterols can decrease the amount of Wnt components in Schwann cells. Therefore, the amount of β-catenin recruited at the level of MPZ and PMP22 promoters is the result of a balance between Wnt and oxysterol pathways. (B) When Wnt ligands are at high amounts, Wnt/β-catenin pathway becomes predominant provoking the increase of β-catenin recruitment on myelin gene promoters. This will allow the expression of myelin genes at high levels. (C) When oxysterols reach high concentrations, they repress myelin gene expression by means of LXRs and decrease the amounts of Wnt/β-catenin components in Schwann cells. In this case, the oxysterol/LXR pathway becomes predominant.

Mpz and *Pmp22* transcripts (Tawk et al., 2011). So, these results show, for the first time, the implication of LXR and oxysterols in the negative regulation of peripheral myelin gene expression, *Pmp22* and *Mpz*.

Usually, liganded LXR activate transcription of genes after binding of coactivators, whereas here, LXR ligands repress myelin genes. Thus, the inhibition of peripheral myelin genes by oxysterols does not implicate the classical mechanism of action of oxysterols. It is a novel mechanism, and Makoukji et al. showed that it is involves the inhibition of the canonical Wnt/β-catenin signaling pathway. In fact, the transcripts of the major components of the canonical Wnt/β-catenin signaling pathway (i.e., Dvl, TCF/LEF) are inhibited by oxysterols. β-catenin transcript and protein are also decreased by oxysterols in Schwann cells, and on the other hand, enhanced in the sciatic nerves of LXRα/β$^{-/-}$ animals. Moreover, after 24 hours of treatment with oxysterols, confocal microscopy showed that β-catenin was not detected in Schwann cell nuclei, and ChIP analysis showed that β-catenin recruitment at the level of *MPZ* and *PMP22* promoters was reduced (Tawk et al., 2011). The reduction of b-catenin could be explained by a potential recruitment by LXRs, since TCF/LEF binding sites are close to LXR response elements in *MPZ* and *PMP22* promoters (Figure 17.9).

Conclusion

To conclude, oxysterols, due to their inhibitory action on the peripheral myelin genes *PMP22* and *MPZ*, could provide a new approach for the treatment of demyelinating diseases such as Charcot-Marie-Tooth type one. The use of oxysterols would be beneficial, especially during the early phase of the disease, characterized by an upregulation of the expression of *PMP22* gene, which may lead to nerve demyelination and axonal loss.

Acknowledgements

I would like to thank Nadine J. Makhoul for a critical reading of the manuscript.

References

Banchs, I., C. Casasnovas, A. Alberti, L. De Jorge, M. Povedano, J. Montero, J.A. Martinez-Matos and V. Volpini. 2009. Diagnosis of Charcot-Marie-Tooth disease. *J Biomed Biotechnol.* 985415.
Bienz, M. 1998. TCF: transcriptional activator or repressor? *Curr Opin Cell Biol.* 10: 366–72.
Bird, T.D. 1993. Hereditary Neuropathy with Liability to Pressure Palsies.
Boerkoel, C.F., H. Takashima, C.A. Garcia, R.K. Olney, J. Johnson, K. Berry, P. Russo, S. Kennedy, A.S. Teebi, M. Scavina, L.L. Williams, P. Mancias, I.J. Butler, K. Krajewski, M. Shy and J.R. Lupski. 2002. Charcot-Marie-Tooth disease and related neuropathies: mutation distribution and genotype-phenotype correlation. *Ann Neurol.* 51: 190–201.

Chance, P.F. 2006. Inherited focal, episodic neuropathies: hereditary neuropathy with liability to pressure palsies and hereditary neuralgic amyotrophy. *Neuromolecular Med*. 8: 159–74.

Chen, G., J. Fernandez, S. Mische and A.J. Courey. 1999. A functional interaction between the histone deacetylase Rpd3 and the corepressor groucho in Drosophila development. *Genes Dev*. 13: 2218–30.

Chen, J.D. and R.M. Evans. 1995. A transcriptional co-repressor that interacts with nuclear hormone receptors. *Nature*. 377: 454–7.

Desarnaud, F., S. Bidichandani, P.I. Patel, E.E. Baulieu and M. Schumacher. 2000. Glucocorticosteroids stimulate the activity of the promoters of peripheral myelin protein-22 and protein zero genes in Schwann cells. *Brain Res*. 865: 12–6.

Eylar, E.H., K. Uyemura, S.W. Brostoff, K. Kitamura, A. Ishaque and S. Greenfield. 1979. Proposed nomenclature for PNS myelin proteins. *Neurochem Res*. 4: 289–93.

Gabreels-Festen, A.A., P.A. Bolhuis, J.E. Hoogendijk, L.J. Valentijn, E.J. Eshuis and F.J. Gabreels. 1995. Charcot-Marie-Tooth disease type 1A: morphological phenotype of the 17p duplication versus PMP22 point mutations. *Acta Neuropathol*. 90: 645–9.

Gabriel, J.M., B. Erne, D. Pareyson, A. Sghirlanzoni, F. Taroni and A.J. Steck. 1997. Gene dosage effects in hereditary peripheral neuropathy. Expression of peripheral myelin protein 22 in Charcot-Marie-Tooth disease type 1A and hereditary neuropathy with liability to pressure palsies nerve biopsies. *Neurology*. 49: 1635–40.

Giese, K.P., R. Martini, G. Lemke, P. Soriano and M. Schachner. 1992. Mouse P0 gene disruption leads to hypomyelination, abnormal expression of recognition molecules, and degeneration of myelin and axons. *Cell*. 71: 565–76.

Grandis, M., M. Leandri, T. Vigo, M. Cilli, M.W. Sereda, G. Gherardi, L. Benedetti, G. Mancardi, M. Abbruzzese, K.A. Nave, L. Nobbio and A. Schenone. 2004. Early abnormalities in sciatic nerve function and structure in a rat model of Charcot-Marie-Tooth type 1A disease. *Exp Neurol*. 190: 213–23.

Greenfield, S., S. Brostoff, E.H. Eylar and P. Morell. 1973. Protein composition of myelin of the peripheral nervous system. *J Neurochem*. 20: 1207–16.

Hanemann, C.O., G. Stoll, D. D'Urso, W. Fricke, J.J. Martin, C. Van Broeckhoven, G.L. Mancardi, I. Bartke and H.W. Muller. 1994. Peripheral myelin protein-22 expression in Charcot-Marie-Tooth disease type 1a sural nerve biopsies. *J Neurosci Res*. 37: 654–9.

Harding, A.E. and P.K. Thomas. 1980. The clinical features of hereditary motor and sensory neuropathy types I and II. *Brain*. 103: 259–80.

Horlein, A.J., A.M. Naar, T. Heinzel, J. Torchia, B. Gloss, R. Kurokawa, A. Ryan, Y. Kamei, M. Soderstrom, C.K. Glass and M.G. Rosenfeld. 1995. Ligand-independent repression by the thyroid hormone receptor mediated by a nuclear receptor co-repressor. *Nature*. 377: 397–404.

Huuskonen, J., P.E. Fielding and C.J. Fielding. 2004. Role of p160 coactivator complex in the activation of liver X receptor. *Arterioscler Thromb Vasc Biol*. 24: 703–8.

Katona, I., X. Wu, S.M. Feely, S. Sottile, C.E. Siskind, L.J. Miller, M.E. Shy and J. Li. 2009. PMP22 expression in dermal nerve myelin from patients with CMT1A. *Brain*. 132: 1734–40.

Lai, S.L., A.J. Chien and R.T. Moon. 2009. Wnt/Fz signaling and the cytoskeleton: potential roles in tumorigenesis. *Cell Res*. 19: 532–45.

Lee, Y.C., M.H. Chang and K.P. Lin. 2008. Charcot-Marie-Tooth disease. *Acta Neurol Taiwan*. 17: 203–13.

Lupski, J.R., R.M. de Oca-Luna, S. Slaugenhaupt, L. Pentao, V. Guzzetta, B.J. Trask, O. Saucedo-Cardenas, D.F. Barker, J.M. Killian, C.A. Garcia, A. Chakravarti and P.I. Patel. 1991. DNA duplication associated with Charcot-Marie-Tooth disease type 1A. *Cell*. 66: 219–32.

Lutjohann, D., A. Papassotiropoulos, I. Bjorkhem, S. Locatelli, M. Bagli, R.D. Oehring, U. Schlegel, F. Jessen, M.L. Rao, K. von Bergmann and R. Heun. 2000. Plasma 24S-hydroxycholesterol (cerebrosterol) is increased in Alzheimer and vascular demented patients. *J Lipid Res*. 41: 195–8.

Madrid, R.E., E. Jaros, M.J. Cullen and W.G. Bradley. 1975. Genetically determined defect of Schwann cell basement membrane in dystrophic mouse. *Nature*. 257: 319–21.

Makoukji, J., M. Belle, D. Meffre, R. Stassart, J. Grenier, G. Shackleford, R. Fledrich, C. Fonte, J. Branchu, M. Goulard, C. de Waele, F. Charbonnier, M.W. Sereda, E.E. Baulieu, M. Schumacher, S. Bernard and C. Massaad. 2012. Lithium enhances remyelination of peripheral nerves. *Proc Natl Acad Sci USA.* 109: 3973–8.

Meyer zu Horste, G., T. Prukop, D. Liebetanz, W. Mobius, K.A. Nave and M.W. Sereda. 2007. Antiprogesterone therapy uncouples axonal loss from demyelination in a transgenic rat model of CMT1A neuropathy. *Ann Neurol.* 61: 61–72.

Nakagawa, M. 2008. [Hereditary neuropathy: recent advance. *Rinsho Shinkeigaku.* 48: 1019–22.

Niemann, A., P. Berger and U. Suter. 2006. Pathomechanisms of mutant proteins in Charcot-Marie-Tooth disease. *Neuromolecular Med.* 8: 217–42.

Nusse, R. 1999. WNT targets. Repression and activation. *Trends Genet.* 15: 1–3.

Oberkofler, H., E. Schraml, F. Krempler and W. Patsch. 2003. Potentiation of liver X receptor transcriptional activity by peroxisome-proliferator-activated receptor gamma co-activator 1 alpha. *Biochem J.* 371: 89–96.

Pareyson, D., M.M. Reilly, A. Schenone, G.M. Fabrizi, T. Cavallaro, L. Santoro, G. Vita, A. Quattrone, L. Padua, F. Gemignani, F. Visioli, M. Laura, D. Radice, D. Calabrese, R.A. Hughes and A. Solari. 2011. Ascorbic acid in Charcot-Marie-Tooth disease type 1A (CMT-TRIAAL and CMT-TRAUK): a double-blind randomised trial. *Lancet Neurol.* 10: 320–8.

Pareyson, D., V. Scaioli and M. Laura. 2006. Clinical and electrophysiological aspects of Charcot-Marie-Tooth disease. *Neuromolecular Med.* 8: 3–22.

Reilly, M.M. 2005. Axonal Charcot-Marie-Tooth disease: the fog is slowly lifting! *Neurology.* 65: 186–7.

Reilly, M.M., M.E. Shy, F. Muntoni and D. Pareyson. 2010. 168th ENMC International Workshop: outcome measures and clinical trials in Charcot-Marie-Tooth disease (CMT). *Neuromuscul Disord.* 20: 839–46.

Sahenk, Z., H.N. Nagaraja, B.S. McCracken, W.M. King, M.L. Freimer, J.M. Cedarbaum and J.R. Mendell. 2005. NT-3 promotes nerve regeneration and sensory improvement in CMT1A mouse models and in patients. *Neurology.* 65: 681–9.

Suter, U. and S.S. Scherer. 2003. Disease mechanisms in inherited neuropathies. *Nat Rev Neurosci.* 4: 714–26.

Syroid, D.E., P.R. Maycox, P.G. Burrola, N. Liu, D. Wen, K.F. Lee, G. Lemke and T.J. Kilpatrick. 1996. Cell death in the Schwann cell lineage and its regulation by neuregulin. *Proc Natl Acad Sci USA.* 93: 9229–34.

Tawk, M., J. Makoukji, M. Belle, C. Fonte, A. Trousson, T. Hawkins, H. Li, S. Ghandour, M. Schumacher and C. Massaad. 2011. Wnt/beta-catenin signaling is an essential and direct driver of myelin gene expression and myelinogenesis. *J Neurosci.* 31: 3729–42.

Teunissen, C.E., C.D. Dijkstra, C.H. Polman, E.L. Hoogervorst, K. von Bergmann and D. Lutjohann. 2003. Decreased levels of the brain specific 24S-hydroxycholesterol and cholesterol precursors in serum of multiple sclerosis patients. *Neurosci Lett.* 347: 159–62.

Tyson, J., S. Malcolm, P.K. Thomas and A.E. Harding. 1996. Deletions of chromosome 17p11.2 in multifocal neuropathies. *Ann Neurol.* 39: 180–6.

Vallat, J.M., P. Sindou, P.M. Preux, F. Tabaraud, A.M. Milor, P. Couratier, E. LeGuern and A. Brice. 1996. Ultrastructural PMP22 expression in inherited demyelinating neuropathies. *Ann Neurol.* 39: 813–7.

van der Kogel, A.J. 1977. Radiation-induced nerve root degeneration and hypertrophic neuropathy in the lumbosacral spinal cord of rats: the relation with changes in aging rats. *Acta Neuropathol.* 39: 139–45.

Yoshikawa, H., T. Nishimura, Y. Nakatsuji, H. Fujimura, M. Himoro, K. Hayasaka, S. Sakoda and T. Yanagihara. 1994. Elevated expression of messenger RNA for peripheral myelin protein 22 in biopsied peripheral nerves of patients with Charcot-Marie-Tooth disease type 1A. *Ann Neurol.* 35: 445–50.

Zelcer, N. and P. Tontonoz. 2006. Liver X receptors as integrators of metabolic and inflammatory signaling. *J Clin Invest.* 116: 607–14.

18

Alpha-Synuclein and 14-3-3 Proteins as Biomarkers of Neurodegenerative Diseases

Molly Foote,[1] Kourtney Graham[2] and Yi Zhou[1,2],*

INTRODUCTION

Alpha-synuclein and the family of 14-3-3 proteins are both highly expressed in the brain and share a structural and functional homology (Ostrerova, 1999). Based on previous clinical and basic research, both alpha-synuclein and 14-3-3 proteins are associated with the pathogenesis of several neurodegenerative diseases including Parkinson's disease, Alzheimer's, Amyotrophic Lateral Sclerosis, etc. In particular, alpha-synuclein and 14-3-3 are major components of Lewy bodies, which are abnormal protein aggregates that develop within the nerve cells of patients with Parkinson's disease and other neurodegenerative diseases. Along these lines, many studies have investigated the use of alpha-synuclein and/or 14-3-3 proteins as potential biomarkers for the differential diagnosis and progression of various neurological diseases.

[1] Department of Biomedical Sciences, Florida State University 1115 West Call Street, College of Medicine, Tallahassee, FL 32306.
Email: molly.foote@med.fsu.edu
[2] Program in Neuroscience, Florida State University 1115 West Call Street, College of Medicine, Tallahassee, FL 32306.
Email: Kourtney.graham@med.fsu.edu
* Corresponding author: yi.zhou@med.fsu.edu

Alpha-Synuclein

Alpha-synuclein is a member of the synuclein family of proteins that is abundantly expressed in the nervous system. All three members of the synuclein family, including alpha, beta and gamma-synuclein, are predominantly neuronal proteins that have been shown to preferentially localize to presynaptic terminals under physiological conditions (George, 2002). These proteins are all composed of three distinct regions: (1) the amino terminus, which is involved in the formation of α-helical structures and also contains apolipoprotein lipid-binding motifs; (2) a central hydrophobic region, termed the NAC (non-AB component), which allows for β-sheet folding; and (3) a carboxyl terminus that is highly negatively charged and primarily unstructured (Maroteaux et al., 1988). What sets alpha-synuclein apart from the rest of the synuclein family is the amino acid sequence divergence in its NAC region. The structure of alpha-synuclein was first determined after cloning of the gene SNCA that encodes for a 140 amino acid protein, which, in aqueous solutions, does not have a definitive structure and is therefore referred to as a "natively unfolded protein" (Stefanis, 2012). However, alpha-synuclein does have the ability to form α-helical structures upon binding to negatively charged lipids, such as phospholipids on cellular membranes, and β-sheet structures in prolonged periods of incubation (Stefanis, 2012). In addition, recent studies suggest that the helical form of alpha-synuclein its native conformation and that destabilization of this structure is what leads to alpha-synuclein misfolding in synucleinopathies (Bartels et al., 2011).

In the nervous system, immunohistochemistry of brain sections, utilizing alpha-synuclein antibodies, revealed a punctate neuropil staining pattern, including presynaptic terminal staining (Huang et al., 2011; Kahle, 2008). In the presynaptic terminals, alpha-synuclein has been shown to be within close proximity of synaptic vesicles, suggesting a possible role for alpha-synuclein in synaptic vesicle regulation (Barbour et al., 2008). The expression of alpha-synuclein is induced during neuronal development, post establishment of synaptic connections and prior to induction of proteins involved in synaptic structure and function, conferring a role for alpha-synuclein in synaptic development and transmission (Withers et al., 1997). Consistent with this hypothesis, mice null for CSP-alpha, a presynaptic protein, exhibit synaptic degeneration that was shown to be regulated by alpha-synuclein (Chandra et al., 2005). This study suggests that alpha-synuclein, through a chaperone-like function, may act, together with CSP-alpha, in the assembly of the SNARE complex, possibly by binding to synaptotagmin-2 (Burre et al., 2010).

Additionally, there has been considerable research into the role of alpha-synuclein in neurodegenerative diseases, particularly regarding the aggregation of alpha-synuclein proteins. Wild-type as well as disease-related mutants of alpha-synuclein form amyloid-like fibrils upon prolonged incubation in solution (Conway et al., 2000). Such fibrils, or protofibrils, gradually become insoluble and coalesce into fibrils (Stefanis, 2012). This cascade of events, starting from the unfolded protein and progressing into the mature fibril formation is known as alpha-synuclein aggregation. Such aggregates form the basis of mature Lewy bodies and Lewy neurites present in synuclein-related diseases referred to as synucleinopathies, including Parkinson's disease, dementia with Lewy bodies, and multiple system atrophy (Kalia et al., 2013). Concurrently, aggregation of two or more alpha-synuclein monomers leads to the formation of soluble oligomeric species. Initial characterization of alpha-synuclein oligomers came from *in vitro* experiments in which recombinant alpha-synuclein was found to spontaneously aggregate in a concentration-dependent manner (Conway et al., 1998). Subsequent experiments conducted in cellular system as well as *in vivo* further support this observation that the presence of soluble alpha-synuclein oligomers generally correlates with neurotoxicity as seen in various synucleinopathies, such as Parkinson's disease (Gosavi et al., 2002; Lo Bianco et al., 2002).

Alpha-synuclein and Parkinson's Disease

Parkinson's Disease (PD) is a neurodegenerative disease characterized by a loss of dopaminergic neurons in the Substantia Nigra *pars compacta* and the presence of Lewy bodies (Cookson and Bandmann, 2010). The first specific genetic association of PD was reported in an Italian family, in which a study identified a G209A substitution of the SNCA gene encoding for alpha-synuclein (Polymeropoulos et al., 1997). The discovery of the genetic link to PD led to histopathological analyses of brain sections from PD patients using specific antibodies against alpha-synuclein (Stefanis, 2012). Based on these histological studies, alpha-synuclein appeared to be abundantly expressed within Lewy bodies, thus, linking alpha-synuclein to the formation of Lewy bodies in PD (Spillantini et al., 1997). Further studies have shown that abnormal deposits of alpha-synuclein in neuritic processes are the driving force in PD pathogenesis and that this deposition occurs early in PD disease progression (Stefanis, 2012). Additionally, alpha-synuclein-null mice exhibit deficits in the dopaminergic system, as evidenced by a reduction in the number of postnatal dopaminergic neurons and a decrease in dopaminergic terminals (Al-Wandi et al., 2010; Robertson et al., 2004). However, the precise role of alpha-synuclein in PD pathogenesis remains unclear.

Alpha-Synuclein as a Biochemical Biomarker

Parkinson's Disease

Currently, most neurological diseases are diagnosed based on the presence of physical symptoms such as motor or cognitive deficiencies. Parkinson's disease is no exception, with the majority of diagnoses based on primary motor features including resting tremor, rigidity, bradykinesia, and loss of postural reflexes (Lang and Lozano, 1998). In order to diagnose these diseases earlier and more efficiently, recent efforts have been exerted to identify biomarkers from various sources. In PD, one of the main potential biomarkers being investigated is alpha-synuclein because of its potential involvement in the pathogenesis of this disease and its presence in different peripheral tissues and fluids (Table 18.1).

Alpha-synuclein has been detected in blood, saliva and cerebrospinal fluid (CSF) as well as peripheral tissues such as the intestine (El-Agnaf et al., 2006; Forsyth et al., 2011; Hong et al., 2010). Most PD biomarker studies, however, have focused on alterations of alpha-synuclein levels in the CSF, because the majority of alpha-synuclein proteins typically present in the CSF

Table 18.1. Alpha-synuclein as a biomarker of neurodegenerative diseases.

Disease	Biomarker Evidence	Reference
PD	Reduced alpha-synuclein levels in the CSF	Tokuda et al., 2006
	Elevated oligomeric alpha-synuclein levels in the CSF	Tokuda et al., 2010
	Elevated ratio of phosphorylated alpha-synuclein to alpha-synuclein in the CSF	Wang et al., 2012
	Reduced alpha-synuclein levels in blood plasma	El-Agnaf et al., 2006
	Elevated phosphorylated alpha-synuclein levels in blood plasma	Foulds et al., 2011
	Reduced alpha-synuclein levels in saliva	Devic et al., 2011
	Elevated alpha-synuclein levels in intestinal tissue	Shannon et al., 2012
	Increased permeability of alpha-synuclein in intestinal tissue	Forsyth et al., 2011
DLB	Reduced alpha-synuclein levels in CSF	Hall et al., 2012
MSA	Reduced alpha-synuclein levels in CSF	Hall et al., 2012
AD	Elevated alpha-synuclein levels in CSF	Korff et al., 2013
Gaucher's Disease	Elevated levels of alpha-synuclein dimers	Argyriou et al., 2012
MS	Elevated levels of alpha-synuclein in inflammatory cells	Lu et al., 2009

Abbreviations: PD, Parkinson's disease; CSF, cerebrospinal fluid; DLB, Dementia with Lewy Body disease; MSA, multiple system atrophy; AD, Alzheimer's disease; MS, multiple sclerosis are derived from brain and spinal cord neurons (Mollenhauer et al., 2012). Many of these studies, with a few exceptions, appear to show a reduction of total alpha-synuclein levels in the CSF of PD patients compared with controls. However, findings from more recent and further comprehensive studies suggest that the use of alpha-synuclein as an individual CSF biomarker may not be sufficient for PD diagnosis due to a low sensitivity and specificity (Mollenhauer et al., 2011). In addition, the level of alpha-synuclein in the CSF does not seem to correlate with the disease progression of PD (van Dijk et al., 2013). On the other hand, analysis of alpha-synuclein, together with other aggregate-prone proteins, such as tau, in the CSF may be helpful in distinguishing PD from other neurodegenerative diseases (Mollenhauer et al., 2011).

Furthermore, examining the presence of various alpha-synuclein species, such as phosphorylated or oligomeric forms, in the CSF may prove to be more suitable biomarkers for PD diagnosis. In fact, compared to total alpha-synuclein levels, oligomeric alpha-synuclein was found to be elevated in the CSF of PD patients (Tokuda et al., 2010). If the toxicity of alpha-synuclein oligomers results in PD-related neuronal death, then the presence of these oligomers is likely a useful marker for identifying at-risk persons before clinical prognosis, as well as for monitoring disease progression in PD patients (Kalia et al., 2013). Additionally, the ratio of alpha-synuclein oligomers to total alpha-synuclein content in the CSF may be considered a more efficient biomarker for PD diagnosis (Tokuda et al., 2010). Alternatively, posttranslational modification of alpha-synuclein may also be a useful target for PD diagnosis (Foulds et al., 2012). Based on a recent study, assessing the ratio between 129-phosphorylated and total alpha-synuclein contents in the CSF improved differential diagnosis of PD from other synucleinopathies (Wang et al., 2012).

In addition to the CSF studies, the presence of alpha-synuclein in other bodily fluids has also been explored as potential biomarkers for PD diagnosis. One of the routes is through the use of blood plasma, in which one study identified a decrease in total alpha-synuclein levels in PD patients (El-Agnaf et al., 2006). In a recent study, plasma alpha-synuclein levels were found to be decreased in male PD patients, but this reduction was not observed in female PD patients (Caranci et al., 2013). On the other hand, another study suggests that the level of phospho-alpha-synuclein is elevated in the plasma of PD patients, whereas plasma contents of other alpha-synuclein species remain unchanged (Foulds et al., 2011). Taken together, alterations of alpha-synuclein levels in blood plasma may be a useful biomarker for PD diagnosis, but more studies with larger samples sizes are needed to further support this hypothesis.

Another alternative approach is the use of salvia to measure changes in alpha-synuclein levels for PD diagnosis. In a postmortem study, alpha-synuclein immunohistochemistry revealed Lewy pathology in the human submandibular gland, a salivary gland, from PD patients, suggesting there may be salivary content alterations associated with PD (Del Tredici, 2010). More recently, reduced levels of alpha-synuclein were detected in the saliva of a PD patient (Devic et al., 2011). However, this is just a preliminary study, and further investigations are needed to examine whether alterations of alpha-synuclein levels in the saliva may potentially be used as a biomarker for PD diagnosis.

One last potential route of alpha-synuclein detection is through the use of gut or intestinal tissues. Recent studies have identified the presence of alpha-synuclein and Lewy bodies within the gut of patients with PD. Interestingly, the observation was based on biopsies obtained several years prior to the onset of PD symptoms, suggesting that the intestine might be a site of early detection for PD prognosis (Shannon et al., 2012). Furthermore, tests on intestinal permeability in PD patients have determined a correlation between increased intestinal permeability in PD and the presence of intestinal alpha-synuclein (Forsyth et al., 2011). Therefore, the detection of alpha-synuclein in intestinal tissue may be a promising biomarker, particularly for early detection and progression of the disease.

Dementia with Lewy Bodies

Dementia with Lewy Bodies (DLB) is a type of dementia closely associated with both Parkinson's (PD) and Alzheimer's disease (AD) and is characterized by the presence of Lewy bodies, or aggregates of alpha-synuclein and ubiquitin proteins in neurons (Kasuga et al., 2012). Due to its close association with both PD and AD, DLB exhibits features of both diseases, including loss of dopamine-producing and acetylcholine-producing neurons. Therefore, measures have been taken to better characterize and differentiate this disease from other related diseases. One such study investigates the utility of different biomarkers within the CSF (Hall et al., 2012). They determined that alpha-synuclein levels in the CSF were reduced in PD, DLB, and other synucleinopathies but not in AD patients. In addition, this study suggests that ascertainment of alpha-synuclein levels along with established AD biomarkers, such as tau, improves the differential diagnosis of AD vs. DLB and PD (Hall et al., 2012). However, they did not find a way to discriminate PD from DLB, and further investigation of novel biomarkers is needed to differentially diagnose these diseases.

Multiple System Atrophy

Multiple System Atrophy (MSA) is an atypical Parkinsonian disorder characterized by the degeneration of nerve cells in specific areas of the brain as well as the presence of abundant glial cytoplasmic inclusions that are mainly composed of alpha-synuclein (Wenning et al., 2004). Despite recent efforts, alpha-synuclein has not been proven to be a specific biomarker for MSA diagnosis. In the CSF, there is a decrease in alpha-synuclein levels in MSA, but this reduction is also seen in other synucleinopathies. Interestingly, a recent study identified an increase in CSF levels of neurofilament light chain specific to atypical Parkinsonian disorders (Hall et al., 2012). Thus, combined analysis of alpha-synuclein and neurofilament light chain may provide a way to discriminate atypical from typical Parkinsonian disorders. The challenge, however, lies in finding suitable biomarkers to differentially diagnose MSA and other atypical Parkinsonian disorders.

Other Neurodegenerative Diseases

Alpha-synuclein may also be used as a potential biomarker in other neurodegenerative diseases such as AD, inflammatory diseases including Multiple Sclerosis (MS), or lysosomal storage diseases such as Gaucher's disease. Some studies have shown higher levels of alpha-synuclein proteins in the CSF of AD patients (Korff et al., 2013), but more comprehensive investigations are needed to determine whether CSF alpha-synuclein could be used as a biomarker for AD diagnosis or progression. Regarding Gaucher's disease and MS, the utility of alpha-synuclein as a biomarker has yet to be investigated. However, recent evidence suggests that alpha-synuclein may play a significant role in the pathology of both diseases. Gaucher's disease is a lysosomal storage disease characterized by fatty lipid accumulation in cells. The acid β-glucosidase mutation that causes Gauchers disease is associated with a five-fold increased risk of developing PD (Grabowski, 2008), a connection that may be explained by the observation that alpha-synuclein dimerization is increased in Gaucher's disease patients (Argyriou et al., 2012). Multiple Sclerosis (MS) is an inflammatory disease that attacks the myelin of neurons in the nervous system, leading to widespread atrophy of the brain and spinal cord. Based on one study, alpha-synuclein immunoreactivity is specifically increased in the inflammatory active cells, including neurons, microglia and oligodendrocytes (Lu et al., 2009). Thus, changes in the level and/or oligomerization of alpha-synuclein proteins may provide potential biomarkers for Gaucher's disease and MS.

14-3-3 Proteins

The 14-3-3 proteins are a family of homologous proteins comprised of seven mammalian isoforms (β, γ, ϵ, η, ζ, σ, and τ/θ) (Rosenquist et al., 2000). They exist as homo- and heterodimers with each 14-3-3 monomer consisting of nine α-helices organized in an antiparallel array (Liu et al., 1995). 14-3-3 proteins function by binding to their target proteins that contain specific pSer/pThr motifs (Muslin et al., 1996). To date, the 14-3-3 family of proteins has been shown to interact with over 200 proteins, and are involved in the regulation of a wide range of biological processes.

While being ubiquitously expressed, 14-3-3 proteins are highly concentrated in the brain, making up about 1% of the total soluble proteins (Berg et al., 2003a). In the nerve cells, 14-3-3 proteins are present in the cytoplasmic compartment, intracellular organelles and plasma membrane (Martin et al., 1994). Certain 14-3-3 isoforms are particularly enriched at the synapse and implicated in the regulation of synaptic transmission and plasticity (Berg et al., 2003a).

While their precise role in the nervous system is not fully understood, 14-3-3 proteins have been linked to a number of neurodegenerative diseases (Foote and Zhou, 2012). Firstly, the presence of 14-3-3 proteins in the CSF is thought to be associated with neuronal death and used as a biomarker for multiple neurodegenerative diseases (Table 18.2). Secondly, 14-3-3 proteins are often detected as a component of inclusion bodies in certain neurodegenerative disorders. Lastly, mutations of different 14-3-3 isoforms have been identified in certain populations of patients with neurodegenerative diseases. Therefore, 14-3-3 proteins have been investigated as a potential biomarker for various neurological disorders.

14-3-3 as a Biochemical Biomarker

Prion and Prion-like Neurological Disorders

Creutzfeldt-Jakob Disease

Creutzfeldt-Jakob Disease (CJD) is a neurodegenerative disease that results from the accumulation of prion proteins in the brain, and belongs to a family of human transmissible spongiform encephalopathies or prion diseases (Spero and Lazibat, 2010). Affecting about one person per million worldwide, CJD occurs in inherited, acquired or infected, and sporadic or unknown cause forms (Collinge et al., 1996). The World Health Organization (WHO) has defined the diagnostic criteria for CJD as a combination of EEG recordings, spongiform changes in brain biopsy, and positive detection of

Table 18.2. 14-3-3 as a biomarker of neurodegenerative diseases.

Disease	Biomarker Evidence	Reference
Prion diseases		
CJD	Elevated 14-3-3 levels in CSF	Hsich et al., 1996
GSS	14-3-3 colocalizes to PrP amyloid plagues	Di Fede et al., 2007
	Elevated 14-3-3 levels in CSF	Ladogana et al., 2009
HIV-associated dementia	Elevated 14-3-3 levels in CSF	Miller et al., 2000
Influenza-associated encephalopathy	Elevated 14-3-3 levels in CSF	Fujii et al., 2006
Herpes simplex encephalitis	Elevated 14-3-3 levels in CSF	Hsich et al., 1996
VGKCE	Elevated 14-3-3 levels in CSF	Geschwind et al., 2008
Inflammatory diseases		
MS	Elevated 14-3-3 levels in CSF	Satoh et al., 1999
	Upregulated 14-3-3 expression in astrocytes and oligodendrocytes of demyelinating brain lesions	Satoh et al., 2004; Kawamoto et al., 2004a
GBS	Elevated 14-3-3 levels in CSF	Satoh et al., 1999
Transverse Myelitis	Elevated 14-3-3 levels in CSF	Irani and Kerr, 2000
Meningitis	Elevated 14-3-3 levels in CSF	Bonora et al., 2003
Other Diseases		
AD	14-3-3 colocalizes to the NFTs	Lee, 1995
	Elevated 14-3-3 levels in CSF	Tschampa et al., 2001
Cerebellar Ataxia	Elevated 14-3-3 levels in CSF	Collins et al., 2000
SCA1	14-3-3 colocalizes to the intranuclear inclusion-like structures	Umahara et al., 2007
	14-3-3 interacts with ataxin1, promoting its neurotoxicity	Chen et al., 2003
PD	14-3-3 colocalizes to Lewy bodies in PD brains	Berg et al., 2003b
	14-3-3 interacts with PD-associated proteins (parkin, alpha-synuclein, LRRK2)	Nichols et al., 2010; Sato et al., 2006; Shirakashi et al., 2006
DLB	14-3-3 colocalizes to Lewy bodies in DLB brains	Kawamoto et al., 2002
	Elevated 14-3-3 levels in CSF	Collins et al., 2000
ALS	14-3-3 interacts with neurofilament mRNA	Ge et al., 2007
	14-3-3 colocalizes to LBHI in ALS brains	Kawamoto et al., 2004b
	Elevated 14-3-3 mRNA in lumbar spinal cord	Malaspina et al., 2000

Abbreviations: CJD, Creutzfeldt-Jakob disease; GSS, Gertsmann-Stäussler-Scheinker disease; VGKCE, Voltage-gated potassium channel encephalopathy; MS, Multiple Sclerosis; GBS, Guillian-Barré syndrome; AD, Alzheimer's disease; SC1, Spinocerebellar ataxia type 1; PD, Parkinson's disease; DLB, Dementia with Lewy Body disease; ALS, Amyotrophic Lateral Sclerosis; CSF, cerebrospinal fluid; NFT, neurofibrillary tangles; LBHI, Lewy body-like hyaline inclusions

CJD biomarkers in the CSF (Weber et al., 1997). Due to its rapid disease progression, there is a great need for practical and reliable premortem tests for CJD. Early evidence for 14-3-3 as a biomarker of CJD came from a study of 71 CJD patients, in which 14-3-3 was detected in the CSF of about 96% of the patients (Hsich et al., 1996). This observation of elevated 14-3-3 proteins in the CSF of patients with CJD was subsequently confirmed by several other groups (Collins et al., 2000; Lemstra et al., 2000; Shiga et al., 2004; Van Everbroeck et al., 2002). In line with these findings, the Quality Standards Subcommittee of the American Academy of Neurology revised their CJD diagnosis parameters, to include the presence of 14-3-3 in the CSF as a useful tool for confirming or rejecting the diagnosis of CJD (Knopman et al., 2001). To date, elevated levels of 14-3-3 proteins in the CSF have been documented within several CJD populations, including Japanese, German, Portuguese, Chinese and Canadian groups (Baldeiras et al., 2009; Coulthart et al., 2011; Hsich et al., 1996; Tschampa et al., 2005; Wang et al., 2010).

CJD occurs as different subtypes based on the underlying cause of the disease, therefore detection of 14-3-3 in the CSF may vary depending on the particular CJD subtypes. In one study, the detection of 14-3-3 in the CSF had higher sensitivity in patients with the molecular features of the classic sporadic form of CJD (sCJD) compared to other non-classical CJD subtypes (Castellani et al., 2004). Another study confirmed that the CSF 14-3-3 levels were elevated in eight patients diagnosed with MM2-type sCJD (Hamaguchi et al., 2005).Consistently, a more recent study showed that the presence of 14-3-3 in the CSF gave the best test specificity (96%) and sensitivity (84%) for sCJD (Bahl et al., 2009). On the other hand, in a study of a population of familial CJD with V180I mutation, CSF testing revealed the absence of 14-3-3 (Mutsukura et al., 2009). In another study, the CSF was tested in 1,859 patients with sporadic, genetic, iatrogenic, and variant CJD cases, in which the presence of 14-3-3 in the CSF was highest in sporadic cases of CJD compared with other CJD subtypes (Sanchez-Juan et al., 2006). However, a different study comparing CSF content between genetic (E200K mutation) and sporadic CJD groups found consistent levels of 14-3-3 (Meiner et al., 2011). Collectively, 14-3-3 appears to be a useful general diagnostic tool for the various subtypes of CJD, with a particularly higher diagnostic accuracy for sporadic cases of CJD.

Another potential use of 14-3-3 in diagnosis of CJD is to examine the presence of different 14-3-3 isoforms in the CSF. Earlier studies identified the presence of 14-3-3γ, -β, -ε, -η and -θ isoforms in the CSF of CJD patients (Lee and Harrington, 1997; Wiltfang et al., 1999). Subsequent studies have confirmed the presence of 14-3-3γ and -β in the CSF of CJD patients (Green et al., 2002; Matsui et al., 2011; Yutzy et al., 2007). However, there have been conflicting reports regarding elevated levels of the ε, η and θ isoforms (Di Fede et al., 2007; Yutzy et al., 2007). Evidence from these studies suggest that certain 14-3-3 isoforms may be more accurate diagnostic tools for CJD, such as 14-3-3γ and -β.

Additionally, the presence of 14-3-3 in the CSF of patients with CJD may be a marker of disease onset and progression. In fact, multiple studies have identified a positive correlation between 14-3-3 detection and early stage of CJD (Geschwind et al., 2003; Van Everbroeck et al., 2003). A more recent study found that CSF biomarkers for sCJD, including 14-3-3, are highly sensitive in the early stages within 6 weeks of onset (Pennington et al., 2009). Consistently, 14-3-3 was detected around the onset of neurological symptoms in an animal model of variant CJD (Yutzy et al., 2007). On the other hand, lower 14-3-3 levels were detected in patients with a longer disease duration, and 14-3-3 concentrations were lowest at either disease onset or the end stage of the disease (Van Everbroeck et al., 2003). Similarly, a later study showed that only 76% of patients tested positive for CSF 14-3-3 in sporadic CJD (MV2) cases that have a long disease duration (Krasnianski et al., 2006). In a Portuguese CJD population, younger patients with long disease duration were found to have decreased 14-3-3 levels in the CSF (Baldeiras et al., 2009). However, in another study of CJD progression, 14-3-3 was detected in all five cases of sporadic CJD, regardless of illness duration (Appleby et al., 2009). Taken together, the amount of 14-3-3 in the CSF of patients with CJD likely correlates to the pathological process and accompanying neuronal damage (Higuma et al., 2013).

Gerstmann-Stäussler-Scheinker Disease

Gerstmann-Stäussler-Scheinker disease (GSS) is a rare autosomal dominant, neurodegenerative disease associated with point mutations of the prion protein gene (Di Fede et al., 2007). In a case study for late-onset GSS, examination of the CSF content of the patient revealed elevated levels of 14-3-3 proteins, providing early evidence for 14-3-3 as a potential GSS biomarker (Takase et al., 2001). In another study, 14-3-3ε was detected in the brain tissue of patients with GSS, specifically within PrP amyloid plaques (Di Fede et al., 2007). However, a later study only detected the presence of 14-3-3 in the CSF of 10% of patients (Ladogana et al., 2009). Similarly, only 11% sensitivity for CSF 14-3-3 was determined by a more recent report (Sano et al., 2013). In addition, 14-3-3 was not detected in the CSF of a GSS patient who bears a rare D202N prion protein mutation and was diagnosed with atypical Parkinsonism (Plate et al., 2013). Based on these results, it is unlikely that the CSF 14-3-3 can be used as a general biomarker for the diagnosis of GSS.

HIV-Associated Dementia

HIV-associated dementia is a neurocognitive disorder considered to be a form of metabolic encephalopathy (Wakabayashi et al., 2001). Early evidence for 14-3-3's potential biomarker role in HIV-associated dementia stems from a case study in which 14-3-3 was detected in the CSF of three out of the 17 patients (Miller et al., 2000). A subsequent study detected the presence of 14-3-3ε, -γ, and -ζ isoforms in the CSF of HIV patients with AIDS dementia complex, or cytomegalovirus encephalitis (Wakabayashi et al., 2001). Interestingly, 14-3-3 was not present in HIV patients that did not have neurological symptoms. In a study using the SIV/macaque model of HIV CNS disease, CSF levels of 14-3-3 were found to be increased in animals starting from infection and continuing until death (Helke et al., 2005). Based on these studies, HIV-associated dementia may result in elevated 14-3-3 protein levels in the CSF. However, more studies with larger sample sizes are needed to validate positive detection of 14-3-3 in the CSF as a true biomarker of this neurocognitive disorder.

Viral Encephalitis

Influenza-associated encephalopathy is a type of viral encephalitis. A study of patients with influenza-associated encephalopathy assessed CSF content and all four patients tested positive for 14-3-3 proteins (Fujii et al., 2006). In a more recent study, six different 14-3-3 isoforms were detected in the CSF of all patients (n = 3) with influenza-associated encephalitis (Fujii et al., 2013). Results from these studies suggest that the presence of 14-3-3 in the CSF may be used as a rapid diagnostic marker of influenza-associated encephalopathy.

Cytomegalovirus, or herpes simplex, encephalitis is another example of viral encephalitis. In a study, almost 92% of patients with herpes simplex encephalitis tested positive for 14-3-3 in their CSF (Hsich et al., 1996). However, a false-positive for 14-3-3 in the CSF was reported in another case study of herpes simplex encephalitis (Zerr et al., 1998). In subsequent studies, 14-3-3η was identified in the CSF of a patient with both herpes simplex encephalitis and dementia (Wiltfang et al., 1999), but no CSF 14-3-3 was detected in patients with cytomegalovirus encephalitis (Wakabayashi et al., 2001). Based on these conflicting reports, 14-3-3's presence in the CSF is probably not a suitable biomarker for herpes simplex encephalitis.

Limbic Encephalitis

Limbic encephalitis is a type of encephalitis caused by auto-immunity which varies with each patient, making accurate differential diagnosis difficult

(Matsui et al., 2011). Voltage-gated potassium channel encephalopathy (VGKCE) is a form of limbic encephalitis (Geschwind et al., 2008). In a case study, patients who were initially thought to have CJD but actually had VGKCE were examined for their CSF content (Geschwind et al., 2008). Analysis revealed that five of the eight patients had increased CSF 14-3-3 protein levels. In another study, 14-3-3γ was detected in the CSF of a patient with limbic encephalitis (Matsui et al., 2011). Limbic encephalitis has similar clinical presentation and CSF biomarkers as that of other neurological diseases, including CJD and other encephalopathies. Therefore, the presence of CSF 14-3-3 protein may provide a clue to the patient's diagnosis, but may not be a useful differential diagnostic tool for limbic encephalitis.

Inflammatory Neurological Diseases

Multiple Sclerosis

Multiple Sclerosis (MS) is a chronic inflammatory demyelinating disease, most common among young adults with an onset around age 30 (Teunissen et al., 2005). This neurological disease typically begins with an Initial Demyelination Event (IDE), followed by a Clinically Isolated Syndrome (CIS), and then progression to MS. Biomarkers are used to help predict and monitor the neurological decline in people with MS. The first evidence for 14-3-3 proteins as a potential biomarker of MS came from a study which identified the presence of 14-3-3 in the CSF of a group of patients with neurodegenerative disorders, including MS (Satoh et al., 1999). This was confirmed in a subsequent case study in which 13% of patients with MS (n = 38) tested positive for 14-3-3 in their CSF (Martinez-Yelamos et al., 2001). Similarly, another study found 14-3-3 in the CSF of about 22% of MS patients, but not in the control patients (Bartosik-Psujek and Archelos, 2004). In addition, 14-3-3 has been identified in the CSF of MS patients with early and late stages of disease progression (Fiorini et al., 2007; Gajofatto et al., 2010; Menon et al., 2011). Taken together, these studies support the presence of 14-3-3 in the CSF as a potential biomarker for MS.

As MS is a degenerative disease with multiple stages of disease progression, the presence of 14-3-3 in the CSF may be altered based on the disease duration and prognosis. Lower CSF levels of 14-3-3 have been observed in early stages of MS patients, while later stages of MS have been associated with increased levels of 14-3-3 in the CSF (Frederiksen et al., 2012; Satoh et al., 2003). Consistently, one study following CIS patients throughout disease progression found that the presence of 14-3-3 in the CSF was suggestive of a prognostic marker for a more rapid conversion to clinically diagnosed MS (Martinez-Yelamos et al., 2004). In a more recent study, CSF biomarkers were assessed in 46 CIS cases, in which CSF samples were obtained at different time points (Gajofatto et al., 2013). Interestingly, during the acute phase, about 71% of CSF from patients tested positive for 14-3-3, compared to about 44% of patients who only tested positive for 14-3-3 during remission. However, this particular study did not identify a correlation between the level of 14-3-3 in the CSF and the different disease stages of MS. Nevertheless, 14-3-3 in the CSF is a promising marker for MS disease progression.

The presence of 14-3-3 in the CSF of MS patients may suggest a role for 14-3-3 proteins in the pathogenesis of this neuroinflammatory disease. In a postmortem study, 14-3-3 expression was found to be upregulated in astrocytes and oligodendrocytes within demyelinating lesions of brain tissue from MS patients (Kawamoto et al., 2004a). This observation was confirmed in a subsequent immunohistochemical study utilizing a panel of 14-3-3 isoform-specific antibodies (Satoh et al., 2004). Several 14-3-3 isoforms, including β, ε, ζ, η and σ, were found to be highly expressed in the reactive astrocytes of chronic demyelinating lesions in brain tissue from MS patients. In addition, they demonstrated that 14-3-3 interacts with vimentin and glial fibrillary acidic protein in cultured human astrocytes, suggesting that 14-3-3 may play a role in the intermediate filament network of reactive astrocytes. Therefore, the presence of 14-3-3 in the CSF of MS patients may be a result of axonal damage and/or inflammation in the brain.

Guillain-Barré Syndrome

Guillain-Barré Syndrome (GBS) is an acute inflammatory polyneuropathy with various subtypes based on the underlying cause of demyelination (Brettschneider et al., 2009). This disruption in the myelin results in proteins being leaked into the CSF, which may be used as a diagnostic tool. Initial work analyzing the CSF content of patients with GBS did not identify the presence of 14-3-3 proteins in the CSF (Satoh et al., 1999). However, in a more recent study, approximately 76% of patients diagnosed with GBS tested positive for 14-3-3 proteins in the CSF (Bersano et al., 2006). Additionally, another study reported that 14-3-3 was upregulated in the CSF of GBS patients compared with controls (Fiorini et al., 2007). Thus, evidence from more recent studies suggests that 14-3-3 detection in the CSF may be a potential biomarker for the diagnosis of GBS.

Transverse Myelitis

Transverse myelitis is a heterogeneous inflammatory disorder of the spinal cord (Pittock and Lucchinetti, 2006). An early study identified the presence of 14-3-3 in the CSF of patients diagnosed with acute transverse myelitis

(Irani and Kerr, 2000). This finding was confirmed in multiple case studies, in which elevated levels of 14-3-3 were detected in the CSF of idiopathic acute transverse myelitis patients (Finsterer and Voigtlander, 2002; Gajofatto et al., 2013). In another study, 14-3-3 levels were found to be elevated in the CSF of 10% of patients diagnosed with both transverse myelitis and MS disorders, but was not detected in patients diagnosed with transverse myelitis alone (de Seze et al., 2002). Based on these conflicting reports, more extensive studies are needed to further investigate the potential use of 14-3-3 as a biomarker for transverse myelitis.

Meningitis

Meningitis is the inflammation of the membranes surrounding the brain and spinal cord, caused by viruses, bacteria or other microorganisms (Vandemark, 2013). In a study of 12 patients with purulent bacterial meningitis, 14-3-3 was present in the CSF of all patients upon disease onset (Bonora et al., 2003). Similar results were obtained in a more recent study conducted in 29 patients with acute bacterial meningitis (Lu et al., 2008). Interestingly, in both studies, all patients who recovered eventually cleared 14-3-3 from their CSF, whereas 14-3-3 was still present in the patients who died (Bonora et al., 2003; Lu et al., 2008). In another study, 14-3-3γ levels were found to be elevated early in disease progression of patients diagnosed with HIV-negative cryptococcal meningitis, a type of fungal meningitis that has a longer disease duration than bacterial meningitis (Chang et al., 2008). Taken together, elevated levels of 14-3-3 in the CSF may provide a useful biomarker for meningitis diagnosis.

Other Neurodegenerative Diseases

Alzheimer's Disease

Alzheimer's Disease (AD) is a neurodegenerative disease with progressive dementia characterized by amyloid plaques and neurofibrillary tangles (NFTs) (Lee, 1995). 14-3-3 proteins were first implicated in AD pathogenesis based on the finding that they colocalize with the NFTs (Layfield et al., 1996). Subsequently, the presence of 14-3-3 proteins in the CSF has been investigated by several groups to determine their potential as a diagnostic biomarker of AD. Originally, only baseline 14-3-3η levels were detected in AD patients with dementia compared with healthy controls (Wiltfang et al., 1999). However, a following case study positively identified 14-3-3 proteins in the CSF from two out of nineteen AD patients with rapid progressive dementia (Tschampa et al., 2001). Similarly, 14-3-3 was detected in the

CSF in two out of forty-five patients suspected of CJD, who were later diagnosed with AD (Van Everbroeck et al., 2004). More recently, another group characterized CSF biomarkers of patients with rapidly progressing Alzheimer's dementia and detected 14-3-3 in the CSF of 31% of cases (Schmidt et al., 2010). Additionally, 14-3-3γ was found to be increased in the CSF of patients present with Alzheimer's related dementia (Matsui et al., 2011). Conversely, a study examined the CSF content in a group of 39 patients with probable AD diagnosis and all tested negative for 14-3-3 in their CSF (Bahl et al., 2009). Taken together, these studies suggest that CSF 14-3-3 is unlikely a general biomarker for AD diagnosis, and future studies may be aimed to investigate its potential use for AD subtype differentiation.

Cerebellar Disorders

Cerebellar ataxia is a form of ataxia resulting from lesions to the cerebellum and is associated with other neurological disorders, including CJD (Laverse et al., 2009). The first evidence indicating 14-3-3 proteins as a potential biomarker of cerebellar ataxia came from a study in which 14-3-3 was detected in the CSF of a patient with slowly progressive cerebellar ataxia (Collins et al., 2000). Subsequent case reports confirmed the presence of elevated 14-3-3 proteins in the CSF from patients with either bilateral middle cerebellar peduncle lesions or cerebellar ataxia with CJD-like symptoms (Chang et al., 2007; Nishida et al., 2003). In more recent case studies, 14-3-3 was detected in the CSF of both an elderly patient with rapidly progressive cerebellar syndrome and children with cerebellar disorders (Fujii et al., 2013; Panegyres and Graves, 2012). Particularly, findings from certain studies indicated a potential use for the 14-3-3ε isoform as a biomarker for cerebellar diseases (Fujii et al., 2013). Therefore, the presence of 14-3-3 in the CSF of cerebellar ataxia patients may be a helpful clue for diagnosis of this disease. However, as elevated CSF 14-3-3 is observed in both CJD and cerebellar ataxia, this alone may not be sufficient for differential diagnosis of these two neurodegenerative diseases.

In addition to being a potential biomarker for cerebellar ataxia, 14-3-3 proteins also play a role in Spinocerebellar ataxia pathogenesis. Spinocerebellar ataxia type 1 (SCA1) is a progressive, degenerative neurological disorder caused by a CAG repeat expansion resulting in abnormally long polyglutamate tract in the ataxin-1 protein that accumulates and leads to neurodegeneration (Umahara and Uchihara, 2010). 14-3-3 binds to and stabilizes the mutant ataxin-1 protein, thus promoting its neurotoxicity (Chen et al., 2003). In a postmortem study, immunolocalization of 14-3-3 proteins were analyzed in samples from SCA1 brains (Umahara et al., 2007). From this study, 14-3-3 proteins were found to colocalize to the intranuclear inclusion-like structures with isoform-specific expression patterns in different regions of the brain. In a mouse model of SCA1, Sca1 (154Q/+) mice were bred with 14-3-3ε haploinsufficient mice which have reduced 14-3-3ε expression. These Sca1/14-3-3ε$^{+/-}$ mice were shown to have decreased incorporation of the expanded ataxin-1 into its large toxic complexes (Jafar-Nejad et al., 2011). Taken together, these studies suggest that this ataxin1/14-3-3 interaction may be a potential therapeutic target for disease intervention in cerebellar ataxia (Umahara and Uchihara, 2010).

Parkinson's Disease

Parkinson's Disease (PD) is a motor system disorder resulting from the loss of dopaminergic neurons and associated with the presence of Lewy bodies (Cookson and Bandmann, 2010). 14-3-3's involvement in PD stemmed from postmortem studies which identified the colocalization of 14-3-3 proteins to the Lewy bodies in PD brain tissue (Berg et al., 2003b; Kawamoto et al., 2002; Ubl et al., 2002). Additionally, 14-3-3 has been shown to interact with and regulate several PD-associated proteins including parkin, α-synuclein, and LRRK2 (Nichols et al., 2010; Sato et al., 2006; Shirakashi et al., 2006). In a study using a mouse model of PD, transgenic mice overexpressing human A53T-alpha-synuclein, proteomic analysis of striatal tissue revealed an upregulation of 14-3-3 isoforms-ε and -η, but a downregulation of 14-3-3γ. Here, this increased expression of 14-3-3ε was also correlated with increased mRNA levels (Kurz et al., 2012). On the other hand, from a recent study, elevated 14-3-3 levels in the CSF have not been observed in PD patients (Plate et al., 2013), suggesting that 14-3-3 may not be a suitable biomarker for PD diagnosis.

Dementia with Lewy Body Disease

Dementia with Lewy Body disease (DLB) is a type of dementia characterized by the presence of Lewy bodies in the brain (Collins et al., 2000). Similar to what was observed in PD, 14-3-3 proteins colocalize with the Lewy bodies in the brains of DLB patients (Kawamoto et al., 2002). In an earlier report, 14-3-3 was detected in the CSF of a patient with DLB (Collins et al., 2000). Subsequent studies have confirmed the presence of elevated 14-3-3 levels in the CSF of a small number of DLB cases (Momjian-Mayor et al., 2006; Van Everbroeck et al., 2004). In contrast, a recent study analyzed the CSF content of DLB patients with variable progressive forms, but 14-3-3 was not detected in the CSF of any DLB patients (Gaig et al., 2011). Collectively, while 14-3-3 proteins may play a role in Lewy body formation and DLB pathogenesis, they do not appear to be a biochemical biomarker of DLB.

Amyotrophic Lateral Sclerosis

Amyotrophic Lateral Sclerosis (ALS) is a heterogeneous neurodegenerative disease classically associated with specific motor neuron degeneration and, more recently, recognized for its degeneration of the frontal and temporal lobes (Strong, 2010). 14-3-3 has been implicated in the pathogenesis of ALS based on two lines of evidence. Firstly, 14-3-3 was identified as a potential binding partner of neurofilament mRNA and regulates its stabilization (Ge et al., 2007; Volkening et al., 2009). Secondly, 14-3-3 was found to colocalize to the Lewy Body-like Hyaline Inclusions (LBHI) present in certain sporadic ALS cases (Kawamoto et al., 2005; Kawamoto et al., 2004b). In addition, 14-3-3θ (YWHAQ) mRNA was found to be elevated in the lumbar spinal cord of patients with sporadic ALS, but no comprehensive study has been carried out to investigate the potential use of 14-3-3 as a biomarker of ALS diagnosis (Malaspina et al., 2000).

Conclusion

Alpha-synuclein is involved in a variety of neuronal functions including neuronal development and synaptic vesicle regulation. In addition, alpha-synuclein's role in protein aggregation has been implicated in a number of neurodegenerative diseases, or synucleinopathies. Therefore, further understanding of the neuropathological alterations that lead to alpha-synuclein-mediated aggregation may provide more potential biomarkers for the differentiation and diagnosis of diseases such as Parkinson's disease.

The family of 14-3-3 proteins share similar structure and neuronal functions to alpha-synuclein. 14-3-3 proteins have been implicated in the pathogenesis of a variety of neurological disorders based on their role in neuronal development, synaptic transmission, aggresome formation, and presence in bodily fluids. Alone, the presence of 14-3-3 proteins in the CSF may not serve as a disease biomarker, but rather a general marker of neuronal degeneration. Future studies to investigate the mechanism underlying altered 14-3-3 expression in neurodegenerative diseases may lead to the identification of more reliable biomarkers for disease diagnosis and differentiation.

References

Al-Wandi, A., N. Ninkina, S. Millership, S.J. Williamson, P.A. Jones and V.L. Buchman. 2010. Absence of alpha-synuclein affects dopamine metabolism and synaptic markers in the striatum of aging mice. *Neurobiol Aging.* 31: 796–804.

Appleby, B.S., K.K. Appleby, B.J. Crain, C.U. Onyike, M.T. Wallin and P.V. Rabins. 2009. Characteristics of established and proposed sporadic Creutzfeldt-Jakob disease variants. *Arch Neurol.* 66: 208–15.

Argyriou, A., G. Dermentzaki, T. Papasilekas, M. Moraitou, E. Stamboulis, K. Vekrellis, H. Michelakakis and L. Stefanis. 2012. Increased dimerization of alpha-synuclein in erythrocytes in Gaucher disease and aging. *Neurosci Lett*. 528: 205–9.

Bahl, J.M., N.H. Heegaard, G. Falkenhorst, H. Laursen, H. Hogenhaven, K. Molbak, C. Jespersgaard, L. Hougs, G. Waldemar, P. Johannsen and M. Christiansen. 2009. The diagnostic efficiency of biomarkers in sporadic Creutzfeldt-Jakob disease compared to Alzheimer's disease. *Neurobiol Aging*. 30: 1834–41.

Baldeiras, I.E., M.H. Ribeiro, P. Pacheco, A. Machado, I. Santana, L. Cunha and C.R. Oliveira. 2009. Diagnostic value of CSF protein profile in a Portuguese population of sCJD patients. *J Neurol*. 256: 1540–50.

Barbour, R., K. Kling, J.P. Anderson, K. Banducci, T. Cole, L. Diep, M. Fox, J.M. Goldstein, F. Soriano, P. Seubert and T.J. Chilcote. 2008. Red blood cells are the major source of alpha-synuclein in blood. *Neurodegener Dis*. 5: 55–9.

Bartels, T., J.G. Choi and D.J. Selkoe. 2011. alpha-Synuclein occurs physiologically as a helically folded tetramer that resists aggregation. *Nature*. 477: 107–10.

Bartosik-Psujek, H. and J.J. Archelos. 2004. Tau protein and 14-3-3 are elevated in the cerebrospinal fluid of patients with multiple sclerosis and correlate with intrathecal synthesis of IgG. *J Neurol*. 251: 414–20.

Berg, D., C. Holzmann and O. Riess. 2003a. 14-3-3 proteins in the nervous system. *Nat Rev Neurosci*. 4: 752–62.

Berg, D., O. Riess and A. Bornemann. 2003b. Specification of 14-3-3 proteins in Lewy bodies. *Ann Neurol*. 54: 135.

Bersano, A., M. Fiorini, S. Allaria, G. Zanusso, E. Fasoli, M. Gelati, H. Monaco, G. Squintani, S. Monaco and E. Nobile-Orazio. 2006. Detection of CSF 14-3-3 protein in Guillain-Barre syndrome. *Neurology*. 67: 2211–6.

Bonora, S., G. Zanusso, R. Raiteri, S. Monaco, A. Rossati, S. Ferrari, M. Boffito, S. Audagnotto, A. Sinicco, N. Rizzuto, E. Concia and G. Di Perri. 2003. Clearance of 14-3-3 protein from cerebrospinal fluid heralds the resolution of bacterial meningitis. *Clin Infect Dis*. 36: 1492–5.

Brettschneider, J., A. Petzold, S. Sussmuth and H. Tumani. 2009. Cerebrospinal fluid biomarkers in Guillain-Barre syndrome—where do we stand? *J Neurol*. 256: 3–12.

Burre, J., M. Sharma, T. Tsetsenis, V. Buchman, M.R. Etherton and T.C. Sudhof. 2010. Alpha-synuclein promotes SNARE-complex assembly *in vivo* and *in vitro*. *Science*. 329: 1663–7.

Caranci, G., P. Piscopo, R. Rivabene, A. Traficante, B. Riozzi, A.E. Castellano, S. Ruggieri, N. Vanacore and A. Confaloni. 2013. Gender differences in Parkinson's disease: focus on plasma alpha-synuclein. *J Neural Transm*. 120: 1209–15.

Castellani, R.J., M. Colucci, Z. Xie, W. Zou, C. Li, P. Parchi, S. Capellari, M. Pastore, M.H. Rahbar, S.G. Chen and P. Gambetti. 2004. Sensitivity of 14-3-3 protein test varies in subtypes of sporadic Creutzfeldt-Jakob disease. *Neurology*. 63: 436–42.

Chandra, S., G. Gallardo, R. Fernandez-Chacon, O.M. Schluter and T.C. Sudhof. 2005. Alpha-synuclein cooperates with CSPalpha in preventing neurodegeneration. *Cell*. 123: 383–96.

Chang, C.C., S.D. Eggers, J.K. Johnson, A. Haman, B.L. Miller and M.D. Geschwind. 2007. Anti-GAD antibody cerebellar ataxia mimicking Creutzfeldt-Jakob disease. *Clin Neurol Neurosurg*. 109: 54–7.

Chang, W.N., C.H. Lu, C.R. Huang, Y.C. Chuang, N.W. Tsai, S.F. Chen, C.C. Chang and H.C. Wang. 2008. Cerebrospinal fluid 14-3-3-gamma protein level in eight HIV-negative cryptococcal meningitis adults. *Eur J Neurol*. 15: 428–30.

Chen, H.K., P. Fernandez-Funez, S.F. Acevedo, Y.C. Lam, M.D. Kaytor, M.H. Fernandez, A. Aitken, E.M. Skoulakis, H.T. Orr, J. Botas and H.Y. Zoghbi. 2003. Interaction of Akt-phosphorylated ataxin-1 with 14-3-3 mediates neurodegeneration in spinocerebellar ataxia type 1. *Cell*. 113: 457–68.

Collinge, J., K.C. Sidle, J. Meads, J. Ironside and A.F. Hill. 1996. Molecular analysis of prion strain variation and the aetiology of 'new variant' CJD. *Nature*. 383: 685–90.

Collins, S., A. Boyd, A. Fletcher, M. Gonzales, C.A. McLean, K. Byron and C.L. Masters. 2000. Creutzfeldt-Jakob disease: diagnostic utility of 14-3-3 protein immunodetection in cerebrospinal fluid. *J Clin Neurosci*. 7: 203–8.

Conway, K.A., J.D. Harper and P.T. Lansbury. 1998. Accelerated *in vitro* fibril formation by a mutant alpha-synuclein linked to early-onset Parkinson disease. *Nat Med*. 4: 1318–20.

Conway, K.A., S.J. Lee, J.C. Rochet, T.T. Ding, R.E. Williamson and P.T. Lansbury, Jr. 2000. Acceleration of oligomerization, not fibrillization, is a shared property of both alpha-synuclein mutations linked to early-onset Parkinson's disease: implications for pathogenesis and therapy. *Proc Natl Acad Sci USA*. 97: 571–6.

Cookson, M.R. and O. Bandmann. 2010. Parkinson's disease: insights from pathways. *Hum Mol Genet*. 19: R21–7.

Coulthart, M.B., G.H. Jansen, E. Olsen, D.L. Godal, T. Connolly, B.C. Choi, Z. Wang and N.R. Cashman. 2011. Diagnostic accuracy of cerebrospinal fluid protein markers for sporadic Creutzfeldt-Jakob disease in Canada: a 6-year prospective study. *BMC Neurol*. 11: 133.

de Seze, J., K. Peoc'h, D. Ferriby, T. Stojkovic, J.L. Laplanche and P. Vermersch. 2002. 14-3-3 Protein in the cerebrospinal fluid of patients with acute transverse myelitis and multiple sclerosis. *J Neurol*. 249: 626–7.

Devic, I., H. Hwang, J.S. Edgar, K. Izutsu, R. Presland, C. Pan, D.R. Goodlett, Y. Wang, J. Armaly, V. Tumas, C.P. Zabetian, J.B. Leverenz, M. Shi and J. Zhang. 2011. Salivary alpha-synuclein and DJ-1: potential biomarkers for Parkinson's disease. *Brain*. 134: e178.

Di Fede, G., G. Giaccone, L. Limido, M. Mangieri, S. Suardi, G. Puoti, M. Morbin, G. Mazzoleni, B. Ghetti and F. Tagliavini. 2007. The epsilon isoform of 14-3-3 protein is a component of the prion protein amyloid deposits of Gerstmann-Straussler-Scheinker disease. *J Neuropathol Exp Neurol*. 66: 124–30.

El-Agnaf, O.M., S.A. Salem, K.E. Paleologou, M.D. Curran, M.J. Gibson, J.A. Court, M.G. Schlossmacher and D. Allsop. 2006. Detection of oligomeric forms of alpha-synuclein protein in human plasma as a potential biomarker for Parkinson's disease. *FASEB J*. 20: 419–25.

Finsterer, J. and T. Voigtlander. 2002. Elevated 14-3-3 protein and axonal loss in immunoglobulin-responsive, idiopathic acute transverse myelitis. *Clin Neurol Neurosurg*. 105: 18–22.

Fiorini, M., G. Zanusso, M.D. Benedetti, P.G. Righetti and S. Monaco. 2007. Cerebrospinal fluid biomarkers in clinically isolated syndromes and multiple sclerosis. *Proteomics Clin Appl*. 1: 963–71.

Foote, M. and Y. Zhou. 2012. 14-3-3 proteins in neurological disorders. *Int J Biochem Mol Biol*. 3: 152–64.

Forsyth, C.B., K.M. Shannon, J.H. Kordower, R.M. Voigt, M. Shaikh, J.A. Jaglin, J.D. Estes, H.B. Dodiya and A. Keshavarzian. 2011. Increased intestinal permeability correlates with sigmoid mucosa alpha-synuclein staining and endotoxin exposure markers in early Parkinson's disease. *PLoS One*. 6: e28032.

Foulds, P., D.M. Mann and D. Allsop. 2012. Phosphorylated alpha-synuclein as a potential biomarker for Parkinson's disease and related disorders. *Expert Rev Mol Diagn*. 12: 115–7.

Foulds, P.G., J.D. Mitchell, A. Parker, R. Turner, G. Green, P. Diggle, M. Hasegawa, M. Taylor, D. Mann and D. Allsop. 2011. Phosphorylated alpha-synuclein can be detected in blood plasma and is potentially a useful biomarker for Parkinson's disease. *FASEB J*. 25: 4127–37.

Frederiksen, J., K. Kristensen, J.M. Bahl and M. Christiansen. 2012. Tau protein: a possible prognostic factor in optic neuritis and multiple sclerosis. *Mult Scler*. 18: 592–9.

Fujii, K., Y. Tanabe, H. Uchikawa, K. Kobayashi, H. Kubota, J. Takanashi and Y. Kohno. 2006. 14-3-3 protein detection in the cerebrospinal fluid of patients with influenza-associated encephalopathy. *J Child Neurol*. 21: 562–5.

Fujii, K., H. Uchikawa, Y. Tanabe, T. Omata, I. Nonaka and Y. Kohno. 2013. 14-3-3 proteins, particularly of the epsilon isoform, are detectable in cerebrospinal fluids of cerebellar diseases in children. *Brain Dev*. 35: 555–60.

Gaig, C., F. Valldeoriola, E. Gelpi, M. Ezquerra, S. Llufriu, M. Buongiorno, M.J. Rey, M.J. Marti, F. Graus and E. Tolosa. 2011. Rapidly progressive diffuse Lewy body disease. *Mov Disord.* 26: 1316–23.

Gajofatto, A., M. Bongianni, G. Zanusso, M.R. Bianchi, M. Turatti, M.D. Benedetti and S. Monaco. 2013. Clinical and biomarker assessment of demyelinating events suggesting multiple sclerosis. *Acta Neurol Scand.*

Gajofatto, A., S. Monaco, M. Fiorini, G. Zanusso, M. Vedovello, F. Rossi, M. Turatti and M.D. Benedetti. 2010. Assessment of outcome predictors in first-episode acute myelitis: a retrospective study of 53 cases. *Arch Neurol.* 67: 724–30.

Ge, W.W., K. Volkening, C. Leystra-Lantz, H. Jaffe and M.J. Strong. 2007. 14-3-3 protein binds to the low molecular weight neurofilament (NFL) mRNA 3′ UTR. *Mol Cell Neurosci.* 34: 80–7.

George, J.M. 2002. The synucleins. *Genome Biol.* 3: REVIEWS3002.

Geschwind, M.D., J. Martindale, D. Miller, S.J. DeArmond, J. Uyehara-Lock, D. Gaskin, J.H. Kramer, N.M. Barbaro and B.L. Miller. 2003. Challenging the clinical utility of the 14-3-3 protein for the diagnosis of sporadic Creutzfeldt-Jakob disease. *Arch Neurol.* 60: 813–6.

Geschwind, M.D., K.M. Tan, V.A. Lennon, R.F. Barajas, Jr., A. Haman, C.J. Klein, S.A. Josephson and S.J. Pittock. 2008. Voltage-gated potassium channel autoimmunity mimicking creutzfeldt-jakob disease. *Arch Neurol.* 65: 1341–6.

Gosavi, N., H.J. Lee, J.S. Lee, S. Patel and S.J. Lee. 2002. Golgi fragmentation occurs in the cells with prefibrillar alpha-synuclein aggregates and precedes the formation of fibrillar inclusion. *J Biol Chem.* 277: 48984–92.

Grabowski, G.A. 2008. Phenotype, diagnosis, and treatment of Gaucher's disease. *Lancet.* 372: 1263–71.

Green, A.J., S. Ramljak, W.E. Muller, R.S. Knight and H.C. Schroder. 2002. 14-3-3 in the cerebrospinal fluid of patients with variant and sporadic Creutzfeldt-Jakob disease measured using capture assay able to detect low levels of 14-3-3 protein. *Neurosci Lett.* 324: 57–60.

Hall, S., A. Ohrfelt, R. Constantinescu, U. Andreasson, Y. Surova, F. Bostrom, C. Nilsson, W. Hakan, H. Decraemer, K. Nagga, L. Minthon, E. Londos, E. Vanmechelen, B. Holmberg, H. Zetterberg, K. Blennow and O. Hansson. 2012. Accuracy of a panel of 5 cerebrospinal fluid biomarkers in the differential diagnosis of patients with dementia and/or parkinsonian disorders. *Arch Neurol.* 69: 1445–52.

Hamaguchi, T., T. Kitamoto, T. Sato, H. Mizusawa, Y. Nakamura, M. Noguchi, Y. Furukawa, C. Ishida, I. Kuji, K. Mitani, S. Murayama, T. Kohriyama, S. Katayama, M. Yamashita, T. Yamamoto, F. Udaka, A. Kawakami, Y. Ihara, T. Nishinaka, S. Kuroda, N. Suzuki, Y. Shiga, H. Arai, M. Maruyama and M. Yamada. 2005. Clinical diagnosis of MM2-type sporadic Creutzfeldt-Jakob disease. *Neurology.* 64: 643–8.

Helke, K.L., S.E. Queen, P.M. Tarwater, J. Turchan-Cholewo, A. Nath, M.C. Zink, D.N. Irani and J.L. Mankowski. 2005. 14-3-3 protein in CSF: an early predictor of SIV CNS disease. *J Neuropathol Exp Neurol.* 64: 202–8.

Higuma, M., N. Sanjo, K. Satoh, Y. Shiga, K. Sakai, I. Nozaki, T. Hamaguchi, Y. Nakamura, T. Kitamoto, S. Shirabe, S. Murayama, M. Yamada, J. Tateishi and H. Mizusawa. 2013. Relationships between clinicopathological features and cerebrospinal fluid biomarkers in Japanese patients with genetic prion diseases. *PLoS One.* 8: e60003.

Hong, Z., M. Shi, K.A. Chung, J.F. Quinn, E.R. Peskind, D. Galasko, J. Jankovic, C.P. Zabetian, J.B. Leverenz, G. Baird, T.J. Montine, A.M. Hancock, H. Hwang, C. Pan, J. Bradner, U.J. Kang, P.H. Jensen and J. Zhang. 2010. DJ-1 and alpha-synuclein in human cerebrospinal fluid as biomarkers of Parkinson's disease. *Brain.* 133: 713–26.

Hsich, G., K. Kenney, C.J. Gibbs, K.H. Lee and M.G. Harrington. 1996. The 14-3-3 brain protein in cerebrospinal fluid as a marker for transmissible spongiform encephalopathies. *N Engl J Med.* 335: 924–30.

Huang, Z., Z. Xu, Y. Wu and Y. Zhou. 2011. Determining nuclear localization of alpha-synuclein in mouse brains. *Neuroscience.* 199: 318–32.

Irani, D.N. and D.A. Kerr. 2000. 14-3-3 protein in the cerebrospinal fluid of patients with acute transverse myelitis. *Lancet*. 355: 901.

Jafar-Nejad, P., C.S. Ward, R. Richman, H.T. Orr and H.Y. Zoghbi. 2011. Regional rescue of spinocerebellar ataxia type 1 phenotypes by 14-3-3epsilon haploinsufficiency in mice underscores complex pathogenicity in neurodegeneration. *Proc Natl Acad Sci USA*. 108: 2142–7.

Kahle, P.J. 2008. alpha-Synucleinopathy models and human neuropathology: similarities and differences. *Acta Neuropathol*. 115: 87–95.

Kalia, L.V., S.K. Kalia, P.J. McLean, A.M. Lozano and A.E. Lang. 2013. alpha-Synuclein oligomers and clinical implications for Parkinson disease. *Ann Neurol*. 73: 155–69.

Kasuga, K., M. Nishizawa and T. Ikeuchi. 2012. alpha-Synuclein as CSF and Blood Biomarker of Dementia with Lewy Bodies. *Int J Alzheimers Dis*. 437025.

Kawamoto, Y., I. Akiguchi, H. Fujimura, Y. Shirakashi, Y. Honjo and S. Sakoda. 2005. 14-3-3 proteins in Lewy body-like hyaline inclusions in a patient with familial amyotrophic lateral sclerosis with a two-base pair deletion in the Cu/Zn superoxide dismutase (SOD1) gene. *Acta Neuropathol*. 110: 203–4.

Kawamoto, Y., I. Akiguchi, G.G. Kovacs, H. Flicker and H. Budka. 2004a. Increased 14-3-3 immunoreactivity in glial elements in patients with multiple sclerosis. *Acta Neuropathol*. 107: 137–43.

Kawamoto, Y., I. Akiguchi, S. Nakamura and H. Budka. 2004b. 14-3-3 proteins in Lewy body-like hyaline inclusions in patients with sporadic amyotrophic lateral sclerosis. *Acta Neuropathol*. 108: 531–7.

Kawamoto, Y., I. Akiguchi, S. Nakamura, Y. Honjyo, H. Shibasaki and H. Budka. 2002. 14-3-3 proteins in Lewy bodies in Parkinson disease and diffuse Lewy body disease brains. *J Neuropathol Exp Neurol*. 61: 245–53.

Knopman, D.S., S.T. DeKosky, J.L. Cummings, H. Chui, J. Corey-Bloom, N. Relkin, G.W. Small, B. Miller and J.C. Stevens. 2001. Practice parameter: diagnosis of dementia (an evidence-based review). Report of the Quality Standards Subcommittee of the American Academy of Neurology. *Neurology*. 56: 1143–53.

Korff, A., C. Liu, C. Ginghina, M. Shi and J. Zhang. 2013. alpha-Synuclein in Cerebrospinal Fluid of Alzheimer's Disease and Mild Cognitive Impairment. *J Alzheimers Dis*. 36: 679–88.

Krasnianski, A., W.J. Schulz-Schaeffer, K. Kallenberg, B. Meissner, D.A. Collie, S. Roeber, M. Bartl, U. Heinemann, D. Varges, H.A. Kretzschmar and I. Zerr. 2006. Clinical findings and diagnostic tests in the MV2 subtype of sporadic CJD. *Brain*. 129: 2288–96.

Kurz, A., C. May, O. Schmidt, T. Muller, C. Stephan, H.E. Meyer, S. Gispert, G. Auburger and K. Marcus. 2012. A53T-alpha-synuclein-overexpression in the mouse nigrostriatal pathway leads to early increase of 14-3-3 epsilon and late increase of GFAP. *J Neural Transm*. 119: 297–312.

Ladogana, A., P. Sanchez-Juan, E. Mitrova, A. Green, N. Cuadrado-Corrales, R. Sanchez-Valle, S. Koscova, A. Aguzzi, T. Sklaviadis, J. Kulczycki, J. Gawinecka, A. Saiz, M. Calero, C.M. van Duijn, M. Pocchiari, R. Knight and I. Zerr. 2009. Cerebrospinal fluid biomarkers in human genetic transmissible spongiform encephalopathies. *J Neurol*. 256: 1620–8.

Lang, A.E. and A.M. Lozano. 1998. Parkinson's disease. First of two parts. *N Engl J Med*. 339: 1044–53.

Laverse, E., S. Shah and M. Mavra. 2009. Sporadic Creutzfeldt-Jakob disease: early signs and pre-mortem diagnosis. *BMJ Case Rep*.

Layfield, R., J. Fergusson, A. Aitken, J. Lowe, M. Landon and R.J. Mayer. 1996. Neurofibrillary tangles of Alzheimer's disease brains contain 14-3-3 proteins. *Neurosci Lett*. 209: 57–60.

Lee, K.H. and M.G. Harrington. 1997. The assay development of a molecular marker for transmissible spongiform encephalopathies. *Electrophoresis*. 18: 502–6.

Lee, V.M. 1995. Disruption of the cytoskeleton in Alzheimer's disease. *Curr Opin Neurobiol*. 5: 663–8.

Lemstra, A.W., M.T. van Meegen, J.P. Vreyling, P.H. Meijerink, G.H. Jansen, S. Bulk, F. Baas and W.A. van Gool. 2000. 14-3-3 testing in diagnosing Creutzfeldt-Jakob disease: a prospective study in 112 patients. *Neurology.* 55: 514–6.

Liu, D., J. Bienkowska, C. Petosa, R.J. Collier, H. Fu and R. Liddington. 1995. Crystal structure of the zeta isoform of the 14-3-3 protein. *Nature.* 376: 191–4.

Lo Bianco, C., J.L. Ridet, B.L. Schneider, N. Deglon and P. Aebischer. 2002. alpha -Synucleinopathy and selective dopaminergic neuron loss in a rat lentiviral-based model of Parkinson's disease. *Proc Natl Acad Sci USA.* 99: 10813–8.

Lu, C.H., W.N. Chang, H.W. Chang, K.J. Chung, H.C. Tsai, H.C. Wang, S.S. Chen, Y.C. Chuang, C.R. Huang, N.W. Tsai and Y.F. Chiang. 2008. The value of serial cerebrospinal fluid 14-3-3 protein levels in adult community-acquired bacterial meningitis. *QJM.* 101: 225–30.

Lu, J.Q., Y. Fan, A.P. Mitha, R. Bell, L. Metz, G.R. Moore and V.W. Yong. 2009. Association of alpha-synuclein immunoreactivity with inflammatory activity in multiple sclerosis lesions. *J Neuropathol Exp Neurol.* 68: 179–89.

Malaspina, A., N. Kaushik and J. de Belleroche. 2000. A 14-3-3 mRNA is up-regulated in amyotrophic lateral sclerosis spinal cord. *J Neurochem.* 75: 2511–20.

Maroteaux, L., J.T. Campanelli and R.H. Scheller. 1988. Synuclein: a neuron-specific protein localized to the nucleus and presynaptic nerve terminal. *J Neurosci.* 8: 2804–15.

Martin, H., J. Rostas, Y. Patel and A. Aitken. 1994. Subcellular localisation of 14-3-3 isoforms in rat brain using specific antibodies. *J Neurochem.* 63: 2259–65.

Martinez-Yelamos, A., A. Rovira, R. Sanchez-Valle, S. Martinez-Yelamos, M. Tintore, Y. Blanco, F. Graus, X. Montalban, T. Arbizu and A. Saiz. 2004. CSF 14-3-3 protein assay and MRI as prognostic markers in patients with a clinically isolated syndrome suggestive of MS. *J Neurol.* 251: 1278–9.

Martinez-Yelamos, A., A. Saiz, R. Sanchez-Valle, V. Casado, J.M. Ramon, F. Graus and T. Arbizu. 2001. 14-3-3 protein in the CSF as prognostic marker in early multiple sclerosis. *Neurology.* 57: 722–4.

Matsui, Y., K. Satoh, T. Miyazaki, S. Shirabe, R. Atarashi, K. Mutsukura, A. Satoh, Y. Kataoka and N. Nishida. 2011. High sensitivity of an ELISA kit for detection of the gamma-isoform of 14-3-3 proteins: usefulness in laboratory diagnosis of human prion disease. *BMC Neurol.* 11: 120.

Meiner, Z., E. Kahana, F. Baitcher, A.D. Korczyn, J. Chapman, O.S. Cohen, R. Milo, J. Aharon-Perez, O. Abramsky, R. Gabizon and H. Rosenmann. 2011. Tau and 14-3-3 of genetic and sporadic Creutzfeldt-Jakob disease patients in Israel. *J Neurol.* 258: 255–62.

Menon, K.N., D.L. Steer, M. Short, S. Petratos, I. Smith and C.C. Bernard. 2011. A novel unbiased proteomic approach to detect the reactivity of cerebrospinal fluid in neurological diseases. *Mol Cell Proteomics.* 10: M110 000042.

Miller, R.F., A.J. Green, G. Giovannoni and E.J. Thompson. 2000. Detection of 14-3-3 brain protein in cerebrospinal fluid of HIV infected patients. *Sex Transm Infect.* 76: 408.

Mollenhauer, B., J.J. Locascio, W. Schulz-Schaeffer, F. Sixel-Doring, C. Trenkwalder and M.G. Schlossmacher. 2011. alpha-Synuclein and tau concentrations in cerebrospinal fluid of patients presenting with parkinsonism: a cohort study. *Lancet Neurol.* 10: 230–40.

Mollenhauer, B., E. Trautmann, B. Otte, J. Ng, A. Spreer, P. Lange, F. Sixel-Doring, M. Hakimi, J.P. Vonsattel, R. Nussbaum, C. Trenkwalder and M.G. Schlossmacher. 2012. alpha-Synuclein in human cerebrospinal fluid is principally derived from neurons of the central nervous system. *J Neural Transm.* 119: 739–46.

Momjian-Mayor, I., G.P. Pizzolato, K. Burkhardt, T. Landis, A. Coeytaux and P.R. Burkhard. 2006. Fulminant Lewy body disease. *Mov Disord.* 21: 1748–51.

Muslin, A.J., J.W. Tanner, P.M. Allen and A.S. Shaw. 1996. Interaction of 14-3-3 with signaling proteins is mediated by the recognition of phosphoserine. *Cell.* 84: 889–97.

Mutsukura, K., K. Satoh, S. Shirabe, I. Tomita, T. Fukutome, M. Morikawa, M. Iseki, K. Sasaki, Y. Shiaga, T. Kitamoto and K. Eguchi. 2009. Familial Creutzfeldt-Jakob disease with a V180I mutation: comparative analysis with pathological findings and diffusion-weighted images. *Dement Geriatr Cogn Disord.* 28: 550–7.

Nichols, R.J., N. Dzamko, N.A. Morrice, D.G. Campbell, M. Deak, A. Ordureau, T. Macartney, Y. Tong, J. Shen, A.R. Prescott and D.R. Alessi. 2010. 14-3-3 binding to LRRK2 is disrupted by multiple Parkinson's disease-associated mutations and regulates cytoplasmic localization. *Biochem J.* 430: 393–404.

Nishida, T., A.M. Tokumaru, K. Doh-Ura, A. Hirata, K. Motoyoshi and K. Kamakura. 2003. Probable sporadic Creutzfeldt-Jakob disease with valine homozygosity at codon 129 and bilateral middle cerebellar peduncle lesions. *Intern Med.* 42: 199–202.

Ostrerova, N. et al. 1999. alpha-Synuclein shares physical and functional homology with 14-3-3 proteins. *J Neurosci.* 19(14): 5782–91.

Panegyres, P.K. and A. Graves. 2012. Anti-Yo and anti-glutamic acid decarboxylase antibodies presenting in carcinoma of the uterus with paraneoplastic cerebellar degeneration: a case report. *J Med Case Rep.* 6: 155.

Pennington, C., G. Chohan, J. Mackenzie, M. Andrews, R. Will, R. Knight and A. Green. 2009. The role of cerebrospinal fluid proteins as early diagnostic markers for sporadic Creutzfeldt-Jakob disease. *Neurosci Lett.* 455: 56–9.

Pittock, S.J. and C.F. Lucchinetti. 2006. Inflammatory transverse myelitis: evolving concepts. *Curr Opin Neurol.* 19: 362–8.

Plate, A., J. Benninghoff, G.H. Jansen, E. Wlasich, S. Eigenbrod, A. Drzezga, N.L. Jansen, H.A. Kretzschmar, K. Botzel, D. Rujescu and A. Danek. 2013. Atypical parkinsonism due to a D202N Gerstmann-Straussler-Scheinker prion protein mutation: first *in vivo* diagnosed case. *Mov Disord.* 28: 241–4.

Polymeropoulos, M.H., C. Lavedan, E. Leroy, S.E. Ide, A. Dehejia, A. Dutra, B. Pike, H. Root, J. Rubenstein, R. Boyer, E.S. Stenroos, S. Chandrasekharappa, A. Athanassiadou, T. Papapetropoulos, W.G. Johnson, A.M. Lazzarini, R.C. Duvoisin, G. Di Iorio, L.I. Golbe and R.L. Nussbaum. 1997. Mutation in the alpha-synuclein gene identified in families with Parkinson's disease. *Science.* 276: 2045–7.

Robertson, D.C., O. Schmidt, N. Ninkina, P.A. Jones, J. Sharkey and V.L. Buchman. 2004. Developmental loss and resistance to MPTP toxicity of dopaminergic neurones in substantia nigra pars compacta of gamma-synuclein, alpha-synuclein and double alpha/gamma-synuclein null mutant mice. *J Neurochem.* 89: 1126–36.

Rosenquist, M., P. Sehnke, R.J. Ferl, M. Sommarin and C. Larsson. 2000. Evolution of the 14-3-3 protein family: does the large number of isoforms in multicellular organisms reflect functional specificity? *J Mol Evol.* 51: 446–58.

Sanchez-Juan, P., A. Green, A. Ladogana, N. Cuadrado-Corrales, R. Saanchez-Valle, E. Mitrovaa, K. Stoeck, T. Sklaviadis, J. Kulczycki, K. Hess, M. Bodemer, D. Slivarichova, A. Saiz, M. Calero, L. Ingrosso, R. Knight, A.C. Janssens, C.M. van Duijn and I. Zerr. 2006. CSF tests in the differential diagnosis of Creutzfeldt-Jakob disease. *Neurology.* 67: 637–43.

Sano, K., K. Satoh, R. Atarashi, H. Takashima, Y. Iwasaki, M. Yoshida, N. Sanjo, H. Murai, H. Mizusawa, M. Schmitz, I. Zerr, Y.S. Kim and N. Nishida. 2013. Early detection of abnormal prion protein in genetic human prion diseases now possible using real-time QUIC assay. *PLoS One.* 8: e54915.

Sato, S., T. Chiba, E. Sakata, K. Kato, Y. Mizuno, N. Hattori and K. Tanaka. 2006. 14-3-3eta is a novel regulator of parkin ubiquitin ligase. *EMBO J.* 25: 211–21.

Satoh, J., K. Kurohara, M. Yukitake and Y. Kuroda. 1999. The 14-3-3 protein detectable in the cerebrospinal fluid of patients with prion-unrelated neurological diseases is expressed constitutively in neurons and glial cells in culture. *Eur Neurol.* 41: 216–25.

Satoh, J., T. Yamamura and K. Arima. 2004. The 14-3-3 protein epsilon isoform expressed in reactive astrocytes in demyelinating lesions of multiple sclerosis binds to vimentin and glial fibrillary acidic protein in cultured human astrocytes. *Am J Pathol.* 165: 577–92.

Satoh, J., M. Yukitake, K. Kurohara, H. Takashima and Y. Kuroda. 2003. Detection of the 14-3-3 protein in the cerebrospinal fluid of Japanese multiple sclerosis patients presenting with severe myelitis. *J Neurol Sci.* 212: 11–20.

Schmidt, C., K. Redyk, B. Meissner, L. Krack, N. von Ahsen, S. Roeber, H. Kretzschmar and I. Zerr. 2010. Clinical features of rapidly progressive Alzheimer's disease. *Dement Geriatr Cogn Disord*. 29: 371–8.

Shannon, K.M., A. Keshavarzian, H.B. Dodiya, S. Jakate and J.H. Kordower. 2012. Is alpha-synuclein in the colon a biomarker for premotor Parkinson's disease? Evidence from 3 cases. *Mov Disord*. 27: 716–9.

Shiga, Y., K. Miyazawa, S. Sato, R. Fukushima, S. Shibuya, Y. Sato, H. Konno, K. Doh-ura, S. Mugikura, H. Tamura, S. Higano, S. Takahashi and Y. Itoyama. 2004. Diffusion-weighted MRI abnormalities as an early diagnostic marker for Creutzfeldt-Jakob disease. *Neurology*. 63: 443–9.

Shirakashi, Y., Y. Kawamoto, H. Tomimoto, R. Takahashi and M. Ihara. 2006. alpha-Synuclein is colocalized with 14-3-3 and synphilin-1 in A53T transgenic mice. *Acta Neuropathol*. 112: 681–9.

Spero, M. and I. Lazibat. 2010. Creutzfeldt-Jakob disease: case report and review of the literature. *Acta Clin Croat*. 49: 181–7.

Spillantini, M.G., M.L. Schmidt, V.M. Lee, J.Q. Trojanowski, R. Jakes and M. Goedert. 1997. Alpha-synuclein in Lewy bodies. *Nature*. 388: 839–40.

Stefanis, L. 2012. alpha-Synuclein in Parkinson's disease. *Cold Spring Harb Perspect Med*. 2: a009399.

Strong, M.J. 2010. The evidence for altered RNA metabolism in amyotrophic lateral sclerosis (ALS). *J Neurol Sci*. 288: 1–12.

Takase, K., H. Furuya, H. Murai, T. Yamada, Y. Oh-yagi, K. Doh-ura, T. Iwaki, S. Tobimatsu and J. Kira. 2001. A case of Gerstmann-Straussler-Scheinker syndrome (GSS) with late onset—a haplotype analysis of Glu219Lys polymorphism in PrP gene. *Rinsho Shinkeigaku*. 41: 318–21.

Teunissen, C.E., C. Dijkstra and C. Polman. 2005. Biological markers in CSF and blood for axonal degeneration in multiple sclerosis. *Lancet Neurol*. 4: 32–41.

Tokuda, T., M.M. Qureshi, M.T. Ardah, S. Varghese, S.A. Shehab, T. Kasai, N. Ishigami, A. Tamaoka, M. Nakagawa and O.M. El-Agnaf. 2010. Detection of elevated levels of alpha-synuclein oligomers in CSF from patients with Parkinson disease. *Neurology*. 75: 1766–72.

Tschampa, H.J., K. Kallenberg, H. Urbach, B. Meissner, C. Nicolay, H.A. Kretzschmar, M. Knauth and I. Zerr. 2005. MRI in the diagnosis of sporadic Creutzfeldt-Jakob disease: a study on inter-observer agreement. *Brain*. 128: 2026–33.

Tschampa, H.J., M. Neumann, I. Zerr, K. Henkel, A. Schroter, W.J. Schulz-Schaeffer, B.J. Steinhoff, H.A. Kretzschmar and S. Poser. 2001. Patients with Alzheimer's disease and dementia with Lewy bodies mistaken for Creutzfeldt-Jakob disease. *J Neurol Neurosurg Psychiatry*. 71: 33–9.

Ubl, A., D. Berg, C. Holzmann, R. Kruger, K. Berger, T. Arzberger, A. Bornemann and O. Riess. 2002. 14-3-3 protein is a component of Lewy bodies in Parkinson's disease-mutation analysis and association studies of 14-3-3 eta. *Brain Res Mol Brain Res*. 108: 33–9.

Umahara, T. and T. Uchihara. 2010. 14-3-3 proteins and spinocerebellar ataxia type 1: from molecular interaction to human neuropathology. *Cerebellum*. 9: 183–9.

Umahara, T., T. Uchihara, S. Yagishita, A. Nakamura, K. Tsuchiya, and T. Iwamoto. 2007. Intranuclear immunolocalization of 14-3-3 protein isoforms in brains with spinocerebellar ataxia type 1. *Neurosci Lett*. 414: 130–5.

van Dijk, K.D., M. Bidinosti, A. Weiss, P. Raijmakers, H.W. Berendse and W.D. van de Berg. 2013. Reduced alpha-synuclein levels in cerebrospinal fluid in Parkinson's disease are unrelated to clinical and imaging measures of disease severity. *Eur J Neurol*.

Van Everbroeck, B., I. Dobbeleir, M. De Waele, P. De Deyn, J.J. Martin and P. Cras. 2004. Differential diagnosis of 201 possible Creutzfeldt-Jakob disease patients. *J Neurol*. 251: 298–304.

Van Everbroeck, B., A.J. Green, E. Vanmechelen, H. Vanderstichele, P. Pals, R. Sanchez-Valle, N.C. Corrales, J.J. Martin and P. Cras. 2002. Phosphorylated tau in cerebrospinal fluid as a marker for Creutzfeldt-Jakob disease. *J Neurol Neurosurg Psychiatry*. 73: 79–81.

Van Everbroeck, B., S. Quoilin, J. Boons, J.J. Martin and P. Cras. 2003. A prospective study of CSF markers in 250 patients with possible Creutzfeldt-Jakob disease. *J Neurol Neurosurg Psychiatry*. 74: 1210–4.

Vandemark, M. 2013. Acute bacterial meningitis: current review and treatment update. *Crit Care Nurs Clin North Am*. 25: 351–61.

Volkening, K., C. Leystra-Lantz, W. Yang, H. Jaffee and M.J. Strong. 2009. Tar DNA binding protein of 43 kDa (TDP-43), 14-3-3 proteins and copper/zinc superoxide dismutase (SOD1) interact to modulate NFL mRNA stability. Implications for altered RNA processing in amyotrophic lateral sclerosis (ALS). *Brain Res*. 1305: 168–82.

Wakabayashi, H., M. Yano, N. Tachikawa, S. Oka, M. Maeda and H. Kido. 2001. Increased concentrations of 14-3-3 epsilon, gamma and zeta isoforms in cerebrospinal fluid of AIDS patients with neuronal destruction. *Clin Chim Acta*. 312: 97–105.

Wang, G.R., C. Gao, Q. Shi, W. Zhou, J.M. Chen, C.F. Dong, S. Shi, X. Wang, Y. Wei, H.Y. Jiang, J. Han and X.P. Dong. 2010. Elevated levels of tau protein in cerebrospinal fluid of patients with probable Creutzfeldt-Jakob disease. *Am J Med Sci*. 340: 291–5.

Wang, Y., M. Shi, K.A. Chung, C.P. Zabetian, J.B. Leverenz, D. Berg, K. Srulijes, J.Q. Trojanowski, V.M. Lee, A.D. Siderowf, H. Hurtig, I. Litvan, M.C. Schiess, E.R. Peskind, M. Masuda, M. Hasegawa, X. Lin, C. Pan, D. Galasko, D.S. Goldstein, P.H. Jensen, H. Yang, K.C. Cain and J. Zhang. 2012. Phosphorylated alpha-synuclein in Parkinson's disease. *Sci Transl Med*. 4: 121ra20.

Weber, T., M. Otto, M. Bodemer and I. Zerr. 1997. Diagnosis of Creutzfeldt-Jakob disease and related human spongiform encephalopathies. *Biomed Pharmacother*. 51: 381–7.

Wenning, G.K., C. Colosimo, F. Geser and W. Poewe. 2004. Multiple system atrophy. *Lancet Neurol*. 3: 93–103.

Wiltfang, J., M. Otto, H.C. Baxter, M. Bodemer, P. Steinacker, E. Bahn, I. Zerr, J. Kornhuber, H.A. Kretzschmar, S. Poser, E. Ruther and A. Aitken. 1999. Isoform pattern of 14-3-3 proteins in the cerebrospinal fluid of patients with Creutzfeldt-Jakob disease. *J Neurochem*. 73: 2485–90.

Withers, G.S., J.M. George, G.A. Banker and D.F. Clayton. 1997. Delayed localization of synelfin (synuclein, NACP) to presynaptic terminals in cultured rat hippocampal neurons. *Brain Res Dev Brain Res*. 99: 87–94.

Yutzy, B., E. Holznagel, C. Coulibaly, A. Stuke, U. Hahmann, J.P. Deslys, G. Hunsmann and J. Lower. 2007. Time-course studies of 14-3-3 protein isoforms in cerebrospinal fluid and brain of primates after oral or intracerebral infection with bovine spongiform encephalopathy agent. *J Gen Virol*. 88: 3469–78.

Zerr, I., M. Bodemer, O. Gefeller, M. Otto, S. Poser, J. Wiltfang, O. Windl, H.A. Kretzschmar and T. Weber. 1998. Detection of 14-3-3 protein in the cerebrospinal fluid supports the diagnosis of Creutzfeldt-Jakob disease. *Ann Neurol*. 43: 32–40.

19

Transcranial Magnetic Stimulation and Deep Brain Stimulation in Neuropsychiatric Disorders: New Dimension of Therapeutic Markers in Psychiatry

Tarek H. Mouhieddine,[1] Wassim Abou-Kheir,[2] Muhieddine M. Itani,[3] Amaly Nokkari[4] and Ziad Nahas[5,]*

INTRODUCTION

Neuropsychiatric disorders are widely prevalent and lead to the disability of millions worldwide with three disorders listed by the World Health Organization (WHO) among the top 10 causes of disability worldwide. The main contributors of the chronicity and unresponsiveness to treatment of

[1] Faculty of Medicine, American University of Beirut Medical Center, Beirut, Lebanon.
  Postal Address: American University of Beirut, P.O. Box 6539, Riad El Solh, Beirut, Lebanon.
  Email: tarek.mouhieddine@gmail.com
[2] Department of Anatomy, Cell Biology and Physiology, American University of Beirut, Beirut, Lebanon.
  Email: wa12@aub.edu.lb
[3] Faculty of Medicine, Saint George University of London, Nicosia, Cyprus.
  Email: muhieddineitani@gmail.com
[4] Department of Biochemistry and Molecular Genetics, American University of Beirut, Beirut, Lebanon.
  Email: amn17@aub.edu.lb
[5] Department of Psychiatry, American University of Beirut Medical Center, Beirut, Lebanon.
* Corresponding author: zn17@aub.edu.lb

these disorders are our incomplete understanding of the pathophysiology of the diseases, our limited therapeutic arsenal and our phenomenological approach in diagnosis (Nahas, 2010). Several biomarkers using neuroimaging techniques, biochemical and genetic tests have been identified, but most lack exact sensitivity or specificity. The field of neuromodulation is advancing at a rapid pace. Brain stimulation techniques are promising for treatments of many neuropsychiatric diseases including depression, schizophrenia, Parkinson's Disease (PD), Obsessive-Compulsive Disorder (OCD) and Tourette's Syndrome (TS). They also aid in the development of other new treatments that incorporate neurogenesis via stem cells, viral vectors and transgenic cells. Although much is known about their action on brain networks, their exact therapeutic mechanism is yet to be agreed upon. This chapter focuses on two specific techniques at the opposite spectrum of invasiveness: Deep Brain Stimulation (DBS) and Transcranial Magnetic Stimulation (TMS).

Somatic interventions for treating mental illnesses date back to ancient times. In the last two decades, investigators have moved away from cross-sectional studies towards an integrated multimodal approach with neuroimaging and genetic studies (Nahas, 2010). Both the anterior and midlateral prefrontal cortices of the brain play diverse but corresponding roles in regulating the mood (Nahas et al., 2010). The frontal pole (connected to anterior cingulate cortex, posterior cingulate, precuneus, dorsolateral prefrontal and orbitofrontal cortex) integrates the control of memory, emotions, motivation, self-awareness and reference while the ventral and dorsal midlateral frontal cortex (bidirectionally connected to multimodal temporal regions and paralimbic cortices) monitor information in the working memory and participate in active judgments, respectively (Nahas et al., 2010). As for the lateral frontal zones, they monitor, assess, verify and organize information and attend to emotional stimuli (Nahas et al., 2010).

The major receptive portion of the basal ganglia is the striatum, which receives input from the substantia nigra (SN), intralaminar nuclei of the thalamus and neocortex (DeLong et al., 1986). The putamen is somatotopically organized and can be divided into two main regions: sensorimotor and associative (Künzle, 1978; Crutcher and DeLong, 1984). The putamen mainly receives the motor, premotor and somatosensory projections from the cortex, while the caudate mainly receives projections of the association cortices (DeLong et al., 1986). The major sources of the basal ganglia output are the SN and the globus pallidus (GP), which is also divided into external (GPe) and internal (GPi) segments (DeLong et al., 1985). The SN and GP further project into the thalamus (Asanuma et al., 1983; DeVito and Anderson, 1982) and they act as funnels for segregated parallel cortico-subcortical loops that control different body parts and other complex functions related to association areas (DeLong et al., 1986).

All these findings of anatomical connections and physiological functions have made these regions the targets of many brain stimulation therapies.

Many groundbreaking techniques in the field of brain stimulation are making their way into clinical practice (George et al., 2007). They widely differ in their invasiveness, stimulation parameters, efficacy and potential side effects. Yet, mechanistically, they appear to adaptively modulate frontal and limbic brain regions over time, similar to what is observed with medications and cognitive behavioral therapy (Nahas et al., 2007).

Transcranial Magnetic Stimulation

One of the least invasive techniques is the transcranial magnetic stimulation, which introduces a flow of electrical current throughout the superficial cortex via alternating magnetic fields applied at the scalp (George et al., 2002, 2007). In addition to its therapeutic role, TMS is widely used as a tool to study brain-behavior relationships, and map the brain's function and its connectivity (George et al., 2010). Over time, different sham methods have been developed to optimally test TMS under rigorous double blind conditions, including a coil without a magnetic field but which mimics the popping sounds of the real TMS, and a coil which in addition to the sound, induces scalp muscle contractions to mimic the sensation of the real TMS (Nahas, 2008).

Successive stimuli administration is referred to as repetitive TMS (rTMS) (George et al., 2002). TMS is able to modulate anatomically and functionally related cortical and subcortical brain regions. TMS thus reaches deep brain structures transynaptically (Nahas, 2008). It is applied sub-convulsively on awake patients, without the use of general anesthesia but with medical supervision (Nahas, 2008). With fast stimulation, it poses a small risk of inducing seizures especially in patients with multiple sclerosis, epilepsy or stroke. It has induced seizures in over 13 reported cases of healthy individuals (Anderson et al., 2006). Where to stimulate, how to stimulate and what the underlying brain activity is at the time of stimulation constitute critical parameters to be considered when using TMS. There are several methodologies for applying TMS, which involve different dosages, intensities, frequencies, treatment durations, pulse durations, different coil positioning and different areas of application depending on its intended use (George et al., 2007; Johnson et al., 2013).

By convention, slow rTMS is when stimuli are repeated at a rate less than or equal to 1 Hz and fast rTMS is when it is more than 1 Hz, sometimes reaching 30 Hz above which poses a risk of inflicting seizures (George et al., 2002). Changing the parameters of TMS leads to the excitation of different structures (axons or cell bodies) and neuronal sets (cortical neurons, U fibers or interneurons) (Borckardt et al., 2006c; George et al., 2002). Nevertheless, it is generally accepted that TMS leads to neuronal "excitation... where the negative-going first spatial derivative of the induced electric field parallel to the long axis of the nerve (i.e., the axon) peaks" (George and Belmaker, 2000). However, it was noted that fast rTMS over the motor cortex does lead to increased excitability of the corticospinal tract while long slow rTMS pulses of around 1 Hz would reduce excitability (Chae et al., 2004). These are also associated with changes in blood flow with increasing intensity (George et al., 2002). Furthermore, more stimuli per session and longer courses of treatment may lead to more pronounced and possibly lasting modulation (Nahas, 2008). TMS can also lead to the release of glutamate or gamma aminobutyric acid (GABA) which may mediate its observed therapeutic effects.

It was shown that after daily fast left rTMS, there is an increase in Fractional Anisotropy (FA) values in the prefrontal white matter of the left and right hemispheres, but higher in the left hemisphere (Kozel et al., 2011). This could mean that rTMS does not cause structural damage (decrease in FA) but could affect white matter organization positively (Kozel et al., 2011) and putatively increase targeted functional connectivity (Kozel et al., 2011). Another study, which showed structural enhancement, by May et al. (2007) who found an increase in gray matter in the region stimulated by rTMS following 5 days of 1 Hz and 110% motor threshold (MT) stimulation of the superior temporal cortex. They even countered the convention of the nervous system being a slow evolving system by detecting rapidly adjusting structural cortical plasticity within 1 week of rTMS stimulation (May et al., 2007).

TMS has been mainly studied for the acute treatment of mild to moderate Treatment-Resistant Depression (TRD) (Nahas, 2010). The Food and Drug Administration (FDA) approved the usage of TMS on adult patients who have major depression disorder on a condition that they have failed to improve with one conventional antidepressant treatment at the minimum effective dosage or more (Nahas and Anderson, 2011; George et al., 2010). In addition, TMS has been investigated to treat schizophrenia, auditory hallucinations and chronic pain (George et al., 2007) and it has even been observed to reduce impulsivity and cravings in smokers even when presented with smoking cues (Li et al., 2013b).

Even though TMS was well tolerated among the patients, with overall low attrition rates (<5%), some patients abandon treatment due to mild pain or discomfort in the scalp (Li et al., 2013b). The pain was likely caused by stimulation of nociceptors in the scalp, periosteum and meninges (Borckardt et al., 2006b). Local anesthetic creams and thin foam pads are sometimes used to reduce skin and scalp stimulation pain (Borckardt et al., 2006b) but with limited success. Moreover, Lidocaine can be injected along with epinephrine to reduce scalp and deeper nociceptive pain (periosteum rather than meninges) and cause vasoconstriction to reduce Lidocaine absorbance and prolong analgesic effect (Borckardt et al., 2006b). However, this method is not clinically practical and has been shown to cause some hypersensitivity to touch in a few individuals and thus may not be a very viable method of anesthesia (Borckardt et al., 2006b).

On a different level, TMS and all other brain stimulation techniques may also be used as techniques for certain experiments, rather than treatments. Li et al. (2010) for example, used TMS along with fMRI on 30 healthy men to induce neuronal activation and then followed the deactivation effect of Lamotrigine (LTG) and Valproic Acid (VPA). Thus, TMS with fMRI promises a valuable methodology for targeting different brain regions and observing the effect of a wide spectrum of drugs in different neuropsychiatric diseases. Another important technique is paired-pulse TMS that involves sending two pulses to the same region with different intensities and interpulse intervals (milliseconds long) (George et al., 2002). According to the intensity and interval, the pulses may enhance or inhibit one another and the brain's excitatory and inhibitory systems could be assessed in different diseases before and after the administration of drugs or treatments (George et al., 2002). TMS is also used to study the effect of lesions in certain brain regions, for it is able to inflict virtual lesions in Broca's area for example, and cause speech arrest (Epstein et al., 1996).

TMS Motor Threshold and Coil Placement

There are very few areas in the brain that can give reliable immediate and observable phenomena when stimulated via TMS. These include the primary motor and visual cortex, and expressive language areas. It is thus the convention before starting TMS, to first locate the left primary motor cortex and then measure the resting Motor Threshold, i.e., the slightest amount of TMS machine power needed to induce a Motor-Evoked Potential (MEP) of about 50 µV in the relaxed contralateral abductor pollicis brevis muscle of the individual in at least 50% of the trials (Herbsman et al., 2009b; Borckardt et al., 2006a). This differs from one individual to another. To reach the MT, an individual is stimulated with a sub-threshold TMS intensity and the intensity is then increased in 5% increments for around 10 to 20 trials to finally reach the MT (Borckardt et al., 2006a). Another method suggested by Rothwell et al. (1999) is to start from supra-threshold intensity and then decrease it in increments of 2 or 5% until motor responses are no longer observed. Mills and Nithi (1997) have also suggested a method in which the MT is represented as an arithmetic mean of the lowest intensity that produces 10 MEPs in 10 trials and the highest intensity that does not produce any MEPs in 10 trials. However, this approach needs more than 50 stimuli over several minutes in order to get the MT (Mishory et al., 2004) and that is why other more efficient techniques were created.

One of those techniques is maximum likelihood parameter estimation by sequential testing (ML-PEST), which creates a curve from data point pairs (intensity of stimulus and success/failure) and feeds them into the computer (Borckardt et al., 2006a). The software determines the stimulus intensity that should be used in the consecutive trials and finally calculates the MT from the curve's midpoint (Borckardt et al., 2006a). Nonetheless, Mishory et al. (2004) did not find a significant difference between the MT values procured through ML-PEST and conventional methods when electromyography, rather than visible detection, was used to identify the MEP. Nevertheless, this could have been due to the fact that the parameters used in that study were increments of 5% and the machines back then were not as sensitive as current ones. Finally, there is the non-parametric simple adaptive PEST (SA-PEST), which starts with a stimulus that is presumably close to the threshold and the stimulus intensity is then increased gradually to search for the correct cortical region that yields an observable MEP (Borckardt et al., 2006a). When comparing the two PEST procedures, it was found that the SA-PEST, even though it needs an average of 3.48 more trials than the ML-PEST, provides a more accurate MT value by 1.41% (Borckardt et al., 2006a). MT is then used to define the pulse intensity needed to stimulate the brain and which mostly ranges between 80 and 130% of MT (Johnson et al., 2013). MT differs a lot from one individual to another and increases linearly with age and skull to cortex distance, but is rather constant in a given person (Herbsman et al., 2009b). However, it has been shown that stimulating at 120% MT covers more than 90% of patients of all ages (i.e., with different levels of atrophy) (Johnson et al., 2013).

The excitation of large structures, like the pyramidal tract neurons, is influenced by trajectory non-uniformities and bends, while short structures like cell bodies, dendrites and interneurons are influenced by the orientation of the electric field and the nerve fiber (Amassian et al., 1998, 1992; Maccabee et al., 1993; Abdeen and Stuchly, 1994). Moreover, the different MEPs that are observed among different individuals are a result of the variable orientations of different cortical fiber groups and the different magnetic field orientations, which employ variable action potentials (Chen et al., 2003). Different current orientation leads to the production of different waves such as the antero-posterior current that excites cortical motor neurons, inducing I-waves, while a latero-medial current induces D-waves (Chen et al., 2003). In a study by Herbsman et al. (2009b) they found that the skull to cortex distance account for 69% of the variance of MT among patients and when combined with motor bundle fiber orientation (cortico-pyramidal diffusion tensor imaging tract) they account for 80% of the variance.

Placing the TMS coil is primarily dependent on the cortical region intended to be modulated. For motor electrophysiology studies, it is placed over the motor cortex. For auditory hallucination or tinnitus, it is placed over the temporal lobe. For depression, its primary clinical application, TMS targets the left dorsolateral prefrontal cortex (DLPFC). This is because, in many of the early works, the left DLPFC has revealed dysregulation in depressed patients (Herbsman et al., 2009a). The prefrontal cortex (PFC) has also shown rich connections with structures of the limbic system that cannot be directly stimulated with TMS because they lie in deeper regions of the brain (Herbsman et al., 2009a). TMS stimulation of the PFC has also rendered an increase in dopamine level in the mesolimbic system and modulated the GABA(b) receptor (Daskalakis et al., 2006). To target the DLPFC, many studies used the "5 cm rule" to probabilistically position the TMS over Brodmann areas (BA) 9 and/or 46 (Herbsman et al., 2009a). This is done by moving 5 cm anteriorly in a parasagittal line from the site of the motor cortex (Johnson et al., 2013). However, individual differences (brain size, motor localization and scalp-to-cortex distance) have rendered that method inapplicable in some patients. One study conducted by Herwig et al. (2001) have showed that the "5 cm rule" positions the TMS over the premotor cortex (BA 6) in 32% of patients, in addition to other areas such as the frontal eye fields (BA 8) and the DLPFC that are superior and posterior to BA 9 (Herbsman et al., 2009a). That is why Johnson et al. (2013) devised a new method called the "5 cm + 1 cm" (or "6 cm rule") which theoretically should eliminate the high percentage of miss-localization, and they applied it on 185 participants (Johnson et al., 2013). Both rules would lead to variable positioning in different patients, but in the "6 cm rule", the range of variability is pushed anteriorly, decreasing stimulation of BA 6 from 9 to 2% and increasing anterior stimulation of BA 46 from 3 to 16% (Johnson et al., 2013). This forward shift could also lead to better treatment responses as was mentioned in a study by Herbsman et al. (2009a) who concluded that using the "5 cm rule" and stimulating more laterally and anteriorly leads to better treatment responses.

Correct brain targeting is continuously proving to be an influential parameter on TMS results, where the electrode location is correlated with the antidepressant effect of the treatment (Herbsman et al., 2009a). That is not surprising given that the cortical region to be excited via TMS is less than 1 cm in diameter, leading to the placement of fiducials at the borders of two adjacent BAs (Herbsman et al., 2009a). According to Herbsman and Nahas (2012) the field would benefit from individualized techniques to target brain regions more accurately. Such techniques would include virtual stereotaxic systems and high resolution T1 scans of the head that require expensive and complex computer hardware and software and accurate optical sensing equipment (Herbsman and Nahas, 2012). Herbsman et al.

(2009a) replicated and improved the work of Andoh et al. (2009), where they simplified the triangulation technique of Andoh et al. (2009) by using only two electrical wires to localize their target on the brain, instead of three. This technique involved an MRI and software-guided triangulation based technique using two electrical wires to localize a fixed cortical coordinate, with an error that is less than 1 cm away from the target.

TMS, Depression and Biomarkers

TMS is most widely used for the treatment of Major Depressive Disorder (MDD) that is both incapacitating, costly and its patients rarely improve (George et al., 2010). TRD is a disease that is very common and highly morbid and is inflicting the health care system with increasing costs (Nahas and Anderson, 2011). Globally, there are around 350 million people from all ages suffering from some form of depression with about 10–50% or less of them receiving adequate treatment (WHO). TRD is a serious episode of depression, which has not improved, even with the administration of two different trials of adequately prescribed medication (Fava, 2003; Berlim and Turecki, 2007; Berlim et al., 2008).

Many studies have tried to find the most effective way of treatment by stimulating different brain regions and using different parameters of stimulation and by improving the sham treatment to be as close to the real TMS as possible. It was found that stimulating the border between BA 9 and BA 46 would lead to a 69% drop in the Hamilton Depression Rating Scale (HDRS) by the end of the third week (Fitzgerald et al., 2009). One study by George et al. (2010) applied daily left prefrontal TMS on 190 MDD patients with a high intensity of 120% MT and a high pulse number (3000 stimuli per session) and used an MRI guided method to place the equipment on the skull and an improved active sham rTMS, which greatly mimicked the real rTMS in terms of somatosensory experience. Studies have shown that patients who have failed to improve on two previous medication trials have a probability of less than 20% to improve with another trial, and those who failed three trials have a probability of 10–20% to improve (Nierenberg et al., 2006; McGrath et al., 2006; Wisniewski et al., 2007; Fava et al., 2006). The remission rate in the blinded phase of the trial by George et al. (2010) was 14.1% by 3–5 weeks, for their patients who failed to improve on three previous medication trials. So, even the time needed for improvement of patients was better in this study compared to others where antidepressant TMS treatment was administered for only 2 weeks, while medications was for 6–8 weeks and electroconvulsive therapy for 3–4 weeks (George et al., 2010). Furthermore, they noticed that most of the remitters in this study had a lower degree of treatment resistance (George et al., 2010).

Some biomarkers (peripheral and central) were found to have some relevance to TRD. The peripheral biomarkers include cytokines and steroids: the action of suppression of dexamethasone on Tumor Necrosis Factor α (TNF-α) decreases in TRD patients as compared to healthy subjects, and a combination of Brain Derived Neurotrophic Factor (BDNF) and 5-HT1A (a subtype of the 5-HT receptor that binds serotonin) genetic polymorphisms were also found to be linked to TRD (Smith, 2013). On the other hand, central biomarkers include the CSF: increased levels of dopamine metabolites and homovanillic acid were found in patients that suffer from TRD (Smith, 2013). Structural magnetic resonance imaging, a central imaging biomarker, includes some observations such as gray matter reduction in the left temporal cortex (Shah et al., 1998), enlarged cerebroventricles, decrease in the size of gray matter in many regions of both hemispheres (Shah et al., 2002), and some white matter hyperintensities in subcortical areas (Smith, 2013). Functional magnetic resonance imaging also revealed some interesting biomarkers; it has been shown that people suffering from TRD have lower functional connectivity in areas of the brain such as the precuneus cortex, right insula, anterior cingulate cortex and left amygdala (Li et al., 2013a).

In a study on TRD patients, McDonald et al. (2011) applied high frequency rTMS, or fast left rTMS, at 10 Hz and 120% MT and for 180,000 pulses throughout the 12 weeks. Patients who did not respond to the treatment were switched to a low frequency stimulation of the DLPFC (slow right rTMS) at 1 Hz. Within 4 weeks, there was some difference in response but it was not until the 6th week of treatment that they found significance between active rTMS and sham rTMS. The remission rate reached 20% for the active treatment compared to 14.2% for those who received a sham treatment followed by an active rTMS treatment. Nevertheless, what was intriguing was that the remission rate for patients undergoing slow right rTMS was 26% and unlike fast left rTMS patients, did not show a correlation between medication resistance and remission rate. However, this could be explained by the fact that the patients who responded to slow right rTMS were already responding to the fast left rTMS but needed additional treatments and more time to respond (McDonald et al., 2011). A complementary explanation would be that as a patient receives more active treatments, the patient's treatment resistance becomes less important in determining the response (Avery et al., 2008).

In addition to the above studies, an open study by Hadley et al. (2011) identified that a more aggressive rTMS treatment leads to better results. In their study, TRD patients' suicidal ideation decreased with increasing dosage of rTMS (Hadley et al., 2011). Out of the 19 depressed patients, 66% achieved remission, and had a better quality of life, social life, emotional well-being and energy.

TMS and Schizophrenia

Schizophrenia is one of the major psychiatric illnesses, affecting almost 1% of the worldwide population, manifesting in late adolescents/young adults (Zaidi et al., 2010) with an overall cost of US$ 62.7 billion in the USA in 2002 (Bonilha et al., 2009). The disease debilitates its patients from knowing what is real from what is not, thinking clearly, having normal emotional responses, and acting appropriately in social scenes, among many other things (Zaidi et al., 2010). Studies show that it is directly related to cardiovascular health as it increases mortality rates by 2–3 folds (Zaidi et al., 2010).

Patients with schizophrenia show structural abnormalities in the brain, including ventricular enlargement and progressive cortical atrophy, in addition to gray matter atrophy in the prefrontal cortex, leading to negative symptoms and insight reduction (Bonilha et al., 2009). Moreover, gray matter loss takes place in the temporal and frontal parietal regions leading to reality distortion, and superior temporal gyrus leading to auditory and visual induced potentials and negative and psychotic symptoms (Bonilha et al., 2009). Very few biomarkers for schizophrenia can be found in the literature; yet, the following were established as biomarkers for this disease. The biomarkers include mediators of the N-methyl-D-aspartate (NMDA) pathway, which is one of the most postulated theories of schizophrenia, such as homocysteine, serine and glycine (Neeman et al., 2005), and regulatory chemicals which are involved in Hypothalamic Pituitary-Adrenal (HPA) axis region, such as cortisol and dehydroepiandrosterone (DHEA). There are also inflammatory biomarkers such as, Nuclear Factor-kappa B (NF-kB), TNF-α, and selectins, factors involved in growth and development such as BDNF (Ritsner et al., 2004; Gama et al., 2007; Song et al., 2009), and finally, genetic factors such as, the schizophrenia candidate gene, NRG1 (Zaidi et al., 2010).

As for the treatment, researchers targeted specifics symptoms of schizophrenia (paranoia, auditory hallucinations, attention problems, etc....) by targeting relevant functional networks with a specific cortical representation (George et al., 2009). Out of the positive symptoms of schizophrenia, only auditory hallucinations were reduced via slow frequency TMS of the temporoparietal junction, while studies about negative symptoms treatment showed variable results (George et al., 2009).

TMS and Obsessive Compulsive Disorder

Obsessive–compulsive disorder is a debilitating neuropsychiatric disease with a 2–3% worldwide occurrence (Robins et al., 1984). "Obsessions are recurrent intrusive thoughts that enter into the stream of consciousness and tend to be unpleasant, distressing and difficult to suppress; compulsions are rituals (physical or mental) performed either in response to obsessions or according to rigid rules" (Chamberlain and Menzies, 2009).

Early genetic studies suggest that the disorder is correlated with chromosome 9 (Hanna et al., 2002; Willour et al., 2004). More recent studies have found that the disorder is located among chromosomes 1, 3, 6, 7 and 15 (Shugart et al., 2006). It was found, via neuroimaging, that there is an increased metabolic activity in the orbitofrontal cortex (OFC) and striatal regions, and, in some studies, a size reduction in the OFC (Cottraux et al., 1996; Rauch et al., 1994; Atmaca et al., 2006, 2007; Kang et al., 2004). It was also noted that reduced gray matter was observed in the bilateral medial prefrontal cortex, left inferior frontal cortex and left DLPFC (van den Heuvel et al., 2009).

An association between the brain region and an aspect of the psychiatric symptoms was also noticed, which was brought to light by a few studies and was probably one of the most direct biomarkers to this disorder (Menzies et al., 2007). "For example, symmetry/ordering was associated with lower global grey and white matter volumes... lower right motor cortex, left insula and left parietal cortex grey matter volume... while contamination/washing with lower grey matter volumes in the bilateral caudate... and harm/checking with lower bilateral temporal lobe grey matter" (Chamberlain and Menzies, 2009).

TMS was investigated to treat OCD where one study (Benjamin et al., 1997) found positive results when stimulating the right lateral prefrontal cortex and another study (Mantovani et al., 2006) found positive results in stimulating the supplemental motor area. Likewise, 1 HZ rTMS over the primary motor area in patients suffering from tic disorder, including TS in some of them, decreased tic frequency (Karp et al., 1997). Another study by Chae et al. (2004) applied rTMS either over the left motor cortex or left PFC for 5 days and found that there is an improvement in tic and OCD symptoms in eight patients suffering from TS where the presence of OCD significantly affected responsivity to TMS.

TMS and Other Neuropsychiatric Disorders

Even though TMS was mostly applied on patients with depression, many studies have tried using it for the treatment of other diseases and had variable results. One of the target diseases was mania in a study performed by Grisaru et al. (1998) who applied TMS for 10 consecutive days on 16 patients with mania and found that those being treated on the right side were responding more than those treated on the left prefrontal. Moreover,

Kaptsan et al. (2003) tried to replicate those results but failed. Therefore treating mania with TMS is still a controversy. Using TMS for the treatment of PD is also inconclusive since one study by Boylan et al. (2001) showed that stimulating the supplementary motor area worsens PD symptoms while another study by Shimamoto et al. (2001) showed that low frequency rTMS on the frontal areas of the brain improves the symptoms. Even with all these discrepancies, it does seem logical to apply TMS to treat PD since it has been shown to affect dopamine activity in the caudate after PFC stimulation (Strafella et al., 2001). In addition to that, it was used for the treatment of posttraumatic stress disorder (PTSD) with variable results and panic disorder with no significant effect (George et al., 2009).

TMS and Pain

Even though the motor cortex was the main target for pain management, imaging studies have shown that PFC stimulation could induce changes in the OFC, cingulate gyrus, hippocampus and insula, which are deep limbic regions that regulate pain and mood (Borckardt et al., 2007; George et al., 2002). The pain inhibitory effect of the PFC can be explained by the fact that there is a neuronal circuitry between the PFC and periaqueductal gray and nucleus cuneiformis (Borckardt et al., 2007). Three days of fast rTMS (20Hz) has also been effective in treating chronic neuropathic pain with a sustained analgesic effect of 15 days, while slow rTMS (1 Hz) did not provide this analgesic effect (Khedr et al., 2005; André-Obadia et al., 2006; Saitoh et al., 2007). Borckardt et al. (2006c) have applied prefrontal rTMS for 20 minutes on patients following gastric bypass surgery and found that it reduced the amount of morphine needed by 40% for a duration of about 24 hours. It has also been found that prefrontal fast rTMS can decrease headache pain and migraine frequency while right prefrontal slow rTMS can treat fibromyalgia (Sampson et al., 2006; Brighina et al., 2004). Since chronic back pain has been correlated with a reduction in the prefrontal cortex gray matter (Apkarian et al., 2004), prefrontal TMS can probably decrease that pain by inducing an increase in gray matter or stimulating limbic structures. It has been shown that rTMS could increase the mechanical and pain threshold in patients (Borckardt et al., 2009). Another explanation for TMS-induced pain reduction is that the pain caused by rTMS treatment is activating anti-nociceptive processes which in turn are releasing endogenous opioids (Borckardt et al., 2007). However, in all studies concerning pain, the effect of TMS was generally short lived with some studies encountering shorter effects than others especially if pain was not the primary complaint of the patient (Borckardt et al., 2009). A recent study by Taylor et al. (2013) found that naloxone, a u-opioid antagonist, was able to block the analgesic effect produced by a 10 Hz rTMS of BA 9 of the DLPFC. This showed that the analgesic effect augmented via left DLPFC rTMS is the result of a top-down opioidergic analgesia.

Of importance, TMS-induced discomfort at the site of stimulation has been shown to decrease pain where in a study on medication-free unipolar depressed patients, left prefrontal rTMS painfulness decreased by 48% throughout 3 weeks, independently of mood changes (Anderson et al., 2009). However, more studies are needed to attribute the pain reduction to cortical brain or scalp receptor stimulation (Anderson et al., 2009).

TMS as an Adjunctive Therapy

Many studies question the possibility of using TMS as a maintenance or adjunctive therapy on patients who have undergone an acute phase of TMS treatment. It is important to have an adjunctive TMS since the Sequence Treatment Alternatives to Relative Depression (STAR*D) Study stated that 40.1% of remitting patients who have previously failed one antidepressant treatment may relapse after an average of 4.1 months (Rush et al., 2006). In a study done by Li et al. (2004) three out of seven bipolar depression patients remained free from acute depression for 1 year after going through an acute TMS treatment followed by a year-long maintenance TMS treatment and medication, showing that TMS can be an effective adjunctive therapy. Another study by O'Reardon et al. (2005) showed a remission rate of about 70% for 6 weeks to 6 years. Moreover, a study performed by Fitzgerald et al. (2006) on 10 depressed TMS responders showed a relapse occurring 6 to 11.6 months after remission while Demitras-Tatlidede et al. (2008) studied 16 TRD patients in which 50% of them sustained a response to continuous courses of TMS where the effect of each course lasted for about 5 months. Furthermore, Cohen et al. (2009) followed up with 204 patients who underwent TMS and found that their remission rates were 75.3, 60, 42.7 and 22.6% at months 2, 3, 4 and 6 respectively.

A study by Janicak et al. (2010) on the durability of TMS benefit on pharmaco-resistant major depression revealed that out of 99 patients, 10% relapsed, 38.4% experienced worsening of conditions and 84.2% returned to remission. Comparable results were found by Mantovani et al. (2012) who examined the antidepressant effect of TMS on 37 out of 61 remitters for 3 months and found that 85% retained remission while 4% partially responded and 2% relapsed at about 7.2 ± 3.3 weeks but regained remission at 11 ± 2 weeks after the start of the follow-up. The overall relapse rate was 13.7% and the remission rate was 85%. It has also been shown that five sessions of daily TMS every fifth week would keep a patient in a depression-free state for a duration of 12 months (Mantovani et al., 2012).

Despite TMS acute therapeutic efficacy, some patients relapse while on medication maintenance and require another round of TMS sessions. A few studies suggest that in such cases, responses to TMS are high (Fitzgerald et al., 2006), implying that patients do not build up a tolerance against continuous administration of rTMS (Hadley et al., 2011). Furthermore, even in patients who received very aggressive treatment (180,000 pulses in 2 weeks) and had a vagus nerve stimulation electrode implanted in them, there were no adverse side effects (Hadley et al., 2011). Some studies also tried to identify certain factors that would predict the time of remission maintenance and it was found that having an acute response to therapy, low treatment resistance, young age and low number of treatments serve as a good prognosis for remission, but these were not replicated in other studies (Janicak et al., 2010). And it was found that a combination of medications and adjunctive TMS treatment is a good method for relapse prevention (Janicak et al., 2010).

TMS Feasibility

Another aspect of treatment that not so many studies have addressed is the cost-effectiveness of TMS. Simpson et al. (2009) quantified the Incremental Cost-Effectiveness Ratio (ICER)—a measurement of treatment benefit—of TMS per QALY (Quality-Adjusted Life Year) where an ICER below US$50,000 is considered highly cost effective (Simpson et al., 2009). Moreover, the WHO suggested that the limit for a treatment to be cost effective is three times the country's GDP per capita, meaning that the upper limit for the USA is $140,000. With a base cost of US$300 per TMS session, they found that the ICER of TMS to be US$34,999 per QALY, which is less than the range of US$50,000–$100,000 per QALY that has been proposed as a standard to pay for a new treatment (Simpson et al., 2009). What is even better is that the ICER would be further reduced to US$6667 per QALY if work productivity including reduced caregiver expenses and clinical recovery were taken into consideration (Simpson et al., 2009).

Deep Brain Stimulation

One of the most invasive techniques of brain stimulation is the deep brain stimulation, which have been applied on over 80,000 people worldwide with a mortality rate of 0–0.4% and intracranial bleeding, infection, seizures, transient confusion and skin complication in 0.4–5, 1.7, 1.5, 15.6 and 25% of patients respectively (Müller et al., 2013). DBS constitutes a multi-contact electrode implanted directly into the subcortical regions of the brain and connected to a pacemaker-like generator that is subcutaneously placed in

Table 19.1. Summary of the studies on TMS and DBS that are discussed in this chapter.

Author(s)	Brain Stimulation	Aim	Subject	Experiment (Parameters)	Result	Comment
Abelson et al. (2005)	DBS	Investigate the efficacy of DBS in treating refractory OCD	4 patients with refractory OCD	Bilateral lead implantation in the anterior limbs of their internal capsules at 130 Hz.	DBS was well tolerated and there were benefits to mood, anxiety, and OCD symptoms in one patient and moderate benefit in a second patient.	DBS has shown positive signs in treating refractory OCD but more investigation is still needed.
Anderson et al. (2009)	rTMS	Investigate whether rTMS procedural pain changes over time	20 patients with unipolar depression and who were non-responders to at least 3 weeks of rTMS treatment	3 weeks (15 sessions) of open-label rTMS with assessment of retrospective pain ratings and state emotional factors.	rTMS painfulness decreased by 48% from baseline, where it decreased greatly in the first few days and then steadily at an average of 2.11 points/session.	Procedural pain decreased irrespective of emotional changes.
Andoh et al. (2009)	TMS	Improved triangulation-based MRI guided method for coil positioning depending on individual MR data	10 healthy controls	Evaluated the spatial accuracy and reproducibility using functional MR activations of 2 targets in the motor and parietal cortices.	The assessment was according to the Euclidean distance (Dm) between the thumb motor target and the coil position that elicits thumb MEP with TMS. Accuracy was Dm: 10 ± 3 mm, reproducibility was Dm: 6.7 ± 1.4 mm and repositioning in the motor cortex was Dm 6.0 ± 3.2 mm in the parietal cortex.	This method proved to be a good method for effective positioning of TMS coil.
Boon et al. (2007)	DBS	Evaluate long term efficacy of MTL DBS in patients with MTL epilepsy	12 patients with refractory MTL epilepsy	Invasive video-EEG monitoring and evaluation for ictal-onset localization, subsequent stimulation, side effects and changes in seizure frequency.	After 12–52 months, 1/10 patients was seizure free (>1 year), 1/10 patients had a >90% seizure frequency reduction; 5/10 patients had ≥50% seizure-frequency reduction, 2/10 patients had	The potential efficacy of DBS in MTL epilepsy patients should be further studied.

Borckardt et al. (2009)	rTMS	Investigating the effect of left fast PFC rTMS on chronic neuropathic pain	4 patients with chronic neuropathic pain	Each patient underwent 3 real and 3 sham 20-minute sessions of 10 Hz left PFC rTMS for 3 weeks.	30–49% seizure-frequency reduction 1 was a nonresponder. No side effects were reported. Real TMS was associated with decrease in average daily pain in 3/4 participants independently of changes in mood in two participants.	rTMS of the PFC may manage certain neuropathic pain syndromes.
Borckardt et al. (2007)	rTMS	Assess the effect of rTMS on thermal pain threshold	20 healthy adults	Participants were blindly assigned either real or sham rTMS. Thermal pain threshold was assessed before and after 15 min of left PFC rTMS (10 Hz, 100% MT, 2 sec on/60 sec off, 300 pulses total).	Participants undergoing real rTMS experienced an increase in thermal pain threshold.	rTMS over the left PFC is a promising technique for pain management.
Borckardt et al. (2006c)	rTMS	Assessing different methods of reducing pain during rTMS	7 healthy adults	Participants rated the unpleasantness of left PFC rTMS before and after (1) application of anesthetics cream, (2) lidocaine scalp injection, (3) lidocaine and epinephrine scalp injection, and (4) with or without thin foam sheets between the scalp and coil.	Anesthetic injections were associated with a decrease in pain, whereas the cream had no effect. The foam sheets had little effect on decreasing pain.	Further research is needed to assess the different techniques for pain reduction.
Borckardt et al. (2006b)	rTMS	Investigating five different PEST methods in MT estimation	Computer simulation models	For each method, the mean number of TMS trials needed to reach the MT estimate and the mean accuracy of the estimates, were recorded.	A simple nonparametric PEST provided the most accurate MT estimates, but took a bit longer than maximum likelihood PEST.	The pros and cons of each PEST technique were mentioned.

Table 19.1. contd....

Table 19.1. contd.

Author(s)	Brain Stimulation	Aim	Subject	Experiment (Parameters)	Result	Comment
Borckardt et al. (2006d)	rTMS	Investigating the effect of postoperative rTMS on analgesia use	20 gastric bypass surgery patients	20 min of active or sham left PFC rTMS applied directly after surgery while tracking the analgesia pump use. Patients rated pain and mood twice per day.	40% reduction in morphine use 44 hrs after surgery, in real rTMS group.	Results suggest a potential use of rTMS in restricting morphine use.
Chae et al. (2004)	rTMS	Assessing the safety of rTMS in Tourette's Syndrome	8 TS patients	rTMS at 110% MT over left PFC (twice) or left motor cortex (twice), using either 1 Hz or 15 Hz TMS, or sham TMS (once).	Tic symptoms improved over 1 week.	rTMS may be a safe and effective treatment for TS.
Encinas et al. (2010)	DBS	Study the effect of HF stimulation of the ATN, on neurogenesis in the DG	Age-matched 13-week-old C57BL/6 mice and aged-matched 2-month-old Nestin-CFPnuc transgenic animals	Changes in certain classes of neural progenitors were quantified after ATN-DBS.	Increase in symmetric division of a specific class of DG neural progenitors, which later led to an increase in number of new neurons.	Stimuli of different types can converge on the same DG neurogenic center. Hippocampal neurogenesis can be an indicator of the limbic circuitry activation via DBS.
Ewing et al. (2013)	DBS	Effect of ventral hippocampus DBS on schizophrenic rodents	Rodent model of schizophrenia (MAM-E17)	Bipolar stimulating electrodes were bilaterally implanted into the ventral hippocampus. HF stimulation was delivered and spectral analysis of amplitude of auditory evoked potential and resting state local field potentials were examined.	Rats exhibited changes in certain components of the auditory evoked potential in the ventral hippocampus, infralimbic cortex, mediodorsal thalamic nucleus and the nucleus accumbens in the left hemisphere. DBS reversed the evoked deficits to levels similar to those observed in control animals while it had no effect on control animals.	DBS normalizes deficits of auditory evoked potentials generation.

Fisher et al. (2010)	DBS	Assess the efficacy of DBS of the Anterior Nucleus of Thalamus in treating epilepsy	110 patients that are 18–65 years old with partial seizures (at least 6 per month, but no more than 10 per day)	A blinded phase where half of the patients had their DBS electrodes on at 5 V and the other half had their DBS electrodes off. This was followed by an open label phase where all patients had their devices turned on for 9 months.	Long-term follow-up yielded a 41% reduction in median seizure frequency at 13 months and a 56% reduction at 25 months.	Some patients might benefit more than others from this treatment. Side effects included transient paresthesia, depression and memory impairment.
Fytagoridis et al. (2013)	DBS	Investigate the spatial distribution and stimulation-induced side effects of PSA	28 patients with essential tremor	The side effects were evaluated 1 year after DBS surgery.	The study showed that similar responses (muscular affection and cerebellar symptoms) can be induced from different points in the PSA.	The side effects couldn't be attributed to specific anatomical locations in the PSA.
George et al. (2010)	rTMS	Assess rTMS LPFC in treatming MDD	190 outpatients with MDD and not taking medication	Daily rTMS to LPFC at 120% MT (10 Hz, 4 sec train and 26 sec intertrain interval) for 37.5 min (3000 pulses per session) using a figure-eight solid-core coil. Sham rTMS with a similar coil and scalp electrodes that delivered the same somatosensory sensations.	Attaining remission was 4.2 times more likely in patients with active treatment than sham.	Daily rTMS showed positive results in treating MDD.
Greenberg et al. (2010)	DBS	Investigate the effect of VC/VS DBS on OCD symptoms	26 OCD patients from different centers	VC/VS DBS stimulation at 100-130 Hz.	Reduction of OCD symptoms and improvement of function in two thirds of patients.	VC/VS DBS is a promising therapuetic technique for the treatment of OCD and it would be more effective at a more posterior target.

Table 19.1. contd....

Table 19.1. contd.

Author(s)	Brain Stimulation	Aim	Subject	Experiment (Parameters)	Result	Comment
Hadley et al. (2011)	rTMS	Assess tolerability, safety and effectiveness of high doses of daily left PFC rTMS as adjunctive therapy for TRD	19 TRD patients	18 months of daily left PFC rTMS at 120% MT, 10 Hz, 5 sec on/10 sec for a mean of 6800 stimuli/session with all patients continuing their antidepressant medication.	High rTMS doses were well tolerated, and depression, suicidal ideation, quality of life, and social and physical functioning were all improved within 1–2 weeks.	High doses of daily left PFC rTMS can be safely used as an adjunctive therapy for TRD.
Herbsman et al. (2009a)	rTMS	Assess the relation between variable coil placement and rTMS antidepressant efficacy	54 depressed subjects	Used MRI scans to image fiducial placement relative to anatomy and employed an imaging-processing algorithm to define the stimulated cortical region.	Patients with electrodes placed more anteriorly and laterally were more likely to respond.	After placing the electrode using the 5 cm rule, stimulating more anteriorly and laterally improved response rates.
Herbsman et al. (2009b)	TMS	Investigate relation between different structural parameters of the corticospinal tract and TMS MT	17 subjects, with and without schizophrenia	Measured skull-to-cortex over the left motor region and obtained structural and diffusion tensor MRI scans and measured the FA and principal diffusion direction of certain regions in the corticospinal tract following TMS.	FA, age and schizophrenia status were not correlated to MT while anterior–posterior trajectory of principle diffusion direction and skull-to-cortex distance were highly correlated with MT.	Skull to cortex distance and motor bundle fiber orientation, are responsible for 80% of the variance seen in TMS MT.
Janicak et al. (2010)	TMS	Assessing the durability of TMS effect on MDD patients who had an acute response to TMS	301 MDD patients	301 patients assigned to sham or active 6-week TMS; non responders enrolled in an extra 6-week study and patients with partial response (n = 142) were tapered off TMS over 3 weeks, while taking maintenance antidepressant	120 patients agreed to 24 week follow-up. 99 were active responders and 21 sham responders. 10/99 patients relapsed. 38/99 had worsening of symptoms and 32/38 reachieved benefit from adjunctive TMS.	TMS effects are durable and TMS can be used as an intermittent rescue strategy from relapse.

Kozel et al. (2011)	rTMS	Assess the effect of daily left HF rTMS on FA of PFC white matter	8 TRD outpatients	monotherapy. Long-term durability was assessed for 24 weeks where TMS was readministered if patients experienced symptom worsening. After 4 to 6 weeks of daily rTMS, mean FA levels were assessed and compared between active rTMS and sham rTMS for the right and left PFC white matter.	There was a significant mean increase in left PFC white matter among the active rTMS participants.	rTMS causes no damage to white matter and diffusion tensor imaging can help increase the understanding of the underlying therapuetic mechanism of rTMS.
Leone et al. (2013)	DBS	Assess hypothalamic stimulation in drug-resistant CH patients	19 patients with chronic CH and who were unresponsive to prophylactics	DBS of the posterior inferior hypothalamic area ipsilateral to the pain with a follow-up for 6–12 years.	6/17 patients are almost pain-free; another 6 experience episodic CH with interspersed remissions. 5 patients did not improve (4 had bilateral CH, and 3 developed tolerance). Adverse events include electrode displacement, electrode malpositioning, infection, third ventricle hemorrhage, slight one-sided muscle weakness and seizure.	Stimulation can be tolerated for years, pain-free condition can be maintained even with DBS turned off, tolerance can occur and bilateral chronic CH responds poorly to hypothalamic DBS.
Mallet et al. (2008)	DBS	Assess the efficacy of subthalamic nucleus DBS in treating severe OCD	18 OCD patients from 10 academic centers	DBS stimulation of the subthalamic nucleus at 130 Hz for two 3 month periods.	Symptoms became less severe and levels of functioning increased but active stimulation did not affect depression, anxiety and the neuropsychological status of patients.	There were 15 serious adverse events, including permanent neurological sequelae, and 23 non-serious events.

Table 19.1. contd....

Table 19.1. contd.

Author(s)	Brain Stimulation	Aim	Subject	Experiment (Parameters)	Result	Comment
Malone et al. (2009)	DBS	Investigated the efficacy of VC/VS DBS for the treatment of refractory depression.	15 patients with chronic, severe, highly refractory depression	Electrodes were implanted bilaterally in the VC/VS region. All patients received continuous stimulation at 100 or 130 Hz and were followed for 6 months –4 years.	Responder rates with the HDRS were 40% at 6 months and 53.3% at the last follow-up while remission rates were 20% at 6 months and 40% at the last follow-up.	VC/VS is a promising technique for the treatment of refractory major depression.
Marshall et al. (2012)	DBS	Evaluating alternating verbal fluency measures in PD patients who underwent STN-DBS	23 STN-DBS patients and 20 non-surgical PD patients	Neuropsychological assessment, at base-line and at 6-month follow-up for cued and uncued intradimensional and extradimensional alternating fluency measures.	STN-DBS patients demonstrated a greater decline on the cued and uncued fluency tasks compared to the non-surgical patients.	Changes in alternating fluency (decline in information processing) are not due to PD progression only but to STN-DBS too.
May et al. (2007)	rTMS	Study brain alterations and neuroplasticity following rTMS	36 healthy subjects	Stimulation to the superior temporal cortex for 5 days at 1 Hz and at 110% MT.	Macroscopic cortical changes in gray matter of the auditory cortex.	The study suggests the existence of fast adjusting cortical plasticity in adult humans unlike slow mechanisms of neuronal and glial cell genesis.
McCairn et al. (2013)	DBS	Investigate the neuronal mechanism of the therapuetic effect of GPi-HF-DBS	Nonhuman primate (Macaca fuscata) that exhibitsf BG-meditated motor tics	HF-GPi-DBS with alternating on and off conditions with a time interval of 30 sec.	During on-stimulation, there was a reduction in tic-related electromyograph amplitude. DBS stimulation induced patterns of inhibition and excitation in GP segments. Neuronal firing rates and behavior modulation were	Gpi DBS may suppress motor tics through temporal locking with the pulse.

related to the decrease in tic-related phasic activity in pallidal cells. Firing rate alterations were associated with the cells' activity locking with the DBS pulse.

Author	Technique	Objective	Sample	Method	Results
McDonald et al. (2011)	TMS	Assess efficacy of increasing number of fast left rTMS to achieve remission in TRD and determine if non-remitters remit via slow right rTMS	141 TRD patients who failed to respond to an rTMS sham-controlled trial	141 patients underwent fast left rTMS (10 Hz, 120% MT over the left DLPFC) for 3–6 weeks and those who did not remit underwent slow right rTMS (1 Hz, 120% MT over the right DLPFC) for an additional 4 weeks.	43 patients remitted after 26 active treatments (90,000 pulses) and 26% of patients who didn't remit via fast left, remitted during slow right treatment. To achieve remission from TRD, a high number of rTMS stimulation is needed and TRD nonresponders to fast left rTMS might remit to slow right rTMS or more rTMS stimulations.
Min et al. (2012)	DBS	Trace DBS-induced neuronal network activation using fMRI	11 normal pigs	STN DBS in 7 pigs and entopeduncular nucleus (EN) DBS (non-primate analog of the primate Gpi) in 4 pigs.	STN and EN DBS increased BOLD activation in the ipsilateral sensorimotor network and each activated brain area showed a group of network connections. DBS may have modulatory effects on areas of higher cognitive and emotional processing in addition to motor areas.
Morrell (2011)	DBS	Assess safety and effectiveness of cortical open loop DBS as an adjuvent for partial onset seizures in adults with refractory epilepsy.	191 patients with medically intractable partial epilepsy	Either subdural or depth electrodes were implanted at 1 or 2 seizure foci to deliver an electrical stimulation to counteract any abnormality that begins. There was a 3-month blinded phase where half of the patients received stimulation, followed by an 84-week open label phase, in which all stimulators were turned on.	In the long term follow-up 46% of patients showed a 50% reduction in the mean seizure frequency. There were no side effects and the active treatment group had greater improvements than the sham group in the domains of visuospatial ability, verbal functioning, and memory.

Table 19.1. contd....

Table 19.1. contd.

Author(s)	Brain Stimulation	Aim	Subject	Experiment (Parameters)	Result	Comment
Schlaepfer et al. (2013)	DBS	Assess safety and efficacy of DBS to supero-lateral branch of medial forebrain bundle in treating refractory depression	7 patients with highly refractory depression	Patients underwent stimulation for 12-33 weeks at 130 Hz.	Onset of antidepressant efficacy was fast, and most of the patients responded at lower stimulation intensities compared to previous studies and there were 4 remitters.	Bilaterally stimulating the supero-lateral branch of the medial forebrain bundle seems an effective method for treating TRD.
Sillay et al. (2013)	DBS	Quantify electrode displacement and brain shift	12 patients with movement disorders who underwent DBS surgery	One post-operative MRI and one time-delayed CT was obtained for each patient to measure electrode deviation.	The greatest shift was in the rostral direction, reaching an average of 1.41 mm.	Further studies should be done to assess the effects of brain shift on rigid and non-rigid devices implanted while in the supine or semi-sitting position.
Sutton et al. (2013a)	DBS	Investigate if elevated potassium level can elicit the same DBS therapuetic effects	Hemiparkinsonian rats - Male Sprague-Dawley (SD) rats	20 mM KCl injection into the (SNr) or the subthalamic nucleus followed by limb-use asymmetry test and self-adjusting stepping test.	KCL injections led to depolarization and a reduction in SNr neuronal activity and SNr beta-frequency oscillations that are associated with akinesia/bradykinesia in PD patients/animals.	Elevated potassium level improved forelimb akinesia, and decreased and desynchronized SNr neuronal activity and SNr beta frequency, respectively. This ionic mechanism may provide an insight on the mode of action of DBS.
Sutton et al. (2013b)	DBS	Assess whether SNr DBS improves motor function in hemiparkinsonian rats	Hemiparkinsonian rats - Male Sprague-Dawley (SD) rats	SNr-HFS (150 Hz) followed by stepping and limb-use asymmetry tests.	After stimulation, rats touched equally with both forepaws. SNr beta oscillations decreased and	Improvement of forelimb akinesia probably due to the

				SNr neuronal spiking activity decreased while ventromedial thalamus neuronal spiking activity increasedt	decrease of SNr neuronal activity and SNr beta oscillations, and increased activity of ventromedial thalamic neurons, meaning that SNr DBS may treat motor symptoms.
Taylor et al. (2013)	rTMS	Investigate the method of augmenting the analgesic effect of rTMS	14 healthy subjects	Stimulation of the BA9 of the left DLPFC at 10 Hz.	The analgesic effect augmented via left DLPFC rTMS is the result of a top-down opioidergic analgesia.
				10 ml of intravenous naloxone (0.1 mg/kg) was able to block the analgesic effect of rTMS.	
Toda et al. (2008)	DBS	Examine DBS effect on hippocampal neurogenesis	Adult male Sprague-Dawley rats	ATN-DBS follcwed by BrdU-positive cells counting in the DG. Fate of cells was investigated via staining for doublecortin, NeuN, and GFAP. Number of DG BrdU-positive cells was compared between animals treated with corticosterone (hippocampal neurogenesis suppressor) and sham surgery, vehicle and sham surgery, or corticosterone and ATN-DBS.	HF ATN-DBS increases hippocampal neurogenesis and counteracts neurogenesis suppression and may be associated with enhanced behavioral performance.
				2-3 fold increase in the number of DG BrdU-positive cells where most of those cells assumed a neuronal cell fate. Corticosterone decreased the number of those cells but ATN-DBS reversed it.	
Wolz et al. (2012)	DBS	Study the effect of STN DBS on non-motor symptoms in PD	34 PD patients	Patients underwent bilateral STN-DBS and the frequency and severity of 10 non-motor symptoms were assessed.	STN DBS doesn't affect frequency of non-motor symptoms nor worsens them but may improve psychiatric symptoms.
				The most prominent symptoms were fatigue (85% of patients), problems with concentration/ attention (71%) and inner restlessness (53%), but symptoms improved in some patients.	

the sub-clavicular region (George et al., 2007; Nahas and Anderson, 2011; Cenci et al., 2011). The electrode has four contact points with a variable gap between the points, depending on the type of electrode being used (Min et al., 2012), and applies a high frequency stimulation of 130–180 Hz (Sutton et al., 2013a), amplitudes of 2–5 V and rectangular impulses of 60–120 μsec (Müller et al., 2013).

In North America, Medtronic 3387 electrode with a gap of 1.5 mm for DBS implantation is often used (Min et al., 2012). The electrode is placed in a flexible insulating carrier and is driven through an insertion tube via a stiff stylet and once the target is reached, they fix the electrode at the burr hole and disconnect the stylet after removing the insertion tube (Sillay et al., 2013). Accurate targeting is one of the main concerns of DBS because the treatment efficacy is highly dependent upon it. Thus, scientists constantly work on improving the way they target brain regions by using and enhancing imaging techniques preoperatively or intra-operatively and even using microelectrode recording for more information during intraoperative imaging (Sillay et al., 2013). Postoperative imaging has also been used to check if the electrodes were correctly implanted and if there are any complications that may have resulted from the surgery (Sillay et al., 2013). DBS is being mostly used nowadays for end-stage PD and some other psychiatric disorders (Benazzouz and Hallett, 2000; Piasecki and Jefferson, 2004) including OCD for which it is FDA approved (Nahas, 2010), depression and TS which are revealing promising results (Nahas, 2010).

In the treatment of movement disorders, DBS targets include the GPi, the thalamic nucleus: ventralis intermedius (Vim), and the subthalamic nucleus (STN) that is associated with a more severe cognitive behavior (Min et al., 2012). Vim DBS is used to reduce tremors and not the more severe PD, and its therapeutic effects can last for as long as 10 years but it does cause complications such as dysarthria in about 20% of patients (Chao et al., 2007). It is even said that a good effect is established when stimulating the Posterior Subthalamic Area (PSA), especially the prelemniscal radiation and the zona incerta (Fytagoridis et al., 2013). GPi and STN are the most common DBS targets since they are important relay regions in the basal ganglia-thalamocortical connections and project to limbic, sensorimotor and associative cortical areas (Min et al., 2012). STN stimulation improves the Unified Parkinson's Disease Rating Scale (UPDRS) by 49% compared to 37% improvement due to GPi stimulation (Chao et al., 2007).

Actually, STN stimulation was more effective in treating akinesia, tremors and reducing the need for L-Dopa drugs while GPi stimulation was more effective in treating dyskinesia (Min et al., 2012). Even though DBS was mostly used to treat motor symptoms, many studies are studying its effect on non-motor symptoms such as Nazzaro et al. (2011) who found that there is a decrease in non-motor symptoms after 1 year of bilateral STN DBS.

Moreover, high frequency DBS has been used to treat TRD by stimulating the caudate nucleus, subgenual cingulate, Anterior Thalamic Nucleus (ATN), nucleus accumbens (NAc) and the anterior limb of the internal capsule (Nahas and Anderson, 2011; Nahas et al., 2010). Furthermore, Positron Emission Tomography (PET) and fMRI have been used to detect DBS induced changes and it was found that DBS causes ipsilateral deactivation or activation of the primary sensorimotor cortex, supplementary motor area, premotor cortex, basal ganglia, DLPFC, superior PFC, medial PFC, caudate, anterior thalamus, anterior cingulate, brainstem and contralateral cerebellum (Min et al., 2012; Stefurak et al., 2003).

DBS Side Effects

DBS parameters can be easily manipulated throughout a single fMRI session that allows one to change and test the stimulus intensity on the subcortical-cortical pathway (Min et al., 2012). DBS, like all other brain stimulation techniques, can cause psychiatric and cognitive side effects due to the current spread into various brain structures surrounding the electrode implantation site (Min et al., 2012). Thus, the observed unwanted psychiatric symptoms may be due to accidentally stimulating specific brain regions rather than side effects of the DBS treatment itself (Müller et al., 2013). Some of the DBS side effects include suicide, with a rate of 0.16–0.32% (Müller et al., 2013), depression, euphoria, hypomania, hypersexuality and impulsivity and which are not surprising given all the interconnections with the DBS stimulated areas (Min et al., 2012). DBS of the PSA for example, can induce dystonic posturing at a low amplitude of 2.6 V, dysarthria due to cerebellothalamic fiber stimulation medial to the STN, and paresthesias at a low amplitude and that has no somatotopic pattern of incidence (Fytagoridis et al., 2013).

Wolz et al. (2012) explain that side effects may be due to electrode misplacement, where implanting the electrode in the SN, ventral to the STN, would cause depression. Yet, placing the electrode in its right position in the STN would decrease fatigue, anxiety, restlessness, concentration problems, dysphagia and bladder urgency (Wolz et al., 2012). Not to mention that different brain regions exhibit different risk severities upon stimulation where PSA, when stimulated, shows variable side effects at lower stimulation intensity and with less treatment impedance than Vim (Fytagoridis et al., 2013). So, to obtain the maximum benefit from a DBS treatment, the current should be focused on the target area with only minimal charge spread to other regions, which would be responsible for the differential target-specific hostile effects (Kringelbach et al., 2007; Montgomery and Gale, 2008). One way of preventing current spread while keeping a high charge density is by minimizing the voltage (1 volt) and increasing the pulse duration (500 μ sec) (Kringelbach et al., 2007; Lozano et al., 2002). Accordingly, in order to further decrease unwanted responses, in DBS for motor diseases, the electrodes should be placed in a way that takes into consideration the consequences of stimulating non-motor cortical areas, and the opposite would be true for treating non-motor diseases such as OCD and depression (Min et al., 2012).

A final important consequence of DBS surgery is brain shift, which is the distortion and/or displacement of the brain laterally and caudally with respect to the burr hole due to gravity, pneumocephalus, CSF egress and mechanical forces produced by the electrode implantation (Sillay et al., 2013). However this shift is reversible, where it starts as soon as the skull and dura are opened to reach a climax during surgery and then resolves days or weeks postoperatively (Sillay et al., 2013). Unfortunately, even when the pneumocephalus and the brain shift resolve, a secondary shift of more than 3 mm of the implanted electrode will remain, resulting in reduced efficacy of neurostimulation (Petersen et al., 2010).

DBS Mode of Action

STN and GPi stimulation leads to the activation of ipsilateral primary motor and somatosensory cortex, premotor cortex, DLPFC, head of caudate, insular cortex and anterior cingulate cortex, meaning that these regions contribute in the therapeutic effect of PD and other movement disorders (Grill et al., 2004; McIntyre et al., 2004a,b). Furthermore, STN is smaller than GPi and more sensitive to stimulation intensity, meaning that GPi requires higher intensities than STN to produce the same therapeutic effect (Min et al., 2012). However, STN stimulation does produce more activation in the putamen and caudate than GPi stimulation, and that may explain why there are more STN rather than GPi DBS patients who are able to reduce their medication dosages (Study-Group, 2001). Moreover, in a study by Min et al. (2010) STN DBS led to seven clusters of functionally defined regions (sensorimotor, cognitive/emotional, and basal ganglia-thalamocortical) while GPi DBS led to five clusters without any known neural-network pattern. In the end, even though STN and GPi stimulation activates a common sensorimotor circuit, each one of them also activates a distinct neural circuit (Min et al., 2012).

DBS activates efferent axons and depolarizes neurons by inactivating voltage-dependent ion channels where a 150 Hz stimulation of the entopeduncular nucleus for 10 seconds increases the concentration of potassium to 18 mM for over a minute through neuronal rather than glial sources (Sutton et al., 2013a). In order to make sure if the increase in potassium concentration is what is mediating the therapeutic effects of DBS,

Sutton et al. (2013a) injected the substantia nigra reticulata (SNr) and STN of hemiparkinsonian rats with 20 mM of KCl and found an improvement in contralateral forelimb akinesia correlated with increased neuronal activity. However, when 10 mM of KCl were injected no improvement was observed and when 50 mM of KCl were injected, contralateral forelimb dyskinesia was observed. On the other hand, when they decreased the volume of injected KCl from 10 uL to 5 uL, no improvement was observed and that could be because there is a "critical mass" of SNr that should be modulated or there are other fiber tracts or nuclei also stimulated (Sutton et al., 2013a). Holthoff and Witte (2000) have also found that a 2 second 50 Hz stimulation leads to a wide dispersion of potassium ions reaching 900 μm away from the electrode (Holthoff and Witte, 2000). So the potassium ions can inhibit SNr by activating GABAergic afferent neurons or activating GPe fibers through glutamatergic efferents. This can inhibit STN which in turn reduces its excitation of SNr, causing a decrease in beta oscillations and akinesia (Sutton et al., 2013a,b). Nevertheless, one should be careful when stimulating the SNr since it has been shown that medial SNr stimulation leads to hypomania due to its connections with the caudate and limbic ventromedial STN (Sutton et al., 2013b). Furthermore, the internal capsule could also be stimulated due to a current spread which would also mediate other consequences (Sutton et al., 2013b).

STN DBS treatment has been associated with mood and motor regulation where right DBS induced acute depressive dysphoria while left DBS improved motor symptoms (Stefurak et al., 2003). It is important to note that left DBS decreased sensorimotor and supplementary motor cortex activity and increased the activity of brain regions that control motor activities such as the motor and premotor areas, putamen, ventrolateral thalamus and cerebellum (Min et al., 2012). On the other hand, right DBS showed the same changes in activation but to a lesser extent in the motor areas (Min et al., 2012). STN DBS has also been associated with an increase in fludeoxyglucose (FDG) uptake in the right dorsal anterior cingulate and DLPFC, which in turn leads to a decrease in verbal fluency (Kalbe et al., 2009). A study by Marshall et al. (2012) has shown that after 6 months of STN DBS, patients face a decline in cued phonemic/phonemic fluency and uncued phonemic/semantic fluency due to changes in frontostriatal functioning that is worse than in PD patients. In that study, PD patients could not generate strategies they internally organized but they became better with practice, and while STN DBS patients did not benefit at all from practice to use the cues to perform better, bilateral STN DBS patients were able to rely more on cues for tasks with increasing demands. Moreover, anteromedial STN stimulation both reduced motor symptoms and produced a hypomanic state due to limbic and association cortex activation (Benarroch, 2008; Mallet et al., 2007).

Several studies have been done to assess the effects of STN stimulation in particular and they found that it evokes excitatory postsynaptic potentials in the dopaminergic neurons of substantia nigra pars compacta (SNRc) and increases dopamine release in rats (Lee et al., 2003, 2004, 2006). Similar results were also observed in STN DBS in pigs where increase in dopamine release was also dependent on the intensity and frequency of the stimulation (Shon et al., 2010). However, the increase in dopamine release in humans could produce behavioral difficulties like impulse control disorders (Broen et al., 2011). One example is a PD patient who underwent ventral STN DBS, which evoked mania and thus had to resort to dorsolateral STN stimulation to reverse the side effects (Ulla et al., 2006). Chopra et al. (2011) also had another PD patient who suffered from a voltage-dependent mania upon STN DBS stimulation. However, since PET studies on PD patients showed that there are no changes in dopamine release in PD patients undergoing DBS (Strafella et al., 2003), it is difficult to assert the exact role of dopamine in DBS (Min et al., 2012). Moreover, the serotonin transmission system is impaired in PD patients and so, if the SNRc is stimulated via DBS, due to current spread for example, this could further impair the serotonin neurotransmission and thus worsen the PD patient's condition (Tan et al., 2013). The serotonin transmission impairment could be via modulating the neural connections (efferents and afferents) between SNRc and the dorsal raphe nucleus (DRN), which in turn is rich in serotonin producing neurons (Tan et al., 2013).

DBS, Parkinson's Disease and Biomarkers

Parkinson's disease is a neurodegenerative disease with a prevalence of 1% in the population of people aged 60 and above and 4% of people aged 80 and above (Benito-León et al., 1998). It is the second-most common chronic neurodegenerative disorder, which is characterized by dopaminergic nigrostriatal neuronal loss in the SNRc area, and the formation of intracytoplasmic inclusions called Lewy Bodies (Michell et al., 2004). Some of the major clinical symptoms include rigidity, progressive tremor and bradykinesia (Michell et al., 2004). Loss of neurons beyond the dopaminergic level does in fact occur and may grow to cause problems such as cognitive deficits (Michell et al., 2004). There is a major difficulty in the early diagnosis of PD since it is mimicked by many other similar diseases, such as progressive supranuclear palsy and Multiple System Atrophy (MSA) (McGhee et al., 2013). In addition to that, depression in PD has been attributed to a decrease in OFC and cingulate white matter (Kostic et al., 2010) while apathy was correlated with gray matter reduction in the cingulate and insular cortex (Reijnders et al., 2010).

A lot of functional neuroimaging research has been done to study the pathophysiology of PD with particular interest in the role of non-motor circuits (Min et al., 2012). Neuroimaging techniques has helped scientists figure out the reasons behind some PD symptoms such as the decrease in verbal learning which has been attributed to a metabolic decrease in dorsal anterior cingulate and left OFC (Min et al., 2012). In functional imaging, and through the use of PET and Single-Photon Emission Computed Tomography (SPECT), the following biomarkers can be detected: a clear decline in the function of neurotransmitter, changes in metabolic activity detected through the use of 18F-FDG, development of new ligands, and protein accumulation such as beta-amyloid and alpha-synuclein (Zhuang et al., 2001).

Through the use of transcranial ultrasound, the SN is found to be bilaterally increased (Michell et al., 2004). However, using cardiac metaiodobenzylguanidine (MIBG) scintigraphy, researchers were able to detect the a formation of Lewy bodies spread through the autonomic nervous system (McGhee et al., 2013). Also, it was noted that cardiac to mediastinal MIBG uptake ratio might help in the identification of idiopathic PD (Michell et al., 2004).

Several biomarkers exist in clinical procedures which include affective and psychological tests, such as depression rating scales, and PD's specific depression attributes such as feelings of self-reproach (Huber et al., 1990) and anxiety-related depression (Hoogendijk et al., 1998), tests of motor performance, such as high-level motor control, and finally clinical neurophysiology, such as tremor, bradykinesia, and rigidity (Valls-sole and Valldeoriola, 2002). Alternatively some biomarkers are observed in olfaction, such as loss of smell detection due to olfactory bulb neurodegeneration (Michell et al., 2004), and vision, such as abnormal color vision and sensitivity in contrast (Buttner et al., 1995).

In serum, there is a mitochondrial-complex I reduction in PD patients' SNbrain region as well as in platelets (Michell et al., 2004). Furthermore, oxidative stress biomarkers can be used in this case since they were found to be at different levels. These markers include malondi-aldehyde, superoxide radicals (Michell et al., 2004), 8-hydroxy-20-deoxy-guanosine from oxidized DNA and 8-hydroxyguanosine from RNA oxidation (Kikuchi et al., 2002), the coenzyme Q10 redox ratio (Götz et al., 2000). It was also noticed that the level in dopaminergic transporter immunoreactivity is reduced in lymphocytes of PD patients (Caronti et al., 2001). In the CSF, a significant increase was noticed in levels of reactive oxygen species malondialdehyde (Michell et al., 2004), and a decrease was found in the levels of b-phenylethylamine (Zhou et al., 1997).

Even though L-DOPA is considered the best choice for PD treatment, one of its drawbacks is that patients' response to treatment decreases as the disease progresses, where they would start needing higher dosages and more doses per day (Cenci et al., 2011). That usually happens in 40% of patients after 4–6 years of treatment (Ahlskog and Muenter, 2001) and in 90% of patients after 10 years of therapy (Mazzella et al., 2005), not to mention that they would also experience dyskinesia and motor fluctuations (Cenci et al., 2011). These motor complications or "wearing-off phenomenon" is due to dysfunctional plastic changes taking place in the patients' brains (Linazasoro, 2005). Furthermore, L-DOPA is incapable of improving the symptoms of advanced PD, such as cardiovascular, gastrointestinal and urological symptoms, mood shifts, sleep problems, cognitive deficits (Chaudhuri and Schapira, 2009) and neurodegeneration (Schapira, 2004). It is because of all of this that scientists headed towards other ways of treating PD and found positive results when using injections or continuous subcutaneous apomorphine infusion, non-ergolinic dopamine receptor agonist transdermal patches and most importantly, DBS, whose efficacy is still being maximized (Cenci et al., 2011).

DBS has proven to be better than pharmacological therapy, with STN DBS being superior to GPi DBS. On the other hand, what makes DBS a second option is that it is an invasive intervention with a higher risk of complications (Cenci et al., 2011). STN stimulation improves UPDRS motor scores by 50–52% and reduces L-DOPA dosages by 50–60% (Kleiner-Fisman et al., 2006) and dyskinesias by 54–74% (Bötzel et al., 2006) while GPi stimulation improves the score by 33% (Stefani et al., 2007) without decreasing the L-DOPA dosage (Anderson et al., 2005) and greatly reduces dyskinesia for at least 3–4 years (Moro et al., 2010). Some of the risks of DBS that render it unfavorable to patients include the risk of infections, intracerebral bleeding and electrode misplacement which occur in about 4% of patients with a mortality of 0.4 and 1% long-term morbidity (Voges et al., 2007). Even though STN and GPi both have the same surgical risks, STN stimulation induces more cognitive side effects in older patients and increases the risk of suicide by 15 folds during the first year following the DBS operation (Cenci et al., 2011). Yet, STN DBS is always preferred over GPi DBS except in cases where dyskinesia is the main problem or when the aim is to decrease cognitive side effects in the elderly (Cenci et al., 2011). Because of all of the above complications, scientists are working on novel methods of treating PD, one of which is gene therapy where they can either implant genetically modified cells or deliver genes to already existing cells via a viral vector (Cenci et al., 2011). The delivered gene would aim at restoring the enzymatic machinery of producing either L-DOPA or dopamine (Cenci et al., 2011). They are even targeting non-dopaminergic systems such as the glutamate receptor (Cenci and Lundblad, 2006) and serotonin receptors (Svenningsson et al., 2002; Zhang et al., 2008) which are involved in L-DOPA induced dyskinesia and modulating signaling pathways of dopamine uptake and conversion downstream of dopamine receptors, respectively. Even adrenergic and adenosine receptors and opioids have found their way into future non-dopaminergic treatments of PD (Cenci et al., 2011).

DBS and Obsessive Compulsive Disorder

For the treatment of OCD, open studies were done to assess the benefit of DBS on different brain regions, which led to the US FDA approval under a Humanitarian Device Exemption (HDE). In an 8-year study by Greenberg et al. (2010) highly treatment resistant OCD was treated via DBS applied to the ventral anterior limb of the internal capsule and the adjoining ventral striatum (VC/VS). The study yielded a clinically significant reduction in OCD symptoms and an improvement in function in about two-thirds of patients. At first, they implanted the electrodes in an anterior-posterior location based on anterior capsulotomy lesions. On the other hand, when they placed the electrodes more posteriorly at the junction of the anterior capsule, anterior commissure and posterior ventral striatum, more effective results were achieved. DBS was well tolerated and the side effects were transient and included exacerbations of psychiatric symptoms. Noteworthy, when DBS was interrupted, there was a greater worsening in mood symptoms than in compulsions and obsessions. In another study by Abelson et al. (2005) DBS was bilaterally applied more anteriorly in the internal capsule of four OCD patients that were incompliant with treatments. Patients tolerated DBS with some side effects such as psychosocial stress, however, only one patient had a significant improvement in anxiety, mood and OCD symptoms while another patient had a moderate benefit during follow-up. A double-blind placebo controlled study of VC/VS DBS for OCD is currently ongoing.

In a case report by Aouizerate et al. (2004), DBS of the ventral caudate nucleus was applied on a patient with intractable severe OCD and associated major depression. Remission from depression and anxiety took place 6 months after the start of DBS treatment, while the remission of OCD began after 12 months without any neuropsychological deterioration. The stimulation of the STN has been previously prescribed for movement disorders, however, a study by Mallet et al. (2008) investigated the efficacy of stimulating it for the treatment of OCD. The 10-month study on patients with refractory OCD involved eight patients undergoing STN DBS followed by sham stimulation and another group of eight patients that underwent sham stimulation followed by active stimulation. Symptoms of OCD became less severe and levels of functioning increased following active stimulation, compared to sham stimulation. However, active stimulation did not affect depression, anxiety and the neuropsychological status of patients. Unfortunately, DBS of the STN was associated with 15 serious adverse events, including permanent neurological sequelae, and 23 non-serious events.

DBS and Chronic Cluster Headache

Messina et al. (2012) have used DBS to stimulate the right posterior hypothalamus of an angiomyolipoma patient suffering from chronic Cluster Headache (CH), which is a disease of unknown etiology but which may be caused by lesions irritating the trigeminal nerve or malformations and neoplasms disturbing the cranio-cervical region. After suffering from 60 attacks per day, the patient's symptoms improved to one attack every 2 months but when the electrode was removed due to sepsis, the initial symptoms returned. The mode of action of DBS may have been inhibiting hypothalamic neurons, increasing trigeminal pain threshold or regulating complex pain-matrix brain regions (Messina et al., 2012).

In another study on CH patients by Leone et al. (2013), after 6–12 years of hypothalamic DBS stimulation, 70% of patients amended with a maintained improvement and most of those who did not improve had a bilateral CH indicating that bilateral CH predicts a bad prognosis. However, electrode implantation at the hypothalamus (near the lateral ventricle wall) has a potential risk of hemorrhage and that is why Seijo et al. (2011) moved the DBS electrode away from the wall and towards the lateral hypothalamus, which in turn is implicated in pain modulation. CH attacks and hypothalamic stimulation have been correlated with an increased blood flow in the insula, anterior cingulate and frontal lobe (Leone et al., 2013). This leads to a long-term potentiation and CH chronicization, which is interrupted by DBS stimulation, turning the chronic CH to an episodic one (Leone et al., 2013). It is interesting that DBS is able to decrease the pain of CH while in other cases of STN DBS, patients feel pain in the off state which may be due to the hypodopaminergic state that STN DBS induces (Nègre-Pagès et al., 2008). It is also advisable to start using greater occipital nerve stimulation before hypothalamic DBS because it is a less invasive/ risky technique which has shown positive results in treating CH (Leone et al., 2013).

DBS and Tourette's Syndrome

Tourette's syndrome is a neuropsychiatric disorder, characterized by the presence of both vocal and motor tics. "A tic is a sudden, rapid, recurrent, non-rhythmic, stereotyped motor movement or vocalization (DSM-IV) such as excessive eye blinking, nose twitching, head jerks and tensing the abdomen, whereas throat clearing, coughing and sniffing are the most prevalent vocalizations" (Hoekstra et al., 2002). The tic onset is mostly between 2 and 15 years of age, but as the patients get older, their "symptoms tend to decrease in intensity and show less variation in both severity and type of tics" (Hoekstra et al., 2002).

Scientists are still struggling to understand the pathophysiology behind TS but it is hypothesized that the brain structures involved are the small basal ganglia and/or hyperactive motor cortex and/or reticular activating system and/or dysfunctional regulating cortical regions (Chae et al., 2004). They also noticed that children with TS have a small PFC but those whose tics start improving while growing up have larger PFC than those whose tics do not improve (Chae et al., 2004). Thus it can be concluded that the PFC can regulate the motor cortex, which in turn controls the tics (Chae et al., 2004). Peterson et al. (1998) have previously found that as TS patients are trying to suppress the tics, there is an activation in the superior temporal, frontal and anterior cingulate cortices, meaning that an underactive PFC may be the reason behind TS.

Furthermore, a recent study on primates used 30 seconds of high frequency DBS to treat TS, which is caused by abnormalities in the basal ganglia and GABAergic networks and activity changes in the GPi and GPe (McCairn et al., 2013). However, the best location for DBS stimulation is still undetermined but potential targets include GPi, GPe, STN and thalamus (McCairn et al., 2013). As for the DBS mechanism, the stimulation pulse is supposedly temporally locking cellular activity and causing a pattern of excitation and inhibition in the posterior region of GP and reducing the number of its tic-responding cells (McCairn et al., 2013). McCairn and Turner (2009) have found that DBS induces a collision-like phenomenon and an antidromic-like short latency responses that are generating a fixed latency spike and without any significant latency shifts. In addition to that, there is a reduction in the firing rate of a subset of neurons, which may be due to refractory period induction, striatum GABAergic afferent neuron and collateral axon activation while latency increase (3–5 ms) may be due to STN glutamatergic and pedunculopontine nucleus activation (McCairn et al., 2013). The increased latency could also be explained by neuronal plasticity (McCairn and Turner, 2009) and some interfering neuronal pathways such as the corticostriatal, thalamstriatal, corticosubthalamic and brainstem networks (DeLong and Wichmann, 2007).

DBS and Epilepsy

Medial Temporal Lobe (MTL) epilepsy has been treated by hippocampectomy or modified temporal lobectomy or amygdalohippocampectomy with a

60–75% success rate and side effects such as verbal memory retardation after operating in the left hemisphere and visual-spatial learning retardation after operating in the right hemisphere (Boon et al., 2007). On the other hand, DBS induces EEG synchronization when applied at a low frequency and desynchronization at a high frequency, where desynchronization is the therapeutic effect of DBS on epilepsy (Boon et al., 2007). Osorio et al. (2001) used DBS on epilepsy patients with a pulse width of 0.45 mseconds and an increasing frequency of 130 Hz to 200 Hz so that they can stimulate a larger volume of epileptic tissue in patients suffering from seizures after 1 year of constant DBS. Yet, some hypothesize that the effect induced by DBS is due to an accidental lesion in the MTL region by the electrode rather than a stimulatory current that is inhibiting the structures that induce, sustain and propagate epileptic activity (Boon et al., 2007).

Lately, two large studies were conducted to assess the efficacy of DBS in treating epilepsy. The electrical Stimulation of the Anterior Nucleus of Thalamus for Epilepsy (SANTE) trial, was a large, double-blind, multicenter, randomized trial that was finished by 2010 (Fisher et al., 2010). It examined the effects of bilateral DBS on the anterior nucleus of the thalamus in 110 patients with intractable epilepsy. A 3-month "blinded" phase was then initiated, in which the patients were randomized to either a stimulation group or a no-stimulation group. At the end of that phase, the stimulation group had 40.4% decline in median seizure frequency compared to a 14.5% decrease in the no-stimulation group. Noteworthy, there was a significant difference in the median seizure decrease between the two groups (44.2% in the stimulation group vs. 21.8% in the no-stimulation group) in patients with seizures originating from one or both temporal lobes. However, this was not the case in patients with seizures originating from the parietal, frontal or occipital lobe. A 9-month open label phase was then commenced, in which all the patients had their stimulators turned on. Only 81 patients were followed up by the end of 25 months. The trial was assessed according to the monthly seizure rate, Quality of Life in Epilepsy Scale, Liverpool Seizure Severity Scale, and neuropsychological assessment. Long-term follow-up yielded a 41% reduction in median seizure frequency at 13 months and a 56% reduction at 25 months. Nonetheless, nine patients had an increase in seizure frequency at 25 months. Furthermore, this trial showed that some patients might benefit more than others from this treatment, since there were 14 patients that remained seizure free for at least 6 months. The most prominent side effects of this trial were paresthesia, depression and memory impairment, but which were mostly transient events that resolved later on.

A more recent study (RNS System, Neuropace) was published in 2011, and it investigated a new development in DBS that is called "open loop" or responsive cortical stimulation (Morrell, 2011). Unlike the closed-loop traditional DBS which involves continuous stimulation of a target, responsive stimulation electrodes have both a stimulation and electrocorticographic activity detection function. Whenever there is an abnormal activity in the brain, the subdural or depth electrodes deliver an electrical stimulation to counteract the abnormality. The study by Morrell (2011), was a large, double-blind, multicenter, randomized, trial on 191 patients with refractory partial-onset seizures. Either subdural or depth electrodes were implanted at one or two seizure foci. The patients were then randomized into either a treatment group or a sham-stimulation group. After that, the treatment group underwent stimulation while the sham group did not, during a 3-month blinded phase. This was followed by an 84-week open label phase, in which the sham group stimulators were turned on. The trial was assessed according to the seizure frequency reduction, neuropsychiatric end points and quality of life measures. During the blinded phase, the treatment group had a 37.9% decrease in mean seizure frequency while the sham group had a 17.3% decrease. Out of the 191 patients, 102 were followed up for 2 years and 46% of them showed at least a 50% reduction in their mean seizure frequency. Nevertheless, the active treatment group had greater improvements than the sham group in the domains of visuospatial ability, verbal functioning and memory. Finally, there were no significant side effects in both groups.

DBS and Depression

In a study conducted by Mayberg et al. (2005) on six treatment-resistant patients with major depression, DBS was bilaterally implanted in the white matter fibers connecting to BA 25, and function was focally ablated using generators. After 6 months of stimulation and taking medication, four out of six patients responded to the treatment and three of the responders qualified for remission. Usually, the remission rates of DBS range between 20 to 35% at the end of a 6-month open-label follow-up (Nahas et al., 2010). In 2009, there was another study conducted by Malone et al. (2009) that investigated the efficacy of VC/VS DBS in the treatment of TRD. Electrodes were bilaterally placed in the VC/VS of 15 patients with chronic, severe, highly refractory depression. All patients received continuous stimulation and significant improvement was noticed throughout the follow-ups that lasted between 6 months and more than 4 years. Mean HDRS scores decreased from 33.1 at baseline to 17.5 at 6 months and 14.3 at the last follow-up. Similar improvements were seen with the Global Assessment of Function Scale (GAF) and the Montgomery-Asberg Depression Rating Scale (MADRS). Responder rates with the HDRS were 40% at 6 months and 53.3% at the last follow-up while remission rates were 20% at 6 months and 40% at the last follow-up.

In a pilot study by Schlaepfer et al. (2013) the safety and efficacy of DBS to the supero-lateral branch of the medial forebrain bundle was assessed in seven TRD patients. The assessment of the trial was according to the MADRS, GAF, quality of life and neuropsychological tests. At the end of 12–33 weeks, six out of the seven patients attained a rapid response in which the mean MADRS of all the patients decreased by 50% on the 7th day of stimulation and GAF improved from serious to mild impairment. Interestingly, all patients showed an increase in appetite while only four patients were considered remitters. Side effects included strabismus, one small intracranial bleeding and infection, but which could be resolved. One interesting event was that a high proportion of the sample responded at lower stimulation intensities compared to previous studies. Unfortunately, to this date, the two large industries that sponsored double-blind placebo controlled studies of DBS for TRD targeting the VC/VS or BA 25 have been halted after a futility analysis. Additional understanding of DBS mechanisms of action and optimal study designs are needed.

DBS and Other Neuropsychiatric Disorders

Kefalopoulou et al. (2013) have also tried high frequency (130 Hz) DBS treatment on two patients with chorea-acanthocytosis (ChAc), which is defined by orofacial dyskinesia, chorea and caudate nucleus atrophy. Both patients showed marked improvement and 5 months after the removal of the DBS hardware, one of the patients' chorea was completely resolved due to a "pallidotomy-like" left intracerebral abscess, which unfortunately also further compromised his dysarthric speech (Kefalopoulou et al., 2013).

In schizophrenia, patients suffer from deficits in responses to and processing of auditory information via the hippocampus, which may be the result of the hyper-dopaminergic state that is found in schizophrenics (Adler et al., 1986; Bickford-Wimer et al., 1990). Ewing and Grace (2013) applied ventral hippocampal DBS on methylazoxymethanol (MAM)-treated rats, which modulated the disrupted mediodorsal thalamic nucleus, medial PFC, ventral pallidum and dopamine system. So the inhibited hippocampus was leading to a feed-forward excitation of neural networks to many parts of the brain that are supposedly causing the schizophrenic symptoms, and stimulating the CA3, CA1 and subiculum of the ventral hippocampus inhibits that feed forward excitation (Ewing and Grace, 2013). Even though schizophrenics have a significantly reduced amount of gray matter in the left hemisphere, there was no significant difference between bilateral and unilateral DBS (Ewing and Grace, 2013). However, another hypothesis states that positive effects of DBS in schizophrenics may not be due to a modulation of dopamine level after all, but due to boosting interneuron function by increasing parvalbumin (Ewing and Grace, 2013).

The NAc has also been a target of DBS because of its implication in the reward system of the brain and its correlation with addiction and other psychiatric disorders (Müller et al., 2013). In addiction and some psychiatric disorders, the ratio of D1 to D2 receptors in the NAc is in disequilibrium and NMDA receptors are activated via D1 receptors (Lee et al., 2011). Many studies have been done to assess the effect of DBS on addicted treatment-resistant patients with positive results in studies on smoking, alcohol and heroin addiction with abstinence or decrease in consumption that lasted for years (Kuhn et al., 2009; Müller et al., 2013; Valencia-Alfonso et al., 2012).

DBS, Neurogenesis and Stem Cells

DBS has also shown positive results in neurogenesis where Encinas et al. (2011) found that high frequency stimulation of the ATN increases the number of cells by 69% in the subgranular zone of the dentate gyrus in adult mice brains. What was exactly affected was the number of dividing Amplifying Neural Progenitor (ANP) cells and not the rate of cell division of ANP or quiescent progenitor (QNP) cells (Encinas et al., 2011). Moreover, it was hypothesized that QNP cells underwent a rapid asymmetric division into ANP before going back to their basal state (Encinas et al., 2011). Those ANPs would later turn into post-mitotic precursors (neuroblast type I and II and immature neurons), which in turn would give rise to new neurons in the granular layer of the neocortex (Encinas et al., 2011). The evidence at hand points out that the hippocampal ANP change their differentiation programs from a glial cell fate towards a differentiating neuron by reducing Hes1 and Id2 gene expression and increasing the expression of the NeuroD gene which positively regulates neuronal differentiation (Toda et al., 2008). This remains a good question to evaluate and see whether DBS can increase neurogenesis by 2–3 folds in humans just like it did in mice (Toda et al., 2008)?

Astrocytes have been interesting players in the therapeutic effects of DBS. It is known that stimulated astrocytes participate in neural signaling by assembling into networks that propagate calcium for communication and forming a three-way synapse with neuronal synapses (Vedam-Mai et al., 2012). It was also found that astrocytes near the implanted electrode undergo a change in function and can become stem cells that produce astrocytes, oligodendrocytes and neurons in response to a depolarizing high frequency stimulation (Vedam-Mai et al., 2012). Astrocytes can release gliotransmitters such as D-serine, glutamate that stimulates synaptic activity (Tawfik et al., 2011) and ATP (Bekar et al., 2008), which will be converted into adenosine that will, in turn, bind to the inhibiting A1 brain receptors (Dunwiddie and Masino, 2001) and open postsynaptic potassium (Trussell and Jackson, 1985) and presynaptic calcium channels (Macdonald et al., 1986). Reactive astrocytes can also change cerebral blood flow and down regulate glutamine synthase, which causes a decrease in GABA that leads to hyperactivity (Vedam-Mai et al., 2012). Furthermore, around the electrode, there would be some neuronal loss, scar tissue, microbleeding remnants, neuronal migration aided by astrocytes and a reactive gliosis, which is an astrocyte hypertrophy due to increased production of glial fibrillary acidic protein, nestin, vimentin and synemin (Pekny and Nilsson, 2005). The increase in cell density around the electrode can be due to the proliferation of progenitor cells, which have been derived from astrocytes (Vedam-Mai et al., 2012).

Inducing the formation of stem cells via DBS is a very intriguing topic that needs further investigation because of its great potential in offering effective treatment for neuropsychiatric disorders and even neurodegenerative diseases. The mechanism of turning normal somatic cells into stem cells is yet to be determined. There are many possibilities and pathways in which DBS can lead to the formation of stem cells. DBS, for example, increases potassium concentration in the extracellular matrix, so that may give a hint on where to start looking. Many questions to be answered can arise here, such as: is it the potassium or the positive charge that is behind the effect of DBS? If another positive ion was used in *in vitro* experiments, would it yield the same results? What is the transduction pathway that is being induced to create stem cells? What is the new expression level profile of cellular proteins in the new stem cells after DBS treatment? Animal models should be used in order to study the vast number of questions in this new field. Researchers can use a pharmacological approach in order to block specific receptors during DBS treatment so that they could have an idea about which receptors are being employed in the therapeutic effect. Viral vectors and transgenic cells can also be used to study genetic changes, their effects on the treatment and if they are actually affected by DBS.

Future Directions and Conclusion

Neuropsychiatric disorders have urged scientists and medical researchers to study several methods of brain stimulation, which have been discussed above. As we learn more about the pathophysiology of psychiatric disorders a lot of new techniques are emerging such as spinal cord stimulation, simple transcutaneous electrical nerve stimulation and electro-acupuncture (Nahas and Anderson, 2011). More studies are being conducted to evaluate the older techniques and the newly emerging techniques so as to find the most effective treatment for psychiatric disorders by combining several techniques together or manipulating their parameters to increase efficacy. And in addition to stimulatory techniques, future studies should work on intricate processes such as apoptosis, inflammatory cascades, metabolism and cerebral blood flow in the microvasculature. All of these could also be targeted via techniques that are still under development such as the use of transgenic cells or viral vectors or even stem cell transplantation.

References

Abdeen, M. and M. Stuchly. 1994. Modeling of magnetic field stimulation of bent neurons. *IEEE Trans Biomed Eng*. 41: 1092–1095.

Abelson, J., G. Curtis, O. Sagher, R. Albucher, M. Harrigan, S. Taylor, B. Martis and B. Giordani. 2005. Deep brain stimulation for refractory obsessive-compulsive disorder. *Biol Psychiatry*. 57: 510–6.

Adler, L., G. Rose and R. Freedman. 1986. Neurophysiological studies of sensory gating in rats: effects of amphetamine, phencyclidine, and haloperidol. *Biol Psychiatry*. 21: 787–98.

Ahlskog, J.E. and M. Muenter. 2001. Frequency of levodopa-related dyskinesias and motor fluctuations as estimated from the cumulative literature. *Mov Disord*. 16: 448–58. doi:10.1016/j.clineuro.2013.07.014.

Amassian, V., R. Cracco and P. Maccabee. 1998. Basic Mechanisms of Magnetic Coil Excitation of Nervous System in Humans and Monkeys and Their Applications. *New Orleans: McGregor and Warner*. 10–17.

Amassian, V., L. Eberle, P. Maccabee and R. Cracco. 1992. Modelling magnetic coil excitation of human cerebral cortex with a pe-ripheral nerve immersed in a brain-shaped volume conductor: The significance of fiber bending in excitation. *Electroencephalogr Clin Neurophysiol*. 82: 291–301.

Anderson, B., A. Mishory, Z. Nahas, J.J. Borckardt, K. Yamanaka, K. Rastogi and M.S. George. 2006. Tolerability and safety of high daily doses of repetitive transcranial magnetic stimulation in healthy young men. *J ECT*. 22: 49–53.

Anderson, B.S., K. Kavanagh, J.J. Borckardt, Z.H. Nahas, S. Kose, S.H. Lisanby, W.M. McDonald, D. Avery, H.A. Sackeim and M.S. George. 2009. Decreasing procedural pain over time of left prefrontal rTMS for depression: initial results from the open-label phase of a multi-site trial (OPT-TMS). *Brain Stimul*. 2: 88–92. doi:10.1016/j.brs.2008.09.001.

Anderson, V.C., K.J. Burchiel, P. Hogarth, J. Favre and J.P. Hammerstad. 2005. Pallidal vs subthalamic nucleus deep brain stimulation in Parkinson disease. *Arch Neurol*. 62: 554–60. doi:10.1001/archneur.62.4.554.

Andoh, J., D. Riviere, J.-F. Mangin, E. Artiges, Y. Cointepas, D. Grevent, M.-L. Paillère-Martinot, J.-L. Martinot and A. Cachia. 2009. A triangulation-based magnetic resonance image-guided method for transcranial magnetic stimulation coil positioning. *Brain Stimul*. 2: 123–31. doi:10.1016/j.brs.2008.10.002.

André-Obadia, N., R. Peyron, P. Mertens, F. Mauguière, B. Laurent and L. Garcia-Larrea. 2006. Transcranial magnetic stimulation for pain control. Double-blind study of different frequencies against placebo, and correlation with motor cortex stimulation efficacy. *Clin Neurophysiol*. 117: 1536–44. doi:10.1016/j.clinph.2006.03.025.

Aouizerate, B., E. Cuny, C. Martin-Guehl, D. Guehl, H. Amieva, A. Benazzouz, C. Fabrigoule, M. Allard, A. Rougier, B. Bioulac, J. Tignol and P. Burbaud. 2004. Deep brain stimulation of the ventral caudate nucleus in the treatment of obsessive-compulsive disorder and major depression. Case report. *J Neurosurg*. 101: 682–6.

Apkarian, A.V., Y. Sosa, S. Sonty, R.M. Levy, R.N. Harden, T.B. Parrish and D.R. Gitelman. 2004. Chronic back pain is associated with decreased prefrontal and thalamic gray matter density. *J Neurosci: The Official Journal of the Society for Neuroscience*. 24: 10410–5. doi:10.1523/JNEUROSCI.2541-04.2004.

Asanuma, C., W. Thach and E. Jones. 1983. Distribution of cerebellar terminations and their relation to other afferent terminations in the ventral lateral thalamic region of the monkey. *Brain Res*. 286: 237–65.

Atmaca, M., B.H. Yildirim, B.H. Ozdemir, B.A. Aydin, A.E. Tezcan and A.S. Ozler. 2006. Volumetric MRI assessment of brain regions in patients with refractory obsessive-compulsive disorder. *Prog Neuropsychopharmacol Biol Psychiatry*. 30: 1051–7.

Atmaca, M., H. Yildirim, H. Ozdemir, E. Tezcan and P. AK. 2007. Volumetric MRI study of key brain regions implicated in obsessive-compulsive disorder. *Prog Neuropsychopharmacol Biol Psychiatry*. 31: 46–52.

Avery, D., K. Isenberg, S. Sampson, P. Janicak, S. Lisanby, D. Maixner, C. Loo, M. Thase, M. Demitrack and M. George. 2008. Transcranial magnetic stimulation in the acute treatment of major depressive disorder: clinical response in an open-label extension trial. *J Clin Psychiatry*. 69: 441–51. doi:10.1017/S1092852913000357.

Bekar, L., W. Libionka, G.-F. Tian, Q. Xu, A. Torres, X. Wang, D. Lovatt, E. Williams, T. Takano, J. Schnermann, R. Bakos and M. Nedergaard. 2008. Adenosine is crucial for deep brain stimulation-mediated attenuation of tremor. *Nat Med*. 14: 75–80. doi:10.1038/nm1693.

Benarroch, E.E. 2008. Subthalamic nucleus and its connections: anatomic substrate for the network effects of deep brain stimulation. *Neurology*. 70: 1991–1995.

Benazzouz, A. and M. Hallett. 2000. Mechanism of action of deep brain stimulation. *Neurology*. 55: 13–16. doi:10.1002/mds.25214.

Benito-León, J., J. Porta-Etessam and F. Bermejo. 1998. Epidemiology of Parkinson disease. *Neurología (Barcelona, Spain)*. 13 Suppl 1: 2–9.

Benjamin, J., T.E. Schlaepfer, M. Altemus, E.M. Wassermann, R.M. Post and D.L. Murphy. 1997. Effect of Prefrontal Repetitive Transcranial Magnetic Stimulation in Obsessive-Compulsive Disorder : A Preliminary Study. *Am J Psychiatry*. 154: 867–869.

Berlim, M.T., M.P. Fleck and G. Turecki. 2008. Current trends in the assessment and somatic treatment of resistant/refractory major depression: an overview. *Ann Med*. 40: 149–59. doi:10.1080/07853890701769728.

Berlim, M.T. and G. Turecki. 2007. Treatment-Resistant Refractory Major Depression: A Review of Current Concepts and Methods. *Can J Psychiatry*. 52: 46–54.

Bickford-Wimer, P., H. Nagamoto, R. Johnson, L. Adler, M. Egan, G. Rose and R. Freedman. 1990. Auditory sensory gating in hippocampal neurons: a model system in the rat. *Biol Psychiatry*. 27: 183–92.

Bonilha, L., C. Molnar, M.D. Horner, B. Anderson, M.S. George and Z. Nahas. 2009. Neurocognitive deficits and prefrontal cortical atrophy in patients with schizophrenia. *Schizophr Res*. 101: 142–151. doi:10.1016/j.schres.2007.11.023. Neurocognitive.

Boon, P., K. Vonck, V. De Herdt, A. Van Dycke, M. Goethals, M. Van Zandijcke, T. De Smedt, I. Dewaele, R. Achten, F. Dewaele, J. Caemaert and D. Van Roost. 2007. Deep Brain Stimulation in Patients with Refractory Temporal Lobe Epilepsy. *Epilepsia*. 48: 1551–1560. doi:10.1111/j.1528-1167.2007.01005.x.

Borckardt, J.J., Z. Nahas, J. Koola and M.S. George. 2006a. Estimating resting motor thresholds in transcranial magnetic stimulation research and practice: a computer simulation evaluation of best methods. *J ECT*. 22: 169–75. doi:10.1097/01.yct.0000235923.52741.72.

Borckardt, J.J., A.R. Smith, K. Hutcheson, K. Johnson, Z. Nahas, B. Anderson, M.B. Schneider, S.T. Reeves and M.S. George. 2006b. Reducing pain and unpleasantness during repetitive transcranial magnetic stimulation. *J ECT*. 22: 259–64. doi:10.1097/01.yct.0000244248.40662.9a.

Borckardt, J.J., A.R. Smith, S.T. Reeves, A. Madan, N. Shelley, R. Branham, Z. Nahas and M.S. George. 2009. A pilot study investigating the effects of fast left prefrontal rTMS on chronic neuropathic pain. *Pain Med*. 10: 840–9. doi:10.1111/j.1526-4637.2009.00657.x.

Borckardt, J.J., A.R. Smith, S.T. Reeves, M. Weinstein, F.A. Kozel, Z. Nahas, N. Shelley, R.K. Branham, K.J. Thomas and M.S. George. 2007. Fifteen minutes of left prefrontal repetitive transcranial magnetic stimulation acutely increases thermal pain thresholds in healthy adults. *Pain Res Manag.* 12: 287–90.

Borckardt, J.J., M. Weinstein, Z. Nahas and M.S. George. 2006c. Postoperative Left Prefrontal Repetitive Transcranial Magnetic Stimulation Reduces Patient-controlled Analgesia Use. *Anesthesiology.* 105: 557–562.

Bötzel, K., D. Ph, C. Daniels, A. Deutschländer, D. Gruber, W. Hamel, J. Herzog, D. Lorenz, S. Lorenzl, H.M. Mehdorn, J.R. Moringlane, W. Oertel, M.O. Pinsker, H. Reichmann and A. Reu. 2006. A Randomized Trial of Deep-Brain Stimulation for Parkinson's Disease. *N Engl J Med.* 355: 896–908.

Boylan, L.S., S.L. Pullman, S.H. Lisanby, K.E. Spicknall and H.A. Sackeim. 2001. Repetitive transcranial magnetic stimulation to SMA worsens complex movements in Parkinson's disease. *Clin Neurophysiol.* 112: 259–264.

Brighina, F., A. Piazza, G. Vitello, A. Aloisio, A. Palermo, O. Daniele and B. Fierro. 2004. rTMS of the prefrontal cortex in the treatment of chronic migraine: a pilot study. *J Neurol Sci.* 227: 67–71. doi:10.1016/j.jns.2004.08.008.

Broen, M., A. Duits, V. Visser-Vandewalle, Y. Temel and A. Winogrodzka. 2011. Impulse control and related disorders in Parkinson's disease patients treated with bilateral subthalamic nucleus stimulation: a review. *Parkinsonism Relat Disord.* 17: 413–7. doi:10.1016/j.parkreldis.2011.02.013.

Buttner, T., W. Kuhn, T. Muller, T. Patzold, K. Heidbrink and H. Przuntek. 1995. Distorted color discriminationin "de nova" parkinsonian patients. *Neurology.* 386–387.

Caronti, B., G. Antonini, C. Calderaro, S. Ruggieri, G. Palladini, F.E. Pontieri and C. Colosimo. 2001. Dopamine transporter immunoreactivity in peripheral blood lymphocytes in Parkinson's disease Short Communication. *J Neural Transm.* 108: 803–807.

Cenci, M.A. and M. Lundblad. 2006. Post- versus presynaptic plasticity in L-DOPA-induced dyskinesia. *J Neurochem.* 99: 381–92. doi:10.1111/j.1471-4159.2006.04124.x.

Cenci, M.A., K.E. Ohlin and P. Odin. 2011. Current Options and Future Possibilities for the Treatment of Dyskinesia and Motor Fluctuations in Parkinson's Disease. *CNS Neurol Disord Drug Targets.* 2: 670–684.

Chae, J.-H., Z. Nahas, E. Wassermann, X. Li, G. Sethuraman, D. Gilbert, F.R. Sallee and M.S. George. 2004. A Pilot Safety Study of Repetitive Transcranial Magnetic Stimulation (rTMS) in Tourette's Syndrome. *Cog Behav Neurol.* 17: 109–117.

Chamberlain, S.R. and L. Menzies. 2009. Endophenotypes of obsessive-compulsive disorder: rationale, evidence and future potential. *Expert Rev Neurother.* 9: 1133–46. doi:10.1586/ern.09.36.

Chao, Y., L. Gang, Z.L. Na, W.Y. Ming, W.S. Zhong and W.S. Mian. 2007. Surgical Management of Parkinson ' s Disease : Update and Review. *Interv Neuroradiol.* 13: 359–368.

Chaudhuri, K.R. and A.H.V. Schapira. 2009. Non-motor symptoms of Parkinson's disease: dopaminergic pathophysiology and treatment. *Lancet Neurol.* 8: 464–74. doi:10.1016/S1474-4422(09)70068-7.

Chen, R., D. Yung and J. Li. 2003. Organization of ipsilateral excitatory and inhibitory pathways in the human motor cortex. *J Neurophysiol.* 89: 1256–1264.

Chopra, A., S.J. Tye, K.H. Lee, J. Matsumoto, B. Klassen, A.C. Adams, M. Stead, S. Sampson, B.A. Kall and M.A. Frye. 2011. Voltage-dependent mania after subthalamic nucleus deep brain stimulation in Parkinson's disease: a case report. *Biol Psychiatry.* 70: e5–7. doi:10.1016/j.biopsych.2010.12.035.

Cohen, R., P. Boggio and F. Fregni. 2009. Risk factors for relapse after remission with repetitive transcranial magnetic stimulation for the treatment of depression. *Depress Anxiety.* 688: 682–688. doi:10.1002/da.20486.

Cottraux, J., D. Gérard, L. Cinotti, J. Froment, M. Deiber, D. Le Bars, G. Galy, P. Millet, C. Labbé, F. Lavenne, M. Bouvard and F. Mauguière. 1996. A controlled positron emission tomography study of obsessive and neutral auditory stimulation in obsessive-compulsive disorder with checking rituals. *Psychiatry Res.* 60: 101–12.

Crutcher, M. and M. DeLong. 1984. Single cell studies of the primate putamen. I. Functional organization. *Exp Brain Res.* 53: 233–43.

Daskalakis, Z.J., B. Möller, B.K. Christensen, P.B. Fitzgerald, C. Gunraj and R. Chen. 2006. The effects of repetitive transcranial magnetic stimulation on cortical inhibition in healthy human subjects. *Exp Brain Res.* 174: 403–12. doi:10.1007/s00221-006-0472-0.

DeLong, M., G. Alexander, S. Mitchell and R. Richardson. 1986. The contribution of basal ganglia to limb control. *Prog Brain Res.* 64: 161–74.

DeLong, M., M. Crutcher and A. Georgopoulos. 1985. Primate globus pallidus and subthalamic nucleus: functional organization. *J Neurophysiol.* 53: 530–43.

DeLong, M.R. and T. Wichmann. 2007. Circuits and circuit disorders of the basal ganglia. *Arch Neurol.* 64: 20–4. doi:10.1001/archneur.64.1.20.

Demirtas-Tatlidede, A., D. Mechanic-Hamilton, D.Z. Press, C. Pearlman, W.M. Stern, M. Thall and A. Pascual-Leone. 2008. An open-label, prospective study of repetitive transcranial magnetic stimulation (rTMS) in the long-term treatment of refractory depression: reproducibility and duration of the antidepressant effect in medication-free patients. *J Clin Psychiatry.* 69: 930–4.

DeVito, J. and M. Anderson. 1982. An autoradiographic study of efferent connections of the globus pallidus in Macaca mulatta. *Exp Brain Res.* 46: 107–17.

Dunwiddie, T. V. and S.A. Masino. 2001. The Role and Regulation of Adenosine in the Central Nervous System. *Annu Rev Neurosci.* 24: 31–55.

Encinas, J.M., C. Hamani, A.M. Lozano and G. Enikolopov. 2011. Neurogenic hippocampal targets of deep brain stimulation. *J Comp Neurol.* 519: 6–20. doi:10.1002/cne.22503.

Epstein, C., J. Lah, K. Meador, J. Weissman, L. Gaitan and B. Dihenia. 1996. Optimum stimulus parameters for lateralized suppression of speech with magnetic brain stimulation. *Neurology.* 47: 1590–3.

Ewing, S.G. and A.A. Grace. 2013. Deep brain stimulation of the ventral hippocampus restores deficits in processing of auditory evoked potentials in a rodent developmental disruption model of schizophrenia. *Schizophr Res.* 143: 377–383. doi:10.1016/j.schres.2012.11.023.

Fava, M. 2003. Diagnosis and definition of treatment-resistant depression. *Biol Psychiatry.* 53: 649–659. doi:10.1016/S0006-3223(03)00231-2.

Fava, M., A. Rush, S. Wisniewski, A. Nierenberg, J. Alpert, P. McGrath, M. Thase, D. Warden, M. Biggs, J. Luther, G. Niederehe, L. Ritz and M. Trivedi. 2006. A comparison of mirtazapine and nortriptyline following two consecutive failed medication treatments for depressed outpatients: a STAR*D report. *Am J Psychiatry.* 163: 1161–1172.

Fisher, R., V. Salanova, T. Witt, R. Worth, T. Henry, R. Gross, K. Oommen, I. Osorio, J. Nazzaro, D. Labar, M. Kaplitt, M. Sperling, E. Sandok, J. Neal, A. Handforth, J. Stern, A. DeSalles, S. Chung, A. Shetter, D. Bergen, R. Bakay, J. Henderson, J. French, G. Baltuch, W. Rosenfeld, A. Youkilis, W. Marks, P. Garcia, N. Barbaro, N. Fountain, C. Bazil, R. Goodman, G. McKhann, K. Babu Krishnamurthy, S. Papavassiliou, C. Epstein, J. Pollard, L. Tonder, J. Grebin, R. Coffey and N. Graves. 2010. Electrical stimulation of the anterior nucleus of thalamus for treatment of refractory epilepsy. *Epilepsia.* 51: 899–908. doi:10.1111/j.1528-1167.2010.02536.x.

Fitzgerald, P.B., J. Benitez, A.R. De Castella, T.L. Brown, Z. Jeff Daskalakis and J. Kulkarni. 2006. Naturalistic study of the use of transcranial magnetic stimulation in the treatment of depressive relapse. *Aust N Z J Psychiatry.* 40: 764–768. doi:10.1080/j.1440-1614.2006.01881.x.

Fitzgerald, P.B., K. Hoy, S. McQueen, J.J. Maller, S. Herring, R. Segrave, M. Bailey, G. Been, J. Kulkarni and Z.J. Daskalakis. 2009. A randomized trial of rTMS targeted with MRI based neuro-navigation in treatment-resistant depression. *Neuropsychopharmacology.* 34: 1255–62. doi:10.1038/npp.2008.233.

Fytagoridis, A., M. Åström, K. Wårdell and P. Blomstedt. 2013. Stimulation-induced side effects in the posterior subthalamic area : Distribution, characteristics and visualization. *Clin Neurol Neurosurg.* 115: 65–71. doi:10.1016/j.clineuro.2012.04.015.

Gama, C.S., A.C. Andreazza, M. Kunz, M. Berk, P.S. Belmonte-de-Abreu and F. Kapczinski. 2007. Serum levels of brain-derived neurotrophic factor in patients with schizophrenia and bipolar disorder. *Neurosci Lett.* 420: 45–8. doi:10.1016/j.neulet.2007.04.001.

George, M. and R. Belmaker. 2000. Transcranial Stimulation in Neuropsychiatry. 205. American Psychiatric Press, Washington DC.

George, M.S., S.H. Lisanby, D. Avery, W.M. Mcdonald and V. Durkalski. 2010. Daily Left Prefrontal Transcranial Magnetic Stimulation Therapy for Major Depressive Disorder. *Arch Gen Psychiatry.* 67: 507–516.

George, M.S., Z. Nahas, J.J. Borckardt, B. Anderson, M.J. Foust, C. Burns, S. Kose and E.B. Short. 2007. Brain stimulation for the treatment of psychiatric disorders. *Curr Opin Psychiatry.* 20: 250–4; discussion 247–9. doi:10.1097/YCO.0b013e3280ad4698.

George, M.S., Z. Nahas, F.A. Kozel, X. Li, S. Denslow, K. Yamanaka, A. Mishory, M.J. Foust and D.E. Bohning. 2002. Mechanisms and state of the art of transcranial magnetic stimulation. *J ECT.* 18: 170–81.

George, M.S., F. Padberg, T.E. Schlaepfer, J.P. O'Reardon, P.B. Fitzgerald, Z.H. Nahas and M.A. Marcolin. 2009. Controversy: Repetitive transcranial magnetic stimulation or transcranial direct current stimulation shows efficacy in treating psychiatric diseases (depression, mania, schizophrenia, obsessive-compulsive disorder, panic, posttraumatic stress disorder). *Brain Stimul.* 2: 14–21. doi:10.1016/j.brs.2008.06.001.

Götz, M.E., A. Gerstner, R. Harth, A. Dirr, B. Janetzky, W. Kuhn, P. Riederer and M. Gerlach. 2000. Altered redox state of platelet coenzyme Q 10 in Parkinson's disease. *Dialogues Clin Neurosci.* 107: 41–48.

Greenberg, B.D., L.A. Gabriels, D.A. Malone, A.R. Rezai, G.M. Friehs, M.S. Okun, N.A. Shapira, K.D. Foote, P.R. Cosyns, C.S. Kubu, P.F. Malloy, S.P. Salloway, J.E. Giftakis, M.T. Rise, A.G. Machado, K.B. Baker, P.H. Stypulkowski, W.K. Goodman, S.A. Rasmussen and B.J. Nuttin. 2010. Deep brain stimulation of the ventral internal capsule/ventral striatum for obsessive-compulsive disorder: worldwide experience. *Mol Psychiatry.* 15: 64–79. doi:10.1038/mp.2008.55.

Grill, W.M., C.A.A.N. Snyder and S. Miocinovic. 2004. Deep brain stimulation creates an informational lesion of the stimulated nucleus. *NeuroReport.* 15: 1137–1140. doi:10.1097/01.wnr.0000125783.35268.9f.

Grisaru, N., B. Chudakov, Y. Yaroslavsky and R. Belmaker. 1998. Transcranial Magnetic Stimulation in Mania : A Controlled Study. *Am J Psychiatry.* 155: 1608–1610.

Hadley, D., B.S. Anderson, J.J. Borckardt, A. Arana, X. Li, Z. Nahas and M.S. George. 2011. Safety, tolerability, and effectiveness of high doses of adjunctive daily left prefrontal repetitive transcranial magnetic stimulation for treatment-resistant depression in a clinical setting. *J ECT.* 27: 18–25. doi:10.1097/YCT.0b013e3181ce1a8c.

Hanna, G.L., J. Veenstra-VanderWeele, N.J. Cox, M. Boehnke, J.A. Himle, G.C. Curtis, B.L. Leventhal and E.H. Cook. 2002. Genome-wide linkage analysis of families with obsessive-compulsive disorder ascertained through pediatric probands. *Am J Med Genet.* 114: 541–52. doi:10.1002/ajmg.10519.

Herbsman, T., D. Avery, D. Ramsey, P. Holtzheimer, C. Wadjik, F. Hardaway, D. Haynor, M.S. George and Z. Nahas. 2009a. More lateral and anterior prefrontal coil location is associated with better repetitive transcranial magnetic stimulation antidepressant response. *Biol Psychiatry.* 66: 509–15. doi:10.1016/j.biopsych.2009.04.034.

Herbsman, T., L. Forster, C. Molnar, R. Dougherty, D. Christie, J. Koola, D. Ramsey, P.S. Morgan, D.E. Bohning, M.S. George and Z. Nahas. 2009b. Motor threshold in transcranial magnetic stimulation: the impact of white matter fiber orientation and skull-to-cortex distance. *Hum Brain Mapp.* 30: 2044–55. doi:10.1002/hbm.20649.

Herbsman, T. and Z. Nahas. 2012. Anatomically based targeting of prefrontal cortex for rTMS. *Brain Stimul.* 4: 300–302. doi:10.1016/j.brs.2011.01.004. Anatomically.

Herwig, U., F. Padberg, J. Unger, M. Spitzer and C. Schönfeldt-Lecuona. 2001. Transcranial magnetic stimulation in therapy studies: examination of the reliability of "standard" coil positioning by neuronavigation. *Biol Psychiatry*. 50: 58–61.

Van den Heuvel, O.A., P.L. Remijnse, D. Mataix-Cols, H. Vrenken, H.J. Groenewegen, H.B.M. Uylings, A.J.L.M. van Balkom and D.J. Veltman. 2009. The major symptom dimensions of obsessive-compulsive disorder are mediated by partially distinct neural systems. *Brain*. 132:853–68. doi:10.1093/brain/awn267.

Hoekstra, P.J., C.G.M. Kallenberg, J. Korf and R.B. Minderaa. 2002. Is Tourette's syndrome an autoimmune disease? *Mol Psychiatry*. 7: 437–45. doi:10.1038/sj.mp.4000972.

Holthoff, K. and O.W. Witte. 2000. Directed Spatial Potassium Redistribution in Rat Neocortex. *Glia*. 292: 288–292.

Hoogendijk, W.J., I.E. Sommer, G. Tissingh, D.J. Deeg and E.C. Wolters. 1998. Depression in Parkinson's disease. The impact of symptom overlap on prevalence. *Psychosomatics*. 39: 416–21. doi:10.1016/S0033-3182(98)71300-3.

Huber, S.J., D.L. Freidenberg, G.W. Paulson, E.C. Shuttleworth and J.A. Christy. 1990. The pattern of depressive symptoms varies with progression of Parkinson's disease. *J Neurol Neurosurg Psychiatry*. 53: 275–278. doi:10.1136/jnnp.53.4.275.

Janicak, P.G., Z. Nahas, S.H. Lisanby, H.B. Solvason, S.M. Sampson, W.M. McDonald, L.B. Marangell, P. Rosenquist, W.V. McCall, J. Kimball, J.P. O'Reardon, C. Loo, M.H. Husain, A. Krystal, W. Gilmer, S.M. Dowd, M.A. Demitrack and A.F. Schatzberg. 2010. Durability of clinical benefit with transcranial magnetic stimulation (TMS) in the treatment of pharmacoresistant major depression: assessment of relapse during a 6-month, multisite, open-label study. *Brain Stimul*. 3: 187–99. doi:10.1016/j.brs.2010.07.003.

Johnson, K.A., M. Baig, D. Ramsey, S.H. Lisanby, D. Avery, W.M. McDonald, X. Li, E.R. Bernhardt, D.R. Haynor, P.E. Holtzheimer, H.A. Sackeim, M.S. George and Z. Nahas. 2013. Prefrontal rTMS for treating depression: location and intensity results from the OPT-TMS multi-site clinical trial. *Brain Stimul*. 6: 108–17. doi:10.1016/j.brs.2012.02.003.

Kalbe, E., J. Voges, T. Weber, M. Haarer, S. Baudrexel, J.C. Klein, J. Kessler, V. Sturm, W.D. Heiss and R. Hilker. 2009. Frontal FDG-PET activity correlates with cognitive outcome after STN-DBS in Parkinson disease. *Neurology*. 72: 42–9. doi:10.1212/01.wnl.0000338536.31388.f0.

Kang, D., J. Kim, J. Choi, Y. Kim, C. Kim, T. Youn, M. Han, K. Chang and J. Kwon. 2004. Volumetric investigation of the frontal-subcortical circuitry in patients with obsessive-compulsive disorder. *J Neuropsychiatry Clin Neurosci*. 16: 342–9.

Kaptsan, A., Y. Yaroslavsky, J. Applebaum, B. Rh and N. Grisaru. 2003. Right prefrontal TMS versus sham treatment of mania: a controlled study. *Bipolar Disorder*. 5: 36–39.

Karp, B., E. Wassermann and S. Porter. 1997. Transcranial magnetic stimulation acutely decreases motor tics (abstract). *Neurology*. 48: A397.

Kefalopoulou, Z., L. Zrinzo, I.A. Kailash, T. Foltynie and P. Jarman. 2013. Deep brain stimulation as a treatment for chorea-acanthocytosis. *J Neurol*. 303–305. doi:10.1007/s00415-012-6714-0.

Khedr, E.M., H. Kotb, N.F. Kamel, M.A. Ahmed, R. Sadek and J.C. Rothwell. 2005. Longlasting antalgic effects of daily sessions of repetitive transcranial magnetic stimulation in central and peripheral neuropathic pain. *J Neurol Neurosurg Psychiatry*. 76: 833–8. doi:10.1136/jnnp.2004.055806.

Kikuchi, A., A. Takeda, H. Onodera, T. Kimpara, K. Hisanaga, N. Sato, A. Nunomura, R.J. Castellani, G. Perry, M.A. Smith and Y. Itoyama. 2002. Systemic increase of oxidative nucleic acid damage in Parkinson's disease and multiple system atrophy. *Neurobiol Dis*. 9: 244–8. doi:10.1006/nbdi.2002.0466.

Kleiner-Fisman, G., J. Herzog, D.N. Fisman, F. Tamma, K.E. Lyons, R. Pahwa, A.E. Lang and G. Deuschl. 2006. Subthalamic nucleus deep brain stimulation: summary and meta-analysis of outcomes. *Mov Disord*. 21 Suppl 1: 290–304. doi:10.1002/mds.20962.

Kostic, V.S., I. Petrovic and M. Jec. 2010. Regional patterns of brain tissue loss associated with depression in Parkinson disease. *Neurology*. 75: 857–863.

Kozel, F.A., K.A. Johnson, Z. Nahas, P.A. Nakonezny, P.S. Morgan, B.S. Anderson, S. Kose, X. Li, K.O. Lim, M.H. Trivedi and M.S. George. 2011. Fractional anisotropy changes after several weeks of daily left high-frequency repetitive transcranial magnetic stimulation of the prefrontal cortex to treat major depression. *J ECT*. 27: 5–10. doi:10.1097/YCT.0b013e3181e6317d.

Kringelbach, M.L., N. Jenkinson, S.L.F. Owen and T.Z. Aziz. 2007. Translational principles of deep brain stimulation. *Nat Rev Neurosci*. 8: 623–35. doi:10.1038/nrn2196.

Kuhn, J., R. Bauer, S. Pohl, D. Lenartz, W. Huff, E. Kim, J. Klosterkoetter and V. Sturm. 2009. Observations on unaided smoking cessation after deep brain stimulation of the nucleus accumbens. *Eur Addict Res*. 15: 196–201.

Künzle, H. 1978. An autoradiographic analysis of the efferent connections from premotor and adjacent prefrontal regions (areas 6 and 9) in macaca fascicularis. *Brain Behav Evol*. 15: 185–234.

Lee, D., S. Ahn, Y. Shim, W. Koh, I. Shim and E. Choe. 2011. Interactions of Dopamine D1 and N-methyl-D-Aspartate Receptors are Required for Acute Cocaine-Evoked Nitric Oxide Efflux in the Dorsal Striatum. *Exp Neurobiol*. 20: 116–22.

Lee, K.H., C.D. Blaha, B.T. Harris, S. Cooper, F.L. Hitti, J.C. Leiter, D.W. Roberts and U. Kim. 2006. Dopamine efflux in the rat striatum evoked by electrical stimulation of the subthalamic nucleus: potential mechanism of action in Parkinson's disease. *Eur J Neurosci*. 23: 1005–14. doi:10.1111/j.1460-9568.2006.04638.x.

Lee, K.H., S.-Y. Chang, D.W. Roberts and U. Kim. 2004. Neurotransmitter release from high-frequency stimulation of the subthalamic nucleus. *J Neurosurg*. 101: 511–7. doi:10.3171/jns.2004.101.3.0511.

Lee, K.H., D.W. Roberts and U. Kim. 2003. Effect of High-Frequency Stimulation of the Subthalamic Nucleus on Subthalamic Neurons: An Intracellular Study. *Stereotact Funct Neurosurg*. 80: 32–36. doi:10.1159/000075157.

Leone, M., A. Franzini, A. Proietti and G. Bussone. 2013. Success, failure, and putative mechanisms in hypothalamic stimulation for drug-resistant chronic cluster headache. *Pain*. 154: 89–94. doi:10.1016/j.pain.2012.09.011.

Li, B., L. Liu, K.J. Friston, H. Shen, L. Wang, L.-L. Zeng and D. Hu. 2013a. A treatment-resistant default mode subnetwork in major depression. *Biol Psychiatry*. 74: 48–54. doi:10.1016/j.biopsych.2012.11.007.

Li, X., K.J. Hartwell, M. Owens, T. Lematty, J.J. Borckardt, C.A. Hanlon, K.T. Brady and M.S. George. 2013b. Repetitive transcranial magnetic stimulation of the dorsolateral prefrontal cortex reduces nicotine cue craving. *Biol Psychiatry*. 73: 714–20. doi:10.1016/j.biopsych.2013.01.003.

Li, X., Z. Nahas, B. Anderson, F.A. Kozel and M.S. George. 2004. Can left prefrontal rTMS be used as a maintenance treatment for bipolar depression? *Depress Anxiety*. 20: 98–100. doi:10.1002/da.20027.

Li, X., R. Ricci, C.H. Large, B. Anderson, Z. Nahas, D.E. Bohning and M.S. George. 2010. Interleaved transcranial magnetic stimulation and fMRI suggests that lamotrigine and valproic acid have different effects on corticolimbic activity. *Psychopharmacology*. 209: 233–44. doi:10.1007/s00213-010-1786-y.

Linazasoro, G. 2005. New ideas on the origin of L-dopa-induced dyskinesias: age, genes and neural plasticity. *Trends Pharmacol Sci*. 26: 391–7. doi:10.1016/j.tips.2005.06.007.

Lozano, A.M., J. Dostrovsky, R. Chen and P. Ashby. 2002. Review Deep brain stimulation for Parkinson's disease : disrupting the disruption. *Lancet Neurol*. 1: 225–231.

Maccabee, P., V. Amassian, L. Eberle and R. Cracco. 1993. Magnetic coil stimulation of straight and bent amphibian and mammalian peripheral nerve *in vitro*: Locus of excitation. *J Physiol*. 460: 201–219.

Macdonald, R.L., J.H. Skerritt and M.A. Werz. 1986. Adenosine Agonists Reduce Voltage-Dependent Calcium Conductance of Mouse Sensory Neurones in Cell Culture. *J Physiol*. 370: 75–90.

Mallet, L., M. Polosan, N. Jaafari, N. Baup, M. Welter, D. Fontaine, S. du Montcel, J. Yelnik, I. Chéreau, C. Arbus, S. Raoul, B. Aouizerate, P. Damier, S. Chabardès, V. Czernecki, C. Ardouin, M. Krebs, E. Bardinet, P. Chaynes, P. Burbaud, P. Cornu, P. Derost, T. Bougerol,

B. Bataille, V. Mattei, D. Dormont, B. Devaux, M. Vérin, J. Houeto, P. Pollak, A. Benabid, Y. Agid, P. Krack, B. Millet, A. Pelissolo and STOC-Study-Group. 2008. Subthalamic nucleus stimulation in severe obsessive-compulsive disorder. *N Engl J Med.* 359: 2121–34.

Mallet, L., M. Schüpbach, K. N'Diaye, P. Remy, E. Bardinet, V. Czernecki, M.-L. Welter, A. Pelissolo, M. Ruberg, Y. Agid and J. Yelnik. 2007. Stimulation of subterritories of the subthalamic nucleus reveals its role in the integration of the emotional and motor aspects of behavior. *Proc Natl Acad Sci USA.* 104: 10661–6. doi:10.1073/pnas.0610849104.

Malone, D.A., D.D. Dougherty, A.R. Rezai, L.L. Carpenter, M. Gerhard, E.N. Eskandar, S.L. Rauch, S.A. Rasmussen, A.G. Machado, C.S. Kubu, A.R. Tyrka, L.H. Price, P.H. Stypulkowski, E. Jonathon, M.T. Rise, P.F. Malloy, S.P. Salloway and B.D. Greenberg. 2009. Deep Brain Stimulation of the Ventral Capsule/Ventral Striatum for Treatment-Resistant Depression. *Biol Psychiatry.* 65: 267–275. doi:10.1016/j.biopsych.2008.08.029.Deep.

Mantovani, A., S.H. Lisanby, F. Pieraccini, M. Ulivelli, P. Castrogiovanni and S. Rossi. 2006. Repetitive transcranial magnetic stimulation (rTMS) in the treatment of obsessive-compulsive disorder (OCD) and Tourette's syndrome (TS). *Int J Neuropsychopharmacol.* 9: 95–100. doi:10.1017/S1461145705005729.

Mantovani, A., M. Pavlicova, D. Avery, Z. Nahas, W.M. McDonald, C.D. Wajdik, P.E. Holtzheimer, M.S. George, H.A. Sackeim and S.H. Lisanby. 2012. Long-term efficacy of repeated daily prefrontal transcranial magnetic stimulation (TMS) in treatment-resistant depression. *Depress Anxiety.* 29: 883–90. doi:10.1002/da.21967.

Marshall, D.F., A.M. Strutt, A.E. Williams, R.K. Simpson, J. Jankovic and M.K. York. 2012. Alternating verbal fluency performance following bilateral subthalamic nucleus deep brain stimulation for Parkinson's disease. *Eur J Neurol.* 19: 1525–1531. doi:10.1111/j.1468-1331.2012.03759.x.

May, A., G. Hajak, S. Gänssbauer, T. Steffens, B. Langguth, T. Kleinjung and P. Eichhammer. 2007. Structural brain alterations following 5 days of intervention: dynamic aspects of neuroplasticity. *Cereb Cortex.* 17: 205–10. doi:10.1093/cercor/bhj138.

Mayberg, H.S., A.M. Lozano, V. Voon, H.E. McNeely, D. Seminowicz, C. Hamani, J.M. Schwalb and S.H. Kennedy. 2005. Deep brain stimulation for treatment-resistant depression. *Neuron.* 45: 651–60. doi:10.1016/j.neuron.2005.02.014.

Mazzella, L., M.D. Yahr, L. Marinelli, N. Huang, E. Moshier and A. Di Rocco. 2005. Dyskinesias predict the onset of motor response fluctuations in patients with Parkinson's disease on L-dopa monotherapy. *Parkinsonism Relat Disord.* 11: 151–5. doi:10.1016/j.parkreldis.2004.10.002.

McCairn, K.W., A. Iriki and M. Isoda. 2013. Deep Brain Stimulation Reduces Tic-Related Neural Activity via Temporal Locking with Stimulus Pulses. *J Neurosci.* 33: 6581–6593. doi:10.1523/JNEUROSCI.4874-12.2013.

McCairn, K.W. and R.S. Turner. 2009. Deep brain stimulation of the globus pallidus internus in the parkinsonian primate: local entrainment and suppression of low-frequency oscillations. *J neurophysiol.* 101: 1941–60. doi:10.1152/jn.91092.2008.

McDonald, W.M., V. Durkalski, E.R. Ball, P.E. Holtzheimer, M. Pavlicova, S.H. Lisanby, D. Avery, B.S. Anderson, Z. Nahas, P. Zarkowski, H.A. Sackeim and M.S. George. 2011. Improving the antidepressant efficacy of transcranial magnetic stimulation: maximizing the number of stimulations and treatment location in treatment-resistant depression. *Depress Anxiety.* 28: 973–80. doi:10.1002/da.20885.

McGhee, D.J.M., P.L. Royle, P.A. Thompson, D.E. Wright, J.P. Zajicek and C.E. Counsell. 2013. A systematic review of biomarkers for disease progression in Parkinson's disease. *BMC Neurology.* 13: 35. doi:10.1186/1471-2377-13-35.

McGrath, P., J. Stewart, M. Fava, M. Trivedi, S. Wisniewski, A. Nierenberg, M. Thase, L. Davis, M. Biggs, K. Shores-Wilson, J. Luther, G. Niederehe, D. Warden and A. Rush. 2006. Tranylcypromine versus venlafaxine plus mirtazapine following three failed antidepressant medication trials for depression: a STAR*D report. *Am J Psychiatry.* 163: 1531–1541, 1666.

McIntyre, C.C., W.M. Grill, D.L. Sherman and N. V Thakor. 2004a. Cellular effects of deep brain stimulation: model-based analysis of activation and inhibition. *J Neurophysiol.* 91: 1457–69. doi:10.1152/jn.00989.2003.

McIntyre, C.C., S. Mori, D.L. Sherman, N.V. Thakor and J.L. Vitek. 2004b. Electric field and stimulating influence generated by deep brain stimulation of the subthalamic nucleus. *Clin Neurophysiol.* 115: 589–95. doi:10.1016/j.clinph.2003.10.033.

Menzies, L., S. Achard, S.R. Chamberlain, N. Fineberg, C.-H. Chen, N. del Campo, B.J. Sahakian, T.W. Robbins and E. Bullmore. 2007. Neurocognitive endophenotypes of obsessive-compulsive disorder. *Brain.* 130: 3223–36. doi:10.1093/brain/awm205.

Messina, G., M. Rizzi, R. Cordella, A. Caraceni, E. Zecca, G. Bussone, A. Franzini and M. Leone. 2012. Secondary chronic cluster headache treated by posterior hypothalamic deep brain stimulation : First reported case. *Cephalgia.* 33: 136–138. doi:10.1177/0333102412468675.

Michell, A.W., S.J.G. Lewis, T. Foltynie and R.A. Barker. 2004. Biomarkers and Parkinson's disease. *Brain.* 127: 1693–705. doi:10.1093/brain/awh198.

Mills, K.R. and K.A. Nithi. 1997. Corticomotor threshold to magnetic stimulation: normal values and repeatability. *Muscle Nerve.* 570–576.

Min, H., S. Hwang, M.P. Marsh, I. Kim, E. Knight, B. Striemer, J.P. Felmlee, K.M. Welker, C.D. Blaha, S. Chang, K.E. Bennet and K.H. Lee. 2012. Deep brain stimulation induces BOLD activation in motor and non-motor networks: An fMRI comparison study of STN and EN/GPi DBS in large animals. *NeuroImage.* 63: 1408–1420. doi:10.1016/j.neuroimage.2012.08.006.

Mishory, A., C. Molnar, F.A. Kozel and M.S. George. 2004. The Maximum-likelihood Strategy for Determining Transcranial Magnetic Stimulation Motor Threshold, Using Parameter Estimation by Sequential Testing Is Faster Than Conventional Methods With Similar Precision. *J ECT.* 20: 160–165.

Montgomery, E.B. and J.T. Gale. 2008. Mechanisms of action of deep brain stimulation(DBS). *Neurosci Biobehav Rev.* 32: 388–407. doi:10.1016/j.neubiorev.2007.06.003.

Moro, E., A.M. Lozano, P. Pollak, Y. Agid, S. Rehncrona, J. Volkmann, J. Kulisevsky, J A. Obeso, A. Albanese, M.I. Hariz, N.P. Quinn, J.D. Speelman, A.L. Benabid, V. Fraix, A. Mendes, M.-L. Welter, J.-L. Houeto, P. Cornu, D. Dormont, A.L. Tornqvist, R. Ekberg, A. Schnitzler, L. Timmermann, L. Wojtecki, A. Gironell, M.C. Rodriguez-Oroz, J. Guridi, A.R. Bentivoglio, M.F. Contarino, L. Romito, M. Scerrati, M. Janssens and A.E. Lang. 2010. Long-term results of a multicenter study on subthalamic and pallidal stimulation in Parkinson's disease. *Mov Disord.* 25: 578–86. doi:10.1002/mds.22735.

Morrell, M.J. 2011. Responsive cortical stimulation for the treatment of medically intractable partial epilepsy. *Neurology.* 77: 1295–304. doi:10.1212/WNL.0b013e3182302056.

Müller, U.J., J. Voges, J. Steiner, I. Galazky, H.-J. Heinze, M. Möller, J. Pisapia, C. Halpern, A. Caplan, B. Bogerts and J. Kuhn5. 2013. Deep brain stimulation of the nucleus accumbens for the treatment of addiction. *Ann NY Acad Sci.* 1282: 119–128. doi:10.1111/j.1749-6632.2012.06834.x.

Nahas, Z. 2008. Transcranial Magnetic Stimulation for Treating Psychiatric Conditions: What Have We Learned So Far? *Can J Psychiatry.* 53(9): 553–4.

Nahas, Z. 2010. The frontiers in brain imaging and neuromodulation: a new challenge. *Front Psychiatry.* 1: 25. doi:10.3389/fpsyt.2010.00025.

Nahas, Z. and B.S. Anderson. 2011. Brain stimulation therapies for mood disorders: the continued necessity of electroconvulsive therapy. *J Am Psychiatr Nurses Assoc.* 17: 214–6. doi:10.1177/1078390311409037.

Nahas, Z., B.S. Anderson, J. Borckardt, A.B. Arana, M.S. George, S.T. Reeves and I. Takacs. 2010. Bilateral epidural prefrontal cortical stimulation for treatment-resistant depression. *Biol Psychiatry.* 67: 101–9. doi:10.1016/j.biopsych.2009.08.021.

Nahas, Z., C. Teneback, J.-H. Chae, Q. Mu, C. Molnar, F.A. Kozel, J. Walker, B. Anderson, J. Koola, S. Kose, M. Lomarev, D.E. Bohning and M.S. George. 2007. Serial vagus nerve stimulation functional MRI in treatment-resistant depression. *Neuropsychopharmacology.* 32: 1649–60. doi:10.1038/sj.npp.1301288.

Nazzaro, J.M., R. Pahwa and K.E. Lyons. 2011. The impact of bilateral subthalamic stimulation on non-motor symptoms of Parkinson's disease. *Parkinsonism Relat Disord*. 17: 606–9. doi:10.1016/j.parkreldis.2011.05.009.

Neeman, G., M. Blanaru, B. Bloch, I. Kremer, M. Ermilov, D.C. Javitt, D. Ph and U. Herescolevy. 2005. Relation of Plasma Glycine, Serine, and Homocysteine Levels to Schizophrenia Symptoms and Medication Type. *Am J Psychiatry*. 162: 1738–1740.

Nègre-Pagès, L., W. Regragui, D. Bouhassira, H. Grandjean and O. Rascol. 2008. Chronic pain in Parkinson's disease: the cross-sectional French DoPaMiP survey. *Mov Disord*. 23: 1361–9. doi:10.1002/mds.22142.

Nierenberg, A., M. Fava, M. Trivedi, S. Wisniewski, M. Thase, P. McGrath, J. Alpert, D. Warden, J. Luther, G. Niederehe, B. Lebowitz, K. Shores-Wilson and A. Rush. 2006. A comparison of lithium and T(3) augmentation following two failed medication treatments for depression: a STAR*D report. *Am J Psychiatry*. 163: 1519– 1530, 1665.

O'Reardon, J., K. Blumner, A. Peshek, R. Pradilla and P. Pimiento. 2005. Long-term maintenance therapy for major depressive disorder with rTMS. *J Clin Psychiatry*. 66: 1524–8.

Osorio, I., M.G. Frei, B.F. Manly, S. Sunderam, N.C. Bhavaraju and S.B. Wilkinson. 2001. An introduction to contingent (closed-loop) brain electrical stimulation for seizure blockage, to ultra-short-term clinical trials, and to multidimensional statistical analysis of therapeutic efficacy. *J Clin Neurophysiol*. 18: 533–44.

Pekny, M. and M. Nilsson. 2005. Astrocyte activation and reactive gliosis. *Glia*. 50: 427–34. doi:10.1002/glia.20207.

Petersen, E.A., E.M. Holl, I. Martinez-Torres, T. Foltynie, P. Limousin, M.I. Hariz and L. Zrinzo. 2010. Minimizing brain shift in stereotactic functional neurosurgery. *Neurosurgery*. 67: ons 213–21; discussion ons221. doi:10.1227/01.NEU.0000380991.23444.08.

Peterson, B., P. Skudlarski, A. Anderson, H. Zhang, J. Gatenby, C. Lacadie, J. Leckman and J. Gore. 1998. A functional magnetic resonance imaging study of tic suppression in Tourette's syndrome. *Arch Gen Psychiatry*. 55: 326–33.

Piasecki, S. and J. Jefferson. 2004. Psychiatric complications of deep brain stimulation for Parkinson's disease. *J Clin Psychiatry*. 65: 845–849. doi:10.1136/jnnp-2012-302387.

Rauch, S., M. Jenike, N. Alpert, L. Baer, H. Breiter, C. Savage and A. Fischman. 1994. Regional cerebral blood flow measured during symptom provocation in obsessive-compulsive disorder using oxygen 15-labeled carbon dioxide and positron emission tomography. *Arch Gen Psychiatry*. 51: 62–70.

Reijnders, J.S.A.M., B. Scholtissen, W.E.J. Weber, P. Aalten, F.R.J. Verhey and A.F.G. Leentjens. 2010. Neuroanatomical correlates of apathy in Parkinson's disease: A magnetic resonance imaging study using voxel-based morphometry. *Mov Disord*. 25: 2318–25. doi:10.1002/mds.23268.

Ritsner, M., R. Maayan, A. Gibel, R.D. Strous, I. Modai and A. Weizman. 2004. Elevation of the cortisol/dehydroepiandrosterone ratio in schizophrenia patients. *Eur Neuropsychopharmacol*. 14: 267–73. doi:10.1016/j.euroneuro.2003.09.003.

Robins, L.N., J.E. Helzer, M.M. Weissman, H. Orvaschel, E. Gruenberg, J.D. Burke and D.A. Regier. 1984. Lifetime prevalence of specific psychiatric disorders in three sites. *Arch Gen Psychiatry*. 41: 949–58.

Rothwell, J.C., M. Hallett, A. Berardelli, A. Eisen, P. Rossini and W. Paulus. 1999. Magnetic stimulation: motor evoked potentials. The International Federation of Clinical Neurophysiology. *Electroencephalogr Clin Neurophysiol*. 52: 97–103. doi:10.1016/j.parkreldis.2013.07.017.

Rush, A.J., M.H. Trivedi, S.R. Wisniewski, A.A. Nierenberg, J.W. Stewart, D. Warden, G. Niederehe, M.E. Thase, P.W. Lavori, B.D. Lebowitz, P.J. McGrath, J.F. Rosenbaum, H.A. Sackeim, D.J. Kupfer, J. Luther and M. Fava. 2006. Acute and longer-term outcomes in depressed outpatients requiring one or several treatment steps: a STAR*D report. *Am J Psychiatry*. 163: 1905–17. doi:10.1176/appi.ajp.163.11.1905.

Saitoh, Y., A. Hirayama, H. Kishima, T. Shimokawa, S. Oshino, M. Hirata, N. Tani, A. Kato and T. Yoshimine. 2007. Reduction of intractable deafferentation pain due to spinal cord or peripheral lesion by high-frequency repetitive transcranial magnetic stimulation of the primary motor cortex. *J Neurosurg.* 107: 555–9. doi:10.3171/JNS-07/09/0555.

Sampson, S.M., J.D. Rome and T.A. Rummans. 2006. Slow-Frequency rTMS Reduces Fibromyalgia Pain. *Pain Med.* 7: 115–118.

Schapira, A.H.V. 2004. Neuroprotection in Parkinson Disease Mysteries, Myths, and Misconceptions. *JAMA.* 291: 358–364.

Schlaepfer, T.E., B.H. Bewernick, S. Kayser, B. Mädler and V.A. Coenen. 2013. Rapid effects of deep brain stimulation for treatment-resistant major depression. *Biol Psychiatry.* 73: 1204–12. doi:10.1016/j.biopsych.2013.01.034.

Seijo, F., A. Saiz, B. Lozano, E. Santamarta, M. Alvarez-Vega, E. Seijo, R. Fernández de León, F. Fernández-González and J. Pascual. 2011. Neuromodulation of the posterolateral hypothalamus for the treatment of chronic refractory cluster headache: Experience in five patients with a modified anatomical target. *Cephalalgia.* 31: 1634–41. doi:10.1177/0333102411430264.

Shah, P.J., K.P. Ebmeier, M.F. Glabus and G.M. Goodwin. 1998. Cortical grey matter reductions associated with treatment-resistant chronic unipolar depression. Controlled magnetic resonance imaging study. *Br J Psychiatry.* 172: 527–32.

Shah, P.J., M.F. Glabus, G.M. Goodwin and K.P. Ebmeier. 2002. Chronic, treatment-resistant depression and right fronto-striatal atrophy. *Br J Psychiatry.* 180: 434–40.

Shimamoto, H., K. Takasaki, M. Shigemori, T. Imaizumi, M. Ayabe and H. Shoji. 2001. Therapeutic effect and mechanism of repetitive transcranial magnetic stimulation in Parkinson's disease. *J Neurol.* 248: 48–53.

Shon, Y.-M., K.H. Lee, S.J. Goerss, I.Y. Kim, C. Kimble, J.J. Van Gompel, K. Bennet, C.D. Blaha and S.-Y. Chang. 2010. High frequency stimulation of the subthalamic nucleus evokes striatal dopamine release in a large animal model of human DBS neurosurgery. *Neurosci Lett.* 475: 136–40. doi:10.1016/j.neulet.2010.03.060.

Shugart, Y.Y., J. Samuels, V.L. Willour, M.A. Grados, B.D. Greenberg, J.A. Knowles, J.T. McCracken, S.L. Rauch, D.L. Murphy, Y. Wang, A. Pinto, A.J. Fyer, J. Piacentini, D.L. Pauls, B. Cullen, J. Page, S.A. Rasmussen, O.J. Bienvenu, R. Hoehn-Saric, D. Valle, K.-Y. Liang, M.A. Riddle and G. Nestadt. 2006. Genomewide linkage scan for obsessive-compulsive disorder: evidence for susceptibility loci on chromosomes 3q, 7p, 1q, 15q, and 6q. *Mol Psychiatry.* 11: 763–70. doi:10.1038/sj.mp.4001847.

Sillay, K.A., L.M. Kumbier, C. Ross, M. Brady, A. Alexander, A. Gupta, N. Adluru, G.S. Miranpuri and J.C. Williams. 2013. Perioperative Brain Shift and Deep Brain Stimulating Electrode Deformation Analysis : Implications for rigid and non-rigid devices. *Ann Biomed Eng.* 41: 293–304. doi:10.1007/s10439-012-0650-0.

Simpson, K.N., M.J. Welch, F.A. Kozel, M.A. Demitrack and Z. Nahas. 2009. Cost-effectiveness of transcranial magnetic stimulation in the treatment of major depression: a health economics analysis. *Adv Ther.* 26: 346–68. doi:10.1007/s12325-009-0013-x.

Smith, D.F. 2013. Quest for biomarkers of treatment-resistant depression: shifting the paradigm toward risk. *Front Psychiatry.* 4: 57. doi:10.3389/fpsyt.2013.00057.

Song, X.-Q., L.-X. Lv, W.-Q. Li, Y.-H. Hao and J.-P. Zhao. 2009. The interaction of nuclear factor-kappa B and cytokines is associated with schizophrenia. *Biol Psychiatry.* 65: 481–8. doi:10.1016/j.biopsych.2008.10.018.

Stefani, A., A.M. Lozano, A. Peppe, P. Stanzione, S. Galati, D. Tropepi, M. Pierantozzi, L. Brusa, E. Scarnati and P. Mazzone. 2007. Bilateral deep brain stimulation of the pedunculopontine and subthalamic nuclei in severe Parkinson's disease. *Brain.* 130: 1596–607. doi:10.1093/brain/awl346.

Stefurak, T., D. Mikulis, H. Mayberg, A.E. Lang, S. Hevenor, P. Pahapill, J. Saint-Cyr and A. Lozano. 2003. Deep brain stimulation for Parkinson's disease dissociates mood and motor circuits: a functional MRI case study. *Mov Disord.* 18: 1508–16. doi:10.1002/mds.10593.

Strafella, A.P., J. Barrett and A. Dagher. 2001. Repetitive Transcranial Magnetic Stimulation of the Human Prefrontal Cortex Induces Dopamine Release in the Caudate Nucleus. *J Neurosci.* 21: 1–4.

Strafella, A.P., A.F. Sadikot and A. Dagher. 2003. Subthalamic deep brain stimulation does not induce striatal dopamine release in Parkinson's disease. *NeuroReport*. 14: 1287–1289. doi:10.1097/00001756-200307010-00020.

Study-Group. 2001. Deep-brain stimulation of the subthalamic nucleus or the pars interna of the globus pallidus in Parkinson's disease. *NJEM*. 345: 956–963.

Sutton, A.C., W. Yu, M.E. Calos, L.E. Mueller, M. Berk, J. Shim, E.S. Molho, J.M. Brotchie, P.L. Carlen and D.S. Shin. 2013a. Elevated potassium provides an ionic mechanism for deep brain stimulation in the hemiparkinsonian rat. *Eur J Neurosci*. 37: 231–241. doi:10.1111/ejn.12040.

Sutton, A.C., W. Yu, M.E. Calos, A.B. Smith, E.S. Molho, J.G. Pilitsis, J.M. Brotchie, S. Damian, A. Ramirez-zamora and D.S. Shin. 2013b. Deep brain stimulation of the substantia nigra pars reticulata improves forelimb akinesia in the hemiparkinsonian rat. *J Neurophysiol*. 109: 363–374. doi:10.1152/jn.00311.2012.

Svenningsson, P., E.T. Tzavara, F. Liu, A.A. Fienberg, G.G. Nomikos and P. Greengard. 2002. DARPP-32 mediates serotonergic neurotransmission in the forebrain. *Proc Natl Acad Sci USA*. 99: 3188–93. doi:10.1073/pnas.052712699.

Tan, S.K.H., H. Hartung, S. Schievink, T. Sharp and Y. Temel. 2013. High-Frequency Stimulation of the Substantia Nigra Induces Serotonin-Dependent Depression-Like Behavior in Animal Models. *BPS*. 73: 1–3. doi:10.1016/j.biopsych.2012.07.032.

Tawfik, V.L., D. Ph, S. Chang, F.L. Hitti, W. Roberts, J.C. Leiter, S. Jovanovic and K.H. Lee. 2011. Deep Brain Stimulation Results in Local Glutamate and Adenosine Release: Investigation into the Role of Astrocytes. *Neurosurgery*. 67: 367–375. doi:10.1227/01.NEU.0000371988.73620.4C.Deep.

Taylor, J.J., J.J. Borckardt, M. Canterberry, X. Li, C.A. Hanlon, T.R. Brown and M.S. George. 2013. Naloxone-reversible modulation of pain circuitry by left prefrontal rTMS. *Neuropsychopharmacology*. 38: 1189–97. doi:10.1038/npp.2013.13.

Toda, H., C. Hamani, A.P. Fawcett, W.D. Hutchison and A.M. Lozano. 2008. The regulation of adult rodent hippocampal neurogenesis by deep brain stimulation. *Journal of Neurosurgery*. 108: 132–8. doi:10.3171/JNS/2008/108/01/0132.

Trussell, L.O. and M.B. Jackson. 1985. Adenosine-activated potassium conductance in cultured striatal neurons. *Proc Natl Acad Sci USA*. 82: 4857–61.

Ulla, M., S. Thobois, J.-J. Lemaire, A. Schmitt, P. Derost, E. Broussolle, P.-M. Llorca and F. Durif. 2006. Manic behaviour induced by deep-brain stimulation in Parkinson's disease: evidence of substantia nigra implication? *J Neurol Neurosurg Psychiatry*. 77: 1363–6. doi:10.1136/jnnp.2006.096628.

Valencia-Alfonso, C., J. Luigjes, R. Smolders, M. Cohen, N. Levar, A. Mazaheri, P. van den Munckhof, P. Schuurman, W. van den Brink and D. Denys. 2012. Effective deep brain stimulation in heroin addiction: a case report with complementary intracranial electroencephalogram. *Biol Psychiatry*. 71.

Valls-sole, J. and F. Valldeoriola. 2002. Neurophysiological correlate of clinical signs in Parkinson's disease. *Clin Neurophysiol*. 113: 792–805.

Vedam-Mai, V., E.Y. van Battum, W. Kamphuis, M.G.P. Feenstra, D. Denys, B.A. Reynolds, M.S. Okun and E.M. Hol. 2012. Deep brain stimulation and the role of astrocytes. *Mol Psychiatry*. 17: 124–31, 115. doi:10.1038/mp.2011.61.

Voges, J., R. Hilker, K. Bötzel, K.L. Kiening, M. Kloss, A. Kupsch, A. Schnitzler, G.-H. Schneider, U. Steude, G. Deuschl and M.O. Pinsker. 2007. Thirty days complication rate following surgery performed for deep-brain-stimulation. *Mov Disord*. 22: 1486–9. doi:10.1002/mds.21481.

Wei, X.F. and W.M. Grill. 2005. Current density distributions, field distributions and impedance analysis of segmented deep brain stimulation electrodes. *J Neural Eng*. 2: 139–47. doi:10.1088/1741-2560/2/4/010.

Willour, V.L., Y. Yao Shugart, J. Samuels, M. Grados, B. Cullen, O.J. Bienvenu, Y. Wang, K.-Y. Liang, D. Valle, R. Hoehn-Saric, M. Riddle and G. Nestadt. 2004. Replication study supports evidence for linkage to 9p24 in obsessive-compulsive disorder. *Am J Hum Genet.* 75: 508–13. doi:10.1086/423899.

Wisniewski, S., M. Fava, M. Trivedi, M. Thase, D. Warden, G. Niederehe, E. Friedman, M. Biggs, H. Sackeim, K. Shores-Wilson, P. McGrath, P. Lavori, S. Miyahara and A. Rush. 2007. Acceptability of second step treatments to depressed outpatients: a STAR*D report. *Am J Psychiatry.* 164: 753–760.

Wolz, M., J. Hauschild, M. Fauser, L. Klingelhöfer, H. Reichmann and A. Storch. 2012. Immediate effects of deep brain stimulation of the subthalamic nucleus on nonmotor symptoms in Parkinson's disease. *Parkinsonism Relat Disord.* 18: 994–997. doi:10.1016/j. parkreldis.2012.05.011.

Zaidi, S.M.A., A.L. Bikak and Rameez-ul-Hassan. 2010. Improving schizophrenia diagnosis through biomarkers : an upcoming prospect. *J Pak Med Assoc.* 60: 595–596.

Zhang, X., P.E. Andren, P. Greengard and P. Svenningsson. 2008. Evidence for a role of the 5-HT1B receptor and its adaptor protein, p11, in L-DOPA treatment of an animal model of Parkinsonism. *PNAS.* 105: 2163–2168.

Zhou, G., H. Shoji, S. Yamada and T. Matsuishi. 1997. Decreased-phenylethylamine in CSF in Parkinson,s disease. *J Neurol Neurosurg Psychiatry.* 63: 754–758.

Zhuang, Z., M. Kung, C. Hou, K. Plo, T.L. Gur, J.Q. Trojanowski, V.M. Lee and H.F. Kung. 2001. IBOX(2-(4'-dimethylaminophenyl)-6-iodobenzoxazole): a ligand for imaging amyloid plaques in the brain. *Nucl Med Biol.* 28: 887–894.

20

Biomarkers of Nerve Regeneration in Peripheral Nerve Injuries: An Emerging Field

Karim A. Sarhane,[1] *Chris Cashman,*[2,3] *Kellin Krick,*[4]
Sami H. Tuffaha,[1] *Justin M. Broyles,*[1] *Saami Khalifian,*[1]
Mohammed Alrakan,[1] *Pablo Baltodano,*[1] *Zuhaib Ibrahim*[1]
and Gerald Brandacher[1,*]

INTRODUCTION

Peripheral nerve injuries constitute a frequent and disabling condition, with an estimated incidence reaching 23 per 100,000 persons per year in developed countries, not accounting for non traumatic cases (i.e., nerve damage secondary to abdominal or pelvic surgeries) and for lesions not treated at health facilities (Taylor et al., 2008; Asplund et al., 2009). Despite great advances in microsurgical techniques, nerve repair continues to be suboptimal, and full functional recovery is seldom achieved.

[1] Department of Plastic and Reconstructive Surgery, Johns Hopkins University School of Medicine, Baltimore, MD, USA.
[2] Department of Neurology, Johns Hopkins University School of Medicine, Baltimore, MD, USA.
[3] Department of Neuroscience, Johns Hopkins University School of Medicine, Baltimore, MD, USA.
[4] Department of Materials Science and Engineering, Johns Hopkins University Whiting School of Engineering, Baltimore, MD, USA.
* Corresponding author: brandacher@jhmi.edu

A significant amount of research is being invested in devising neuroregenerative therapies aiming at increasing the rate of peripheral axonal growth. Despite encouraging results in animal models using various growth factors, neurohormones and cell-based therapies, clinical trials in humans have had variable success (Rodrigues et al., 2012). One of these reasons lies in the very slow regenerative process in humans (as opposed to rodents) (Buchthal and Kühl, 1979; Dolenc and Janko, 1976). Furthermore, this process cannot be either measured nor reliably followed, and there are no objective means to assess the effectiveness of the different nerve regenerative therapies in humans. There is thus a critical need for a modality that measures and follows the growth of peripheral axons in humans, and evaluates the effectiveness of various therapeutic interventions aiming at speeding nerve regeneration; ideally this modality is simple, does not involve complicated procedures, and is sensitive enough to detect specific regenerative cues allowing targeted and timely therapeutic strategies.

In this regard, biomarkers have a great potential in objectively assessing the evolution of physiological processes (such as nerve regeneration), and in evaluating the effect of various agents (or growth factors) in modulating such processes. They play a crucial role in the successful translation of findings from experimental studies in animals to therapeutic benefits in humans. Unfortunately, much of the work in this field has been done in characterizing biomarkers that impede neuroregeneration, and not those that promote it (Olsson et al., 2011; Lovestone, 2010). Biomarkers of peripheral nerve regeneration is an emerging area of research and holds great promise in improving the interventions aimed at enhancing axonal growth.

The purpose of this chapter is to explore the role of biomarkers in advancing the field of peripheral nerve regeneration. We will first revisit the integral role biomarkers play in translational medicine and the development of new therapeutic strategies. Then, we present potential biomarkers for peripheral nerve regeneration that might fuel novel treatment modalities. Finally, we discuss the feasibility and applicability of neuroregenerative biomarkers in the field of Vascularized Composite Allotransplantation (VCA), the main interest of our research laboratory.

Definition and Types of Biomarkers

A biomarker is a measurable and quantifiable biological molecule that serves as an index for specific physiologic/pathologic processes, or for a particular pharmacologic response to a treatment modality (Biomarkers and surrogate endpoints: preferred definitions and conceptual framework, 2001). In an era dominated by high-throughput molecular technologies, the term biomarker typically refers to molecular biomarkers (i.e., any cellular change at the nuclear or mitochondrial DNA, messenger or micro RNA or protein level). Simple molecules such as carbohydrates (e.g., glycosylated hemoglobin) (Lyons and Basu, 2012), lipids (e.g., fatty acids) (Mohanty et al., 2013) or peptides (e.g., albumin) (Lu et al., 2012) can also be considered as biomarkers. More complex categories include whole cells (e.g., T-cells) (Ferran et al., 2013). Biomarkers are usually assayed in blood or other body fluids (CSF, urine, sputum…). A new generation of biomarkers based on imaging techniques (CT, MRI, PET, and SPECT technologies) is evolving, especially in the field of neurodegenerative disorders. By visualizing changes in the brain or spinal cord specific for certain pathologies, these can be used as guides for treatment (Willmann et al., 2008).

Three broad categories of biomarkers exist as shown in Figure 20.1: (1) Biomarkers following a physiological or pathological process over time; most of these parallel changes seen during clinical assessments, and can thus be used to monitor and follow-up diseases. (2) Biomarkers reflecting physiological responses to pharmacological interventions; such biomarkers aid in the discovery of new drugs and can potentially uncover novel targets

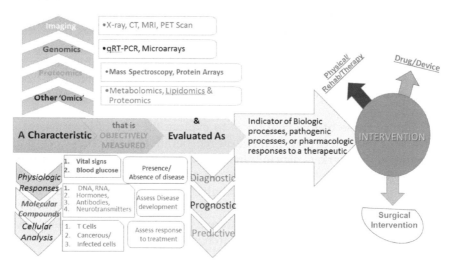

Figure 20.1. Definition and Types of Biomarkers. Biomarker terminology was introduced in 1989 as a Medical Subject Heading (MeSH) term: "measurable and quantifiable biological parameters (e.g., specific enzyme concentration, specific hormone concentration, specific gene phenotype distribution in a population, presence of biological substances) which serve as indices for health- and physiology-related assessments, such as disease risk, psychiatric disorders, environmental exposure and its effects, disease diagnosis, metabolic processes, substance abuse, pregnancy, cell line development, epidemiologic studies, etc." In 2001, an NIH working group standardized the definition of a biomarker as "a characteristic that is objectively measured and evaluated as an indicator of normal biological processes, pathogenic processes, or pharmacologic responses to a therapeutic intervention" and defined types of biomarkers.

for intervention. (3) Biomarkers representing clinical endpoints potentially serving as alternative endpoint assessment tools; such biomarkers characterize the final outcome of an intervention and can thus be used as clinical endpoints in trials evaluating safety and effectiveness of various therapies. It is worth adding that these are most useful when providing information about the outcome of interest significantly earlier than the gold standard clinical measurement (Biomarkers and surrogate endpoints: preferred definitions and conceptual framework, 2001; Fox and Growdon, 2004; De Gruttola et al., 2001).

Biomarkers in Translational Medicine

The identification and characterization of various Schwann Cells (SCs) trophic factors as the main neuroregenerative substrates after peripheral nerve injury (Zhang et al., 2000; Meyer et al., 1992; Funakoshi et al., 1993; Boyd and Gordon, 2003) has driven many efforts in search for suitable biomarkers for neuroregeneration. While Nerve Growth Factor (NGF) and other molecules released by SCs may potentiate axonal regeneration by upregulating Neural Cell Adhesion Molecules (NCAMs) expression (Seilheimer and Schachner, 1987; Friedlander et al., 1986), other aspects of this multifaceted regenerative process, namely the molecular and cellular changes in the distal portion of the nerve that impede regeneration, need to be unraveled. In order to develop this comprehensive approach for demystifying nerve regeneration mechanisms, biomarkers assessing injury severity, demonstrating changes with regeneration progression, and predicting functional outcomes in longitudinal studies are needed. These hold great translational potential as their use in clinical trials will help in identifying appropriate drugs, drug dosing, drug efficacy and in improving safety assessments. The pathway to biomarker translation is however a long process and carries multiple intricacies. It can be divided into three phases as described in Figure 20.2 (Rai, 2007):

1. Biomarker discovery is the first phase. It entails using high-throughput computational analysis and bioinformatics algorithms to search for molecular (or other biological) candidates in tissue samples; this is followed by validation of potential biomarkers using a different set of samples. This first phase can be divided into multiple stages: Study conception and design, samples' collection (from control and experimental tissue), separation and quantification of samples' analytes, identification of differences between analyzed samples, verification of differences (using a different algorithm than that used for discovery), and validation of biomarker candidates in larger studies (using the same algorithm for discovery but different samples)

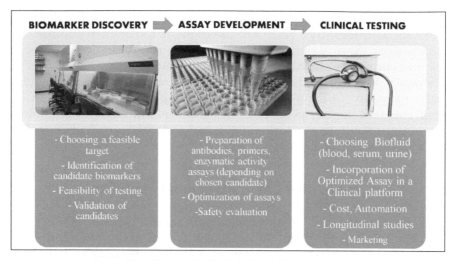

Figure 20.2. Phases for Clinical Translation of Biomarkers.

(Le-Niculescu et al., 2013). Ideally, it is recommended to start analyzing tissues that are as close to the site of injury as possible; for instance, in addition to the axotomized proximal and distal stumps, analyzing tissue from the innervated muscle and skin or even from fluid released close to the site of injury would be of potential interest. Serum/plasma or urine samples might also be valuable. The concentration of a biomarker derived from the injured tissue will decrease as one moves further away from the site of injury. Thus, it will be highest in the tissue of interest, and becomes lower in interstitial fluid, less in blood, and might even be undetectable in urine. When the ultimate goal is a biomarker with direct translational potential, blood-based samples (serum or plasma) are the most favorable. Serum is stored in most hospital clinical laboratories, and the assays to analyze it for various analytes are widely available (Aziz, 1998).

2. The second phase entails developing a simple and readily available assay that reliably detect identified candidate biomarkers, and can be easily implemented in a clinical setting. This phase is also divided into multiple stages: Analyte of interest characterization, selection of most appropriate assay type (e.g., qRT-PCR, immunohistochemistry, enzymatic activity assays, etc…), acquisition/preparation of reagents (e.g., primer probes, antibodies, etc…), protocol development/optimization (initial testing on positive controls), clinical validation on experimental samples, availability of calibrators, and quality control of materials and reagents (Drake et al., 2004). Quality control, certified reference and linearity samples should also be tested to define limits,

and to assess the measurement accuracy and trueness of the devised assay system (Vitzthum et al., 2005). It is worth noting that since biomarkers derived from nerve injury site might be present at very low concentrations in serum, sensitivity of the optimized assay is critical.

3. The third phase is clinical translation. Success of this phase is dependent on the clinical feasibility/applicability of the developed assay; ideally, this new assay would make use of an already established clinical test, with the aim of using it as a platform. This expedites the implementation of the in-house developed test. It is therefore crucial for basic scientists to be familiar with the various assays widely used in today's clinical laboratories (Burtis et al., 2001). Technologies like genotyping, gene expression, proteomics, immunohistochemistry, and *in situ* hybridization are now able to identify key molecules, and are thus becoming more and more important tools for biomarkers (Sarhane, 2012). Other considerations include cost and possibilities for automation (Rai, 2007). Finally, once the appropriate assay format has been confirmed, commercialization of this biomarker clinical test is needed. An assay is considered to be a medical device, and therefore needs to be registered with the US FDA (FDA, 2010).

Biomarkers in Peripheral Nerve Regeneration

The hallmark of the peripheral nervous system is its ability to regenerate following injury, a process which depends on complex cell–cell and cell–extracellular matrix interactions (Geuna et al., 2009). These interactions involve a myriad of messenger molecules (Fu and Gordon, 1997), which could be of great potential interest for biomarker discovery.

Biomarker discovery involves an in-depth comparison of the transcriptome, metabolome and proteome of tissues differing in one or several physiologic processes (Eugster et al., 2013). In the case of nerve regeneration, the molecular fingerprint of tissue implicated in nerve regeneration is compared to that of control tissue, with the goal of identifying expression differences, and a subset of molecules specific for nerve regeneration. A sequence of molecular responses takes place in response to injury for the successful nerve regeneration and recovery of function. These molecular processes can be mapped to three different sites illustrated in Figure 20.3: Spinal cord cell bodies, site of axotomy, innervated end organ.

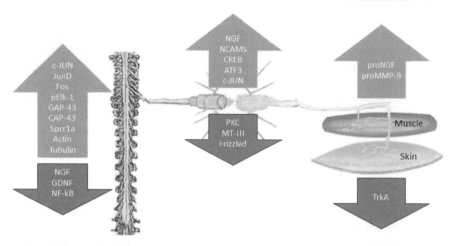

Figure 20.3. Candidate Biomarkers at three different Locations. pElk-1: phosphorylated Elk-1; GAP-43: Growth Associated Protein 43; Sprr1a: Small Proline-rich Repeat Protein 1A; NGF: Neural Growth Factor; NF-kB: Nuclear Factor Kappa Beta; CREB: cAMP Responsive Element binding Protein; ATF-3: Activating Transcription Factor-3; PKC: protein kinase C; MT-III: Metallothionein III; proMMP9: pro Matrix MetalloProteinase-9; TrKA: Neurotrophic Tyrosine Kinase Antigen.

Color image of this figure appears in the color plate section at the end of the book.

Site of Axotomy

After nerve injury, calcium and sodium ions enter the axoplasm through the lesioned plasma membranes, and generate high frequency burst of action potentials at the injured site (Makwana and Raivich, 2005; Navarro et al., 2007). This first signal promotes an influx of additional calcium through voltage-dependent ion channels, which activates several protein kinase pathways, including: calcium/calmodulin dependent kinase 2 (CMAK2), Protein Kinase A (PKA), Protein Kinase C (PKC), and Mitogen-Activated Protein Kinase (MAPK), such as Erk1 and Erk2, c-jun N-terminal kinase (JNK) and P38 kinase (Makwana and Raivich, 2005; Raivich and Makwana, 2007). The change in activity of transcription factors, mediators of gene expression, is considered the downstream event of these axotomy-activated protein kinases (Dahlin et al., 2009b). The activation of transcription factor cAMP responsive element binding protein (CREB) has been demonstrated in early stages after injury (Miletic et al., 2004b). Phosphorylation of CREB can be mediated by multiple protein kinase pathways through activation of tyrosine kinase eceptors (Trk) (Miletic et al., 2004a). Two other transcription factors induced by nerve injury are activating transcription factor 3 (ATF3) and c-Jun. c-Jun upregulation and phosphorylation is induced by activated JNK, leading to the formation of activating protein 1 (AP-1) complexes.

JNK pathways and the Erk1/2 pathways also show cross-talk coordinated by MEKK1 in PC12 cells (Waetzig and Herdegen, 2005). Novel modalities tracing the expression changes of these two key molecules hold the potential for devising biomarkers estimating neuroregenerative potential.

c-Jun

c-Jun is the fundamental component of the Activating Protein 1 (AP-1) complex (Karin, 1995), and is one of the targets of the stress activated c-Jun N-terminal Linase (JNK), which catalyzes its phosphorylation (Neumann et al., 2002; Nilsson et al., 2005; Perlson et al., 2004). Conditional knockout of c-Jun (Raivich et al., 2004) as well as pharmacological inhibition of JNK (Lindwall et al., 2004) were shown to impede nerve regeneration. Thus, the JNK family of kinases is required for successful regeneration of peripheral neurons, and more specifically for sensory axons (Middlemas et al., 2003; Qiu et al., 2002a; Perlson et al., 2004). As such, they represent attractive targeted therapy for sensory reinnervation.

AFT3

ATF3 is a member of the ATF/CREB transcription factor family (Qiu et al., 2002b; Raivich et al., 2004). It is rapidly upregulated by a variety of signals post-injury, including agents triggering the JNK signaling cascade (Seijffers et al., 2006, 2007; Sung et al., 2001; Tanabe et al., 2003). Inhibition of JNK reduces ATF3 protein levels (Lindwall et al., 2004), which in turn inhibits nerve regeneration. Importantly, ectopic expression of ATF3 has the potential of promoting neurite outgrowth of peripheral neurons, possibly through an increase in the intrinsic growth capacity of neurons (Wong and Oblinger, 1991; Woolf et al., 1990). The presence of these factors over time can probably modulate the rate of axonal growth. This is very crucial when deciding on the optimal timing for nerve repair. Thus, if the levels of nerve regeneration molecules (JNK, c-Jun, and ATF3) can be selectively modified following peripheral nerve injury, an increase in neuritogenesis might be obtained. Such targets for nerve regeneration also serve as biomarkers indices for axonal growth.

The changes in the gene expression that promote nerve regeneration could also be induced by extrinsic factors, released by non-neuronal cells (Zigmond, 2012). In fact, for optimal effects of the intrinsic growth capacity of peripheral nerve regeneration, an adequate environment favorable for axonal growth is a must. Such an environment is provided primarily by SCs (Gordon, 2009). These ensheath and myelinate growing axons providing them with an effective scaffold (basal lamina) that promotes their extension. More specifically, following injury, SCs de-differentiate and aid in the clearing of damaged debris, while during regeneration they act as guides for sprouting axons (Mirsky and Jessen, 1996). Different molecular processes govern these two phases and might be used to indicate the timing of nerve regeneration, also with potential targeted and timely therapeutic interventions. During the regenerative process SCs upregulate several specific genes involved in guiding axonal sprouts by Schwann cell-axon attachment (Martini et al., 1994). For instance, the extracellular matrix molecule laminin, which is produced by SCs, plays a significant role during regeneration. Laminin receptors, such as integrins, are expressed on growing axons, which supports regeneration. In mouse knockout models of laminin, axonal regeneration is significantly impaired (Zhang and Ambron, 2000). Searching for sensitive modalities to detect the level of laminin in the distal stump might provide useful information on the growth potential of axons.

During regeneration the axonal sprouts grow down the distal nerve segment to finally reinnervate their correct targets. Axons must make correct discriminatory choices in order to reinnervate the correct target tissue (muscle versus skin), and during this process they are often misrouted (Lee and Farel, 1988). In order to develop therapeutic strategies to improve both rate and accuracy of target reinnervation, we need biomarkers that can clarify both the molecular events influencing the intrinsic growth capacity as well as axonal discrimination of the extrinsic cues encountered by the axon along the regenerative pathway (Dahlin et al., 2009a). It is worth noting that these signal transduction mechanisms do not occur only in neurons and axons, but also in all types of growing cells in both a temporal and spatial resolution (Massing et al., 2010). Hence, lessons learnt from other regenerative processes might be applied to the context of nerve regeneration.

Spinal Cord Site: The Cell Body Reaction

The activated proteins from the injured site incorporate into the retrograde transport system for trafficking back to the cell body where they induce several signaling pathways genes in the neurons (Hanz and Fainzilber, 2006). This pathway is termed "positive injury signals", i.e., retrograde transport of "harmful" substrates from the site of injury to the cell body. Another pathway implicated in inducing molecular changes in the neurons is termed "negative injury signals", i.e., loss of trophic retrograde transport from end organs (Gordon et al., 1991); examples include loss of NGF (Fitzgerald et al., 1985; Schicho et al., 1999; Verge et al., 1995) and glial-derived neurotrophic factor support (Bennett et al., 1998). The latter pathway (negative signaling) seems to be more important, as pharmacological inhibition of retrograde axonal transport with microtubule-binding proteins, such as colchicine or vinblastine, mimics many of the neuronal and perineuronal responses of neurons to nerve transection (Aldskogius and Svensson, 1988; Landmesser and Pilar, 1974; Leah et al., 1991; Woolf et al., 1990). Furthermore, neuronal death is thought to be caused by severe loss of retrograde trophic support from target tissues (Lieberman, 1974; Pettmann and Henderson, 1998), and that can be reduced by exogenous trophic agents (Li et al., 1994). Other neuronal changes are caused by interruption of retrograde transport of unknown molecules (Verge et al., 1990). Clearly, more research into the implications of these retrograde reactions on serum/plasma global protein expression changes might offer new insights in assessing nerve injury severity and thus potentially predicting functional outcomes.

Collectively, all injury signals conveyed to the cell body are known to induce the "Cell Body Reaction", which consists of changes in transcription, translation, and posttranslational mechanisms (Sterman and Delannoy, 1985). Microarray analysis of these neurons has revealed injury-induced regulation of genes encoding neurotrophin receptors, cytoskeletal proteins, cell adhesion and guidance molecules, trophic factors and receptors, cytokines, neuropeptides and neurotransmitter synthesizing enzymes, ion channels, and membrane transporters transcription factors (Curtis, 1998; Drysdale et al., 1996; Gunstream et al., 1995; Gupta et al., 1996; Cummins and Waxman, 1997; Verge et al., 1995; Zhang et al., 1997; Hökfelt et al., 1994). As a result of these expression changes, nerve cell bodies shift their physiology from synaptic transmission and maintenance of structure to the formation of new growth cones and elongation of the regenerating axon (McQuarrie and Grafstein, 1973; Richardson and Issa, 1984; Skene, 1989; Lieberman, 1971).

Several kinases are activated in cell bodies in response to an injury event. Among these are the MAP kinases Erk1 and Erk2 (Obata et al., 2004), and Jnk (Kenney and Kocsis, 1998). Interestingly, Jnk activation in Dorsal Root Ganglion (DRG) neurons was shown to be dependent on the distance of the site of axotomy (Kenney and Kocsis, 1998). Axotomy may also induce elevated transcription and translation of kinases in cell bodies, as exemplified in the case of Janus Kinases (JAK) (Rajan et al., 1995; Yao et al., 1997), and regeneration-enhancing changes in intracellular cAMP levels (Neumann et al., 2002; Qiu et al., 2002b). Downstream events include upregulation or activation of transcription factors. For example, lesion-induced Jnk induces upregulation and phosphorylation of the transcription factors c-Jun, JunD and Fos and also translocation of ATF3 into the nucleus, leading to formation of complexes with DNA binding activity (Kenney and Kocsis, 1998; Lindwall and Kanje, 2005; Lindwall et al., 2004; Seijffers et al., 2006). Activated retrogradely transported Erk induces phosphorylation of the ETS domain transcription factor Elk-1 in DRG neuronal cell bodies (Perlson et al., 2005). Other transcription factors, including STAT3, P311, Sox11 and C/

EBPb, were also found to be upregulated and activated following injury (Nadeau et al., 2005; Schwaiger et al., 2000), while NF-kB in contrast is decreased (Povelones et al., 1997). These alterations in transcription factor activity result in changes of gene expression in the injured neuron, with upregulation of a growing list of "regeneration-associated genes". So far the main studies of this aspect of neuronal growth programs have followed prior analyses of proteins associated with axonal growth in embryos, such as GAP-43, CAP-43, Sprr1a, actin, tubulin, assorted neuropeptides and growth factors (Hanz and Fainzilber, 2006). Some attention has also been directed to decreases in the expression of developmentally regulated proteins such as the neurofilament and neurotransmitter systems (Fawcett and Keynes, 1990; Bonilla et al., 2002; Caroni, 1998; Goldberg, 2003). Evidently, additional research is needed to define the pertinent elements of the neuronal regeneration network.

Innervated End-Organ Site

As outlined above, blood-based biomarkers are the most favorable for clinical translation. However, in certain instances when skin biopsies (of the innervated limb for example) are taken to assess other parameters (skin rejection), a minute portion (1 mm) can be assayed for some molecules known to be dysregulated in response to denervation/innervation. Effectively, some epidermal and dermal proteins (namely Neurotrophic Tyrosine Kinase Antigen [TrkA]) decrease in response to de-innervation, and others (proNGF, pro Matrix MetalloProteinase-9 [proMMP-9]) increase with recovery (Peleshok and Ribeiro-da-Silva, 2012). These can be measured on simple skin biopsies, and may reflect the underlying neuroregenerative processes.

Histologic Parameters of Neuroregeneration

Histopathologic changes remain by and large the most currently used predictors of peripheral nerve damage and regeneration (Vleggeert-Lankamp, 2007; Castro et al., 2008). It is true that with the advent of "omics" technologies all focus has centered on molecular pathways, however it should be remembered that adequate knowledge of nerve histology is a prerequisite for peripheral nerve regeneration research. Here we outline the salient morphological changes occurring in the nerve cell body and proximal segment after nerve damage and regeneration.

Cell Body

As early as several hours after the injury, changes are observed in the nerve cell body. The series of morphological changes is known as chromatolysis; this includes cell body and nucleolar swelling and nuclear eccentricity (Goldstein et al., 1987). All of these changes involve an alteration of the metabolic machinery from being primarily concerned with transmitting nerve impulses to fabricating structural components for reconstruction of the injured nerve (Ducker et al., 1969; Lieberman, 1971). The neurons switch from a "signaling mode" to a "growing mode" (Fu and Gordon, 1997), and protein synthesis switches from neurotransmitter-related substances to those required for axonal reconstruction (Müller and Stoll, 1998). Metabolic changes include altered synthesis of many neuropeptides (Hökfelt et al., 1994) and changes in synthesis of cytoskeletal proteins (Fornaro et al., 2008; Tetzlaff et al., 1988) and growth-associated proteins (Schreyer and Skene, 1991).

Proximal Segment

In the proximal segment, axons degenerate for some distance back from the site of injury, leaving the corresponding endoneurial tubes (the basal laminae of the Schwann cell) behind as empty cylinders. This retrograde degeneration may extend over one or several internodal segments, the length depending on the severity of the lesion (Cajal, 1928). Within hours after injury, the axon in the proximal segment produces a great number of collateral and terminal sprouts that advance distally along the tube on the inside of the basal lamina (Fawcett and Keynes, 1990; Mira, 1984). The terminal sprouts arise from the tip of the remaining axon. Within hours of axotomy, small axoplasmic outgrowths have been observed from axoplasmic tips (Zelená et al., 1968). This first wave of sprouts is followed by a second wave, appearing within the first 2 days (Mira, 1984; Grafstein and McQuarrie, 1978; Cajal, 1928). Early sprouts can apparently degenerate before the definitive sprouting phase occurs. The time required for the definite sprouts to appear has been called the "initial delay" (Sunderland, 1978).

Distal Segment

Following nerve injury, Wallerian degeneration is triggered, a process necessary for axons in the distal stump undergoing demyelination to be cleared of myelin debris for subsequent sprouting of uninjured axons (Gaudet et al., 2011). In Wallerian degeneration the primary histological change involves physical fragmentation of both axons and myelin, a process that begins within hours of injury (Waller, 1850). Ultrastructurally, both neurotubules and neurofilaments become disarrayed, and axonal contour becomes irregular, due to varicose swellings. By 48 to 96 hours post injury, axonal continuity is lost and conduction of impulses is no longer possible (Sunderland and Bradley, 1950). Myelin disintegration lags slightly behind that of axons but is well advanced by 36 to 48 hours (Esposito et al., 2002). Mast cells in the immediate vicinity undergo rapid degranulation (Olsson, 1967). This degranulation contributes to the pain elicited by NGF (Woolf et al., 1996; Lewin et al., 1994) and to other pathological processes in peripheral nerves (Brosman et al., 1985; Dines and Powell, 1997; Zochodne et al., 1994). Macrophages are recruited to the site of the lesion during the first week post injury (Perry et al., 1987), contributing to lysis and phagocytosis of myelin and subsequent Schwann cell proliferation (Beuche and Friede, 1984). Immunohistochemical stains for macrophages highlight an influx of macrophages that enter the space formerly occupied by the nerve fiber and clear away cellular debris. These debris may be seen on routine histologic sections and occupies the center of phagocytizing cells (Daniel and Strich, 1969). Both cell types (mast cells and macrophages) secrete mitogens and growth factors which also play a role in axonal regeneration and remyelination (De Vries, 1993; Reynolds and Woolf, 1993).

Biomarkers in the Field of Vascularized Composite Allotransplantation

Vascularized Composite Allotransplantation is an emerging field in transplant surgery. It consists of transplanting composite allografts of various tissue types (skin, muscle, bone, tendon, nerves, vessels, bone marrow). This is performed in patients suffering from complex tissue injuries for the restoration of form and function when conventional reconstructive options are not possible. Over the past decade, VCA has become the preeminent reconstructive modality with over 200 procedures performed worldwide including hand, forearm, arm, partial facial tissue and full face, abdominal wall, larynx, trachea, vascularized bone and joint, tongue or even uterus and penis (Schneeberger et al., 2011; Swearingen et al., 2008). Widespread expansion of VCA clinical indications is dependent on further breakthroughs in immunomodulation and graft tolerance strategies (Sarhane et al., 2013). Just as important to the ultimate success of hand and face transplantation is improving nerve functional outcomes. The unique requirement for nerve regeneration over long distances to regain full motor and sensory function significantly hampers expansion of VCA for upper arm or lower extremity transplantation, and therefore, its full utilization as a reconstructive modality for devastating limb loss. Modalities for enhancing nerve regeneration in VCA specifically are needed. Progress in this field would also benefit from adequate biomarkers that can reliably monitor the process of regeneration and indicate the effectiveness of various neuroregenerative therapies.

Our group has a long track record in exploring novel modalities for promoting immune tolerance and enhancing nerve regeneration after VCA. We have recently found that intraneural injection of chondroitinase in a rat orthotopic hind limb transplant model enhances nerve regeneration (Tuffaha et al., 2011). In this study, the mechanism underlying the neurotherapeutic effect was decreased in chondroitin sulfate proteoglycans in the extracellular matrix of donor's nerve. Effectively, it is known that the presence of chondroitin sulfate proteoglycans in distal nerve stumps inhibits regenerating axons (Carulli et al., 2005); blood-based biomarkers capable of assessing the amount of chondroitin sulfate proteoglycans in the nerves of hand or face donors might be indicative of the ability of recipient's nerves to effectively regenerate into the distal stump. This can further guide targeted therapy.

In another study by our group (report in preparation), intraneural and systemic delivery of Mesenchymal Stem Cells (MSCs) was shown to promote axonal sprouting in a rat othotopic hind limb transplant model. Preliminary data from another group seemed to show the same positive results of MSCs on functional outcomes (Fitzpatrick et al., 2013). Mechanistically, we found that this effect might be mediated by a decrease in collagen deposition distal to the repair site. Effectively, scar tissue between coapted nerve stumps was frequently demonstrated to impede axonal growth (Boedts, 1987; Ngeow, 2010; Karim A. Sarhane et al., 2014), and MSCs were shown in other studies to have anti-inflammatory properties (MacFarlane et al., 2013). Hence, biomarkers reflecting the degree of fibrosis and inflammation at the repair site might be useful in predicting the effectiveness of nerve regeneration after VCA.

In addition to assessing the amount of inhibitory cues in the donor's nerve (distalstump) and the degree of scarring at the repair site, a greater understanding of the interaction between the recipient's immune system and the unique nerve regenerative mechanisms in VCA is essential when working towards discovering novel biomarkers for nerve regeneration. This interaction can pose great challenges to effective nerve regeneration, but might also confer unique immunologic benefits that might allow the prediction and promotion of nerve regeneration. On the one hand, SCs, by virtue of their MHC-I and MHC-II expression, are highly immunogenic (Yang et al., 2013; Yu et al., 1990; Gulati and Cole, 1990), and thus boost the alloimmune reaction at the regenerating axonal front which might impede regeneration. Moreover, growing recipient's axons will be surrounded by immunogenic cells all along their pathway into the donor's nerve. Unless a state of donor-specific tolerance or immune quiescence is reached, this highly immunogenic pathway might hamper their growth.

On another hand, SCs, being exceedingly antigenic, might be rapidly eliminated by the host immune system, and subsequently replaced by recipient's own SCs. Effectively it is hypothesized that acute rejection episodes, by destroying some of the donor's SCs, might benefit nerve regeneration by allowing host SCs migration. This phenomenon of host SCs migration and repopulation of the donor's nerve remains to be elucidated (Mackinnon et al., 2001). Considerations include mitogenic cytokines released by regenerating axons as well as those secreted by infiltrating alloimmune cells, and their effects on the molecular and cellular dynamics underlying recipient's SCs migration. Ultimately, the entire length of the donor's nerve might be repopulated with host SCs. When this repopulation occurs (several months post-transplant), the level of immunosuppressive medication that patients usually receive would have been tapered down; although skin rejection and serum level of donor specific antibodies can still indicate the safety of such drug tapering, a potential *de novo* proliferation of donor SCs and re-invasion of the growing recipient's axons (triggered by drug tapering) might occur. Should this occur with growing axons getting re-populated by donor SCs after minimization of immunosuppression, debilitating conduction block might ensue. Lessons learned from nerve allografts indicate that nerve conduction will only be supported long-term if regenerated host axons were myelinated by host SCs that have migrated into the graft (Glaus et al., 2011). Whether this is also the case in VCA remains to be elucidated.

Other considerations include the use of highly potent immunosuppression in VCA to prevent skin rejection. It is not clear whether this strong immunosuppression would permit the elimination of donor SCs and the subsequent migration of host SCs. But still, benefits of such strong immunosuppression exist. In fact, some agents are favorable for nerve regeneration, namely tacrolimus (Konofaos and Terzis, 2013). Tacrolimus has been shown to have neuroprotective and neurotrophic actions in experimental models, increasing neurite elongation and accelerating the rate of nerve regeneration *in vitro* and *in vivo* (Doolabh and Mackinnon, 1999; Gold et al., 1995; Katsube et al., 1998; Lee et al., 2000; Wang et al., 2002). This effect is believed to be mediated by prevention of the dephosphorylation/ inactivation of growth-associated protein 43 (GAP-43) (Lyons et al., 1994) and the TGF-β1 pathway (Fansa et al., 1999). Biomarkers reflecting the status of these two pathways might be of benefit in this setting.

In sum, knowledge in the field of peripheral nerve regeneration in VCA, particularly the molecular and cellular changes underlying SCs' immunologic and myelinating dynamics, is limited. Greater understanding of such mechanisms is needed. Many of the unique aspects of VCA (namely immunologic factors) need to be taken into account when attempting to translate the biomarkers candidate identified in peripheral nerve regeneration studies to VCA. Further research in this field is practically feasible, since VCA patients undergo frequent blood draws and skin biopsies; these can be assayed for potential nerve regeneration cues. Along these lines, it is also equally important to develop adequate animal models that reliably replicate the complex nerve regenerative scenario of VCA.

Conclusion

Biomarkers play a critical role in patient management and in pharmaceutical trials as diagnostic, prognostic, predictive and drug efficacy indicators. The overall expectation of a peripheral nerve regeneration biomarker is to enhance the ability of clinicians to optimally manage their patients (see Table 20.1). Such biomarkers need to be able to assess the severity of axonal damage, the likelihood of regeneration and functional recovery, and the effectiveness of various therapeutic strategies. In order to be of clinical value biomarkers need also to be accurate, reproducible (obtained in a standardized fashion), easy to interpret, and most importantly possess a high sensitivity and specificity for the outcome they are meant to estimate (Morrow et al., 2003; Zolg and Langen, 2004).

Currently, with the advent of new and improved genomic and proteomic technologies coupled with advances in bioinformatics tools, there is a rising activity towards identifying novel biomarkers for peripheral nerve regeneration. Research in this area needs to focus on developing a better understanding of the post-injury mechanisms underlying SCs atrophy and

Table 20.1. Desirable Features of Peripheral Nerve Regeneration Biomarkers.

Screening	Diagnostic	Prognostic
Defines extent of injury (motor, sensory)	Specific for nerve injury	Adds to known prognostic measures
Rapid test	Diagnostic cutoff well defined	Fluctuations alter management plan
Inexpensive	Cost-effective	Affects choice of therapy
Readily available	Long half-life	Monitors recovery progression

loss of basal lamina, as well as the principles that govern axonal growth and transport. More insight into the molecular changes in the distal stump that impede regeneration is also crucial (Höke, 2006). In a VCA setting, the various immunomodulatory and immunosuppressive protocols need to be added to this complex picture. The ultimate goal is to discover simple biomarkers (blood or skin biopsy based) that can guide novel strategies aiming at promoting both donor-specific tolerance and nerve regeneration after VCA.

References

Aldskogius, H. and M. Svensson. 1988. Effect on the rat hypoglossal nucleus of vinblastine and colchicine applied to the intact or transected hypoglossal nerve. *Exp Neurol.* 99: 461–73.

Asplund, M., M. Nilsson, A. Jacobsson and H. von Holst. 2009. Incidence of traumatic peripheral nerve injuries and amputations in Sweden between 1998 and 2006. *Neuroepidemiology.* 32: 217–28. doi: 10.1159/000197900.

Aziz, K.J. 1998. Tumor markers: reclassification and new approaches to evaluation. *Adv Clin Chem.* 33: 169–99.

Bennett, D.L., G.J. Michael, N. Ramachandran, J.B. Munson, S. Averill, Q. Yan, S.B. McMahon and J.V. Priestley. 1998. A distinct subgroup of small DRG cells express GDNF receptor components and GDNF is protective for these neurons after nerve injury. *J Neurosci.* 18: 3059–72.

Biomarkers and surrogate endpoints: preferred definitions and conceptual framework. 2001. *Clin Pharmacol Ther.* 69: 89–95. doi: 10.1067/mcp.2001.113989.

Boedts, D. 1987. A comparative experimental study on nerve repair. *Arch Otorhinolaryngol.* 244: 1–6.

Bonilla, I.E., K. Tanabe and S.M. Strittmatter. 2002. Small proline-rich repeat protein 1A is expressed by axotomized neurons and promotes axonal outgrowth. *J Neurosci.* 22: 1303–15.

Boyd, J.G. and T. Gordon. 2003. Neurotrophic factors and their receptors in axonal regeneration and functional recovery after peripheral nerve injury. *Mol Neurobiol.* 27: 277–324. doi: 10.1385/MN: 27: 3: 277.

Brosman, C.F., W.D. Lyman, F.A. Tansey and T.H. Carter. 1985. Quantitation of mast cells in experimental allergic neuritis. *J Neuropathol Exp Neurol.* 44: 196–203.

Buchthal, F. and V. Kühl. 1979. Nerve conduction, tactile sensibility, and the electromyogram after suture or compression of peripheral nerve: a longitudinal study in man. *J Neurol Neurosurg Psychiatry.* 42: 436–51.

Burtis, C., E. Ashwood and N. Tietz. 2001. Fundamentals of Clinical Chemistry. WB Saunders Company, PA, USA.

Cajal, R. 1928. Degeneration and Regeneration of the Nervous System. Oxford University Press, London.

Caroni, P. 1998. Neuro-regeneration: plasticity for repair and adaptation. *Essays Biochem.* 33: 53–64.

Carulli, D., T. Laabs, H.M. Geller and J.W. Fawcett. 2005. Chondroitin sulfate proteoglycans in neural development and regeneration. *Curr Opin Neurobiol.* 15: 116–20. doi: 10.1016/j.conb.2005.01.014.

Castro, J., P. Negredo and C. Avendaño. 2008. Fiber composition of the rat sciatic nerve and its modification during regeneration through a sieve electrode. *Brain Res.* 1190: 65–77. doi: 10.1016/j.brainres.2007.11.028.

Cummins, T.R. and S.G. Waxman. 1997. Downregulation of tetrodotoxin-resistant sodium currents and upregulation of a rapidly repriming tetrodotoxin-sensitive sodium current in small spinal sensory neurons after nerve injury. *J Neurosci.* 17: 3503–14.

Curtis, R. 1998. Neuronal Injury Increases Retrograde Axonal Transport of the Neurotrophins to Spinal Sensory Neurons and Motor Neurons via Multiple Receptor Mechanisms. *Mol Cell Neurosci.* 12: 105–118. doi: 10.1006/mcne.1998.0704.

Dahlin, L., F. Johansson, C. Lindwall and M. Kanje. 2009a. Future Perspective In Peripheral Nerve Reconstruction. *In:* Essays on Peripheral Nerve Repair and Regeneration. S. Geuna, P. Tos and I. Battiston, editors. Academic Press. 507–530.

Dahlin, L., F. Johansson, C. Lindwall and M. Kanje. 2009b. Chapter 28: Future perspective in peripheral nerve reconstruction. *Int Rev Neurobiol.* 87: 507–30. doi: 10.1016/S0074-7742(09)87028-1.

Daniel, P.M. and S.J. Strich. 1969. Histological observations on Wallerian degeneration in the spinal cord of the baboon, Papio papio. *Acta Neuropathol.* 12: 314–328. doi: 10.1007/BF00809128.

Dines, K.C. and H.C. Powell. 1997. Mast cell interactions with the nervous system: relationship to mechanisms of disease. *J Neuropathol Exp Neurol.* 56: 627–40.

Dolenc, V. and M. Janko. 1976. Nerve regeneration following primary repair. *Acta Neurochir. (Wien).* 34: 223–34.

Doolabh, V.B. and S.E. Mackinnon. 1999. FK506 accelerates functional recovery following nerve grafting in a rat model. *Plast Reconstr Surg.* 103: 1928–36.

Drake, R., L. Cazares, A. Corsica, G. Malik, E. Schwegler, A. Libby, G. Wright, B. Adam and O. Semmes. 2004. Quality Control, Preparation, and Protein Stability Issues for Blood Serum and Plasma Used In Biomarker Discovery and Proteomic Profiling Assays. *BioProcess J.* 3: 45–50.

Drysdale, B.E., D.L. Howard and R.J. Johnson. 1996. Identification of a lipopolysaccharide inducible transcription factor in murine macrophages. *Mol Immunol.* 33: 989–98.

Ducker, T.B., L.G. Kempe and G.J. Hayes. 1969. The metabolic background for peripheral nerve surgery. *J Neurosurg.* 30: 270–80. doi: 10.3171/jns.1969.30.3part1.0270.

Esposito, B., A. De Santis, R. Monteforte and G.C. Baccari. 2002. Mast cells in Wallerian degeneration: morphologic and ultrastructural changes. *J Comp Neurol.* 445: 199–210.

Eugster, P.J., G. Glauser and J.-L. Wolfender. 2013. Strategies in Biomarker Discovery. Peak Annotation by MS and Targeted LC-MS Micro-Fractionation for *De Novo* Structure Identification by Micro-NMR. *Methods Mol Biol.* 1055: 267–89. doi: 10.1007/978-1-62703-577-4_19.

Fansa, H., G. Keilhoff, S. Altmann, K. Plogmeier, G. Wolf and W. Schneider. 1999. The effect of the immunosuppressant FK 506 on peripheral nerve regeneration following nerve grafting. *J Hand Surg Br.* 24: 38–42.

Fawcett, J.W. and R.J. Keynes. 1990. Peripheral nerve regeneration. *Annu Rev Neurosci.* 13: 43–60. doi: 10.1146/annurev.ne.13.030190.000355.

FDA. 2010. Center for Biologics Evaluation and Research. *Food Drug Adm.*

Ferran, M., E.R. Romeu, C. Rincón, M. Sagristà, A.M. Giménez Arnau, A. Celada, R.M. Pujol, P. Holló, H. Jókai and L.F. Santamaria-Babí. 2013. Circulating CLA+ T lymphocytes as peripheral cell biomarkers in T-cell-mediated skin diseases. *Exp Dermatol.* 22: 439–42. doi: 10.1111/exd.12154.

Fitzgerald, M., P.D. Wall, M. Goedert and P.C. Emson. 1985. Nerve growth factor counteracts the neurophysiological and neurochemical effects of chronic sciatic nerve section. *Brain Res.* 332: 131–41.

Fitzpatrick, E., M. DeHart, J. Tercero, T. Brown and S. Salgar. 2013. Bone marrow derived mesenchymal stem cell therapy to improve nerve regeneration and function in a rat limb transplant model (P2170)—Fitzpatrick et al. 190 (1001): 69.23—The Journal of Immunology. *J Immunol.* 190.

Fornaro, M., J.M. Lee, S. Raimondo, S. Nicolino, S. Geuna and M. Giacobini-Robecchi. 2008. Neuronal intermediate filament expression in rat dorsal root ganglia sensory neurons: an *in vivo* and *in vitro* study. *Neuroscience.* 153: 1153–63. doi: 10.1016/j.neuroscience.2008.02.080.

Fox, N. and J.H. Growdon. 2004. Biomarkers and surrogates. *NeuroRX.* 1: 181–181. doi: 10.1602/neurorx.1.2.181.

Friedlander, D.R., M. Grumet and G.M. Edelman. 1986. Nerve growth factor enhances expression of neuron-glia cell adhesion molecule in PC12 cells. *J Cell Biol.* 102: 413–9.

Fu, S.Y. and T. Gordon. 1997. The cellular and molecular basis of peripheral nerve regeneration. *Mol Neurobiol.* 14: 67–116. doi: 10.1007/BF02740621.

Funakoshi, H., J. Frisén, G. Barbany, T. Timmusk, O. Zachrisson, V.M. Verge and H. Persson. 1993. Differential expression of mRNAs for neurotrophins and their receptors after axotomy of the sciatic nerve. *J Cell Biol.* 123: 455–65.

Gaudet, A.D., P.G. Popovich and M.S. Ramer. 2011. Wallerian degeneration: gaining perspective on inflammatory events after peripheral nerve injury. *J Neuroinflammation.* 8: 110. doi: 10.1186/1742-2094-8-110.

Geuna, S., S. Raimondo, G. Ronchi, F. Di Scipio, P. Tos, K. Czaja and M. Fornaro. 2009. Chapter 3: Histology of the peripheral nerve and changes occurring during nerve regeneration. *Int Rev Neurobiol.* 87: 27–46. doi: 10.1016/S0074-7742(09)87003-7.

Glaus, S.W., P.J. Johnson and S.E. Mackinnon. 2011. Clinical strategies to enhance nerve regeneration in composite tissue allotransplantation. *Hand Clin.* 27: 495–509, ix. doi: 10.1016/j.hcl.2011.07.002.

Gold, B.G., K. Katoh and T. Storm-Dickerson. 1995. The immunosuppressant FK506 increases the rate of axonal regeneration in rat sciatic nerve. *J Neurosci.* 15: 7509–16.

Goldberg, J.L. 2003. How does an axon grow? *Genes Dev.* 17: 941–58. doi: 10.1101/gad.1062303.

Goldstein, M.E., H.S. Cooper, J. Bruce, M.J. Carden, V.M. Lee and W.W. Schlaepfer. 1987. Phosphorylation of neurofilament proteins and chromatolysis following transection of rat sciatic nerve. *J Neurosci.* 7: 1586–94.

Gordon, T. 2009. The role of neurotrophic factors in nerve regeneration. *Neurosurg Focus.* 26: E3. doi: 10.3171/FOC.2009.26.2.E3.

Gordon, T., J. Gillespie, R. Orozco and L. Davis. 1991. Axotomy-induced changes in rabbit hindlimb nerves and the effects of chronic electrical stimulation. *J Neurosci.* 11: 2157–69.

Grafstein, B. and I. McQuarrie. 1978. Role of the nerve cell body in axonal regeneration. *In:* Neuronal Plasticity. C. Cotman, editor. Raven Press, New York. 155–195.

De Gruttola, V.G., P. Clax, D.L. DeMets, G.J. Downing, S.S. Ellenberg, L. Friedman, M.H. Gail, R. Prentice, J. Wittes and S.L. Zeger. 2001. Considerations in the evaluation of surrogate endpoints in clinical trials. summary of a National Institutes of Health workshop. *Control Clin Trials.* 22: 485–502.

Gulati, A.K. and G.P. Cole. 1990. Nerve graft immunogenicity as a factor determining axonal regeneration in the rat. *J Neurosurg.* 72: 114–22. doi: 10.3171/jns.1990.72.1.0114.

Gunstream, J.D., G.A. Castro and E.T. Walters. 1995. Retrograde transport of plasticity signals in Aplysia sensory neurons following axonal injury. *J Neurosci.* 15: 439–48.

Gupta, S., T. Barrett, A.J. Whitmarsh, J. Cavanagh, H.K. Sluss, B. Dérijard and R.J. Davis. 1996. Selective interaction of JNK protein kinase isoforms with transcription factors. *EMBO J.* 15: 2760–70.

Hanz, S. and M. Fainzilber. 2006. Retrograde signaling in injured nerve—the axon reaction revisited. *J Neurochem.* 99: 13–9. doi: 10.1111/j.1471-4159.2006.04089.x.

Höke, A. 2006. Mechanisms of Disease: what factors limit the success of peripheral nerve regeneration in humans? *Nat Clin Pract Neurol.* 2: 448–54. doi: 10.1038/ncpneuro0262.

Hökfelt, T., X. Zhang and Z. Wiesenfeld-Hallin. 1994. Messenger plasticity in primary sensory neurons following axotomy and its functional implications. *Trends Neurosci.* 17: 22–30.

Karin, M. 1995. The regulation of AP-1 activity by mitogen-activated protein kinases. *J Biol Chem.* 270: 16483–6.

Katsube, K., K. Doi, T. Fukumoto, Y. Fujikura, M. Shigetomi and S. Kawai. 1998. Successful nerve regeneration and persistence of donor cells after a limited course of immunosuppression in rat peripheral nerve allografts. *Transplantation.* 66: 772–7.

Kenney, A.M. and J.D. Kocsis. 1998. Peripheral axotomy induces long-term c-Jun amino-terminal kinase-1 activation and activator protein-1 binding activity by c-Jun and junD in adult rat dorsal root ganglia In vivo. *J Neurosci.* 18: 1318–28.

Konofaos, P. and J.K. Terzis. 2013. FK506 and nerve regeneration: past, present and future. *J Reconstr Microsurg.* 29: 141–8. doi: 10.1055/s-0032-1333314.

Landmesser, L. and G. Pilar. 1974. Synaptic transmission and cell death during normal ganglionic development. *J Physiol.* 241: 737–49.

Leah, J.D., T. Herdegen and R. Bravo. 1991. Selective expression of Jun proteins following axotomy and axonal transport block in peripheral nerves in the rat: evidence for a role in the regeneration process. *Brain Res.* 566: 198–207.

Lee, M., V.B. Doolabh, S.E. Mackinnon and S. Jost. 2000. FK506 promotes functional recovery in crushed rat sciatic nerve. *Muscle Nerve.* 23: 633–40.

Lee, M.T. and P.B. Farel. 1988. Guidance of regenerating motor axons in larval and juvenile bullfrogs. *J Neurosci.* 8: 2430–7.

Le-Niculescu, H., D.F. Levey, M. Ayalew, L. Palmer, L.M. Gavrin, N. Jain, E. Winiger, S. Bhosrekar, G. Shankar, M. Radel, E. Bellanger, H. Duckworth, K. Olesek, J. Vergo, R. Schweitzer, M. Yard, A. Ballew, A. Shekhar, G.E. Sandusky, N.J. Schork, S.M. Kurian, D.R. Salomon and A.B. Niculescu. 2013. Discovery and validation of blood biomarkers for suicidality. *Mol Psychiatry.* doi: 10.1038/mp.2013.95.

Lewin, G.R., A. Rueff and L.M. Mendell. 1994. Peripheral and central mechanisms of NGF-induced hyperalgesia. *Eur J Neurosci.* 6: 1903–12.

Li, L., R.W. Oppenheim, M. Lei and L.J. Houenou. 1994. Neurotrophic agents prevent motoneuron death following sciatic nerve section in the neonatal mouse. *J Neurobiol.* 25: 759–66. doi: 10.1002/neu.480250702.

Lieberman, A.R. 1971. The axon reaction: a review of the principal features of perikaryal responses to axon injury. *Int Rev Neurobiol.* 14: 49–124.

Lieberman, A.R. 1974. Some factors affecting retrograde neuronal responses to axonal lesions. *In:* Essays on the nervous system. R. Bellairs and E. Gray, editors. Oxford, Clareddon. 71–105.

Lindwall, C., L. Dahlin, G. Lundborg and M. Kanje. 2004. Inhibition of c-Jun phosphorylation reduces axonal outgrowth of adult rat nodose ganglia and dorsal root ganglia sensory neurons. *Mol Cell Neurosci.* 27: 267–79. doi: 10.1016/j.mcn.2004.07.001.

Lindwall, C. and M. Kanje. 2005. The role of p-c-Jun in survival and outgrowth of developing sensory neurons. *Neuroreport.* 16: 1655–9.

Lovestone, S. 2010. Searching for biomarkers in neurodegeneration. *Nat Med.* 16: 1371–2. doi: 10.1038/nm1210-1371b.

Lu, J., Y. Huang, Y. Wang, Y. Li, Y. Zhang, J. Wu, F. Zhao, S. Meng, X. Yu, Q. Ma, M. Song, N. Chang, A.H. Bittles and W. Wang. 2012. Profiling plasma peptides for the identification of potential ageing biomarkers in Chinese Han adults. *PLoS One.* 7: e39726. doi: 10.1371/journal.pone.0039726.

Lyons, T.J. and A. Basu. 2012. Biomarkers in diabetes: hemoglobin A1c, vascular and tissue markers. *Transl Res.* 159: 303–12. doi: 10.1016/j.trsl.2012.01.009.

Lyons, W.E., E.B. George, T.M. Dawson, J.P. Steiner and S.H. Snyder. 1994. Immunosuppressant FK506 promotes neurite outgrowth in cultures of PC12 cells and sensory ganglia. *Proc Natl Acad Sci USA.* 91: 3191–5.

MacFarlane, R.J., S.M. Graham, P.S.E. Davies, N. Korres, H. Tsouchnica, M. Heliotis, A. Mantalaris and E. Tsiridis. 2013. Anti-inflammatory role and immunomodulation of mesenchymal stem cells in systemic joint diseases: potential for treatment. *Expert Opin Ther Targets*. 17: 243–54.

Mackinnon, S.E., V.B. Doolabh, C.B. Novak and E.P. Trulock. 2001. Clinical outcome following nerve allograft transplantation. *Plast Reconstr Surg*. 107: 1419–29.

Makwana, M. and G. Raivich. 2005. Molecular mechanisms in successful peripheral regeneration. *FEBS J*. 272: 2628–38. doi: 10.1111/j.1742-4658.2005.04699.x.

Martini, R., M. Schachner and T.M. Brushart. 1994. The L2/HNK-1 carbohydrate is preferentially expressed by previously motor axon-associated Schwann cells in reinnervated peripheral nerves. *J Neurosci*. 14: 7180–91.

Massing, M., G. Robinson, C. Marx, O. Alzate and R. Madison. 2010. Applications of Proteomics to Nerve Regeneration Research. *In:* Neuroproteomics. O. Alzate, editor. CRC Press, FL.

McQuarrie, I.G. and B. Grafstein. 1973. Axon outgrowth enhanced by a previous nerve injury. *Arch Neurol*. 29: 53–5.

Meyer, M., I. Matsuoka, C. Wetmore, L. Olson and H. Thoenen. 1992. Enhanced synthesis of brain-derived neurotrophic factor in the lesioned peripheral nerve: different mechanisms are responsible for the regulation of BDNF and NGF mRNA. *J Cell Biol*. 119: 45–54.

Middlemas, A., J.-D. Delcroix, N.M. Sayers, D.R. Tomlinson and P. Fernyhough. 2003. Enhanced activation of axonally transported stress-activated protein kinases in peripheral nerve in diabetic neuropathy is prevented by neurotrophin-3. *Brain*. 126: 1671–82. doi: 10.1093/brain/awg150.

Miletic, G., E.N. Hanson and V. Miletic. 2004a. Brain-derived neurotrophic factor-elicited or sciatic ligation-associated phosphorylation of cyclic AMP response element binding protein in the rat spinal dorsal horn is reduced by block of tyrosine kinase receptors. *Neurosci Lett*. 361: 269–71. doi: 10.1016/j.neulet.2003.12.029.

Miletic, G., E.N. Hanson, C.A. Savagian and V. Miletic. 2004b. Protein kinase A contributes to sciatic ligation-associated early activation of cyclic AMP response element binding protein in the rat spinal dorsal horn. *Neurosci. Lett*. 360: 149–52. doi: 10.1016/j.neulet.2004.02.060.

Mira, J.C. 1984. Effects of repeated denervation on muscle reinnervation. *Clin Plast Surg*. 11: 31–8.

Mirsky, R. and K.R. Jessen. 1996. Schwann cell development, differentiation and myelination. *Curr Opin Neurobiol*. 6: 89–96.

Mohanty, B.P., S. Bhattacharjee, P. Paria, A. Mahanty and A.P. Sharma. 2013. Lipid biomarkers of lens aging. *Appl Biochem Biotechnol*. 169: 192–200. doi: 10.1007/s12010-012-9963-6.

Morrow, D.A., J.A. de Lemos, M.S. Sabatine and E.M. Antman. 2003. The search for a biomarker of cardiac ischemia. *Clin Chem*. 49: 537–9.

Müller, H.W. and G. Stoll. 1998. Nerve injury and regeneration: basic insights and therapeutic interventions. *Curr Opin Neurol*. 11: 557–62.

Nadeau, S., P. Hein, K.J.L. Fernandes, A.C. Peterson and F.D. Miller. 2005. A transcriptional role for C/EBP beta in the neuronal response to axonal injury. *Mol Cell Neurosci*. 29: 525–35. doi: 10.1016/j.mcn.2005.04.004.

Neumann, S., F. Bradke, M. Tessier-Lavigne and A.I. Basbaum. 2002. Regeneration of sensory axons within the injured spinal cord induced by intraganglionic cAMP elevation. *Neuron*. 34: 885–93.

Ngeow, W.C. 2010. Scar less: a review of methods of scar reduction at sites of peripheral nerve repair. *Oral Surg Oral Med Oral Pathol Oral Radiol Endod*. 109: 357–66. doi: 10.1016/j.tripleo.2009.06.030.

Nilsson, A., L. Dahlin, G. Lundborg and M. Kanje. 2005. Graft repair of a peripheral nerve without the sacrifice of a healthy donor nerve by the use of acutely dissociated autologous Schwann cells. *Scand J Plast Reconstr Surg Hand Surg.* 39: 1–6. doi: 10.1080/02844310410017979.

Obata, K., H. Yamanaka, Y. Dai, T. Mizushima, T. Fukuoka, A. Tokunaga and K. Noguchi. 2004. Differential activation of MAPK in injured and uninjured DRG neurons following chronic constriction injury of the sciatic nerve in rats. *Eur J Neurosci.* 20: 2881–95. doi: 10.1111/j.1460-9568.2004.03754.x.

Olsson, B., H. Zetterberg, H. Hampel and K. Blennow. 2011. Biomarker-based dissection of neurodegenerative diseases. *Prog Neurobiol.* 95: 520–34. doi: 10.1016/j.pneurobio.2011.04.006.

Olsson, Y. 1967. Degranulation of mast cells in peripheral nerve injuries. *Acta Neurol Scand.* 43: 365–74.

Peleshok, J.C. and A. Ribeiro-da-Silva. 2012. Neurotrophic factor changes in the rat thick skin following chronic constriction injury of the sciatic nerve. *Mol Pain.* 8: 1. doi: 10.1186/1744-8069-8-1.

Perlson, E., S. Hanz, K. Ben-Yaakov, Y. Segal-Ruder, R. Seger and M. Fainzilber. 2005. Vimentin-dependent spatial translocation of an activated MAP kinase in injured nerve. *Neuron.* 45: 715–26. doi: 10.1016/j.neuron.2005.01.023.

Perlson, E., K.F. Medzihradszky, Z. Darula, D.W. Munno, N.I. Syed, A.L. Burlingame and M. Fainzilber. 2004. Differential proteomics reveals multiple components in retrogradely transported axoplasm after nerve injury. *Mol Cell Proteomics.* 3: 510–20. doi: 10.1074/mcp.M400004-MCP200.

Pettmann, B. and C.E. Henderson. 1998. Neuronal cell death. *Neuron.* 20: 633–47.

Povelones, M., K. Tran, D. Thanos and R.T. Ambron. 1997. An NF-kappaB-like transcription factor in axoplasm is rapidly inactivated after nerve injury in Aplysia. *J Neurosci.* 17: 4915–20.

Qiu, J., D. Cai, H. Dai, M. McAtee, P.N. Hoffman, B.S. Bregman and M.T. Filbin. 2002a. Spinal axon regeneration induced by elevation of cyclic AMP. *Neuron.* 34: 895–903.

Qiu, J., D. Cai and M.T. Filbin. 2002b. A role for cAMP in regeneration during development and after injury. *Prog Brain Res.* 137: 381–7.

Rai, A.J. 2007. Biomarkers in translational research: focus on discovery, development and translation of protein biomarkers to clinical immunoassays. *Expert Rev Mol Diagn.* 7: 545–53. doi: 10.1586/14737159.7.5.545.

Raivich, G., M. Bohatschek, C. Da Costa, O. Iwata, M. Galiano, M. Hristova, A.S. Nateri, M. Makwana, L. Riera-Sans, D.P. Wolfer, H.-P. Lipp, A. Aguzzi, E.F. Wagner and A. Behrens. 2004. The AP-1 transcription factor c-Jun is required for efficient axonal regeneration. *Neuron.* 43: 57–67. doi: 10.1016/j.neuron.2004.06.005.

Raivich, G. and M. Makwana. 2007. The making of successful axonal regeneration: genes, molecules and signal transduction pathways. *Brain Res Rev.* 53: 287–311. doi: 10.1016/j.brainresrev.2006.09.005.

Rajan, P., C.L. Stewart and J.S. Fink. 1995. LIF-mediated activation of STAT proteins after neuronal injury *in vivo*. *Neuroreport.* 6: 2240–4.

Richardson, P.M. and V.M. Issa. 1984. Peripheral injury enhances central regeneration of primary sensory neurones. *Nature.* 309: 791–3.

Rodrigues, M.C.O., A.A. Rodrigues, L.E. Glover, J. Voltarelli and C. V Borlongan. 2012. Peripheral nerve repair with cultured schwann cells: getting closer to the clinics. *Scientific World Journal.* 2012: 413091.

Sarhane, K.A. 2012. Tissue CLN3: A potential biomarker for breast cancer? Thesis Dissertation (M.Sc), American University of Beirut. http://hdl.handle.net/10938/9318.

Sarhane, K.A., Z. Ibrahim, A.A.L. Barone, D.S. Cooney, W.P.A. Lee and G. Brandacher. 2013. Minimization of Immunosuppression and Tolerance Induction in Reconstructive Transplantation. *Curr Surg Reports*. 1: 40–46.

Sarhane, K.A., Z. Ibrahim, K. Krick, R. Martin, B. Pan, S. Tuffaha, J. Broyles, C. Cashman, R. Mi, M. Alrakan, S. Tehrani, L. Goldberg, H.-Q. Mao, W.A. Lee and G. Brandacher. 2014. Selectively Permeable Nanofiber Constructs to Prevent Inflammatory Scarring And Enhance Nerve Regeneration in Peripheral Nerve Injury. *Plast Reconstr Surg*. 133: 16–17. doi: 10.1097/01.prs.0000444952.84840.10.

Schicho, R., G. Skofitsch and J. Donnerer. 1999. Regenerative effect of human recombinant NGF on capsaicin-lesioned sensory neurons in the adult rat. *Brain Res*. 815: 60–9.

Schneeberger, S., L. Landin, J. Jableki, P. Butler, C. Hoehnke, G. Brandacher and E. Morelon. 2011. Achievements and challenges in composite tissue allotransplantation. *Transpl Int*. 24: 760–9. doi: 10.1111/j.1432-2277.2011.01261.x.

Schreyer, D.J. and J.H. Skene. 1991. Fate of GAP-43 in ascending spinal axons of DRG neurons after peripheral nerve injury: delayed accumulation and correlation with regenerative potential. *J Neurosci*. 11: 3738–51.

Schwaiger, F.W., G. Hager, A.B. Schmitt, A. Horvat, R. Streif, C. Spitzer, S. Gamal, S. Breuer, G.A. Brook, W. Nacimiento and G.W. Kreutzberg. 2000. Peripheral but not central axotomy induces changes in Janus kinases (JAK) and signal transducers and activators of transcription (STAT). *Eur J Neurosci*. 12: 1165–76.

Seijffers, R., A.J. Allchorne and C.J. Woolf. 2006. The transcription factor ATF-3 promotes neurite outgrowth. *Mol Cell Neurosci*. 32: 143–54. doi: 10.1016/j.mcn.2006.03.005.

Seijffers, R., C.D. Mills and C.J. Woolf. 2007. ATF3 increases the intrinsic growth state of DRG neurons to enhance peripheral nerve regeneration. *J Neurosci*. 27: 7911–20. doi: 10.1523/JNEUROSCI.5313-06.2007.

Seilheimer, B. and M. Schachner. 1987. Regulation of neural cell adhesion molecule expression on cultured mouse Schwann cells by nerve growth factor. *EMBO J*. 6: 1611–6.

Skene, J.H. 1989. Axonal growth-associated proteins. *Annu Rev Neurosci*. 12: 127–56. doi: 10.1146/annurev.ne.12.030189.001015.

Sterman, A.B. and M.R. Delannoy. 1985. Cell body responses to axonal injury: traumatic axotomy versus toxic neuropathy. *Exp Neurol*. 89: 408–19.

Sunderland, S. 1987. Nerves and Nerve Injuries. 2nd ed. Churchill Livingstone, Edinburgh.

Sunderland, S. and K.C. Bradley. 1950. Endoneurial tube shrinkage in the distal segment of a severed nerve. *J Comp Neurol*. 93: 411–20.

Sung, Y.J., M. Povelones and R.T. Ambron. 2001. RISK-1: a novel MAPK homologue in axoplasm that is activated and retrogradely transported after nerve injury. *J Neurobiol*. 47: 67–79.

Swearingen, B., K. Ravindra, H. Xu, S. Wu, W.C. Breidenbach and S.T. Ildstad. 2008. Science of composite tissue allotransplantation. *Transplantation*. 86: 627–35. doi: 10.1097/TP.0b013e318184ca6a.

Tanabe, K., I. Bonilla, J.A. Winkles and S.M. Strittmatter. 2003. Fibroblast growth factor-inducible-14 is induced in axotomized neurons and promotes neurite outgrowth. *J Neurosci*. 23: 9675–86.

Taylor, C., D. Braza, J. Rice and T. Dillingham. 2008. The incidence of peripheral nerve injury in extremity trauma. *Am J Phys Med Rehabil*. 87: 381–385.

Tetzlaff, W., M.A. Bisby and G.W. Kreutzberg. 1988. Changes in cytoskeletal proteins in the rat facial nucleus following axotomy. *J Neurosci*. 8: 3181–9.

Tuffaha, S., M. Quigley, T. Ng, V.S. Gorantla, J.T. Shores, B. Pulikkottil, C. Shestak, G. Brandacher and W.P.A. Lee. 2011. The effect of chondroitinase on nerve regeneration following composite tissue allotransplantation. *J Hand Surg Am*. 36: 1447–52. doi: 10.1016/j.jhsa.2011.06.007.

Verge, V.M., P.M. Richardson, Z. Wiesenfeld-Hallin and T. Hökfelt. 1995. Differential influence of nerve growth factor on neuropeptide expression *in vivo*: a novel role in peptide suppression in adult sensory neurons. *J Neurosci.* 15: 2081–96.

Verge, V.M., W. Tetzlaff, P.M. Richardson and M.A. Bisby. 1990. Correlation between GAP43 and nerve growth factor receptors in rat sensory neurons. *J Neurosci.* 10: 926–34.

Vitzthum, F., F. Behrens, N.L. Anderson and J.H. Shaw. 2005. Proteomics: from basic research to diagnostic application. A review of requirements & needs. *J Proteome Res.* 4: 1086–97. doi: 10.1021/pr050080b.

Vleggeert-Lankamp, C.L.A.M. 2007. The role of evaluation methods in the assessment of peripheral nerve regeneration through synthetic conduits: a systematic review. Laboratory investigation. *J Neurosurg.* 107: 1168–89. doi: 10.3171/JNS-07/12/1168.

Waetzig, V. and T. Herdegen. 2005. MEKK1 controls neurite regrowth after experimental injury by balancing ERK1/2 and JNK2 signaling. *Mol Cell Neurosci.* 30: 67–78. doi: 10.1016/j.mcn.2005.06.001.

Waller, A. 1850. Experiments on the section of the glossopharyngeal and hypoglossal nerves of the frog, and observations of the alterations produced thereby in the structure of their primitive fibers. *Phil Trans Roy Soc.* 140: 423–429.

Wang, J., X. Liu, J. Zhu, W. Li and J. Xiang. 2002. Study *in vitro* of populating autogenous Schwann cells into chemical extracted allogenous nerve. *Chin J Traumatol.* 5: 326–8.

Willmann, J.K., N. van Bruggen, L.M. Dinkelborg and S.S. Gambhir. 2008. Molecular imaging in drug development. *Nat Rev Drug Discov.* 7: 591–607. doi: 10.1038/nrd2290.

Wong, J. and M.M. Oblinger. 1991. NGF rescues substance P expression but not neurofilament or tubulin gene expression in axotomized sensory neurons. *J Neurosci.* 11: 543–52.

Woolf, C.J., Q.P. Ma, A. Allchorne and S. Poole. 1996. Peripheral cell types contributing to the hyperalgesic action of nerve growth factor in inflammation. *J Neurosci.* 16: 2716–23.

Woolf, C.J., M.L. Reynolds, C. Molander, C. O'Brien, R.M. Lindsay and L.I. Benowitz. 1990. The growth-associated protein GAP-43 appears in dorsal root ganglion cells and in the dorsal horn of the rat spinal cord following peripheral nerve injury. *Neuroscience.* 34: 465–78.

Yang, Y., W. Dai, Z. Chen, Z. Yan, Z. Yao and C. Zhang. 2013. Downregulating immunogenicity of Schwann cells via inhibiting a potential target of class II transactivator (CIITA) gene. *Biosci Trends.* 7: 50–5.

Yao, G.L., H. Kato, M. Khalil, S. Kiryu and H. Kiyama. 1997. Selective upregulation of cytokine receptor subchain and their intracellular signalling molecules after peripheral nerve injury. *Eur J Neurosci.* 9: 1047–54.

Yu, L.T., A. Rostami, W.K. Silvers, D. Larossa and W.F. Hickey. 1990. Expression of major histocompatibility complex antigens on inflammatory peripheral nerve lesions. *J Neuroimmunol.* 30: 121–8.

Zelená, J., L. Lubińska and E. Gutmann. 1968. Accumulation of organelles at the ends of interrupted axons. *Z Zellforsch Mikrosk Anat.* 91: 200–19.

Zhang, J.M., D.F. Donnelly, X.J. Song and R.H. Lamotte. 1997. Axotomy increases the excitability of dorsal root ganglion cells with unmyelinated axons. *J Neurophysiol.* 78: 2790–4.

Zhang, J.Y., X.G. Luo, C.J. Xian, Z.H. Liu and X.F. Zhou. 2000. Endogenous BDNF is required for myelination and regeneration of injured sciatic nerve in rodents. *Eur J Neurosci.* 12: 4171–80.

Zhang, X.P. and R.T. Ambron. 2000. Positive injury signals induce growth and prolong survival in Aplysia neurons. *J Neurobiol.* 45: 84–94.

Zigmond, R.E. 2012. Cytokines that promote nerve regeneration. *Exp Neurol.* 238: 101–6. doi: 10.1016/j.expneurol.2012.08.017.

Zochodne, D.W., C. Nguyen and K.A. Sharkey. 1994. Accumulation and degranulation of mast cells in experimental neuromas. *Neurosci Lett.* 182: 3–6.

Zolg, J.W. and H. Langen. 2004. How industry is approaching the search for new diagnostic markers and biomarkers. *Mol Cell Proteomics.* 3: 345–54. doi: 10.1074/mcp.M400007-MCP200.

21

Biomarkers of Drug Abuse-induced Brain Changes: Role of Microglia in Alcohol-induced Neurotoxicity

Harris Bell-Temin,[1] Bin Liu,[2] Ping Zhang[2] and Stanley M. Stevens, Jr.[1,]*

INTRODUCTION

Overconsumption of alcohol can have a significant negative impact on the brain, manifesting in neurological dysfunctions including cognitive impairment and dementia (Bates et al., 2002; Nixon, 2006). These adverse effects have been attributed to ethanol's effects on neurotransmission, synaptic plasticity and the structural integrity of the neuronal network (Harper, 1998; Nixon, 2006). The mechanisms through which ethanol acts upon these processes, however, are not clearly understood. Some studies have suggested that ethanol and its metabolites including acetaldehyde and acetate are directly neurotoxic; others suggest that a hyperglutamatergic

[1] Department of Cell Biology, Microbiology, and Molecular Biology, University of South Florida, 4202 E. Fowler Ave., Tampa, FL 33620.

[2] Department of Pharmacodynamics, University of Florida, 1600 SW Archer Rd., Gainesville, FL 32610.

* Corresponding author: smstevens@usf.edu

state is triggered by ethanol (Moykkynen and Korpi, 2012; Zou and Crews, 2005). However, a growing body of evidence posits the role of the major types of non-neuronal cells in the brain, glia, in alcohol-induced neurotoxicity through oxidative stress (Blanco and Guerri, 2007a; Blanco et al., 2004; Blanco et al., 2005; Boyadjieva and Sarkar, 2013b; Crews et al., 2006a; Crews et al., 2006b; Fernandez-Lizarbe et al., 2013; He and Crews, 2008; Lee et al., 2004; Qin and Crews, 2012b; Qin et al., 2008). Neuroinflammation is caused through the activation of glia, astrocytes and microglia, the resident immune cells found in the central nervous system (Liu, 2006). Activated glia, particularly microglia, produce a variety of pro-inflammatory cytokines and neurotoxic factors such as Reactive Oxygen Species (ROS), Reactive Nitrogen Species (RNS) and lipid mediators (David and Kroner, 2011; Liu et al., 2003; Liu and Hong, 2003b). If uncontrolled, sustained activation of microglia could lead to functional impairment and eventually structural damage to surrounding neuronal tissue (Block et al., 2007; Qin et al., 2007). Microglia, however, are not limited to morphing into an aggressive neurotoxic posture when activated; activated microglia can also transform into a variety of non-neurotoxic phenotypes to perform other immune-related functions which may be triggered when exposed to ethanol (Colton, 2009; Czapiga and Colton, 2003; Kigerl et al., 2009; Martinez et al., 2006; Martinez et al., 2009; Martinez et al., 2008; Mills, 2012).

An increasing body of literature is being amassed implicating activated microglia in response to chronic ethanol exposure in the process of neurodegeneration (Blanco and Guerri, 2007a; Blanco et al., 2004; Blanco et al., 2005; Boyadjieva and Sarkar, 2013b; Crews et al., 2006a; Crews et al., 2006b; Fernandez-Lizarbe et al., 2013; He and Crews, 2008; Lee et al., 2004; Qin and Crews, 2012b; Qin et al., 2008). In this chapter, we will attempt to summarize the current literature on the potential relationship between alcohol neurotoxicity and neuroinflammation, with a particular emphasis on the role of microglia and related activation markers and/or pathways that could serve as biomarkers of alcohol-induced brain changes (Fig. 21.1). Additionally, we will discuss the various methodologies and models available for the study of ethanol neurotoxicity and neuroinflammation.

Microglia and Neurotoxicity

Microglia comprise approximately 12% of the cells of the brain, varying in density by region but predominating in the hippocampus and substantianigra (Block et al., 2007). Microglia function similarly to macrophages outside of the brain, and share many of the same activation/responsiveness states and signal transduction pathways. Microglia can respond to pathogens, toxins or to injury, radiation or immune response (Colton, 2009). During a bacterial

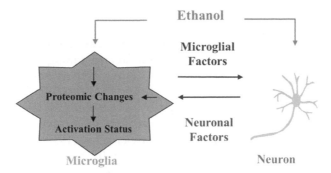

Figure 21.1. Identification of Biomarkers Associated with Ethanol-induced Microglial Response. Ethanol exposure induces an activation response in microglia associated with a specific proteome profile. Ethanol-induced neuronal injury results in modulation of microglial activation phenotype depending on dose and exposure time. Significant progress is being made using global-scale expression profiling methods such as proteomics to gain a better understanding of the molecular markers associated with ethanol-induced microglial activation and role in ethanol neurotoxicity.

Color image of this figure appears in the color plate section at the end of the book.

infection, microglia reduce the pathogenical load using a magnitude of responses: they induce pro-inflammatory cytokines for recruitment of additional immune responsive cells from within the brain and from across the blood-brain barrier, produce ROS and RNS to target the pathogen, and remove cellular debris following infection. As such, neuroinflammatory microglia are a hallmark of multiple neurodegenerative disorders and are often defined by their cytotoxic functions (Banati et al., 1993; Block and Hong, 2005).

The role of microglia in the monitoring and response of pathogens and injury in the brain has become more complex in recent years, taking into consideration the quiescent, pre-inflammatory and post-inflammatory phases of microglia response (Saijo and Glass, 2011). Originally microglia were only thought of as a neurotoxic responder to pathogens, injury, and disease while our understanding has swelled to include several different anti-inflammatory states (Kreutzberg, 1996). In a review by Graeber and Streit (2010) the presence of activated microglia was not found to correlate with inflammation state and concludes that a negative connotation has been attached to microglial activity; that it leads only to inflammation, subsuming the possibility of a neuroprotective role for its cytotoxic activities, and ignoring the vast complexity and shades of microglial responses. In a study of Alzheimer's disease, where microglial involvement and neuroinflammation have been implicated, multiple states of microglial activation have been observed, often simultaneously (Colton and Wilcock, 2010). In this and other cases, after performing their pro-inflammatory tasks, microglia morph into a healing response followed by a return to a quiescent state using anti-inflammatory cytokines that often follow the neurotoxic state (Block et al., 2007; He and Crews, 2008). A similar cycle of an initial neurotoxic response in order to remove dead and damaged cells followed by an anti-inflammatory cessation of neurotoxic activity has been found in response to spinal cord injury as well (Kigerl et al., 2009).

Quiescent microglia are typified by a ramified morphology as the microglia actively monitor the environment for cues for brain injury, foreign pathogens or immunological stimuli. Raivich et al. (1999a) described five stages of microglial activation; resting (stage 0), alert (stage 1), homing (stage 2), phagocytic (stage 3a) and bystander activation (stage 3b). During the alert stage, microglia are highly responsive to immunoglobulins and increase surface expression of ligands involved in adhesion such as CD11b. In addition, microglia in the alert stage begin to produce Tumor Growth Factor β (TGFβ), a neuroprotective factor (Raivich et al., 1999b). The third stage of activation, dubbed the homing stage as described by Raivich et al. (1999a) is characterized by a further increase in expression of integrins, leading to changes in cell mobility and adhesion. Useful in response to injury, homing stage microglia increase expression of colony stimulating factor 1 receptor (CSF1R) in addition to the neuroprotective and pro-healing cytokine interleukin-10 (IL-10) (Raivich et al., 1999b). The fourth and phagocytotic stage is characterized by microglial response to neuronal cell death to trigger phagocytotic behavior, removing disconnected and severed axons and myelin debris. This response includes the production of pro-inflammatory cytokines IL-6 and Tumor Necrosis Factor alpha (TNF-α), as well as the marker of phagocytosis CD68. The final phase of microglial response, the bystander activation stage, involves the production of inflammatory cytokines such as interferon gamma (IFN-γ) that stimulate additional microglial activation to respond to continued inflammatory triggers. If these triggers are absent, the cells can return to their quiescent state in response to anti-inflammatory cytokines (Raivich et al., 1999a).

A mix of microglia in all of these stages exists at any given time, and the proportions of microglia in various stages vary due to inflammatory trigger, severity of injury and brain region (Harting et al., 2008; Lai and Todd, 2008). In response to a bacterial infection, microglia can respond directly to the lipopolysaccharide component in the outer membrane of Gram negative bacteria or to IFN-γ. This phase, known as the M1 subtype of microglial activation and similar to immune response-mediated activation which could include stages 2–3 of Raivich's spectrum of microglial activation, leads to the production of RNS, specifically nitric oxide by inducible nitric oxide synthase, iNOS, and ROS by a plasma membrane NADPH complex. IFN-γ leads to upregulation of iNOS through a JAK/STAT mediated pathway while LPS upregulates iNOS ultimately through Nuclear Factor kappa B (NFκB).

During M1 phase, microglia and macrophages produce a number of cytokines depending on the species, such as the pro-inflammatory interleukins IL-1β, IL-6, IL-12, IL-15 and TNF-α (Gomez-Nicola et al., 2010). In addition, a number of pro-inflammatory chemokines, chemotactic cytokines, are produced. These include the β-chemokines, members of the CC chemokine group, typified by two or more adjacent cysteine residues at the amino terminus, including CCL8, CCL 17, CCL19and CCL20 (Aravalli et al., 2005; Gomez-Nicola et al., 2010; Martinez et al., 2006; Martinez et al., 2009; Martinez et al., 2008; Rom et al., 2010; Terao et al., 2009). Members of the CXC chemokine group, typified by two cysteine residues separated by one amino acid, CXCL9, CXCL10, CXCL11 and CXCL13 are also produced in microglia (Carter et al., 2007; Martinez et al., 2006; Martinez et al., 2009; Martinez et al., 2008; Rainey-Barger et al., 2011; Rock et al., 2005). M1 microglia have multiple surface markers, such as the cluster of differentiation receptors CD86 and the major histocompatibility class 2 receptors (MHCII). Additionally upregulated proteins in M1 phase include proteins involved in the microglia's antioxidant response to its own ROS production such as cyclooxygenase-2 (COX-2), sequestosome, and hemeoxygenase. Microglia also produce chondroitin sulfate proteoglycan (CSPG) which can stimulate additional activation by binding to the CD44 receptor, aiding in healing, but later becomes the predominant proteoglycan in glial scar formation (Rolls et al., 2008). Through ROS and RNS production and phagocytosis, M1 microglia are aggressively neurotoxic from their response to invading pathogens and damaged tissue.

As pro-inflammatory neurotoxic factors are produced by M1 microglia, or as neurons are directly damaged by glutamate or other direct insults, neurons produce a variety of microglial activators. These include laminin, MMP3, α-synuclein, and neuromelanin (Block et al., 2007). In turn, these byproducts of neuronal damage can activate additional microglia in a process known as reactive microgliosis. Additional LPS stimulation after first treatment will provoke a stronger response than the first treatment, and *in utero* LPS treatment has led to a progressive LPS response in adult mice (Cunningham et al., 2005; Ling et al., 2006). These additional microglia, if unchecked by anti-inflammatory neuropeptides or cytokines can lead to severe and progressive neuronal damage.

Microglia can be restrained from over-activation through the involvement of anti-inflammatory cytokines. These cytokines can not only stop the production of pro-inflammatory products such as nitric oxide, but also produce several alternative microglial phenotypes to perform different immune-related activities. These phenotypes play a homeostatic role, returning the active and neurotoxic microglia to its resting state, limiting additional RNS, ROS and pro-inflammatory cytokine production. In response to the anti-inflammatory cytokines IL-4 and/or IL-13, microglia adopt an M2a phenotype, characterized by tissue repair and growth stimulation. This phase is often found in response to axonal injury. During this phase, microglia and macrophages produce anti-inflammatory cytokines such as the interleukins IL-10, IL-1Ra, an IL-1 receptor antagonist, and Tumor Growth Factor β (TGFβ). M2a microglia and macrophages produce a variety of chemokines including CCL13, CCL14, CCL17, CCL18, CCL22, CCL23, CCL24 and CCL26 (Martinez et al., 2006; Martinez et al., 2009; Martinez et al., 2008). Surface markers and receptors of this phase include CD163, CD204 and CD206. Important cytokine markers of this phase in murine microglia are YM1, a lectin with chemotactic activity, and Fizz1 (Raes et al., 2002). Tissue repair and growth stimulation are promoted through the actions of various chitinases, fibrinogenic and coagulation factors. Most important to the anti-inflammatory, and M1 phase-countering, activity of the M2a phase is the production of arginase-1. Arginase-1 converts arginine into ornithine. Arginine also is the source of nitric oxide for inducible Nitric Oxide Synthase (iNOS), which converts arginine to citrulline via an Nω-hydroxy-L-arginine intermediate rather than ornithine. iNOS in turn produces both nitric oxide and NADPH which can then be used as a substrate for the ROS-producing NADPH oxidase complex. By starving iNOS of its nitrogen source, arginase-1 clamps down on the production of neurotoxic reactive species by M1-phase microglia (Pesce et al., 2009; Xu et al., 2003).

Microglia that are stimulated with IL-1β and LPS concurrently or IgA immune complexes that bind to FcγR receptors adopt the M2b polarization. This stage is a combined M1 and M2a state, with a significant reduction in both IL-12 production from M1-phase microglia and arginase-1 production, but high levels of pro-inflammatory cytokines such as IL-6, IL-1β and TNFα in addition to IL-10 production. As such, this stage has a pro- and anti-inflammatory function. The cluster of differentiation marker CD86 is highly specific to this stage of microglial activation (Mosser, 2003). The stage also is typified by the production of the chemokines CCL1, CCL20, CXCL1, CXCL2 and CXCL3 (Akimoto et al., 2013; Johnson et al., 2011; Martinez et al., 2006; Martinez et al., 2009; Martinez et al., 2008; Shiratori et al., 2010; Terao et al., 2009).

Microglia and macrophages can be stimulated to adopt the M2c phenotype through treatment with anti-inflammatory IL-10, TGFβ, and glucocorticoids. M2c microglia serve a debris scavenging and pro-healing function in the brain, as additional arginase-1 is produced as well as the markers CD163, CD204, and CD206, as are seen in the M2a phenotype. The two phenotypes are differentiated by the production of the chemokine CCL16 and the lack of fibrinogenic and coagulation factors (Martinez et al., 2009; Martinez et al., 2008). IL-10 and TGFβ are both produced by M2a microglia, demonstrating that microglia do not appear in their various activation stages in a binary fashion; microglial phenotype is fluid, and is more often than not characterized by a mix of M1 and M2 states. The quantity and type of microglia in each state is informative towards the assessment of overall microglial response (Olah et al., 2011; Schwartz et al., 2006).

Methods of Studying Microglial Response

Several avenues are available for the study of microglia. The most direct is the use of *ex vivo* microglia isolated from the subject post-treatment. Multiple methods of isolation are available. One method involves the use of a Percoll gradient for discontinuous density gradient centrifugation as the microglia appear between the 30 and 37% Percoll layers and all myelinated cells and astrocytes appear in lower layer transitions (Cardona et al., 2006; Frank et al., 2006). Other methods include the use of marker antibodies bound to magnetic beads for capture, which appears to offer higher yields and purities at an increased cost (Gordon et al., 2011). Primary cell culture as well as immortalized cell lines are also available for the study of microglia, in addition to differentiated monocytes in order to mimic human microglia (Etemad et al., 2012; Leone et al., 2006).

Primary Microglia Cell Culture

The most common method of studying microglia in the laboratory is the use of primary cell cultures.Microglia are easily isolated from mixed glial cultures from neonatal mice or rats using agitation to pull the gently anchored microglia off of the astrocytic basal layer (Liu and Hong, 2003a). Primary microglia have been found to produce many of the pro-inflammatory cytokines as described above and will release nitric oxide in response to INF-γ, TNF-α or through other pro-inflammatory classical activators such as LPS (Zhang et al., 2010). Primary microglia have also been shown to be capable of induction into alternative activation states (Colton et al., 2006; Lee et al., 2002; Lyons et al., 2007).

Immortalized Microglial Cell Lines

Several immortalized cell lines that include retroviral immortalized cells or cells resulting from spontaneous immortalization are available. Rat HAPI (Highly Aggressively Proliferating Immortalized) microglial cells were first isolated in 2001 by Cheepsunthorn et al. (2001). HAPI cells stain positive for microglial markers isolectin B4, OX42 and GLUT5 and do not exhibit immunoreactivity for astrocytic or neuronal markers. HAPI cells express Macrophage Inhibitory Factor (MIF), TNF-α and demonstrate a nitric oxide production response when stimulated by LPS (Bell-Temin et al., 2013). A proteomic catalog of over 3,000 proteins was produced for HAPI cells which showed enrichment in proteins for cell proliferation and survival along with proteins involved in microglial action (Bell-Temin et al., 2012). First isolated in 1989 by Righi et al. (1989) the rat N9 microglial cell line is a retrovirally immortalized cell line which produce IL-6, TNF-α, and IL-1 upon stimulation with LPS. In addition, the cells possess purinergic receptors involved in IL-1β release in response to ATP stimulation (Ferrari et al., 1996).

One of the widely used microglia cell lines is the murine BV2 microglial cell line. BV2 cells are purified by primary murine microglia obtained from the agitation method of isolation that were immortalized using a v-raf/v-myc-containing retrovirus. BV2 cells stain positive for MAC-1 and MAC-2 antigens and were used by Wilcock in 2012 to successfully stimulate the M1, M2a, M2b, and M2c polarizations of microglia. When treated with INF-γ to stimulate an M1 response, BV2 cells showed a significant increase in IL-1β, IL-6, TNF-α, and IL-12 release when compared to untreated cells. Treatment with a combination of IL-4 and IL-13 to stimulate an M2a response yielded an increase in arginase-1 by >45 fold, as well as significant increases in CHI3L3/YM1 (chitinase 3-like 3), an extracellular matrix remodeler, IL-1Ra (IL-1 receptor antagonist), and MRC1 (mannose receptor 1) which mediates endocytosis. Treatment with immune complexes of mouse IgA bound to amyloid-β plaques caused the mixed M2b phenotype, demonstrating significant increases in expression of CD86, IL-10, and FCGR1/2, the IgG responsive FCγ receptors 1 and 2. Finally, direct exposure to IL-10 led to the M2c polarization and significant upregulation of the anti-inflammatory cytokine TGFβ and SPHK1, sphingosine kinase 1, which downregulates production of TNFα through phosphorylation.

As with all immortalized cell lines, BV2 cells and the others mentioned vary significantly from primary cell culture in terms of proliferation, adherence and homogeneity of physical appearance. In addition, primary cell cultures offer more similar protein expression profiles to *ex vivo* microglia than immortalized cells. In a comparison among LPS-stimulated *ex vivo* microglia, neonatal brain-derived primary microglia, and the mouse microglia cell line BV2, primary microglia showed an upregulation of 28 out of the 29 LPS-responsive cytokine, chemokines and receptor proteins from the *ex vivo* microglia, as compared to 12 out of 29 for the BV2 microglia cells (Henn et al., 2009). The magnitude of response, however, was exceedingly large in the primary microglia, whereas the protein expression profile of BV2 cells was far closer to that of the *ex vivo* microglia.

Human Microglia and the Perils of Extrapolation

As with all animal models, the conclusions reached while using rat or mouse models, whether cell culture or *in vivo*, cannot be assumed to match the human response with certitude. Due to restrictions on human primary microglia, several groups have used results from differentiated human monocytes to model microglial action (Etemad et al., 2012; Leone et al., 2006; Martinez et al., 2006). Monocytes derived from CD34$^+$ myeloid progenitor cells in the bone marrow treated with M-CSF produce a shift towards the macrophagic M2 transcriptome. These newly differentiated cells can be stimulated towards an M1 phenotype through treatment with pro-inflammatory triggers, LPS and INFγ, and towards an M2a phenotype using IL-4 and IL-13 treatment (Martinez et al., 2006; Scotton et al., 2005). The M1 phase is enriched with pro-inflammatory cytokines, chemokines, and NADPH oxidase for the production of neurotoxic ROS. Notably absent, or produced at very low level, is iNOS or its end product NO; and in the IL-4 and IL-13 treatments no arginase-1 can be detected, nor its end-product ornithine. This result and others have led to the belief that human microglia might not rely on iNOS and NO as part of its complement of cytotoxic chemicals, and subsequently would have no need for arginase-1 production in order to remove the pool of arginine available for conversion to NO and L-citrulline. The differences between human and rodent microglial and macrophagic activity which include differences in signaling pathways and receptors (MacMicking et al., 1997; Mestas and Hughes, 2004; Schneemann and Schoeden, 2007; Scotton et al., 2005), demonstrate the possible limited translational potential of mouse and rat models to human microglial function.

Fortunately for the study of microglia and macrophagic function, there are over 250 publications illustrating an increase in NO production from macrophages from patient blood samples for a variety of diseases (Fang and Nathan, 2007). Some concerns may arise when that information is interpreted in the context of brain-derived microglia, which have significant differences in phenotype, morphology and function from blood-derived macrophages; however, some important work has been performed in human fetal microglia (Lee et al., 2002). Ding et al. (1997) demonstrated an increase in NO production in primary mixed glia and microglia prepared from human fetal tissue in response to INF-γ. Significantly, not only was an increase in NO production confirmed by the Griess reagent assay, an increase in nitrotyrosine, a protein modification formed in the presence of peroxynitrite (ONOO⁻), was detected as well. The upregulation of iNOS in human fetal microglia was detected by Northern blot and confirmed by *in situ* mRNA hybridization and immunocytochemistry. In addition to success in primary cell culture, the use of immortalized human microglia cell lines HMO6 and C13NJ has been reported (Nagai et al., 2001). In HMO6 cells, treatment with the activator lysophosphatidylcholine led to the induction of IL-1β and iNOS (Sheikh et al., 2009). In C13NJ cells, treatment with amyloid-β peptide resulted in the production of NO as well (Liew et al., 2010).

Ethanol and Neurotoxicity

In order to understand the potential role of microglia in alcohol-induced neurotoxicity, it is important to first explore the impact of alcohol overconsumption on the functional and structural integrity of the brain. Chronic alcohol use is a vicious cycle as initial damage of brain tissue caused by alcohol abuse is thought to play a crucial role in the development of a chronic Alcohol Use Disorder (AUD), which in turn causes additional neurotoxicity (Crews and Boettiger, 2009; Crews et al., 2004; Koob and Le Moal, 1997). Specifically, ethanol-induced neurotoxicity results in functional changes in learning, memory, and mood as well as the induction of seizures (Nixon, 2006). Binge drinking has been shown to induce clear changes in behavior due to alcohol-induced changes in the brain that promote the development of alcoholism. These include loss in executive functions and olfactory impairment found in humans and adult rats (Kesslak et al., 1991; Obernier et al., 2002; Townshend and Duka, 2002). Other changes involve anxiety due to dysfunctional amygdala (Holmes et al., 2012; McCown and Breese, 1990). Further studies have investigated the link of neuroinflammation to the neurodegenerative effects of alcohol (Blanco and Guerri, 2007a; Blanco et al., 2004; Blanco et al., 2005; Boyadjieva and Sarkar, 2013b; Crews et al., 2006a; Crews et al., 2006b; Fernandez-Lizarbe et al., 2013; He and Crews, 2008; Lee et al., 2004; Qin and Crews, 2012b; Qin et al., 2008).

Of interest is the fact that much of the reduction in brain volume due to alcohol exposure appears to be reversible (Crews et al., 2005; Harper, 1998; Nixon, 2006). This result leaves open the enticing possibility of pathways that can undo some of the neuronal loss linked to AUDs. Human hippocampus dentate gyrus granule cells are known to undergo neurogenesis in adults contributing to learning and memory (van Praag et al., 2002). Nixon and Crews observed in 2004 that during alcohol intoxication, rats show a ~50% decrease in progenitor cells in the dentate gyrus and there is a gradual accumulation of these cells until there is a four-fold increase of progenitor cells seven days following ethanol exposure. In a study, Nixon et al. (2008) reported that on day 2 of abstinence following alcohol binge in rats that a large number of microglia, identified using the microglia-specific marker Iba-1, were detected in the hippocampus and cortex, outside the areas where neural progenitor cell production is known to take place and survived for 2 months. The results are similar to microglial proliferation response to a partially deafferented hippocampus, suggesting that the ethanol-induced microglial response is similar to that in response to brain injury, as the new microglia help recover lost brain volume following neuronal death(Fagan and Gage, 1994; Gall et al., 1979). Earlier work by Riikonen et al. (2002) observed an increase in the number of microglia in cerebellar anterior vermis in intermittent ethanol-exposed rats prior to cerebellar atrophy. They found that the increase in microglial abundance was greater in the intermittently exposed animals than the chronically exposed, leading to the possibility that ethanol withdrawal effects on the CNS rather than direct ethanol intoxication might have resulted in the additional microglial proliferation.

It has been known for over half a century that structural and functional changes occur in the human brain due to excessive alcohol use (Courville, 1955). Cirrhosis of the liver causes vitamin B1 deficiency, leading to severe brain damage in the form of Wernicke-Korsakoff syndrome, the most severe form of the progressive spectrum of alcohol-induced dementia. Pfefferbaum et al. (1992) found that alcoholics without Wernicke-Korsakoff symptoms have decreased volumes of both gray and white matter in the brain, especially in the cortex, with no differentiation of loss by cortical region. Also of interest was that brains of older subjects were found to be far more vulnerable to loss when compared to non-drinkers of the same age. Other studies have found that the superior frontal cortex is vulnerable and the primary motor cortex is affected at a lower level than average (Kril and Harper, 1989). In a study, Kubota et al. (2001) used MRI to image the brains of those classified as abstainers, light, moderate and heavy drinkers, and concluded that the contribution of frontal lobe shrinking incidence was 11.3%.

Specific neuronal loss beyond overall volume changes has been documented in the cerebral cortex, hypothalamus and cerebellum of alcoholics (Harper, 1998). Using an ethanol binge model to simulate AUD-induced neurodegeneration, Obernier et al. (2002) demonstrated significant necrotic cell death in multiple regions due to dark cell degeneration in the agranular insular cortex, anterior piriform cortex, perirhinal cortex, lateral entorhinal cortex and the temporal dentate gyrus. In the developing rat brain, ethanol has been shown to cause widespread apoptotic neurodegeneration in the rat forebrain by blocking NMDA glutamate receptors while activating GABA(A) receptors (Ikonomidou et al., 2000). Although the damage caused by ethanol on the brain at the organ and cellular level has been firmly established, the mode of action that causes neurotoxicity and neurodegeneration is not clearly understood.

Oxidative Stress, Neuroinflammation and Neurodegeneration

Certain reports suggest that oxidative stress, particularly that produced as part of an inflammatory response by astrocytes and microglia, leads to neuroinflammation that could cause alcohol-induced neurodegeneration. Ethanol has been found to inhibit astrocyte proliferation in a culture system of *ex vivo* human astrocytes derived from cerebrum (Kane et al., 1996). Knapp and Crews (1999) found an increase in COX-2 during acute and chronic ethanol treatment of rats. An even greater increase in COX-2 expression was seen during the withdrawal phase post-ethanol treatment. Blanco et al. (2004) found, in cultured astrocytes, an increase in COX-2 and also iNOS in response to ethanol treatment through the actions of the transcription regulator NF-κB. Additionally, an increase in iNOS, COX-2 and IL-1β was found in both rat cerebral cortex and cultured astrocytes (Valles et al., 2004).

Crews and Nixon (2009) have demonstrated that a four-day alcohol binge model in rats mimics the cognitive deficits found in humans with neurodegeneration due to AUDs. They suggest that neuronal death is related to oxidative stress in addition to an increase in NF-κB and downstream pro-inflammatory cytokines and a decrease in cAMP-response element-binding protein (CREB), a protein that has a vital role in neuronal plasticity and long-term memory formation. In addition, increases in the NADPH oxidase subunit gp91phox and p67phox were observed in a study by the Crews laboratory, suggesting the involvement of ROS in ethanol neurotoxicity (Qin et al., 2008).

Ethanol and Classical Microglial Activation

When used concomitantly in treatment, ethanol and classical activators such as LPS and INFγ cause changes to normal activation response in both *in vivo* and *in vitro* models. In a 2008 study by Qin et al. (2008) treatment with LPS and ethanol caused an exaggerated response in CCL2 (or MCP-1) and IL-1β response. This work was continued in a study using intragastric ethanol dosing in mice; Qin and Crews (2012a) note a significant increase in TNFα and IL-6 expression of a similar magnitude to the mRNA TLR3 activator polyinosinic: polycytidylic acid (poly I:C), and a significant increase in TNFα, IL-6, and IL-1β production in the concomitant administration of poly I:C and ethanol. In a study by Lee et al. (2004) pretreatment with ethanol caused a decrease in NO production in LPS-stimulated BV2 microglia but had no effect on INFγ-induced NO production. Ethanol appeared to be inhibiting p300, a co-activator of NF-κB, selectively affecting the LPS pathway of NO production while leaving the INF-γ pathway unchanged.

In a study using primary rat microglia cultures, Aroor and Baker (1998) found that ethanol treatment led to a markedly lower level of phagocytosis of heavy-labeled *E. coli* cultures over a 7-day treatment. When treated with ethanol, the cells showed no significant change in ROS production. In addition, treatment with a known microglial activator phorbolmyristate acetate (PMA) and ethanol led to a significantly decreased level of superoxide production when compared to treatment with PMA only and to the control baseline. Whether this reduction in ROS production, in response to the combined treatment of PMA and ethanol, was due to a potential over-activation of microglia and subsequent reduction in microglial activity remains to be determined (Liu et al., 2001). Interestingly, Colton et al. (1998) found that ethanol treatment caused an increase in ROS production in cultured neonatal hamster microglia as compared to the resting state but no priming effect on ROS production when the microglia were exposed to the activator poly I:C. Exposure to concomitant treatment with ethanol and poly I:C led to a decrease in nitric oxide production similar to that seen in the study by Lee et al. (2004); again, either this reduction reflected a compromised cell viability or a convergence of multiple pathways remains to be determined.

Work by the group of Blanco and Guerri (2007b) posits that ethanol has a pro-inflammatory effect on quiescent microglia. They state that ethanol can cause a microglial activation affect by perturbations in membrane fluidity and recruitment of the activating receptor TLR4 to lipid rafts. The group first reported an increase in iNOS, nitric oxide, and COX2 in the murine macrophage line RAW 264.7, and were able to ablate the effect using anti-TLR4 antibody pretreatment (Fernandez-Lizarbe et al., 2008). They followed on this work using primary rat and mouse microglia in response to ethanol stimulation, finding increased phagocytosis, in contrast to the previously mentioned finding of Aroor et al. (1998) and increases in TNF-α and IL-1β and nitric oxide (Fernandez-Lizarbe et al., 2009). Recently, they have posited that the ethanol-induced activation response they observed is due to an increase in the association of TLR4 and TLR2 activating receptor heterodimer (Fernandez-Lizarbe et al., 2013). Knocking out TLR2 or TLR4 mitigates ethanol's activating effects.

Ethanol's Effect on ROS Production in Microglia

Several studies have investigated the link of ethanol treatment to the production of ROS. In an article, Crews et al. (2006a) used intragastric ethanol treatment on rats prior to immunohistochemical investigation. They found, as had been discussed previously, ethanol exposure caused corticolimbic brain damage. Butylatedhydroxytoluene (BHT) was also administered in addition to three other antioxidants and only the BHT reversed the brain damage. It was noted that BHT had no effect on NFκB binding and microglial involvement was demonstrated by morphological changes of the ethanol-treated microglia and expression of the microglial activation marker, OX42 (Crews et al., 2006a).

In a study by Qin and Crews (2012b) the authors posit a direct link between NADPH oxidase and ROS to alcohol-induced microglial activation and neurotoxicity. Using intragastric ethanol treatment of mice rather than a binge model, their results show an increase in markers of neuronal death in the cortex and dentate gyrus as revealed by Fluoro-Jade B (a marker of degenerating neurons)-positive cells co-localized with the neuronal marker Neu-N in the mice and in the orbitofrontal cortex of human postmortem alcoholics compared to moderate drinkers (Qin and Crews, 2012b). Activation of glia was demonstrated based on observed morphology of the microglia and astrocytes using Iba-1 and GFAP immunohistochemistry, respectively. Also observed were increased levels of ROS in the brains of ethanol-treated, which, based on immunofluorescence studies were attributed to mainly microglial and neuronal cells.

Boyadjieva and Sarkar (2013b) have used supernatant from ethanol-treated microglia to apply to a culture of hypothalamic neurons to study the role of microglia in ethanol neurotoxicity. Exposure of neuronal cultures with culture medium from ethanol-treated microglia lead to a concentration-dependent increase in ROS and RNS production, as well as a reduction in cellular antioxidants glutathione, superoxide dismutase and catalase and finally neuronal apoptosis. Increases in the control group of no neuronal treatment and another control group of conditioned microglia media without ethanol treatment were also observed. The same group later demonstrated the neuroprotective effects of neurotrophic factor BDNF or dibutyryl cyclic AMP in this model system (Boyadjieva and Sarkar, 2013a).

Ethanol and Pro-inflammatory Cytokines

In addition to ROS production, several papers have shown that ethanol treatment leads to the production of pro-inflammatory cytokines *in vivo* and *in vitro*. Boyadjieva and Sarkar (2010) found that ethanol treatment of primary microglia led to significant increases in TNFα, CCL3, CXCL2 and IL-6. As discussed previously, TNFα and IL-6 are upregulated in the M1 and M2b phenotypes while CXCL2 could be related to the M2b phenotype, although CCL3 has been observed during a lipid A-triggered M1 response in human monocyte-derived macrophage cultures (Cassol et al., 2009; Owen et al., 2011; Wilcock, 2012).

Increased expression of certain cytokines have also been detected in postmortem analysis of human alcoholic brains (He and Crews, 2008). In mice, gene deletion of the chemokines Ccr2, Ccl2 in females, and Ccl3 led to lower preference and higher taste aversion in mutant mice, showing a role of the immune chemokine network in the motivational effects of alcohol (Blednov et al., 2005). Recent work shows transgenic mice expressing high levels of CCL2 are resistant to the effects of ethanol on synaptic plasticity (Bray et al., 2013). In a study, He and Crews (2008) report an increase in expression of CCL2 (MCP-1) in the ventral tegmental area, substantia nigra, hippocampus and amygdala. This result was in addition to increased ionized calcium binding adaptor protein-1 (Iba-1) and glucose transporter-5 (GluT$_5$) staining, specifically in the cingulate cortex, suggesting changes in microglia activation. Microglia are implicated in the process of neuroinflammation due to increase in microglial cell numbers and increased CCL2 production.

Ethanol and Microglial Partial Activation

Saito et al. (2010) demonstrated that microglia are activated following ethanol-induced modification of tau that leads to neuronal injury. Microglia were observed to phagocytize the cellular debris as a reactive microgliosis response to the neuronal apoptosis. Furthermore, several recent studies suggest that acute ethanol treatment of microglia may not lead to an M1 polarization or to immune response-mediated activation but rather to an earlier stage on Raivich's spectrum of activation, resulting in a greater ability in microglia to respond to insult or pathogens. Using an AUD 4-day binge model in adolescent rats, McClain et al. (2011) noted significant morphological changes in hippocampal microglia consistent with microglial activation, as had been reported previously (Nixon and Crews, 2004). In addition, a number of dividing cells were seen in the hippocampus 2 days post-treatment, as discussed previously (van Praag et al., 2002). No increase in the M1-phase markers CD86, MHC-II, or TNFα was observed, suggesting a partial activation of microglia without advancement to the neurotoxic M1 phase (McClain et al., 2011). Marshall et al. (2013) followed up on this work in an AUD 4-day binge model of adult rats to control for age-related differences, again demonstrating morphological changes consistent with microglial activation without CD86 or MHC-II upregulation. In addition, no increase was seen in TNF-α or IL-6, but a small increase in both IL-10 and TGFβ were observed; both of which relate to cellular growth and repair following damage and also to the earlier stages of Raivich's spectrum of microglial activation, the alert and homing stages, respectively (Marshall et al., 2013).

Bell-Temin et al. (2013) found similar results using a proteome-wide analysis of the effect of ethanol on rat immortalized HAPI microglial cells. Using a stable isotope labeling by amino acids in cell culture (SILAC) workflow shown in Fig. 21.2, the expression of over 3,000 proteins were

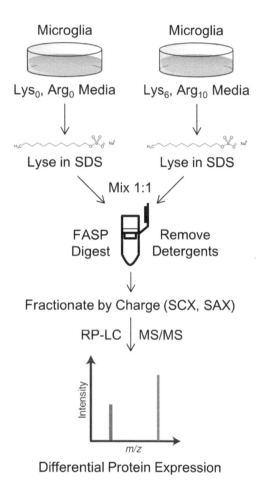

Figure 21.2. Basic Metabolic Labeling Workflow. Microglia incorporate light labeled Lysine$_0$ and Arginine$_0$, or heavy labeled Lysine$_6$ and Arginine$_{10}$ in Stable Isotope Labeling by Amino Acids in Cell Culture (SILAC) media (Ong et al., 2002). Cells are lysed in a high detergent lysis buffer for best protein solvation prior to digestion and detergent removal using the FASP method (Wisniewski et al., 2009). Samples are fractionated by charge using strong cation exchange (SCX) or strong anion exchange (SAX) to reduce complexity before tandem reverse phase liquid chromatography mass spectrometric analysis. The peptide pairs are separated by the mass difference of the labeled amino acids divided by the peptide charge. The intensity differential of the pair provides relative quantification of the protein.

Color image of this figure appears in the color plate section at the end of the book.

identified and quantified between the control and ethanol-treated HAPI microglial cells. No detectable changes in expression levels were found for iNOS, NFκB, TNFα, or IL-6 in response to acute ethanol treatment, although all those factors were highly upregulated in microglial cells stimulated with LPS. Additionally, ethanol-treated HAPI microglial cells exhibited increased expression of proteins involved in the homing stage of Raivich's activation spectrum including colony stimulating factor 1 receptor (CSF1R) and its regulator PU.1. Also observed were increases in the lectin binding proteins CD9 and LGALS3BP which are involved in cell motility, an observation consistent with the findings from McClain and Marshall's *in vivo* studies (Marshall et al., 2013; McClain et al., 2011).

Additional large-scale studies support the hypothesis that ethanol exposure may modulate relatively specific pathways in microglia, or other cells types in the brain. In a study by Osterndorff-Kahanek et al. (2013) bulk gene expression profiles in mice following ethanol and LPS treatments were analyzed. The study found modest changes in microglial cell-type specific gene expression in mice in their chronic, intermittent, or drinking in the dark (DID) alcohol administration models, while demonstrating major increases in microglial activation-related pathways in the LPS-treated animals (Osterndorff-Kahanek et al., 2013).

Closing Remarks

Much progress has been made in the study of the role of neuroinflammation, especially microglia in alcohol-induced neurotoxicity, owing mostly to the effort of pioneers in the field. A large body of literature has helped to define a potentially neurotoxic role of microglia and related biomarkers in ethanol toxicity. However, due to the dynamic and multi-faceted functions of microglia, our understanding of the involvement of microglia in ethanol neurotoxicity may be on an evolving course. We may come to realize that microglia will exhibit a full spectrum of functions in responding to the presence of increased levels of ethanol in the CNS. Advances in techniques for both global and targeted pathway analysis should enable us to define a perhaps more dynamic, yet complex role for microglia in maintaining the homeostasis of the CNS and responding to the "invasion" of agents such as ethanol, albeit a seemingly very simple molecule, that in and of itself can elicit a full spectrum of physiological and pathological responses in the brain. Ultimately, a full understanding of the roles of microglia in the interaction between ethanol and neurons as well as other types of cells in the brain should help us identify and manage adverse effects caused by alcohol overconsumption in humans.

While numerous studies evaluate microglial activation using a handful of known markers, exciting progress has been made using "-omics" based approaches to identify entire pathways or protein interaction networks that could be altered and consequently serve as a class of biomarkers associated with the impact of ethanol on the brain, particularly in the process of neuroinflammation. Recent work employing large-scale quantitative proteomic and transcriptomic analyses show a modulation of microglial activation from the resting stage to the homing stage after acute ethanol exposure where cytokines and iNOS expression that are solely attributed to the M1 phase of activation are notably absent (Bell-Temin et al., 2013; Osterndorff-Kahanek et al., 2013). Nonetheless, the molecular mechanisms of ethanol-induced microglial activation in the context of the central nervous system as well as chronic ethanol exposure are unclear, and additional experimentation employing co-culture methods of microglia and ethanol, using state-of-the-art isotopic labeling techniques for proteins, ROS, and RNS may lead to a better understanding of the role of microglia in ethanol-induced neurodegeneration.

Acknowledgement

The authors acknowledge the support of their work by the National Institutes of Health, National Institute on Alcohol Abuse and Alcoholism, grant AA021245.

References

Akimoto, N., M. Ifuku, Y. Mori and M. Noda. 2013. Effects of chemokine (C-C motif) ligand 1 on microglial function. *Biochem Biophys Res Commun.* 436: 455–461.

Aravalli, R.N., S. Hu, T.N. Rowen, J.M. Palmquist and J.R. Lokensgard. 2005. Cutting edge: TLR2-mediated proinflammatory cytokine and chemokine production by microglial cells in response to herpes simplex virus. *J Immunol.* 175: 4189–4193.

Aroor, A.R. and R.C. Baker. 1998. Ethanol inhibition of phagocytosis and superoxide anion production by microglia. *Alcohol.* 15: 277–280.

Banati, R.B., J. Gehrmann, P. Schubert and G.W. Kreutzberg. 1993. Cytotoxicity of microglia. *Glia.* 7: 111–118.

Bates, M.E., S.C. Bowden and D. Barry. 2002. Neurocognitive impairment associated with alcohol use disorders: implications for treatment. *Experimental and Clinical Psychopharmacology.* 10: 193–212.

Bell-Temin, H., D.S. Barber, P. Zhang, B. Liu and S.M. Stevens, Jr. 2012. Proteomic analysis of rat microglia establishes a high-confidence reference data set of over 3000 proteins. *Proteomics.* 12: 246–250.

Bell-Temin, H., P. Zhang, D. Chaput, M.A. King, M. You, B. Liu and S.M. Stevens, Jr. 2013. Quantitative proteomic characterization of ethanol-responsive pathways in rat microglial cells. *J Proteome Res.* 12: 2067–2077.

Blanco, A.M. and C. Guerri. 2007. Ethanol intake enhances inflammatory mediators in brain: role of glial cells and TLR4/IL-1RI receptors. *Frontiers in Bioscience: A Journal and Virtual Library*. 12: 2616–2630.

Blanco, A.M., M. Pascual, S.L. Valles and C. Guerri. 2004. Ethanol-induced iNOS and COX-2 expression in cultured astrocytes via NF-kappa B. *Neuroreport*. 15: 681–685.

Blanco, A.M., S.L. Valles, M. Pascual and C. Guerri. 2005. Involvement of TLR4/type I IL-1 receptor signaling in the induction of inflammatory mediators and cell death induced by ethanol in cultured astrocytes. *J Immunol*. 175: 6893–6899.

Blednov, Y.A., S.E. Bergeson, D. Walker, V.M. Ferreira, W.A. Kuziel and R.A. Harris. 2005. Perturbation of chemokine networks by gene deletion alters the reinforcing actions of ethanol. *Behav Brain Res*. 165: 110–125.

Block, M.L. and J.S. Hong. 2005. Microglia and inflammation-mediated neurodegeneration: multiple triggers with a common mechanism. *Prog Neurobiol*. 76: 77–98.

Block, M.L., L. Zecca and J.S. Hong. 2007. Microglia-mediated neurotoxicity: uncovering the molecular mechanisms. *Nat Rev Neurosci*. 8: 57–69.

Boyadjieva, N.I. and D.K. Sarkar. 2010. Role of microglia in ethanol's apoptotic action on hypothalamic neuronal cells in primary cultures. *Alcohol Clin Exp Res*. 34: 1835–1842.

Boyadjieva, N.I. and D.K. Sarkar. 2013a. Cyclic adenosine monophosphate and brain-derived neurotrophic factor decreased oxidative stress and apoptosis in developing hypothalamic neuronal cells: role of microglia. *Alcohol Clin Exp Res*. 37: 1370–1379.

Boyadjieva, N.I. and D.K. Sarkar. 2013b. Microglia play a role in ethanol-induced oxidative stress and apoptosis in developing hypothalamic neurons. *Alcohol Clin Exp Res*. 37: 252–262.

Bray, J.G., K.C. Reyes, A.J. Roberts, R.M. Ransohoff and D.L. Gruol. 2013. Synaptic plasticity in the hippocampus shows resistance to acute ethanol exposure in transgenic mice with astrocyte-targeted enhanced CCL2 expression. *Neuropharmacology*. 67: 115–125.

Cardona, A.E., D. Huang, M.E. Sasse and R.M. Ransohoff. 2006. Isolation of murine microglial cells for RNA analysis or flow cytometry. *Nat Protoc*. 1: 1947–1951.

Carter, S.L., M. Muller, P.M. Manders and I.L. Campbell. 2007. Induction of the genes for Cxcl9 and Cxcl10 is dependent on IFN-gamma but shows differential cellular expression in experimental autoimmune encephalomyelitis and by astrocytes and microglia *in vitro*. *Glia*. 55: 1728–1739.

Cassol, E., L. Cassetta, C. Rizzi, M. Alfano and G. Poli. 2009. M1 and M2a polarization of human monocyte-derived macrophages inhibits HIV-1 replication by distinct mechanisms. *J Immunol*. 182: 6237–6246.

Cheepsunthorn, P., L. Radov, S. Menzies, J. Reid and J.R. Connor. 2001. Characterization of a novel brain-derived microglial cell line isolated from neonatal rat brain. *Glia*. 35: 53–62.

Colton, C. and D.M. Wilcock. 2010. Assessing activation states in microglia. *CNS Neurol Disord Drug Targets*. 9: 174–191.

Colton, C.A. 2009. Heterogeneity of microglial activation in the innate immune response in the brain. *J Neuroimmune Pharmacol*. 4: 399–418.

Colton, C.A., R.T. Mott, H. Sharpe, Q. Xu, W.E. Van Nostrand and M.P. Vitek. 2006. Expression profiles for macrophage alternative activation genes in AD and in mouse models of AD. *J Neuroinflammation*. 3: 27.

Colton, C.A., J. Snell-Callanan and O.N. Chernyshev. 1998. Ethanol induced changes in superoxide anion and nitric oxide in cultured microglia. *Alcohol Clin Exp Res*. 22: 710–716.

Courville, C.B. 1955. Effects of alcohol on the nervous system of man. San Lucas Press, Los Angeles. 102 p.

Crews, F., K. Nixon, D. Kim, J. Joseph, B. Shukitt-Hale, L. Qin and J. Zou. 2006a. BHT blocks NF-kappaB activation and ethanol-induced brain damage. *Alcohol Clin Exp Res*. 30: 1938–1949.

Crews, F.T., R. Bechara, L.A. Brown, D.M. Guidot, P. Mandrekar, S. Oak, L. Qin, G. Szabo, M. Wheeler and J. Zou. 2006b. Cytokines and alcohol. *Alcohol Clin Exp Res*. 30: 720–730.

Crews, F.T. and C.A. Boettiger. 2009. Impulsivity, frontal lobes and risk for addiction. *Pharmacol Biochem Behav*. 93: 237–247.

Crews, F.T., T. Buckley, P.R. Dodd, G. Ende, N. Foley, C. Harper, J. He, D. Innes, W. Loh el, A. Pfefferbaum, J. Zou and E.V. Sullivan. 2005. Alcoholic neurobiology: changes in dependence and recovery. *Alcohol Clin Exp Res.* 29: 1504–1513.

Crews, F.T., M.A. Collins, C. Dlugos, J. Littleton, L. Wilkins, E.J. Neafsey, R. Pentney, L.D. Snell, B. Tabakoff, J. Zou and A. Noronha. 2004. Alcohol-induced neurodegeneration: when, where and why? *Alcohol Clin Exp Res.* 28: 350–364.

Crews, F.T. and K. Nixon. 2009. Mechanisms of neurodegeneration and regeneration in alcoholism. *Alcohol Alcohol.* 44: 115–127.

Cunningham, C., D.C. Wilcockson, S. Campion, K. Lunnon and V.H. Perry. 2005. Central and systemic endotoxin challenges exacerbate the local inflammatory response and increase neuronal death during chronic neurodegeneration. *J Neurosci.* 25: 9275–9284.

Czapiga, M. and C.A. Colton. 2003. Microglial function in human APOE3 and APOE4 transgenic mice: altered arginine transport. *J Neuroimmunol.* 134: 44–51.

David, S. and A. Kroner. 2011. Repertoire of microglial and macrophage responses after spinal cord injury. *Nat Rev Neurosci.* 12: 388–399.

Ding, M., B.A. St Pierre, J.F. Parkinson, P. Medberry, J.L. Wong, N.E. Rogers, L.J. Ignarro and J.E. Merrill. 1997. Inducible nitric-oxide synthase and nitric oxide production in human fetal astrocytes and microglia. A kinetic analysis. *J Biol Chem.* 272: 11327–11335.

Dutta, G., D.S. Barber, P. Zhang, N.J. Doperalski and B. Liu. 2012. Involvement of dopaminergic neuronal cystatin C in neuronal injury-induced microglial activation and neurotoxicity. *J Neurochem.* 122: 752–763.

Etemad, S., R.M. Zamin, M.J. Ruitenberg and L. Filgueira. 2012. A novel *in vitro* human microglia model: characterization of human monocyte-derived microglia. *J Neurosci Methods.* 209: 79–89.

Fagan, A.M. and F.H. Gage. 1994. Mechanisms of sprouting in the adult central nervous system: cellular responses in areas of terminal degeneration and reinnervation in the rat hippocampus. *Neuroscience.* 58: 705–725.

Fang, F.C. and C.F. Nathan. 2007. Man is not a mouse: reply. *J Leukoc Biol.* 81: 580.

Fernandez-Lizarbe, S., J. Montesinos and C. Guerri. 2013. Ethanol induces TLR4/TLR2 association, triggering an inflammatory response in microglial cells. *J Neurochem.* 126: 261–273.

Fernandez-Lizarbe, S., M. Pascual, M.S. Gascon, A. Blanco and C. Guerri. 2008. Lipid rafts regulate ethanol-induced activation of TLR4 signaling in murine macrophages. *Mol Immunol.* 45: 2007–2016.

Fernandez-Lizarbe, S., M. Pascual and C. Guerri. 2009. Critical role of TLR4 response in the activation of microglia induced by ethanol. *J Immunol.* 183: 4733–4744.

Ferrari, D., M. Villalba, P. Chiozzi, S. Falzoni, P. Ricciardi-Castagnoli and F. Di Virgilio. 1996. Mouse microglial cells express a plasma membrane pore gated by extracellular ATP. *J Immunol.* 156: 1531–1539.

Frank, M.G., J.L. Wieseler-Frank, L.R. Watkins and S.F. Maier. 2006. Rapid isolation of highly enriched and quiescent microglia from adult rat hippocampus: immunophenotypic and functional characteristics. *J Neurosci Methods.* 151: 121–130.

Gall, C., G. Rose and G. Lynch. 1979. Proliferative and migratory activity of glial cells in the partially deafferented hippocampus. *J Comp Neurol.* 183: 539–549.

Gomez-Nicola, D., B. Valle-Argos and M. Nieto-Sampedro. 2010. Blockade of IL-15 activity inhibits microglial activation through the NFkappaB, p38, and ERK1/2 pathways, reducing cytokine and chemokine release. *Glia.* 58: 264–276.

Gordon, R., C.E. Hogan, M.L. Neal, V. Anantharam, A.G. Kanthasamy and A. Kanthasamy. 2011. A simple magnetic separation method for high-yield isolation of pure primary microglia. *J Neurosci Methods.* 194: 287–296.

Graeber, M.B. and W.J. Streit. 2010. Microglia: biology and pathology. *Acta Neuropathol.* 119: 89–105.

Harper, C. 1998. The neuropathology of alcohol-specific brain damage, or does alcohol damage the brain? *J Neuropathol Exp Neurol.* 57: 101–110.

Harting, M.T., F. Jimenez, S.D. Adams, D.W. Mercer and C.S. Cox, Jr. 2008. Acute, regional inflammatory response after traumatic brain injury: Implications for cellular therapy. *Surgery.* 144: 803–813.

He, J. and F.T. Crews. 2008. Increased MCP-1 and microglia in various regions of the human alcoholic brain. *Exp Neurol.* 210: 349–358.

Henn, A., S. Lund, M. Hedtjarn, A. Schrattenholz, P. Porzgen and M. Leist. 2009. The suitability of BV2 cells as alternative model system for primary microglia cultures or for animal experiments examining brain inflammation. *Altex.* 26: 83–94.

Holmes, A., P.J. Fitzgerald, K.P. MacPherson, L. DeBrouse, G. Colacicco, S.M. Flynn, S. Masneuf, K.E. Pleil, C. Li, C.A. Marcinkiewcz, T.L. Kash, O. Gunduz-Cinar and M. Camp. 2012. Chronic alcohol remodels prefrontal neurons and disrupts NMDAR-mediated fear extinction encoding. *Nat Neurosci.* 15: 1359–1361.

Ikonomidou, C., P. Bittigau, M.J. Ishimaru, D.F. Wozniak, C. Koch, K. Genz, M.T. Price, V. Stefovska, F. Horster, T. Tenkova, K. Dikranian and J.W. Olney. 2000. Ethanol-induced apoptotic neurodegeneration and fetal alcohol syndrome. *Science.* 287: 1056–1060.

Johnson, E.A., T.L. Dao, M.A. Guignet, C.E. Geddes, A.I. Koemeter-Cox and R.K. Kan. 2011. Increased expression of the chemokines CXCL1 and MIP-1alpha by resident brain cells precedes neutrophil infiltration in the brain following prolonged soman-induced status epilepticus in rats. *J Neuroinflammation.* 8: 41.

Kane, C.J., A. Berry, F.A. Boop and D.L. Davies. 1996. Proliferation of astroglia from the adult human cerebrum is inhibited by ethanol *in vitro. Brain Res.* 731: 39–44.

Kesslak, J.P., B.F. Profitt and P. Criswell. 1991. Olfactory function in chronic alcoholics. *Percept Mot Skills.* 73: 551–554.

Kigerl, K.A., J.C. Gensel, D.P. Ankeny, J.K. Alexander, D.J. Donnelly and P.G. Popovich. 2009. Identification of two distinct macrophage subsets with divergent effects causing either neurotoxicity or regeneration in the injured mouse spinal cord. *J Neurosci.* 29: 13435–13444.

Knapp, D.J. and F.T. Crews. 1999. Induction of cyclooxygenase-2 in brain during acute and chronic ethanol treatment and ethanol withdrawal. *Alcohol Clin Exp Res.* 23: 633–643.

Koob, G.F. and M. Le Moal. 1997. Drug abuse: hedonic homeostatic dysregulation. *Science.* 278: 52–58.

Kreutzberg, G.W. 1996. Microglia: a sensor for pathological events in the CNS. *Trends Neurosci.* 19: 312–318.

Kril, J.J. and C.G. Harper. 1989. Neuronal counts from four cortical regions of alcoholic brains. *Acta Neuropathol.* 79: 200–204.

Kubota, M., S. Nakazaki, S. Hirai, N. Saeki, A. Yamaura and T. Kusaka. 2001. Alcohol consumption and frontal lobe shrinkage: study of 1432 non-alcoholic subjects. *J Neurol Neurosurg Psychiatry.* 71: 104–106.

Lai, A.Y. and K.G. Todd. 2008. Differential regulation of trophic and proinflammatory microglial effectors is dependent on severity of neuronal injury. *Glia.* 56: 259–270.

Lee, H., J. Jeong, E. Son, A. Mosa, G.J. Cho, W.S. Choi, J.H. Ha, I.K. Kim, M.G. Lee, C.Y. Kim and K. Suk. 2004. Ethanol selectively modulates inflammatory activation signaling of brain microglia. *J Neuroimmunol.* 156: 88–95.

Lee, Y.B., A. Nagai and S.U. Kim. 2002. Cytokines, chemokines, and cytokine receptors in human microglia. *J Neurosci Res.* 69: 94–103.

Leone, C., G. Le Pavec, W. Meme, F. Porcheray, B. Samah, D. Dormont and G. Gras. 2006. Characterization of human monocyte-derived microglia-like cells. *Glia.* 54: 183–192.

Liew, Y.F., C.T. Huang, S.S. Chou, Y.C. Kuo, S.H. Chou, J.Y. Leu, W.F. Tzeng, S.J. Wang, M.C. Tang and R.F. Huang. 2010. The isolated and combined effects of folic acid and synthetic bioactive compounds against Abeta(25-35)-induced toxicity in human microglial cells. *Molecules.* 15: 1632–1644.

Ling, Z., Y. Zhu, C. Tong, J.A. Snyder, J.W. Lipton and P.M. Carvey. 2006. Progressive dopamine neuron loss following supra-nigral lipopolysaccharide (LPS) infusion into rats exposed to LPS prenatally. *Exp Neurol.* 199: 499–512.

Liu, B. 2006. Modulation of microglial pro-inflammatory and neurotoxic activity for the treatment of Parkinson's disease. *Aaps J.* 8: E606–621.

Liu, B., H.M. Gao and J.S. Hong. 2003. Parkinson's disease and exposure to infectious agents and pesticides and the occurrence of brain injuries: role of neuroinflammation. *Environ Health Perspect.* 111: 1065–1073.

Liu, B. and J.S. Hong. 2003a. Primary rat mesencephalic neuron-glia, neuron-enriched, microglia-enriched, and astroglia-enriched cultures. *Methods Mol Med.* 79: 387–395.

Liu, B. and J.S. Hong. 2003b. Role of microglia in inflammation-mediated neurodegenerative diseases: mechanisms and strategies for therapeutic intervention. *J Pharmacol Exp Ther.* 304: 1–7.

Liu, B., K. Wang, H.M. Gao, B. Mandavilli, J.Y. Wang and J.S. Hong. 2001. Molecular consequences of activated microglia in the brain: overactivation induces apoptosis. *J Neurochem.* 77: 182–189.

Lyons, A., R.J. Griffin, C.E. Costelloe, R.M. Clarke and M.A. Lynch. 2007. IL-4 attenuates the neuroinflammation induced by amyloid-beta *in vivo* and *in vitro*. *J Neurochem.* 101: 771–781.

MacMicking, J., Q.W. Xie and C. Nathan. 1997. Nitric oxide and macrophage function. *Annu Rev Immunol.* 15: 323–350.

Marshall, S.A., J.A. McClain, M.L. Kelso, D.M. Hopkins, J.R. Pauly and K. Nixon. 2013. Microglial activation is not equivalent to neuroinflammation in alcohol-induced neurodegeneration: The importance of microglia phenotype. *Neurobiol Dis.* 54: 239–251.

Martinez, F.O., S. Gordon, M. Locati and A. Mantovani. 2006. Transcriptional profiling of the human monocyte-to-macrophage differentiation and polarization: new molecules and patterns of gene expression. *J Immunol.* 177: 7303–7311.

Martinez, F.O., L. Helming and S. Gordon. 2009. Alternative activation of macrophages: an immunologic functional perspective. *Annu Rev Immunol.* 27: 451–483.

Martinez, F.O., A. Sica, A. Mantovani and M. Locati. 2008. Macrophage activation and polarization. *Frontiers in Bioscience: A Journal and Virtual Library.* 13: 453–461.

McClain, J.A., S.A. Morris, M.A. Deeny, S.A. Marshall, D.M. Hayes, Z.M. Kiser and K. Nixon. 2011. Adolescent binge alcohol exposure induces long-lasting partial activation of microglia. *Brain Behav Immun.* 25 Suppl 1: S120–128.

McCown, T.J. and G.R. Breese. 1990. Multiple withdrawals from chronic ethanol "kindles" inferior collicular seizure activity: evidence for kindling of seizures associated with alcoholism. *Alcohol Clin Exp Res.* 14: 394–399.

Mestas, J. and C.C. Hughes. 2004. Of mice and not men: differences between mouse and human immunology. *J Immunol.* 172: 2731–2738.

Mills, C.D. 2012. M1 and M2 macrophages: oracles of health and disease. *Crit Rev Immunol.* 32: 463–488.

Mosser, D.M. 2003. The many faces of macrophage activation. *J Leukoc Biol.* 73: 209–212.

Moykkynen, T. and E.R. Korpi. 2012. Acute effects of ethanol on glutamate receptors. *Basic Clin Pharmacol Toxicol.* 111: 4–13.

Nagai, A., E. Nakagawa, K. Hatori, H.B. Choi, J.G. McLarnon, M.A. Lee and S.U. Kim. 2001. Generation and characterization of immortalized human microglial cell lines: expression of cytokines and chemokines. *Neurobiol Dis.* 8: 1057–1068.

Nixon, K. 2006. Alcohol and adult neurogenesis: roles in neurodegeneration and recovery in chronic alcoholism. *Hippocampus.* 16: 287–295.

Nixon, K. and F.T. Crews. 2004. Temporally specific burst in cell proliferation increases hippocampal neurogenesis in protracted abstinence from alcohol. *J Neurosci.* 24: 9714–9722.

Nixon, K., D.H. Kim, E.N. Potts, J. He and F.T. Crews. 2008. Distinct cell proliferation events during abstinence after alcohol dependence: microglia proliferation precedes neurogenesis. *Neurobiol Dis.* 31: 218–229.

Obernier, J.A., T.W. Bouldin and F.T. Crews. 2002. Binge ethanol exposure in adult rats causes necrotic cell death. *Alcohol Clin Exp Res.* 26: 547–557.

Olah, M., K. Biber, J. Vinet and H.W. Boddeke. 2011. Microglia phenotype diversity. *CNS Neurol Disord Drug Targets.* 10: 108–118.

Ong, S.E., B. Blagoev, I. Kratchmarova, D.B. Kristensen, H. Steen, A. Pandey and M. Mann. 2002. Stable isotope labeling by amino acids in cell culture, SILAC, as a simple and accurate approach to expression proteomics. *Mol Cell Proteomics*. 1: 376–386.

Osterndorff-Kahanek, E., I. Ponomarev, Y.A. Blednov and R.A. Harris. 2013. Gene expression in brain and liver produced by three different regimens of alcohol consumption in mice: comparison with immune activation. *PLoS One*. 8: e59870.

Owen, J.L., M.F. Criscitiello, S. Libreros, R. Garcia-Areas, K. Guthrie, M. Torroella-Kouri and V. Iragavarapu-Charyulu. 2011. Expression of the inflammatory chemokines CCL2, CCL5 and CXCL2 and the receptors CCR1-3 and CXCR2 in T lymphocytes from mammary tumor-bearing mice. *Cell Immunol*. 270: 172–182.

Pesce, J.T., T.R. Ramalingam, M.M. Mentink-Kane, M.S. Wilson, K.C. El Kasmi, A.M. Smith, R.W. Thompson, A.W. Cheever, P.J. Murray and T.A. Wynn. 2009. Arginase-1-expressing macrophages suppress Th2 cytokine-driven inflammation and fibrosis. *PLoS Pathog*. 5: e1000371.

Pfefferbaum, A., K.O. Lim, R.B. Zipursky, D.H. Mathalon, M.J. Rosenbloom, B. Lane, C.N. Ha and E.V. Sullivan. 1992. Brain gray and white matter volume loss accelerates with aging in chronic alcoholics: a quantitative MRI study. *Alcohol Clin Exp Res*. 16: 1078–1089.

Qin, L. and F.T. Crews. 2012a. Chronic ethanol increases systemic TLR3 agonist-induced neuroinflammation and neurodegeneration. *J Neuroinflammation*. 9: 130.

Qin, L. and F.T. Crews. 2012b. NADPH oxidase and reactive oxygen species contribute to alcohol-induced microglial activation and neurodegeneration. *J Neuroinflammation*. 9: 5.

Qin, L., J. He, R.N. Hanes, O. Pluzarev, J.S. Hong and F.T. Crews. 2008. Increased systemic and brain cytokine production and neuroinflammation by endotoxin following ethanol treatment. *J Neuroinflammation*. 5: 10.

Qin, L., X. Wu, M.L. Block, Y. Liu, G.R. Breese, J.S. Hong, D.J. Knapp and F.T. Crews. 2007. Systemic LPS causes chronic neuroinflammation and progressive neurodegeneration. *Glia*. 55: 453–462.

Raes, G., P. De Baetselier, W. Noel, A. Beschin, F. Brombacher and G. Hassanzadeh Gh. 2002. Differential expression of FIZZ1 and Ym1 in alternatively versus classically activated macrophages. *J Leukoc Biol*. 71: 597–602.

Rainey-Barger, E.K., J.M. Rumble, S.J. Lalor, N. Esen, B.M. Segal and D.N. Irani. 2011. The lymphoid chemokine, CXCL13, is dispensable for the initial recruitment of B cells to the acutely inflamed central nervous system. *Brain Behav Immun*. 25: 922–931.

Raivich, G., M. Bohatschek, C.U. Kloss, A. Werner, L.L. Jones and G.W. Kreutzberg. 1999a. Neuroglial activation repertoire in the injured brain: graded response, molecular mechanisms and cues to physiological function. *Brain Res Brain Res Rev*. 30: 77–105.

Raivich, G., L.L. Jones, A. Werner, H. Bluthmann, T. Doetschmann and G.W. Kreutzberg. 1999b. Molecular signals for glial activation: pro- and anti-inflammatory cytokines in the injured brain. *Acta Neurochir Suppl*. 73: 21–30.

Righi, M., L. Mori, G. Delibero, M. Sironi, A. Biondi, A. Mantovani, S.D. Donini and P. Ricciardicastagnoli. 1989. Monokine Production by Microglial Cell Clones. *Eur J Immunol*. 19: 1443–1448.

Riikonen, J., P. Jaatinen, J. Rintala, I. Porsti, K. Karjala and A. Hervonen. 2002. Intermittent ethanol exposure increases the number of cerebellar microglia. *Alcohol Alcohol*. 37: 421–426.

Rock, R.B., S. Hu, A. Deshpande, S. Munir, B.J. May, C.A. Baker, P.K. Peterson and V. Kapur. 2005. Transcriptional response of human microglial cells to interferon-gamma. *Genes Immun*. 6: 712–719.

Rolls, A., R. Shechter, A. London, Y. Segev, J. Jacob-Hirsch, N. Amariglio, G. Rechavi and M. Schwartz. 2008. Two faces of chondroitin sulfate proteoglycan in spinal cord repair: a role in microglia/macrophage activation. *PLoS Med*. 5: e171.

Rom, S., I. Rom, G. Passiatore, M. Pacifici, S. Radhakrishnan, L. Del Valle, S. Pina-Oviedo, K. Khalili, D. Eletto and F. Peruzzi. 2010. CCL8/MCP-2 is a target for mir-146a in HIV-1-infected human microglial cells. *Faseb J*. 24: 2292–2300.

Saijo, K. and C.K. Glass. 2011. Microglial cell origin and phenotypes in health and disease. *Nat Rev Immunol.* 11: 775–787.

Saito, M., G. Chakraborty, R.F. Mao, S.M. Paik and C. Vadasz. 2010. Tau phosphorylation and cleavage in ethanol-induced neurodegeneration in the developing mouse brain. *Neurochem Res.* 35: 651–659.

Schneemann, M. and G. Schoeden. 2007. Macrophage biology and immunology: man is not a mouse. *J Leukoc Biol.* 81: 579; discussion 580.

Schwartz, M., O. Butovsky, W. Bruck and U.K. Hanisch. 2006. Microglial phenotype: is the commitment reversible? *Trends Neurosci.* 29: 68–74.

Scotton, C.J., F.O. Martinez, M.J. Smelt, M. Sironi, M. Locati, A. Mantovani and S. Sozzani. 2005. Transcriptional profiling reveals complex regulation of the monocyte IL-1 beta system by IL-13. *J Immunol.* 174: 834–845.

Sheikh, A.M., A. Nagai, J.K. Ryu, J.G. McLarnon, S.U. Kim and J. Masuda. 2009. Lysophosphatidylcholine induces glial cell activation: role of rho kinase. *Glia.* 57: 898–907.

Shiratori, M., H. Tozaki-Saitoh, M. Yoshitake, M. Tsuda and K. Inoue. 2010. P2X7 receptor activation induces CXCL2 production in microglia through NFAT and PKC/MAPK pathways. *J Neurochem.* 114: 810–819.

Terao, Y., H. Ohta, A. Oda, Y. Nakagaito, Y. Kiyota and Y. Shintani. 2009. Macrophage inflammatory protein-3alpha plays a key role in the inflammatory cascade in rat focal cerebral ischemia. *Neurosci Res.* 64: 75–82.

Townshend, J.M. and T. Duka. 2002. Patterns of alcohol drinking in a population of young social drinkers: a comparison of questionnaire and diary measures. *Alcohol Alcohol.* 37: 187–192.

Valles, S.L., A.M. Blanco, M. Pascual and C. Guerri. 2004. Chronic ethanol treatment enhances inflammatory mediators and cell death in the brain and in astrocytes. *Brain Pathol.* 14: 365–371.

van Praag, H., A.F. Schinder, B.R. Christie, N. Toni, T.D. Palmer and F.H. Gage. 2002. Functional neurogenesis in the adult hippocampus. *Nature.* 415: 1030–1034.

Wilcock, D.M. 2012. A changing perspective on the role of neuroinflammation in Alzheimer's disease. *Int J Alzheimers Dis.* 2012: 495243.

Wisniewski, J.R., A. Zougman, N. Nagaraj and M. Mann. 2009. Universal sample preparation method for proteome analysis. *Nat Methods.* 6: 359–362.

Xu, L., B. Hilliard, R.J. Carmody, G. Tsabary, H. Shin, D.W. Christianson and Y.H. Chen. 2003. Arginase and autoimmune inflammation in the central nervous system. *Immunology.* 110: 141–148.

Zhang, P., K.M. Lokuta, D.E. Turner and B. Liu. 2010. Synergistic dopaminergic neurotoxicity of manganese and lipopolysaccharide: differential involvement of microglia and astroglia. *J Neurochem.* 112: 434–443.

Zou, J.Y. and F.T. Crews. 2005. TNF alpha potentiates glutamate neurotoxicity by inhibiting glutamate uptake in organotypic brain slice cultures: neuroprotection by NF kappa B inhibition. *Brain Res.* 1034: 11–24.

Index

About the Editors

Kevin K.W. Wang, PhD

<u>Institution: University of Florida, Gainesville, FL</u>

Dr. Wang is currently at the Departments of Psychiatry and Neuroscience of University of Florida McKnight Brain Institute as the Executive Director of the Center for Neuroproteomics and Biomarkers Research/Associate Professor and Chief of Translational Research. Dr. Kevin Wang is internationally recognized for his original contributions to the fields of traumatic brain injury (TBI)-linked proteolytic enzymes, therapeutic targets, neuroproteomics/systems biology, biomarker discovery and validation. The clinical diagnostic utility for two TBI protein biomarkers during the acute phase of brain injury has now been confirmed in peer-reviewed journals. These TBI diagnostic biomarker tests are now moving forward to FDA-approval seeking pivotal clinical study. His current research directions include studying mechanisms for CNS injury and substance abuse-induced brain perturbation using systems biology approach. He published more than 200 peer-reviewed papers, reviews and book chapters and co-authored eight US patents and co-edited four books. Dr. Wang is also past President (2011–12) and current Councilor (2013–present) of the National Neurotrauma Society (USA).

Zhiqun Zhang, PhD

<u>Institution: University of Florida, Gainesville, FL</u>

Dr. Zhang is an Adjunct Assistant Professor at the University of Florida, Department of Psychaitry. Her research focus is on biomarkers identification and therapeutic strategies in the area of organ injury by applying cutting-edge technologies, such proteomics, genomics and systems biology. She obtained her MD in Nephrology from Nanjing

University with focus on the epidemiologic and etiologic studies on organ injury and autoimmune diseases. In 2007, she obtained my PhD from the Department of Neuroscience at University of Florida with focusing on biomarkers identification in the area of traumatic brain injury and systematic study on the signal pathways in neuronal injury and regeneration.

Firas Hosni Kobeissy, PhD

Institution: University of Florida, Gainesville, FL

Dr. Kobeissy is a Research Assistant Professor at the Department of Psychiatry, College of Medicine; University of Florida, Gainesville, FL. A native of Lebanon, Dr. Kobeissy graduated from the American University of Beirut. He also holds a Master's in *Molecular Genetics* from Bowling Green State University, Ohio. He obtained his PhD from the University of Florida in the area of Neuroscience. His current research overlaps the fields of neuroscience and psychiatry with a focus on drug abuse neurotoxicity and traumatic brain injury neuroproteomics. Dr. Kobeissy has authored more than 45 peer-reviewed scientific papers and 18 book chapters. He holds five patents in the areas of inflammation, drug abuse, and traumatic brain injury biomarkers. Dr. Kobeissy is the editor of the book: Psychiatric Disorders: Methods and Protocols (Springer, Humana Press, 2013). Dr. Kobeissy serves as an editorial member on several journals related to proteomics and is a member of the VA RRD grant review panel. He is also an actively participating member at the Center of Neuroproteomics and Biomarker Research and at the Center for Traumatic Brain Injury Studies at the University of Florida McKnight Brain Institute.

Color Plate Section

Chapter 1

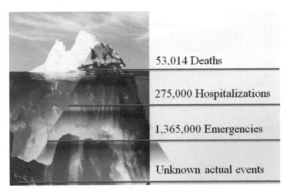

53,014 Deaths	
275,000 Hospitalizations	
1,365,000 Emergencies	
Unknown actual events	

Figure 1.1. Deaths from TBI are only the tip of the iceberg compared to the actual incidence of mild TBI not drawn to medical attention.

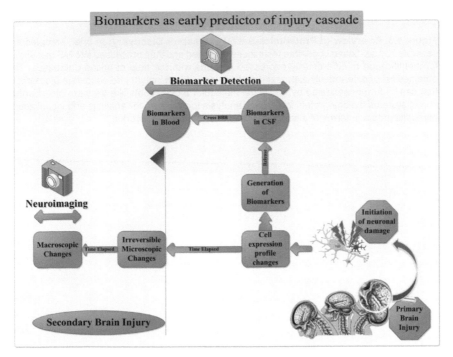

Figure 1.2. The advantage of tissue specific biomarker discovery over current imaging and diagnostic tools is that it can allow for the detection of injury early on before irreversible damage to the brain tissue ensues.

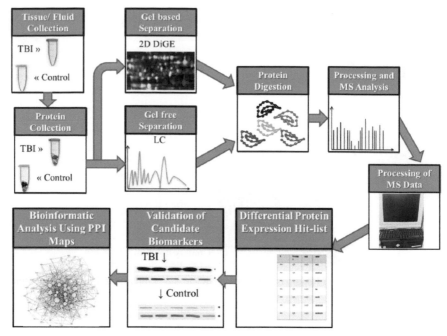

Figure 1.3. Overview of Proteomics-Based Biomarker Discovery: In brief, samples are separated by gel-based or gel-free techniques, digested and then processed into MS that allows for identification of differentially expressed proteins with the help of online databases. The obtained list of differentially expressed proteins provide an insight into candidate biomarkers that can be further validated by traditional molecular biology tools like Western blot. Further use of systems biology protein interaction analysis tools allows for analysis and visualization of resulting data in terms of interacting proteins and protein networks.

Chapter 3

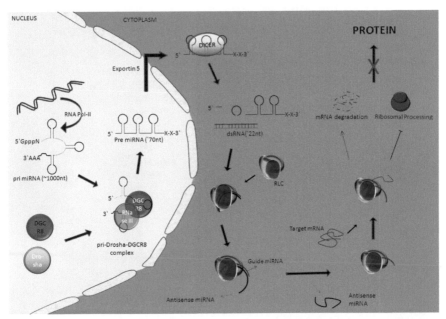

Figure 3.1. MicroRNA biogenesis and functions. MicroRNA (miRNA) genes arc transcribed by RNA polymerase II enzyme into long primary miRNAs (pri-miRNA). The pri-miRNA is processed within the nucleus by RNase III enzyme Drosha and the dsRNA binding protein DiGeorge Syndrome Critical Region 8 (DGCR8), into precursor miRNA (Pre-miRNA). Pre-miRNA is then transported to cytoplasm by Exportin-5. In the cytoplasm, the pre-miRNA is cleaved by Dicer and form ~22 bp miRNA duplex. The strand that is less thermodynamically stable at 5' end (guide strand) gets incorporated into RNA induced silencing complex (RISC) whereas the other strand (passenger strand) in most cases is degraded. The mRNA gets degraded when a perfect match between the 2–8 nt from the miRNA 5'end (seed sequence) and 3'UTR target sequence is available. Without a perfect match, mRNA gets either destabilized or repressed.

Chapter 4

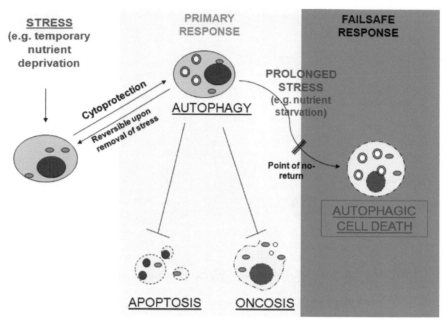

Figure 4.4. From autophagy to autophagic cell death.

Chapter 5

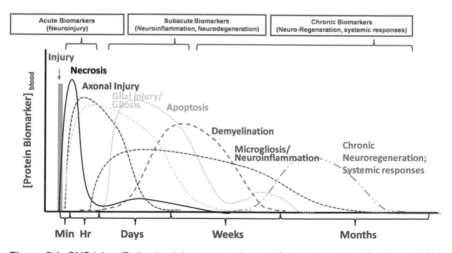

Figure 5.1. CNS Injury Pathophysiology—a continuum of biomarkers. Modified from (118).

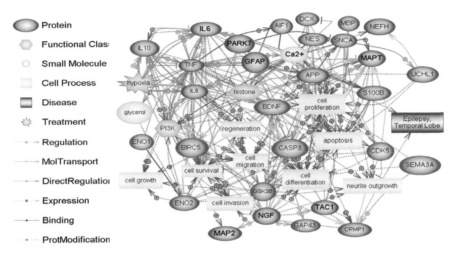

Figure 5.2. Systems Biology mapping of interactome of various TBI biomarker types.

Chapter 7

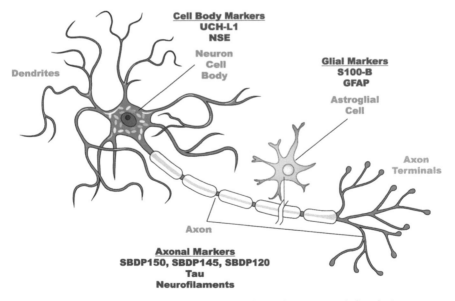

Figure 7.1. Neuroanatomic areas where biomarkers are most abundant.

Chapter 11

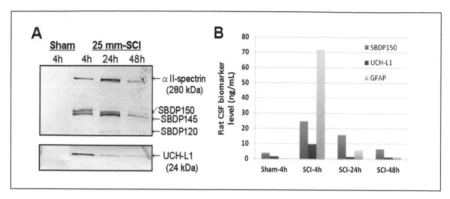

Figure 11.2. αII-Spectrin-BDPs, UCH-L1 , GFAP biomarker release into rat CSF after experimental spinal cord injury. Biomarkers were detected by (A) immunblotting showing increase release of αII-spectrin breakdown products SBDP150, SBDP145, SBDP120 and UCH-L1 at 4, 24 and 48 hours after injury, and (B) ELISA showing increase CSF levels of SBDP150, UCH-L1 and GFAP at 4, 24 and 48 hours after SCI. Results shown are representative of two experiments.

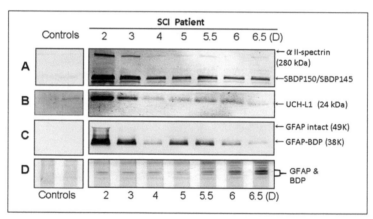

Figure 11.3. Time course of SBDP150, UCH-L1 , GFAP biomarker release into human biofluid after spinal cord injury. Biomarkers were detected by immunblotting (A) CSF αII-spectrin and SBDPs, (B) CSF UCH-L1, (C) CSF GFAP and BDP and (D) serum autoantibody to GFAP and BDP in one SCI patient (various time points-2 day to 6.5 days post-injury) in comparison to biofluid samples from two controls.

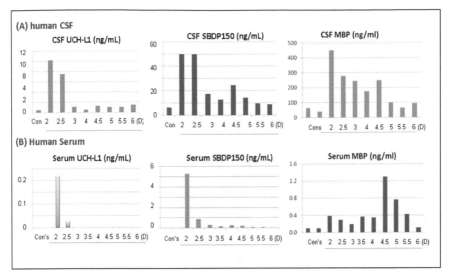

Figure 11.4. SCI biomarker levels in human SCI CSF (A) and serum samples (B) by swELISA. In comparison to CSF control and normal serum control, in one SCI human patient, Both CSF and serum UCH-L1 and SBDP145 levels are elevated in day 2 after SCI. In contrast, CSF MBP elevation peaks in day 2 but also sustained to 4 days, while serum MBP elevation peaks at day 4–5.

Chapter 17

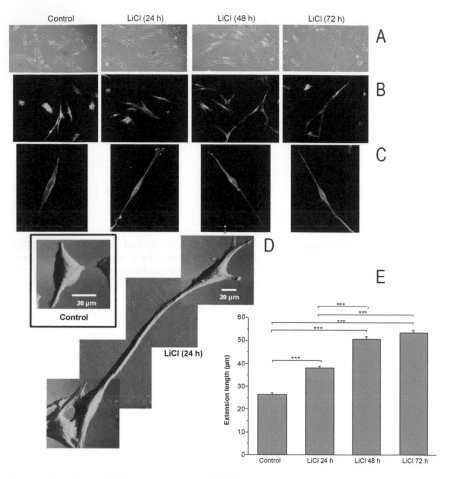

Figure 17.4. Effect of LiCl on the shape of MSC80 cells and on the localization of β-catenin in MSC80 cells (Tawk et al., 2011). MSC80 cells were incubated in the absence (control) or in the presence of LiCl (10 mM) during 24, 48 and/or 72 hours and then monitored by light microscopy (A), confocal microscopy to localize b-catenin (B;C) and atomic force microscopy (D). Scale bar = 20 μm. (B;C) Antibodies directed against b-catenin were added to the cells overnight. (E) Extension lengths of MSC80 cells. Data are presented as mean ± S.E.M. of three independent experiments and a minimum of 100 cells per experiment. Kruskal-Wallis one-way ANOVA with Bonferroni, Scheffe, Tukey and Sidak comparison tests were used to determine the statistical significance between data, ***P<0.001.

Chapter 20

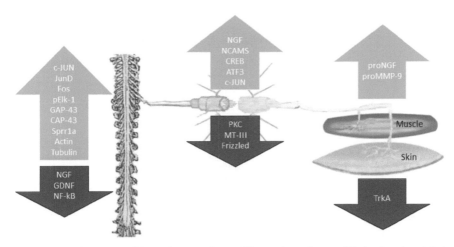

Figure 20.3. Candidate Biomarkers at three different Locations. pElk-1: phosphorylated Elk-1; GAP-43: Growth Associated Protein 43; Sprr1a: Small Proline-rich Repeat Protein 1A; NGF: Neural Growth Factor; NF-kB: Nuclear Factor Kappa Beta; CREB: cAMP Responsive Element binding Protein; ATF-3: Activating Transcription Factor-3; PKC: protein kinase C; MT-III: Metallothionein III; proMMP9: pro Matrix MetalloProteinase-9; TrKA: Neurotrophic Tyrosine Kinase Antigen.

Chapter 21

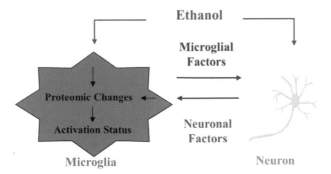

Figure 21.1. Identification of Biomarkers Associated with Ethanol-induced Microglial Response. Ethanol exposure induces an activation response in microglia associated with a specific proteome profile. Ethanol-induced neuronal injury results in modulation of microglial activation phenotype depending on dose and exposure time. Significant progress is being made using global-scale expression profiling methods such as proteomics to gain a better understanding of the molecular markers associated with ethanol-induced microglial activation and role in ethanol neurotoxicity.

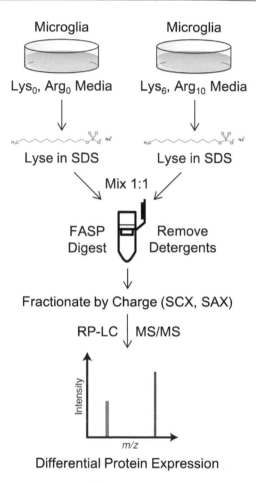

Figure 21.2. Basic Metabolic Labeling Workflow. Microglia incorporate light labeled Lysine$_0$ and Arginine$_0$, or heavy labeled Lysine$_6$ and Arginine$_{10}$ in Stable Isotope Labeling by Amino Acids in Cell Culture (SILAC) media (Ong et al., 2002). Cells are lysed in a high detergent lysis buffer for best protein solvation prior to digestion and detergent removal using the FASP method (Wisniewski et al., 2009). Samples are fractionated by charge using strong cation exchange (SCX) or strong anion exchange (SAX) to reduce complexity before tandem reverse phase liquid chromatography mass spectrometric analysis. The peptide pairs are separated by the mass difference of the labeled amino acids divided by the peptide charge. The intensity differential of the pair provides relative quantification of the protein.